STUDENT SOLUTIONS MANUAL
MARK McCOMBS

PRECALCULUS
SEVENTH EDITION

SULLIVAN

PEARSON

Prentice
Hall

Upper Saddle River, NJ 07458

Editor-in-Chief: Sally Yagan
Senior Acquisitions Editor: Eric Frank
Project Manager: Dawn Murrin
Executive Managing Editor: Vince O'Brien
Production Editor: Allyson Kloss
Supplement Cover Manager: Paul Gourhan
Supplement Cover Designer: Joanne Alexandris
Manufacturing Buyer: Ilene Kahn

© 2005 Pearson Education, Inc.
Pearson Prentice Hall
Pearson Education, Inc.
Upper Saddle River, NJ 07458

The author and publisher of this book have used their best efforts in preparing this book. These efforts include the development, research, and testing of the theories and programs to determine their effectiveness. The author and publisher make no warranty of any kind, expressed or implied, with regard to these programs or the documentation contained in this book. The author and publisher shall not be liable in any event for incidental or consequential damages in connection with, or arising out of, the furnishing, performance, or use of these programs.

Printed in the United States of America

10   9   8   7   6   5   4   3   2

ISBN 0-13-143136-6

Pearson Education Ltd., *London*
Pearson Education Australia Pty. Ltd., *Sydney*
Pearson Education Singapore, Pte. Ltd.
Pearson Education North Asia Ltd., *Hong Kong*
Pearson Education Canada, Inc., *Toronto*
Pearson Educación de Mexico, S.A. de C.V.
Pearson Education—Japan, *Tokyo*
Pearson Education Malaysia, Pte. Ltd.

Contents

Chapter 5   Trigonometric Functions

Chapter 6   Analytic Trigonometry

Chapter 7   Applications of Trigonometric Functions

Chapter 8   Polar Coordinates; Vectors

Chapter 9   The Conics

Chapter 10   Systems of Equations and Inequalities

Chapter 11   Sequences; Induction; The Binomial Theorem

Chapter 12   Counting and Probability

Chapter 13   A Preview of Calculus: The Limit, Derivative, and Integral of a Function

Appendix A   Review

Appendix B Graphing Utilities

Preface

The <u>Student Solutions Manual</u> to accompany <u>Precalculus, 7th Edition</u> by Michael Sullivan contains detailed solutions to all of the odd-numbered problems in the textbook. TI-83 graphing calculator screens have been included to demonstrate the use of the graphics calculator in solving and in checking solutions to the problems where requested. Every attempt has been made to make this manual as error free as possible. If you have suggestions, error corrections, or comments please feel free to write to me about them.

A number of people need to be recognized for their contributions in the preparation of this manual. Thanks go to Sally Yagan and Dawn Murrin at Prentice Hall. Thanks also to Cindy Trimble for her meticulous error-checking of the solutions.

I especially wish to thank my mother, Sarah, and my brothers, Kirk and Doug, for their unwavering support and encouragement.

Finally, I want to thank my wife for helping me endure the long hours of editing the manuscript. Thank you, Tate, I love you that much.

Mark A. McCombs
Department of Mathematics
Campus Box 3250
University of North Carolina at Chapel Hill
Chapel Hill, NC 27599
mccombs@math.unc.edu

Graphs

1.1 Rectangular Coordinates

1. 0

3. $d = |5-(-3)| = |5+3| = |8| = 8$

5. quadrants

7. x

9. False

11. (a) Quadrant II
    (b) Positive x-axis
    (c) Quadrant III
    (d) Quadrant I
    (e) Negative y-axis
    (f) Quadrant IV

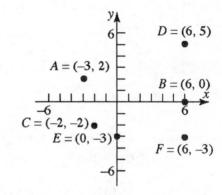

13. The points will be on a vertical line that is two units to the right of the y-axis.

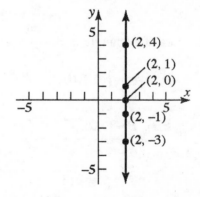

15. $d(P_1, P_2) = \sqrt{(2-0)^2 + (1-0)^2} = \sqrt{2^2 + 1^2} = \sqrt{4+1} = \sqrt{5}$

17. $d(P_1, P_2) = \sqrt{(-2-1)^2 + (2-1)^2} = \sqrt{(-3)^2 + 1^2} = \sqrt{9+1} = \sqrt{10}$

19. $d(P_1, P_2) = \sqrt{(5-3)^2 + (4-(-4))^2} = \sqrt{2^2 + 8^2} = \sqrt{4+64} = \sqrt{68} = 2\sqrt{17}$

21. $d(P_1, P_2) = \sqrt{(6-(-3))^2 + (0-2)^2} = \sqrt{9^2 + (-2)^2} = \sqrt{81+4} = \sqrt{85}$

23. $d(P_1,P_2) = \sqrt{(6-4)^2 + (4-(-3))^2} = \sqrt{2^2 + 7^2} = \sqrt{4+49} = \sqrt{53}$

25. $d(P_1,P_2) = \sqrt{(2.3-(-0.2))^2 + (1.1-0.3)^2} = \sqrt{(2.5)^2 + (0.8)^2}$
$\qquad\qquad = \sqrt{6.25+0.64} = \sqrt{6.89}$

27. $d(P_1,P_2) = \sqrt{(0-a)^2 + (0-b)^2} = \sqrt{a^2 + b^2}$

29. $A = (-2,5), \ B = (1,3), \ C = (-1,0)$
$\quad d(A,B) = \sqrt{(1-(-2))^2 + (3-5)^2} = \sqrt{3^2 + (-2)^2}$
$\qquad\qquad = \sqrt{9+4} = \sqrt{13}$
$\quad d(B,C) = \sqrt{(-1-1)^2 + (0-3)^2} = \sqrt{(-2)^2 + (-3)^2}$
$\qquad\qquad = \sqrt{4+9} = \sqrt{13}$
$\quad d(A,C) = \sqrt{(-1-(-2))^2 + (0-5)^2} = \sqrt{1^2 + (-5)^2}$
$\qquad\qquad = \sqrt{1+25} = \sqrt{26}$

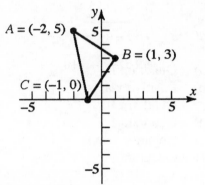

Verifying that $\triangle ABC$ is a right triangle by the Pythagorean Theorem:
$$[d(A,B)]^2 + [d(B,C)]^2 = [d(A,C)]^2$$
$$\left(\sqrt{13}\right)^2 + \left(\sqrt{13}\right)^2 = \left(\sqrt{26}\right)^2$$
$$13+13 = 26 \Rightarrow 26 = 26$$
The area of a triangle is $A = \dfrac{1}{2} \cdot bh$. Here,

$$A = \frac{1}{2} \cdot [d(A,B)] \cdot [d(B,C)] = \frac{1}{2} \cdot \sqrt{13} \cdot \sqrt{13} = \frac{1}{2} \cdot 13 = \frac{13}{2} \text{ square units}$$

31. $A = (-5,3), \ B = (6,0), \ C = (5,5)$
$\quad d(A,B) = \sqrt{(6-(-5))^2 + (0-3)^2} = \sqrt{11^2 + (-3)^2}$
$\qquad\qquad = \sqrt{121+9} = \sqrt{130}$
$\quad d(B,C) = \sqrt{(5-6)^2 + (5-0)^2} = \sqrt{(-1)^2 + 5^2}$
$\qquad\qquad = \sqrt{1+25} = \sqrt{26}$
$\quad d(A,C) = \sqrt{(5-(-5))^2 + (5-3)^2} = \sqrt{10^2 + 2^2}$
$\qquad\qquad = \sqrt{100+4} = \sqrt{104}$

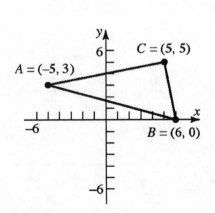

Verifying that $\triangle ABC$ is a right triangle by the Pythagorean Theorem:
$$[d(A,C)]^2 + [d(B,C)]^2 = [d(A,B)]^2$$
$$\left(\sqrt{104}\right)^2 + \left(\sqrt{26}\right)^2 = \left(\sqrt{130}\right)^2$$
$$104 + 26 = 130 \Rightarrow 130 = 130$$

The area of a triangle is $A = \frac{1}{2} \cdot bh.$ Here,
$$A = \frac{1}{2} \cdot [d(A,C)] \cdot [d(B,C)] = \frac{1}{2} \cdot \sqrt{104} \cdot \sqrt{26} = \frac{1}{2} \cdot \sqrt{2704} = \frac{1}{2} \cdot 52 = 26 \text{ square units}$$

33. $A = (4,-3),\; B = (0,-3),\; C = (4,2)$

$d(A,B) = \sqrt{(0-4)^2 + (-3-(-3))^2} = \sqrt{(-4)^2 + 0^2}$
$\qquad = \sqrt{16+0} = \sqrt{16} = 4$

$d(B,C) = \sqrt{(4-0)^2 + (2-(-3))^2} = \sqrt{4^2 + 5^2}$
$\qquad = \sqrt{16+25} = \sqrt{41}$

$d(A,C) = \sqrt{(4-4)^2 + (2-(-3))^2} = \sqrt{0^2 + 5^2}$
$\qquad = \sqrt{0+25} = \sqrt{25} = 5$

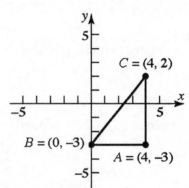

Verifying that $\triangle ABC$ is a right triangle by the Pythagorean Theorem:
$$[d(A,B)]^2 + [d(A,C)]^2 = [d(B,C)]^2$$
$$4^2 + 5^2 = \left(\sqrt{41}\right)^2 \Rightarrow 16 + 25 = 41 \Rightarrow 41 = 41$$

The area of a triangle is $A = \frac{1}{2} \cdot bh.$ Here,
$$A = \frac{1}{2} \cdot [d(A,B)] \cdot [d(A,C)] = \frac{1}{2} \cdot 4 \cdot 5 = 10 \text{ square units}$$

35.   All points having an x-coordinate of 2 are of the form $(2, y).$ Those which are 5 units from $(-2, -1)$ are:
$$\sqrt{(2-(-2))^2 + (y-(-1))^2} = 5 \Rightarrow \sqrt{4^2 + (y+1)^2} = 5$$
$$\text{Squaring both sides: } 4^2 + (y+1)^2 = 25$$
$$16 + y^2 + 2y + 1 = 25$$
$$y^2 + 2y - 8 = 0$$
$$(y+4)(y-2) = 0 \Rightarrow y = -4 \text{ or } y = 2$$
Therefore, the points are $(2, -4)$ and $(2, 2).$

37.   All points on the x-axis are of the form $(x, 0).$ Those which are 5 units from $(4, -3)$ are:
$$\sqrt{(x-4)^2 + (0-(-3))^2} = 5 \Rightarrow \sqrt{(x-4)^2 + 3^2} = 5$$
$$\text{Squaring both sides: } (x-4)^2 + 9 = 25$$

$$x^2 - 8x + 16 + 9 = 25$$
$$x^2 - 8x = 0$$
$$x(x - 8) = 0 \Rightarrow x = 0 \text{ or } x = 8$$

Therefore, the points are $(0, 0)$ and $(8, 0)$.

39. The coordinates of the midpoint are:

$$(x, y) = \left(\frac{x_1 + x_2}{2}, \frac{y_1 + y_2}{2}\right) = \left(\frac{5 + 3}{2}, \frac{-4 + 2}{2}\right) = \left(\frac{8}{2}, \frac{-2}{2}\right) = (4, -1)$$

41. The coordinates of the midpoint are:

$$(x, y) = \left(\frac{x_1 + x_2}{2}, \frac{y_1 + y_2}{2}\right) = \left(\frac{-3 + 6}{2}, \frac{2 + 0}{2}\right) = \left(\frac{3}{2}, \frac{2}{2}\right) = \left(\frac{3}{2}, 1\right)$$

43. The coordinates of the midpoint are:

$$(x, y) = \left(\frac{x_1 + x_2}{2}, \frac{y_1 + y_2}{2}\right) = \left(\frac{4 + 6}{2}, \frac{-3 + 1}{2}\right) = \left(\frac{10}{2}, \frac{-2}{2}\right) = (5, -1)$$

45. The coordinates of the midpoint are:

$$(x, y) = \left(\frac{x_1 + x_2}{2}, \frac{y_1 + y_2}{2}\right) = \left(\frac{-0.2 + 2.3}{2}, \frac{0.3 + 1.1}{2}\right) = \left(\frac{2.1}{2}, \frac{1.4}{2}\right) = (1.05, 0.7)$$

47. The coordinates of the midpoint are:

$$(x, y) = \left(\frac{x_1 + x_2}{2}, \frac{y_1 + y_2}{2}\right) = \left(\frac{a + 0}{2}, \frac{b + 0}{2}\right) = \left(\frac{a}{2}, \frac{b}{2}\right)$$

49. The midpoint of AB is: $D = \left(\frac{0 + 0}{2}, \frac{0 + 6}{2}\right) = (0, 3)$

The midpoint of AC is: $E = \left(\frac{0 + 4}{2}, \frac{0 + 4}{2}\right) = (2, 2)$

The midpoint of BC is: $F = \left(\frac{0 + 4}{2}, \frac{6 + 4}{2}\right) = (2, 5)$

$$d(C, D) = \sqrt{(0 - 4)^2 + (3 - 4)^2} = \sqrt{(-4)^2 + (-1)^2} = \sqrt{16 + 1} = \sqrt{17}$$

$$d(B, E) = \sqrt{(2 - 0)^2 + (2 - 6)^2} = \sqrt{2^2 + (-4)^2} = \sqrt{4 + 16} = \sqrt{20} = 2\sqrt{5}$$

$$d(A, F) = \sqrt{(2 - 0)^2 + (5 - 0)^2} = \sqrt{2^2 + 5^2} = \sqrt{4 + 25} = \sqrt{29}$$

51. $d(P_1, P_2) = \sqrt{(-4 - 2)^2 + (1 - 1)^2} = \sqrt{(-6)^2 + 0^2} = \sqrt{36} = 6$

$d(P_2, P_3) = \sqrt{(-4 - (-4))^2 + (-3 - 1)^2} = \sqrt{0^2 + (-4)^2} = \sqrt{16} = 4$

$d(P_1, P_3) = \sqrt{(-4 - 2)^2 + (-3 - 1)^2} = \sqrt{(-6)^2 + (-4)^2} = \sqrt{36 + 16} = \sqrt{52} = 2\sqrt{13}$

Since $[d(P_1, P_2)]^2 + [d(P_2, P_3)]^2 = [d(P_1, P_3)]^2$, the triangle is a right triangle.

53. $d(P_1,P_2) = \sqrt{(0-(-2))^2 + (7-(-1))^2} = \sqrt{2^2 + 8^2} = \sqrt{4+64} = \sqrt{68} = 2\sqrt{17}$

$d(P_2,P_3) = \sqrt{(3-0)^2 + (2-7)^2} = \sqrt{3^2 + (-5)^2} = \sqrt{9+25} = \sqrt{34}$

$d(P_1,P_3) = \sqrt{(3-(-2))^2 + (2-(-1))^2} = \sqrt{5^2 + 3^2} = \sqrt{25+9} = \sqrt{34}$

Since $d(P_2,P_3) = d(P_1,P_3)$, the triangle is isosceles.

Since $[d(P_1,P_3)]^2 + [d(P_2,P_3)]^2 = [d(P_1,P_2)]^2$, the triangle is also a right triangle.

Therefore, the triangle is an isosceles right triangle.

55. $P_1 = (1,3); \ P_2 = (5,15)$

$\begin{aligned} d(P_1,P_2) &= \sqrt{(5-1)^2 + (15-3)^2} \\ &= \sqrt{(4)^2 + (12)^2} \\ &= \sqrt{16+144} \\ &= \sqrt{160} = 4\sqrt{10} \end{aligned}$

57. $P_1 = (-4,6); \ P_2 = (4,-8)$

$\begin{aligned} d(P_1,P_2) &= \sqrt{(4-(-4))^2 + (-8-6)^2} \\ &= \sqrt{(8)^2 + (-14)^2} \\ &= \sqrt{64+196} \\ &= \sqrt{260} = 2\sqrt{65} \end{aligned}$

59. Plot the vertices of the square at $(0, 0)$, $(0, s)$, (s, s), and $(s, 0)$.

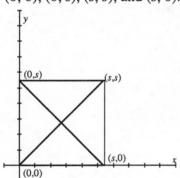

Find the midpoints of the diagonals.

$M_1 = \left(\dfrac{0+s}{2}, \dfrac{0+s}{2}\right) = \left(\dfrac{s}{2}, \dfrac{s}{2}\right)$

$M_2 = \left(\dfrac{0+s}{2}, \dfrac{s+0}{2}\right) = \left(\dfrac{s}{2}, \dfrac{s}{2}\right)$

Since the coordinates of the midpoints are the same, the diagonals of a square intersect at their midpoints.

61. Using the Pythagorean Theorem:

$90^2 + 90^2 = d^2$

$8100 + 8100 = d^2$

$16,200 = d^2$

$d = \sqrt{16,200} = 90\sqrt{2} \approx 127.28 \text{ feet}$

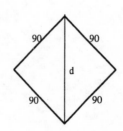

63. (a)   First: (90, 0), Second: (90, 90)
           Third: (0, 90)

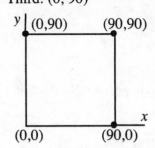

   (b) Using the distance formula:
        $$d = \sqrt{(310-90)^2 + (15-90)^2}$$
        $$= \sqrt{220^2 + (-75)^2}$$
        $$= \sqrt{54,025} \approx 232.4 \text{ feet}$$

   (c) Using the distance formula:
        $$d = \sqrt{(300-0)^2 + (300-90)^2}$$
        $$= \sqrt{300^2 + 210^2}$$
        $$= \sqrt{134,100} \approx 366.2 \text{ feet}$$

65. The Intrepid heading east moves a distance $30t$ after t
     hours. The truck heading south moves a distance $40t$
     after t hours. Their distance apart after t hours is:
     $$d = \sqrt{(30t)^2 + (40t)^2}$$
     $$= \sqrt{900t^2 + 1600t^2}$$
     $$= \sqrt{2500t^2}$$
     $$= 50t$$

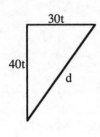

Chapter 1

Graphs

1.2 Graphs of Equations; Circles

1. add, 4

3. intercepts

5. (3, –4)

7. True

9.
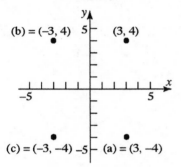

11.
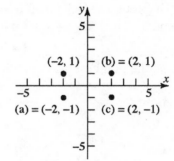

13.
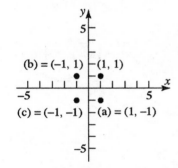

15.
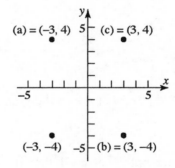

17.
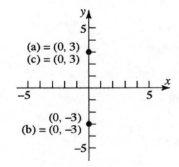

19. (a) (–1, 0), (1, 0)

(b) symmetric with respect to the *x*-axis, *y*-axis, and origin

21. (a) $\left(-\dfrac{\pi}{2},0\right),\left(\dfrac{\pi}{2},0\right),(0,1)$ (b) symmetric with respect to the y-axis

23. (a) $(0,0)$ (b) symmetric with respect to the x-axis

25. (a) $(1,0)$ (b) not symmetric with respect to x-axis, y-axis, or origin

27. (a) $(-1,0),(1,0),(0,-1)$ (b) symmetric with respect to the y-axis

29. (a) none   (b) symmetric with respect to the origin

31. $y = x^4 - \sqrt{x}$

$0 = 0^4 - \sqrt{0}$ $1 = 1^4 - \sqrt{1}$ $0 = (-1)^4 - \sqrt{-1}$

$0 = 0$ $1 \neq 0$ $0 \neq 1 - \sqrt{-1}$

$(0, 0)$ is on the graph of the equation.

33. $y^2 = x^2 + 9$

$3^2 = 0^2 + 9$ $0^2 = 3^2 + 9$ $0^2 = (-3)^2 + 9$

$9 = 9$ $0 \neq 18$ $0 \neq 18$

$(0, 3)$ is on the graph of the equation.

35. $x^2 + y^2 = 4$

$0^2 + 2^2 = 4$ $(-2)^2 + 2^2 = 4$ $\left(\sqrt{2}\right)^2 + \left(\sqrt{2}\right)^2 = 4$

$4 = 4$ $8 \neq 4$ $4 = 4$

$(0, 2)$ and $\left(\sqrt{2}, \sqrt{2}\right)$ are on the graph of the equation.

37. $x^2 = y$

y - intercept : Let $x = 0$, then $0^2 = y \Rightarrow y = 0$ $(0,0)$

x - intercept : Let $y = 0$, then $x^2 = 0 \Rightarrow x = 0$ $(0,0)$

Test for symmetry:

x - axis : Replace y by $-y$: $x^2 = -y$, which is not equivalent to $x^2 = y$.

y - axis : Replace x by $-x$: $(-x)^2 = y$ or $x^2 = y$, which is equivalent to $x^2 = y$.

Origin : Replace x by $-x$ and y by $-y$: $(-x)^2 = -y$ or $x^2 = -y$,

which is not equivalent to $x^2 = y$.

Therefore, the graph is symmetric with respect to the y - axis .

39. $y = 3x$

y - intercept : Let $x = 0$, then $y = 3 \cdot 0 = 0$ $(0,0)$

x - intercept : Let $y = 0$, then $3x = 0 \Rightarrow x = 0$ $(0,0)$

Test for symmetry:

x - axis: Replace y by $-y$: $-y = 3x$, which is not equivalent to $y = 3x$.

y - axis: Replace x by $-x$: $y = 3(-x)$ or $y = -3x$,

which is not equivalent to $y = 3x$.

Origin: Replace x by $-x$ and y by $-y$: $-y = 3(-x)$ or $y = 3x$,

which is equivalent to $y = 3x$.

Therefore, the graph is symmetric with respect to the origin.

41. $x^2 + y - 9 = 0$

y - intercept: Let $x = 0$, then $0 + y - 9 = 0 \Rightarrow y = 9$ $(0,9)$

x - intercept: Let $y = 0$, then $x^2 - 9 = 0 \Rightarrow x = \pm 3$ $(-3,0),(3,0)$

Test for symmetry:

x - axis: Replace y by $-y$: $x^2 + (-y) - 9 = 0$ or $x^2 - y - 9 = 0$,

which is not equivalent to $x^2 + y - 9 = 0$.

y - axis: Replace x by $-x$: $(-x)^2 + y - 9 = 0$ or $x^2 + y - 9 = 0$,

which is equivalent to $x^2 + y - 9 = 0$.

Origin: Replace x by $-x$ and y by $-y$: $(-x)^2 + (-y) - 9 = 0$ or $x^2 - y - 9 = 0$,

which is not equivalent to $x^2 + y - 9 = 0$.

Therefore, the graph is symmetric with respect to the y-axis.

43. $9x^2 + 4y^2 = 36$

y - intercept: Let $x = 0$, then $4y^2 = 36 \Rightarrow y^2 = 9 \Rightarrow y = \pm 3$ $(0,-3),(0,3)$

x - intercept: Let $y = 0$, then $9x^2 = 36 \Rightarrow x^2 = 4 \Rightarrow x = \pm 2$ $(-2,0),(2,0)$

Test for symmetry:

x - axis: Replace y by $-y$: $9x^2 + 4(-y)^2 = 36$ or $9x^2 + 4y^2 = 36$,

which is equivalent to $9x^2 + 4y^2 = 36$.

y - axis: Replace x by $-x$: $9(-x)^2 + 4y^2 = 36$ or $9x^2 + 4y^2 = 36$,

which is equivalent to $9x^2 + 4y^2 = 36$.

Origin: Replace x by $-x$ and y by $-y$: $9(-x)^2 + 4(-y)^2 = 36$ or $9x^2 + 4y^2 = 36$,

which is equivalent to $9x^2 + 4y^2 = 36$.

Therefore, the graph is symmetric with respect to the x-axis, the y-axis, and the origin.

45. $y = x^3 - 27$

y - intercept: Let $x = 0$, then $y = 0^3 - 27 \Rightarrow y = -27$ $(0,-27)$

x - intercept: Let $y = 0$, then $0 = x^3 - 27 \Rightarrow x^3 = 27 \Rightarrow x = 3$ $(3,0)$

Test for symmetry:

x - axis: Replace y by $-y$: $-y = x^3 - 27$, which is not equivalent to $y = x^3 - 27$.

y - axis: Replace x by $-x$: $y = (-x)^3 - 27$ or $y = -x^3 - 27$,

which is not equivalent to $y = x^3 - 27$.

Origin: Replace x by $-x$ and y by $-y$: $-y = (-x)^3 - 27$ or

$y = x^3 + 27$, which is not equivalent to $y = x^3 - 27$.

Therefore, the graph is not symmetric with respect to the x-axis, the y-axis, or the origin.

47. $y = x^2 - 3x - 4$

y - intercept : Let $x = 0$, then $y = 0^2 - 3(0) - 4 \Rightarrow y = -4$ $(0, -4)$

x - intercept : Let $y = 0$, then $0 = x^2 - 3x - 4 \Rightarrow (x - 4)(x + 1) = 0 \Rightarrow x = 4, x = -1$ $(4, 0), (-1, 0)$

Test for symmetry:

x - axis: Replace y by $-y$: $-y = x^2 - 3x - 4$, which is not

equivalent to $y = x^2 - 3x - 4$.

y - axis: Replace x by $-x$: $y = (-x)^2 - 3(-x) - 4$ or $y = x^2 + 3x - 4$,

which is not equivalent to $y = x^2 - 3x - 4$.

Origin: Replace x by $-x$ and y by $-y$: $-y = (-x)^2 - 3(-x) - 4$ or

$y = -x^2 - 3x + 4$, which is not equivalent to $y = x^2 - 3x - 4$.

Therefore, the graph is not symmetric with respect to the x-axis, the y-axis, or the origin.

49. $y = \dfrac{3x}{x^2 + 9}$

y - intercept : Let $x = 0$, then $y = \dfrac{0}{0 + 9} = 0$ $(0, 0)$

x - intercept : Let $y = 0$, then $0 = \dfrac{3x}{x^2 + 9} \Rightarrow 3x = 0 \Rightarrow x = 0$ $(0, 0)$

Test for symmetry:

x - axis: Replace y by $-y$: $-y = \dfrac{3x}{x^2 + 9}$, which is not

equivalent to $y = \dfrac{3x}{x^2 + 9}$.

y - axis: Replace x by $-x$: $y = \dfrac{3(-x)}{(-x)^2 + 9}$ or $y = \dfrac{-3x}{x^2 + 9}$,

which is not equivalent to $y = \dfrac{3x}{x^2 + 9}$.

Origin: Replace x by $-x$ and y by $-y$: $-y = \dfrac{-3x}{(-x)^2 + 9}$ or

$y = \dfrac{3x}{x^2 + 9}$, which is equivalent to $y = \dfrac{3x}{x^2 + 9}$.

Therefore, the graph is symmetric with respect to the origin.

51. $y = \dfrac{-x^3}{x^2 - 9}$

y-intercept: Let $x = 0$, then $y = \dfrac{0}{-9} = 0$ $(0,0)$

x-intercept: Let $y = 0$, then $0 = \dfrac{-x^3}{x^2 - 9} \Rightarrow -x^3 = 0 \Rightarrow x = 0$ $(0,0)$

Test for symmetry:

x-axis: Replace y by $-y$: $-y = \dfrac{-x^3}{x^2 - 9}$, which is not equivalent to $y = \dfrac{-x^3}{x^2 - 9}$.

y-axis: Replace x by $-x$: $y = \dfrac{-(-x)^3}{(-x)^2 - 9}$ or $y = \dfrac{x^3}{x^2 - 9}$,

which is not equivalent to $y = \dfrac{-x^3}{x^2 - 9}$.

Origin: Replace x by $-x$ and y by $-y$: $-y = \dfrac{-(-x)^3}{(-x)^2 - 9}$ or

$-y = \dfrac{x^3}{x^2 - 9}$, which is equivalent to $y = \dfrac{-x^3}{x^2 - 9}$.

Therefore, the graph is symmetric with respect to the origin.

53. $y = x^3$

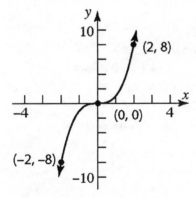

55. $y = \sqrt{x}$

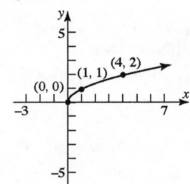

57. $y = 3x + 5$

$2 = 3a + 5 \Rightarrow 3a = -3 \Rightarrow a = -1$

59. $2x + 3y = 6$

$2a + 3b = 6 \Rightarrow b = 2 - \dfrac{2}{3}a$

61. Center = (2, 1)
     Radius = distance from (0,1) to (2,1)
$$= \sqrt{(2-0)^2 + (1-1)^2}$$
$$= \sqrt{4} = 2$$

$$(x-2)^2 + (y-1)^2 = 4$$

63. Center = midpoint of (1,2) and (4,2)
$$= \left(\frac{1+4}{2}, \frac{2+2}{2}\right) = \left(\frac{5}{2}, 2\right)$$
Radius = distance from $\left(\frac{5}{2}, 2\right)$ to (4,2)
$$= \sqrt{\left(4 - \frac{5}{2}\right)^2 + (2-2)^2}$$
$$= \sqrt{\frac{9}{4}} = \frac{3}{2}$$

$$\left(x - \frac{5}{2}\right)^2 + (y-2)^2 = \frac{9}{4}$$

65. $(x-h)^2 + (y-k)^2 = r^2$
     $(x-0)^2 + (y-0)^2 = 2^2$
     $x^2 + y^2 = 4$
     General form: $x^2 + y^2 - 4 = 0$

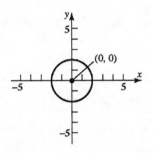

67. $(x-h)^2 + (y-k)^2 = r^2$
     $(x-1)^2 + (y-(-1))^2 = 1^2$
     $(x-1)^2 + (y+1)^2 = 1$
     General form:
     $x^2 - 2x + 1 + y^2 + 2y + 1 = 1$
     $x^2 + y^2 - 2x + 2y + 1 = 0$

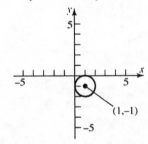

69. $(x-h)^2 + (y-k)^2 = r^2$
     $(x-0)^2 + (y-2)^2 = 2^2$
     $x^2 + (y-2)^2 = 4$
     General form:
     $x^2 + y^2 - 4y + 4 = 4$
     $x^2 + y^2 - 4y = 0$

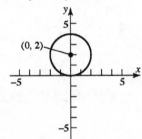

71. $(x - h)^2 + (y - k)^2 = r^2$
       $(x - 4)^2 + (y - (-3))^2 = 5^2$
          $(x - 4)^2 + (y + 3)^2 = 25$
   General form:
   $x^2 - 8x + 16 + y^2 + 6y + 9 = 25$
          $x^2 + y^2 - 8x + 6y = 0$

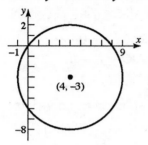

73. $(x - h)^2 + (y - k)^2 = r^2$
       $(x - (-3))^2 + (y - (-6))^2 = 6^2$
            $(x + 3)^2 + (y + 6)^2 = 36$
   General form:
   $x^2 + 6x + 9 + y^2 + 12y + 36 = 36$
         $x^2 + y^2 + 6x + 12y + 9 = 0$

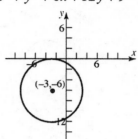

75. $(x - h)^2 + (y - k)^2 = r^2$
       $(x - 0)^2 + (y - (-3))^2 = 3^2$
            $x^2 + (y + 3)^2 = 9$
   General form:
   $x^2 + y^2 + 6y + 9 = 9$
      $x^2 + y^2 + 6y = 0$

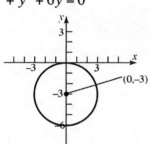

77. $x^2 + y^2 = 4$
      $x^2 + y^2 = 2^2$
   (a) Center : $(0,0)$; Radius $= 2$
   (b)

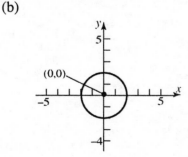

(c) x-intercepts: $y = 0$
              $x^2 + 0 = 4$
                 $x = \pm 2$
                 $(-2,0),(2,0)$
   y-intercepts: $x = 0$
              $0 + y^2 = 4$
                 $y = \pm 2$
                 $(0,-2),(0,2)$

79. $2(x-3)^2 + 2y^2 = 8$

$(x-3)^2 + y^2 = 4$

$(x-3)^2 + y^2 = 2^2$

(a) Center: $(3,0)$; Radius $= 2$

(b)

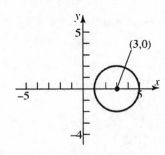

(c) x-intercepts: $y = 0$

$(x-3)^2 + 0 = 4$

$(x-3)^2 = 4$

$x - 3 = \pm 2$

$x = 5, x = 1$

$(1,0),(5,0)$

y-intercepts: $x = 0$

$9 + y^2 = 4$

$y^2 = -5$

no solution $\Rightarrow$ no y-intercepts

81. $x^2 + y^2 + 4x - 4y - 1 = 0$

$x^2 + 4x + y^2 - 4y = 1$

$(x^2 + 4x + 4) + (y^2 - 4y + 4) = 1 + 4 + 4$

$(x+2)^2 + (y-2)^2 = 3^2$

(a) Center: $(-2,2)$; Radius $= 3$

(b)

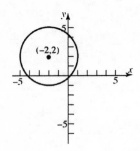

(c) x-intercepts: $y = 0$

$(x+2)^2 + 4 = 9$

$(x+2)^2 = 5$

$x + 2 = \pm\sqrt{5}$

$x = \sqrt{5} - 2, x = -\sqrt{5} - 2$

$\left(-\sqrt{5} - 2, 0\right), \left(\sqrt{5} - 2, 0\right)$

y-intercepts: $x = 0$

$4 + (y-2)^2 = 9$

$(y-2)^2 = 5$

$y - 2 = \pm\sqrt{5}$

$y = \sqrt{5} + 2, y = -\sqrt{5} + 2$

$\left(0, -\sqrt{5} + 2\right), \left(0, \sqrt{5} + 2\right)$

83.
$$x^2 + y^2 - 2x + 4y - 4 = 0$$
$$x^2 - 2x + y^2 + 4y = 4$$
$$(x^2 - 2x + 1) + (y^2 + 4y + 4) = 4 + 1 + 4$$
$$(x - 1)^2 + (y + 2)^2 = 3^2$$

(a) Center: $(1, -2)$; Radius $= 3$

(b)

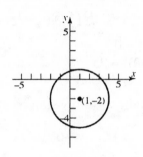

(c) x-intercepts: $y = 0$
$$(x - 1)^2 + 4 = 9$$
$$(x - 1)^2 = 5$$
$$x + 1 = \pm\sqrt{5}$$
$$x = \sqrt{5} - 1, x = -\sqrt{5} - 1$$
$$\left(-\sqrt{5} - 1, 0\right), \left(\sqrt{5} - 1, 0\right)$$

y-intercepts: $x = 0$
$$1 + (y + 2)^2 = 9$$
$$(y + 2)^2 = 8$$
$$y + 2 = \pm\sqrt{8}$$
$$y = \sqrt{8} - 2, y = -\sqrt{8} - 2$$
$$\left(0, -\sqrt{8} - 2\right), \left(0, \sqrt{8} - 2\right)$$

85.
$$x^2 + y^2 - x + 2y + 1 = 0$$
$$x^2 - x + y^2 + 2y = -1$$
$$\left(x^2 - x + \frac{1}{4}\right) + (y^2 + 2y + 1) = -1 + \frac{1}{4} + 1$$
$$\left(x - \frac{1}{2}\right)^2 + (y + 1)^2 = \left(\frac{1}{2}\right)^2$$

(a) Center: $\left(\frac{1}{2}, -1\right)$; Radius $= \frac{1}{2}$

(b)

(c) x-intercepts: $y = 0$
$$\left(x - \frac{1}{2}\right)^2 + 1 = \frac{1}{4}$$
$$\left(x - \frac{1}{2}\right)^2 = -\frac{3}{4}$$

no solution $\Rightarrow$ no x - intercepts

y-intercepts: $x = 0$
$$\frac{1}{4} + (y + 1)^2 = \frac{1}{4}$$
$$(y + 1)^2 = 0$$
$$y + 1 = 0$$
$$y = -1$$
$$(0, -1)$$

87. $2x^2 + 2y^2 - 12x + 8y - 24 = 0$
$$x^2 + y^2 - 6x + 4y = 12$$
$$x^2 - 6x + y^2 + 4y = 12$$
$$(x^2 - 6x + 9) + (y^2 + 4y + 4) = 12 + 9 + 4$$
$$(x - 3)^2 + (y + 2)^2 = 5^2$$
(a) Center: $(3, -2)$; Radius $= 5$
(b)

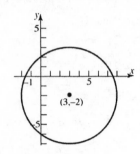

(c) x-intercepts: $y = 0$
$$(x - 3)^2 + 4 = 25$$
$$(x - 3)^2 = 21$$
$$x - 3 = \pm\sqrt{21}$$
$$x = \sqrt{21} + 3, x = -\sqrt{21} + 3$$
$$\left(-\sqrt{21} + 3, 0\right), \left(\sqrt{21} + 3, 0\right)$$

y-intercepts: $x = 0$
$$9 + (y + 2)^2 = 25$$
$$(y + 2)^2 = 16$$
$$y + 2 = \pm 4$$
$$y = 2, y = -6$$
$$(0, -6), (0, 2)$$

89.   Center at $(0,0)$; containing point $(-3, 2)$.
$$r = \sqrt{(-3 - 0)^2 + (2 - 0)^2} = \sqrt{9 + 4} = \sqrt{13}$$
Equation:
$$(x - 0)^2 + (y - 0)^2 = \left(\sqrt{13}\right)^2$$
$$x^2 + y^2 = 13$$

91.   Center at $(2,3)$; tangent to the x-axis.
$r = 3$
Equation:
$$(x - 2)^2 + (y - 3)^2 = 3^2$$
$$x^2 - 4x + 4 + y^2 - 6y + 9 = 9$$
$$x^2 + y^2 - 4x - 6y + 4 = 0$$

93.   Endpoints of a diameter are $(1,4)$ and $(-3,2)$.
The center is at the midpoint of that diameter:
Center: $\left(\dfrac{1 + (-3)}{2}, \dfrac{4 + 2}{2}\right) = (-1, 3)$
Radius: $r = \sqrt{(1 - (-1))^2 + (4 - 3)^2} = \sqrt{4 + 1} = \sqrt{5}$
Equation: $\left(x - (-1)\right)^2 + (y - 3)^2 = \left(\sqrt{5}\right)^2$
$$x^2 + 2x + 1 + y^2 - 6y + 9 = 5$$
$$x^2 + y^2 + 2x - 6y + 5 = 0$$

95.   (c)

97.   (b)

99.   (b), (c), (e) and (g)

101. $x^2 + y^2 + 2x + 4y - 4091 = 0$
$$x^2 + 2x + y^2 + 4y - 4091 = 0$$
$$x^2 + 2x + 1 + y^2 + 4y + 4 = 4091 + 5$$
$$(x + 1)^2 + (y + 2)^2 = 4096$$
The circle representing Earth has center $(-1, -2)$ and radius $= \sqrt{4096} = 64$
So the radius of the satellite's orbit is $64 + 0.6 = 64.6$ units.

The equation of the orbit is

$$(x+1)^2 + (y+2)^2 = (64.6)^2$$

$$x^2 + y^2 + 2x + 4y - 4168.16 = 0$$

103–105.   Answers will vary

Graphs

1.3 Lines

1. undefined, 0

3. $y = b$, y-intercept

5. False

7. $m_1 = m_2$, y-intercepts, $m_1 \cdot m_2 = -1$

9. $-\dfrac{1}{2}$

11. (a) Slope $= \dfrac{1-0}{2-0} = \dfrac{1}{2}$

    (b) If x increases by 2 units, y will increase by 1 unit.

13. (a) Slope $= \dfrac{1-2}{1-(-2)} = -\dfrac{1}{3}$

    (b) If x increases by 3 units, y will decrease by 1 unit.

15. (x_1, y_1) (x_2, y_2)
    $(2,3)$ $(4,0)$

    Slope $= \dfrac{y_2 - y_1}{x_2 - x_1} = \dfrac{0-3}{4-2} = -\dfrac{3}{2}$

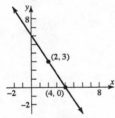

17. (x_1, y_1) (x_2, y_2)
    $(-2,3)$ $(2,1)$

    Slope $= \dfrac{y_2 - y_1}{x_2 - x_1} = \dfrac{1-3}{2-(-2)} = \dfrac{-2}{4} = -\dfrac{1}{2}$

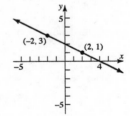

19. (x_1, y_1) (x_2, y_2)
    $(-3,-1)$ $(2,-1)$

    Slope $= \dfrac{y_2 - y_1}{x_2 - x_1} = \dfrac{-1-(-1)}{2-(-3)} = \dfrac{0}{5} = 0$

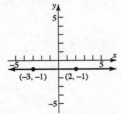

21. (x_1, y_1) (x_2, y_2)
    $(-1, 2)$ $(-1, -2)$

    Slope $= \dfrac{y_2 - y_1}{x_2 - x_1} = \dfrac{-2 - 2}{-1 - (-1)} = \dfrac{-4}{0}$

    Slope is undefined.

    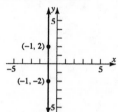

23.

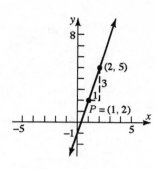

25.

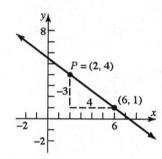

27.

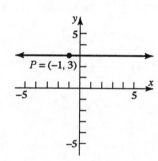

29.

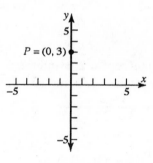

31. $(0,0)$ and $(2,1)$ are points on the line.

    Slope $= \dfrac{1 - 0}{2 - 0} = \dfrac{1}{2}$

    y-intercept is 0; using $y = mx + b$:

    $y = \dfrac{1}{2}x + 0$

    $2y = x$

    $0 = x - 2y$

    $x - 2y = 0$ or $y = \dfrac{1}{2}x$

33. $(-2, 2)$ and $(1, 1)$ are points on the line.

    Slope $= \dfrac{1 - 2}{1 - (-2)} = \dfrac{-1}{3} = -\dfrac{1}{3}$

    Using $y - y_1 = m(x - x_1)$

    $y - 1 = -\dfrac{1}{3}(x - 1)$

    $y - 1 = -\dfrac{1}{3}x + \dfrac{1}{3}$

    $y = -\dfrac{1}{3}x + \dfrac{4}{3}$

    $x + 3y = 4$ or $y = -\dfrac{1}{3}x + \dfrac{4}{3}$

35. $y - y_1 = m(x - x_1)$, $m = 2$
    $y - 3 = 2(x - 3)$
    $y - 3 = 2x - 6$
    $y = 2x - 3$
    $2x - y = 3$ or $y = 2x - 3$

37. $y - y_1 = m(x - x_1), \ m = -\dfrac{1}{2}$

$$y - 2 = -\frac{1}{2}(x - 1)$$

$$y - 2 = -\frac{1}{2}x + \frac{1}{2}$$

$$y = -\frac{1}{2}x + \frac{5}{2}$$

$$x + 2y = 5 \ \text{ or } \ y = -\frac{1}{2}x + \frac{5}{2}$$

39. Slope = 3; containing (–2,3)

$$y - y_1 = m(x - x_1)$$

$$y - 3 = 3(x - (-2))$$

$$y - 3 = 3x + 6$$

$$y = 3x + 9$$

$$3x - y = -9 \ \text{ or } \ y = 3x + 9$$

41. Slope = $-\dfrac{2}{3}$; containing (1,–1)

$$y - y_1 = m(x - x_1)$$

$$y - (-1) = -\frac{2}{3}(x - 1)$$

$$y + 1 = -\frac{2}{3}x + \frac{2}{3}$$

$$y = -\frac{2}{3}x - \frac{1}{3}$$

$$2x + 3y = -1 \ \text{ or } \ y = -\frac{2}{3}x - \frac{1}{3}$$

43. Containing (1,3) and (–1,2)

$$m = \frac{2 - 3}{-1 - 1} = \frac{-1}{-2} = \frac{1}{2}$$

$$y - y_1 = m(x - x_1)$$

$$y - 3 = \frac{1}{2}(x - 1)$$

$$y - 3 = \frac{1}{2}x - \frac{1}{2}$$

$$y = \frac{1}{2}x + \frac{5}{2}$$

$$x - 2y = -5 \ \text{ or } \ y = \frac{1}{2}x + \frac{5}{2}$$

45. Slope = –3; y-intercept =3

$$y = mx + b$$

$$y = -3x + 3$$

$$3x + y = 3 \ \text{ or } \ y = -3x + 3$$

47. x-intercept = 2; y-intercept = –1

Points are (2,0) and (0,–1)

$$m = \frac{-1 - 0}{0 - 2} = \frac{-1}{-2} = \frac{1}{2}$$

$$y = mx + b$$

$$y = \frac{1}{2}x - 1$$

$$x - 2y = 2 \ \text{ or } \ y = \frac{1}{2}x - 1$$

49. Slope undefined; passing through (2,4)
This is a vertical line.

$x = 2$
No slope-intercept form.

51. Horizontal; containing the point (–3,2)
slope = 0
$y = 2$

53. Parallel to $y = 2x$; Slope = 2
Containing (–1,2)

$$y - y_1 = m(x - x_1)$$

$$y - 2 = 2(x - (-1))$$

$$y - 2 = 2x + 2$$

$$y = 2x + 4$$

$$2x - y = -4 \ \text{ or } \ y = 2x + 4$$

55. Parallel to $2x - y = -2$; Slope = 2
Containing (0,0)

$$y - y_1 = m(x - x_1)$$

$$y - 0 = 2(x - 0)$$

$$y = 2x$$

$$2x - y = 0 \ \text{ or } \ y = 2x$$

57. Parallel to $x = 5$
Containing $(4,2)$
This is a vertical line.
$x = 4$

No slope - intercept form.

59. Perpendicular to $y = \dfrac{1}{2}x + 4$
Slope of perpendicular $= -2$
Containing $(1,-2)$
$$y - y_1 = m(x - x_1)$$
$$y - (-2) = -2(x - 1)$$
$$y + 2 = -2x + 2$$
$$y = -2x$$
$$2x + y = 0 \ \text{ or } \ y = -2x$$

61. Perpendicular to $2x + y = 2$
$2x + y = 2 \Rightarrow y = -2x + 2$

Slope of perpendicular $= \dfrac{1}{2}$

Containing $(-3,0)$

$$y - y_1 = m(x - x_1)$$
$$y - 0 = \frac{1}{2}(x - (-3)) \Rightarrow y = \frac{1}{2}x + \frac{3}{2}$$
$$x - 2y = -3 \ \text{ or } \ y = \frac{1}{2}x + \frac{3}{2}$$

63. Perpendicular to $x = 8$
Slope of perpendicular $= 0$
$\Rightarrow$ horizontal line
Containing the point $(3,4)$
$y = 4$

65. $y = 2x + 3$
Slope $= 2$
y-intercept $= 3$

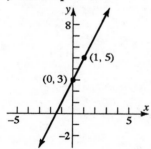

67. $\dfrac{1}{2}y = x - 1$
$y = 2x - 2$
Slope $= 2$
y-intercept $= -2$

69. $y = \dfrac{1}{2}x + 2$

Slope $= \dfrac{1}{2}$

y-intercept $= 2$

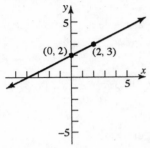

71. $x + 2y = 4$

$2y = -x + 4 \Rightarrow y = -\dfrac{1}{2}x + 2$

Slope $= -\dfrac{1}{2}$

y-intercept $= 2$

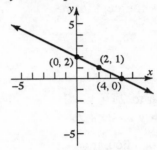

73. $2x - 3y = 6$

$-3y = -2x + 6 \Rightarrow y = \dfrac{2}{3}x - 2$

Slope $= \dfrac{2}{3}$

y-intercept $= -2$

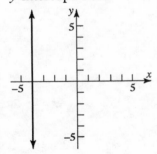

75. $x + y = 1$

$y = -x + 1$

Slope $= -1$

y-intercept $= 1$

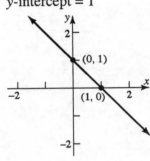

77. $x = -4$

Slope is undefined

y-intercept: none

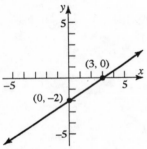

79. $y = 5$

Slope $= 0$

y-intercept $= 5$

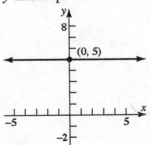

81. $y - x = 0$
$y = x$
Slope = 1
y-intercept = 0

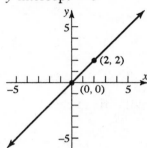

83. $2y - 3x = 0$
$2y = 3x \Rightarrow y = \dfrac{3}{2}x$

Slope $= \dfrac{3}{2}$; y-intercept = 0

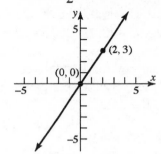

85. $y = 2x - 3 \Rightarrow$ slope = 2
$y = 2x + 4 \Rightarrow$ slope = 2
The lines are parallel.

87. $y = 4x + 5 \Rightarrow$ slope = 4
$y = -4x + 2 \Rightarrow$ slope = -4
The lines are neither parallel nor perpendicular.

89. Consider the points $A(-2,5)$, $B(1,3)$ and $C(-1,0)$

slope of $\overline{AB} = \dfrac{3-5}{1-(-2)} = -\dfrac{2}{3}$; slope of $\overline{AC} = \dfrac{0-5}{-1-(-2)} = -5$; slope of $\overline{BC} = \dfrac{0-3}{-1-1} = \dfrac{3}{2}$

Therefore, $\triangle ABC$ has a right angle at vertex B since

slope $\overline{AB} = -\dfrac{2}{3}$ and slope $\overline{BC} = \dfrac{3}{2} \Rightarrow \overline{AB}$ is perpendicular to $\overline{BC}$.

91. Consider the points $A(-1,0)$, $B(2,3)$, $C(1,-2)$ and $D(4,1)$

slope of $\overline{AB} = \dfrac{3-0}{2-(-1)} = 1$; slope of $\overline{CD} = \dfrac{1-(-2)}{4-1} = 1$

slope of $\overline{AC} = \dfrac{-2-0}{1-(-1)} = -1$; slope of $\overline{BD} = \dfrac{1-3}{4-2} = -1$

Therefore, the quadrilateral $ACDB$ is a parallelogram since
slope $\overline{AB} = 1$ and slope $\overline{CD} = 1 \Rightarrow \overline{AB}$ is parallel to $\overline{CD}$

slope $\overline{AC} = -1$ and slope $\overline{BD} = -1 \Rightarrow \overline{AC}$ is parallel to $\overline{BD}$
Furthermore,
slope $\overline{AB} = 1$ and slope $\overline{BD} = -1 \Rightarrow \overline{AB}$ is perpendicular to $\overline{BD}$

slope $\overline{AC} = -1$ and slope $\overline{CD} = 1 \Rightarrow \overline{AC}$ is perpendicular to $\overline{CD}$
So the quadrilateral $ACDB$ is a rectangle.

93. Slope = 1; y-intercept = 2
$y = x + 2$ or $x - y = -2$

95. Slope $= -\dfrac{1}{3}$; y-intercept = 1

$y = -\dfrac{1}{3}x + 1$ or $x + 3y = 3$

97. The equation of the x-axis is $y = 0$. (The slope is 0 and the y-intercept is 0.)

99. Let x = number of miles driven, and let C = cost in dollars.
Total cost = (cost per mile)(number of miles) + fixed cost
$C = 0.07x + 29$
When $x = 110$, $C = (0.07)(110) + 29 = \$36.70$.
When $x = 230$, $C = (0.07)(230) + 29 = \$45.10$.

101. Let x = number newspapers delivered, and let C = cost in dollars.
Total cost = (delivery cost per paper)(number of papers delivered) + fixed cost
$C = 0.53x + 1{,}070{,}000$

103. (a) $C = 0.08275x + 7.58$

(b)
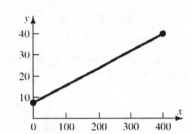

(c) For 100 kWh,
$C = 0.08275(100) + 7.58 = \$15.86$

(d) For 300 kWh,
$C = 0.08725(300) + 7.58 = \$32.41$

(e) For each usage increase of 1 kWh, the monthly charge increases by 8.27 cents.

105. $(°C, °F) = (0, 32);\quad (°C, °F) = (100, 212)$

$\text{slope} = \dfrac{212 - 32}{100 - 0} = \dfrac{180}{100} = \dfrac{9}{5}$

$°F - 32 = \dfrac{9}{5}(°C - 0)$

$°F - 32 = \dfrac{9}{5}(°C)$

$°C = \dfrac{5}{9}(°F - 32)$

If $°F = 70$, then

$°C = \dfrac{5}{9}(70 - 32) = \dfrac{5}{9}(38)$

$°C \approx 21°$

107. (a) Let x = number of boxes sold, A = money, in dollars, spent on advertising.
We have the points $(x_1, A_1) = (100{,}000, 40{,}000); (x_2, A_2) = (200{,}000, 60{,}000)$

$\text{slope} = \dfrac{60{,}000 - 40{,}000}{200{,}000 - 100{,}000} = \dfrac{20{,}000}{100{,}000} = \dfrac{1}{5}$

$A - 40{,}000 = \dfrac{1}{5}(x - 100{,}000)$

$A - 40{,}000 = \dfrac{1}{5}x - 20{,}000$

$A = \dfrac{1}{5}x + 20{,}000$

(b) If $x = 300{,}000$, then $A = \dfrac{1}{5}(300{,}000) + 20{,}000 = \$80{,}000$

(c) To increase the number of boxes sold by 5 units, the amount spent on advertising needs to increase by 1 dollar.

109. (b), (c), (e) and (g)   111.   (c)

113. (a)   Since the tangent line intersects the circle in a single point, there is exactly one point on the line $y = mx + b$ such that $x^2 + y^2 = r^2$. That is, $x^2 + (mx + b)^2 = r^2$ has exactly one solution.

$$x^2 + (mx + b)^2 = r^2$$
$$x^2 + m^2 x^2 + 2bmx + b^2 = r^2$$
$$(1 + m^2)x^2 + 2bmx + b^2 - r^2 = 0$$

There is one solution if and only if the discriminant is zero.

$$(2bm)^2 - 4(1 + m^2)(b^2 - r^2) = 0$$
$$4b^2 m^2 - 4b^2 + 4r^2 - 4b^2 m^2 + 4m^2 r^2 = 0$$
$$-4b^2 + 4r^2 + 4m^2 r^2 = 0$$
$$-b^2 + r^2 + m^2 r^2 = 0$$
$$r^2(1 + m^2) = b^2$$

(b)   Use the quadratic formula, knowing that the discriminant is zero:

$$x = \frac{-2bm}{2(1 + m^2)} = \frac{-bm}{\dfrac{b^2}{r^2}} = \frac{-bmr^2}{b^2} = \frac{-mr^2}{b}$$

$$y = m\left(\frac{-mr^2}{b}\right) + b = \frac{-m^2 r^2}{b} + b = \frac{-m^2 r^2 + b^2}{b} = \frac{r^2}{b}$$

(c)   The slope of the tangent line is m.
The slope of the line joining the point of tangency and the center is:

$$\frac{\dfrac{r^2}{b} - 0}{\dfrac{-mr^2}{b} - 0} = \frac{r^2}{b} \cdot \frac{b}{-mr^2} = -\frac{1}{m}$$

115. $$x^2 + y^2 - 4x + 6y + 4 = 0$$
$$(x^2 - 4x + 4) + (y^2 + 6y + 9) = -4 + 4 + 9$$
$$(x - 2)^2 + (y + 3)^2 = 9$$

Center: $(2, -3)$

Slope from center to $\left(3, 2\sqrt{2} - 3\right)$ is $\dfrac{2\sqrt{2} - 3 - (-3)}{3 - 2} = \dfrac{2\sqrt{2}}{1} = 2\sqrt{2}$

Slope of the tangent line is: $\dfrac{-1}{2\sqrt{2}} = -\dfrac{\sqrt{2}}{4}$

Equation of the tangent line:

$$y - \left(2\sqrt{2} - 3\right) = -\frac{\sqrt{2}}{4}(x - 3)$$

$$y - 2\sqrt{2} + 3 = -\frac{\sqrt{2}}{4}x + \frac{3\sqrt{2}}{4}$$

$$4y - 8\sqrt{2} + 12 = -\sqrt{2}x + 3\sqrt{2}$$

$$\sqrt{2}x + 4y = 11\sqrt{2} - 12$$

117. Find the centers of the two circles:

$$x^2 + y^2 - 4x + 6y + 4 = 0$$

$$(x^2 - 4x + 4) + (y^2 + 6y + 9) = -4 + 4 + 9$$

$$(x - 2)^2 + (y + 3)^2 = 9 \qquad \text{Center: } (2, -3)$$

$$x^2 + y^2 + 6x + 4y + 9 = 0$$

$$(x^2 + 6x + 9) + (y^2 + 4y + 4) = -9 + 9 + 4$$

$$(x + 3)^2 + (y + 2)^2 = 4 \qquad \text{Center: } (-3, -2)$$

Find the slope of the line containing the centers:

$$m = \frac{-2 - (-3)}{-3 - 2} = -\frac{1}{5}$$

Find the equation of the line containing the centers:

$$y + 3 = -\frac{1}{5}(x - 2)$$

$$5y + 15 = -x + 2$$

$$x + 5y = -13$$

119. $2x - y = C$

Graph the lines:

$$2x - y = 4 \Rightarrow y = 2x + 4$$

$$2x - y = 0 \Rightarrow y = 2x$$

$$2x - y = 2 \Rightarrow y = 2x - 2$$

All the lines have the same slope, 2.
The lines are parallel.

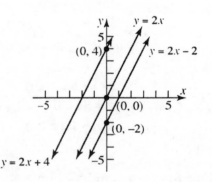

121. Consider the following diagram:

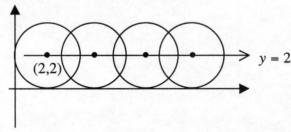

The path of the center of the circle is the line $y = 2$.

123.   Answers will vary.

125.   Not every line has two distinct intercepts since a horizontal line might not touch the x-axis and a vertical line might not touch the y-axis. Also, a non-vertical, non-horizontal line that passes through the origin will have only one intercept, $(0, 0)$.
A line must have at least one intercept since a vertical line always crosses the x-axis, a horizontal line always crosses the y-axis and a non-vertical, non-horizontal line always crosses both axes.

127.   Two lines with the same non-zero x-intercept and the same y-intercept must have the same slope and therefore must be represented by equivalent equations.

129.   Two lines that have the same y-intercept but different slopes can only have the same x-intercept if the y-intercept is zero.
Assume Line 1 has equation $y = m_1 x + b$ and Line 2 has equation $y = m_2 x + b$,

Line 1 has x-intercept $-\dfrac{b}{m_1}$ and y-intercept b.

Line 2 has x-intercept $-\dfrac{b}{m_2}$ and y-intercept b.

Assume also that Line 1 and Line 2 have unequal slopes, that is $m_1 \neq m_2$.

If the lines have the same x-intercept, then $-\dfrac{b}{m_1} = -\dfrac{b}{m_2}$.

$$-\frac{b}{m_1} = -\frac{b}{m_2}$$
$$-m_2 b = -m_1 b$$
$$-m_2 b + m_1 b = 0$$
$$\text{But } -m_2 b + m_1 b = 0 \Rightarrow b(m_1 - m_2) = 0$$
$$\Rightarrow b = 0$$
$$\text{or } m_1 - m_2 = 0 \Rightarrow m_1 = m_2$$

Since we are assuming that $m_1 \neq m_2$, the only way that the two lines can have the same x-intercept is if $b = 0$.

131.   Answers will vary.

Graphs

1.4   Scatter Diagrams; Linear Curve Fitting

1.   scatter diagram   3.   Linear, $m > 0$

5.   Linear, $m < 0$ 7.   Nonlinear

9.   (a), (c)

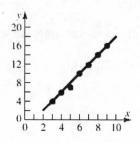

(d)   Using the LINear REGresssion program, the line of best fit is:
$$y = 2.0357x - 2.3571$$

(e)

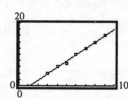

(b)   Answers will vary. We select (3,4) and (9,16). The slope of the line containing these points is: $m = \dfrac{16-4}{9-3} = \dfrac{12}{6} = 2$.

The equation of the line is:
$$y - y_1 = m(x - x_1)$$
$$y - 4 = 2(x - 3)$$
$$y - 4 = 2x - 6$$
$$y = 2x - 2$$

11.   (a), (c)

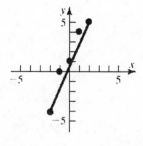

(d)   Using the LINear REGresssion program, the line of best fit is:
$$y = 2.2x + 1.2$$

(e)

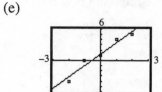

(b)   Answers will vary. We select (–2,–4) and (2,5). The slope of the line containing

these points is: $m = \dfrac{5-(-4)}{2-(-2)} = \dfrac{9}{4}$.

The equation of the line is:

$$y - y_1 = m(x - x_1)$$

$$y - (-4) = \frac{9}{4}(x - (-2))$$

$$y + 4 = \frac{9}{4}x + \frac{9}{2}$$

$$y = \frac{9}{4}x + \frac{1}{2}$$

13.   (a), (c)

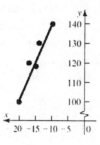

(d)   Using the LINear REGresssion program, the line of best fit is:

$$y = 3.8613x + 180.2920$$

(e)

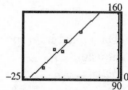

(b)   Answers will vary. We select (–20,100) and (–10,140). The slope of the line

containing these points is: $m = \dfrac{140 - 100}{-10 - (-20)} = \dfrac{40}{10} = 4$.

The equation of the line is:

$$y - y_1 = m(x - x_1)$$

$$y - 100 = 4(x - (-20))$$

$$y - 100 = 4x + 80$$

$$y = 4x + 180$$

15. (a)

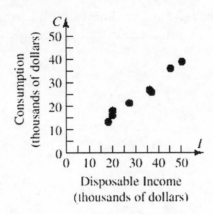

(b) Using points $(20,16)$ and $(50,39)$,

$$\text{slope} = \frac{39-16}{50-20} = \frac{23}{30}.$$

The point-slope formula yields

$$C - 16 = \frac{23}{30}(I - 20)$$

$$C = \frac{23}{30}I - \frac{460}{30} + 16$$

$$C \approx 0.77I + 0.67$$

(c) As disposable income increases by $1000, personal consumption expenditure increases by about $770.

(d) $C = 0.77(42) + 0.67 = \$33.01$

A family with disposable income of $42,000 has personal consumption expenditure of about $33,010.

(e) $C = 0.7549I + 0.6266$

17. (a) (Data used in graphs is in thousands.)

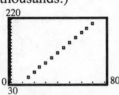

(b) $L = 2.9814I - 0.0761$

(c)

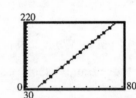

(d) As annual income increases by $1000, the loan amount increases by about $2981.40.

(e) $L = 2.9814(42) - 0.0761 \approx 125.143$

An individual with an annual income of $42,000 would qualify for a loan of about $125,143.

19. (a)

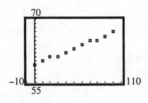

(b) $T = 0.0782h + 59.0909$

(c)

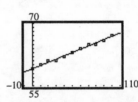

(d) As relative humidity increases by 1%, the apparent temperature increases by $0.0782°$.

(e) $T = 0.0782(75) + 59.0909 \approx 64.96$

A relative humidity of 75% would give an apparent temperature of $65°$.

Graphs

1.R Chapter Review

1. $(0,0), (4,2)$

    (a) distance $= \sqrt{(4-0)^2 + (2-0)^2}$
    $$= \sqrt{16+4} = \sqrt{20} = 2\sqrt{5}$$

    (b) midpoint $= \left(\dfrac{0+4}{2}, \dfrac{0+2}{2}\right) = \left(\dfrac{4}{2}, \dfrac{2}{2}\right) = (2,1)$

    (c) slope $= \dfrac{\Delta y}{\Delta x} = \dfrac{2-0}{4-0} = \dfrac{2}{4} = \dfrac{1}{2}$

    (d) When x increases by 2 units, y increases by 1 unit.

3. $(1,-1), (-2,3)$

    (a) distance $= \sqrt{(-2-1)^2 + (3-(-1))^2}$
    $$= \sqrt{9+16} = \sqrt{25} = 5$$

    (b) midpoint $= \left(\dfrac{1+(-2)}{2}, \dfrac{-1+3}{2}\right)$
    $$= \left(\dfrac{-1}{2}, \dfrac{2}{2}\right) = \left(-\dfrac{1}{2}, 1\right)$$

    (c) slope $= \dfrac{\Delta y}{\Delta x} = \dfrac{3-(-1)}{-2-1} = \dfrac{4}{-3} = -\dfrac{4}{3}$

    (d) When x increases by 3 units, y decreases by 4 units.

5. $(4,-4), (4,8)$

    (a) distance $= \sqrt{(4-4)^2 + (8-(-4))^2}$
    $$= \sqrt{0+144} = \sqrt{144} = 12$$

    (b) midpoint $= \left(\dfrac{4+4}{2}, \dfrac{-4+8}{2}\right)$
    $$= \left(\dfrac{8}{2}, \dfrac{4}{2}\right) = (4,2)$$

    (c) slope $= \dfrac{\Delta y}{\Delta x} = \dfrac{8-(-4)}{4-4} = \dfrac{12}{0}$, undefined

    (d) Undefined slope means the points lie on a vertical line.

7. $y = x^2 + 4$

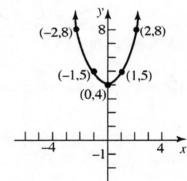

9. $2x = 3y^2$

x-intercept(s): $2x = 3(0)^2 \Rightarrow 2x = 0 \Rightarrow x = 0 \Rightarrow (0,0)$

y-intercept(s): $2(0) = 3y^2 \Rightarrow 0 = 3y^2 \Rightarrow y = 0 \Rightarrow (0,0)$

Test for symmetry:

    x-axis: Replace y by $-y$: $2x = 3(-y)^2$ or $2x = 3y^2$, which is

                        equivalent to $2x = 3y^2$.

    y-axis: Replace x by $-x$: $2(-x) = 3y^2$ or $-2x = 3y^2$,

                        which is not equivalent to $2x = 3y^2$.

    Origin: Replace x by $-x$ and y by $-y$: $2(-x) = 3(-y)^2$ or

                        $-2x = 3y^2$, which is not equivalent to $2x = 3y^2$.

Therefore, the graph is symmetric with respect to the x-axis.

11. $x^2 + 4y^2 = 16$

x-intercept(s): $x^2 + 4(0)^2 = 16 \Rightarrow x^2 = 16 \Rightarrow x = \pm 4 \Rightarrow (-4,0),(4,0)$

y-intercept(s): $(0)^2 + 4y^2 = 16 \Rightarrow 4y^2 = 16 \Rightarrow y^2 = 4 \Rightarrow y = \pm 2 \Rightarrow (0,-2),(0,2)$

Test for symmetry:

    x-axis: Replace y by $-y$: $x^2 + 4(-y)^2 = 16$ or $x^2 + 4y^2 = 16$,

                        which is equivalent to $x^2 + 4y^2 = 16$.

    y-axis: Replace x by $-x$: $(-x)^2 + 4y^2 = 16$ or $x^2 + 4y^2 = 16$,

                        which is equivalent to $x^2 + 4y^2 = 16$.

    Origin: Replace x by $-x$ and y by $-y$: $(-x)^2 + 4(-y)^2 = 16$ or $x^2 + 4y^2 = 16$,

                        which is equivalent to $x^2 + 4y^2 = 16$.

Therefore, the graph is symmetric with respect to the x-axis, the y-axis, and the origin.

13. $y = x^4 + 2x^2 + 1$

x-intercept(s): $0 = x^4 + 2x^2 + 1 \Rightarrow 0 = (x^2 + 1)(x^2 + 1) \Rightarrow x^2 + 1 = 0 \Rightarrow x^2 = -1$

                  $\Rightarrow$ no solution $\Rightarrow$ no x-intercepts

y-intercept(s): $y = (0)^4 + 2(0)^2 + 1 = 1 \Rightarrow (0,1)$

Test for symmetry:

    x-axis: Replace y by $-y$: $-y = x^4 + 2x^2 + 1$,

                        which is not equivalent to $y = x^4 + 2x^2 + 1$.

    y-axis: Replace x by $-x$: $y = (-x)^4 + 2(-x)^2 + 1$ or $y = x^4 + 2x^2 + 1$,

                        which is equivalent to $y = x^4 + 2x^2 + 1$.

    Origin: Replace x by $-x$ and y by $-y$: $-y = (-x)^4 + 2(-x)^2 + 1$ or $-y = x^4 + 2x^2 + 1$,

                        which is not equivalent to $y = x^4 + 2x^2 + 1$.

Therefore, the graph is symmetric with respect to the y-axis.

15. $x^2 + x + y^2 + 2y = 0$

    x-intercept(s): $x^2 + x + (0)^2 + 2(0) = 0 \Rightarrow x^2 + x = 0 \Rightarrow x(x+1) = 0$

        $x = 0, x = -1 \Rightarrow (-1,0),(0,0)$

    y-intercept(s): $(0)^2 + 0 + y^2 + 2y = 0 \Rightarrow y^2 + 2y = 0 \Rightarrow y(y+2) = 0$

        $y = 0, y = -2 \Rightarrow (0,-2),(0,0)$

    Test for symmetry:

       x-axis: Replace y by $-y$: $x^2 + x + (-y)^2 + 2(-y) = 0$ or $x^2 + x + y^2 - 2y = 0$,

              which is not equivalent to $x^2 + x + y^2 + 2y = 0$.

       y-axis: Replace x by $-x$: $(-x)^2 + (-x) + y^2 + 2y = 0$ or $x^2 - x + y^2 + 2y = 0$,

              which is not equivalent to $x^2 + x + y^2 + 2y = 0$.

      Origin: Replace x by $-x$ and y by $-y$: $(-x)^2 + (-x) + (-y)^2 + 2(-y) = 0$ or

            $x^2 - x + y^2 - 2y = 0$, which is not equivalent to

            $x^2 + x + y^2 + 2y = 0$.

    Therefore, the graph is not symmetric with respect to the x-axis, the y-axis, or the origin.

17. $(x-h)^2 + (y-k)^2 = r^2$

    $\left(x-(-2)\right)^2 + (y-3)^2 = 4^2$

      $(x+2)^2 + (y-3)^2 = 16$

19. $(x-h)^2 + (y-k)^2 = r^2$

    $\left(x-(-1)\right)^2 + \left(y-(-2)\right)^2 = 1^2$

      $(x+1)^2 + (y+2)^2 = 1$

21. $x^2 + (y-1)^2 = 4$

    $x^2 + (y-1)^2 = 2^2$

    Center: $(0,1)$

    Radius $= 2$

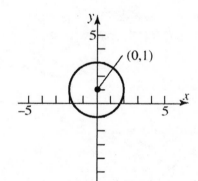

23. $x^2 + y^2 - 2x + 4y - 4 = 0$

        $x^2 - 2x + y^2 + 4y = 4$

    $(x^2 - 2x + 1) + (y^2 + 4y + 4) = 4 + 1 + 4$

        $(x-1)^2 + (y+2)^2 = 3^2$

    Center: $(1,-2)$ Radius $= 3$

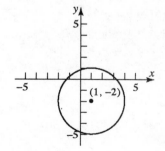

25.
$$3x^2 + 3y^2 - 6x + 12y = 0$$
$$x^2 + y^2 - 2x + 4y = 0$$
$$x^2 - 2x + y^2 + 4y = 0$$
$$(x^2 - 2x + 1) + (y^2 + 4y + 4) = 1 + 4$$
$$(x-1)^2 + (y+2)^2 = \left(\sqrt{5}\right)^2$$
Center: $(1,-2)$ Radius $= \sqrt{5}$

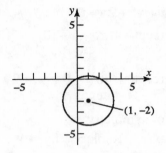

27. Slope $= -2$; containing $(3,-1)$
$$y - y_1 = m(x - x_1)$$
$$y - (-1) = -2(x - 3)$$
$$y + 1 = -2x + 6$$
$$y = -2x + 5$$
$$2x + y = 5 \text{ or } y = -2x + 5$$

29. vertical; containing $(-3,4)$
$$x = -3$$
No slope - intercept form.

31. y-intercept $= -2$; containing $(5,-3)$
Points are $(5,-3)$ and $(0,-2)$
$$m = \frac{-2-(-3)}{0-5} = \frac{1}{-5} = -\frac{1}{5}$$
$$y = mx + b$$
$$y = -\frac{1}{5}x - 2$$
$$x + 5y = -10 \text{ or } y = -\frac{1}{5}x - 2$$

33. Parallel to $2x - 3y = -4$
$$2x - 3y = -4 \Rightarrow y = \frac{2}{3}x + \frac{4}{3}$$
Slope $= \frac{2}{3}$; containing $(-5,3)$
$$y - y_1 = m(x - x_1)$$
$$y - 3 = \frac{2}{3}(x - (-5))$$
$$y - 3 = \frac{2}{3}x + \frac{10}{3}$$
$$y = \frac{2}{3}x + \frac{19}{3}$$
$$2x - 3y = -19 \text{ or } y = \frac{2}{3}x + \frac{19}{3}$$

35. Perpendicular to $x + y = 2$
$$x + y = 2 \Rightarrow y = -x + 2$$
Containing $(4,-3)$
Slope of perpendicular $= 1$
$$y - y_1 = m(x - x_1)$$
$$y - (-3) = 1(x - 4)$$
$$y + 3 = x - 4$$
$$y = x - 7$$
$$x - y = 7 \text{ or } y = x - 7$$

37. $4x - 5y = -20$
$$-5y = -4x - 20 \Rightarrow y = \frac{4}{5}x + 4$$
slope $= \frac{4}{5}$; y-intercept $= 4$

39. $\dfrac{1}{2}x - \dfrac{1}{3}y = -\dfrac{1}{6}$

$-\dfrac{1}{3}y = -\dfrac{1}{2}x - \dfrac{1}{6} \Rightarrow y = \dfrac{3}{2}x + \dfrac{1}{2}$

slope $= \dfrac{3}{2}$; y-intercept $= \dfrac{1}{2}$

41. slope $= \dfrac{2}{3}$, containing the point $(1,2)$

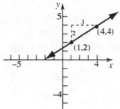

43. Given the points $A = (-2,0)$, $B = (-4,4)$, $C = (8,5)$.

(a) Find the distance between each pair of points.

$d_{A,B} = \sqrt{(-4-(-2))^2 + (4-0)^2} = \sqrt{4+16} = \sqrt{20} = 2\sqrt{5}$

$d_{B,C} = \sqrt{(8-(-4))^2 + (5-4)^2} = \sqrt{144+1} = \sqrt{145}$

$d_{A,C} = \sqrt{(8-(-2))^2 + (5-0)^2} = \sqrt{100+25} = \sqrt{125} = 5\sqrt{5}$

$\left(\sqrt{20}\right)^2 + \left(\sqrt{125}\right)^2 = \left(\sqrt{145}\right)^2 \Rightarrow 20+125 = 145 \Rightarrow 145 = 145$

The Pythagorean theorem is satisfied, so this is a right triangle.

(b) Find the slopes:

$m_{AB} = \dfrac{4-0}{-4-(-2)} = \dfrac{4}{-2} = -2; \; m_{BC} = \dfrac{5-4}{8-(-4)} = \dfrac{1}{12}$

$m_{AC} = \dfrac{5-0}{8-(-2)} = \dfrac{5}{10} = \dfrac{1}{2}$

$m_{AB} \cdot m_{AC} = -2 \cdot \dfrac{1}{2} = -1$

Since the product of the slopes is -1, two sides of the triangle are perpendicular and the triangle is a right triangle.

45. slope of $\overline{AB} = \dfrac{1-5}{6-2} = -1$; slope of $\overline{AC} = \dfrac{-1-5}{8-2} = -1$; slope of $\overline{BC} = \dfrac{-1-1}{8-6} = -1$

Since all the slopes are equal, the points lie on a line.

47. Answers will vary.

49. Set the axes so that the field's maximum dimension is along the x-axis.

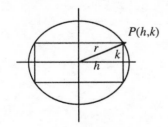

Let $2h = $ width, $2k = $ height, therefore the point farthest from the origin has coordinates $P(h,k)$. So the distance from the origin to point P is

$r = \sqrt{h^2 + k^2} = $ the radius of the circle .

Using 1 sprinkler arm:

If we place the sprinkler at the origin, we get a circle with equation $x^2 + y^2 = r^2$, where

$r = \sqrt{h^2 + k^2}$. So how much excess land is being watered ? The area of the field $= A_F = 4h$

The area of the circular water pattern $A_C = \pi r^2 = \pi\left(\sqrt{h^2 + k^2}\right)^2 = \pi\left(h^2 + k^2\right)$.

Therefore the amount of excess land being watered $= A_C - A_F = \pi\left(h^2 + k^2\right) - 4hk$.

Using 2 sprinkler arms:

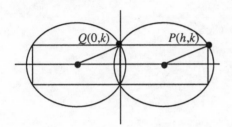

We want to place the sprinklers so that they overlap as little as possible while watering the entire field. The equation of the circle with center on the positive x-axis that passes through the point $P(h,k)$

and $Q(0,k)$ is $\left(x - \dfrac{h}{2}\right)^2 + y^2 = \dfrac{1}{4}h^2 + k^2$, since the

center is $\left(\dfrac{h}{2},0\right)$, and the radius is $\sqrt{\dfrac{1}{4}h^2 + k^2}$.

The case for the other side is similar. Thus, the sprinklers should have their centers at

$\left(-\dfrac{h}{2},0\right)$ and $\left(\dfrac{h}{2},0\right)$, with the arm lengths set at $\sqrt{\dfrac{1}{4}h^2 + k^2}$.

Each sprinkler waters the same area, so the total area watered is

$A_C = 2\pi r^2 = 2\pi\left(\sqrt{\dfrac{1}{4}h^2 + k^2}\right)^2 = 2\pi\left(\dfrac{1}{4}h^2 + k^2\right)$. The amount of excess land being watered is

$A_C - A_F = 2\pi\left(\dfrac{1}{4}h^2 + k^2\right) - 4hk$.

Comparison:

In order to determine when to switch from 1 sprinkler to 2 sprinklers, we want to determine when 2 sprinklers water less excess land than 1 sprinkler waters.

That is, we want to solve: $2\pi\left(\dfrac{1}{4}h^2 + k^2\right) - 4hk < \pi\left(h^2 + k^2\right) - 4hk$.

$$2\pi\left(\frac{1}{4}h^2 + k^2\right) - 4hk < \pi\left(h^2 + k^2\right) - 4hk$$

$$\frac{\pi}{2}h^2 + 2\pi k^2 - 4hk < \pi h^2 + \pi k^2 - 4hk$$

$$\frac{\pi}{2}h^2 + 2\pi k^2 < \pi h^2 + \pi k^2$$

$$\frac{1}{2}h^2 + 2k^2 < h^2 + k^2$$

$$k^2 < \frac{1}{2}h^2$$

$$k < \sqrt{\frac{1}{2}}h \Rightarrow h > \sqrt{2}k$$

So 2 sprinklers is the better choice when the longer dimension of the rectangle exceeds the shorter dimension by a factor of more than $\sqrt{2} \approx 1.414$.

Chapter 2

Functions and Their Graphs

2.1 Functions

1. $(-1,3)$

3. We must not allow the denominator to be 0.
   $x + 4 \neq 0 \Rightarrow x \neq -4$; Domain: $\{x | x \neq -4\}$.

5. independent, dependent

7. The intersection of the intervals $[0,7]$ and $[-2,5]$, i.e., $[0,5]$.

9. $(g - f)(x)$

11. True

13. True

15. Function
    Domain: {Dad, Colleen, Kaleigh, Marissa}
    Range: {Jan. 8, Mar. 15, Sept. 17}

17. Not a function

19. Not a function

21. Function
    Domain: {1, 2, 3, 4}
    Range: {3}

23. Not a function

25. Function
    Domain: {−2, −1, 0, 1}
    Range: {4, 1, 0}

27. $f(x) = 3x^2 + 2x - 4$

    (a) $f(0) = 3(0)^2 + 2(0) - 4 = -4$

    (b) $f(1) = 3(1)^2 + 2(1) - 4 = 3 + 2 - 4 = 1$

    (c) $f(-1) = 3(-1)^2 + 2(-1) - 4 = 3 - 2 - 4 = -3$

    (d) $f(-x) = 3(-x)^2 + 2(-x) - 4 = 3x^2 - 2x - 4$

    (e) $-f(x) = -(3x^2 + 2x - 4) = -3x^2 - 2x + 4$

    (f) $f(x+1) = 3(x+1)^2 + 2(x+1) - 4 = 3(x^2 + 2x + 1) + 2x + 2 - 4$
    $$= 3x^2 + 6x + 3 + 2x + 2 - 4$$
    $$= 3x^2 + 8x + 1$$

(g) $f(2x) = 3(2x)^2 + 2(2x) - 4 = 12x^2 + 4x - 4$

(h) $f(x+h) = 3(x+h)^2 + 2(x+h) - 4 = 3(x^2 + 2xh + h^2) + 2x + 2h - 4$
$$= 3x^2 + 6xh + 3h^2 + 2x + 2h - 4$$

29. $f(x) = \dfrac{x}{x^2+1}$

(a) $f(0) = \dfrac{0}{0^2+1} = \dfrac{0}{1} = 0$

(b) $f(1) = \dfrac{1}{1^2+1} = \dfrac{1}{2}$

(c) $f(-1) = \dfrac{-1}{(-1)^2+1} = \dfrac{-1}{1+1} = -\dfrac{1}{2}$

(d) $f(-x) = \dfrac{-x}{(-x)^2+1} = \dfrac{-x}{x^2+1}$

(e) $-f(x) = -\left(\dfrac{x}{x^2+1}\right) = \dfrac{-x}{x^2+1}$

(f) $f(x+1) = \dfrac{x+1}{(x+1)^2+1} = \dfrac{x+1}{x^2+2x+1+1} = \dfrac{x+1}{x^2+2x+2}$

(g) $f(2x) = \dfrac{2x}{(2x)^2+1} = \dfrac{2x}{4x^2+1}$

(h) $f(x+h) = \dfrac{x+h}{(x+h)^2+1} = \dfrac{x+h}{x^2+2xh+h^2+1}$

31. $f(x) = |x| + 4$

(a) $f(0) = |0| + 4 = 0 + 4 = 4$

(b) $f(1) = |1| + 4 = 1 + 4 = 5$

(c) $f(-1) = |-1| + 4 = 1 + 4 = 5$

(d) $f(-x) = |-x| + 4 = |x| + 4$

(e) $-f(x) = -(|x| + 4) = -|x| - 4$

(f) $f(x+1) = |x+1| + 4$

(g) $f(2x) = |2x| + 4 = 2|x| + 4$

(h) $f(x+h) = |x+h| + 4$

33. $f(x) = \dfrac{2x+1}{3x-5}$

(a) $f(0) = \dfrac{2(0)+1}{3(0)-5} = \dfrac{0+1}{0-5} = -\dfrac{1}{5}$

(b) $f(1) = \dfrac{2(1)+1}{3(1)-5} = \dfrac{2+1}{3-5} = \dfrac{3}{-2} = -\dfrac{3}{2}$

(c) $f(-1) = \dfrac{2(-1)+1}{3(-1)-5} = \dfrac{-2+1}{-3-5} = \dfrac{-1}{-8} = \dfrac{1}{8}$

(d) $f(-x) = \dfrac{2(-x)+1}{3(-x)-5} = \dfrac{-2x+1}{-3x-5} = \dfrac{2x-1}{3x+5}$

(e) $-f(x) = -\left(\dfrac{2x+1}{3x-5}\right) = \dfrac{-2x-1}{3x-5}$

(f) $f(x+1) = \dfrac{2(x+1)+1}{3(x+1)-5} = \dfrac{2x+2+1}{3x+3-5} = \dfrac{2x+3}{3x-2}$

(g) $f(2x) = \dfrac{2(2x)+1}{3(2x)-5} = \dfrac{4x+1}{6x-5}$

(h) $f(x+h) = \dfrac{2(x+h)+1}{3(x+h)-5} = \dfrac{2x+2h+1}{3x+3h-5}$

35.   Graph $y = x^2$. The graph passes the vertical line test. Thus, the equation represents a function.

37.   Graph $y = \dfrac{1}{x}$. The graph passes the vertical line test. Thus, the equation represents a function.

39. $y^2 = 4 - x^2$
Solve for y: $y = \pm\sqrt{4 - x^2}$
For $x = 0$, $y = \pm 2$. Thus, $(0, 2)$ and $(0, -2)$ are on the graph. This is not a function, since a distinct x corresponds to two different y's.

41. $x = y^2$
Solve for y: $y = \pm\sqrt{x}$
For $x = 1$, $y = \pm 1$. Thus, $(1, 1)$ and $(1, -1)$ are on the graph. This is not a function, since a distinct x corresponds to two different y's.

43.   Graph $y = 2x^2 - 3x + 4$. The graph passes the vertical line test. Thus, the equation represents a function.

45. $2x^2 + 3y^2 = 1$
Solve for y:

$2x^2 + 3y^2 = 1 \Rightarrow 3y^2 = 1 - 2x^2 \Rightarrow y^2 = \dfrac{1-2x^2}{3} \Rightarrow y = \pm\sqrt{\dfrac{1-2x^2}{3}}$

For $x = 0$, $y = \pm\sqrt{\dfrac{1}{3}}$. Thus, $\left(0, \sqrt{\dfrac{1}{3}}\right)$ and $\left(0, -\sqrt{\dfrac{1}{3}}\right)$ are on the graph. This is not a function, since a distinct x corresponds to two different y's.

47. $f(x) = -5x + 4$
Domain: $\{x \mid x \text{ is any real number}\}$

49. $f(x) = \dfrac{x}{x^2+1}$
Domain: $\{x \mid x \text{ is any real number}\}$

51.　$g(x) = \dfrac{x}{x^2 - 16}$

$x^2 - 16 \neq 0$

$x^2 \neq 16 \Rightarrow x \neq \pm 4$

Domain: $\left\{ x \mid x \neq -4,\ x \neq 4 \right\}$

53.　$F(x) = \dfrac{x - 2}{x^3 + x}$

$x^3 + x \neq 0$

$x(x^2 + 1) \neq 0$

$x \neq 0,\quad x^2 \neq -1$

Domain: $\left\{ x \mid x \neq 0 \right\}$

55.　$h(x) = \sqrt{3x - 12}$

$3x - 12 \geq 0$

$3x \geq 12$

$x \geq 4$

Domain: $\left\{ x \mid x \geq 4 \right\}$

57.　$f(x) = \dfrac{4}{\sqrt{x - 9}}$

$x - 9 > 0$

$x > 9$

Domain: $\left\{ x \mid x > 9 \right\}$

59.　$p(x) = \sqrt{\dfrac{2}{x - 1}} = \dfrac{\sqrt{2}}{\sqrt{x - 1}}$

$x - 1 > 0$

$x > 1$

Domain: $\left\{ x \mid x > 1 \right\}$

61.　$f(x) = 3x + 4 \qquad g(x) = 2x - 3$

(a)　$(f + g)(x) = 3x + 4 + 2x - 3 = 5x + 1$

The domain is $\left\{ x \mid x \text{ is any real number} \right\}$.

(b)　$(f - g)(x) = (3x + 4) - (2x - 3) = 3x + 4 - 2x + 3 = x + 7$

The domain is $\left\{ x \mid x \text{ is any real number} \right\}$.

(c)　$(f \cdot g)(x) = (3x + 4)(2x - 3) = 6x^2 - 9x + 8x - 12 = 6x^2 - x - 12$

The domain is $\left\{ x \mid x \text{ is any real number} \right\}$.

(d)　$\left(\dfrac{f}{g} \right)(x) = \dfrac{3x + 4}{2x - 3}$

$2x - 3 \neq 0$

$2x \neq 3 \Rightarrow x \neq \dfrac{3}{2}$

The domain is $\left\{ x \mid x \neq \dfrac{3}{2} \right\}$.

63.　$f(x) = x - 1 \qquad g(x) = 2x^2$

(a)　$(f + g)(x) = x - 1 + 2x^2 = 2x^2 + x - 1$

The domain is $\left\{ x \mid x \text{ is any real number} \right\}$.

(b)　$(f - g)(x) = (x - 1) - (2x^2) = x - 1 - 2x^2 = -2x^2 + x - 1$

The domain is $\left\{ x \mid x \text{ is any real number} \right\}$.

(c) $(f \cdot g)(x) = (x-1)(2x^2) = 2x^3 - 2x^2$
The domain is $\{x \mid x \text{ is any real number}\}$.

(d) $\left(\dfrac{f}{g}\right)(x) = \dfrac{x-1}{2x^2}$
The domain is $\{x \mid x \neq 0\}$.

65. $f(x) = \sqrt{x}$ $g(x) = 3x - 5$
(a) $(f + g)(x) = \sqrt{x} + 3x - 5$
The domain is $\{x \mid x \geq 0\}$.

(b) $(f - g)(x) = \sqrt{x} - (3x - 5) = \sqrt{x} - 3x + 5$
The domain is $\{x \mid x \geq 0\}$.

(c) $(f \cdot g)(x) = \sqrt{x}(3x - 5) = 3x\sqrt{x} - 5\sqrt{x}$
The domain is $\{x \mid x \geq 0\}$.

(d) $\left(\dfrac{f}{g}\right)(x) = \dfrac{\sqrt{x}}{3x - 5}$
$x \geq 0$ and $3x - 5 \neq 0$

$$3x \neq 5 \Rightarrow x \neq \dfrac{5}{3}$$

The domain is $\left\{x \mid x \geq 0 \text{ and } x \neq \dfrac{5}{3}\right\}$.

67. $f(x) = 1 + \dfrac{1}{x}$ $g(x) = \dfrac{1}{x}$
(a) $(f + g)(x) = 1 + \dfrac{1}{x} + \dfrac{1}{x} = 1 + \dfrac{2}{x}$
The domain is $\{x \mid x \neq 0\}$.

(b) $(f - g)(x) = 1 + \dfrac{1}{x} - \dfrac{1}{x} = 1$
The domain is $\{x \mid x \neq 0\}$.

(c) $(f \cdot g)(x) = \left(1 + \dfrac{1}{x}\right)\dfrac{1}{x} = \dfrac{1}{x} + \dfrac{1}{x^2}$
The domain is $\{x \mid x \neq 0\}$.

(d) $\left(\dfrac{f}{g}\right)(x) = \dfrac{1 + \dfrac{1}{x}}{\dfrac{1}{x}} = \dfrac{\dfrac{x+1}{x}}{\dfrac{1}{x}} = \dfrac{x+1}{x} \cdot \dfrac{x}{1} = x + 1$
The domain is $\{x \mid x \neq 0\}$.

69. $f(x) = \dfrac{2x+3}{3x-2}$ $g(x) = \dfrac{4x}{3x-2}$

   (a) $(f+g)(x) = \dfrac{2x+3}{3x-2} + \dfrac{4x}{3x-2} = \dfrac{2x+3+4x}{3x-2} = \dfrac{6x+3}{3x-2}$

   $3x - 2 \neq 0$

   $3x \neq 2 \Rightarrow x \neq \dfrac{2}{3}$

   The domain is $\left\{ x \middle| x \neq \dfrac{2}{3} \right\}$.

   (b) $(f-g)(x) = \dfrac{2x+3}{3x-2} - \dfrac{4x}{3x-2} = \dfrac{2x+3-4x}{3x-2} = \dfrac{-2x+3}{3x-2}$

   $3x - 2 \neq 0$

   $3x \neq 2 \Rightarrow x \neq \dfrac{2}{3}$

   The domain is $\left\{ x \middle| x \neq \dfrac{2}{3} \right\}$.

   (c) $(f \cdot g)(x) = \left(\dfrac{2x+3}{3x-2} \right)\left(\dfrac{4x}{3x-2} \right) = \dfrac{8x^2 + 12x}{(3x-2)^2}$

   $3x - 2 \neq 0$

   $3x \neq 2 \Rightarrow x \neq \dfrac{2}{3}$

   The domain is $\left\{ x \middle| x \neq \dfrac{2}{3} \right\}$.

   (d) $\left(\dfrac{f}{g} \right)(x) = \dfrac{\dfrac{2x+3}{3x-2}}{\dfrac{4x}{3x-2}} = \dfrac{2x+3}{3x-2} \cdot \dfrac{3x-2}{4x} = \dfrac{2x+3}{4x}$

   $3x - 2 \neq 0 \quad \text{and} \quad x \neq 0$

   $3x \neq 2$

   $x \neq \dfrac{2}{3}$

   The domain is $\left\{ x \middle| x \neq \dfrac{2}{3} \text{ and } x \neq 0 \right\}$.

71. $f(x) = 3x + 1$ $(f+g)(x) = 6 - \dfrac{1}{2}x$

   $6 - \dfrac{1}{2}x = 3x + 1 + g(x)$

   $5 - \dfrac{7}{2}x = g(x) \Rightarrow g(x) = 5 - \dfrac{7}{2}x$

73. $f(x) = 4x + 3$

$$\frac{f(x+h) - f(x)}{h} = \frac{4(x+h) + 3 - 4x - 3}{h}$$

$$= \frac{4x + 4h + 3 - 4x - 3}{h} = \frac{4h}{h} = 4$$

75. $f(x) = x^2 - x + 4$

$$\frac{f(x+h) - f(x)}{h} = \frac{(x+h)^2 - (x+h) + 4 - (x^2 - x + 4)}{h}$$

$$= \frac{x^2 + 2xh + h^2 - x - h + 4 - x^2 + x - 4}{h}$$

$$= \frac{2xh + h^2 - h}{h}$$

$$= 2x + h - 1$$

77. $f(x) = x^3 - 2$

$$\frac{f(x+h) - f(x)}{h} = \frac{(x+h)^3 - 2 - (x^3 - 2)}{h}$$

$$= \frac{x^3 + 3x^2h + 3xh^2 + h^3 - 2 - x^3 + 2}{h}$$

$$= \frac{3x^2h + 3xh^2 + h^3}{h}$$

$$= 3x^2 + 3xh + h^2$$

79. $f(x) = 2x^3 + Ax^2 + 4x - 5$ and $f(2) = 5$

$f(2) = 2(2)^3 + A(2)^2 + 4(2) - 5$

$5 = 16 + 4A + 8 - 5 \Rightarrow 5 = 4A + 19$

$-14 = 4A \Rightarrow A = -\dfrac{7}{2}$

81. $f(x) = \dfrac{3x + 8}{2x - A}$ and $f(0) = 2$

$f(0) = \dfrac{3(0) + 8}{2(0) - A}$

$2 = \dfrac{8}{-A} \Rightarrow -2A = 8 \Rightarrow A = -4$

83. $f(x) = \dfrac{2x - A}{x - 3}$ and $f(4) = 0$

$f(4) = \dfrac{2(4) - A}{4 - 3}$

$0 = \dfrac{8 - A}{1}$

$0 = 8 - A$

$A = 8$

f is undefined when $x = 3$.

85. Let x represent the length of the rectangle.

Then $\dfrac{x}{2}$ represents the width of the rectangle, since the length is twice the width.

The function for the area is: $A(x) = x \cdot \dfrac{x}{2} = \dfrac{x^2}{2} = \dfrac{1}{2}x^2$

87. Let x represent the number of hours worked.
The function for the gross salary is: $G(x) = 10x$

89. (a) $H(1) = 20 - 4.9(1)^2 = 20 - 4.9 = 15.1$ meters

$H(1.1) = 20 - 4.9(1.1)^2 = 20 - 4.9(1.21) = 20 - 5.929 = 14.071$ meters

$H(1.2) = 20 - 4.9(1.2)^2 = 20 - 4.9(1.44) = 20 - 7.056 = 12.944$ meters

$H(1.3) = 20 - 4.9(1.3)^2 = 20 - 4.9(1.69) = 20 - 8.281 = 11.719$ meters

(b) $H(x) = 15$ $\qquad\qquad$ $H(x) = 10$ $\qquad\qquad$ $H(x) = 5$

$15 = 20 - 4.9x^2$ $\qquad$ $10 = 20 - 4.9x^2$ $\qquad$ $5 = 20 - 4.9x^2$

$-5 = -4.9x^2$ $\qquad\qquad$ $-10 = -4.9x^2$ $\qquad\quad$ $-15 = -4.9x^2$

$x^2 \approx 1.0204$ $\qquad\qquad$ $x^2 \approx 2.0408$ $\qquad\qquad$ $x^2 \approx 3.0612$

$x \approx 1.01$ seconds $\qquad$ $x \approx 1.43$ seconds $\qquad$ $x \approx 1.75$ seconds

(c) $H(x) = 0$

$0 = 20 - 4.9x^2$

$-20 = -4.9x^2$

$x^2 \approx 4.0816$

$x \approx 2.02$ seconds

91. $C(x) = 100 + \dfrac{x}{10} + \dfrac{36{,}000}{x}$

(a) $C(500) = 100 + \dfrac{500}{10} + \dfrac{36{,}000}{500} = 100 + 50 + 72 = \$222$

(b) $C(450) = 100 + \dfrac{450}{10} + \dfrac{36{,}000}{450} = 100 + 45 + 80 = \$225$

(c) $C(600) = 100 + \dfrac{600}{10} + \dfrac{36{,}000}{600} = 100 + 60 + 60 = \$220$

(d) $C(400) = 100 + \dfrac{400}{10} + \dfrac{36{,}000}{400} = 100 + 40 + 90 = \$230$

93. $R(x) = \left(\dfrac{L}{P}\right)(x) = \dfrac{L(x)}{P(x)}$ $\qquad\qquad$ 95. $H(x) = (P \cdot I)(x) = P(x) \cdot I(x)$

97. (a) $h(x) = 2x$

$h(a + b) = 2(a + b) = 2a + 2b = h(a) + h(b)$

$h(x) = 2x$ has the property.

(b) $g(x) = x^2$

$g(a+b) = (a+b)^2 = a^2 + 2ab + b^2 \neq a^2 + b^2 = g(a) + g(b)$

$g(x) = x^2$ does not have the property.

(c) $F(x) = 5x - 2$

$F(a+b) = 5(a+b) - 2 = 5a + 5b - 2 \neq 5a - 2 + 5b - 2 = F(a) + F(b)$

$F(x) = 5x - 2$ does not have the property.

(d) $G(x) = \dfrac{1}{x}$

$G(a+b) = \dfrac{1}{a+b} \neq \dfrac{1}{a} + \dfrac{1}{b} = G(a) + G(b)$

$G(x) = \dfrac{1}{x}$ does not have the property.

99. Answers will vary.

Functions and Their Graphs

2.2   The Graph of a Function

1. $x^2 + 4y^2 = 16$

     x-intercepts: $x^2 + 4(0)^2 = 16 \Rightarrow x^2 = 16 \Rightarrow x = \pm 4 \Rightarrow (-4,0),(4,0)$

     y-intercepts: $(0)^2 + 4y^2 = 16 \Rightarrow 4y^2 = 16 \Rightarrow y^2 = 4 \Rightarrow y = \pm 2 \Rightarrow (0,-2),(0,2)$

3.   vertical

5. $f(x) = ax^2 + 4$

     $a(-1)^2 + 4 = 2 \Rightarrow a = -2$

7.   False

9.   (a) $f(0) = 3$ since $(0,3)$ is on the graph.

          $f(-6) = -3$ since $(-6,-3)$ is on the graph.

     (b) $f(6) = 0$ since $(6,0)$ is on the graph.

          $f(11) = 1$ since $(11,1)$ is on the graph.

     (c) $f(3)$ is positive since $f(3) \approx 3.7$.

     (d) $f(-4)$ is negative since $f(-4) \approx -1$.

     (e) $f(x) = 0$ when $x = -3$, $x = 6$, and $x = 10$.

     (f) $f(x) > 0$ when $-3 < x < 6$, and $10 < x \le 11$.

     (g)   The domain of f is $\{x | -6 \le x \le 11\}$ or $[-6,11]$

     (h)   The range of f is $\{y | -3 \le y \le 4\}$ or $[-3, 4]$

     (i)   The x-intercepts are $(-3, 0)$, $(6, 0)$, and $(11, 0)$.

     (j)   The y-intercept is $(0, 3)$

     (k)   The line $y = \dfrac{1}{2}$ intersects the graph 3 times.

     (l)   The line $x = 5$ intersects the graph 1 time

     (m) $f(x) = 3$ when $x = 0$ and $x = 4$.

     (n) $f(x) = -2$ when $x = -5$ and $x = 8$.

11.   Not a function since vertical lines will intersect the graph in more than one point.

13.   Function   (a)   Domain: $\{x | -\pi \le x \le \pi\}$;   Range: $\{y | -1 \le y \le 1\}$

               (b)   intercepts: $\left(-\dfrac{\pi}{2},0\right)$, $\left(\dfrac{\pi}{2},0\right)$, $(0,1)$

15.   Not a function since vertical lines will intersect the graph in more than one point.

17. Function   (a) Domain: $\{x\,|\,x > 0\}$;   Range: $\{y\,|\,y \text{ is any real number}\}$
                 (b) intercepts: $(1, 0)$

19. Function   (a) Domain: $\{x\,|\,x \text{ is any real number}\}$;   Range: $\{y\,|\,y \le 2\}$
                 (b) intercepts: $(-3,0), (3,0), (0,2)$

21. Function   (a) Domain: $\{x\,|\,x \text{ is any real number}\}$;   Range: $\{y\,|\,y \ge -3\}$
                 (b) intercepts: $(1,0), (3,0), (0,9)$

23. $f(x) = 2x^2 - x - 1$

(a) $f(-1) = 2(-1)^2 - (-1) - 1 = 2$ $(-1,2)$ is on the graph of f.

(b) $f(-2) = 2(-2)^2 - (-2) - 1 = 9$ $(-2,9)$ is on the graph of f.

(c) Solve for x:
$$-1 = 2x^2 - x - 1 \Rightarrow 0 = 2x^2 - x$$
$$0 = x(2x - 1) \Rightarrow x = 0, x = \frac{1}{2}$$
$(0, -1)$ and $\left(\dfrac{1}{2}, -1\right)$ are on the graph of f.

(d) The domain of f is: $\{x\,|\,x \text{ is any real number}\}$.

(e) x-intercepts:
$$f(x) = 0 \Rightarrow 2x^2 - x - 1 = 0$$
$$(2x + 1)(x - 1) = 0 \Rightarrow x = -\frac{1}{2}, x = 1$$
$\left(-\dfrac{1}{2}, 0\right)$ and $(1,0)$

(f) y-intercept: $f(0) = 2(0)^2 - 0 - 1 = -1 \Rightarrow (0,-1)$

25. $f(x) = \dfrac{x+2}{x-6}$

(a) $f(3) = \dfrac{3+2}{3-6} = -\dfrac{5}{3} \ne 14$ $(3,14)$ is not on the graph of f.

(b) $f(4) = \dfrac{4+2}{4-6} = \dfrac{6}{-2} = -3$ $(4,-3)$ is on the graph of f.

(c) Solve for x:
$$2 = \frac{x+2}{x-6}$$
$$2x - 12 = x + 2$$
$$x = 14$$
$(14, 2)$ is a point on the graph of f.

(d) The domain of f is: $\{x\,|\,x \ne 6\}$.

(e) x-intercepts:

$$f(x) = 0 \Rightarrow \frac{x+2}{x-6} = 0$$

$$x + 2 = 0 \Rightarrow x = -2 \Rightarrow (-2, 0)$$

(f) y-intercept: $f(0) = \dfrac{0+2}{0-6} = -\dfrac{1}{3} \Rightarrow \left(0, -\dfrac{1}{3}\right)$

27. $f(x) = \dfrac{2x^2}{x^4 + 1}$

(a) $f(-1) = \dfrac{2(-1)^2}{(-1)^4 + 1} = \dfrac{2}{2} = 1$ $(-1, 1)$ is on the graph of f.

(b) $f(2) = \dfrac{2(2)^2}{(2)^4 + 1} = \dfrac{8}{17}$ $\left(2, \dfrac{8}{17}\right)$ is on the graph of f.

(c)   Solve for x:

$$1 = \frac{2x^2}{x^4 + 1}$$

$$x^4 + 1 = 2x^2$$

$$x^4 - 2x^2 + 1 = 0$$

$$(x^2 - 1)^2 = 0$$

$$x^2 - 1 = 0 \Rightarrow x = \pm 1$$

$(1, 1)$ and $(-1, 1)$ are on the graph of f.

(d)   The domain of f is: $\{x \mid x \text{ is any real number}\}$.

(e) x-intercept:

$$f(x) = 0 \Rightarrow \frac{2x^2}{x^4 + 1} = 0$$

$$2x^2 = 0 \Rightarrow x = 0 \Rightarrow (0, 0)$$

(f) y-intercept: $f(0) = \dfrac{2(0)^2}{0^4 + 1} = \dfrac{0}{0 + 1} = 0 \Rightarrow (0, 0)$

29. $h(x) = \dfrac{-32x^2}{130^2} + x$

(a) $h(100) = \dfrac{-32(100)^2}{130^2} + 100 = \dfrac{-320,000}{16,900} + 100 \approx -18.93 + 100 \approx 81.07 \text{ feet}$

(b) $h(300) = \dfrac{-32(300)^2}{130^2} + 300 = \dfrac{-2,880,000}{16,900} + 300 \approx -170.41 + 300 \approx 129.59 \text{ feet}$

(c) $h(500) = \dfrac{-32(500)^2}{130^2} + 500 = \dfrac{-8,000,000}{16,900} + 500 \approx -473.37 + 500 \approx 26.63 \text{ feet}$

(d)   Solving $h(x) = \dfrac{-32x^2}{130^2} + x = 0$

$\dfrac{-32x^2}{130^2} + x = 0 \Rightarrow x\left(\dfrac{-32x}{130^2} + 1\right) = 0 \Rightarrow x = 0$ or $\dfrac{-32x}{130^2} + 1 = 0$

$\dfrac{-32x}{130^2} + 1 = 0 \Rightarrow 1 = \dfrac{32x}{130^2} \Rightarrow 130^2 = 32x \Rightarrow x = \dfrac{130^2}{32} = 528.125$ feet

Therefore, the golf ball travels 528.125 feet.

(e) $y_1 = \dfrac{-32x^2}{130^2} + x$

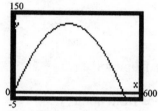

(f)   Use INTERSECT on the graphs of $y_1 = \dfrac{-32x^2}{130^2} + x$ and $y_2 = 90$.

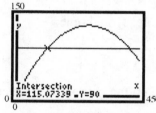   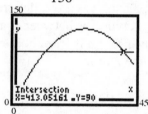

The ball reaches a height of 90 feet twice. The first time is when the ball has traveled approximately 115 feet, and the second time is when the ball has traveled approximately 413 feet.

(g)   The ball travels approximately 275 feet before it reaches its maximum height of approximately 131.8 feet.

X	Y1
0	0
25	23.817
50	45.266
75	64.349
100	81.065
125	95.414
150	107.4

X=0

(h)   The ball travels approximately 264 feet before it reaches its maximum height of approximately 132.03 feet.

X	Y1
263.5	132.03
264	132.03
264.5	132.03
265	132.03
265.5	132.03
266	132.02
266.5	132.02

X=264

31.   (a)   III   (b)   IV   (c)   I   (d)   V   (e)   II

33.

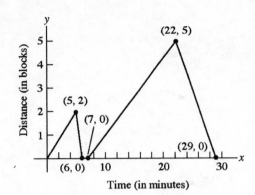

35. (a) 2 hours elapsed; Kevin was between 0 and 3 miles from home.
     (b) 0.5 hours elapsed; Kevin was 3 miles from home.
     (c) 0.3 hours elapsed; Kevin was between 0 and 3 miles from home.
     (d) 0.2 hours elapsed; Kevin was at home.
     (e) 0.9 hours elapsed; Kevin was between 0 and 2.8 miles from home.
     (f) 0.3 hours elapsed; Kevin was 2.8 miles from home.
     (g) 1.1 hours elapsed; Kevin was between 0 and 2.8 miles from home.
     (h) The farthest distance Kevin is from home is 3 miles.
     (i) Kevin returned home 2 times.

37. Points of the form $(5, y)$ and of the form $(x, 0)$ cannot be on the graph of the function.

39. A function may have any number of x-intercepts; it can have only one y-intercept.

41. The only such function is $f(x) = 0$.

Chapter 2

Functions and Their Graphs

2.3 Properties of Functions

1. $2 \le x \le 5$

3. $y = 2x^2 + 3$
   Test for symmetry:

   x-axis: Replace y by $-y$: $-y = 2x^2 + 3 \Rightarrow y = -2x^2 - 3$,

   which is not equivalent to $y = 2x^2 + 3$.

   y-axis: Replace x by $-x$: $y = 2(-x)^2 + 3 = 2x^2 + 3$,

   which equivalent to $y = 2x^2 + 3$.

   Origin: Replace x by $-x$ and y by $-y$: $-y = 2(-x)^2 + 3 \Rightarrow y = -2x^2 - 3$,

   which is not equivalent to $y = 2x^2 + 3$.

   Therefore, the graph is symmetric with respect to the y-axis.

5. $y = x^2 - 9$

   y-intercept: Let $x = 0$, then $y = 0^2 - 9 = -9$ $(0, -9)$

   x-intercepts: Let $y = 0$, then $0 = x^2 - 9 \Rightarrow x^2 = 9 \Rightarrow x = \pm 3$ $(-3, 0), (3, 0)$

7. even, odd

9. True

11. Yes

13. No, it only increases on (5, 10).

15. f is increasing on the intervals $(-8, -2)$, $(0, 2)$, $(5, \infty)$.

17. Yes. The local maximum at $x = 2$ is 10.

19. f has local maxima at $x = -2$ and $x = 2$. The local maxima are 6 and 10, respectively.

21. (a) Intercepts: $(-2, 0)$, $(2, 0)$, and $(0, 3)$.
    (b) Domain: $\{x \mid -4 \le x \le 4\}$; Range: $\{y \mid 0 \le y \le 3\}$.
    (c) Increasing: $(-2, 0)$ and $(2, 4)$; Decreasing: $(-4, -2)$ and $(0, 2)$.
    (d) Since the graph is symmetric with respect to the y-axis, the function is <u>even</u>.

23. (a) Intercepts: $(0,1)$.
    (b) Domain: $\{x \mid x \text{ is any real number}\}$; Range: $\{y \mid y > 0\}$.
    (c) Increasing: $(-\infty, \infty)$; Decreasing: never.
    (d) Since the graph is not symmetric with respect to the y-axis or the origin, the function is <u>neither</u> even nor odd.

25. (a) Intercepts: $(-\pi, 0)$, $(\pi, 0)$, and $(0, 0)$.
    (b) Domain: $\{x \mid -\pi \le x \le \pi\}$; Range: $\{y \mid -1 \le y \le 1\}$.
    (c) Increasing: $\left(-\dfrac{\pi}{2}, \dfrac{\pi}{2}\right)$; Decreasing: $\left(-\pi, -\dfrac{\pi}{2}\right)$ and $\left(\dfrac{\pi}{2}, \pi\right)$.
    (d) Since the graph is symmetric with respect to the origin, the function is <u>odd</u>.

27. (a) Intercepts: $\left(\dfrac{1}{2}, 0\right)$, $\left(\dfrac{5}{2}, 0\right)$, and $\left(0, \dfrac{1}{2}\right)$.
    (b) Domain: $\{x \mid -3 \le x \le 3\}$; Range: $\{y \mid -1 \le y \le 2\}$.
    (c) Increasing: $(2, 3)$; Decreasing: $(-1, 1)$; Constant: $(-3, -1)$ and $(1, 2)$
    (d) Since the graph is not symmetric with respect to the y-axis or the origin, the function is <u>neither</u> even nor odd.

29. (a) f has a local maximum of 3 at $x = 0$.
    (b) f has a local minimum of 0 at both $x = -2$ and $x = 2$.

31. (a) f has a local maximum of 1 at $x = \dfrac{\pi}{2}$.
    (b) f has a local minimum of -1 at $x = -\dfrac{\pi}{2}$.

33. $f(x) = -2x^2 + 4$
    (a) Average rate of change of f from $x = 0$ to $x = 2$:
$$\frac{f(2) - f(0)}{2 - 0} = \frac{\left(-2(2)^2 + 4\right) - \left(-2(0)^2 + 4\right)}{2} = \frac{(-4) - (4)}{2} = \frac{-8}{2} = -4$$
    (b) Average rate of change of f from $x = 1$ to $x = 3$:
$$\frac{f(3) - f(1)}{3 - 1} = \frac{\left(-2(3)^2 + 4\right) - \left(-2(1)^2 + 4\right)}{2} = \frac{(-14) - (2)}{2} = \frac{-16}{2} = -8$$
    (c) Average rate of change of f from $x = 1$ to $x = 4$:
$$\frac{f(4) - f(1)}{4 - 1} = \frac{\left(-2(4)^2 + 4\right) - \left(-2(1)^2 + 4\right)}{3} = \frac{(-28) - (2)}{3} = \frac{-30}{3} = -10$$

35. $f(x) = 5x$

(a) $\dfrac{f(x) - f(1)}{x - 1} = \dfrac{5x - 5}{x - 1} = \dfrac{5(x - 1)}{x - 1} = 5$

(b) $\dfrac{f(2) - f(1)}{2 - 1} = 5$

(c) Slope = 5; Containing $(1, 5)$:
$y - 5 = 5(x - 1)$
$y - 5 = 5x - 5 \Rightarrow y = 5x$

37. $f(x) = 1 - 3x$

(a) $\dfrac{f(x) - f(1)}{x - 1} = \dfrac{1 - 3x - (-2)}{x - 1}$

$= \dfrac{-3x + 3}{x - 1} = \dfrac{-3(x - 1)}{x - 1} = -3$

(b) $\dfrac{f(2) - f(1)}{2 - 1} = -3$

(c) Slope = -3; Containing $(1, -2)$:
$y - (-2) = -3(x - 1)$
$y + 2 = -3x + 3 \Rightarrow y = -3x + 1$

39. $f(x) = x^2 - 2x$

(a) $\dfrac{f(x) - f(1)}{x - 1} = \dfrac{x^2 - 2x - (-1)}{x - 1}$

$= \dfrac{x^2 - 2x + 1}{x - 1} = \dfrac{(x - 1)^2}{x - 1}$

$= x - 1$

(b) $\dfrac{f(2) - f(1)}{2 - 1} = 2 - 1 = 1$

(c) Slope = 1; Containing $(1, -1)$:
$y - (-1) = 1(x - 1)$
$y + 1 = x - 1 \Rightarrow y = x - 2$

41. $f(x) = x^3 - x$

(a) $\dfrac{f(x) - f(1)}{x - 1} = \dfrac{x^3 - x - 0}{x - 1} = \dfrac{x^3 - x}{x - 1}$

$= \dfrac{x(x - 1)(x + 1)}{x - 1} = x^2 + x$

(b) $\dfrac{f(2) - f(1)}{2 - 1} = 2^2 + 2 = 6$

(c) Slope = 6; Containing $(1, 0)$:
$y - 0 = 6(x - 1) \Rightarrow y = 6x - 6$

43. $f(x) = \dfrac{2}{x + 1}$

(a) $\dfrac{f(x) - f(1)}{x - 1} = \dfrac{\dfrac{2}{x + 1} - \dfrac{2}{2}}{x - 1} = \dfrac{\dfrac{2}{x + 1} - 1}{x - 1}$

$= \dfrac{\dfrac{2 - x - 1}{x + 1}}{x - 1} = \dfrac{1 - x}{(x - 1)(x + 1)}$

$= \dfrac{-1}{x + 1}$

(b) $\dfrac{f(2) - f(1)}{2 - 1} = \dfrac{-1}{2 + 1} = -\dfrac{1}{3}$

(c) Slope $= -\dfrac{1}{3}$; Containing $(1, 1)$:

$y - 1 = -\dfrac{1}{3}(x - 1)$

$y - 1 = -\dfrac{1}{3}x + \dfrac{1}{3} \Rightarrow y = -\dfrac{1}{3}x + \dfrac{4}{3}$

45. $f(x) = \sqrt{x}$

(a) $\dfrac{f(x) - f(1)}{x - 1} = \dfrac{\sqrt{x} - 1}{x - 1}$

(b) $\dfrac{f(2) - f(1)}{2 - 1} = \dfrac{\sqrt{2} - 1}{2 - 1} = \sqrt{2} - 1$

(c) Slope $= \sqrt{2} - 1$; Containing $(1, 1)$:
$y - 1 = \left(\sqrt{2} - 1\right)(x - 1)$
$y - 1 = \left(\sqrt{2} - 1\right)x - \left(\sqrt{2} - 1\right)$
$y = \left(\sqrt{2} - 1\right)x - \sqrt{2} + 2$

47. $f(x) = 4x^3$

$\qquad f(-x) = 4(-x)^3 = -4x^3$

f is odd.

49. $g(x) = -3x^2 - 5$

$\qquad g(-x) = -3(-x)^2 - 5 = -3x^2 - 5$

g is even.

51. $F(x) = \sqrt[3]{x}$

$\qquad F(-x) = \sqrt[3]{-x} = -\sqrt[3]{x}$

F is odd.

53. $f(x) = x + |x|$

$\qquad f(-x) = -x + |-x| = -x + |x|$

f is neither even nor odd.

55. $g(x) = \dfrac{1}{x^2}$

$\qquad g(-x) = \dfrac{1}{(-x)^2} = \dfrac{1}{x^2}$

g is even.

57. $h(x) = \dfrac{-x^3}{3x^2 - 9}$

$\qquad h(-x) = \dfrac{-(-x)^3}{3(-x)^2 - 9} = \dfrac{x^3}{3x^2 - 9}$

h is odd.

59. $f(x) = x^3 - 3x + 2$ on the interval $(-2,2)$

Use MAXIMUM and MINIMUM on the graph of $y_1 = x^3 - 3x + 2$.

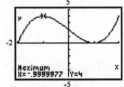

local maximum at: $(-1,4)$;   local minimum at: $(1,0)$

f is increasing on: $(-2,-1)$ and $(1,2)$; f is decreasing on: $(-1,1)$

61. $f(x) = x^5 - x^3$ on the interval $(-2,2)$

Use MAXIMUM and MINIMUM on the graph of $y_1 = x^5 - x^3$.

local maximum at: $(-0.77,0.19)$;   local minimum at: $(0.77,-0.19)$

f is increasing on: $(-2,-0.77)$ and $(0.77,2)$; f is decreasing on: $(-0.77,0.77)$

63. $f(x) = -0.2x^3 - 0.6x^2 + 4x - 6$ on the interval $(-6,4)$

Use MAXIMUM and MINIMUM on the graph of $y_1 = -0.2x^3 - 0.6x^2 + 4x - 6$.

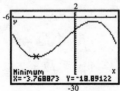

local maximum at: $(1.77,-1.91)$;   local minimum at: $(-3.77,-18.89)$

f is increasing on: $(-3.77,1.77)$; f is decreasing on: $(-6,-3.77)$ and $(1.77,4)$

65. $f(x) = 0.25x^4 + 0.3x^3 - 0.9x^2 + 3$ on the interval $(-3,2)$

Use MAXIMUM and MINIMUM on the graph of $y_1 = 0.25x^4 + 0.3x^3 - 0.9x^2 + 3$.

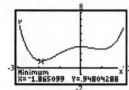 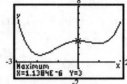 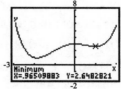

local maximum at: $(0,3)$; local minimum at: $(-1.87, 0.95)$, $(0.97, 2.65)$

f is increasing on: $(-1.87, 0)$ and $(0.97, 2)$; f is decreasing on: $(-3, -1.87)$ and $(0, 0.97)$

67. (a) length $= 24 - 2x$
     width $= 24 - 2x$
     height $= x$
     $V(x) = x(24 - 2x)(24 - 2x) = x(24 - 2x)^2$

     (b) $V(3) = 3(24 - 2(3))^2 = 3(18)^2 = 3(324) = 972$ cu. in.

     (c) $V(10) = 3(24 - 2(10))^2 = 3(4)^2 = 3(16) = 48$ cu. in.

     (d) $y_1 = x(24 - 2x)^2$ Use MAXIMUM.
                                              The volume is largest when $x = 4$ inches.

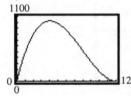

69. (a) $y_1 = -16x^2 + 80x + 6$

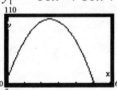

     (b) Use MAXIMUM. The maximum height occurs when $t = 2.5$ seconds.

     (c) From the graph, the maximum height is 106 feet.

71. $f(x) = x^2$

     (a) Average rate of change of f from $x = 0$ to $x = 1$:
     $$\frac{f(1) - f(0)}{1 - 0} = \frac{1^2 - 0^2}{1} = \frac{1}{1} = 1$$

     (b) Average rate of change of f from $x = 0$ to $x = 0.5$:
     $$\frac{f(0.5) - f(0)}{0.5 - 0} = \frac{(0.5)^2 - 0^2}{0.5} = \frac{0.25}{0.5} = 0.5$$

     (c) Average rate of change of f from $x = 0$ to $x = 0.1$:
     $$\frac{f(0.1) - f(0)}{0.1 - 0} = \frac{(0.1)^2 - 0^2}{0.1} = \frac{0.01}{0.1} = 0.1$$

(d)   Average rate of change of f from $x = 0$ to $x = 0.01$:

$$\frac{f(0.01) - f(0)}{0.01 - 0} = \frac{(0.01)^2 - 0^2}{0.01} = \frac{0.0001}{0.01} = 0.01$$

(e)   Average rate of change of f from $x = 0$ to $x = 0.001$:

$$\frac{f(0.001) - f(0)}{0.001 - 0} = \frac{(0.001)^2 - 0^2}{0.001} = \frac{0.000001}{0.001} = 0.001$$

(f)   Graphing the secant lines:

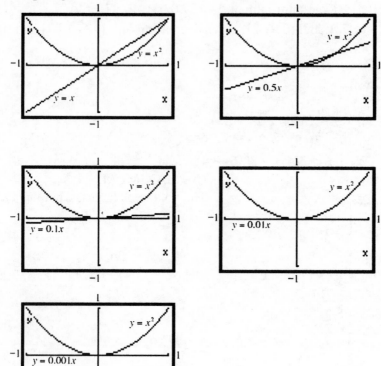

(g)   The secant lines are beginning to look more and more like the tangent line to the graph of f at the point where $x = 0$.

(h)   The slopes of the secant lines are getting smaller and smaller. They seem to be approaching the number zero.

73.   Answers will vary.

75.   One, at most, because if f is increasing it could only cross the x-axis at most one time. The graph of f could not "turn" and cross it again or it would start to decrease.

77.   The only such function is $f(x) = 0$.

Chapter 2

Functions and Their Graphs

2.4   Library of Functions; Piecewise-Defined Functions

1. $y = \sqrt{x}$

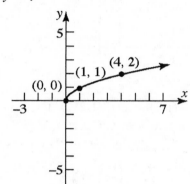

3. $y = x^3 - 8$

y - intercept :   Let $x = 0$, then $y = (0)^3 - 8 = -8$ $(0, -8)$

x - intercept :   Let $y = 0$, then $0 = x^3 - 8 \Rightarrow x^3 = 8 \Rightarrow x = 2$ $(2, 0)$

5.   piecewise-defined

7.   False

9.   C

11.   E

13.   B

15.   F

17. $f(x) = x$

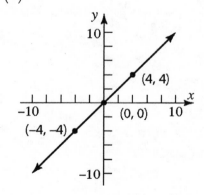

19. $f(x) = x^3$

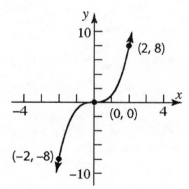

21. $f(x) = \dfrac{1}{x}$

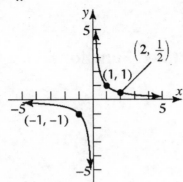

23. $f(x) = \sqrt[3]{x}$

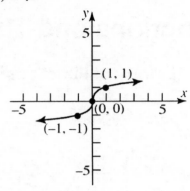

25. (a) $f(-2) = (-2)^2 = 4$
    (b) $f(0) = 2$
    (c) $f(2) = 2(2) + 1 = 5$

27. (a) $f(1.2) = \text{int}(2(1.2)) = \text{int}(2.4) = 2$
    (b) $f(1.6) = \text{int}(2(1.6)) = \text{int}(3.2) = 3$
    (c) $f(-1.8) = \text{int}(2(-1.8)) = \text{int}(-3.6) = -$

29. $f(x) = \begin{cases} 2x & \text{if } x \neq 0 \\ 1 & \text{if } x = 0 \end{cases}$

    (a) Domain: $\{x \,|\, x \text{ is any real number}\}$

    (b) x-intercept: none
        y-intercept: $(0,1)$

    (c)

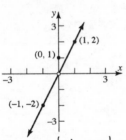

    (d) Range: $\{y \,|\, y \neq 0\}$

31. $f(x) = \begin{cases} -2x + 3 & \text{if } x < 1 \\ 3x - 2 & \text{if } x \geq 1 \end{cases}$

    (a) Domain: $\{x \,|\, x \text{ is any real number}\}$

    (b) x-intercept: none
        y-intercept: $(0,3)$

    (c)

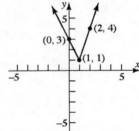

    (d) Range: $\{y \,|\, y \geq 1\}$

33. $f(x) = \begin{cases} x+3 & \text{if } -2 \le x < 1 \\ 5 & \text{if } x = 1 \\ -x+2 & \text{if } x > 1 \end{cases}$

(a) Domain: $\{x \mid x \ge -2\}$

(b) x-intercept: $(2, 0)$
     y-intercept: $(0, 3)$

(c)

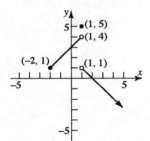

(d) Range: $\{y \mid y < 4\}$ and $y = 5$

35. $f(x) = \begin{cases} 1+x & \text{if } x < 0 \\ x^2 & \text{if } x \ge 0 \end{cases}$

(a) Domain: $\{x \mid x \text{ is any real number}\}$

(b) x-intercepts: $(-1,0), (0,0)$
     y-intercept: $(0,0)$

(c)

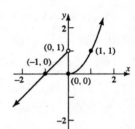

(d) Range: $\{y \mid y \text{ is any real number}\}$

37. $f(x) = \begin{cases} |x| & \text{if } -2 \le x < 0 \\ 1 & \text{if } x = 0 \\ x^3 & \text{if } x > 0 \end{cases}$

(a) Domain: $\{x \mid x \ge -2\}$

(b) x-intercept: none
     y-intercept: $(0, 1)$

(c)

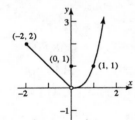

(d) Range: $\{y \mid y > 0\}$

39. $f(x) = 2\text{int}(x)$

(a) Domain: $\{x \mid x \text{ is any real number}\}$

(b) x-intercepts: all ordered pairs
     $(x,0)$ when $0 \le x < 1$.
     y-intercept: $(0,0)$

(c)

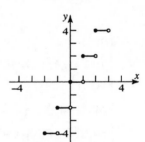

(d) Range: $\{y \mid y \text{ is an even integer}\}$

41. $f(x) = \begin{cases} -x & \text{if } -1 \le x \le 0 \\ \dfrac{1}{2}x & \text{if } 0 < x \le 2 \end{cases}$

43. $f(x) = \begin{cases} -x & \text{if } x \le 0 \\ -x+2 & \text{if } 0 < x \le 2 \end{cases}$

45. $C = \begin{cases} 39.99 & \text{if } 0 < x \le 350 \\ 0.25x - 47.51 & \text{if } x > 350 \end{cases}$

(a) $C(200) = \$39.99$

(b) $C(365) = 0.25(365) - 47.51 = \$43.74$

(c) $C(351) = 0.25(351) - 47.51 = \$40.24$

47.   (a)   Charge for 50 therms: $C = 9.45 + 0.6338(50) + 0.3128(50) = \$59.33$

      (b)   Charge for 500 therms:
            $C = 9.45 + 0.36375(50) + 0.11445(450) + 0.6338(500) = \$396.04$

      (c)   The monthly charge function:

$$C = \begin{cases} 9.45 + 0.36375x + 0.6338x & \text{for } 0 \le x \le 50 \\ 9.45 + 0.36375(50) + 0.11445(x - 50) + 0.6338x & \text{for } x > 50 \end{cases}$$

$$= \begin{cases} 9.45 + 0.99755x & \text{for } 0 \le x \le 50 \\ 9.45 + 18.1875 + 0.11445x - 5.7225 + 0.6338x & \text{for } x > 50 \end{cases}$$

$$= \begin{cases} 9.45 + 0.99755x & \text{for } 0 \le x \le 50 \\ 21.915 + 0.74825x & \text{for } x > 50 \end{cases}$$

      (d)   Graphing:

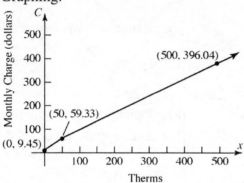

49.   For schedule X:

$$f(x) = \begin{cases} 0.10x & \text{if } 0 < x \le 7000 \\ 700 + 0.15(x - 7000) & \text{if } 7000 < x \le 28400 \\ 3910 + 0.25(x - 28,400) & \text{if } 28,400 < x \le 68,800 \\ 14,010 + 0.28(x - 68,800) & \text{if } 68,800 < x \le 143,500 \\ 34,926 + 0.33(x - 143,500) & \text{if } 143,500 < x \le 311,950 \\ 90,514.50 + 0.35(x - 311,950) & \text{if } x > 311,950 \end{cases}$$

51.   (a)   Let x represent the number of miles and C be the cost of transportation.

$$C(x) = \begin{cases} 0.50x & \text{if } 0 \le x \le 100 \\ 0.50(100) + 0.40(x - 100) & \text{if } 100 < x \le 400 \\ 0.50(100) + 0.40(300) + 0.25(x - 400) & \text{if } 400 < x \le 800 \\ 0.50(100) + 0.40(300) + 0.25(400) + 0(x - 800) & \text{if } 800 < x \le 960 \end{cases}$$

$$C(x) = \begin{cases} 0.50x & \text{if } 0 \le x \le 100 \\ 10 + 0.40x & \text{if } 100 < x \le 400 \\ 70 + 0.25x & \text{if } 400 < x \le 800 \\ 270 & \text{if } 800 < x \le 960 \end{cases}$$

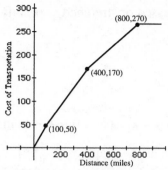

(b)   For hauls between 100 and 400 miles the cost is: $C(x) = 10 + 0.40x$.

(c)   For hauls between 400 and 800 miles the cost is: $C(x) = 70 + 0.25x$.

53.   Let x = the amount of the bill in dollars. The minimum payment due is given by

$$f(x) = \begin{cases} x & \text{if } x < 10 \\ 10 & \text{if } 10 \le x < 500 \\ 30 & \text{if } 500 \le x < 1000 \\ 50 & \text{if } 1000 \le x < 1500 \\ 70 & \text{if } 1500 \le x \end{cases}$$

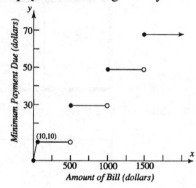

55.   (a) $W = 10°C$

(b) $W = 33 - \dfrac{(10.45 + 10\sqrt{5} - 5)(33 - 10)}{22.04} \approx 3.98°C$

(c) $W = 33 - \dfrac{(10.45 + 10\sqrt{15} - 15)(33 - 10)}{22.04} \approx -2.67°C$

(d) $W = 33 - 1.5958(33 - 10) = -3.7°C$

(e)   When $0 \le v < 1.79$, the wind speed is so small that there is no effect on the temperature.

(f)   For each drop of 1° in temperature, the wind chill factor drops approximately 1.6°C. When the wind speed exceeds 20, there is a constant drop in temperature.

57.   Each graph is that of $y = x^2$, but shifted vertically.

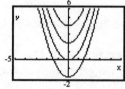

If $y = x^2 + k$, $k > 0$, the shift is up k units; if $y = x^2 + k$, $k < 0$, the shift is down $|k|$ units. The graph of $y = x^2 - 4$ is the same as the graph of $y = x^2$, but shifted down 4 units. The graph of $y = x^2 + 5$ is the graph of $y = x^2$, but shifted up 5 units.

59. Each graph is that of $y = |x|$, but either compressed or stretched vertically.

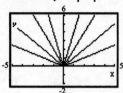

If $y = k|x|$ and $k > 1$, the graph is stretched; if $y = k|x|$ and $0 < k < 1$, the graph is compressed. The graph of $y = \frac{1}{4}|x|$ is the same as the graph of $y = |x|$, but compressed. The graph of $y = 5|x|$ is the same as the graph of $y = |x|$, but stretched.

61. The graph of $y = \sqrt{-x}$ is the reflection about the y-axis of the graph of $y = \sqrt{x}$.

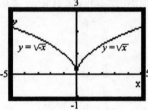

The same type of reflection occurs when graphing $y = 2x + 1$ and $y = 2(-x) + 1$.

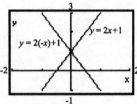

The graph of $y = f(-x)$ is the reflection about the y-axis of the graph of $y = f(x)$.

63. For the graph of $y = x^n$, n a positive even integer, as n increases, the graph of the function is narrower for $|x| > 1$ and flatter for $|x| < 1$.

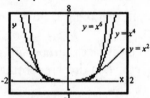

65. $f(x) = \begin{cases} 1 & \text{if } x \text{ is rational} \\ 0 & \text{if } x \text{ is irrational} \end{cases}$ $\{x \mid x \text{ is any real number}\}$ Range $= \{0, 1\}$

y-intercept: $x = 0 \Rightarrow x$ is rational $\Rightarrow y = 1$ So the y-intercept is (0, 1).
x-intercept: $y = 0 \Rightarrow x$ is irrational So the graph has infinitely many x-intercepts, namely, there is an x-intercept at each irrational value of x.
$f(-x) = 1 = f(x)$ when x is rational; $f(-x) = 0 = f(x)$ when x is irrational, So f is even.
The graph of f consists of 2 infinite clusters of distinct points, extending horizontally in both directions.
One cluster is located 1 unit above the x-axis, and the other is located along the x-axis.

Functions and Their Graphs

2.5 Graphing Techniques; Transformations

1. horizontal, right

3. –5, –2, and 2

5. False

7. *B*

9. *H*

11. *I*

13. *L*

15. *F*

17. *G*

19. $y = (x - 4)^3$

21. $y = x^3 + 4$

23. $y = (-x)^3 = -x^3$

25. $y = 4x^3$

27. (1) $y = \sqrt{x} + 2$
    (2) $y = -\left(\sqrt{x} + 2\right)$
    (3) $y = -\left(\sqrt{-x} + 2\right) = -\sqrt{-x} - 2$

29. (1) $y = -\sqrt{x}$
    (2) $y = -\sqrt{x} + 2$
    (3) $y = -\sqrt{x + 3} + 2$

31. c

33. c

35. $f(x) = x^2 - 1$
    Using the graph of $y = x^2$, vertically shift downward 1 unit.

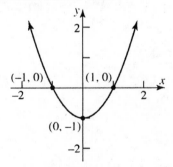

37. $g(x) = x^3 + 1$
    Using the graph of $y = x^3$, vertically shift upward 1 unit.

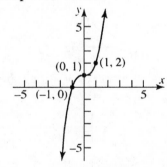

39. $h(x) = \sqrt{x-2}$

Using the graph of $y = \sqrt{x}$, horizontally
shift to the right 2 units.

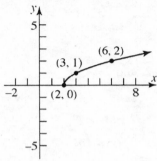

41. $f(x) = (x-1)^3 + 2$

Using the graph of $y = x^3$, horizontally
shift to the right 1 unit, then vertically
shift up 2 units.

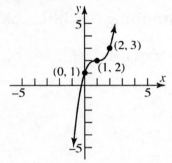

43. $g(x) = 4\sqrt{x}$

Using the graph of $y = \sqrt{x}$, vertically
stretch by a factor of 4.

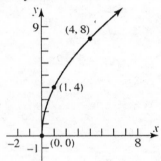

45. $h(x) = \dfrac{1}{2x} = \left(\dfrac{1}{2}\right)\left(\dfrac{1}{x}\right)$

Using the graph of $y = \dfrac{1}{x}$, vertically

compress by a factor of $\dfrac{1}{2}$.

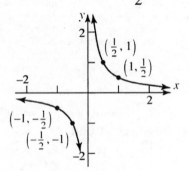

47. $f(x) = -\sqrt[3]{x}$

Reflect the graph of $y = \sqrt[3]{x}$, about the
x-axis.

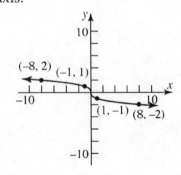

49. $g(x) = \left|-x\right|$

Reflect the graph of $y = |x|$ about the
y-axis.

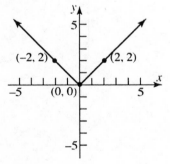

51. $h(x) = -x^3 + 2$

Reflect the graph of $y = x^3$ about the x-axis, then shift vertically upward 2 units.

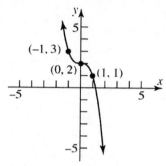

53. $f(x) = 2(x+1)^2 - 3$

Using the graph of $y = x^2$, horizontally shift to the left 1 unit, vertically stretch by a factor of 2, and vertically shift downward 3 units.

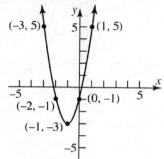

55. $g(x) = \sqrt{x-2} + 1$

Using the graph of $y = \sqrt{x}$, horizontally shift to the right 2 units and vertically shift upward 1 unit.

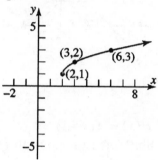

57. $h(x) = \sqrt{-x} - 2$

Reflect the graph of $y = \sqrt{x}$ about the y-axis and vertically shift downward 2 units.

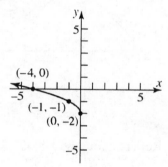

59. $f(x) = -(x+1)^3 - 1$

Using the graph of $y = x^3$, horizontally shift to the left 1 unit, reflect the graph about the x-axis, and vertically shift downward 1 unit.

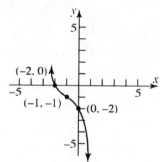

61. $g(x) = 2|1-x| = 2|-(-1+x)| = 2|x-1|$

Using the graph of $y = |x|$, horizontally shift to the right 1 unit, and vertically stretch by a factor or 2.

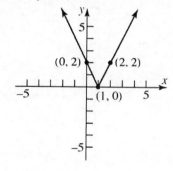

63. $h(x) = 2\,\text{int}(x-1)$
     Using the graph of $y = \text{int}(x)$,
     horizontally shift to the right 1 unit, and
     vertically stretch by a factor of 2.

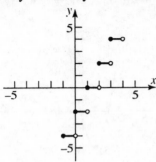

65. (a) $F(x) = f(x) + 3$
           Shift up 3 units.

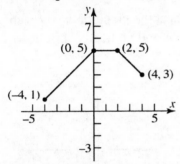

     (b) $G(x) = f(x+2)$
           Shift left 2 units.

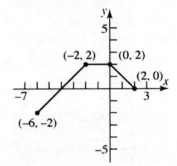

     (c) $P(x) = -f(x)$
           Reflect about the x-axis.

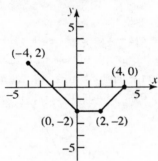

     (d) $H(x) = f(x+1) - 2$
           Shift left 1 unit and shift down 2
           units.

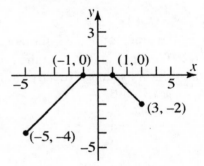

(e) $Q(x) = \dfrac{1}{2} f(x)$

Compress vertically by a factor of $\dfrac{1}{2}$.

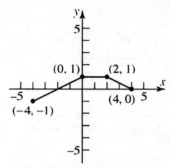

(f) $g(x) = f(-x)$
Reflect about the y-axis.

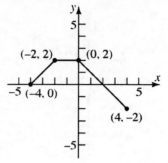

(g) $h(x) = f(2x)$

Compress horizontally by a factor of $\dfrac{1}{2}$.

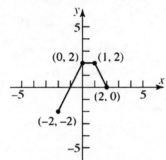

67.   (a) $F(x) = f(x) + 3$
Shift up 3 units.

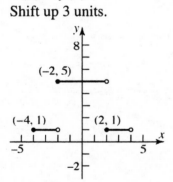

(b) $G(x) = f(x + 2)$
Shift left 2 units.

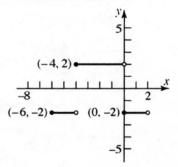

(c) $P(x) = -f(x)$
Reflect about the x-axis.

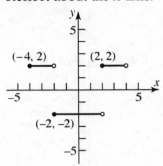

(d) $H(x) = f(x+1) - 2$
Shift left 1 unit and shift down 2 units.

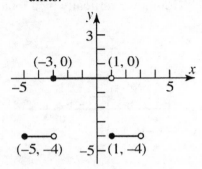

(e) $Q(x) = \frac{1}{2} f(x)$

Compress vertically by a factor of $\frac{1}{2}$.

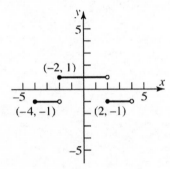

(f) $g(x) = f(-x)$
Reflect about the y-axis.

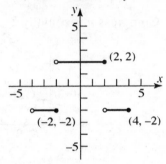

(g) $h(x) = f(2x)$

Compress horizontally by a factor of $\frac{1}{2}$.

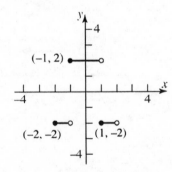

69. (a) $F(x) = f(x) + 3$
Shift up 3 units.

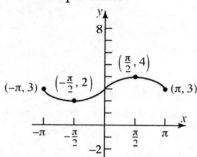

(b) $G(x) = f(x + 2)$
Shift left 2 units.

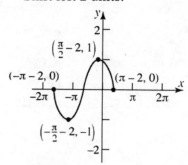

(c) $P(x) = -f(x)$
Reflect about the x-axis.

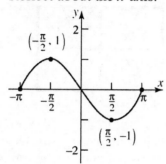

(d) $H(x) = f(x + 1) - 2$
Shift left 1 unit and shift down 2 units.

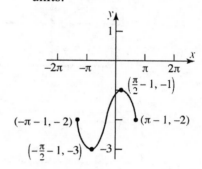

(e) $Q(x) = \dfrac{1}{2} f(x)$

Compress vertically by a factor of $\dfrac{1}{2}$.

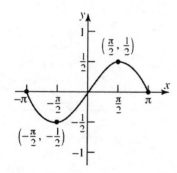

(f) $g(x) = f(-x)$
Reflect about the y-axis.

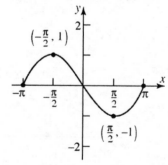

(g) $h(x) = f(2x)$

Compress horizontally by a factor of $\dfrac{1}{2}$.

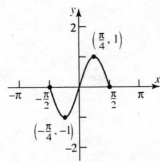

71. (a) $y = |x+1|$

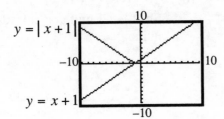

$y = x+1$

(b) $y = |4 - x^2|$

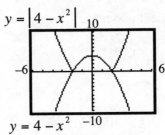

$y = 4 - x^2$

(c) $y = |x^3 + x|$

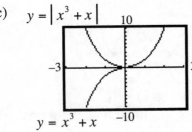

$y = x^3 + x$

(d) Any part of the graph of $y = f(x)$ that lies below the x-axis is reflected about the x-axis to obtain the graph of $y = |f(x)|$.

73. (a) $y = |f(x)|$

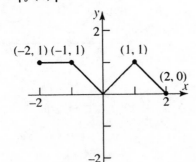

(b) $y = f(|x|)$

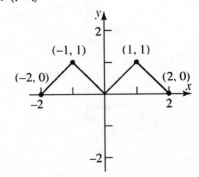

75. $f(x) = x^2 + 2x$
$f(x) = (x^2 + 2x + 1) - 1$
$f(x) = (x + 1)^2 - 1$
Using $f(x) = x^2$, shift left 1 unit and shift down 1 unit.

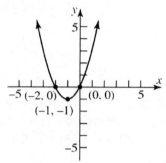

77. $f(x) = x^2 - 8x + 1$
$f(x) = (x^2 - 8x + 16) + 1 - 16$
$f(x) = (x - 4)^2 - 15$
Using $f(x) = x^2$, shift right 4 units and shift down 15 units.

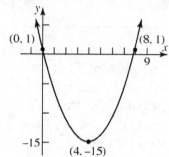

79. $f(x) = x^2 + x + 1$
$f(x) = \left(x^2 + x + \dfrac{1}{4}\right) + 1 - \dfrac{1}{4}$
$f(x) = \left(x + \dfrac{1}{2}\right)^2 + \dfrac{3}{4}$

Using $f(x) = x^2$, shift left $\dfrac{1}{2}$ unit and

shift up $\dfrac{3}{4}$ unit.

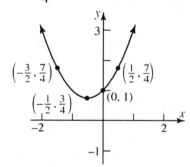

81. $y = (x - c)^2$
If $c = 0$, $y = x^2$.

If $c = 3$, $y = (x - 3)^2$; shift right 3 units.
If $c = -2$, $y = (x + 2)^2$; shift left 2 units.

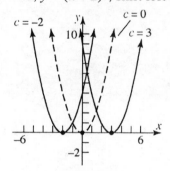

83. $F = \dfrac{9}{5}C + 32$

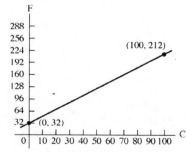

$F = \dfrac{9}{5}(K - 273) + 32$

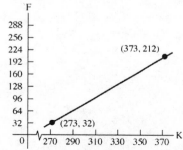

Shift the graph 273 units to the right.

85. (a)

(b) Select the 10% tax since the profits are higher.

(c) The graph of Y1 is obtained by shifting the graph of $p(x)$ vertically down 10,000 units. The graph of Y2 is obtained by multiplying the y-coordinate of the graph of $p(x)$ by 0.9. Thus, Y2 is the graph of $p(x)$ vertically compressed by a factor of 0.9.

(d) Select the 10% tax since the graph of $Y1 = 0.9p(x) \geq Y2 = -0.05x^2 + 100x - 6800$ for all x in the domain.

Chapter 2

Functions and Their Graphs

2.6 Mathematical Models; Constructing Functions

1. $V = \pi r^2 h, \; h = 2r \Rightarrow V(r) = \pi r^2 \cdot (2r) = 2\pi r^3$

3. (a) $R(x) = x\left(-\dfrac{1}{6}x + 100\right) = -\dfrac{1}{6}x^2 + 100x$

    (b) $R(200) = -\dfrac{1}{6}(200)^2 + 100(200) = \dfrac{-20{,}000}{3} + 20{,}000 = \dfrac{40{,}000}{3} \approx \$13{,}333$

    (c)

    (d) $x = 300$ maximizes revenue

         $R(300) = -\dfrac{1}{6}(300)^2 + 100(300) = -15{,}000 + 30{,}000 = \$15{,}000 = $ maximum revenue

    (e) $p = -\dfrac{1}{6}(300) + 100 = -50 + 100 = \$50$ maximizes revenue

5. (a) If $x = -5p + 100$, then $p = \dfrac{100 - x}{5}$. $R(x) = x\left(\dfrac{100 - x}{5}\right) = -\dfrac{1}{5}x^2 + 20x$

    (b) $R(15) = -\dfrac{1}{5}(15)^2 + 20(15) = -45 + 300 = \$255$

    (c)

    (d) $x = 50$ maximizes revenue

         $R(50) = -\dfrac{1}{5}(50)^2 + 20(50) = -500 + 1000 = \$500 = $ maximum revenue

    (e) $p = \dfrac{100 - 50}{5} = \dfrac{50}{5} = \$10$ maximizes revenue

7.   (a)   Let x = width and y = length of the rectangular area.

$$P = 2x + 2y = 400 \Rightarrow y = \frac{400 - 2x}{2} = 200 - x$$

Then $A(x) = (200 - x)x = 200x - x^2 = -x^2 + 200x$

(b)   We need $x \geq 0$ and $y \geq 0 \Rightarrow 200 - x \geq 0 \Rightarrow 200 \geq x$

So the domain of A is $\{x | 0 \leq x \leq 200\}$

(c)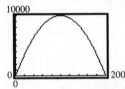

$x = 100$ yards maximizes area

9.   (a)   The distance d from P to the origin is $d = \sqrt{x^2 + y^2}$. Since P is a point on the graph

of $y = x^2 - 8$, we have:

$$d(x) = \sqrt{x^2 + (x^2 - 8)^2} = \sqrt{x^4 - 15x^2 + 64}$$

(b) $d(0) = \sqrt{0^4 - 15(0)^2 + 64} = \sqrt{64} = 8$

(c) $d(1) = \sqrt{(1)^4 - 15(1)^2 + 64} = \sqrt{1 - 15 + 64} = \sqrt{50} = 5\sqrt{2} \approx 7.07$

(d)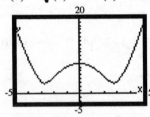

(e) d is smallest when $x \approx -2.74$ and when $x \approx 2.74$.

11. (a)   The distance d from P to the point $(1, 0)$ is $d = \sqrt{(x - 1)^2 + y^2}$. Since P is a point on

the graph of $y = \sqrt{x}$, we have:

$$d(x) = \sqrt{(x - 1)^2 + \left(\sqrt{x}\right)^2} = \sqrt{x^2 - x + 1}$$

(b)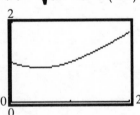

(c) d is smallest when x is 0.50.

13. By definition, a triangle has area $A = \frac{1}{2}bh, b$ = base, h = height. From the figure, we know

that $b = x$ and $h = y$. Expressing the area of the triangle as a function of x, we have:

$$A(x) = \frac{1}{2}xy = \frac{1}{2}x\left(x^3\right) = \frac{1}{2}x^4.$$

15. (a) $A(x) = xy = x(16 - x^2) = -x^3 + 16x$

    (b) Domain: $\{x \mid 0 < x < 4\}$

    (c)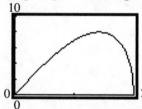

    The area is largest when x is approximately 2.31.

17. (a) $A(x) = (2x)(2y) = 4x(4 - x^2)^{1/2}$

    (b) $p(x) = 2(2x) + 2(2y) = 4x + 4(4 - x^2)^{1/2}$

    (c) Graphing the area equation: (d) Graphing the perimeter equation:

    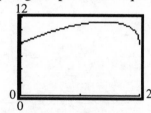

    The area is largest when x is   The perimeter is largest when x is
    approximately 1.41.   approximately 1.41.

19. (a) $C =$ circumference, $TA =$ total area, $r =$ radius, $x =$ side of square

    $$C = 2\pi r = 10 - 4x \quad \Rightarrow \quad r = \frac{5 - 2x}{\pi}$$

    Total Area = area of the square + area of the circle = $x^2 + \pi r^2$

    $$TA(x) = x^2 + \pi \left(\frac{5 - 2x}{\pi}\right)^2 = x^2 + \frac{25 - 20x + 4x^2}{\pi}$$

    (b) Since the lengths must be positive, we have:

    $$10 - 4x > 0 \quad \text{and} \quad x > 0$$
    $$-4x > -10 \quad \text{and} \quad x > 0$$
    $$x < 2.5 \qquad \text{and} \quad x > 0$$

    Domain: $\{x \mid 0 < x < 2.5\}$

    (c)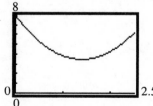

    The total area is smallest when x is approximately 1.40 meters.

21. (a) Since the wire of length x is bent into a circle, the circumference is x. Therefore, $C(x) = x$.

    (b) Since $C = x = 2\pi r$, $r = \dfrac{x}{2\pi}$.

$$A(x) = \pi r^2 = \pi\left(\frac{x}{2\pi}\right)^2 = \frac{x^2}{4\pi}.$$

23. (a) A = area, r = radius; diameter = $2r$ (b) p = perimeter

    $A(r) = (2r)(r) = 2r^2$ $p(r) = 2(2r) + 2r = 6r$

25. Area of the equilateral triangle $= \dfrac{1}{2}x \cdot \dfrac{\sqrt{3}}{2}x = \dfrac{\sqrt{3}}{4}x^2$

    Area of $\dfrac{1}{3}$ of the equilateral triangle $= \dfrac{1}{2}x\sqrt{r^2 - \left(\dfrac{x}{2}\right)^2} = \dfrac{1}{2}x\sqrt{r^2 - \dfrac{x^2}{4}} = \dfrac{1}{3} \cdot \dfrac{\sqrt{3}}{4}x^2$

    Solving for r^2:

$$\frac{1}{2}x\sqrt{r^2 - \frac{x^2}{4}} = \frac{1}{3} \cdot \frac{\sqrt{3}}{4}x^2 \Rightarrow \sqrt{r^2 - \frac{x^2}{4}} = \frac{2}{x} \cdot \frac{\sqrt{3}}{12}x^2$$

$$\sqrt{r^2 - \frac{x^2}{4}} = \frac{\sqrt{3}}{6}x \Rightarrow r^2 - \frac{x^2}{4} = \frac{3}{36}x^2 \Rightarrow r^2 = \frac{x^2}{3}$$

Area inside the circle, but outside the triangle:

$$A(x) = \pi r^2 - \frac{\sqrt{3}}{4}x^2 = \pi\frac{x^2}{3} - \frac{\sqrt{3}}{4}x^2 = \left(\frac{\pi}{3} - \frac{\sqrt{3}}{4}\right)x^2$$

27. (a)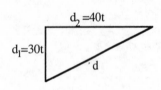

$$d^2 = d_1^2 + d_2^2$$

$$d^2 = (2 - 30t)^2 + (3 - 40t)^2$$

$$d(t) = \sqrt{(2 - 30t)^2 + (3 - 40t)^2}$$

$$= \sqrt{4 - 120t + 900t^2 + 9 - 240t + 1600t^2}$$

$$= \sqrt{2500t^2 - 360t + 13}$$

   (b)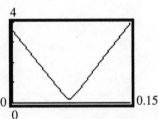

   The distance is smallest at $t \approx 0.072$ hours.

29. r = radius of cylinder, h = height of cylinder, V = volume of cylinder

By similar triangles: $\dfrac{H}{R} = \dfrac{H-h}{r}$

$$Hr = R(H - h)$$

$$Hr = RH - Rh$$

$$Rh = RH - Hr$$

$$h = \frac{RH - Hr}{R} = H - \frac{Hr}{R}$$

$$V = \pi r^2 h = \pi r^2 \left(H - \frac{Hr}{R} \right) = H\pi r^2 \left(1 - \frac{r}{R} \right)$$

31.   (a)   The time on the boat is given by $\dfrac{d_1}{3}$. The time on

land is given by $\dfrac{12 - x}{5}$.

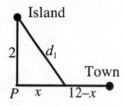

$$d_1 = \sqrt{x^2 + 2^2} = \sqrt{x^2 + 4}$$

The total time for the trip is:

$$T(x) = \frac{12 - x}{5} + \frac{d_1}{3} = \frac{12 - x}{5} + \frac{\sqrt{x^2 + 4}}{3}$$

(b)   Domain: $\left\{ x \mid 0 \le x \le 12 \right\}$

(c) $T(4) = \dfrac{12 - 4}{5} + \dfrac{\sqrt{4^2 + 4}}{3} = \dfrac{8}{5} + \dfrac{\sqrt{20}}{3} \approx 3.09$ hours

(d) $T(8) = \dfrac{12 - 8}{5} + \dfrac{\sqrt{8^2 + 4}}{3} = \dfrac{4}{5} + \dfrac{\sqrt{68}}{3} \approx 3.55$ hours

77

Functions and Their Graphs

2.R Chapter Review

1. This relation represents a function. Domain = $\{-1, 2, 4\}$; Range = $\{0, 3\}$.

3. $f(x) = \dfrac{3x}{x^2 - 1}$

   (a) $f(2) = \dfrac{3(2)}{(2)^2 - 1} = \dfrac{6}{4 - 1} = \dfrac{6}{3} = 2$

   (b) $f(-2) = \dfrac{3(-2)}{(-2)^2 - 1} = \dfrac{-6}{4 - 1} = \dfrac{-6}{3} = -2$

   (c) $f(-x) = \dfrac{3(-x)}{(-x)^2 - 1} = \dfrac{-3x}{x^2 - 1}$

   (d) $-f(x) = -\left(\dfrac{3x}{x^2 - 1}\right) = \dfrac{-3x}{x^2 - 1}$

   (e) $f(x + 2) = \dfrac{3(x + 2)}{(x + 2)^2 - 1} = \dfrac{3x + 6}{x^2 + 4x + 4 - 1} = \dfrac{3x + 6}{x^2 + 4x + 3}$

   (f) $f(2x) = \dfrac{3(2x)}{(2x)^2 - 1} = \dfrac{6x}{4x^2 - 1}$

5. $f(x) = \sqrt{x^2 - 4}$

   (a) $f(2) = \sqrt{2^2 - 4} = \sqrt{4 - 4} = \sqrt{0} = 0$

   (b) $f(-2) = \sqrt{(-2)^2 - 4} = \sqrt{4 - 4} = \sqrt{0} = 0$

   (c) $f(-x) = \sqrt{(-x)^2 - 4} = \sqrt{x^2 - 4}$

   (d) $-f(x) = -\sqrt{x^2 - 4}$

   (e) $f(x + 2) = \sqrt{(x + 2)^2 - 4} = \sqrt{x^2 + 4x + 4 - 4} = \sqrt{x^2 + 4x}$

   (f) $f(2x) = \sqrt{(2x)^2 - 4} = \sqrt{4x^2 - 4} = \sqrt{4(x^2 - 1)} = 2\sqrt{x^2 - 1}$

7. $f(x) = \dfrac{x^2 - 4}{x^2}$

   (a) $f(2) = \dfrac{2^2 - 4}{2^2} = \dfrac{4 - 4}{4} = \dfrac{0}{4} = 0$

   (b) $f(-2) = \dfrac{(-2)^2 - 4}{(-2)^2} = \dfrac{4 - 4}{4} = \dfrac{0}{4} = 0$

   (c) $f(-x) = \dfrac{(-x)^2 - 4}{(-x)^2} = \dfrac{x^2 - 4}{x^2}$

(d) $-f(x) = -\left(\dfrac{x^2 - 4}{x^2}\right) = \dfrac{4 - x^2}{x^2}$

(e) $f(x+2) = \dfrac{(x+2)^2 - 4}{(x+2)^2} = \dfrac{x^2 + 4x + 4 - 4}{x^2 + 4x + 4} = \dfrac{x^2 + 4x}{x^2 + 4x + 4}$

(f) $f(2x) = \dfrac{(2x)^2 - 4}{(2x)^2} = \dfrac{4x^2 - 4}{4x^2} = \dfrac{4(x^2 - 1)}{4x^2} = \dfrac{x^2 - 1}{x^2}$

9. $f(x) = \dfrac{x}{x^2 - 9}$

The denominator cannot be zero:
$$x^2 - 9 \neq 0$$
$$(x+3)(x-3) \neq 0$$
$$x \neq -3 \text{ or } 3$$
Domain: $\{x \mid x \neq -3, \ x \neq 3\}$

11. $f(x) = \sqrt{2 - x}$

The radicand must be non-negative:
$$2 - x \geq 0$$
$$x \leq 2$$
Domain: $\{x \mid x \leq 2\}$ or $(-\infty, 2]$

13. $f(x) = \dfrac{\sqrt{x}}{|x|}$

The radicand must be non-negative and
the denominator cannot be zero: $x > 0$
Domain: $\{x \mid x > 0\}$ or $(0, \infty)$

15. $f(x) = \dfrac{x}{x^2 + 2x - 3}$

The denominator cannot be zero:
$$x^2 + 2x - 3 \neq 0$$
$$(x+3)(x-1) \neq 0$$
$$x \neq -3 \text{ or } 1$$
Domain: $\{x \mid x \neq -3, \ x \neq 1\}$

17. $f(x) = 2 - x \quad g(x) = 3x + 1$

$(f + g)(x) = f(x) + g(x) = 2 - x + 3x + 1 = 2x + 3$

Domain: $\{x \mid x \text{ is any real number}\}$

$(f - g)(x) = f(x) - g(x) = 2 - x - (3x + 1) = 2 - x - 3x - 1 = -4x + 1$

Domain: $\{x \mid x \text{ is any real number}\}$

$(f \cdot g)(x) = f(x) \cdot g(x) = (2 - x)(3x + 1) = 6x + 2 - 3x^2 - x = -3x^2 + 5x + 2$

Domain: $\{x \mid x \text{ is any real number}\}$

$\left(\dfrac{f}{g}\right)(x) = \dfrac{f(x)}{g(x)} = \dfrac{2 - x}{3x + 1}$

$3x + 1 \neq 0$

$3x \neq -1 \Rightarrow x \neq -\dfrac{1}{3}$

Domain: $\left\{ x \mid x \neq -\dfrac{1}{3} \right\}$

19. $f(x) = 3x^2 + x + 1 \qquad g(x) = 3x$

     $(f + g)(x) = f(x) + g(x) = 3x^2 + x + 1 + 3x = 3x^2 + 4x + 1$

        Domain: $\{x \mid x \text{ is any real number}\}$

     $(f - g)(x) = f(x) - g(x) = 3x^2 + x + 1 - 3x = 3x^2 - 2x + 1$

        Domain: $\{x \mid x \text{ is any real number}\}$

     $(f \cdot g)(x) = f(x) \cdot g(x) = \left(3x^2 + x + 1\right)(3x) = 9x^3 + 3x^2 + 3x$

        Domain: $\{x \mid x \text{ is any real number}\}$

     $\left(\dfrac{f}{g}\right)(x) = \dfrac{f(x)}{g(x)} = \dfrac{3x^2 + x + 1}{3x}$

        $3x \neq 0 \Rightarrow x \neq 0$

        Domain: $\{x \mid x \neq 0\}$

21. $f(x) = \dfrac{x+1}{x-1} \qquad g(x) = \dfrac{1}{x}$

     $(f + g)(x) = f(x) + g(x) = \dfrac{x+1}{x-1} + \dfrac{1}{x} = \dfrac{x(x+1) + 1(x-1)}{x(x-1)} = \dfrac{x^2 + x + x - 1}{x(x-1)}$

        $= \dfrac{x^2 + 2x - 1}{x(x-1)}$

        Domain: $\{x \mid x \neq 0, x \neq 1\}$

     $(f - g)(x) = f(x) - g(x) = \dfrac{x+1}{x-1} - \dfrac{1}{x} = \dfrac{x(x+1) - 1(x-1)}{x(x-1)} = \dfrac{x^2 + x - x + 1}{x(x-1)}$

        $= \dfrac{x^2 + 1}{x(x-1)}$

        Domain: $\{x \mid x \neq 0, x \neq 1\}$

     $(f \cdot g)(x) = f(x) \cdot g(x) = \left(\dfrac{x+1}{x-1}\right)\left(\dfrac{1}{x}\right) = \dfrac{x+1}{x(x-1)}$

        Domain: $\{x \mid x \neq 0, x \neq 1\}$

     $\left(\dfrac{f}{g}\right)(x) = \dfrac{f(x)}{g(x)} = \dfrac{\dfrac{x+1}{x-1}}{\dfrac{1}{x}} = \left(\dfrac{x+1}{x-1}\right)\left(\dfrac{x}{1}\right) = \dfrac{x^2 + x}{x-1}$

        Domain: $\{x \mid x \neq 0, x \neq 1\}$

23. $f(x) = -2x^2 + x + 1$

$$\frac{f(x+h) - f(x)}{h} = \frac{-2(x+h)^2 + (x+h) + 1 - \left(-2x^2 + x + 1\right)}{h}$$

$$= \frac{-2\left(x^2 + 2xh + h^2\right) + x + h + 1 + 2x^2 - x - 1}{h}$$

$$= \frac{-2x^2 - 4xh - 2h^2 + x + h + 1 + 2x^2 - x - 1}{h}$$

$$= \frac{-4xh - 2h^2 + h}{h} = \frac{h(-4x - 2h + 1)}{h}$$

$$= -4x - 2h + 1$$

25.   (a)   Domain: $\left\{ x \mid -4 \le x \le 3 \right\}$ Range: $\left\{ y \mid -3 \le y \le 3 \right\}$

(b) x-intercept: $(0,0)$; y-intercept: $(0,0)$

(c) $f(-2) = -1$

(d) $f(x) = -3$ when $x = -4$

(e) $f(x) > 0$ when $0 < x \le 3$

(f)   To graph $y = f(x-3)$, shift the graph of f horizontally 3 units to the right.

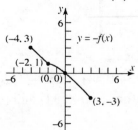

(g)   To graph $y = f\left(\dfrac{1}{2}x\right)$, stretch the graph of f horizontally by a factor of 2.

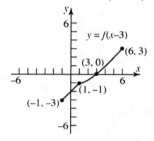

(h)   To graph $y = -f(x)$, reflect the graph of f vertically about the y-axis.

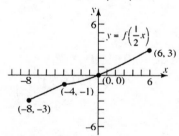

27. (a) Domain: $\{x| -4 \le x \le 4\}$ Range: $\{y| -3 \le y \le 1\}$

   (b) Increasing: $(-4,-1)$ and $(3,4)$;   Decreasing: $(-1, 3)$

   (c) Local minimum $(3,-3)$;   Local maximum $(-1,1)$

   (d) The graph is not symmetric with respect to the x-axis, the y-axis or the origin.

   (e) The function is neither even nor odd.

   (f) x-intercepts: $(-2,0)$, $(0,0)$, $(4,0)$; y-intercept: $(0,0)$

29. $f(x) = x^3 - 4x$

$$f(-x) = (-x)^3 - 4(-x) = -x^3 + 4x = -\left(x^3 - 4x\right) = -f(x) \qquad f \text{ is odd.}$$

31. $h(x) = \dfrac{1}{x^4} + \dfrac{1}{x^2} + 1$

$$h(-x) = \frac{1}{(-x)^4} + \frac{1}{(-x)^2} + 1 = \frac{1}{x^4} + \frac{1}{x^2} + 1 = h(x) \qquad h \text{ is even.}$$

33. $G(x) = 1 - x + x^3$

$$G(-x) = 1 - (-x) + (-x)^3 = 1 + x - x^3 \ne -G(x) \text{ or } G(x)$$

G is neither even nor odd.

35. $f(x) = \dfrac{x}{1+x^2}$

$$f(-x) = \frac{-x}{1+(-x)^2} = \frac{-x}{1+x^2} = -f(x) \quad f \text{ is odd.}$$

37. $f(x) = 2x^3 - 5x + 1$ on the interval $(-3,3)$

Use MAXIMUM and MINIMUM on the graph of $y_1 = 2x^3 - 5x + 1$.

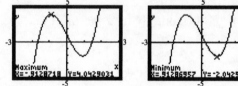

local maximum at: $(-0.91, 4.04)$;   local minimum at: $(0.91, -2.04)$

f is increasing on: $(-3, -0.91)$ and $(0.91, 3)$; f is decreasing on: $(-0.91, 0.91)$

39. $f(x) = 2x^4 - 5x^3 + 2x + 1$ on the interval $(-2,3)$

Use MAXIMUM and MINIMUM on the graph of $y_1 = 2x^4 - 5x^3 + 2x + 1$.

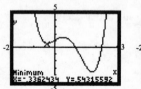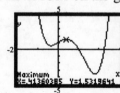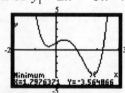

local maximum at: $(0.41, 1.53)$;   local minima at: $(-0.34, 0.54)$, $(1.80, -3.56)$

f is increasing on: $(-0.34, 0.41)$ and $(1.80, 3)$; f is decreasing on: $(-2, -0.34)$ and $(0.41, 1.80)$

41. $f(x) = 8x^2 - x$

(a) $\dfrac{f(2) - f(1)}{2 - 1} = \dfrac{8(2)^2 - 2 - \left(8(1)^2 - 1\right)}{1} = 32 - 2 - (7) = 23$

(b) $\dfrac{f(1) - f(0)}{1 - 0} = \dfrac{8(1)^2 - 1 - \left(8(0)^2 - 0\right)}{1} = 8 - 1 - (0) = 7$

(c) $\dfrac{f(4) - f(2)}{4 - 2} = \dfrac{8(4)^2 - 4 - \left(8(2)^2 - 2\right)}{2} = \dfrac{128 - 4 - (30)}{2} = \dfrac{94}{2} = 47$

43. $f(x) = 2 - 5x$

$\dfrac{f(x) - f(2)}{x - 2} = \dfrac{2 - 5x - (-8)}{x - 2} = \dfrac{-5x + 10}{x - 2} = \dfrac{-5(x - 2)}{x - 2} = -5$

45. $f(x) = 3x - 4x^2$

$\dfrac{f(x) - f(2)}{x - 2} = \dfrac{3x - 4x^2 - (-10)}{x - 2} = \dfrac{-4x^2 + 3x + 10}{x - 2}$

$= \dfrac{-(4x^2 - 3x - 10)}{x - 2} = \dfrac{-(4x + 5)(x - 2)}{x - 2} = -4x - 5$

47.   (b), (c), and (d) pass the Vertical Line Test and are therefore functions.

49. $f(x) = \sqrt[3]{x}$

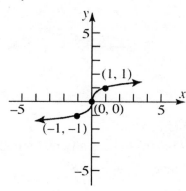

51. $F(x) = |x| - 4$

Using the graph of $y = |x|$, vertically shift the graph downward 4 units.

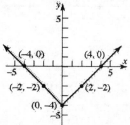

Intercepts: $(-4, 0)$, $(4, 0)$, $(0, -4)$
Domain: $\{x \mid x \text{ is any real number}\}$
Range: $\{y \mid y \geq -4\}$

53. $g(x) = -2|x|$

Reflect the graph of $y = |x|$ about the x-axis and vertically stretch the graph by a factor of 2.

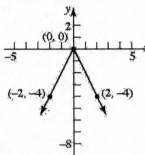

Intercepts: $(0,0)$
Domain: $\{x \mid x \text{ is any real number}\}$
Range: $\{y \mid y \leq 0\}$

55. $h(x) = \sqrt{x-1}$

Using the graph of $y = \sqrt{x}$, horizontally shift the graph to the right 1 unit.

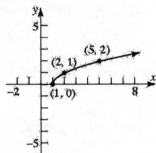

Intercept: $(1,0)$
Domain: $\{x \mid x \geq 1\}$
Range: $\{y \mid y \geq 0\}$

57. $f(x) = \sqrt{1-x} = \sqrt{-1(x-1)}$

Reflect the graph of $y = \sqrt{x}$ about the y-axis and horizontally shift the graph to the right 1 unit..

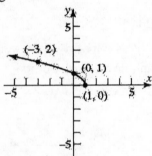

Intercepts: $(1,0)$, $(0,1)$
Domain: $\{x \mid x \leq 1\}$
Range: $\{y \mid y \geq 0\}$

59. $h(x) = (x-1)^2 + 2$

Using the graph of $y = x^2$, horizontally shift the graph to the right 1 unit and vertically shift the graph up 2 units.

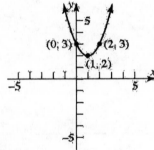

Intercepts: $(0,3)$
Domain: $\{x \mid x \text{ is any real number}\}$
Range: $\{y \mid y \geq 2\}$

61. $g(x) = 3(x-1)^3 + 1$

Using the graph of $y = x^3$, horizontally shift the graph to the right 1 unit vertically stretch the graph by a factor of 3, and vertically shift the graph up 1 unit.

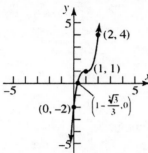

Intercepts: $(0,-2)$, $\left(1 - \dfrac{\sqrt[3]{3}}{3}, 0\right)$

Domain: $\left\{x \mid x \text{ is any real number}\right\}$

Range: $\left\{y \mid y \text{ is any real number}\right\}$

65. $f(x) = \begin{cases} x & \text{if } -4 \le x < 0 \\ 1 & \text{if } x = 0 \\ 3x & \text{if } x > 0 \end{cases}$

(a) Domain: $\left\{x \mid x \ge -4\right\}$

(b) x-intercept: none

  y-intercept: $(0, 1)$

(c)

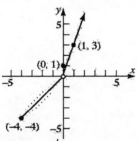

(d) Range: $\left\{y \mid y \ge -4,\ y \ne 0\right\}$

63. $f(x) = \begin{cases} 3x & \text{if } -2 < x \le 1 \\ x+1 & \text{if } x > 1 \end{cases}$

(a) Domain: $\left\{x \mid x > -2\right\}$

(b) x-intercept: $(0,0)$

  y-intercept: $(0,0)$

(c)

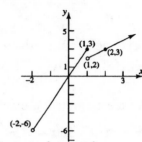

(d) Range: $\left\{y > -6\right\}$

67. $f(4) = -5$ gives the ordered pair $(4,-5)$.

  $f(0) = 3$ gives $(0,3)$.

Finding the slope: $m = \dfrac{3-(-5)}{0-4} = \dfrac{8}{-4} = -2$

Using slope-intercept form: $f(x) = -2x + 3$

69. $f(x) = \dfrac{Ax+5}{6x-2}$ and $f(1) = 4$

$$\dfrac{A(1)+5}{6(1)-2} = 4$$

$$\dfrac{A+5}{4} = 4$$

$$A + 5 = 16$$

$$A = 11$$

71. We have the points $(h_1, T_1) = (0,30)$ and $(h_2, T_2) = (10000,5)$.

$$\text{slope} = \dfrac{\Delta T}{\Delta h} = \dfrac{5-30}{10,000-0} = \dfrac{-25}{10,000} = -0.0025$$

Using the point-slope formula yields

$T - T_1 = m(h - h_1) \Rightarrow T - 30 = -0.0025(h - 0)$

$T - 30 = -0.0025h \Rightarrow T = -0.0025h + 30$

$T(h) = -0.0025h + 30$

73. $S = 4\pi r^2 \Rightarrow r = \sqrt{\dfrac{S}{4\pi}}$　　$V(S) = \dfrac{4}{3}\pi r^3 = \dfrac{4\pi}{3}\left(\sqrt{\dfrac{S}{4\pi}}\right)^3 = \dfrac{4\pi}{3} \cdot \dfrac{S}{4\pi}\sqrt{\dfrac{S}{4\pi}} = \dfrac{S}{6}\sqrt{\dfrac{S}{\pi}}$

$V(2S) = \dfrac{2S}{6}\sqrt{\dfrac{2S}{\pi}} = 2\sqrt{2}\left(\dfrac{S}{6}\sqrt{\dfrac{S}{\pi}}\right)$

The volume is $2\sqrt{2}$ times as large.

75. $S = kxd^3$,　$x = $ width;　$d = $ depth

In the diagram, depth = diameter of the log = 6

$S(x) = kx(6)^3 = 216kx$　　Domain: $\{x \mid 0 \le x \le 6\}$

77. (a) We are given that the volume is 500 centimeters, so we have

$V = \pi r^2 h = 500 \Rightarrow h = \dfrac{500}{\pi r^2}$

Total Cost = cost of top + cost of bottom + cost of body

$= 2(\text{cost of top}) + \text{cost of body}$

$= 2(\text{area of top})(\text{cost per area of top}) + (\text{area of body})(\text{cost per area of body})$

$= 2(\pi r^2)(0.06) + (2\pi rh)(0.04)$

$= 0.12\pi r^2 + 0.08\pi rh = 0.12\pi r^2 + 0.08\pi r\left(\dfrac{500}{\pi r^2}\right)$

$= 0.12\pi r^2 + \dfrac{40}{r}$

$C(r) = 0.12\pi r^2 + \dfrac{40}{r}$

(b) $C(4) = 0.12\pi(4)^2 + \dfrac{40}{4} = 1.92\pi + 10 \approx 16.03$ dollars

(c) $C(8) = 0.12\pi(8)^2 + \dfrac{40}{8} = 7.68\pi + 5 \approx 29.13$ dollars

(d) Graphing:

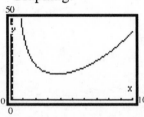

The minimum cost occurs when $r \approx 3.79$.

Chapter 2

Functions and Their Graphs

2.CR   Cumulative Review

1. $-5x + 4 = 0$
   $$-5x = -4$$
   $$x = \frac{-4}{-5} = \frac{4}{5}$$
   The solution set is $\left\{\frac{4}{5}\right\}$.

3. $3x^2 - 5x - 2 = 0$
   $$(3x + 1)(x - 2) = 0 \Rightarrow x = -\frac{1}{3}, x = 2$$
   The solution set is $\left\{-\frac{1}{3}, 2\right\}$.

5. $4x^2 - 2x + 4 = 0 \Rightarrow 2x^2 - x + 2 = 0$
   $$x = \frac{-(-1) \pm \sqrt{(-1)^2 - 4(2)(2)}}{2(2)}$$
   $$= \frac{1 \pm \sqrt{1 - 16}}{4} = \frac{1 \pm \sqrt{-15}}{4}$$
   no real solution

7. $\sqrt[5]{1 - x} = 2$
   $$\left(\sqrt[5]{1 - x}\right)^5 = 2^5$$
   $$1 - x = 32$$
   $$-x = 31$$
   $$x = -31$$
   Check $x = -31$:
   $$\sqrt[5]{1 - (-31)} = 2 \Rightarrow \sqrt[5]{32} = 2 \Rightarrow 2 = 2$$
   The solution set is $\{-31\}$.

9. $4x^2 - 2x + 4 = 0 \Rightarrow 2x^2 - x + 2 = 0$
   $$x = \frac{-(-1) \pm \sqrt{(-1)^2 - 4(2)(2)}}{2(2)}$$
   $$= \frac{1 \pm \sqrt{1 - 16}}{4} = \frac{1 \pm \sqrt{-15}}{4} = \frac{1 \pm \sqrt{15}i}{4}$$
   The solution set is $\left\{\frac{1 - \sqrt{15}i}{4}, \frac{1 + \sqrt{15}i}{4}\right\}$.

11. $-3x + 4y = 12 \Rightarrow 4y = 3x + 12$
    $$y = \frac{3}{4}x + 3$$
    This is a line with slope $\frac{3}{4}$ and y-intercept $(0,3)$.

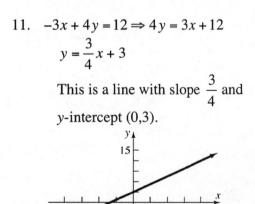

13. $x^2 + y^2 + 2x - 4y + 4 = 0$

$x^2 + 2x + y^2 - 4y = -4$

$(x^2 + 2x + 1) + (y^2 - 4y + 4) = -4 + 1 + 4$

$(x + 1)^2 + (y - 2)^2 = 1$

$(x + 1)^2 + (y - 2)^2 = 1^2$

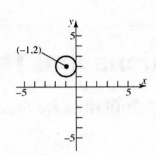

This is a circle with center $(-1, 2)$ and radius 1.

15.   (a)   Domain: $\{x | -4 \le x \le 4\}$ Range: $\{y | -3 \le y \le 3\}$

(b) x-intercept: $(1, 0)$; y-intercept: $(0, -1)$

(c)   The graph is not symmetric to the x-axis, the y-axis or the origin.

(d) $f(2) = 1$

(e) $f(x) = 3$ when $x = 4$

(f) $f(x) < 0$ when $-4 < x < 1$

(g) $y = f(x + 2)$

   Horizontally shift left 2 units

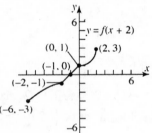

(h) $y = f(-x)$

   Reflect about the y-axis.

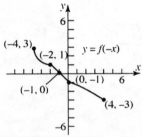

(i) $y = 2f(x)$

   Vertical stretch by a factor of 2.

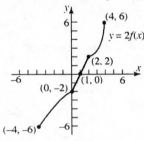

(j)   The function is neither even nor odd.

(k)   Increasing on the interval $(-4, 4)$

(l)   Decreasing: never

(m)   No local maximum. No local minimum.

(n) $\dfrac{f(4) - f(1)}{4 - 1} = \dfrac{3 - 0}{3} = \dfrac{3}{3} = 1$

Polynomial and Rational Functions

3.1 Quadratic Functions and Models

1. x-intercepts: $(-3,0)$, $(3,0)$
   y-intercept: $(0,-9)$

3. $\left(\dfrac{-5}{2}\right)^2 = \dfrac{25}{4}$

5. parabola

7. $-\dfrac{b}{2a}$

9. False

11. C

13. F

15. G

17. H

19. $f(x) = \dfrac{1}{4}x^2$

Using the graph of $y = x^2$, compress vertically by a factor of $\dfrac{1}{4}$.

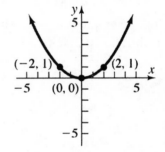

21. $f(x) = \dfrac{1}{4}x^2 - 2$

Using the graph of $y = x^2$, compress vertically by a factor of $\dfrac{1}{4}$, then shift down 2 units.

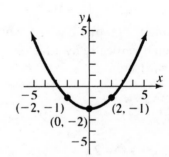

23. $f(x) = \dfrac{1}{4}x^2 + 2$

Using the graph of $y = x^2$, compress vertically by a factor of $\dfrac{1}{4}$, then shift up 2 units.

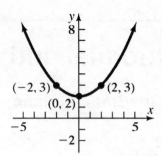

25. $f(x) = \dfrac{1}{4}x^2 + 1$

Using the graph of $y = x^2$, compress vertically by a factor of $\dfrac{1}{4}$, then shift up 1 unit.

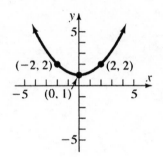

27. $f(x) = x^2 + 4x + 2$
$\quad = \left(x^2 + 4x + 4\right) + 2 - 4 = (x + 2)^2 - 2$

Using the graph of $y = x^2$, shift left 2 units, then shift down 2 units.

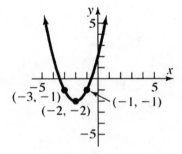

29. $f(x) = 2x^2 - 4x + 1$
$\quad = 2\left(x^2 - 2x + 1\right) + 1 - 2 = 2(x - 1)^2 - 1$

Using the graph of $y = x^2$, shift right 1 unit, stretch vertically by a factor of 2, then shift down 1 unit.

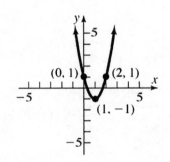

31. $f(x) = -x^2 - 2x$
$\quad = -\left(x^2 + 2x + 1\right) + 1 = -(x + 1)^2 + 1$

Using the graph of $y = x^2$, shift left 1 unit, reflect across the x-axis, then shift up 1 unit.

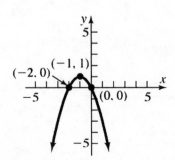

90

33. $f(x) = \dfrac{1}{2}x^2 + x - 1$

$= \dfrac{1}{2}\left(x^2 + 2x + 1\right) - 1 - \dfrac{1}{2}$

$= \dfrac{1}{2}(x + 1)^2 - \dfrac{3}{2}$

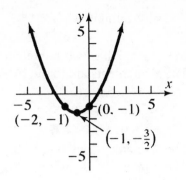

Using the graph of $y = x^2$, shift left 1 unit,

compress vertically by a factor of $\dfrac{1}{2}$, then shift

down $\dfrac{3}{2}$ units.

35. $f(x) = x^2 + 2x$

$a = 1,\ b = 2,\ c = 0$. Since $a = 1 > 0$, the graph opens up.

The x-coordinate of the vertex is $x = -\dfrac{b}{2a} = -\dfrac{2}{2(1)} = -\dfrac{2}{2} = -1$.

The y-coordinate of the vertex is $f\left(-\dfrac{b}{2a}\right) = f(-1) = (-1)^2 + 2(-1) = 1 - 2 = -1$.

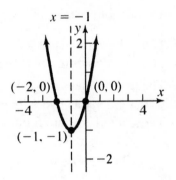

Thus, the vertex is $(-1, -1)$.

The axis of symmetry is the line $x = -1$.

The discriminant is

$b^2 - 4ac = (2)^2 - 4(1)(0) = 4 > 0$,

so the graph has two x-intercepts.

The x-intercepts are found by solving:

$x^2 + 2x = 0$

$x(x + 2) = 0$

$x = 0\ $ or $\ x = -2$

The x-intercepts are -2 and 0.

The y-intercept is $f(0) = 0$.

Domain: All real numbers.

Range: $\left\{y \mid y \ge -1\right\}$.

f is increasing on $(-1, \infty)$

f is decreasing on $(-\infty, -1)$

37. $f(x) = -x^2 - 6x$

$a = -1,\ b = -6,\ c = 0$. Since $a = -1 < 0$, the graph opens down.

The x-coordinate of the vertex is $x = -\dfrac{b}{2a} = -\dfrac{-6}{2(-1)} = \dfrac{6}{-2} = -3$.

The y-coordinate of the vertex is $f\left(-\dfrac{b}{2a}\right) = f(-3) = -(-3)^2 - 6(-3) = -9 + 18 = 9$.

Thus, the vertex is $(-3, 9)$.

The axis of symmetry is the line $x = -3$.

The discriminant is
$$b^2 - 4ac = (-6)^2 - 4(-1)(0) = 36 > 0,$$
so the graph has two x-intercepts.
The x-intercepts are found by solving:
$$-x^2 - 6x = 0$$
$$-x(x + 6) = 0$$
$$x = 0 \text{ or } x = -6$$
The x-intercepts are -6 and 0.
The y-intercept is $f(0) = 0$.
Domain: All real numbers.
Range: $\{y | y \le 9\}$.
f is increasing on $(-\infty, -3)$
f is decreasing on $(-3, \infty)$

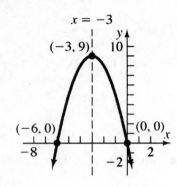

39. $f(x) = 2x^2 - 8x$
$a = 2, \ b = -8, \ c = 0.$ Since $a = 2 > 0,$ the graph opens up.

The x-coordinate of the vertex is $x = -\dfrac{b}{2a} = -\dfrac{-8}{2(2)} = \dfrac{8}{4} = 2.$

The y-coordinate of the vertex is $f\left(-\dfrac{b}{2a}\right) = f(2) = 2(2)^2 - 8(2) = 8 - 16 = -8.$

Thus, the vertex is $(2, -8).$
The axis of symmetry is the line $x = 2.$
The discriminant is $b^2 - 4ac = (-8)^2 - 4(2)(0) = 64 > 0,$
so the graph has two x-intercepts.
The x-intercepts are found by solving:
$$2x^2 - 8x = 0$$
$$2x(x - 4) = 0$$
$$x = 0 \text{ or } x = 4$$
The x-intercepts are 0 and 4.
The y-intercept is $f(0) = 0.$
Domain: All real numbers.
Range: $\{y | y \ge -8\}$.
f is increasing on $(2, \infty)$
f is decreasing on $(-\infty, 2)$

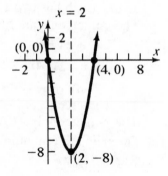

41. $f(x) = x^2 + 2x - 8$
$a = 1, \ b = 2, \ c = -8.$ Since $a = 1 > 0,$ the graph opens up.

The x-coordinate of the vertex is $x = -\dfrac{b}{2a} = -\dfrac{2}{2(1)} = -\dfrac{2}{2} = -1.$

The y-coordinate of the vertex is $f\left(-\dfrac{b}{2a}\right) = f(-1) = (-1)^2 + 2(-1) - 8 = 1 - 2 - 8 = -9.$

Thus, the vertex is $(-1, -9).$
The axis of symmetry is the line $x = -1.$

The discriminant is
$$b^2 - 4ac = 2^2 - 4(1)(-8) = 4 + 32 = 36 > 0,$$
so the graph has two x-intercepts.
The x-intercepts are found by solving:
$$x^2 + 2x - 8 = 0$$
$$(x + 4)(x - 2) = 0$$
$$x = -4 \text{ or } x = 2$$
The x-intercepts are –4 and 2.
The y-intercept is $f(0) = -8$.
Domain: All real numbers.
Range: $\{y | y \geq -9\}$.
f is increasing on $(-1, \infty)$
f is decreasing on $(-\infty, -1)$

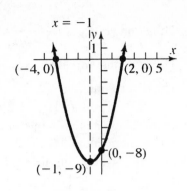

43. $f(x) = x^2 + 2x + 1$
    $a = 1, \ b = 2, \ c = 1.$ Since $a = 1 > 0,$ the graph opens up.

    The x-coordinate of the vertex is $x = -\dfrac{b}{2a} = -\dfrac{2}{2(1)} = -\dfrac{2}{2} = -1.$

    The y-coordinate of the vertex is $f\left(-\dfrac{b}{2a}\right) = f(-1) = (-1)^2 + 2(-1) + 1 = 1 - 2 + 1 = 0.$

    Thus, the vertex is (–1, 0).
    The axis of symmetry is the line $x = -1$.
    The discriminant is
    $$b^2 - 4ac = 2^2 - 4(1)(1) = 4 - 4 = 0,$$
    so the graph has one x-intercept.
    The x-intercept is found by solving:
    $$x^2 + 2x + 1 = 0$$
    $$(x + 1)^2 = 0$$
    $$x = -1$$
    The x-intercept is –1.
    The y-intercept is $f(0) = 1$.
    Domain: All real numbers.
    Range: $\{y | y \geq 0\}$.
    f is increasing on $(-1, \infty)$
    f is decreasing on $(-\infty, -1)$

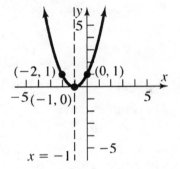

45. $f(x) = 2x^2 - x + 2$
    $a = 2, \ b = -1, \ c = 2.$ Since $a = 2 > 0,$ the graph opens up.

    The x-coordinate of the vertex is $x = -\dfrac{b}{2a} = -\dfrac{-1}{2(2)} = \dfrac{1}{4}.$

    The y-coordinate of the vertex is $f\left(-\dfrac{b}{2a}\right) = f\left(\dfrac{1}{4}\right) = 2\left(\dfrac{1}{4}\right)^2 - \dfrac{1}{4} + 2 = \dfrac{1}{8} - \dfrac{1}{4} + 2 = \dfrac{15}{8}.$

Thus, the vertex is $\left(\frac{1}{4}, \frac{15}{8}\right)$.

The axis of symmetry is the line $x = \frac{1}{4}$.

The discriminant is
$$b^2 - 4ac = (-1)^2 - 4(2)(2) = 1 - 16 = -15,$$
so the graph has no x-intercepts.

The y-intercept is $f(0) = 2$.

Domain: All real numbers.

Range: $\left\{y \middle| y \geq \frac{15}{8}\right\}$.

f is increasing on $\left(\frac{1}{4}, \infty\right)$

f is decreasing on $\left(-\infty, \frac{1}{4}\right)$

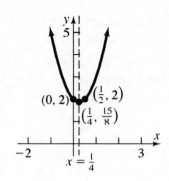

47. $f(x) = -2x^2 + 2x - 3$

$a = -2,\ b = 2,\ c = -3.$ Since $a = -2 < 0,$ the graph opens down.

The x-coordinate of the vertex is $x = -\dfrac{b}{2a} = -\dfrac{2}{2(-2)} = -\dfrac{2}{-4} = \dfrac{1}{2}.$

The y-coordinate of the vertex is $f\left(-\dfrac{b}{2a}\right) = f\left(\dfrac{1}{2}\right) = -2\left(\dfrac{1}{2}\right)^2 + 2\left(\dfrac{1}{2}\right) - 3 = -\dfrac{1}{2} + 1 - 3 = -\dfrac{5}{2}.$

Thus, the vertex is $\left(\dfrac{1}{2}, -\dfrac{5}{2}\right).$

The axis of symmetry is the line $x = \dfrac{1}{2}.$

The discriminant is
$$b^2 - 4ac = 2^2 - 4(-2)(-3) = 4 - 24 = -20,$$
so the graph has no x-intercepts.

The y-intercept is $f(0) = -3.$

Domain: All real numbers.

Range: $\left\{y \middle| y \leq -\dfrac{5}{2}\right\}.$

f is increasing on $\left(-\infty, \dfrac{1}{2}\right)$

f is decreasing on $\left(\dfrac{1}{2}, \infty\right)$

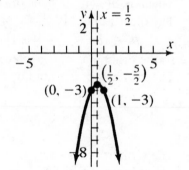

49. $f(x) = 3x^2 + 6x + 2$

$a = 3,\ b = 6,\ c = 2.$ Since $a = 3 > 0,$ the graph opens up.

The x-coordinate of the vertex is $x = -\dfrac{b}{2a} = -\dfrac{6}{2(3)} = -\dfrac{6}{6} = -1.$

The y-coordinate of the vertex is $f\left(-\dfrac{b}{2a}\right) = f(-1) = 3(-1)^2 + 6(-1) + 2 = 3 - 6 + 2 = -1.$

Thus, the vertex is $(-1, -1).$

The axis of symmetry is the line $x = -1$.

The discriminant is

$$b^2 - 4ac = 6^2 - 4(3)(2) = 36 - 24 = 12,$$

so the graph has two x-intercepts.

The x-intercepts are found by solving $3x^2 + 6x + 2 = 0$:

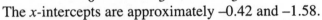

$$x = \frac{-b \pm \sqrt{b^2 - 4ac}}{2a} = \frac{-6 \pm \sqrt{12}}{2(3)}$$

$$= \frac{-6 \pm 2\sqrt{3}}{6} = \frac{-3 \pm \sqrt{3}}{3} \approx \frac{-3 \pm 1.732}{3}$$

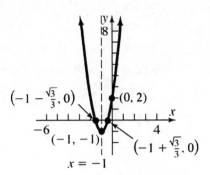

The x-intercepts are approximately -0.42 and -1.58.

The y-intercept is $f(0) = 2$.

Domain: All real numbers.

Range: $\{y | y \geq -1\}$.

f is increasing on $(-1, \infty)$

f is decreasing on $(-\infty, -1)$

51. $f(x) = -4x^2 - 6x + 2$

$a = -4$, $b = -6$, $c = 2$. Since $a = -4 < 0$, the graph opens down.

The x-coordinate of the vertex is $x = -\dfrac{b}{2a} = -\dfrac{-6}{2(-4)} = \dfrac{6}{-8} = -\dfrac{3}{4}$.

The y-coordinate of the vertex is $f\left(-\dfrac{b}{2a}\right) = f\left(-\dfrac{3}{4}\right) = -4\left(-\dfrac{3}{4}\right)^2 - 6\left(-\dfrac{3}{4}\right) + 2 = -\dfrac{9}{4} + \dfrac{9}{2} + 2 = \dfrac{17}{4}$.

Thus, the vertex is $\left(-\dfrac{3}{4}, \dfrac{17}{4}\right)$.

The axis of symmetry is the line $x = -\dfrac{3}{4}$.

The discriminant is $b^2 - 4ac = (-6)^2 - 4(-4)(2) = 36 + 32 = 68$, so the graph has two x-intercepts.

The x-intercepts are found by solving $-4x^2 - 6x + 2 = 0$:

$$x = \frac{-b \pm \sqrt{b^2 - 4ac}}{2a} = \frac{-(-6) \pm \sqrt{68}}{2(-4)} = \frac{6 \pm 2\sqrt{17}}{-8} = \frac{-3 \pm \sqrt{17}}{4} \approx \frac{-3 \pm 4.123}{4}$$

The x-intercepts are approximately -1.78 and 0.28.

The y-intercept is $f(0) = 2$.

Domain: All real numbers.

Range: $\left\{y | y \leq \dfrac{17}{4}\right\}$.

f is increasing on $\left(-\infty, \dfrac{3}{4}\right)$

f is decreasing on $\left(\dfrac{3}{4}, \infty\right)$

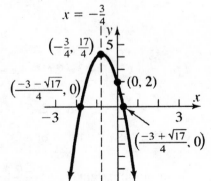

53. Use the form $f(x) = a(x - h)^2 + k$.

The vertex is $(-1, -2)$, so $h = -1$ and $k = -2$.

$f(x) = a(x - (-1))^2 + (-2) = a(x + 1)^2 - 2$.

Since the graph passes through $(0, -1)$, $f(0) = -1$.

$f(x) = a(x + 1)^2 - 2$

$-1 = a(0 + 1)^2 - 2$

$-1 = a - 2$

$1 = a$

$f(x) = (1)(x + 1)^2 - 2 = x^2 + 2x - 1$.

55. Use the form $f(x) = a(x - h)^2 + k$.

The vertex is $(-3, 5)$, so $h = -3$ and $k = 5$.

$f(x) = a(x - (-3))^2 + 5 = a(x + 3)^2 + 5$.

Since the graph passes through $(0, -4)$, $f(0) = -4$.

$f(x) = a(x + 3)^2 + 5$

$-4 = a(0 + 3)^2 + 5$

$-4 = 9a + 5$

$-9 = 9a$

$-1 = a$

$f(x) = (-1)(x + 3)^2 + 5 = -x^2 - 6x - 4$.

57. Use the form $f(x) = a(x - h)^2 + k$.

The vertex is $(1, -3)$, so $h = 1$ and $k = -3$.

$f(x) = a(x - 1)^2 + (-3) = a(x - 1)^2 - 3$.

Since the graph passes through $(3, 5)$, $f(3) = 5$.

$f(x) = a(x - 1)^2 - 3$

$5 = a(3 - 1)^2 - 3$

$5 = 4a - 3$

$8 = 4a$

$2 = a$

$f(x) = 2(x - 1)^2 - 3 = 2x^2 - 4x - 1$.

59. $f(x) = 2x^2 + 12x$, $a = 2$, $b = 12$, $c = 0$. Since $a = 2 > 0$, the graph opens up, so

the vertex is a minimum point. The minimum occurs at $x = -\dfrac{b}{2a} = -\dfrac{12}{2(2)} = -\dfrac{12}{4} = -3$.

The minimum value is $f\left(-\dfrac{b}{2a}\right) = f(-3) = 2(-3)^2 + 12(-3) = 18 - 36 = -18$.

61. $f(x) = 2x^2 + 12x - 3$, $a = 2$, $b = 12$, $c = -3$. Since $a = 2 > 0$, the graph opens up,

so the vertex is a minimum point. The minimum occurs at $x = -\dfrac{b}{2a} = -\dfrac{12}{2(2)} = -\dfrac{12}{4} = -3$.

The minimum value is $f\left(-\dfrac{b}{2a}\right) = f(-3) = 2(-3)^2 + 12(-3) - 3 = 18 - 36 - 3 = -21$.

63. $f(x) = -x^2 + 10x - 4$

$a = -1$, $b = 10$, $c = -4$. Since $a = -1 < 0$, the graph opens down, so the vertex is a

maximum point. The maximum occurs at $x = -\dfrac{b}{2a} = -\dfrac{10}{2(-1)} = -\dfrac{10}{-2} = 5$.

The maximum value is $f\left(-\dfrac{b}{2a}\right) = f(5) = -(5)^2 + 10(5) - 4 = -25 + 50 - 4 = 21$.

65. $f(x) = -3x^2 + 12x + 1$

$a = -3$, $b = 12$, $c = 1$. Since $a = -3 < 0$, the graph opens down, so the vertex is a

maximum point. The maximum occurs at $x = -\dfrac{b}{2a} = -\dfrac{12}{2(-3)} = -\dfrac{12}{-6} = 2$.

The maximum value is $f\left(-\dfrac{b}{2a}\right) = f(2) = -3(2)^2 + 12(2) + 1 = -12 + 24 + 1 = 13$.

67. (a) $f(x) = 1(x - (-3))(x - 1) = 1(x + 3)(x - 1) = 1\left(x^2 + 2x - 3\right) = x^2 + 2x - 3$

$f(x) = 2(x - (-3))(x - 1) = 2(x + 3)(x - 1) = 2\left(x^2 + 2x - 3\right) = 2x^2 + 4x - 6$

$f(x) = -2(x - (-3))(x - 1) = -2(x + 3)(x - 1) = -2\left(x^2 + 2x - 3\right) = -2x^2 - 4x + 6$

$f(x) = 5(x - (-3))(x - 1) = 5(x + 3)(x - 1) = 5\left(x^2 + 2x - 3\right) = 5x^2 + 10x - 15$

(b)   The value of a multiplies the value of the y-intercept by the value of a. The values
of the x-intercepts are not changed.

(c)   The axis of symmetry is unaffected by the value of a.

(d)   The y-coordinate of the vertex is multiplied by the value of a.

(e)   The x-coordinate of the vertex is the midpoint of the x-intercepts.

69. $R(p) = -4p^2 + 4000p$, $a = -4$, $b = 4000$, $c = 0$. Since $a = -4 < 0$, the graph is a
parabola that opens down, so the vertex is a maximum point. The maximum occurs at

$p = -\dfrac{b}{2a} = -\dfrac{4000}{2(-4)} = 500$.

Thus, the unit price should be \$500 for maximum revenue.

The maximum revenue is

$R(500) = -4(500)^2 + 4000(500) = -1{,}000{,}000 + 2{,}000{,}000 = \$1{,}000{,}000$

71. (a) $R(x) = x\left(-\dfrac{1}{6}x + 100\right) = -\dfrac{1}{6}x^2 + 100x$

(b) $R(200) = -\dfrac{1}{6}(200)^2 + 100(200) = -\dfrac{20{,}000}{3} + 20{,}000 = \dfrac{40{,}000}{3} \approx \$13{,}333.33$

(c)   A quantity of $x = -\dfrac{b}{2a} = -\dfrac{100}{2(-1/6)} = -\dfrac{100}{(-1/3)} = 300$ maximizes revenue.

The maximum revenue is $R(300) = -\dfrac{1}{6}(300)^2 + 100(300) = -15{,}000 + 30{,}000 = \$15{,}000$

(d)   The unit price should be $p = -\dfrac{1}{6}(300) + 100 = -50 + 100 = \$50$ to maximize revenue.

73.   (a)   If $x = -5p + 100$, then $p = \dfrac{100 - x}{5}$. $R(x) = x\left(\dfrac{100 - x}{5}\right) = -\dfrac{1}{5}x^2 + 20x$

(b) $R(15) = -\dfrac{1}{5}(15)^2 + 20(15) = -45 + 300 = \$255$

(c)   A quantity of $x = -\dfrac{b}{2a} = -\dfrac{20}{2(-1/5)} = -\dfrac{20}{(-2/5)} = \dfrac{100}{2} = 50$ maximizes revenue.

The maximum revenue is $R(50) = -\dfrac{1}{5}(50)^2 + 20(50) = -500 + 1000 = \$500$.

(d)   The unit price should be $p = \dfrac{100 - 50}{5} = \dfrac{50}{5} = \$10$ to maximize revenue.

75.   (a)   Let w = width and l = length of the rectangular area.

$P = 2w + 2l = 400 \Rightarrow l = \dfrac{400 - 2w}{2} = 200 - w$

Then $A(w) = (200 - w)w = 200w - w^2 = -w^2 + 200w$

(b)   A width of $w = \dfrac{b}{2a} = -\dfrac{200}{2(-1)} = -\dfrac{200}{-2} = 100$ yards maximizes area

(c)   The maximum area is $A(100) = -(100)^2 + 200(100) = -10{,}000 + 20{,}000 = 10{,}000$ sq yds.

77.   Since the width of the plot is x, and the length of the plot is $4000 - x$, the area is

$A(x) = (4000 - 2x)x = 4000x - 2x^2 = -2x^2 + 4000x$.

$x = -\dfrac{b}{2a} = -\dfrac{4000}{2(-2)} = -\dfrac{4000}{-4} = 1000$ maximizes area

$A(1000) = -2(1000)^2 + 4000(1000) = -2{,}000{,}000 + 4{,}000{,}000 = 2{,}000{,}000$

The largest area that can be enclosed is 2,000,000 square meters.

79.   (a) $a = \dfrac{-32}{(50)^2}$, $b = 1$, $c = 200$. The maximum height occurs when

$x = -\dfrac{b}{2a} = -\dfrac{1}{2\left(-\dfrac{32}{(50)^2}\right)} = \dfrac{2500}{64} = 39.0625$ feet from base of the cliff.

(b)   The maximum height is

$h(39.0625) = \dfrac{-32(39.0625)^2}{(50)^2} + 39.0625 + 200 \approx 219.53$ feet.

(c)   Solving $h(x) = 0 \Rightarrow \dfrac{-32x^2}{(50)^2} + x + 200 = 0$.

$$x = \frac{-1 \pm \sqrt{1^2 - 4\left(\dfrac{-32}{(50)^2}\right)(200)}}{2\left(\dfrac{-32}{(50)^2}\right)} = \frac{-1 \pm \sqrt{11.24}}{-0.0256} \Rightarrow x \approx -91.90 \text{ or } x \approx 170.02$$

Since the distance cannot be negative, the projectile strikes the water approximately 170.02 feet from the base of the cliff.

(d)   Graphing $y_1 = \dfrac{-32x^2}{(50)^2} + x + 200$

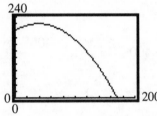

(e)   Solving $h(x) = 100 \Rightarrow \dfrac{-32x^2}{(50)^2} + x + 200 = 100 \Rightarrow \dfrac{-32x^2}{(50)^2} + x + 100 = 0$:

$$x = \frac{-1 \pm \sqrt{1^2 - 4\left(\dfrac{-32}{(50)^2}\right)(100)}}{2\left(\dfrac{-32}{(50)^2}\right)} = \frac{-1 \pm \sqrt{6.12}}{-0.0256} \Rightarrow x \approx -57.57 \text{ or } x \approx 135.70$$

Since the distance cannot be negative, the projectile is 100 feet above the water when it is approximately 135.70 feet from the base of the cliff.

81.   Locate the origin at the point where the cable touches the road. Then the equation of the parabola is of the form: $y = ax^2$, where $a > 0$. Since the point (200, 75) is on the parabola, we can find the constant a:

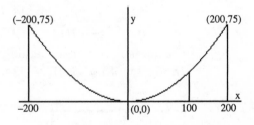

$$75 = a(200)^2 \Rightarrow a = \frac{75}{200^2}$$

When $x = 100$, we have:

$$y = \left(\frac{75}{200^2}\right)(100)^2 = 18.75 \text{ meters}$$

83.   Let x = the depth of the gutter, then the width of the gutter is $12 - 2x$.
Thus, the cross-sectional area of the gutter is $A = x(12 - 2x) = -2x^2 + 12x$.
This equation is a parabola opening down; thus, it has a maximum when
$$x = -\frac{b}{2a} = -\frac{12}{2(-2)} = -\frac{12}{-4} = 3.$$
Thus, a depth of 3 inches produces a maximum cross-sectional area.

85.   Let x = the width of the rectangle or the diameter of the semicircle.
Let y = the length of the rectangle.

The perimeter of each semicircle is $\frac{\pi x}{2}$.

The perimeter of the track is given by: $\frac{\pi x}{2} + \frac{\pi x}{2} + y + y = 400$.

Solving for x:

$$\frac{\pi x}{2} + \frac{\pi x}{2} + y + y = 1500 \Rightarrow \pi x + 2y = 400 \Rightarrow \pi x = 400 - 2y \Rightarrow x = \frac{400 - 2y}{\pi}$$

The area of the rectangle is: $A = xy = \left(\frac{400 - 2y}{\pi}\right)y = -\frac{2}{\pi}y^2 + \frac{400}{\pi}y$

This equation is a parabola opening down; thus, it has a maximum when

$$y = -\frac{b}{2a} = -\frac{\frac{400}{\pi}}{2\left(-\frac{2}{\pi}\right)} = -\frac{400}{-4} = 100. \quad \text{Thus, } x = \frac{400 - 2(100)}{\pi} = \frac{200}{\pi} \approx 63.66.$$

The dimensions for the rectangle with maximum area are $\frac{200}{\pi} \approx 63.66$ meters
by 100 meters.

87.   (a) $a = -1.01, b = 114.3, c = 451.0$ The maximum number of hunters occurs when
        the income level is
        $$x = -\frac{b}{2a} = -\frac{114.3}{2(-1.01)} = -\frac{114.3}{-2.02} \approx 56.584158 \text{ thousand dollars} = \$56,584.16$$
        The number of hunters earning this amount is:
        $H(56.584158) = -1.01(56.584158)^2 + 114.3(56.584158) + 451.0 \approx 3685$ hunters.

    (b)   Graphing: The function H is increasing on $(0, 56)$,

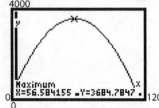

therefore the number of hunters is increasing
for individuals earning between $20,000 and
$40,000.

89. (a) $M(23) = 0.76(23)^2 - 107.00(23) + 3854.18 = 1795.22$ victims.

(b)   Solve for x: $M(x) = 0.76x^2 - 107.00x + 3854.18 = 1456$

$0.76x^2 - 107.00x + 3854.18 = 1456$

$0.76x^2 - 107.00x + 2398.18 = 0$
$a = 0.76, b = -107.00, c = 2398.18$

$$x = \frac{-b \pm \sqrt{b^2 - 4ac}}{2a} = \frac{-(-107) \pm \sqrt{(-107)^2 - 4(0.76)(2398.18)}}{2(0.76)}$$

$$= \frac{107 \pm \sqrt{4158.5328}}{1.52} \approx \frac{107 \pm 64.49}{1.52} \approx 112.82 \text{ or } 27.97$$

Since the model is valid on the interval $20 \le x < 90$, the only solution is $x \approx 27.97$.

(c)   Graphing:

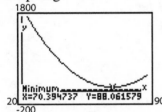

(d)   As age increases between the ages of 20 and 70.39, the number of murder victims decreases. After age 70.39, the number of murder victims increases as age increases.

91. (a)   The data appear to be quadratic with $a < 0$.

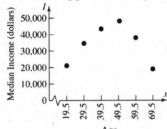

(b) $x = -\dfrac{b}{2a} = -\dfrac{3806}{2(-42.6)} \approx 44.6714$

An individual will earn the most income at an age of approximately 44.7 years.

(c)   The maximum income will be:

$I(44.7) = -42.6(44.7)^2 + 3806(44.7) - 38,526 \approx \$46,483.57$

(d) and (e)   Graphing the quadratic function of best fit:

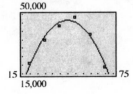

101

93. (a) The data appear to be quadratic with $a < 0$.

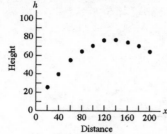

(b) $x = -\dfrac{b}{2a} = -\dfrac{1.03}{2(-0.0037)} \approx 139.19$ feet

The ball travels about 139 feet before reaching its maximum height.

(c) The maximum height will be:

$h(139) = -0.0037(139)^2 + 1.03(139) + 5.7 \approx 77.38$ feet

(d) and (e)   Graphing the quadratic function of best fit:

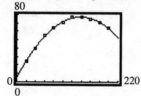

95. We are given: $V(x) = kx(a - x) = -kx^2 + akx$

The reaction rate is a maximum when: $x = -\dfrac{b}{2a} = -\dfrac{ak}{2(-k)} = \dfrac{ak}{2k} = \dfrac{a}{2}$.

97. $f(x) = -5x^2 + 8,\quad h = 1$

Area $= \dfrac{h}{3}\left(2ah^2 + 6c\right) = \dfrac{1}{3}\left(2(-5)(1)^2 + 6(8)\right) = \dfrac{1}{3}(-10 + 48) = \dfrac{38}{3} \approx 12.67$ sq. units.

99. $f(x) = x^2 + 3x + 5,\quad h = 4$

Area $= \dfrac{h}{3}\left(2ah^2 + 6c\right) = \dfrac{4}{3}\left(2(1)(4)^2 + 6(5)\right) = \dfrac{4}{3}(32 + 30) = \dfrac{248}{3} \approx 82.67$ sq. units.

101. Consider the diagram:

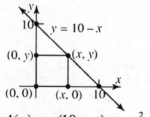

$A(x) = x(10 - x) = -x^2 + 10x$

The area is a maximum when: $x = \dfrac{b}{2a} = -\dfrac{10}{2(-1)} = \dfrac{10}{2} = 5$.

$A(5) = -(5)^2 + 10(5) = -25 + 50 = 25$

The largest area that can be enclosed is 25 square units.

103. Answers will vary.

105. $y = x^2 - 4x + 1$; $y = x^2 + 1$; $y = x^2 + 4x + 1$

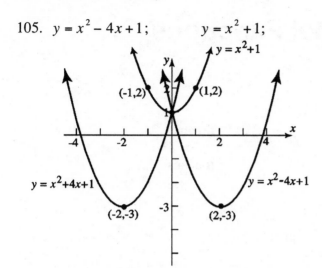

Each member of this family will be a parabola with the following characteristics:

(i) opens upwards since $a > 0$
(ii) y-intercept occurs at $(0, 1)$

107. By completing the square on the quadratic function $f(x) = ax^2 + bx + c$ we obtain the equation $y = a\left(x + \dfrac{b}{2a}\right)^2 + c - \dfrac{b^2}{4a}$. We can then draw the graph by applying transformations to the graph of the basic parabola $y = x^2$, which opens up. When $a > 0$, the basic parabola will either be stretched or compressed vertically.

When $a < 0$, the basic parabola will either be stretched or compressed vertically as well as reflected across the x-axis. Therefore, when $a > 0$, the graph of $f(x) = ax^2 + bx + c$ will open up, and when $a < 0$, the graph of $f(x) = ax^2 + bx + c$ will open down.

Polynomial and Rational Functions

3.2 Polynomial Functions

1. $(2, 0), (-2, 0)$, and $(0, 9)$

3. down, 4

5. smooth, continuous

7. touches

9. False

11. $f(x) = 4x + x^3$ is a polynomial function of degree 3.

13. $g(x) = \dfrac{1-x^2}{2} = \dfrac{1}{2} - \dfrac{1}{2}x^2$ is a polynomial function of degree 2.

15. $f(x) = 1 - \dfrac{1}{x} = 1 - x^{-1}$ is not a polynomial function because it contains a negative exponent.

17. $g(x) = x^{3/2} - x^2 + 2$ is not a polynomial function because it contains a fractional exponent.

19. $F(x) = 5x^4 - \pi x^3 + \dfrac{1}{2}$ is a polynomial function of degree 4.

21. $G(x) = 2(x-1)^2(x^2+1) = 2(x^2 - 2x + 1)(x^2 + 1) = 2(x^4 + x^2 - 2x^3 - 2x + x^2 + 1)$
     $= 2x^4 - 4x^3 + 4x^2 - 4x + 2$ is a polynomial function of degree 4.

23. $f(x) = (x+1)^4$
     Using the graph of $y = x^4$, shift the graph horizontally, to the left 1 unit.

25. $f(x) = x^5 - 3$
     Using the graph of $y = x^5$, shift the graph vertically, down 3 units.

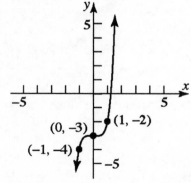

27. $f(x) = \dfrac{1}{2}x^4$

Using the graph of $y = x^4$, compress the graph vertically by a factor of $\dfrac{1}{2}$.

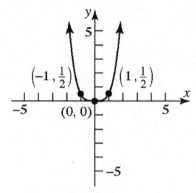

29. $f(x) = -x^5$

Using the graph of $y = x^5$, reflect the graph about the x-axis.

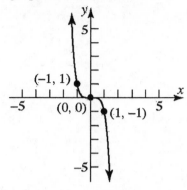

31. $f(x) = (x-1)^5 + 2$

Using the graph of $y = x^5$, shift the graph horizontally, to the right 1 unit, and shift vertically up 2 units.

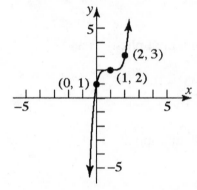

33. $f(x) = 2(x+1)^4 + 1$

Using the graph of $y = x^4$, shift the graph horizontally, to the left 1 unit, stretch vertically by a factor of 2, and shift vertically up 1 unit.

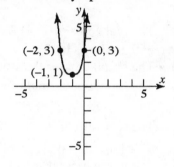

35. $f(x) = 4 - (x-2)^5 = -(x-2)^5 + 4$

Using the graph of $y = x^5$, shift the graph horizontally, to the right 2 units, reflect about the x-axis, and shift vertically up 4 units.

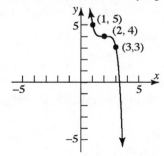

37. $f(x) = a(x - (-1))(x - 1)(x - 3)$

For $a = 1$: $f(x) = (x + 1)(x - 1)(x - 3) = (x^2 - 1)(x - 3) = x^3 - 3x^2 - x + 3$

39. $f(x) = a(x - (-3))(x - 0)(x - 4)$

For $a = 1$: $f(x) = (x + 3)(x)(x - 4) = (x^2 + 3x)(x - 4)$

$$= x^3 - 4x^2 + 3x^2 - 12x$$

$$= x^3 - x^2 - 12x$$

41. $f(x) = a(x - (-4))(x - (-1))(x - 2)(x - 3)$

For $a = 1$: $f(x) = (x + 4)(x + 1)(x - 2)(x - 3)$

$$= (x^2 + 5x + 4)(x^2 - 5x + 6)$$

$$= x^4 - 5x^3 + 6x^2 + 5x^3 - 25x^2 + 30x + 4x^2 - 20x + 24$$

$$= x^4 - 15x^2 + 10x + 24$$

43. $f(x) = a(x - (-1))(x - 3)^2$

For $a = 1$: $f(x) = (x + 1)(x - 3)^2$

$$= (x + 1)(x^2 - 6x + 9)$$

$$= x^3 - 6x^2 + 9x + x^2 - 6x + 9$$

$$= x^3 - 5x^2 + 3x + 9$$

45.   (a)   The real zeros of $f(x) = 3(x - 7)(x + 3)^2$ are: 7, with multiplicity 1; and –3, with multiplicity 2.
      (b)   The graph crosses the x-axis at 7 and touches it at –3.
      (c)   The function resembles $y = 3x^3$ for large values of $|x|$.

47.   (a)   The real zero of $f(x) = 4(x^2 + 1)(x - 2)^3$ is: 2, with multiplicity 3.

$x^2 + 1 = 0$ has no real solution.
      (b)   The graph crosses the x-axis at 2.
      (c)   The function resembles $y = 4x^5$ for large values of $|x|$.

49.   (a)   The real zero of $f(x) = -2\left(x + \frac{1}{2}\right)^2 (x^2 + 4)^2$ is: $-\frac{1}{2}$, with multiplicity 2.

$x^2 + 4 = 0$ has no real solution.
      (b)   The graph touches the x-axis at $-\frac{1}{2}$.
      (c)   The function resembles $y = -2x^6$ for large values of $|x|$.

51.   (a)   The real zeros of $f(x) = (x - 5)^3(x + 4)^2$ are: 5, with multiplicity 3; and –4, with multiplicity 2.
      (b)   The graph crosses the x-axis at 5 and touches it at –4.
      (c)   The function resembles $y = x^5$ for large values of $|x|$.

53. (a) $f(x) = 3(x^2 + 8)(x^2 + 9)^2$ has no real zeros, since $x^2 + 8 = 0$ and $x^2 + 9 = 0$ have no real solutions.
     (b) The graph neither touches nor crosses the x-axis.
     (c) The function resembles $y = 3x^6$ for large values of $|x|$.

55. (a) The real zeros of $f(x) = -2x^2(x^2 - 2)$ are: $-\sqrt{2}$ and $\sqrt{2}$ with multiplicity 1; and 0, with multiplicity 2.
     (b) The graph touches the x-axis at 0 and crosses the x-axis at $-\sqrt{2}$ and $\sqrt{2}$.
     (c) The function resembles $y = -2x^4$ for large values of $|x|$.

57. $f(x) = (x - 1)^2$
     (a) y-intercept: $f(0) = (0 - 1)^2 = 1$
          x-intercept: solve $f(x) = 0$
          $$(x - 1)^2 = 0 \Rightarrow x = 1$$
     (b) The graph touches the x-axis at $x = 1$, since this zero has multiplicity 2.
     (c) The graph of the function resembles $y = x^2$ for large values of $|x|$.
     (d) The graph has at most 1 turning point.
     (e) The x-intercept yields the intervals $(-\infty, 1)$ and $(1, \infty)$.

Interval	$(-\infty, 1)$	$(1, \infty)$
Number Chosen	-1	2
Value of f	$f(-1) = 4$	$f(2) = 1$
Location of Graph	Above x-axis	Above x-axis
Point on Graph	$(-1, 4)$	$(2, 1)$

f is above the x-axis on the intervals $(-\infty, 1)$ and $(1, \infty)$

     (f) Graphing:

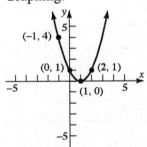

59. $f(x) = x^2(x - 3)$
     (a) y-intercept: $f(0) = 0^2(0 - 3) = 0$
          x-intercept: solve $f(x) = 0$
          $$x^2(x - 3) = 0 \Rightarrow x = 0 \text{ or } x = 3$$
     (b) The graph touches the x-axis at $x = 0$, since this zero has multiplicity 2.
          The graph crosses the x-axis at $x = 3$, since this zero has multiplicity 1.
     (c) The graph of the function resembles $y = x^3$ for large values of $|x|$.
     (d) The graph has at most 2 turning points.
     (e) The x-intercepts yield the intervals $(-\infty, 0)$, $(0, 3)$ and $(3, \infty)$.

	0	3	

Interval	$(-\infty, 0)$	$(0, 3)$	$(3, \infty)$
Number Chosen	-1	2	4
Value of f	$f(-1) = -4$	$f(2) = -4$	$f(4) = 16$
Location of Graph	Below x-axis	Below x-axis	Above x-axis
Point on Graph	$(-1, -4)$	$(2, -4)$	$(4, 16)$

f is below the x-axis on the intervals $(-\infty, 0)$ and $(0, 3)$

f is above the x-axis on the interval $(3, \infty)$

(f)   Graphing:

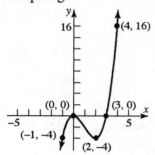

61. $f(x) = 6x^3(x + 4)$

(a) y-intercept: $f(x) = 6(0)^3(0 + 4) = 0$

x-intercept: solve $f(x) = 0$

$$6x^3(x + 4) = 0 \Rightarrow x = 0 \text{ or } x = -4$$

(b)   The graph crosses the x-axis at $x = 0$, since this zero has multiplicity 3.
The graph crosses the x-axis at $x = -4$, since this zero has multiplicity 1.

(c)   The graph of the function resembles $y = 6x^4$ for large values of $|x|$.

(d)   The graph has at most 3 turning points.

(e)   The x-intercepts yield the intervals $(-\infty, -4)$, $(-4, 0)$, and $(0, \infty)$.

	-4	0	

Interval	$(-\infty, -4)$	$(-4, 0)$	$(0, \infty)$
Number Chosen	-5	-2	1
Value of f	$f(-5) = 750$	$f(-2) = -96$	$f(1) = 30$
Location of Graph	Above x-axis	Below x-axis	Above x-axis
Point on Graph	$(-5, 750)$	$(-2, -96)$	$(1, 30)$

f is below the x-axis on the interval $(-4, 0)$

f is above the x-axis on the intervals $(-\infty, -4)$ and $(0, \infty)$

(f)   Graphing:

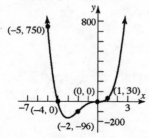

63. $f(x) = -4x^2(x+2)$

(a) y-intercept: $f(0) = -4(0)^2(0+2) = 0$

x-intercept: solve $f(x) = 0$

$$-4x^2(x+2) = 0 \Rightarrow x = 0 \text{ or } x = -2$$

(b) The graph crosses the x-axis at $x = -2$, since this zero has multiplicity 1. The graph touches the x-axis at $x = 0$, since this zero has multiplicity 2.

(c) The graph of the function resembles $y = -4x^3$ for large values of $|x|$.

(d) The graph has at most 2 turning points.

(e) The x-intercepts yield the intervals $(-\infty, -2)$, $(-2, 0)$, and $(0, \infty)$.

Interval	$(-\infty, -2)$	$(-2, 0)$	$(0, \infty)$
Number Chosen	-3	-1	1
Value of f	$f(-3) = 36$	$f(-1) = -4$	$f(1) = -12$
Location of Graph	Above x-axis	Below x-axis	Below x-axis
Point on Graph	(-3, 36)	(-1, -4)	(1, -12)

f is below the x-axis on the intervals $(-2, 0)$ and $(0, \infty)$

f is above the x-axis on the interval $(-\infty, -2)$

(f) Graphing:

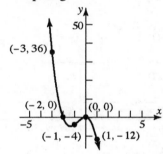

65. $f(x) = (x-1)(x-2)(x+4)$

(a) y-intercept: $f(0) = (0-1)(0-2)(0+4) = 8$

x-intercept: solve $f(x) = 0$

$$(x-1)(x-2)(x+4) = 0 \Rightarrow x = 1 \text{ or } x = 2 \text{ or } x = -4$$

(b) The graph crosses the x-axis at $x = 1$, $x = 2$, and $x = -4$ since each zero has multiplicity 1.

(c) The graph of the function resembles $y = x^3$ for large values of $|x|$.

(d) The graph has at most 2 turning points.

(e) The x-intercepts yield the intervals $(-\infty, -4)$, $(-4, 1)$, $(1, 2)$, and $(2, \infty)$.

Interval	$(-\infty, -4)$	$(-4, 1)$	$(1, 2)$	$(2, \infty)$
Number Chosen	-5	-2	$\frac{3}{2}$	3
Value of f	$f(-5) = -42$	$f(-2) = 24$	$f\left(\frac{3}{2}\right) = -\frac{11}{8}$	$f(3) = 14$
Location of Graph	Below x-axis	Above x-axis	Below x-axis	Above x-axis
Point on Graph	$(-5, -42)$	$(-2. 24)$	$\left(\frac{3}{2}, -\frac{11}{8}\right)$	(3, 14)

f is below the x-axis on the intervals $(-\infty,-4)$ and $(1,2)$

f is above the x-axis on the intervals $(-4,1)$ and $(2,\infty)$

(f)   Graphing:

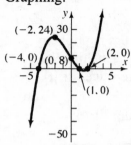

67. $f(x) = 4x - x^3 = x\left(4 - x^2\right) = x(2 + x)(2 - x)$

(a) y-intercept: $f(0) = 4x - x^3 = 4(0) - 0^3 = 0$

x-intercept: solve $f(x) = 0$

$$x(2 + x)(2 - x) = 0 \Rightarrow x = 0 \text{ or } x = -2 \text{ or } x = 2$$

(b)   The graph crosses the x-axis at $x = 0$, $x = -2$ and $x = 2$ since each zero has multiplicity 1.

(c)   The function resembles $y = -x^3$ for large values of $|x|$.

(d)   The graph has, at most, 2 turning points.

(e)   The x-intercepts yield the intervals $(-\infty,-2)$, $(-2,0)$, $(0,2)$ and $(2,\infty)$.

Interval	$(-\infty, -2)$	$(-2, 0)$	$(0, 2)$	$(2, \infty)$
Number Chosen	-3	-1	1	3
Value of f	$f(-3) = 15$	$f(-1) = -3$	$f(1) = 3$	$f(3) = -15$
Location of Graph	Above x-axis	Below x-axis	Above x-axis	Below x-axis
Point on Graph	$(-3, 15)$	$(-1, -3)$	$(1, 3)$	$(3, -15)$

f is below the x-axis on the intervals $(-2,0)$ and $(2,\infty)$

f is above the x-axis on the intervals $(-\infty,-2)$ and $(0,2)$

(f)   Graphing:

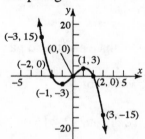

69. $f(x) = x^2(x - 2)(x + 2)$

(a) y-intercept: $f(0) = 0^2(0 - 2)(0 + 2) = 0$

x-intercept: solve $f(x) = 0$

$$x^2(x - 2)(x + 2) = 0 \Rightarrow x = 0 \text{ or } x = 2 \text{ or } x = -2$$

(b)   The graph touches the x-axis at $x = 0$, since this zero has multiplicity 2.

The graph crosses the x-axis at $x = 2$ and $x = -2$, since these zeros have multiplicity 1.

(c)   The graph of the function resembles $y = x^4$ for large values of $|x|$.

(d)   The graph has at most 3 turning points.

(e)   The x-intercepts yield the intervals $(-\infty, -2)$, $(-2, 0)$, $(0, 2)$, and $(2, \infty)$.

Interval	$(-\infty, -2)$	$(-2, 0)$	$(0, 2)$	$(2, \infty)$
Number Chosen	-3	-1	1	3
Value of f	$f(-3) = 45$	$f(-1) = -3$	$f(1) = -3$	$f(3) = 45$
Location of Graph	Above x-axis	Below x-axis	Below x-axis	Above x-axis
Point on Graph	$(-3, 45)$	$(-1, -3)$	$(1, -3)$	$(3, 45)$

f is below the x-axis on the intervals $(-2, 0)$ and $(0, 2)$

f is above the x-axis on the intervals $(-\infty, -2)$ and $(2, \infty)$

(f)   Graphing:

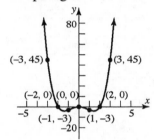

71. $f(x) = (x + 2)^2 (x - 2)^2$

(a) y-intercept: $f(0) = (0 + 2)^2 (0 - 2)^2 = 16$

x-intercept: solve $f(x) = 0$

$$(x + 2)^2 (x - 2)^2 = 0 \Rightarrow x = -2 \text{ or } x = 2$$

(b)   The graph touches the x-axis at $x = -2$ and $x = 2$, since each zero has multiplicity 2.

(c)   The graph of the function resembles $y = x^4$ for large values of $|x|$.

(d)   The graph has at most 3 turning points.

(e)   The x-intercepts yield the intervals $(-\infty, -2)$, $(-2, 2)$, and $(2, \infty)$.

Interval	$(-\infty, -2)$	$(-2, 2)$	$(2, \infty)$
Number Chosen	-3	0	3
Value of f	$f(-3) = 25$	$f(0) = 16$	$f(3) = 25$
Location of Graph	Above x-axis	Above x-axis	Above x-axis
Point on Graph	$(-3, 25)$	$(0, 16)$	$(3, 25)$

f is above the x-axis on the intervals $(-\infty, -2)$, $(-2, 2)$, and $(2, \infty)$.

(f)   Graphing:

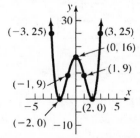

111

73. $f(x) = (x-1)^2(x-3)(x+1)$

   (a) y-intercept: $f(0) = (0-1)^2(0-3)(0+1) = -3$

         x-intercept: solve $f(x) = 0$

         $$(x-1)^2(x-3)(x+1) = 0 \Rightarrow x = 1 \text{ or } x = 3 \text{ or } x = -1$$

   (b)   The graph touches the x-axis at $x = 1$, since this zero has multiplicity 2.
         The graph crosses the x-axis at $x = 3$ and $x = -1$ since each zero has multiplicity 1.

   (c)   The graph of the function resembles $y = x^4$ for large values of $|x|$.

   (d)   The graph has at most 3 turning points.

   (e)   The x-intercepts yield the intervals $(-\infty,-1)$, $(-1,1)$, $(1,3)$, and $(3,\infty)$.

Interval	$(-\infty, -1)$	$(-1, 1)$	$(1, 3)$	$(3, \infty)$
Number Chosen	-2	0	2	4
Value of f	$f(-2) = 45$	$f(0) = -3$	$f(2) = -3$	$f(4) = 45$
Location of Graph	Above x-axis	Below x-axis	Below x-axis	Above x-axis
Point on Graph	$(-2, 45)$	$(0, -3)$	$(2, -3)$	$(4, 45)$

         f is below the x-axis on the intervals $(-1,1)$ and $(1,3)$

         f is above the x-axis on the intervals $(-\infty,-1)$ and $(3,\infty)$

   (f)   Graphing:

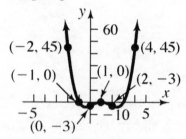

75. $f(x) = (x+2)^2(x-4)^2$

   (a) y-intercept: $f(0) = (0+2)^2(0-4)^2 = 64$

         x-intercept: solve $f(x) = 0$

         $$(x+2)^2(x-4)^2 = 0 \Rightarrow x = -2 \text{ or } x = 4$$

   (b)   The graph touches the x-axis at $x = -2$ and $x = 4$, since each zero has multiplicity 2.

   (c)   The graph of the function resembles $y = x^4$ for large values of $|x|$.

   (d)   The graph has at most 3 turning points.

   (e)   The x-intercepts yield the intervals $(-\infty,-2)$, $(-2,4)$, and $(4,\infty)$.

Interval	$(-\infty, -2)$	$(-2, 4)$	$(4, \infty)$
Number Chosen	-3	0	5
Value of f	$f(-3) = 49$	$f(0) = 64$	$f(5) = 49$
Location of Graph	Above x-axis	Above x-axis	Above x-axis
Point on Graph	(-3, 49)	(0, 64)	(5, 49)

         f is above the x-axis on the intervals $(-\infty,-2)$, $(-2,4)$, and $(4,\infty)$.

(f)   Graphing:

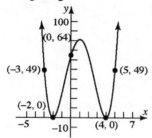

77. $f(x) = x^2(x-2)(x^2+3)$

(a) y-intercept: $f(0) = 0^2(0-2)(0^2+3) = 0$
x-intercept: solve $f(x) = 0$
$$x^2(x-2)(x^2+3) = 0 \Rightarrow x = 0 \text{ or } x = 2$$
$x^2 + 3 = 0$ has no real solution

(b)   The graph touches the x-axis at $x = 0$, since this zero has multiplicity 2.
The graph crosses the x-axis at $x = 2$, since this zero has multiplicity 1.

(c)   The graph of the function resembles $y = x^5$ for large values of $|x|$.

(d)   The graph has at most 4 turning points.

(e)   The x-intercepts yield the intervals $(-\infty,0)$, $(0,2)$ and $(2,\infty)$.

Interval	$(-\infty, 0)$	$(0, 2)$	$(2, \infty)$
Number Chosen	-1	1	3
Value of f	$f(-1) = -12$	$f(1) = -4$	$f(3) = 108$
Location of Graph	Below x-axis	Below x-axis	Above x-axis
Point on Graph	$(-1, -12)$	$(1, -4)$	$(3, 108)$

f is below the x-axis on the intervals $(-\infty,0)$ and $(0,2)$.
f is above the x-axis on the interval $(2,\infty)$.

(f)   Graphing:

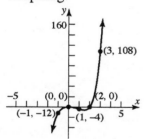

79. $f(x) = -x^2(x^2-1)(x+1) = -x^2(x-1)(x+1)(x+)1 = -x^2(x-1)(x+1)^2$

(a) y-intercept: $f(0) = -0^2(0^2-1)(0+1) = 0$
x-intercept: solve $f(x) = 0$
$$-x^2(x-1)(x+1)^2 = 0 \Rightarrow x = 0 \text{ or } x = 1 \text{ or } x = -1$$

(b)   The graph touches the x-axis at $x = 0$ and $x = -1$, since each zero has multiplicity 2.
The graph crosses the x-axis at $x = 1$, since this zero has multiplicity 1.

(c)   The graph of the function resembles $y = -x^5$ for large values of $|x|$.

(d)   The graph has at most 4 turning points.

(e)   The x-intercepts yield the intervals $(-\infty,-1)$, $(-1,0)$, $(0,1)$, and $(1,\infty)$.

Interval	$(-\infty, -1)$	$(-1, 0)$	$(0, 1)$	$(1, \infty)$
Number Chosen	-2	$-\frac{1}{2}$	$\frac{1}{2}$	2
Value of f	$f(-2) = 12$	$f\left(-\frac{1}{2}\right)=\frac{3}{32}$	$f\left(\frac{1}{2}\right)=\frac{9}{32}$	$f(2) = -36$
Location of Graph	Above x-axis	Above x-axis	Above x-axis	Below x-axis
Point on Graph	$(-2, 12)$	$\left(-\frac{1}{2}, \frac{3}{32}\right)$	$\left(\frac{1}{2}, \frac{9}{32}\right)$	$(2, -36)$

f is below the x-axis on the interval $(1,\infty)$.

f is above the x-axis on the intervals $(-\infty,-1)$, $(-1,0)$, and $(0,1)$.

(f)   Graphing:

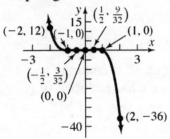

81.   c, e, and f   83.   c and e

85. $f(x) = x^3 + 0.2x^2 - 1.5876x - 0.31752$

(a)   Degree = 3; The graph of the function resembles $y = x^3$ for large values of $|x|$.

(b)   graphing utility

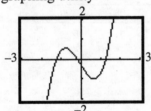

(c) x-intercepts: -1.26, -0.2, 1.26; y-intercept: -0.31752

(d)

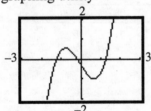

(e)   2 turning points; local maximum: $(-0.80, 0.57)$; local minimum: $(0.66, -0.99)$

(f)   graphing by hand

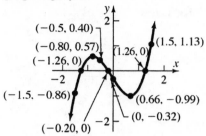

(g)   Domain: $\{x \mid x$ is any real number$\}$; Range: $\{y \mid y$ is any real number$\}$.

(h) f is increasing on $(-\infty, -0.80)$ and $(0.66, \infty)$; f is decreasing on $(-0.80, 0.66)$

$f(x) = x^3 + 2.56x^2 - 3.31x + 0.89$

(a)   Degree = 3; The graph of the function resembles $y = x^3$ for large values of $|x|$.

(b)   graphing utility

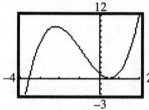

(c) x-intercepts: $-3.56, 0.50$; y-intercept: 0.89

(d)

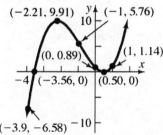

(e)   2 turning points; local maximum: $(-2.21, 9.91)$; local minimum: $(0.50, 0)$

(f)   graphing by hand

(g)   Domain: $\{x \mid x$ is any real number$\}$; Range: $\{y \mid y$ is any real number$\}$.

(h) f is increasing on $(-\infty, -2.21)$ and $(0.50, \infty)$; f is decreasing on $(-2.21, 0.50)$

89. $f(x) = x^4 - 2.5x^2 + 0.5625$

(a) Degree = 4; The graph of the function resembles $y = x^4$ for large values of $|x|$.

(b) graphing utility

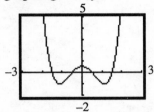

(c) x-intercepts: $-1.50, -0.50, 0.50, 1.50$; y-intercept: 0.5625

(d)

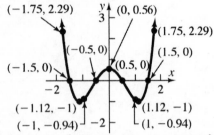

(e) 3 turning points: local maximum: $(0, 0.5625)$; local minima: $(-1.12, -1)$, $(1.12, -1)$

(f) graphing by hand

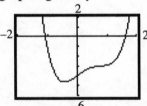

(g) Domain: $\{x \mid x \text{ is any real number}\}$; Range: $\{y \mid y \geq -1\}$.

(h) f is increasing on $(-1.12, 0)$ and $(1.12, \infty)$;
f is decreasing on $(-\infty, -1.12)$ and $(0, 1.12)$

91. $f(x) = 2x^4 - \pi x^3 + \sqrt{5}x - 4$

(a) Degree = 4; The graph of the function resembles $y = 2x^4$ for large values of $|x|$.

(b) graphing utility:

(c) x-intercepts: $-1.07, 1.62$; y-intercept: -4

(d)

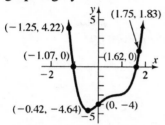

$Y_1 \boxplus 2X^4 - \pi X^3 + \sqrt{(5\ldots}$

(e)   1 turning point; local minimum: $(-0.42, -4.64)$

(f)   graphing by hand

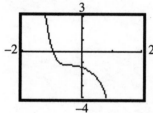

(−1.25, 4.22)
(1.75, 1.83)
(−1.07, 0)
(1.62, 0)
(−0.42, −4.64)
(0, −4)

(g)   Domain: $\{x \mid x \text{ is any real number}\}$; Range: $\{y \mid y \geq -4.62\}$.

(h) f is increasing on $(-0.42, \infty)$; f is decreasing on $(-\infty, -0.42)$

93. $f(x) = -2x^5 - \sqrt{2}x^2 - x - \sqrt{2}$

(a)   Degree = 5; The graph of the function resembles $y = -2x^5$ for large values of $|x|$.

(b)   graphing utility

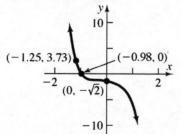

(c) x-intercept: -0.98; y-intercept: $-\sqrt{2} \approx -1.41$

(d)

$Y_1 \boxplus -2X^5 - \sqrt{(2)X^2}\ldots$

(e)   No turning points; No local extrema

(f)   graphing by hand

(−1.25, 3.73)
(−0.98, 0)
(0, −√2)

(g)   Domain: $\{x \mid x \text{ is any real number}\}$; Range: $\{y \mid y \text{ is any real number}\}$.

(h) f is decreasing on $(-\infty, \infty)$

95. (a)   A scatter diagram indicates the data may have a cubic relation.

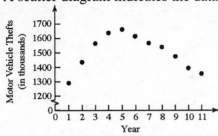

(b) $T(8) = (1.52)(8)^3 - (39.81)(8)^2 + (282.29)(8) + 1035.5 = 1524.22$

The function predicts approximately 1,524,220 thefts in 1994.

(c) and (d)   Graphing the cubic function of best fit:

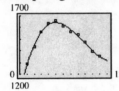

97. (a)   A scatter diagram indicates the data may have a cubic relation.

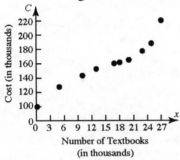

(b)   Average rate of change $= \dfrac{C(13) - C(10)}{13 - 10} = \dfrac{153.5 - 144}{13 - 10} = \dfrac{9.5}{3} \approx \$3.17$ per book

(c)   Average rate of change $= \dfrac{C(20) - C(18)}{20 - 18} = \dfrac{166.3 - 162.6}{20 - 18} = \dfrac{3.7}{2} = \$1.85$ per book

(d) $C(22) = (0.015)(22)^3 - (0.595)(22)^2 + (9.15)(22) + 98.43 = 171.47$

The cost of producing 22,000 texts per week would be about \$171,470.

(e) and (f)   Graphing the cubic function of best fit:

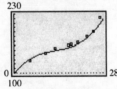

(g)   The y-intercept represents a fixed cost of \$98,000.

99. The graph of a polynomial function will always have a y-intercept since the domain of every polynomial function is the set of real numbers. Therefore $f(0)$ will always produce a y-coordinate on the graph.

A polynomial function might have no x-intercepts. For example $f(x) = x^2 + 1$ has no x-intercepts since the equation $x^2 + 1 = 0$ has no real solutions.

101. Answers will vary.

103. $f(x) = \dfrac{1}{x}$ is smooth but not continuous; $g(x) = |x|$ is continuous but not smooth.

Polynomial and Rational Functions

3.3 Rational Functions I

1. True

3. $y = \dfrac{1}{x}$

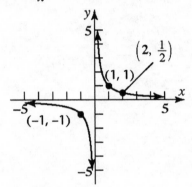

5. $y = 1$

7. proper

9. True

11. In $R(x) = \dfrac{4x}{x-3}$, the denominator, $q(x) = x - 3$, has a zero at 3. Thus, the domain of $R(x)$ is $\left\{x \mid x \neq 3\right\}$.

13. In $H(x) = \dfrac{-4x^2}{(x-2)(x+4)}$, the denominator, $q(x) = (x-2)(x+4)$, has zeros at 2 and –4. Thus, the domain of $H(x)$ is $\left\{x \mid x \neq -4,\, x \neq 2\right\}$.

15. In $F(x) = \dfrac{3x(x-1)}{2x^2 - 5x - 3}$, the denominator, $q(x) = 2x^2 - 5x - 3 = (2x+1)(x-3)$, has zeros at $-\dfrac{1}{2}$ and 3. Thus, the domain of $F(x)$ is $\left\{x \mid x \neq -\dfrac{1}{2}, x \neq 3\right\}$.

17. In $R(x) = \dfrac{x}{x^3 - 8}$, the denominator, $q(x) = x^3 - 8 = (x-2)(x^2 + 2x + 4)$, has a zero at 2. ($x^2 + 2x + 4$ has no real zeros.) Thus, the domain of $R(x)$ is $\left\{x \mid x \neq 2\right\}$.

19.   In $H(x) = \dfrac{3x^2 + x}{x^2 + 4}$, the denominator, $q(x) = x^2 + 4$, has no real zeros. Thus, the domain of

$H(x)$ is $\{x \mid x \text{ is a real number}\}$.

21.   In $R(x) = \dfrac{3(x^2 - x - 6)}{4(x^2 - 9)}$, the denominator, $q(x) = 4(x^2 - 9) = 4(x - 3)(x + 3)$, has zeros at

3 and –3. Thus, the domain of $R(x)$ is $\{x \mid x \neq -3, x \neq 3\}$.

23.   (a)   Domain: $\{x \mid x \neq 2\}$; Range: $\{y \mid y \neq 1\}$

     (b)   Intercept: $(0, 0)$ (c)   Horizontal Asymptote: $y = 1$

     (d)   Vertical Asymptote: $x = 2$ (e)   Oblique Asymptote: none

25.   (a)   Domain: $\{x \mid x \neq 0\}$; Range: all real numbers

     (b)   Intercepts: $(-1, 0), (1, 0)$ (c)   Horizontal Asymptote: none

     (d)   Vertical Asymptote: $x = 0$ (e)   Oblique Asymptote: $y = 2x$

27.   (a)   Domain: $\{x \mid x \neq -2, x \neq 2\}$; Range: $\{y \mid y \leq 0 \text{ or } y > 1\}$

     (b)   Intercept: $(0, 0)$ (c)   Horizontal Asymptote: $y = 1$

     (d)   Vertical Asymptotes: $x = -2, x = 2$ (e)   Oblique Asymptote: none

29. $F(x) = 2 + \dfrac{1}{x}$

Using the function, $y = \dfrac{1}{x}$, shift the graph

vertically 2 units up.

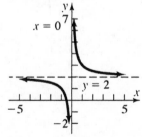

31. $R(x) = \dfrac{1}{(x - 1)^2}$

Using the function, $y = \dfrac{1}{x^2}$, shift the

graph horizontally 1 unit to the right.

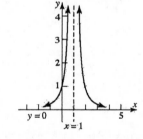

33. $H(x) = \dfrac{-2}{x+1} = -2\left(\dfrac{1}{x+1}\right)$

Using the function $y = \dfrac{1}{x}$, shift the graph horizontally 1 unit to the left, reflect about the x-axis, and stretch vertically by a factor of 2.

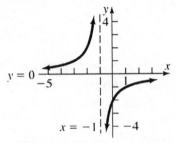

35. $R(x) = \dfrac{-1}{x^2 + 4x + 4} = -\dfrac{1}{(x+2)^2}$

Using the function $y = \dfrac{1}{x^2}$, shift the graph horizontally 2 units to the left, and reflect about the x-axis.

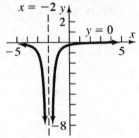

37. $G(x) = 1 + \dfrac{2}{(x-3)^2} = \dfrac{2}{(x-3)^2} + 1$

$= 2\left(\dfrac{1}{(x-3)^2}\right) + 1$

Using the function $y = \dfrac{1}{x^2}$, shift the graph right 3 units, stretch vertically by a factor of 2, and shift vertically 1 unit up.

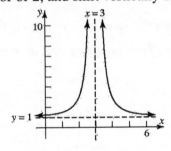

39. $R(x) = \dfrac{x^2 - 4}{x^2} = 1 - \dfrac{4}{x^2} = -4\left(\dfrac{1}{x^2}\right) + 1$

Using the function $y = \dfrac{1}{x^2}$, reflect about the x-axis, stretch vertically by a factor of 4, and shift vertically 1 unit up.

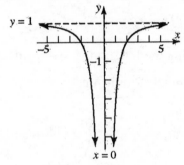

41. $R(x) = \dfrac{3x}{x+4}$

The degree of the numerator, $p(x) = 3x$, is $n = 1$. The degree of the denominator, $q(x) = x + 4$, is $m = 1$. Since $n = m$, the line $y = \dfrac{3}{1} = 3$ is a horizontal asymptote. The denominator is zero at $x = -4$, so $x = -4$ is a vertical asymptote.

43. $H(x) = \dfrac{x^4 + 2x^2 + 1}{x^2 - x + 1}$

The degree of the numerator, $p(x) = x^4 + 2x^2 + 1$, is $n = 4$. The degree of the denominator, $q(x) = x^2 - x + 1$, is $m = 2$. Since $n > m + 1$, there is no horizontal asymptote or oblique asymptote. The denominator has no real zeros, so there is no vertical asymptote.

45. $T(x) = \dfrac{x^3}{x^4 - 1}$

The degree of the numerator, $p(x) = x^3$, is $n = 3$. The degree of the denominator, $q(x) = x^4 - 1$ is $m = 4$. Since $n < m$, the line $y = 0$ is a horizontal asymptote. The denominator is zero at $x = -1$ and $x = 1$, so $x = -1$ and $x = 1$ are vertical asymptotes.

47. $Q(x) = \dfrac{5 - x^2}{3x^4}$

The degree of the numerator, $p(x) = 5 - x^2$, is $n = 2$. The degree of the denominator, $q(x) = 3x^4$ is $m = 4$. Since $n < m$, the line $y = 0$ is a horizontal asymptote. The denominator is zero at $x = 0$, so $x = 0$ is a vertical asymptote.

49. $R(x) = \dfrac{3x^4 + 4}{x^3 + 3x}$

The degree of the numerator, $p(x) = 3x^4 + 4$, is $n = 4$. The degree of the denominator, $q(x) = x^3 + 3x$ is $m = 3$. Since $n = m + 1$, there is an oblique asymptote.
Dividing:

$$
\begin{array}{r}
3x \\
x^3 + 3x \overline{)\,3x^4 \qquad\qquad + 4} \\
\underline{3x^4 \qquad + 9x^2} \\
-9x^2 \qquad + 4
\end{array}
\qquad
R(x) = 3x + \dfrac{-9x^2 + 4}{x^3 + 3x}
$$

Thus, the oblique asymptote is $y = 3x$.
The denominator is zero at $x = 0$, so $x = 0$ is a vertical asymptote.

51. $G(x) = \dfrac{x^3 - 1}{x - x^2}$

The degree of the numerator, $p(x) = x^3 - 1$, is $n = 3$. The degree of the denominator, $q(x) = x - x^2$ is $m = 2$. Since $n = m + 1$, there is an oblique asymptote.
Dividing:

$$
\begin{array}{r}
-x - 1 \\
-x^2 + x \overline{)\,x^3 \qquad\qquad -1} \\
\underline{x^3 - x^2} \\
x^2 \\
\underline{x^2 - x} \\
x - 1
\end{array}
\qquad
G(x) = -x - 1 + \dfrac{x - 1}{x - x^2} = -x - 1 - \dfrac{1}{x},\ x \neq 1
$$

Thus, the oblique asymptote is $y = -x - 1$.
$G(x)$ must be in lowest terms to find the vertical asymptote:

$$
G(x) = \dfrac{x^3 - 1}{x - x^2} = \dfrac{(x - 1)(x^2 + x + 1)}{-x(x - 1)} = \dfrac{x^2 + x + 1}{-x}
$$

The denominator is zero at $x = 0$, so $x = 0$ is a vertical asymptote.

53. $g(h) = \dfrac{3.99 \times 10^{14}}{\left(6.374 \times 10^6 + h\right)^2}$

(a) $g(0) = \dfrac{3.99 \times 10^{14}}{\left(6.374 \times 10^6 + 0\right)^2} \approx 9.8208 \text{ m/s}^2$

(b) $g(443) = \dfrac{3.99 \times 10^{14}}{\left(6.374 \times 10^6 + 443\right)^2} \approx 9.8195 \text{ m/s}^2$

(c) $g(8848) = \dfrac{3.99 \times 10^{14}}{\left(6.374 \times 10^6 + 8848\right)^2} \approx 9.7936 \text{ m/s}^2$

(d) $g(h) = \dfrac{3.99 \times 10^{14}}{\left(6.374 \times 10^6 + h\right)^2} \approx \dfrac{3.99 \times 10^{14}}{h^2} \to 0 \text{ as } h \to \infty$

Thus, $g = 0$ is the horizontal asymptote.

(e) $g(h) = \dfrac{3.99 \times 10^{14}}{\left(6.374 \times 10^6 + h\right)^2} = 0$, to solve this equation would require that

$3.99 \times 10^{14} = 0$, which is impossible. Therefore, there is no height above sea level at which $g = 0$. In other words, there is no point in the entire universe that is unaffected by the Earth's gravity!

55–57.   Answers will vary.

Chapter 3

Polynomial and Rational Functions

3.4 Rational Functions II: Analyzing Graphs

1. False

3. in lowest terms

5. False

In problems 7–43, we will use the terminology: $R(x) = \dfrac{p(x)}{q(x)}$, *where the degree of* $p(x) = n$ *and the degree of* $q(x) = m$.

7. $R(x) = \dfrac{x+1}{x(x+4)}$ $p(x) = x+1$; $q(x) = x(x+4) = x^2 + 4x$; $n = 1$; $m = 2$

   Step 1: Domain: $\{x \mid x \ne -4, \; x \ne 0\}$

   Step 2: (a) The x-intercept is the zero of $p(x)$: -1

   (b) There is no y-intercept; $R(0)$ is not defined, since $q(0) = 0$.

   Step 3: $R(-x) = \dfrac{-x+1}{-x(-x+4)} = \dfrac{-x+1}{x^2 - 4x}$; this is neither $R(x)$ nor $-R(x)$, so there is no symmetry.

   Step 4: $R(x) = \dfrac{x+1}{x(x+4)}$ is in lowest terms.

   The vertical asymptotes are the zeros of $q(x)$: $x = -4$ and $x = 0$.

   Step 5: Since $n < m$, the line $y = 0$ is the horizontal asymptote.

   Solve to find intersection points:

   $$\frac{x+1}{x(x+4)} = 0$$

   $$x + 1 = 0$$

   $$x = -1$$

   $R(x)$ intersects $y = 0$ at $(-1, 0)$.

   Step 6:

Interval	$(-\infty, -4)$	$(-4, -1)$	$(-1, 0)$	$(0, \infty)$
Number Chosen	-5	-2	$-\frac{1}{2}$	1
Value of R	$R(-5) = -\frac{4}{5}$	$R(-2) = \frac{1}{4}$	$R\left(-\frac{1}{2}\right) = -\frac{2}{7}$	$R(1) = \frac{2}{5}$
Location of Graph	Below x-axis	Above x-axis	Below x-axis	Above x-axis
Point on Graph	$\left(-5, -\frac{4}{5}\right)$	$\left(-2, \frac{1}{4}\right)$	$\left(-\frac{1}{2}, -\frac{2}{7}\right)$	$\left(1, \frac{2}{5}\right)$

Step 7:   Graphing

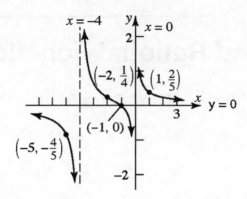

9. $R(x) = \dfrac{3x+3}{2x+4}$ $p(x) = 3x + 3;$ $q(x) = 2x + 4;$ $n = 1;$ $m = 1$

Step 1:   Domain: $\{x \mid x \neq -2\}$

Step 2:   (a)   The x-intercept is the zero of $p(x)$: -1

(b)   The y-intercept is $R(0) = \dfrac{3(0) + 3}{2(0) + 4} = \dfrac{3}{4}$.

Step 3: $R(-x) = \dfrac{3(-x) + 3}{2(-x) + 4} = \dfrac{-3x + 3}{-2x + 4} = \dfrac{3x - 3}{2x - 4}$; this is neither $R(x)$ nor $-R(x)$,

so there is no symmetry.

Step 4: $R(x) = \dfrac{3x+3}{2x+4}$ is in lowest terms.

The vertical asymptote is the zero of $q(x)$: $x = -2$

Step 5:   Since $n = m$, the line $y = \dfrac{3}{2}$ is the horizontal asymptote.

Solve to find intersection points:

$$\frac{3x+3}{2x+4} = \frac{3}{2}$$

$$2(3x + 3) = 3(2x + 4)$$

$$6x + 6 = 6x + 4$$

$$0 \neq 2$$

$R(x)$ does not intersect $y = \dfrac{3}{2}$.

Step 6:

Interval	$(-\infty, -2)$	$(-2, -1)$	$(-1, \infty)$
Number Chosen	-3	$-\frac{3}{2}$	0
Value of R	$R(-3) = 3$	$R\left(-\frac{3}{2}\right) = -\frac{3}{2}$	$R(0) = \frac{3}{4}$
Location of Graph	Above x-axis	Below x-axis	Above x-axis
Point on Graph	$(-3, 3)$	$\left(-\frac{3}{2}, -\frac{3}{2}\right)$	$\left(0, \frac{3}{4}\right)$

Step 7:   Graphing:

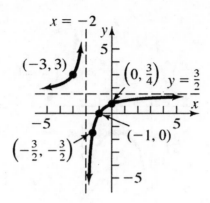

11. $R(x) = \dfrac{3}{x^2 - 4}$ $p(x) = 3;\ q(x) = x^2 - 4;\ n = 0;\ m = 2$

Step 1:   Domain: $\left\{x \mid x \neq -2,\ x \neq 2\right\}$

Step 2:   (a)   There is no x-intercept.

(b)   The y-intercept is $R(0) = \dfrac{3}{0^2 - 4} = \dfrac{3}{-4} = -\dfrac{3}{4}$.

Step 3: $R(-x) = \dfrac{3}{(-x)^2 - 4} = \dfrac{3}{x^2 - 4} = R(x);\ R(x)$ is symmetric with respect to the y-axis.

Step 4: $R(x) = \dfrac{3}{x^2 - 4}$ is in lowest terms.

The vertical asymptotes are the zeros of $q(x)$: $x = -2$ and $x = 2$

Step 5:   Since $n < m$, the line $y = 0$ is the horizontal asymptote.

Solve to find intersection points:

$$\dfrac{3}{x^2 - 4} = 0$$

$$3 = 0\left(x^2 - 4\right)$$

$$3 \neq 0$$

$R(x)$ does not intersect $y = 0$.

Step 6:

Interval	$(-\infty, -2)$	$(-2, 2)$	$(2, \infty)$
Number Chosen	-3	0	3
Value of R	$R(-3) = \frac{3}{5}$	$R(0) = -\frac{3}{4}$	$R(3) = \frac{3}{5}$
Location of Graph	Above x-axis	Below x-axis	Above x-axis
Point on Graph	$\left(-3, \frac{3}{5}\right)$	$\left(0, -\frac{3}{4}\right)$	$\left(3, \frac{3}{5}\right)$

Step 7:   Graphing:

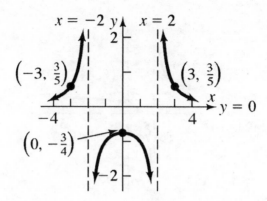

13. $P(x) = \dfrac{x^4 + x^2 + 1}{x^2 - 1}$ $p(x) = x^4 + x^2 + 1;\ q(x) = x^2 - 1;\ n = 4;\ m = 2$

Step 1:   Domain: $\left\{ x \mid x \ne -1,\ x \ne 1 \right\}$

Step 2:   (a)   There is no x-intercept.

(b)   The y-intercept is $P(0) = \dfrac{0^4 + 0^2 + 1}{0^2 - 1} = \dfrac{1}{-1} = -1$.

Step 3: $P(-x) = \dfrac{(-x)^4 + (-x)^2 + 1}{(-x)^2 - 1} = \dfrac{x^4 + x^2 + 1}{x^2 - 1} = P(x)$

$P(x)$ is symmetric with respect to the y-axis.

Step 4: $P(x) = \dfrac{x^4 + x^2 + 1}{x^2 - 1}$ is in lowest terms.

The vertical asymptotes are the zeros of $q(x)$: $x = -1$ and $x = 1$

Step 5:   Since $n > m + 1$, there is no horizontal or oblique asymptote.

Step 6:

Interval	$(-\infty, -1)$	$(-1, 1)$	$(1, \infty)$
Number Chosen	-2	0	2
Value of P	$P(-2) = 7$	$P(0) = -1$	$P(2) = 7$
Location of Graph	Above x-axis	Below x-axis	Above x-axis
Point on Graph	$(-2, 7)$	$(0, -1)$	$(2, 7)$

Step 7:   Graphing:

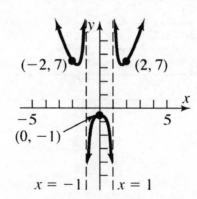

15. $H(x) = \dfrac{x^3 - 1}{x^2 - 9} = \dfrac{(x-1)(x^2 + x + 1)}{(x+3)(x-3)}$ $p(x) = x^3 - 1;\ q(x) = x^2 - 9;\ n = 3;\ m = 2$

Step 1: Domain: $\{x \mid x \neq -3,\ x \neq 3\}$

Step 2: (a)   The x-intercept is the zero of $p(x)$: 1.

(b)   The y-intercept is $H(0) = \dfrac{0^3 - 1}{0^2 - 9} = \dfrac{-1}{-9} = \dfrac{1}{9}$.

Step 3: $H(-x) = \dfrac{(-x)^3 - 1}{(-x)^2 - 9} = \dfrac{-x^3 - 1}{x^2 - 9}$; this is neither $H(x)$ nor $-H(x)$,

so there is no symmetry.

Step 4: $H(x) = \dfrac{x^3 - 1}{x^2 - 9}$ is in lowest terms.

The vertical asymptotes are the zeros of $q(x)$: $x = -3$ and $x = 3$

Step 5: Since $n = m + 1$, there is an oblique asymptote. Dividing:

$$\begin{array}{r} x \phantom{{}^2 + 0x^2 + 0x - 1} \\ x^2 - 9 \overline{)\,x^3 + 0x^2 + 0x - 1} \\ \underline{x^3 - 9x} \\ 9x - 1 \end{array}$$ $\qquad H(x) = x + \dfrac{9x - 1}{x^2 - 9}$

The oblique asymptote is $y = x$.

Solve to find intersection points:

$$\dfrac{x^3 - 1}{x^2 - 9} = x$$

$$x^3 - 1 = x^3 - 9x$$

$$-1 = -9x$$

$$x = \dfrac{1}{9}$$

The oblique asymptote intersects $H(x)$ at $\left(\dfrac{1}{9}, \dfrac{1}{9}\right)$.

Step 6:

Interval	$(-\infty, -3)$	$(-3, 1)$	$(1, 3)$	$(3, \infty)$
Number Chosen	-4	0	2	4
Value of H	$H(-4) \approx -9.3$	$H(0) = \frac{1}{9}$	$H(2) = -1.4$	$H(4) = 9$
Location of Graph	Below x-axis	Above x-axis	Below x-axis	Above x-axis
Point on Graph	$(-4, -9.3)$	$\left(0, \frac{1}{9}\right)$	$(2, -1.4)$	$(4, 9)$

Step 7:   Graphing:

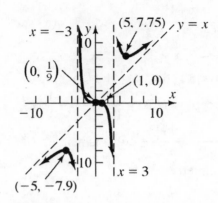

17. $R(x) = \dfrac{x^2}{x^2 + x - 6} = \dfrac{x^2}{(x+3)(x-2)}$ $p(x) = x^2;\; q(x) = x^2 + x - 6;\; n = 2;\; m = 2$

Step 1:   Domain: $\left\{ x \mid x \neq -3,\, x \neq 2 \right\}$

Step 2:   (a)   The x-intercept is the zero of $p(x)$: 0

(b)   The y-intercept is $R(0) = \dfrac{0^2}{0^2 + 0 - 6} = \dfrac{0}{-6} = 0$.

Step 3: $R(-x) = \dfrac{(-x)^2}{(-x)^2 + (-x) - 6} = \dfrac{x^2}{x^2 - x - 6}$; this is neither $R(x)$ nor $-R(x)$,

so there is no symmetry.

Step 4: $R(x) = \dfrac{x^2}{x^2 + x - 6}$ is in lowest terms.

The vertical asymptotes are the zeros of $q(x)$: $x = -3$ and $x = 2$

Step 5:   Since $n = m$, the line $y = 1$ is the horizontal asymptote.

Solve to find intersection points:

$$\frac{x^2}{x^2 + x - 6} = 1$$
$$x^2 = x^2 + x - 6$$
$$0 = x - 6$$
$$x = 6$$

$R(x)$ intersects $y = 1$ at $(6, 1)$.

Step 6:

Interval	$(-\infty, -3)$	$(-3, 0)$	$(0, 2)$	$(2, \infty)$
Number Chosen	-6	-1	1	3
Value of R	$R(-6) = 1.5$	$R(-1) = -\frac{1}{6}$	$R(1) = -0.25$	$R(3) = 1.5$
Location of Graph	Above x-axis	Below x-axis	Below x-axis	Above x-axis
Point on Graph	$(-6, 1.5)$	$\left(-1, -\frac{1}{6}\right)$	$(1, -0.25)$	$(3, 1.5)$

Step 7:   Graphing:

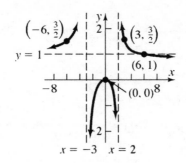

19. $G(x) = \dfrac{x}{x^2 - 4} = \dfrac{x}{(x + 2)(x - 2)}$ $p(x) = x;\ q(x) = x^2 - 4;\ n = 1;\ m = 2$

Step 1:   Domain: $\left\{x \mid x \neq -2,\ x \neq 2\right\}$

Step 2:   (a)   The x-intercept is the zero of $p(x)$: 0

(b)   The y-intercept is $G(0) = \dfrac{0}{0^2 - 4} = \dfrac{0}{-4} = 0$.

Step 3: $G(-x) = \dfrac{-x}{(-x)^2 - 4} = \dfrac{-x}{x^2 - 4} = -G(x)$; $G(x)$ is symmetric with respect to the origin.

Step 4: $G(x) = \dfrac{x}{x^2 - 4}$ is in lowest terms.

The vertical asymptotes are the zeros of $q(x)$: $x = -2$ and $x = 2$

Step 5:   Since $n < m$, the line $y = 0$ is the horizontal asymptote.

Solve to find intersection points:

$$\frac{x}{x^2 - 4} = 0$$

$$x = 0$$

$G(x)$ intersects $y = 0$ at $(0, 0)$.

Step 6:

Interval	$(-\infty, -2)$	$(-2, 0)$	$(0, 2)$	$(2, \infty)$
Number Chosen	-3	-1	1	3
Value of G	$G(-3) = -\frac{3}{5}$	$G(-1) = \frac{1}{3}$	$G(1) = -\frac{1}{3}$	$G(3) = \frac{3}{5}$
Location of Graph	Below x-axis	Above x-axis	Below x-axis	Above x-axis
Point on Graph	$\left(-3, -\frac{3}{5}\right)$	$\left(-1, \frac{1}{3}\right)$	$\left(1, -\frac{1}{3}\right)$	$\left(3, \frac{3}{5}\right)$

Step 7:   Graphing:

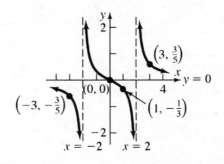

131

21. $R(x) = \dfrac{3}{(x-1)(x^2-4)} = \dfrac{3}{(x-1)(x+2)(x-2)}$ $p(x) = 3;\ q(x) = (x-1)(x^2-4);$
       $n = 0;\ m = 3$

Step 1:   Domain: $\left\{x \mid x \neq -2,\ x \neq 1,\ x \neq 2\right\}$

Step 2:   (a)   There is no x-intercept.

          (b)   The y-intercept is $R(0) = \dfrac{3}{(0-1)(0^2-4)} = \dfrac{3}{4}$.

Step 3: $R(-x) = \dfrac{3}{(-x-1)\left((-x)^2-4\right)} = \dfrac{3}{(-x-1)(x^2-4)}$; this is neither $R(x)$ nor $-R(x)$,

          so there is no symmetry.

Step 4: $R(x) = \dfrac{3}{(x-1)(x^2-4)}$ is in lowest terms.

          The vertical asymptotes are the zeros of $q(x)$: $x = -2$, $x = 1$, and $x = 2$

Step 5:   Since $n < m$, the line $y = 0$ is the horizontal asymptote.

          Solve to find intersection points:

          $$\dfrac{3}{(x-1)(x^2-4)} = 0$$

          $$3 \neq 0$$

          $R(x)$ does not intersect $y = 0$.

Step 6:

Interval	$(-\infty, -2)$	$(-2, 1)$	$(1, 2)$	$(2, \infty)$
Number Chosen	-3	0	1.5	3
Value of R	$R(-3) = -\frac{3}{20}$	$R(0) = \frac{3}{4}$	$R(1.5) = -\frac{24}{7}$	$R(3) = \frac{3}{10}$
Location of Graph	Below x-axis	Above x-axis	Below x-axis	Above x-axis
Point on Graph	$\left(-3, -\frac{3}{20}\right)$	$\left(0, \frac{3}{4}\right)$	$\left(1.5, -\frac{24}{7}\right)$	$\left(3, \frac{3}{10}\right)$

Step 7:   Graphing:

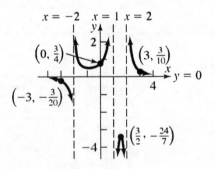

23. $H(x) = \dfrac{4(x^2-1)}{x^4-16} = \dfrac{4(x-1)(x+1)}{(x^2+4)(x+2)(x-2)}$ $p(x) = 4(x^2-1);\ q(x) = x^4-16;$
       $n = 2;\ m = 4$

Step 1:   Domain: $\left\{x \mid x \neq -2,\ x \neq 2\right\}$

Step 2:   (a)   The x-intercepts are the zeros of $p(x)$: -1 and 1

        (b)   The y-intercept is $H(0) = \dfrac{4(0^2 - 1)}{0^4 - 16} = \dfrac{-4}{-16} = \dfrac{1}{4}$.

Step 3: $H(-x) = \dfrac{4\left((-x)^2 - 1\right)}{(-x)^4 - 16} = \dfrac{4\left(x^2 - 1\right)}{x^4 - 16} = H(x)$; $H(x)$ is symmetric with respect to the y-axis.

Step 4: $H(x) = \dfrac{4(x^2 - 1)}{x^4 - 16}$ is in lowest terms.

        The vertical asymptotes are the zeros of $q(x)$: $x = -2$ and $x = 2$

Step 5:   Since $n < m$, the line $y = 0$ is the horizontal asymptote.

        Solve to find intersection points:

$$\frac{4(x^2 - 1)}{x^4 - 16} = 0$$

$$4(x^2 - 1) = 0$$

$$x^2 - 1 = 0$$

$$x = \pm 1$$

        $H(x)$ intersects $y = 0$ at $(-1, 0)$ and $(1, 0)$.

Step 6:

Interval	$(-\infty, -2)$	$(-2, -1)$	$(-1, 1)$	$(1, 2)$	$(2, \infty)$
Number Chosen	-3	-1.5	0	1.5	3
Value of H	$H(-3) \approx 0.49$	$H(-1.5) \approx -0.46$	$H(0) = \frac{1}{4}$	$H(1.5) \approx -0.46$	$H(3) \approx 0.49$
Location of Graph	Above x-axis	Below x-axis	Above x-axis	Below x-axis	Above x-axis
Point on Graph	$(-3, 0.49)$	$(-1.5, -0.46)$	$\left(0, \frac{1}{4}\right)$	$(1.5, -0.46)$	$(3, 0.49)$

Step 7:   Graphing:

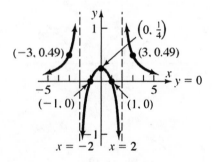

25. $F(x) = \dfrac{x^2 - 3x - 4}{x + 2} = \dfrac{(x + 1)(x - 4)}{x + 2}$ $p(x) = x^2 - 3x - 4$; $q(x) = x + 2$; $n = 2$; $m = 1$

Step 1:   Domain: $\left\{x \mid x \neq -2\right\}$

Step 2:   (a)   The x-intercepts are the zeros of $p(x)$: -1 and 4.

        (b)   The y-intercept is $F(0) = \dfrac{0^2 - 3(0) - 4}{0 + 2} = \dfrac{-4}{2} = -2$.

Step 3: $F(-x) = \dfrac{(-x)^2 - 3(-x) - 4}{-x + 2} = \dfrac{x^2 + 3x - 4}{-x + 2}$; this is neither $F(x)$ nor $-F(x)$, so there is no symmetry.

Step 4: $F(x) = \dfrac{x^2 - 3x - 4}{x + 2}$ is in lowest terms.

The vertical asymptote is the zero of $q(x)$: $x = -2$

Step 5: Since $n = m + 1$, there is an oblique asymptote. Dividing:

$$\begin{array}{r} x - 5 \\ x + 2 \overline{)\, x^2 - 3x - 4} \\ \underline{x^2 + 2x} \\ -5x - 4 \\ \underline{-5x - 10} \\ 6 \end{array} \qquad F(x) = x - 5 + \dfrac{6}{x + 2}$$

The oblique asymptote is $y = x - 5$.

Solve to find intersection points:

$$\frac{x^2 - 3x - 4}{x + 2} = x - 5$$

$$x^2 - 3x - 4 = x^2 - 3x - 10$$

$$-4 \neq -10$$

The oblique asymptote does not intersect $F(x)$.

Step 6:

Interval	$(-\infty, -2)$	$(-2, -1)$	$(-1, 4)$	$(4, \infty)$
Number Chosen	-3	-1.5	0	5
Value of F	$F(-3) = -14$	$F(-1.5) = 5.5$	$F(0) = -2$	$F(5) \approx 0.86$
Location of Graph	Below x-axis	Above x-axis	Below x-axis	Above x-axis
Point on Graph	$(-3, -14)$	$(-1.5, 5.5)$	$(0, -2)$	$(5, 0.86)$

Step 7: Graphing:

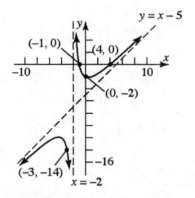

27. $R(x) = \dfrac{x^2 + x - 12}{x - 4} = \dfrac{(x + 4)(x - 3)}{x - 4}$ $p(x) = x^2 + x - 12$; $q(x) = x - 4$; $n = 2$; $m = 1$

Step 1: Domain: $\left\{ x \mid x \neq 4 \right\}$

Step 2: (a)   The x-intercepts are the zeros of $p(x)$: -4 and 3.

  (b)   The y-intercept is $R(0) = \dfrac{0^2 + 0 - 12}{0 - 4} = \dfrac{-12}{-4} = 3$.

Step 3: $R(-x) = \dfrac{(-x)^2 + (-x) - 12}{-x - 4} = \dfrac{x^2 - x - 12}{-x - 4}$; this is neither $R(x)$ nor $-R(x)$,
so there is no symmetry.

Step 4: $R(x) = \dfrac{x^2 + x - 12}{x - 4}$ is in lowest terms.

The vertical asymptote is the zero of $q(x)$: $x = 4$

Step 5:   Since $n = m + 1$, there is an oblique asymptote. Dividing:

$$\begin{array}{r} x + 5 \\ x - 4 \overline{)\, x^2 + x - 12 } \\ \underline{x^2 - 4x } \\ 5x - 12 \\ \underline{5x - 20 } \\ 8 \end{array}$$

$R(x) = x + 5 + \dfrac{8}{x - 4}$

The oblique asymptote is $y = x + 5$.
Solve to find intersection points:

$$\dfrac{x^2 + x - 12}{x - 4} = x + 5$$

$$x^2 + x - 12 = x^2 + x - 20$$

$$-12 \neq -20$$

The oblique asymptote does not intersect $R(x)$.

Step 6:

Interval	$(-\infty, -4)$	$(-4, 3)$	$(3, 4)$	$(4, \infty)$
Number Chosen	-5	0	3.5	5
Value of R	$R(-5) = -\frac{8}{9}$	$R(0) = 3$	$R(3.5) = -7.5$	$R(5) = 18$
Location of Graph	Below x-axis	Above x-axis	Below x-axis	Above x-axis
Point on Graph	$\left(-5, -\frac{8}{9}\right)$	$(0, 3)$	$(3.5, -7.5)$	$(5, 18)$

Step 7:   Graphing:

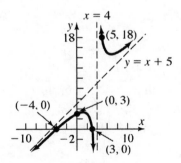

29. $F(x) = \dfrac{x^2 + x - 12}{x + 2} = \dfrac{(x + 4)(x - 3)}{x + 2}$ $p(x) = x^2 + x - 12;\ q(x) = x + 2;\ n = 2;\ m = 1$

Step 1:   Domain: $\{x \mid x \neq -2\}$

Step 2:   (a)   The x-intercepts are the zeros of $p(x)$: -4 and 3.

(b)   The y-intercept is $F(0) = \dfrac{0^2 + 0 - 12}{0 + 2} = \dfrac{-12}{2} = -6$.

Step 3: $F(-x) = \dfrac{(-x)^2 + (-x) - 12}{-x + 2} = \dfrac{x^2 - x - 12}{-x + 2}$; this is neither $F(x)$ nor $-F(x)$, so there is no symmetry.

Step 4: $F(x) = \dfrac{x^2 + x - 12}{x + 2}$ is in lowest terms.

The vertical asymptote is the zero of $q(x)$: $x = -2$

Step 5:   Since $n = m + 1$, there is an oblique asymptote. Dividing:

$$\begin{array}{r}
x - 1 \\
x + 2 \overline{)x^2 + \ x - 12} \\
\underline{x^2 + 2x} \\
- x - 12 \\
\underline{- x - 2} \\
-10
\end{array}$$

$F(x) = x - 1 + \dfrac{-10}{x + 2}$

The oblique asymptote is $y = x - 1$.

Solve to find intersection points:

$$\dfrac{x^2 + x - 12}{x + 2} = x - 1$$

$$x^2 + x - 12 = x^2 + x - 2$$

$$-12 \neq -2$$

The oblique asymptote does not intersect $F(x)$.

Step 6:

Interval	$(-\infty, -4)$	$(-4, -2)$	$(-2, 3)$	$(3, \infty)$
Number Chosen	-5	-3	0	4
Value of F	$F(-5) = -\frac{8}{3}$	$F(-3) = 6$	$F(0) = -6$	$F(4) = \frac{4}{3}$
Location of Graph	Below x-axis	Above x-axis	Below x-axis	Above x-axis
Point on Graph	$\left(-5, -\frac{8}{3}\right)$	$(-3, 6)$	$(0, -6)$	$\left(4, \frac{4}{3}\right)$

Step 7:   Graphing:

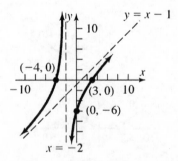

31. $R(x) = \dfrac{x(x - 1)^2}{(x + 3)^3}$ $p(x) = x(x - 1)^2$; $q(x) = (x + 3)^3$; $n = 3$; $m = 3$

Step 1:   Domain: $\left\{x \mid x \neq -3\right\}$

Step 2:   (a)   The x-intercepts are the zeros of $p(x)$: 0 and 1

(b)   The y-intercept is $R(0) = \dfrac{0(0 - 1)^2}{(0 + 3)^3} = \dfrac{0}{27} = 0$.

Step 3: $R(-x) = \dfrac{-x(-x-1)^2}{(-x+3)^3}$; this is neither $R(x)$ nor $-R(x)$, so there is no symmetry.

Step 4: $R(x) = \dfrac{x(x-1)^2}{(x+3)^3}$ is in lowest terms.

The vertical asymptote is the zero of $q(x)$: $x = -3$

Step 5:   Since $n = m$, the line $y = 1$ is the horizontal asymptote.

Solve to find intersection points:

$$\frac{x(x-1)^2}{(x+3)^3} = 1$$

$$x^3 - 2x^2 + x = x^3 + 9x^2 + 27x + 27$$

$$0 = 11x^2 + 26x + 27$$

$$b^2 - 4ac = 26^2 - 4(11)(27) = -512$$

no real solution

$R(x)$ does not intersect $y = 1$.

Step 6:

Interval	$(-\infty, -3)$	$(-3, 0)$	$(0, 1)$	$(1, \infty)$
Number Chosen	-4	-1	$\frac{1}{2}$	2
Value of R	$R(-4) = 100$	$R(-1) = -0.5$	$R\left(\frac{1}{2}\right) \approx 0.003$	$R(2) = 0.016$
Location of Graph	Above x-axis	Below x-axis	Above x-axis	Above x-axis
Point on Graph	$(-4, 100)$	$(-1, -0.5)$	$\left(\frac{1}{2}, 0.003\right)$	$(2, 0.016)$

Step 7:   Graphing:

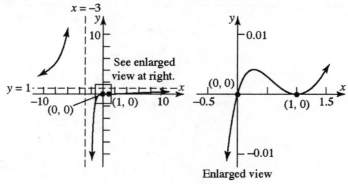

Enlarged view

33. $R(x) = \dfrac{x^2 + x - 12}{x^2 - x - 6} = \dfrac{(x+4)(x-3)}{(x-3)(x+2)} = \dfrac{x+4}{x+2}$ $p(x) = x^2 + x - 12;\ q(x) = x^2 - x - 6;$

$n = 2;\ m = 2$

Step 1:   Domain: $\left\{x \mid x \neq -2,\ x \neq 3\right\}$

Step 2:   (a)   The x-intercept is the zero of $y = x + 4$: -4 ; Note: 3 is not a zero because reduced form must be used to find the zeros.

(b)   The y-intercept is $R(0) = \dfrac{0^2 + 0 - 12}{0^2 - 0 - 6} = \dfrac{-12}{-6} = 2$.

137

Step 3: $R(-x) = \dfrac{(-x)^2 + (-x) - 12}{(-x)^2 - (-x) - 6} = \dfrac{x^2 - x - 12}{x^2 + x - 6}$; this is neither $R(x)$ nor $-R(x)$,

so there is no symmetry.

Step 4: In lowest terms, $R(x) = \dfrac{x+4}{x+2}$, $x \neq 3$.

The vertical asymptote is the zero of $f(x) = x + 2$: $x = -2$; Note: $x = 3$ is not a vertical asymptote because reduced form must be used to find the asymptotes.

The graph has a hole at $\left(3, \dfrac{7}{5}\right)$.

Step 5: Since $n = m$, the line $y = 1$ is the horizontal asymptote.

Solve to find intersection points:

$$\frac{x^2 + x - 12}{x^2 - x - 6} = 1$$
$$x^2 + x - 12 = x^2 - x - 6$$
$$2x = 6$$
$$x = 3$$

$R(x)$ does not intersect $y = 1$ because $R(x)$ is not defined at $x = 3$.

Step 6:

Interval	$(-\infty, -4)$	$(-4, -2)$	$(-2, 3)$	$(3, \infty)$
Number Chosen	-5	-3	0	4
Value of R	$R(-5) = \frac{1}{3}$	$R(-3) = -1$	$R(0) = 2$	$R(4) = \frac{4}{3}$
Location of Graph	Above x-axis	Below x-axis	Above x-axis	Above x-axis
Point on Graph	$\left(-5, \frac{1}{3}\right)$	$(-3, -1)$	$(0, 2)$	$\left(4, \frac{4}{3}\right)$

Step 7: Graphing:

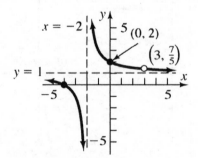

35. $R(x) = \dfrac{6x^2 - 7x - 3}{2x^2 - 7x + 6} = \dfrac{(3x+1)(2x-3)}{(2x-3)(x-2)} = \dfrac{3x+1}{x-2}$ $p(x) = 6x^2 - 7x - 3;$

$q(x) = 2x^2 - 7x + 6;$ $n = 2;$ $m = 2$

Step 1: Domain: $\left\{ x \,\middle|\, x \neq \dfrac{3}{2}, x \neq 2 \right\}$

Step 2: (a) The x-intercept is the zero of $y = 3x + 1$: $-\dfrac{1}{3}$; Note: $x = \dfrac{3}{2}$ is not a zero because reduced form must be used to find the zeros.

(b) The y-intercept is $R(0) = \dfrac{6(0)^2 - 7(0) - 3}{2(0)^2 - 7(0) + 6} = \dfrac{-3}{6} = -\dfrac{1}{2}$.

Step 3:　$R(-x) = \dfrac{6(-x)^2 - 7(-x) - 3}{2(-x)^2 - 7(-x) + 6} = \dfrac{6x^2 + 7x - 3}{2x^2 + 7x + 6}$; this is neither $R(x)$ nor $-R(x)$,

so there is no symmetry.

Step 4:　In lowest terms, $R(x) = \dfrac{3x+1}{x-2}$, $x \neq \dfrac{3}{2}$.

The vertical asymptote is the zero of $f(x) = x - 2$: $x = 2$; Note: $x = \dfrac{3}{2}$ is not a

vertical asymptote because reduced form must be used to find the asymptotes.

The graph has a hole at $\left(\dfrac{3}{2}, -11\right)$.

Step 5:　Since $n = m$, the line $y = 3$ is the horizontal asymptote.

Solve to find intersection points:

$$\dfrac{6x^2 - 7x - 3}{2x^2 - 7x + 6} = 3$$

$$6x^2 - 7x - 3 = 6x^2 - 21x + 18$$

$$14x = 21$$

$$x = \dfrac{3}{2}$$

$R(x)$ does not intersect $y = 3$ because $R(x)$ is not defined at $x = \dfrac{3}{2}$.

Step 6:

Interval	$\left(-\infty, -\frac{1}{3}\right)$	$\left(-\frac{1}{3}, \frac{3}{2}\right)$	$\left(\frac{3}{2}, 2\right)$	$(2, \infty)$
Number Chosen	-1	0	1.7	6
Value of R	$R(-1) = \frac{2}{3}$	$R(0) = -\frac{1}{2}$	$R(1.7) \approx -20.3$	$R(6) = 4.75$
Location of Graph	Above x-axis	Below x-axis	Below x-axis	Above x-axis
Point on Graph	$\left(-1, \frac{2}{3}\right)$	$\left(0, -\frac{1}{2}\right)$	$(1.7, -20.3)$	$(6, 4.75)$

Step 7:　Graphing:

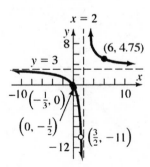

37.　$R(x) = \dfrac{x^2 + 5x + 6}{x + 3} = \dfrac{(x+2)(x+3)}{x+3} = x + 2$　$p(x) = x^2 + 5x + 6$;　$q(x) = x + 3$;

$n = 2$; $m = 1$

Step 1:　Domain: $\left\{x \mid x \neq -3\right\}$

Step 2: (a) The x-intercept is the zero of $y = x + 2$: -2; Note: -3 is not a zero because reduced form must be used to find the zeros.

(b) The y-intercept is $R(0) = \dfrac{0^2 + 5(0) + 6}{0 + 3} = \dfrac{6}{3} = 2$.

Step 3: $R(-x) = \dfrac{(-x)^2 + 5(-x) + 6}{-x + 3} = \dfrac{x^2 - 5x + 6}{-x + 3}$; this is neither $R(x)$ nor $-R(x)$, so there is no symmetry.

Step 4: In lowest terms, $R(x) = x + 2$, $x \neq -3$.

There are no vertical asymptotes. Note: $x = -3$ is not a vertical asymptote because reduced form must be used to find the asymptotes. The graph has a hole at $(-3, -1)$.

Step 5: Since $n = m + 1$ there is an oblique asymptote. The line $y = x + 2$ is the oblique asymptote.

Solve to find intersection points:

$$\frac{x^2 + 5x + 6}{x + 3} = x + 2$$

$$x^2 + 5x + 6 = (x + 2)(x + 3)$$

$$x^2 + 5x + 6 = x^2 + 5x + 6$$

$$0 = 0$$

The oblique asymptote intersects $R(x)$ at every point of the form $(x, x + 2)$ except $(-3, -1)$.

Step 6:

	-3		-2

Interval	$(-\infty, -3)$	$(-3, -2)$	$(-2, \infty)$
Number Chosen	-4	-2.5	0
Value of R	$R(-4) = -2$	$R(-2.5) = -\frac{1}{2}$	$R(0) = 2$
Location of Graph	Below x-axis	Below x-axis	Above x-axis
Point on Graph	$(-4, -2)$	$\left(-2.5, -\frac{1}{2}\right)$	$(0, 2)$

Step 7: Graphing:

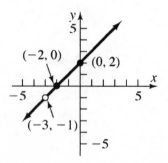

39. $f(x) = x + \dfrac{1}{x} = \dfrac{x^2 + 1}{x}$ $p(x) = x^2 + 1$; $q(x) = x$; $n = 2$; $m = 1$

Step 1: Domain: $\{x \mid x \neq 0\}$

Step 2: (a) There are no x-intercepts since $x^2 + 1 = 0$ has no real solutions

(b) There is no y-intercept because 0 is not in the domain.

Step 3: $f(-x) = \dfrac{(-x)^2 + 1}{-x} = \dfrac{x^2 + 1}{-x} = -f(x)$; The graph of $f(x)$ is symmetric with respect to the origin.

Step 4: $f(x) = \dfrac{x^2 + 1}{x}$ is in lowest terms.

The vertical asymptote is the zero of $q(x)$: $x = 0$

Step 5:   Since $n = m + 1$, there is an oblique asymptote. Dividing:

$$\begin{array}{r} x \\ x \overline{)\, x^2 + 1} \\ \underline{x^2} \\ 1 \end{array}$$

$f(x) = x + \dfrac{1}{x}$

The oblique asymptote is $y = x$.

Solve to find intersection points:

$$\dfrac{x^2 + 1}{x} = x$$

$$x^2 + 1 = x^2$$

$$1 \neq 0$$

The oblique asymptote does not intersect $f(x)$.

Step 6:

Interval	$(-\infty, 0)$	$(0, \infty)$
Number Chosen	-1	1
Value of f	$f(-1) = -2$	$f(1) = 2$
Location of Graph	Below x-axis	Above x-axis
Point on Graph	$(-1, -2)$	$(1, 2)$

Step 7:   Graphing :

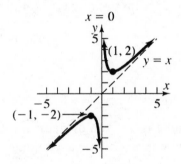

41. $f(x) = x^2 + \dfrac{1}{x} = \dfrac{x^3 + 1}{x}$ $p(x) = x^3 + 1;\ q(x) = x;\ n = 3;\ m = 1$

Step 1:   Domain: $\{x \mid x \neq 0\}$

Step 2:   (a)   The x-intercept is the zero of $p(x)$: -1

(b)   There is no y-intercept because 0 is not in the domain.

Step 3: $f(-x) = \dfrac{(-x)^3 + 1}{-x} = \dfrac{-x^3 + 1}{-x}$; this is neither $f(x)$ nor $-f(x)$, so there is no symmetry.

Step 4: $f(x) = \dfrac{x^3 + 1}{x}$ is in lowest terms.

The vertical asymptote is the zero of $q(x)$: $x = 0$

Step 5:   Since $n > m + 1$, there is no horizontal or oblique asymptote.

Step 6:

	-1	0	
Interval	$(-\infty, -1)$	$(-1, 0)$	$(0, \infty)$
Number Chosen	-2	$-\frac{1}{2}$	1
Value of f	$f(-2) = 3.5$	$f\left(-\frac{1}{2}\right) = -1.75$	$f(1) = 2$
Location of Graph	Above x-axis	Below x-axis	Above x-axis
Point on Graph	$(-2, 3.5)$	$\left(-\frac{1}{2}, -1.75\right)$	$(1, 2)$

Step 7:   Graphing:

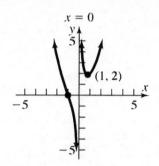

43. $f(x) = x + \dfrac{1}{x^3} = \dfrac{x^4 + 1}{x^3}$ $p(x) = x^4 + 1;\ q(x) = x^3;\ n = 4;\ m = 3$

Step 1:   Domain: $\left\{x \mid x \neq 0\right\}$

Step 2:   (a)   There are no x-intercepts since $x^4 + 1 = 0$ has no real solutions.

         (b)   There is no y-intercept because 0 is not in the domain.

Step 3: $f(-x) = \dfrac{(-x)^4 + 1}{(-x)^3} = \dfrac{x^4 + 1}{-x^3} = -f(x)$; the graph of $f(x)$ is symmetric with respect to

the origin.

Step 4: $f(x) = \dfrac{x^4 + 1}{x^3}$ is in lowest terms.

The vertical asymptote is the zero of $q(x)$: $x = 0$

Step 5:   Since $n = m + 1$, there is an oblique asymptote. Dividing:

$$\begin{array}{r} x \\ x^3 \overline{)\,x^4 + 1} \\ \underline{x^4 } \\ 1 \end{array} \qquad f(x) = x + \dfrac{1}{x^3}$$

The oblique asymptote is $y = x$.

Solve to find intersection points:

$$\frac{x^4 + 1}{x^3} = x$$

$$x^4 + 1 = x^4$$

$$1 \neq 0$$

The oblique asymptote does not intersect $f(x)$.

Step 6:

Interval	$(-\infty, 0)$	$(0, \infty)$
Number Chosen	-1	1
Value of f	$f(-1) = -2$	$f(1) = 2$
Location of Graph	Below x-axis	Above x-axis
Point on Graph	$(-1, -2)$	$(1, 2)$

Step 7:   Graphing:

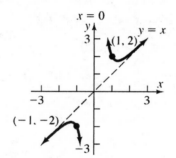

45. $f(x) = \dfrac{x^2}{x^2 - 4}$

47. $f(x) = \dfrac{(x-1)(x-3)\left(x^2 + \dfrac{4}{3}\right)}{(x+1)^2(x-2)^2}$

49.   (a)   The degree of the numerator is 1 and the degree of the denominator is 2. Thus, the horizontal asymptote is $C(t) = 0$. The concentration of the drug decreases to 0 as time increases.

   (b)   Graphing:

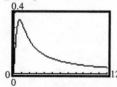

   (c)   Using MAXIMUM, the concentration is highest when $t \approx 0.71$ hours.

51.   (a)   The average cost function is: $\overline{C}(x) = \dfrac{0.2x^3 - 2.3x^2 + 14.3x + 10.2}{x}$

   (b) $\overline{C}(6) = \dfrac{(0.2)(6)^3 - (2.3)(6)^2 + (14.3)(6) + 10.2}{6} = \dfrac{56.4}{6} = 9.4$

   The average cost of producing 6 Cavaliers is $9400 per car.

(c) $\overline{C}(9) = \dfrac{(0.2)(9)^3 - (2.3)(9)^2 + 14.3(9) + 10.2}{9} = \dfrac{98.4}{9} \approx 10.933$

The average cost of producing 9 Cavaliers is $10,933 per car.

(d) Graphing:

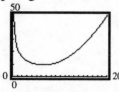

(e) Using MINIMUM, the number of Cavaliers that should be produced to minimize cost is 6.38 cars

(f) The minimum average cost is

$\overline{C}(6.38) = \dfrac{0.2(6.38)^3 - 2.3(6.38)^2 + 14.3(6.38) + 10.2}{6.38} \approx \$9366$ per car.

53. (a) The surface area is the sum of the areas of the six sides.

$$S = xy + xy + xy + xy + x^2 + x^2 = 4xy + 2x^2$$

The volume is $x \cdot x \cdot y = x^2 y = 10,000 \quad \Rightarrow \quad y = \dfrac{10,000}{x^2}$

Thus, $S(x) = 4x\left(\dfrac{10,000}{x^2}\right) + 2x^2 = 2x^2 + \dfrac{40,000}{x} = \dfrac{2x^3 + 40,000}{x}$

(b) Graphing:

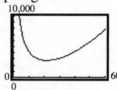

(c) Using MINIMUM, the minimum surface area (amount of cardboard) is about 2785 square inches.

(d) The surface area is a minimum when $x \approx 21.544$.

$y = \dfrac{10,000}{(21.544)^2} \approx 21.545$

The dimensions of the box are: 21.544 in. by 21.544 in. by 21.545 in.

55. (a) $500 = \pi r^2 h \quad \Rightarrow \quad h = \dfrac{500}{\pi r^2}$

$C(r) = 6(2\pi r^2) + 4(2\pi rh) = 12\pi r^2 + 8\pi r\left(\dfrac{500}{\pi r^2}\right) = 12\pi r^2 + \dfrac{4000}{r}$

(b) Graphing:

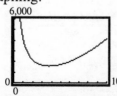

Using MINIMUM, the cost is least for $r \approx 3.76$ cm.

57.

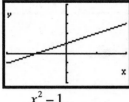 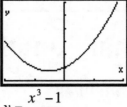 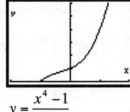 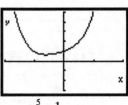

$$y = \frac{x^2 - 1}{x - 1}$$ $$y = \frac{x^3 - 1}{x - 1}$$ $$y = \frac{x^4 - 1}{x - 1}$$ $$y = \frac{x^5 - 1}{x - 1}$$

$x = 1$ is not a vertical asymptote because of the following behavior:

$$y = \frac{x^2 - 1}{x - 1} = \frac{(x + 1)(x - 1)}{x - 1} = x + 1 \text{ when } x \neq 1$$

$$y = \frac{x^3 - 1}{x - 1} = \frac{(x - 1)(x^2 + x + 1)}{x - 1} = x^2 + x + 1 \text{ when } x \neq 1$$

$$y = \frac{x^4 - 1}{x - 1} = \frac{(x^2 + 1)(x^2 - 1)}{x - 1} = \frac{(x^2 + 1)(x - 1)(x + 1)}{x - 1} = x^3 + x^2 + x + 1 \text{ when } x \neq 1$$

$$y = \frac{x^5 - 1}{x - 1} = \frac{(x^4 + x^3 + x^2 + x + 1)(x - 1)}{x - 1} = x^4 + x^3 + x^2 + x + 1 \text{ when } x \neq 1$$

In general, the graph of $y = \dfrac{x^n - 1}{x - 1}$, $n \geq 1$, an integer, will have a "hole" with coordinates $(1, n)$.

59. $f(x) = x + \dfrac{1}{x}, x > 0$

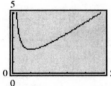

minimum value: 2

61. $f(x) = x^2 + \dfrac{1}{x}, x > 0$

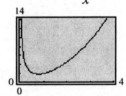

minimum value: ≈ 1.89

63. $f(x) = x + \dfrac{1}{x^3}, x > 0$

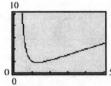

minimum value: ≈ 1.75

65. Answers will vary.

67. Answers will vary, one example is $R(x) = \dfrac{2(x - 3)(x + 2)^2}{(x - 1)^3}$.

Polynomial and Rational Functions

3.5   Polynomial and Rational Inequalities

1. $3 - 4x > 5$

$-4x > 2$

$x < -\dfrac{1}{2}$

The solution set is $\left\{ x \middle| x < -\dfrac{1}{2} \right\}$ or $\left(-\infty, -\dfrac{1}{2} \right)$.

3. $(x - 5)(x + 2) < 0$ $f(x) = (x - 5)(x + 2)$

$x = 5,\ x = -2$ are the zeros of f.

Interval	$(-\infty, -2)$	$(-2, 5)$	$(5, \infty)$
Number Chosen	-3	0	6
Value of f	8	-10	8
Conclusion	Positive	Negative	Positive

The solution set is $\left\{ x \middle| -2 < x < 5 \right\}$.

5. $x^2 - 4x \geq 0$ $f(x) = x^2 - 4x$

$x(x - 4) \geq 0$ $x = 0,\ x = 4$ are the zeros of f.

Interval	$(-\infty, 0)$	$(0, 4)$	$(4, \infty)$
Number Chosen	-1	1	5
Value of f	5	-3	5
Conclusion	Positive	Negative	Positive

The solution set is $\left\{ x \middle| x \leq 0 \text{ or } x \geq 4 \right\}$.

7. $x^2 - 9 < 0$ $f(x) = x^2 - 9$

$(x + 3)(x - 3) < 0$ $x = -3,\ x = 3$ are the zeros of f.

Interval	$(-\infty, -3)$	$(-3, 3)$	$(3, \infty)$
Number Chosen	-4	0	4
Value of f	7	-9	7
Conclusion	Positive	Negative	Positive

The solution set is $\left\{ x \middle| -3 < x < 3 \right\}$.

9.

$$x^2 + x \geq 2 \qquad f(x) = x^2 + x - 2$$
$$x^2 + x - 2 \geq 0 \qquad x = -2, x = 1 \text{ are the zeros of } f.$$
$$(x+2)(x-1) \geq 0$$

Interval	$(-\infty, -2)$	$(-2, 1)$	$(1, \infty)$
Number Chosen	-3	0	2
Value of f	4	-2	4
Conclusion	Positive	Negative	Positive

The solution set is $\left\{ x \mid x \leq -2 \ \text{or} \ x \geq 1 \right\}$.

11.

$$2x^2 \leq 5x + 3 \qquad f(x) = 2x^2 - 5x - 3$$
$$2x^2 - 5x - 3 \leq 0 \qquad x = -\frac{1}{2}, x = 3 \text{ are the zeros of } f.$$
$$(2x+1)(x-3) \leq 0$$

Interval	$\left(-\infty, -\frac{1}{2}\right)$	$\left(-\frac{1}{2}, 3\right)$	$(3, \infty)$
Number Chosen	-1	0	4
Value of f	4	-3	9
Conclusion	Positive	Negative	Positive

The solution set is $\left\{ x \mid -\frac{1}{2} \leq x \leq 3 \right\}$.

13.

$$x(x-7) > 8 \qquad f(x) = x^2 - 7x - 8$$
$$x^2 - 7x > 8 \qquad x = -1, x = 8 \text{ are the zeros of } f.$$
$$x^2 - 7x - 8 > 0$$
$$(x+1)(x-8) > 0$$

Interval	$(-\infty, -1)$	$(-1, 8)$	$(8, \infty)$
Number Chosen	-2	0	9
Value of f	10	-8	10
Conclusion	Positive	Negative	Positive

The solution set is $\left\{ x \mid x < -1 \ \text{or} \ x > 8 \right\}$.

15.

$$4x^2 + 9 < 6x \qquad f(x) = 4x^2 - 6x + 9$$
$$4x^2 - 6x + 9 < 0 \qquad b^2 - 4ac = (-6)^2 - 4(4)(9)$$
$$= 36 - 144$$
$$= -108$$

Since the discriminant is negative, f has no real zeros.

Therefore, f is either always positive or always negative. Choose any value and test.

For $x = 0$, $f(0) = 4(0)^2 - 6(0) + 9 = 9 > 0$. Thus, there is no real solution.

17.
$$6(x^2-1)>5x \qquad f(x)=6x^2-5x-6$$
$$6x^2-6>5x \qquad x=-\frac{2}{3}, x=\frac{3}{2} \text{ are the zeros of } f.$$
$$6x^2-5x-6>0$$
$$(3x+2)(2x-3)>0$$

Interval	$\left(-\infty,-\dfrac{2}{3}\right)$	$\left(-\dfrac{2}{3},\dfrac{3}{2}\right)$	$\left(\dfrac{3}{2},\infty\right)$
Number Chosen	-1	0	2
Value of f	5	-6	8
Conclusion	Positive	Negative	Positive

The solution set is $\left\{ x \middle| x<-\dfrac{2}{3} \text{ or } x>\dfrac{3}{2} \right\}$.

19.
$$(x-1)(x^2+x+4)\geq 0 \qquad f(x)=(x-1)(x^2+x+4)$$
$$x=1 \text{ is the zero of } f. \quad x^2+x+4=0 \text{ has no real solution.}$$

Interval	$(-\infty,1)$	$(1,\infty)$
Number Chosen	0	2
Value of f	-4	10
Conclusion	Negative	Positive

The solution set is $\left\{ x \middle| x\geq 1 \right\}$.

21.
$$(x-1)(x-2)(x-3)\leq 0 \qquad f(x)=(x-1)(x-2)(x-3)$$
$$x=1,\ x=2,\ x=3 \text{ are the zeros of } f.$$

Interval	$(-\infty,1)$	$(1,2)$	$(2,3)$	$(3,\infty)$
Number Chosen	0	1.5	2.5	4
Value of f	-6	0.375	-0.375	6
Conclusion	Negative	Positive	Negative	Positive

The solution set is $\left\{ x \middle| x\leq 1 \text{ or } 2\leq x\leq 3 \right\}$.

23.
$$x^3-2x^2-3x>0 \qquad f(x)=x^3-2x^2-3x$$
$$x(x^2-2x-3)>0 \qquad x=-1,\ x=0,\ x=3 \text{ are the zeros of } f.$$
$$x(x+1)(x-3)>0$$

Interval	$(-\infty,-1)$	$(-1,0)$	$(0,3)$	$(3,\infty)$
Number Chosen	-2	-0.5	1	4
Value of f	-10	0.875	-4	20
Conclusion	Negative	Positive	Negative	Positive

The solution set is $\left\{ x \middle| -1<x<0 \text{ or } x>3 \right\}$.

25.

$$x^4 > x^2$$

$$x^4 - x^2 > 0$$

$$x^2(x^2 - 1) > 0$$

$$x^2(x + 1)(x - 1) > 0$$

$f(x) = x^4 - x^2$

$x = -1, x = 0, x = 1$ are the zeros of f.

Interval	$(-\infty, -1)$	$(-1, 0)$	$(0, 1)$	$(1, \infty)$
Number Chosen	-2	-0.5	0.5	2
Value of f	12	-0.1875	-0.1875	12
Conclusion	Positive	Negative	Negative	Positive

The solution set is $\{x \mid x < -1 \text{ or } x > 1\}$.

27.

$$x^3 \geq 4x^2$$

$$x^3 - 4x^2 \geq 0$$

$$x^2(x - 4) \geq 0$$

$f(x) = x^3 - 4x^2$

$x = 0, x = 4$ are the zeros of f.

Interval	$(-\infty, 0)$	$(0, 4)$	$(4, \infty)$
Number Chosen	-1	1	5
Value of f	-5	-3	25
Conclusion	Negative	Negative	Positive

The solution set is $\{x \mid x \geq 4\}$.

29.

$$x^4 > 1$$

$$x^4 - 1 > 0$$

$$(x^2 + 1)(x^2 - 1) > 0$$

$$(x^2 + 1)(x + 1)(x - 1) > 0$$

$f(x) = x^4 - 1$

$x = -1, x = 1$ are the zeros of f. $x^2 + 1 = 0$ has no real solution.

Interval	$(-\infty, -1)$	$(-1, 1)$	$(1, \infty)$
Number Chosen	-2	0	2
Value of f	15	-1	15
Conclusion	Positive	Negative	Positive

The solution set is $\{x \mid x < -1 \text{ or } x > 1\}$.

31.

$$\frac{x+1}{x-1} > 0$$

$f(x) = \dfrac{x+1}{x-1}$

The zeros and values where f is undefined are $x = -1$ and $x = 1$.

Interval	$(-\infty, -1)$	$(-1, 1)$	$(1, \infty)$
Number Chosen	-2	0	2
Value of f	$\dfrac{1}{3}$	-1	3
Conclusion	Positive	Negative	Positive

The solution set is $\{x \mid x < -1 \text{ or } x > 1\}$.

33.　$\dfrac{(x-1)(x+1)}{x} \le 0$ 　　　$f(x) = \dfrac{(x-1)(x+1)}{x}$

The zeros and values where f is undefined are
$x = -1, x = 0$ and $x = 1$.

Interval	$(-\infty,-1)$	$(-1,0)$	$(0,1)$	$(1,\infty)$
Number Chosen	-2	-0.5	0.5	2
Value of f	-1.5	1.5	-1.5	1.5
Conclusion	Negative	Positive	Negative	Positive

The solution set is $\left\{ x \mid x \le -1 \text{ or } 0 < x \le 1 \right\}$.

35.　$\dfrac{(x-2)^2}{x^2-1} \ge 0$ 　　　$f(x) = \dfrac{(x-2)^2}{x^2-1}$

$\dfrac{(x-2)^2}{(x+1)(x-1)} \ge 0$

The zeros and values where f is undefined are
$x = -1, x = 1$ and $x = 2$.

Interval	$(-\infty,-1)$	$(-1,1)$	$(1,2)$	$(2,\infty)$
Number Chosen	-2	0	1.5	3
Value of f	$\dfrac{16}{3}$	-4	0.2	0.125
Conclusion	Positive	Negative	Positive	Positive

The solution set is $\left\{ x \mid x < -1 \text{ or } x > 1 \right\}$.

37.　$6x - 5 < \dfrac{6}{x}$ 　　　$f(x) = 6x - 5 - \dfrac{6}{x}$

$6x - 5 - \dfrac{6}{x} < 0$

The zeros and values where f is undefined are

$6x - 5 - \dfrac{6}{x} < 0$

$x = -\dfrac{2}{3}, x = 0$ and $x = \dfrac{3}{2}$.

$\dfrac{6x^2 - 5x - 6}{x} < 0$

$\dfrac{(2x-3)(3x+2)}{x} < 0$

Interval	$\left(-\infty,-\dfrac{2}{3}\right)$	$\left(-\dfrac{2}{3},0\right)$	$\left(0,\dfrac{3}{2}\right)$	$\left(\dfrac{3}{2},\infty\right)$
Number Chosen	-1	-0.5	1	2
Value of f	-5	4	-5	4
Conclusion	Negative	Positive	Negative	Positive

The solution set is $\left\{ x \mid x < -\dfrac{2}{3} \text{ or } 0 < x < \dfrac{3}{2} \right\}$.

39.

$$\frac{x+4}{x-2} \le 1$$

$$\frac{x+4}{x-2} - 1 \le 0$$

$$\frac{x+4-(x-2)}{x-2} \le 0$$

$$\frac{6}{x-2} \le 0$$

$f(x) = \frac{x+4}{x-2} - 1$

The value where f is undefined is $x = 2$.

Interval	$(-\infty, 2)$	$(2, \infty)$
Number Chosen	0	3
Value of f	-3	6
Conclusion	Negative	Positive

The solution set is $\left\{ x \mid x < 2 \right\}$.

41.

$$\frac{3x-5}{x+2} \le 2$$

$$\frac{3x-5}{x+2} - 2 \le 0$$

$$\frac{3x-5-2(x+2)}{x+2} \le 0$$

$$\frac{x-9}{x+2} \le 0$$

$f(x) = \frac{3x-5}{x+2} - 2$

The zeros and values where f is undefined are $x = -2$ and $x = 9$.

Interval	$(-\infty, -2)$	$(-2, 9)$	$(9, \infty)$
Number Chosen	-3	0	10
Value of f	12	-4.5	$\dfrac{1}{12}$
Conclusion	Positive	Negative	Positive

The solution set is $\left\{ x \mid -2 < x \le 9 \right\}$.

43.

$$\frac{1}{x-2} < \frac{2}{3x-9}$$

$$\frac{1}{x-2} - \frac{2}{3x-9} < 0$$

$$\frac{3x-9-2(x-2)}{(x-2)(3x-9)} < 0$$

$$\frac{x-5}{(x-2)(3x-9)} < 0$$

$f(x) = \frac{1}{x-2} - \frac{2}{3x-9}$

The zeros and values where f is undefined are $x = 2$, $x = 3$, and $x = 5$.

Interval	$(-\infty, 2)$	$(2, 3)$	$(3, 5)$	$(5, \infty)$
Number Chosen	0	2.5	4	6
Value of f	$-\dfrac{5}{18}$	$\dfrac{10}{3}$	$-\dfrac{1}{6}$	$\dfrac{1}{36}$
Conclusion	Negative	Positive	Negative	Positive

The solution set is $\left\{ x \mid x < 2 \text{ or } 3 < x < 5 \right\}$.

45.

$$\frac{2x+5}{x+1} > \frac{x+1}{x-1}$$

$$\frac{2x+5}{x+1} - \frac{x+1}{x-1} > 0$$

$$\frac{(2x+5)(x-1)-(x+1)(x+1)}{(x+1)(x-1)} > 0$$

$$\frac{2x^2+3x-5-\left(x^2+2x+1\right)}{(x+1)(x-1)} > 0$$

$$\frac{x^2+x-6}{(x+1)(x-1)} > 0$$

$$\frac{(x+3)(x-2)}{(x+1)(x-1)} > 0$$

$$f(x) = \frac{2x+5}{x+1} - \frac{x+1}{x-1}$$

The zeros and values where f is undefined are $x = -3, x = -1, x = 1$ and $x = 2$.

Interval	$(-\infty,-3)$	$(-3,-1)$	$(-1,1)$	$(1,2)$	$(2,\infty)$
Number Chosen	-4	-2	0	1.5	3
Value of f	0.4	$-\dfrac{4}{3}$	6	-1.8	0.75
Conclusion	Positive	Negative	Positive	Negative	Positive

The solution set is $\left\{ x \mid x < -3 \text{ or } -1 < x < 1 \text{ or } x > 2 \right\}$.

47.

$$\frac{x^2(3+x)(x+4)}{(x+5)(x-1)} \geq 0$$

$$f(x) = \frac{x^2(3+x)(x+4)}{(x+5)(x-1)}$$

The zeros and values where f is undefined are $x = -5, x = -4, x = -3, x = 0$ and $x = 1$.

Interval	$(-\infty,-5)$	$(-5,-4)$	$(-4,-3)$	$(-3,0)$	$(0,1)$	$(1,\infty)$
Number Chosen	-6	-4.5	-3.5	-1	0.5	2
Value of f	$\dfrac{216}{7}$	$-\dfrac{243}{44}$	$\dfrac{49}{108}$	-0.75	$-\dfrac{63}{44}$	$\dfrac{120}{7}$
Conclusion	Positive	Negative	Positive	Negative	Negative	Positiv

The solution set is $\left\{ x \mid x < -5 \text{ or } -4 \leq x \leq -3 \text{ or } x > 1 \right\}$.

49.

$$\frac{(3-x)^3(2x+1)}{x^3-1} < 0$$

$$f(x) = \frac{(3-x)^3(2x+1)}{x^3-1}$$

The zeros and values where f is undefined are $x = -\dfrac{1}{2}, x = 1$ and $x =$

Interval	$\left(-\infty,-\dfrac{1}{2}\right)$	$\left(-\dfrac{1}{2},1\right)$	$(1,3)$	$(3,\infty)$
Number Chosen	-1	0	2	4
Value of f	32	-27	$\dfrac{5}{7}$	$-\dfrac{1}{7}$
Conclusion	Positive	Negative	Positive	Negative

The solution set is $\left\{ x \mid -\dfrac{1}{2} < x < 1 \text{ or } x > 3 \right\}$.

51. Let x be the positive number. $f(x) = x^3 - 4x^2$
    Then $x^3 > 4x^2$ The zeros of f are $x = 0$ and $x = 4$.
    $x^3 - 4x^2 > 0$

    $x^2(x - 4) > 0$

Interval	$(-\infty, 0)$	$(0, 4)$	$(4, \infty)$
Number Chosen	-1	1	5
Value of f	-5	-3	25
Conclusion	Negative	Negative	Positive

The solution set is $\{x \mid x > 4\}$. Since x must be positive, all real numbers greater than 4 satisfy the condition.

53. The domain of the expression consists $p(x) = x^2 - 16$
    of all real numbers x for which   The zeros of p are $x = -4$ and $x = 4$.
    $x^2 - 16 \geq 0$

    $(x + 4)(x - 4) \geq 0$

Interval	$(-\infty, -4)$	$(-4, 4)$	$(4, \infty)$
Number Chosen	-5	0	5
Value of p	9	-16	9
Conclusion	Positive	Negative	Positive

The domain of f is $\{x \mid x \leq -4 \text{ or } x \geq 4\}$.

55. The domain of the expression consists $p(x) = \dfrac{x - 2}{x + 4}$
    of all real numbers x for which
    $\dfrac{x - 2}{x + 4} \geq 0$ The zeros and values where p is undefined are
                                               $x = -4$ and $x = 2$.

Interval	$(-\infty, -4)$	$(-4, 2)$	$(2, \infty)$
Number Chosen	-5	0	3
Value of p	7	-0.5	$\dfrac{1}{7}$
Conclusion	Positive	Negative	Positive

The domain of f is $\{x \mid x < -4 \text{ or } x \geq 2\}$.

57. Find the values of t for which $s(t) > 96$. $p(t) = -16t^2 + 80t - 96$
    $80t - 16t^2 > 96$ The zeros of p are $t = 2$ and $t = 3$.

    $-16t^2 + 80t - 96 > 0$

    $-16(t^2 - 5t + 6) > 0 \Rightarrow -16(t - 2)(t - 3) > 0$

Interval	$(-\infty, 2)$	$(2, 3)$	$(3, \infty)$
Number Chosen	1	2.5	4
Value of p	-32	4	-32
Conclusion	Negative	Positive	Negative

The solution set is $\{t \mid 2 < t < 3\}$. The ball is more than 96 feet above the ground for times between 2 and 3 seconds.

59.   Profit = Revenue − Cost = $x(40 − 0.2x) − 32x = 40x − 0.2x^2 − 32x = −0.2x^2 + 8x$
In order to achieve a profit of a least $50, we need to solve the
inequality $−0.2x^2 + 8x \geq 50$.

$−0.2x^2 + 8x \geq 50$

$−0.2x^2 + 8x − 50 \geq 0$
Dividing by $−0.2$ yields
$x^2 − 40x + 250 \leq 0$

$f(x) = x^2 − 40x + 250$
Solving $x^2 − 40x + 250 = 0$:

$$x = \frac{-(-40) \pm \sqrt{(-40)^2 - 4(1)(250)}}{2(1)}$$

$$= \frac{40 \pm \sqrt{600}}{2}$$

$$= \frac{40 \pm 10\sqrt{6}}{2} = 20 \pm 5\sqrt{6}$$

The zeros of f are $x = 20 − 5\sqrt{6} \approx 7.75$ and
$x = 20 + 5\sqrt{6} \approx 32.25$
Also note that x must be positive.

Interval	$\left(0, 20 − 5\sqrt{6}\right)$	$\left(20 − 5\sqrt{6}, 20 + 5\sqrt{6}\right)$	$\left(20 + 5\sqrt{6}, \infty\right)$
Number Chosen	1	8	33
Value of f	211	−6	19
Conclusion	Positive	Negative	Positive

The solution set is $\left\{ x \mid 20 − 5\sqrt{6} < x < 20 + 5\sqrt{6} \right\}$. The profit is at least $50 when at least 8
and no more than 32 watches are sold.

61.   The equation $x^2 + kx + 1 = 0$ has no real solutions whenever the discriminant is less
than zero. Note that $b^2 − 4ac < 0 \Rightarrow k^2 − 4 < 0$.

$k^2 − 4 < 0$

$(k + 2)(k − 2) < 0$

$f(k) = k^2 − 4$
The zeros of f are $k = −2$ and $k = 2$.

Interval	$(-\infty, -2)$	$(-2, 2)$	$(2, \infty)$
Number Chosen	−3	0	3
Value of f	5	−4	5
Conclusion	Positive	Negative	Positive

The solution set is $\left\{ k \mid −2 < k < 2 \right\}$. Therefore the equation $x^2 + kx + 1 = 0$ has no real
solutions whenever $−2 < k < 2$.

63.   Answers will vary,

Polynomial and Rational Functions

3.6   The Real Zeros of a Polynomial Function

1. $f(-1) = 2(-1)^2 - (-1) = 2 + 1 = 3$

3.   Using synthetic division:

$$3\overline{)3 \;\; -5 \;\;\; 0 \;\;\;\;\; 7 \;\; -4}$$
$$ \quad\quad 9 \;\;\; 12 \;\;\; 36 \;\;\; 129$$
$$\overline{ \; 3 \;\;\;\; 4 \;\;\; 12 \;\;\; 43 \;\;\; 125}$$

Quotient: $3x^3 + 4x^2 + 12x + 43$ \qquad Remainder: 125

5.   Remainder, Dividend   7.   –4

9.   False

11. $f(x) = 4x^3 - 3x^2 - 8x + 4; \quad c = 2$
$f(2) = 4(2)^3 - 3(2)^2 - 8(2) + 4 = 32 - 12 - 16 + 4 = 8 \neq 0$
Thus, 2 is not a zero of f and $x - 2$ is not a factor of f.

13. $f(x) = 3x^4 - 6x^3 - 5x + 10; \quad c = 2$
$f(2) = 3(2)^4 - 6(2)^3 - 5(2) + 10 = 48 - 48 - 10 + 10 = 0$
Thus, 2 is a zero of f and $x - 2$ is a factor of f.

15. $f(x) = 3x^6 + 82x^3 + 27; \quad c = -3$
$f(-3) = 3(-3)^6 + 82(-3)^3 + 27 = 2187 - 2214 + 27 = 0$
Thus, –3 is a zero of f and $x + 3$ is a factor of f.

17. $f(x) = 4x^6 - 64x^4 + x^2 - 15; \quad c = -4$
$f(-4) = 4(-4)^6 - 64(-4)^4 + (-4)^2 - 15 = 16,384 - 16,384 + 16 - 15 = 1 \neq 0$
Thus, –4 is not a zero of f and $x + 4$ is not a factor of f.

19. $f(x) = 2x^4 - x^3 + 2x - 1; \quad c = \dfrac{1}{2}$

$f\left(\dfrac{1}{2}\right) = 2\left(\dfrac{1}{2}\right)^4 - \left(\dfrac{1}{2}\right)^3 + 2\left(\dfrac{1}{2}\right) - 1 = \dfrac{1}{8} - \dfrac{1}{8} + 1 - 1 = 0$

Thus, $\dfrac{1}{2}$ is a zero of f and $x - \dfrac{1}{2}$ is a factor of f.

21. $f(x) = -4x^7 + x^3 - x^2 + 2$

The maximum number of zeros is the degree of the polynomial, which is 7.

Examining $f(x) = -4x^7 + x^3 - x^2 + 2$, there are three variations in sign; thus, there are three positive real zeros or there is one positive real zero.

Examining $f(-x) = -4(-x)^7 + (-x)^3 - (-x)^2 + 2 = 4x^7 - x^3 - x^2 + 2$, there are two variations in sign; thus, there are two negative real zeros or no negative real zeros.

23. $f(x) = 2x^6 - 3x^2 - x + 1$

The maximum number of zeros is the degree of the polynomial, which is 6.

Examining $f(x) = 2x^6 - 3x^2 - x + 1$, there are two variations in sign; thus, there are two positive real zeros or no positive real zeros.

Examining $f(-x) = 2(-x)^6 - 3(-x)^2 - (-x) + 1 = 2x^6 - 3x^2 + x + 1$, there are two variations in sign; thus, there are two negative real zeros or no negative real zeros.

25. $f(x) = 3x^3 - 2x^2 + x + 2$

The maximum number of zeros is the degree of the polynomial, which is 3.

Examining $f(x) = 3x^3 - 2x^2 + x + 2$, there are two variations in sign; thus, there are two positive real zeros or no positive real zeros.

Examining $f(-x) = 3(-x)^3 - 2(-x)^2 + (-x) + 2 = -3x^3 - 2x^2 - x + 2$, there is one variation in sign; thus, there is one negative real zero.

27. $f(x) = -x^4 + x^2 - 1$

The maximum number of zeros is the degree of the polynomial, which is 4.

Examining $f(x) = -x^4 + x^2 - 1$, there are two variations in sign; thus, there are two positive real zeros or no positive real zeros.

Examining $f(-x) = -(-x)^4 + (-x)^2 - 1 = -x^4 + x^2 - 1$, there are two variations in sign; thus, there are two negative real zeros or no negative real zeros.

29. $f(x) = x^5 + x^4 + x^2 + x + 1$

The maximum number of zeros is the degree of the polynomial, which is 5.

Examining $f(x) = x^5 + x^4 + x^2 + x + 1$, there are no variations in sign; thus, there are no positive real zeros.

Examining $f(-x) = (-x)^5 + (-x)^4 + (-x)^2 + (-x) + 1 = -x^5 + x^4 + x^2 - x + 1$, there are three variations in sign; thus, there are three negative real zeros or there is one negative real zero.

31. $f(x) = x^6 - 1$

The maximum number of zeros is the degree of the polynomial, which is 6.

Examining $f(x) = x^6 - 1$, there is one variation in sign; thus, there is one positive real zero.

Examining $f(-x) = (-x)^6 - 1 = x^6 - 1$, there is one variation in sign; thus, there is one negative real zero.

33. $f(x) = 3x^4 - 3x^3 + x^2 - x + 1$
    p must be a factor of 1: $p = \pm 1$
    q must be a factor of 3: $q = \pm 1, \pm 3$

    The possible rational zeros are: $\dfrac{p}{q} = \pm 1, \pm \dfrac{1}{3}$

35. $f(x) = x^5 - 6x^2 + 9x - 3$
    p must be a factor of -3: $p = \pm 1, \pm 3$
    q must be a factor of 1: $q = \pm 1$

    The possible rational zeros are: $\dfrac{p}{q} = \pm 1, \pm 3$

37. $f(x) = -4x^3 - x^2 + x + 2$
    p must be a factor of 2: $p = \pm 1, \pm 2$
    q must be a factor of -4: $q = \pm 1, \pm 2, \pm 4$

    The possible rational zeros are: $\dfrac{p}{q} = \pm 1, \pm 2, \pm \dfrac{1}{2}, \pm \dfrac{1}{4}$

39. $f(x) = 6x^4 - x^2 + 9$
    p must be a factor of 9: $p = \pm 1, \pm 3, \pm 9$
    q must be a factor of 6: $q = \pm 1, \pm 2, \pm 3, \pm 6$

    The possible rational zeros are: $\dfrac{p}{q} = \pm 1, \pm \dfrac{1}{2}, \pm \dfrac{1}{3}, \pm \dfrac{1}{6}, \pm 3, \pm \dfrac{3}{2}, \pm 9, \pm \dfrac{9}{2}$

41. $f(x) = 2x^5 - x^3 + 2x^2 + 12$
    p must be a factor of 12: $p = \pm 1, \pm 2, \pm 3, \pm 4, \pm 6, \pm 12$
    q must be a factor of 2: $q = \pm 1, \pm 2$

    The possible rational zeros are: $\dfrac{p}{q} = \pm 1, \pm 2, \pm 4, \pm \dfrac{1}{2}, \pm 3, \pm \dfrac{3}{2}, \pm 6, \pm 12$

43. $f(x) = 6x^4 + 2x^3 - x^2 + 20$
    p must be a factor of 20: $p = \pm 1, \pm 2, \pm 4, \pm 5, \pm 10, \pm 20$
    q must be a factor of 6: $q = \pm 1, \pm 2, \pm 3, \pm 6$
    The possible rational zeros are:

    $\dfrac{p}{q} = \pm 1, \pm 2, \pm \dfrac{1}{2}, \pm \dfrac{1}{3}, \pm \dfrac{2}{3}, \pm \dfrac{1}{6}, \pm 4, \pm \dfrac{4}{3}, \pm 5, \pm \dfrac{5}{2}, \pm \dfrac{5}{3}, \pm \dfrac{5}{6}, \pm 10, \pm \dfrac{10}{3}, \pm 20, \pm \dfrac{20}{3}$

45. $f(x) = x^3 + 2x^2 - 5x - 6$
    Step 1: $f(x)$ has at most 3 real zeros.
    Step 2:   By Descartes' Rule of Signs, there is one positive real zero.
    $f(-x) = (-x)^3 + 2(-x)^2 - 5(-x) - 6 = -x^3 + 2x^2 + 5x - 6$, thus, there are two negative real zeros or no negative real zeros.

Step 3:   Possible rational zeros:
$p = \pm 1, \pm 2, \pm 3, \pm 6; \quad q = \pm 1;$

$$\frac{p}{q} = \pm 1, \pm 2, \pm 3, \pm 6$$

Step 4:   Using synthetic division:
We try $x + 3$:

$$\begin{array}{r|rrrr} -3) & 1 & 2 & -5 & -6 \\ & & -3 & 3 & 6 \\ \hline & 1 & -1 & -2 & 0 \end{array}$$

Since the remainder is 0, $x - (-3) = x + 3$ is a factor. The other factor is the quotient: $x^2 - x - 2$.

Thus, $f(x) = (x + 3)\left(x^2 - x - 2\right) = (x + 3)(x + 1)(x - 2)$.

The real zeros are $-3, -1$, and 2, each of multiplicity 1.

47. $f(x) = 2x^3 - x^2 + 2x - 1$

Step 1: $f(x)$ has at most 3 real zeros.

Step 2:   By Descartes' Rule of Signs, there are three positive real zeros or there is one positive real zero.
$f(-x) = 2(-x)^3 - (-x)^2 + 2(-x) - 1 = -2x^3 - x^2 - 2x - 1$, thus, there are no negative real zeros.

Step 3:   Possible rational zeros:
$p = \pm 1 \quad q = \pm 1, \pm 2$

$$\frac{p}{q} = \pm 1, \pm \frac{1}{2}$$

Step 4:   Using synthetic division:

We try $x - 1$:

$$\begin{array}{r|rrrr} 1) & 2 & -1 & 2 & -1 \\ & & 2 & 1 & 3 \\ \hline & 2 & 1 & 3 & 2 \end{array}$$

$x - 1$ is **not** a factor

We try $x - \dfrac{1}{2}$:

$$\begin{array}{r|rrrr} \frac{1}{2}) & 2 & -1 & 2 & -1 \\ & & 1 & 0 & 1 \\ \hline & 2 & 0 & 2 & 0 \end{array}$$

$x - \dfrac{1}{2}$ is a factor and the quotient is $2x^2 + 2$.

Thus, $f(x) = 2x^3 - x^2 + 2x - 1 = \left(x - \dfrac{1}{2}\right)\left(2x^2 + 2\right)$.

Since $2x^2 + 2 = 0$ has no real solutions, the only real zero is $x = \dfrac{1}{2}$, of multiplicity 1.

49. $f(x) = x^4 + x^2 - 2$

Step 1: $f(x)$ has at most 4 real zeros.

Step 2:   By Descartes' Rule of Signs, there is one positive real zero.
$f(-x) = (-x)^4 + (-x)^2 - 2 = x^4 + x^2 - 2$, thus, there is one negative real zero.

Step 3:   Possible rational zeros:
$$p = \pm1, \pm2; \quad q = \pm1;$$
$$\frac{p}{q} = \pm1, \pm2$$

Step 4:   Using synthetic division:
We try $x + 1$:

$$-1\overline{)\begin{array}{ccccc} 1 & 0 & 1 & 0 & -2 \\ & -1 & 1 & -2 & 2 \\ \hline 1 & -1 & 2 & -2 & 0 \end{array}}$$

Since the remainder is 0, $x - (-1) = x + 1$ is a factor. The other factor is the quotient: $x^3 - x^2 + 2x - 2$.

Thus, $f(x) = (x+1)(x^3 - x^2 + 2x - 2)$. We can factor $x^3 - x^2 + 2x - 2$ by grouping terms:

$$x^3 - x^2 + 2x - 2 = x^2(x-1) + 2(x-1) = (x-1)(x^2 + 2)$$

Thus, $f(x) = (x+1)(x-1)(x^2 + 2)$.

Since $x^2 + 2 = 0$ has no real solutions, the real zeros are -1 and 1, each of multiplicity 1.

51. $f(x) = 4x^4 + 7x^2 - 2$

Step 1: $f(x)$ has at most 4 real zeros.

Step 2:   By Descartes' Rule of Signs, there is one positive real zero.
$f(-x) = 4(-x)^4 + 7(-x)^2 - 2 = 4x^4 + 7x^2 - 2$, thus, there is one negative real zero.

Step 3:   Possible rational zeros:
$$p = \pm1, \pm2; \quad q = \pm1, \pm2, \pm4$$
$$\frac{p}{q} = \pm1, \pm\frac{1}{2}, \pm\frac{1}{4} \pm 2$$

Factoring: $f(x) = 4x^4 + 7x^2 - 2 = (4x^2 - 1)(x^2 + 2) = (2x+1)(2x-1)(x^2 + 2)$.

Since $x^2 + 1 = 0$ has no real solutions, the real zeros are $-\frac{1}{2}$ and $\frac{1}{2}$, each of multiplicity 1.

53. $f(x) = x^4 + x^3 - 3x^2 - x + 2$

Step 1: $f(x)$ has at most 4 real zeros.

Step 2:   By Descartes' Rule of Signs, there are two positive real zeros or no positive real zeros.
$f(-x) = (-x)^4 + (-x)^3 - 3(-x)^2 - (-x) + 2 = x^4 - x^3 - 3x^2 + x + 2$, thus, there are two negative real zeros or no negative real zeros.

Step 3:   Possible rational zeros:
$$p = \pm1, \pm2; \quad q = \pm1;$$
$$\frac{p}{q} = \pm1, \pm2$$

Step 4:   Using synthetic division:

We try $x + 2$:

$$-2\overline{)\begin{array}{rrrrr} 1 & 1 & -3 & -1 & 2 \\ & -2 & 2 & 2 & -2 \\ \hline 1 & -1 & -1 & 1 & 0 \end{array}}$$

$x + 2$ is a factor and

the quotient is $x^3 - x^2 - x + 1$

We try $x + 1$ on $x^3 - x^2 - x + 1$

$$-1\overline{)\begin{array}{rrrr} 1 & -1 & -1 & 1 \\ & -1 & 2 & -1 \\ \hline 1 & -2 & 1 & 0 \end{array}}$$

$x + 1$ is a factor and

the quotient is $x^2 - 2x + 1$

Thus, $f(x) = (x + 2)(x + 1)(x^2 - 2x + 1) = (x + 2)(x + 1)(x - 1)^2$.

The real zeros are $-2, -1$, each of multiplicity 1, and 1, of multiplicity 2.

55. $f(x) = 4x^5 - 8x^4 - x + 2$

Step 1: $f(x)$ has at most 5 real zeros.

Step 2:   By Descartes' Rule of Signs, there are two positive real zeros or no positive real zeros.

$f(-x) = 4(-x)^5 - 8(-x)^4 - (-x) + 2 = -4x^5 - 8x^4 + x + 2$, thus, there is one negative real zero.

Step 3:   Possible rational zeros:

$p = \pm 1, \pm 2; \quad q = \pm 1, \pm 2, \pm 4;$

$\dfrac{p}{q} = \pm 1, \pm 2, \pm \dfrac{1}{2}, \pm \dfrac{1}{4}$

Step 4:   Using synthetic division:

We try $x - 2$:

$$2\overline{)\begin{array}{rrrrrr} 4 & -8 & 0 & 0 & -1 & 2 \\ & 8 & 0 & 0 & 0 & -2 \\ \hline 4 & 0 & 0 & 0 & -1 & 0 \end{array}}$$

Since the remainder is 0, $x - 2$ is a factor. The other factor is the quotient: $4x^4 - 1$.

Thus,

$$f(x) = (x - 2)(4x^4 - 1)$$

$$= (x - 2)(2x^2 - 1)(2x^2 + 1)$$

$$= (x - 2)(\sqrt{2}x - 1)(\sqrt{2}x + 1)(2x^2 + 1)$$

Since $2x^2 + 1 = 0$ has no real solutions, the real zeros are $-\dfrac{\sqrt{2}}{2}, \dfrac{\sqrt{2}}{2}$, and 2, each of multiplicity 1.

57. $x^4 - x^3 + 2x^2 - 4x - 8 = 0$

The solutions of the equation are the zeros of $f(x) = x^4 - x^3 + 2x^2 - 4x - 8$.

Step 1: $f(x)$ has at most 4 real zeros.

Step 2:   By Descartes' Rule of Signs, there are three positive real zeros or there is one positive real zero.

$f(-x) = (-x)^4 - (-x)^3 + 2(-x)^2 - 4(-x) - 8 = x^4 + x^3 + 2x^2 + 4x - 8$, thus, there is one negative real zero.

Step 3:   Possible rational zeros:
$$p = \pm 1, \pm 2, \pm 4, \pm 8; \quad q = \pm 1;$$

$$\frac{p}{q} = \pm 1, \pm 2, \pm 4, \pm 8$$

Step 4:   Using synthetic division:

We try $x + 1$:

$$\begin{array}{r|rrrrr} -1 & 1 & -1 & 2 & -4 & -8 \\ & & -1 & 2 & -4 & 8 \\ \hline & 1 & -2 & 4 & -8 & 0 \end{array}$$

$x + 1$ is a factor and

the quotient is $x^3 - 2x^2 + 4x - 8$

We try $x - 2$ on $x^3 - 2x^2 + 4x - 8$

$$\begin{array}{r|rrrr} 2 & 1 & -2 & 4 & -8 \\ & & 2 & 0 & 8 \\ \hline & 1 & 0 & 4 & 0 \end{array}$$

$x - 2$ **is** a factor and

the quotient is $x^2 + 4$

Thus, $f(x) = (x + 1)(x - 2)(x^2 + 4)$.

Since $x^2 + 4 = 0$ has no real solutions, the solution set is $\{-1, 2\}$.

59. $3x^3 + 4x^2 - 7x + 2 = 0$

The solutions of the equation are the zeros of $f(x) = 3x^3 + 4x^2 - 7x + 2$.

Step 1: $f(x)$ has at most 3 real zeros.

Step 2:   By Descartes' Rule of Signs, there are two positive real zeros or no positive real zeros.

$f(-x) = 3(-x)^3 + 4(-x)^2 - 7(-x) + 2 = -3x^3 + 4x^2 + 7x + 2$;

thus, there is one negative real zero.

Step 3:   Possible rational zeros:
$$p = \pm 1, \pm 2; \quad q = \pm 1, \pm 3$$

$$\frac{p}{q} = \pm 1, \pm 2, \pm \frac{1}{3}, \pm \frac{2}{3}$$

Step 4:   Using synthetic division:

We try $x - \dfrac{2}{3}$:

$$\begin{array}{r|rrrr} \frac{2}{3} & 3 & 4 & -7 & 2 \\ & & 2 & 4 & -2 \\ \hline & 3 & 6 & -3 & 0 \end{array}$$

$x - \dfrac{2}{3}$ is a factor. The other factor is the quotient $3x^2 + 6x - 3$.

Thus, $f(x) = \left(x - \dfrac{2}{3}\right)(3x^2 + 6x - 3) = 3\left(x - \dfrac{2}{3}\right)(x^2 + 2x - 1)$

Using the quadratic formula to solve $x^2 + 2x - 1 = 0$:

$$x = \frac{-2 \pm \sqrt{4 - 4(1)(-1)}}{2(1)} = \frac{-2 \pm \sqrt{8}}{2} = \frac{-2 \pm 2\sqrt{2}}{2} = -1 \pm \sqrt{2}$$

The solution set is $\left\{-1 - \sqrt{2}, \; -1 + \sqrt{2}, \; \dfrac{2}{3}\right\}$.

61. $3x^3 - x^2 - 15x + 5 = 0$
Solving by factoring:
$$x^2(3x - 1) - 5(3x - 1) = 0$$
$$(3x - 1)(x^2 - 5) = 0$$
$$(3x - 1)(x - \sqrt{5})(x + \sqrt{5}) = 0$$
The solution set is $\left\{-\sqrt{5}, \ \sqrt{5}, \ \dfrac{1}{3}\right\}$.

63. $x^4 + 4x^3 + 2x^2 - x + 6 = 0$
The solutions of the equation are the zeros of $f(x) = x^4 + 4x^3 + 2x^2 - x + 6$.
Step 1: $f(x)$ has at most 4 real zeros.
Step 2: By Descartes' Rule of Signs, there are two positive real zeros or no positive real zeros.
$f(-x) = (-x)^4 + 4(-x)^3 + 2(-x)^2 - (-x) + 6 = x^4 - 4x^3 + 2x^2 + x + 6$, thus, there are two negative real zeros or no negative real zeros.
Step 3: Possible rational zeros:
$$p = \pm 1, \pm 2, \pm 3, \pm 6; \quad q = \pm 1;$$
$$\frac{p}{q} = \pm 1, \pm 2, \pm 3, \pm 6$$
Step 4: Using synthetic division:

We try $x + 3$:

$$\begin{array}{r|rrrrr} -3 & 1 & 4 & 2 & -1 & 6 \\ & & -3 & -3 & 3 & -6 \\ \hline & 1 & 1 & -1 & 2 & 0 \end{array}$$

$x + 3$ is a factor and

the quotient is $x^3 + x^2 - x + 2$

We try $x + 2$ on $x^3 + x^2 - x + 2$:

$$\begin{array}{r|rrrr} -2 & 1 & 1 & -1 & 2 \\ & & -2 & 2 & -2 \\ \hline & 1 & -1 & 1 & 0 \end{array}$$

$x + 2$ is a factor and

the quotient is $x^2 - x + 1$

Thus, $f(x) = (x + 3)(x + 2)(x^2 - x + 1)$.
Since $x^2 - x + 1 = 0$ has no real solutions , the solution set is $\{-3, -2\}$.

65. $x^3 - \dfrac{2}{3}x^2 + \dfrac{8}{3}x + 1 = 0 \Rightarrow 3x^3 - 2x^2 + 8x + 3 = 0$
The solutions of the equation are the zeros of $f(x) = 3x^3 - 2x^2 + 8x + 3$.
Step 1: $f(x)$ has at most 3 real zeros.
Step 2: By Descartes' Rule of Signs, there are two positive real zeros or no positive real zeros.
$f(-x) = 3(-x)^3 - 2(-x)^2 + 8(-x) + 3 = -3x^3 - 2x^2 - 8x + 3$, thus, there is one negative real zero.
Step 3: To find the possible rational zeros:
$$p = \pm 1, \pm 3; \quad q = \pm 1, \pm 3$$
$$\frac{p}{q} = \pm 1, \pm 3, \pm \frac{1}{3}$$

Step 4:   Using synthetic division:

We try $x + \dfrac{1}{3}$:

$$-\dfrac{1}{3} \overline{\smash{)}\,3 \quad -2 \quad 8 \quad 3}$$
$$\underline{ -1 \quad 1 \quad -3}$$
$$3 \quad -3 \quad 9 \quad 0$$

$x + \dfrac{1}{3}$ is a factor. The other factor is the quotient: $3x^2 - 3x + 9$.

Thus, $f(x) = \left(x + \dfrac{1}{3}\right)\left(3x^2 - 3x + 9\right) = \left(x + \dfrac{1}{3}\right)(3)\left(x^2 - x + 3\right) = (3x + 1)\left(x^2 - x + 3\right)$.

Since $x^2 - x + 3 = 0$ has no real solutions, the solution set is $\left\{-\dfrac{1}{3}\right\}$.

67. $2x^4 - 19x^3 + 57x^2 - 64x + 20 = 0$

The solutions of the equation are the zeros of $f(x) = 2x^4 - 19x^3 + 57x^2 - 64x + 20$.

Step 1: $f(x)$ has at most 4 real zeros.

Step 2:   By Descartes' Rule of Signs, there are four positive real zeros or two positive real zeros or no positive real zeros.

$$f(-x) = 2(-x)^4 - 19(-x)^3 + 57(-x)^2 - 64(-x) + 20$$
$$= 2x^4 + 19x^3 + 57x^2 + 64x + 20$$

Thus, there are no negative real zeros.

Step 3:   To find the possible rational zeros:

$p = \pm 1, \pm 2, \pm 4, \pm 5, \pm 10, \pm 20;\quad q = \pm 1, \pm 2;$

$\dfrac{p}{q} = \pm 1, \pm \dfrac{1}{2}, \pm 2, \pm 4, \pm 5, \pm \dfrac{5}{2}, \pm 10, \pm 20$

Step 4:   Using synthetic division:

We try $x - 1$:

$$1 \overline{\smash{)}\,2 \quad -19 \quad 57 \quad -64 \quad 20}$$
$$\underline{ 2 \quad -17 \quad 40 \quad -24}$$
$$2 \quad -17 \quad 40 \quad -24 \quad -4$$

$x + 1$ is not a factor

We try $x - \dfrac{1}{2}$:

$$\dfrac{1}{2} \overline{\smash{)}\,2 \quad -19 \quad 57 \quad -64 \quad 20}$$
$$\underline{ 1 \quad -9 \quad 24 \quad -20}$$
$$2 \quad -18 \quad 48 \quad -40 \quad 0$$

$x - \dfrac{1}{2}$ is a factor and

the quotient is $2x^3 - 18x^2 + 48x - 40$

Thus,

$$f(x) = \left(x - \dfrac{1}{2}\right)\left(2x^3 - 18x^2 + 48x - 40\right) = 2\left(x - \dfrac{1}{2}\right)\left(x^3 - 9x^2 + 24x - 20\right)$$

Now try $x - 2$ as a factor of $x^3 - 9x^2 + 24x - 20$.

$$2 \overline{\smash{)}\,1 \quad -9 \quad 24 \quad -20}$$
$$\underline{ 2 \quad -14 \quad 20}$$
$$1 \quad -7 \quad 10 \quad 0$$

$x - 2$ is a factor, and the other factor is the quotient $x^2 - 7x + 10$.

Thus, $x^3 - 9x^2 + 24x - 20 = (x-2)(x^2 - 7x + 10) = (x-2)(x-2)(x-5)$

$f(x) = 2\left(x - \dfrac{1}{2}\right)(x-2)^2(x-2)(x-5)$

The solution set is $\left\{\dfrac{1}{2}, 2, 5\right\}$.

69. $f(x) = x^3 + 2x^2 - 5x - 6 = (x+3)(x+1)(x-2)$

x-intercepts: $-3, -1, 2$;   y-intercept: $f(0) = 0^3 + 2(0)^2 - 5(0) - 6 = -6$;

The graph of f crosses the x-axis at $x = -3, -1$ and 2 since each zero has multiplicity 1.

Interval	$(-\infty, -3)$	$(-3, -1)$	$(-1, 2)$	$(2, \infty)$
Number Chosen	-4	-2	0	3
Value of f	-18	4	-6	24
Location of Graph	Below x-axis	Above x-axis	Below x-axis	Above x-axis
Point on Graph	$(-4, -18)$	$(-2, 4)$	$(0, -6)$	$(3, 24)$

The graph of f is above the x-axis on the intervals $(-3, -1)$ and $(2, \infty)$.

The graph of f is below the x-axis on the intervals $(-\infty, -3)$ and $(-1, -2)$.

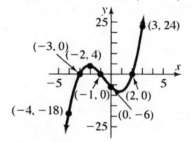

71. $f(x) = 2x^3 - x^2 + 2x - 1 = \left(x - \dfrac{1}{2}\right)(2x^2 + 2)$

x-intercept: $\dfrac{1}{2}$;   y-intercept: $f(0) = 2(0)^3 - 0^2 + 2(0) - 1 = -1$;

The graph of f crosses the x-axis at $x = \dfrac{1}{2}$ since the zero has multiplicity 1.

Interval	$\left(-\infty, \dfrac{1}{2}\right)$	$\left(\dfrac{1}{2}, \infty\right)$
Number Chosen	0	1
Value of f	-1	2
Location of Graph	Below x-axis	Above x-axis
Point on Graph	$(0, -1)$	$(1, 2)$

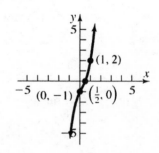

The graph of f is above the x-axis on the interval $\left(\dfrac{1}{2}, \infty\right)$.

The graph of f is below the x-axis on the interval $\left(-\infty, \dfrac{1}{2}\right)$.

73. $f(x) = x^4 + x^2 - 2 = (x+1)(x-1)(x^2+2)$

x-intercepts: $-1, 1$; y-intercept: $f(0) = 0^4 + 0^2 - 2 = -2$;

The graph of f crosses the x-axis at $x = -1$ and 1 since each zero has multiplicity 1.

Interval	$(-\infty, -1)$	$(-1, 1)$	$(1, \infty)$
Number Chosen	-2	0	2
Value of f	18	-2	18
Location of Graph	Above x-axis	Below x-axis	Above x-axis
Point on Graph	$(-2, 18)$	$(0, -2)$	$(2, 18)$

The graph of f is above the x-axis on the intervals $(-\infty, -1)$ and $(1, \infty)$.

The graph of f is below the x-axis on the interval $(-1, -1)$.

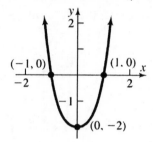

75. $f(x) = 4x^4 + 7x^2 - 2 = (2x+1)(2x-1)(x^2+2)$

x-intercepts: $-\dfrac{1}{2}, \dfrac{1}{2}$; y-intercept: $f(0) = 4(0)^4 + 7(0)^2 - 2 = -2$;

The graph of f crosses the x-axis at $x = -\dfrac{1}{2}$ and $\dfrac{1}{2}$ since each zero has multiplicity 1.

Interval	$\left(-\infty, -\dfrac{1}{2}\right)$	$\left(-\dfrac{1}{2}, \dfrac{1}{2}\right)$	$\left(\dfrac{1}{2}, \infty\right)$
Number Chosen	-1	0	1
Value of f	9	-2	9
Location of Graph	Above x-axis	Below x-axis	Above x-axis
Point on Graph	$(-1, 9)$	$(0, -2)$	$(1, 9)$

The graph of f is above the x-axis on the intervals $\left(-\infty, -\dfrac{1}{2}\right)$ and $\left(\dfrac{1}{2}, \infty\right)$.

The graph of f is below the x-axis on the interval $\left(-\dfrac{1}{2}, \dfrac{1}{2}\right)$.

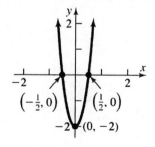

77. $f(x) = x^4 + x^3 - 3x^2 - x + 2 = (x+2)(x+1)(x-1)^2$

   x-intercepts: $-2, -1, 1$;   y-intercept: $f(0) = 0^4 + 0^3 - 3(0)^2 - 0 + 2 = 2$;

   The graph of f crosses the x-axis at $x = -2$ and -1 since each zero has multiplicity 1.

   The graph of f touches the x-axis at $x = 1$ since the zero has multiplicity 2.

Interval	$(-\infty, -2)$	$(-2, -1)$	$(-1, 1)$	$(1, \infty)$
Number Chosen	-3	-1.5	0	2
Value of f	32	-1.5625	2	12
Location of Graph	Above x-axis	Below x-axis	Above x-axis	Above x-axis
Point on Graph	$(-3, 32)$	$(-1.5, -1.5625)$	$(0, 2)$	$(2, 12)$

   The graph of f is above the x-axis on the intervals $(-\infty, -2)$ and $(-1, 1)$ and $(1, \infty)$.

   The graph of f is below the x-axis on the interval $(-2, -1)$.

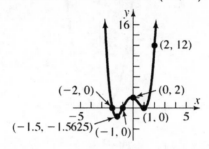

79. $f(x) = 4x^5 - 8x^4 - x + 2 = (x-2)\left(\sqrt{2}x - 1\right)\left(\sqrt{2}x + 1\right)\left(2x^2 + 1\right)$

   x-intercepts: $-\dfrac{\sqrt{2}}{2}, \dfrac{\sqrt{2}}{2}, 2$;   y-intercept: $f(0) = 4(0)^5 - 8(0)^4 - 0 + 2 = 2$;

   The graph of f crosses the x-axis at $x = -\dfrac{\sqrt{2}}{2}$, $x = \dfrac{\sqrt{2}}{2}$ and $x = 2$ since each zero has

   multiplicity 1.

Interval	$\left(-\infty, -\dfrac{\sqrt{2}}{2}\right)$	$\left(-\dfrac{\sqrt{2}}{2}, \dfrac{\sqrt{2}}{2}\right)$	$\left(\dfrac{\sqrt{2}}{2}, 2\right)$	$(2, \infty)$
Number Chosen	-1	0	1	3
Value of f	-9	2	-3	323
Location of Graph	Below x-axis	Above x-axis	Below x-axis	Above x-axis
Point on Graph	$(-1, -9)$	$(0, 2)$	$(1, -3)$	$(3, 323)$

   The graph of f is above the x-axis on the intervals $\left(-\dfrac{\sqrt{2}}{2}, \dfrac{\sqrt{2}}{2}\right)$ and $(2, \infty)$.

   The graph of f is below the x-axis on the intervals $\left(-\infty, -\dfrac{\sqrt{2}}{2}\right)$ and $\left(\dfrac{\sqrt{2}}{2}, 2\right)$.

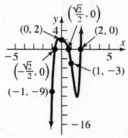

81. $f(x) = x^4 - 3x^2 - 4$

$a_3 = 0, a_2 = -3, a_1 = 0, a_0 = -4$

$\text{Max}\{1, |-4| + |0| + |-3| + |0|\}$ $1 + \text{Max}\{|-4|, |0|, |-3|, |0|\}$

$= \text{Max}\{1, 4 + 0 + 3 + 0\}$ $= 1 + \text{Max}\{4, 0, 3, 0\}$

$= \text{Max}\{1, 7\} = 7$ $= 1 + 4 = 5$

The smaller of the two numbers is 5. Thus, every zero of f lies between -5 and 5.

83. $f(x) = x^4 + x^3 - x - 1$

$a_3 = 1, a_2 = 0, a_1 = -1, a_0 = -1$

$\text{Max}\{1, |-1| + |-1| + |0| + |1|\}$ $1 + \text{Max}\{|-1|, |-1|, |0|, |1|\}$

$= \text{Max}\{1, 1 + 1 + 0 + 1\}$ $= 1 + \text{Max}\{1, 1, 0, 1\}$

$= \text{Max}\{1, 3\} = 3$ $= 1 + 1 = 2$

The smaller of the two numbers is 2. Thus, every zero of f lies between -2 and 2.

85. $f(x) = 3x^4 + 3x^3 - x^2 - 12x - 12 = 3\left(x^4 + x^3 - \dfrac{1}{3}x^2 - 4x - 4\right)$

Note: The leading coefficient must be 1.

$a_3 = 1, a_2 = -\dfrac{1}{3}, a_1 = -4, a_0 = -4$

$\text{Max}\left\{1, |-4| + |-4| + \left|-\dfrac{1}{3}\right| + |1|\right\}$ $1 + \text{Max}\left\{|-4|, |-4|, \left|-\dfrac{1}{3}\right|, |1|\right\}$

$= \text{Max}\left\{1, 4 + 4 + \dfrac{1}{3} + 1\right\}$ $= 1 + \text{Max}\left\{4, 4, \dfrac{1}{3}, 1\right\}$

$= \text{Max}\left\{1, \dfrac{28}{3}\right\} = \dfrac{28}{3}$ $= 1 + 4 = 5$

The smaller of the two numbers is 5. Thus, every zero of f lies between -5 and 5.

87. $f(x) = 4x^5 - x^4 + 2x^3 - 2x^2 + x - 1 = 4\left(x^5 - \dfrac{1}{4}x^4 + \dfrac{1}{2}x^3 - \dfrac{1}{2}x^2 + \dfrac{1}{4}x - \dfrac{1}{4}\right)$

Note: The leading coefficient must be 1.

$a_4 = -\dfrac{1}{4}, a_3 = \dfrac{1}{2}, a_2 = -\dfrac{1}{2}, a_1 = \dfrac{1}{4}, a_0 = -\dfrac{1}{4}$

$\text{Max}\left\{1, \left|-\dfrac{1}{4}\right| + \left|\dfrac{1}{4}\right| + \left|-\dfrac{1}{2}\right| + \left|\dfrac{1}{2}\right| + \left|-\dfrac{1}{4}\right|\right\}$ $1 + \text{Max}\left\{\left|-\dfrac{1}{4}\right|, \left|\dfrac{1}{4}\right|, \left|-\dfrac{1}{2}\right|, \left|\dfrac{1}{2}\right|, \left|-\dfrac{1}{4}\right|\right\}$

$= \text{Max}\left\{1, \dfrac{1}{4} + \dfrac{1}{4} + \dfrac{1}{2} + \dfrac{1}{2} + \dfrac{1}{4}\right\}$ $= 1 + \text{Max}\left\{\dfrac{1}{4}, \dfrac{1}{4}, \dfrac{1}{2}, \dfrac{1}{2}, \dfrac{1}{4}\right\}$

$= \text{Max}\left\{1, \dfrac{7}{4}\right\} = \dfrac{7}{4}$ $= 1 + \dfrac{1}{2} = \dfrac{3}{2}$

The smaller of the two numbers is $\dfrac{3}{2}$. Thus, every zero of f lies between $-\dfrac{3}{2}$ and $\dfrac{3}{2}$.

89. $f(x) = 8x^4 - 2x^2 + 5x - 1;$ $[0, 1]$
$f(0) = -1 < 0$ and $f(1) = 10 > 0$
Since one value is positive and
one value is negative, there is a
zero in the interval.

91. $f(x) = 2x^3 + 6x^2 - 8x + 2;$ $[-5, -4]$
$f(-5) = -58 < 0$ and $f(-4) = 2 > 0$
Since one value is positive and
one value is negative, there is a
zero in the interval.

93. $f(x) = x^5 - x^4 + 7x^3 - 7x^2 - 18x + 18;$ $[1.4, 1.5]$
$f(1.4) = -0.1754 < 0$ and $f(1.5) = 1.4063 > 0$
Since one value is positive and
one value is negative, there is a
zero in the interval.

95. $8x^4 - 2x^2 + 5x - 1 = 0;$ $0 \le r \le 1$
Consider the function $f(x) = 8x^4 - 2x^2 + 5x - 1$
Subdivide the interval [0,1] into 10 equal subintervals:
[0,0.1]; [0.1,0.2]; [0.2,0.3]; [0.3,0.4]; [0.4,0.5]; [0.5,0.6]; [0.6,0.7]; [0.7,0.8];
[0.8,0.9]; [0.9,1]
$f(0) = -1; f(0.1) = -0.5192$
$f(0.1) = -0.5192; f(0.2) = -0.0672$
$f(0.2) = -0.0672; f(0.3) = 0.3848$
So f has a real zero on the interval [0.2,0.3].

Subdivide the interval [0.2,0.3] into 10 equal subintervals:
[0.2,0.21]; [0.21,0.22]; [0.22,0.23]; [0.23,0.24]; [0.24,0.25]; [0.25,0.26];[0.26,0.27];
[0.27,0.28]; [0.28,0.29]; [0.29,0.3]
$f(0.2) = -0.0672; f(0.21) \approx -0.02264$
$f(0.21) \approx -0.02264; f(0.22) \approx 0.0219$
So f has a real zero on the interval [0.21,0.22], therefore $r = 0.21$, correct to two decimal places.

97. $2x^3 + 6x^2 - 8x + 2 = 0;$ $-5 \le r \le -4$
Consider the function $f(x) = 2x^3 + 6x^2 - 8x + 2$
Subdivide the interval [–5, –4] into 10 equal subintervals:
[–5, –4.9]; [–4.9, –4.8]; [–4.8, –4.7]; [–4.7, –4.6]; [–4.6, –4.5]; [–4.5, –4.4]; [–4.4, –4.3];
[–4.3, –4.2]; [–4.2, –4.1]; [–4.1, –4]
$f(-5) = -58; f(-4.9) = -50.038$
$f(-4.9) = -50.038; f(-4.8) = -42.544$
$f(-4.8) = -42.544; f(-4.7) = -35.506$
$f(-4.7) = -35.506; f(-4.6) = -28.912$
$f(-4.6) = -28.912; f(-4.5) = -22.75$
$f(-4.5) = -22.75; f(-4.4) = -17.008$
$f(-4.4) = -17.008; f(-4.3) = -11.674$
$f(-4.3) = -11.674; f(-4.2) = -6.736$

$f(-4.2) = -6.736; f(-4.1) = -2.182$
$f(-4.1) = -2.182; f(-4) = 2$
So f has a real zero on the interval $[-4.1, -4]$.

Subdivide the interval $[-4.1, -4]$ into 10 equal subintervals:
$[-4.1, -4.09]; [-4.09, -4.08]; [-4.08, -4.07]; [-4.07, -4.06]; [-4.06, -4.05]; [-4.05, -4.04];$
$[-4.04, -4.03]; [-4.03, -4.02]; [-4.02, -4.01]; [-4.01, -4]$
$f(-4.1) = -2.182; f(-4.09) \approx -1.7473$
$f(-4.09) \approx -1.7473; f(-4.08) \approx -1.3162$
$f(-4.08) \approx -1.3162; f(-4.07) \approx -0.8889$
$f(-4.07) \approx -0.8889; f(-4.06) \approx -0.4652$
$f(-4.06) \approx -0.4652; f(-4.05) \approx -0.0453$
$f(-4.05) \approx -0.4653; f(-4.04) \approx 0.3711$
So f has a real zero on the interval $[-4.05, -4.04]$, therefore $r = -4.04$,
correct to two decimal places.

99. $f(x) = x^3 + x^2 + x - 4$
  $f(1) = -1; f(2) = 10$ So f has a real zero on the interval $[1,2]$.

  Subdivide the interval $[1,2]$ into 10 equal subintervals:
  $[1,1.1]; [1.1,1.2]; [1.2,1.3]; [1.3,1.4]; [1.4,1.5]; [1.5,1.6]; [1.6,1.7]; [1.7,1.8];$
  $[1.8,1.9]; [1.9,2]$
  $f(1) = -1; f(1.1) = -0.359$
  $f(1.1) = -0.359; f(1.2) = 0.368$ So f has a real zero on the interval $[1.1,1.2]$.

  Subdivide the interval $[1.1,1.2]$ into 10 equal subintervals:
  $[1.1,1.11]; [1.11,1.12]; [1.12,1.13]; [1.13,1.14]; [1.14,1.15]; [1.15,1.16];[1.16,1.17];$
  $[1.17,1.18]; [1.18,1.19]; [1.19,1.2]$
  $f(1.1) = -0.359; f(1.11) \approx -0.2903$
  $f(1.11) \approx -0.2903; f(1.12) \approx -0.2207$
  $f(1.12) \approx -0.2207; f(1.13) \approx -0.1502$
  $f(1.13) \approx -0.1502; f(1.14) \approx -0.0789$
  $f(1.14) \approx -0.0789; f(1.15) \approx -0.0066$
  $f(1.15) \approx -0.0066; f(1.16) \approx 0.0665$
  So f has a real zero on the interval $[1.15,1.16]$, therefore $r = 1.15$, correct
  to two decimal places.

101. $f(x) = 2x^4 - 3x^3 - 4x^2 - 8$
  $f(2) = -16; f(3) = 37$ So f has a real zero on the interval $[2,3]$.
  Subdivide the interval $[2,3]$ into 10 equal subintervals:
  $[2,2.1]; [2.1,2.2]; [2.2,2.3]; [2.3,2.4]; [2.4,2.5]; [2.5,2.6]; [2.6,2.7]; [2.7,2.8];$
  $[2.8,2.9]; [2.9,3]$
  $f(2) = -16; f(2.1) = -14.5268$
  $f(2.1) = -14.5268; f(2.2) = -12.4528$

$f(2.2) = -12.4528; f(2.3) = -9.6928$
$f(2.3) = -9.6928; f(2.4) = -6.1568$
$f(2.4) = -6.1568; f(2.5) = -1.75$
$f(2.5) = -1.75; f(2.6) = 3.6272$

So f has a real zero on the interval $[2.5, 2.6]$.

Subdivide the interval $[2.5, 2.6]$ into 10 equal subintervals:
$[2.5, 2.51]; [2.51, 2.52]; [2.52, 2.53]; [2.53, 2.54]; [2.54, 2.55]; [2.55, 2.56]; [2.56, 2.57];$
$[2.57, 2.58]; [2.58, 2.59]; [2.59, 2.6]$

$f(2.5) = -1.75; f(2.51) \approx -1.2576$
$f(2.51) \approx -1.2576; f(2.52) \approx -0.7555$
$f(2.52) \approx -0.7555; f(2.53) \approx -0.2434$
$f(2.53) \approx -0.2434; f(2.54) \approx 0.2787$

So f has a real zero on the interval $[2.53, 2.54]$, therefore $r = 2.53$, correct to two decimal places.

103. $x - 2$ is a factor of $f(x) = x^3 - kx^2 + kx + 2$ only if the remainder that results when $f(x)$ is divided by $x - 2$ is 0. Dividing, we have:

$$\begin{array}{r|rrrr} 2) & 1 & -k & k & 2 \\ & & 2 & -2k+4 & -2k+8 \\ \hline & 1 & -k+2 & -k+4 & -2k+10 \end{array}$$

Since we want the remainder to equal 0, set the remainder equal to zero and solve:

$-2k + 10 = 0$

$-2k = -10$

$k = 5$

105. By the Remainder Theorem we know that the remainder from synthetic division by c is equal to $f(c)$. Thus the easiest way to find the remainder is to evaluate:

$f(1) = 2(1)^{20} - 8(1)^{10} + 1 - 2 = 2 - 8 + 1 - 2 = -7$

The remainder is -7.

107. We want to prove that $x - c$ is a factor of $x^n - c^n$, for any positive integer n. By the Factor Theorem, $x - c$ will be a factor of $f(x)$ provided $f(c) = 0$. Here, $f(x) = x^n - c^n$, so that $f(c) = c^n - c^n = 0$. Therefore, $x - c$ is a factor of $x^n - c^n$.

109. $x^3 - 8x^2 + 16x - 3 = 0$ has solution $x = 3$, so $x - 3$ is a factor of $f(x) = x^3 - 8x^2 + 16x - 3$. Using synthetic division

$$\begin{array}{r|rrrr} 3) & 1 & -8 & 16 & -3 \\ & & 3 & -15 & 3 \\ \hline & 1 & -5 & 1 & 0 \end{array}$$

Thus, $f(x) = x^3 - 8x^2 + 16x - 3 = (x - 3)(x^2 - 5x + 1)$.

Solving $x^2 - 5x + 1 = 0$:
$$x = \frac{5 \pm \sqrt{25 - 4}}{2} = \frac{5 \pm \sqrt{21}}{2}$$
The sum of these two roots is $\dfrac{5 + \sqrt{21}}{2} + \dfrac{5 - \sqrt{21}}{2} = \dfrac{10}{2} = 5$.

111. $f(x) = 2x^3 + 3x^2 - 6x + 7$

By the Rational Zero Theorem, the only possible rational zeros are: $\dfrac{p}{q} = \pm 1, \pm 7, \pm \dfrac{1}{2}, \pm \dfrac{7}{2}$.

Since $\dfrac{1}{3}$ is not in the list of possible rational zeros, it is not a zero of f.

113. $f(x) = 2x^6 - 5x^4 + x^3 - x + 1$

By the Rational Zero Theorem, the only possible rational zeros are: $\dfrac{p}{q} = \pm 1, \pm \dfrac{1}{2}$.

Since $\dfrac{3}{5}$ is not in the list of possible rational zeros, it is not a zero of $f(x)$.

115. Let x be the length of a side of the original cube.
After removing the 1-inch slice, one dimension will be $x - 1$.
The volume of the new solid will be: $(x - 1) \cdot x \cdot x$.
Solve the volume equation:
$$(x - 1) \cdot x \cdot x = 294$$
$$x^3 - x^2 = 294$$
$$x^3 - x^2 - 294 = 0$$
The solutions to this equation are the same as the real zeros of $f(x) = x^3 - x^2 - 294$.
By Descartes' Rule of Signs, we know that there is one positive real zero.
The possible rational zeros are:
$$p = \pm 1, \pm 2, \pm 3, \pm 6, \pm 7, \pm 14, \pm 21, \pm 42, \pm 49, \pm 98, \pm 147, \pm 294; \quad q = \pm 1$$
The rational zeros are the same as the values for p.
Using synthetic division:

$$7 \overline{)\begin{array}{cccc} 1 & -1 & 0 & -294 \\ & 7 & 42 & 294 \\ \hline 1 & 6 & 42 & 0 \end{array}}$$

7 is a zero, so the length of the edge of the original cube was 7 inches.

117. $f(x) = x^n + a_{n-1}x^{n-1} + a_{n-2}x^{n-2} + \ldots + a_1 x + a_0$; where $a_{n-1}, a_{n-2}, \ldots a_1, a_0$ are integers
If r is a real zero of f, then r is either rational or irrational. We know that the rational roots of f must be of the form $\dfrac{p}{q}$ where p is a divisor of a_0 and q is a divisor of 1. This means that $q = \pm 1$. So if r is rational, then $r = \dfrac{p}{q} = \pm p$. Therefore, r is an integer or r is irrational.

119. (a) $f(x) = 8x^4 - 2x^2 + 5x - 1$ $0 \le r \le 1$
We begin with the interval $[0,1]$.
$f(0) = -1;$ $f(1) = 10$
Let m_i = the midpoint of the interval being considered.
So $m_1 = 0.5$

n	m_n	$f(m_n)$	New interval
1	0.5	$f(0.5) = 1.5 > 0$	$[0, 0.5]$
2	0.25	$f(0.25) = 0.15625 > 0$	$[0, 0.25]$
3	0.125	$f(0.125) \approx -0.4043 < 0$	$[0.125, 0.25]$
4	0.1875	$f(0.1875) \approx -0.1229 < 0$	$[0.1875, 0.25]$
5	0.21875	$f(0.21875) \approx 0.0164 > 0$	$[0.1875, 0.21875]$
6	0.203125	$f(0.203125) \approx -0.0533 < 0$	$[0.203125, 0.21875]$
7	0.2109375	$f(0.2109375) \approx -0.0185 < 0$	$[0.2109375, 0.21875]$

Since the endpoints of the new interval at Step 7 agree to two decimal places,
$r = 0.21$, correct to two decimal places.

(b) $f(x) = x^4 + 8x^3 - x^2 + 2;$ $-1 \le r \le 0$
We begin with the interval $[-1,0]$.
$f(-1) = -6;$ $f(0) = 2$
Let m_i = the midpoint of the interval being considered.
So $m_1 = -0.5$

n	m_n	$f(m_n)$	New interval
1	−0.5	$f(-0.5) = 0.8125 > 0$	$[-1, -0.5]$
2	−0.75	$f(-0.75) \approx -1.6211 < 0$	$[-0.75, -0.5]$
3	−0.625	$f(-0.625) \approx -0.1912 < 0$	$[-0.625, -0.5]$
4	−0.5625	$f(-0.5625) \approx 0.3599 > 0$	$[-0.625, -0.5625]$
5	−0.59375	$f(-0.59375) \approx 0.0972 > 0$	$[-0.625, -0.59375]$
6	−0.609375	$f(-0.609375) \approx -0.0437 < 0$	$[-0.609375, -0.59375]$
7	−0.6015625	$f(-0.6015625) \approx 0.0275 > 0$	$[-0.609375, -0.6015625]$

Since the endpoints of the new interval at Step 7 agree to two decimal places,
$r = -0.60$, correct to two decimal places.

(c) $f(x) = 2x^3 + 6x^2 - 8x + 2;$ $-5 \le r \le -4$
We begin with the interval $[-5,-4]$.
$f(-5) = -58;$ $f(-4) = 2$
Let m_i = the midpoint of the interval being considered.
So $m_1 = -4.5$

n	m_n	$f(m_n)$	New interval
1	–4.5	$f(-4.5) = -22.75 < 0$	[–4.5,–4]
2	–4.25	$f(-4.25) \approx -9.156 < 0$	[–4.25,–4]
3	–4.125	$f(-4.125) \approx -3.2852 < 0$	[–4.125,–4]
4	–4.0625	$f(-4.0625) \approx -0.5708 < 0$	[–4.0625,–4]
5	–4.03125	$f(-4.03125) \approx 0.7324 > 0$	[–4.0625, –4.03125]
6	–4.046875	$f(-4.046875) \approx 0.0852 > 0$	[–4.0625, –4.046875]
7	–4.0546875	$f(-4.0546875) \approx -0.2417 < 0$	[–4.0546875, –4.046875]
8	–4.05078125	$f(-4.05078125) \approx -0.0779 < 0$	[–4.05078125, –4.046875]
9	–4.048828125	$f(-4.048828125) \approx 0.0037 > 0$	[–4.05078125, –4.048828125]
10	–4.0498046875	$f(-4.0498045875) \approx -0.0371 < 0$	[–4.0498046875, –4.048828125]

Since the endpoints of the new interval at Step 10 agree to two decimal places,
$r = -4.05$, correct to two decimal places.

(d) $f(x) = 3x^3 - 10x + 9; \quad -3 \le r \le -2$

We begin with the interval [–3,–2].

$f(-3) = -42; \quad f(-2) = 5$

Let m_i = the midpoint of the interval being considered.

So $m_1 = -2.5$

n	m_n	$f(m_n)$	New interval
1	–2.5	$f(-2.5) = -12.875 < 0$	[–2.5, –2]
2	–2.25	$f(-2.25) \approx -2.6719 < 0$	[–2.25, –2]
3	–2.125	$f(-2.125) \approx 1.4629 > 0$	[–2.25, –2.125]
4	–2.1875	$f(-2.1875) \approx -0.5276 < 0$	[–2.1875, –2.125]
5	–2.15625	$f(-2.15625) \approx 0.4866 > 0$	[–2.1875, –2.15625]
6	–2.171875	$f(-2.171875) \approx -0.0157 < 0$	[–2.171875, –2.15625]
7	–2.1640625	$f(-2.1640625) \approx 0.2366 > 0$	[–2.171875, –2.1640625]
8	–2.16796875	$f(-2.16796875) \approx 0.1108 > 0$	[–2.171875, –2.16796875]
9	–2.169921875	$f(-2.169921875) \approx 0.0476 > 0$	[–2.171875, –2.169921875]
10	–2.1708984375	$f(-2.1708984375) \approx 0.0160 > 0$	[–2.171875, –2.1708984375]

Since the endpoints of the new interval at Step 10 agree to two decimal places,
$r = -2.17$, correct to two decimal places.

(e) $f(x) = x^3 + x^2 + x - 4; \quad 1 \le r \le 2$

We begin with the interval [1,2].

$f(1) = -1; \quad f(2) = 10$

Let m_i = the midpoint of the interval being considered.

So $m_1 = 1.5$

n	m_n	$f(m_n)$	New interval
1	1.5	$f(1.5) = 3.125 > 0$	[1,1.5]
2	1.25	$f(1.25) \approx 0.7656 > 0$	[1,1.25]
3	1.125	$f(1.125) \approx -0.1855 < 0$	[1.125,1.25]
4	1.1875	$f(1.1875) \approx 0.2722 > 0$	[1.125,1.1875]
5	1.15625	$f(1.15625) \approx 0.0390 > 0$	[1.125,1.15625]
6	1.140625	$f(1.140625) \approx -0.0744 < 0$	[1.140625,1.15625]
7	1.1484375	$f(1.1484375) \approx -0.0180 < 0$	[1.1484375,1.15625]
8	1.15234375	$f(1.15234375) \approx 0.0140 > 0$	[1.1484375,1.15625]
9	1.150390625	$f(1.150390625) \approx -0.0038 < 0$	[1.1484375,1.15234375]

Since the endpoints of the new interval at Step 9 agree to two decimal places, $r = 1.15$, correct to two decimal places.

(f) $f(x) = 2x^4 + x^2 - 1; \qquad 0 \le r \le 1$

We begin with the interval [0,1].

$f(0) = -1; \quad f(1) = 2$

Let m_i = the midpoint of the interval being considered.

So $m_1 = 0.5$

n	m_n	$f(m_n)$	New interval
1	0.5	$f(0.5) = -0.625 < 0$	[0.5,1]
2	0.75	$f(0.75) \approx 0.1953 > 0$	[0.5,0.75]
3	0.625	$f(0.625) \approx -0.3042 < 0$	[0.625,0.75]
4	0.6875	$f(0.6875) \approx -0.0805 < 0$	[0.6875,0.75]
5	0.71875	$f(0.71875) \approx 0.0504 > 0$	[0.6875,0.71875]
6	0.703125	$f(0.703125) \approx -0.0168 < 0$	[0.703125,0.71875]
7	0.7109375	$f(0.7109375) \approx 0.0164 > 0$	[0.703125, 0.7109375]
8	0.70703125	$f(0.70703125) \approx -0.0003 < 0$	[0.70703125, 0.7109375]
9	0.708984375	$f(0.708984375) \approx 0.0080 > 0$	[0.70703125, 0.708984375]

Since the endpoints of the new interval at Step 9 agree to two decimal places, $r = 0.70$, correct to two decimal places

(g) $f(x) = 2x^4 - 3x^3 - 4x^2 - 8; \qquad 2 \le r \le 3$

We begin with the interval [2,3]

$f(2) = -16; \quad f(3) = 37$

Let m_i = the midpoint of the interval being considered.

So $m_1 = 2.5$

n	m_n	$f(m_n)$	New interval
1	2.5	$f(2.5) = -1.75 < 0$	[2.5,3]
2	2.75	$f(2.75) \approx 13.7422 > 0$	[2.5,2.75]
3	2.625	$f(2.625) \approx 5.1353 > 0$	[2.5,2.625]
4	2.5625	$f(2.5625) \approx 1.4905 > 0$	[2.5,2.5625]
5	2.53125	$f(2.53125) \approx -0.1787 < 0$	[2.53125,2.5625]
6	2.546875	$f(2.546875) \approx 0.6435 > 0$	[2.53125, 2.546875]
7	2.5390625	$f(2.5390625) \approx 0.2293 > 0$	[2.53125, 2.5390625]

Since the endpoints of the new interval at Step 7 agree to two decimal places, $r = 2.53$, correct to two decimal places.

(h) $f(x) = 3x^3 - 2x^2 - 20;$ $2 \le r \le 3$

We begin with the interval [2,3].

$f(2) = -4;$ $f(3) = 43$

Let m_i = the midpoint of the interval being considered.

So $m_1 = 2.5$

n	m_n	$f(m_n)$	New interval
1	2.5	$f(2.5) = 14.375 > 0$	[2,2.5]
2	2.25	$f(2.25) \approx 4.0469 > 0$	[2,2.25]
3	2.125	$f(2.125) \approx -0.2441 < 0$	[2.125,2.25]
4	2.1875	$f(2.1875) \approx 1.8323 > 0$	[2.125,2.1875]
5	2.15625	$f(2.15625) \approx 0.7771 > 0$	[2.125,2.15625]
6	2.140625	$f(2.140625) \approx 0.2622 > 0$	[2.125, 2.140625]
7	2.1328125	$f(2.1328125) \approx 0.0080 > 0$	[2.125, 2.1328125]
8	2.12890625	$f(2.12890625) \approx -0.1183 < 0$	[2.12890625, 2.1328125]
9	2.130859375	$f(2.130859375) \approx -0.0552 < 0$	[2.130859375, 2.1328125]

Since the endpoints of the new interval at Step 7 agree to two decimal places, $r = 2.13$, correct to two decimal places.

Polynomial and Rational Functions

3.7 Complex Zeros; Fundamental Theorem of Algebra

1. $(3-2i)+(-3+5i) = 3-3-2i+5i = 3i$
   $(3-2i)(-3+5i) = -9+15i+6i-10i^2 = -9+21i-10(-1) = 1+21i$

3. One

5. True

7. Since complex zeros appear in conjugate pairs, $4+i$, the conjugate of $4-i$, is the remaining zero of f.

9. Since complex zeros appear in conjugate pairs, $-i$, the conjugate of i, and $1-i$, the conjugate of $1+i$, are the remaining zeros of f.

11. Since complex zeros appear in conjugate pairs, $-i$, the conjugate of i, and $-2i$, the conjugate of $2i$, are the remaining zeros of f.

13. Since complex zeros appear in conjugate pairs, $-i$, the conjugate of i, is the remaining zero of f.

15. Since complex zeros appear in conjugate pairs, $2-i$, the conjugate of $2+i$, and $-3+i$, the conjugate of $-3-i$, are the remaining zeros of f.

17. Since $3+2i$ is a zero, its conjugate $3-2i$ is also a zero of f.
$$f(x) = (x-4)(x-4)(x-(3+2i))(x-(3-2i))$$
$$= \left(x^2 - 8x + 16\right)((x-3)-2i)((x-3)+2i)$$
$$= \left(x^2 - 8x + 16\right)\left(x^2 - 6x + 9 - 4i^2\right)$$
$$= \left(x^2 - 8x + 16\right)\left(x^2 - 6x + 13\right)$$
$$= x^4 - 6x^3 + 13x^2 - 8x^3 + 48x^2 - 104x + 16x^2 - 96x + 208$$
$$= x^4 - 14x^3 + 77x^2 - 200x + 208$$

19.   Since $-i$ is a zero, its conjugate i is also a zero, and since $1+i$ is a zero, its conjugate $1-i$
      is also a zero of f.
      $$\begin{aligned}
      f(x) &= (x-2)(x+i)(x-i)(x-(1+i))(x-(1-i)) \\
      &= (x-2)(x^2 - i^2)((x-1)-i)((x-1)+i) \\
      &= (x-2)(x^2 + 1)(x^2 - 2x + 1 - i^2) \\
      &= (x^3 - 2x^2 + x - 2)(x^2 - 2x + 2) \\
      &= x^5 - 2x^4 + 2x^3 - 2x^4 + 4x^3 - 4x^2 + x^3 - 2x^2 + 2x - 2x^2 + 4x - 4 \\
      &= x^5 - 4x^4 + 7x^3 - 8x^2 + 6x - 4
      \end{aligned}$$

21.   Since $-i$ is a zero, its conjugate i is also a zero of f.
      $$\begin{aligned}
      f(x) &= (x-3)(x-3)(x+i)(x-i) \\
      &= (x^2 - 6x + 9)(x^2 - i^2) \\
      &= (x^2 - 6x + 9)(x^2 + 1) \\
      &= x^4 + x^2 - 6x^3 - 6x + 9x^2 + 9 \\
      &= x^4 - 6x^3 + 10x^2 - 6x + 9
      \end{aligned}$$

23.   Since $2i$ is a zero, its conjugate $-2i$ is also a zero of f. $x-2i$ and $x+2i$ are factors of f.
      Thus, $(x-2i)(x+2i) = x^2 + 4$ is a factor of f. Using division to find the other factor:

$$
\require{enclose}
\begin{array}{r}
x-4 \\
x^2+4 \enclose{longdiv}{x^3 - 4x^2 + 4x - 16} \\
\underline{x^3 + 4x} \\
-4x^2 - 16 \\
\underline{-4x^2 - 16} \\
\end{array}
$$

$x-4$ is a factor, so the remaining zero is 4.
The zeros of f are $4, 2i, -2i$.

25.   Since $-2i$ is a zero, its conjugate $2i$ is also a zero of f. $x-2i$ and $x+2i$ are factors of f.
      Thus, $(x-2i)(x+2i) = x^2 + 4$ is a factor of f. Using division to find the other factor:

$$
\require{enclose}
\begin{array}{r}
2x^2 + 5x - 3 \\
x^2+4 \enclose{longdiv}{2x^4 + 5x^3 + 5x^2 + 20x - 12} \\
\underline{2x^4 + 8x^2} \\
5x^3 - 3x^2 + 20x \\
\underline{5x^3 + 20x} \\
-3x^2 - 12 \\
\underline{-3x^2 - 12} \\
\end{array}
$$

$2x^2 + 5x - 3 = (2x-1)(x+3)$

The remaining zeros are $\dfrac{1}{2}$ and -3.

The zeros of f are $2i, -2i, -3, \dfrac{1}{2}$.

27. Since $3 - 2i$ is a zero, its conjugate $3 + 2i$ is also a zero of h. $x - (3 - 2i)$ and $x - (3 + 2i)$ are factors of h.

Thus, $(x - (3 - 2i))(x - (3 + 2i)) = ((x - 3) + 2i)((x - 3) - 2i) = x^2 - 6x + 9 - 4i^2 = x^2 - 6x + 13$ is factor of h.

Using division to find the other factor:

$$
\begin{array}{r}
x^2 - 3x - 10 \\
x^2 - 6x + 13\overline{)x^4 - 9x^3 + 21x^2 + 21x - 130} \\
\underline{x^4 - 6x^3 + 13x^2} \\
-3x^3 + 8x^2 + 21x \\
\underline{-3x^3 + 18x^2 - 39x} \\
-10x^2 + 60x - 130 \\
\underline{-10x^2 + 60x - 130}
\end{array}
$$

$x^2 - 3x - 10 = (x + 2)(x - 5)$

The remaining zeros are -2 and 5.

The zeros of h are $3 - 2i$, $3 + 2i$, -2, 5.

29. Since $-4i$ is a zero, its conjugate $4i$ is also a zero of h. $x - 4i$ and $x + 4i$ are factors of h.

Thus, $(x - 4i)(x + 4i) = x^2 + 16$ is a factor of h. Using division to find the other factor:

$$
\begin{array}{r}
3x^3 + 2x^2 - 33x - 22 \\
x^2 + 16\overline{)3x^5 + 2x^4 + 15x^3 + 10x^2 - 528x - 352} \\
\underline{3x^5 + 48x^3} \\
2x^4 - 33x^3 + 10x^2 \\
\underline{2x^4 + 32x^2} \\
-33x^3 - 22x^2 - 528x \\
\underline{-33x^3 - 528x} \\
-22x^2 - 352 \\
\underline{-22x^2 - 352}
\end{array}
$$

$3x^3 + 2x^2 - 33x - 22 = x^2(3x + 2) - 11(3x + 2)$

$ = (3x + 2)(x^2 - 11)$

$ = (3x + 2)\left(x - \sqrt{11}\right)\left(x + \sqrt{11}\right)$

The remaining zeros are $-\dfrac{2}{3}, \sqrt{11},$ and $-\sqrt{11}$.

The zeros of h are $4i, -4i, -\sqrt{11}, \sqrt{11}, -\dfrac{2}{3}$.

31. $f(x) = x^3 - 1 = (x - 1)\left(x^2 + x + 1\right)$ The solutions of $x^2 + x + 1 = 0$ are:

$$x = \frac{-1 \pm \sqrt{1^2 - 4(1)(1)}}{2(1)} = \frac{-1 \pm \sqrt{-3}}{2} = -\frac{1}{2} + \frac{\sqrt{3}}{2}i \text{ and } -\frac{1}{2} - \frac{\sqrt{3}}{2}i$$

The zeros are: $1, -\dfrac{1}{2} + \dfrac{\sqrt{3}}{2}i, -\dfrac{1}{2} - \dfrac{\sqrt{3}}{2}i$.

33. $f(x) = x^3 - 8x^2 + 25x - 26$

Step 1: $f(x)$ has 3 complex zeros.

Step 2:   By Descartes Rule of Signs, there are three positive real zeros or there is one positive real zero.

$f(-x) = (-x)^3 - 8(-x)^2 + 25(-x) - 26 = -x^3 - 8x^2 - 25x - 26$, thus, there are no negative real zeros.

Step 3:   Possible rational zeros:
$p = \pm 1, \pm 2, \pm 13, \pm 26;\quad q = \pm 1;$

$\dfrac{p}{q} = \pm 1, \pm 2, \pm 13, \pm 26$

Step 4:   Using synthetic division:
We try $x - 2$:

$$\begin{array}{r|rrrr} 2 & 1 & -8 & 25 & -26 \\ & & 2 & -12 & 26 \\ \hline & 1 & -6 & 13 & 0 \end{array}$$

$x - 2$ is a factor. The other factor is the quotient: $x^2 - 6x + 13$.

The solutions of $x^2 - 6x + 13 = 0$ are:

$x = \dfrac{-(-6) \pm \sqrt{(-6)^2 - 4(1)(13)}}{2(1)} = \dfrac{6 \pm \sqrt{-16}}{2} = \dfrac{6 \pm 4i}{2} = 3 \pm 2i.$

The zeros are $2,\ 3 - 2i,\ 3 + 2i$.

35. $f(x) = x^4 + 5x^2 + 4 = \left(x^2 + 4\right)\left(x^2 + 1\right) = (x + 2i)(x - 2i)(x + i)(x - i)$

The zeros are: $-2i,\ -i,\ i,\ 2i$.

37. $f(x) = x^4 + 2x^3 + 22x^2 + 50x - 75$

Step 1: $f(x)$ has 4 complex zeros.

Step 2:   By Descartes Rule of Signs, there is 1 positive real zero.

$f(-x) = (-x)^4 + 2(-x)^3 + 22(-x)^2 + 50(-x) - 75$

$= x^4 - 2x^3 + 22x^2 - 50x - 75$

Thus, there are three negative real zeros or there is one negative real zero.

Step 3:   Possible rational zeros:
$p = \pm 1, \pm 3, \pm 5, \pm 15, \pm 25, \pm 75;\quad q = \pm 1;$

$\dfrac{p}{q} = \pm 1, \pm 3, \pm 5, \pm 15, \pm 25, \pm 75$

Step 4:   Using synthetic division:
We try $x + 3$:

$$\begin{array}{r|rrrrr} -3 & 1 & 2 & 22 & 50 & -75 \\ & & -3 & 3 & -75 & 75 \\ \hline & 1 & -1 & 25 & -25 & 0 \end{array}$$

$x + 3$ is a factor. The other factor is the quotient: $x^3 - x^2 + 25x - 25$.

$x^3 - x^2 + 25x - 25 = x^2(x - 1) + 25(x - 1)$

$= (x - 1)\left(x^2 + 25\right) = (x - 1)(x + 5i)(x - 5i)$

The zeros are $-3,\ 1,\ -5i,\ 5i$.

39. $f(x) = 3x^4 - x^3 - 9x^2 + 159x - 52$

Step 1: $f(x)$ has 4 complex zeros.

Step 2:   By Descartes Rule of Signs, there are three positive real zeros or there is one positive real zero.

$$f(-x) = 3(-x)^4 - (-x)^3 - 9(-x)^2 + 159(-x) - 52$$
$$= 3x^4 + x^3 - 9x^2 - 159x - 52$$

thus, there is 1 negative real zero.

Step 3:   Possible rational zeros:

$p = \pm1, \pm2, \pm4, \pm13, \pm26, \pm52;$ $q = \pm1, \pm3;$

$\dfrac{p}{q} = \pm1, \pm2, \pm4, \pm13, \pm26, \pm52, \pm\dfrac{1}{3}, \pm\dfrac{2}{3}, \pm\dfrac{4}{3}, \pm\dfrac{13}{3}, \pm\dfrac{26}{3}, \pm\dfrac{52}{3}$

Step 4:   Using synthetic division:

We try $x + 4$:

$$
\begin{array}{r|rrrrr}
-4 & 3 & -1 & -9 & 159 & -52 \\
   &   & -12 & 52 & -172 & 52 \\
\hline
   & 3 & -13 & 43 & -13 & 0
\end{array}
$$

$x + 4$ is a factor and

the quotient is $3x^3 - 13x^2 + 43x - 13$

We try $x - \dfrac{1}{3}$ on $3x^3 - 13x^2 + 43x - 13$

$$
\begin{array}{r|rrrr}
\frac{1}{3} & 3 & -13 & 43 & -13 \\
   &   & 1 & -4 & 13 \\
\hline
   & 3 & -12 & 39 & 0
\end{array}
$$

$x - \dfrac{1}{3}$ is a factor and

the quotient is $3x^2 - 12x + 39$

$3x^2 - 12x + 39 = 3(x^2 - 4x + 13)$

The solutions of $x^2 - 4x + 13 = 0$ are:

$$x = \frac{-(-4) \pm \sqrt{(-4)^2 - 4(1)(13)}}{2(1)} = \frac{4 \pm \sqrt{-36}}{2} = \frac{4 \pm 6i}{2} = 2 \pm 3i.$$

The zeros are -4, $\dfrac{1}{3}$, $2 - 3i$, $2 + 3i$.

41.   If the coefficients are real numbers and $2 + i$ is a zero, then $2 - i$ would also be a zero. This would then require a polynomial of degree 4.

43.   If the coefficients are real numbers, then complex zeros must appear in conjugate pairs. We have a conjugate pair and one real zero. Thus, there is only one remaining zero, and it must be real because a complex zero would require a pair of complex conjugates.

Polynomial and Rational Functions

3.R   Chapter Review

1. $f(x) = (x-2)^2 + 2$

Using the graph of $y = x^2$, shift right 2 units, then shift up 2 units

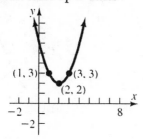

3. $f(x) = -(x-1)^2$

Using the graph of $y = x^2$, shift right 1 unit, then reflect about the x-axis.

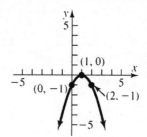

5. $f(x) = (x-1)^2 + 2$

Using the graph of $y = x^2$, shift right 1 unit, then shift up 2 units.

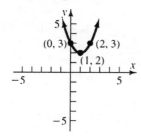

7. $f(x) = (x-2)^2 + 2 = x^2 - 4x + 4 + 2 = x^2 - 4x + 6$

$a = 1, b = -4, c = 6$. Since $a = 1 > 0$, the graph opens up.

The x-coordinate of the vertex is $x = -\dfrac{b}{2a} = -\dfrac{-4}{2(1)} = \dfrac{4}{2} = 2$.

The y-coordinate of the vertex is $f\left(-\dfrac{b}{2a}\right) = f(2) = (2)^2 - 4(2) + 6 = 2$.

Thus, the vertex is $(2, 2)$.
The axis of symmetry is the line $x = 2$.
The discriminant is:

$$b^2 - 4ac = (-4)^2 - 4(1)(6) = -8 < 0,$$

so the graph has no x-intercepts.
The y-intercept is $f(0) = 6$.

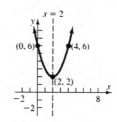

9. $f(x) = \dfrac{1}{4}x^2 - 16$, $a = \dfrac{1}{4}, b = 0, c = -16$. Since $a = \dfrac{1}{4} > 0$, the graph opens up.

The x-coordinate of the vertex is $x = -\dfrac{b}{2a} = -\dfrac{-0}{2\left(\dfrac{1}{4}\right)} = -\dfrac{0}{\dfrac{1}{2}} = 0$.

The y-coordinate of the vertex is $f\left(-\dfrac{b}{2a}\right) = f(0) = \dfrac{1}{4}(0)^2 - 16 = -16$.

Thus, the vertex is $(0, -16)$.

The axis of symmetry is the line $x = 0$.

The discriminant is:

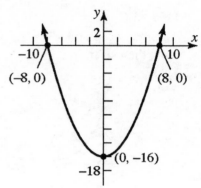

$$b^2 - 4ac = (0)^2 - 4\left(\dfrac{1}{4}\right)(-16) = 16 > 0,$$

so the graph has two x-intercepts.

The x-intercepts are found by solving:

$$\dfrac{1}{4}x^2 - 16 = 0$$

$$x^2 - 64 = 0$$

$$x^2 = 64$$

$$x = 8 \ \text{ or } \ x = -8$$

The x-intercepts are -8 and 8.

The y-intercept is $f(0) = -16$.

11. $f(x) = -4x^2 + 4x$, $a = -4$, $b = 4$, $c = 0$. Since $a = -4 < 0$, the graph opens down.

The x-coordinate of the vertex is $x = -\dfrac{b}{2a} = -\dfrac{4}{2(-4)} = -\dfrac{4}{-8} = \dfrac{1}{2}$.

The y-coordinate of the vertex is $f\left(-\dfrac{b}{2a}\right) = f\left(\dfrac{1}{2}\right) = -4\left(\dfrac{1}{2}\right)^2 + 4\left(\dfrac{1}{2}\right) = -1 + 2 = 1$.

Thus, the vertex is $\left(\dfrac{1}{2}, 1\right)$.

The axis of symmetry is the line $x = \dfrac{1}{2}$.

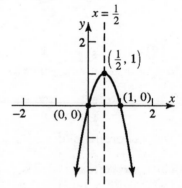

The discriminant is:

$$b^2 - 4ac = 4^2 - 4(-4)(0) = 16 > 0,$$

so the graph has two x-intercepts.

The x-intercepts are found by solving:

$$-4x^2 + 4x = 0$$

$$-4x(x - 1) = 0$$

$$x = 0 \ \text{ or } \ x = 1$$

The x-intercepts are 0 and 1.

The y-intercept is $f(0) = -4(0)^2 + 4(0) = 0$.

13. $f(x) = \dfrac{9}{2}x^2 + 3x + 1$

$a = \dfrac{9}{2}, b = 3, c = 1$. Since $a = \dfrac{9}{2} > 0$, the graph opens up.

The x-coordinate of the vertex is $x = -\dfrac{b}{2a} = -\dfrac{3}{2\left(\dfrac{9}{2}\right)} = -\dfrac{3}{9} = -\dfrac{1}{3}$.

The y-coordinate of the vertex is $f\left(-\dfrac{b}{2a}\right) = f\left(-\dfrac{1}{3}\right) = \dfrac{9}{2}\left(-\dfrac{1}{3}\right)^2 + 3\left(-\dfrac{1}{3}\right) + 1 = \dfrac{1}{2} - 1 + 1 = \dfrac{1}{2}$.

Thus, the vertex is $\left(-\dfrac{1}{3}, \dfrac{1}{2}\right)$.

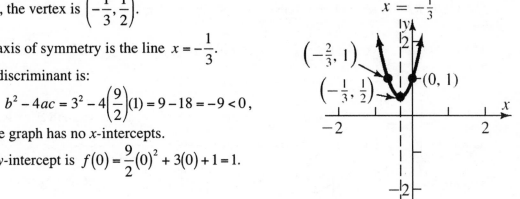

The axis of symmetry is the line $x = -\dfrac{1}{3}$.

The discriminant is:

$$b^2 - 4ac = 3^2 - 4\left(\dfrac{9}{2}\right)(1) = 9 - 18 = -9 < 0,$$

so the graph has no x-intercepts.

The y-intercept is $f(0) = \dfrac{9}{2}(0)^2 + 3(0) + 1 = 1$.

15. $f(x) = 3x^2 + 4x - 1$, $a = 3, b = 4, c = -1$. Since $a = 3 > 0$, the graph opens up.

The x-coordinate of the vertex is $x = -\dfrac{b}{2a} = -\dfrac{4}{2(3)} = -\dfrac{4}{6} = -\dfrac{2}{3}$.

The y-coordinate of the vertex is $f\left(-\dfrac{b}{2a}\right) = f\left(-\dfrac{2}{3}\right) = 3\left(-\dfrac{2}{3}\right)^2 + 4\left(-\dfrac{2}{3}\right) - 1 = \dfrac{4}{3} - \dfrac{8}{3} - 1 = -\dfrac{7}{3}$.

Thus, the vertex is $\left(-\dfrac{2}{3}, -\dfrac{7}{3}\right)$.

The axis of symmetry is the line $x = -\dfrac{2}{3}$.

The discriminant is: $b^2 - 4ac = (4)^2 - 4(3)(-1) = 28 > 0$, so the graph has two x-intercepts.

The x-intercepts are found by solving:

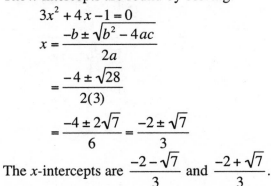

$$3x^2 + 4x - 1 = 0$$

$$x = \dfrac{-b \pm \sqrt{b^2 - 4ac}}{2a}$$

$$= \dfrac{-4 \pm \sqrt{28}}{2(3)}$$

$$= \dfrac{-4 \pm 2\sqrt{7}}{6} = \dfrac{-2 \pm \sqrt{7}}{3}$$

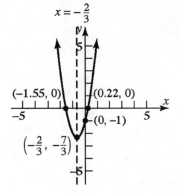

The x-intercepts are $\dfrac{-2 - \sqrt{7}}{3}$ and $\dfrac{-2 + \sqrt{7}}{3}$.

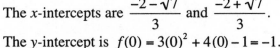

The y-intercept is $f(0) = 3(0)^2 + 4(0) - 1 = -1$.

17. $f(x) = 3x^2 - 6x + 4$

$a = 3, b = -6, c = 4$. Since $a = 3 > 0$, the graph opens up, so the vertex is a minimum

point. The minimum occurs at $x = -\dfrac{b}{2a} = -\dfrac{-6}{2(3)} = \dfrac{6}{6} = 1$.

The minimum value is $f\left(-\dfrac{b}{2a}\right) = f(1) = 3(1)^2 - 6(1) + 4 = 3 - 6 + 4 = 1$.

19. $f(x) = -x^2 + 8x - 4$

$a = -1, b = 8, c = -4$. Since $a = -1 < 0$, the graph opens down, so the vertex is a

maximum point. The maximum occurs at $x = -\dfrac{b}{2a} = -\dfrac{8}{2(-1)} = -\dfrac{8}{-2} = 4$.

The maximum value is $f\left(-\dfrac{b}{2a}\right) = f(4) = -(4)^2 + 8(4) - 4 = -16 + 32 - 4 = 12$.

21. $f(x) = -3x^2 + 12x + 4$

$a = -3, b = 12, c = 4$. Since $a = -3 < 0$, the graph opens down, so the vertex is a

maximum point. The maximum occurs at $x = -\dfrac{b}{2a} = -\dfrac{12}{2(-3)} = -\dfrac{12}{-6} = 2$.

The maximum value is $f\left(-\dfrac{b}{2a}\right) = f(2) = -3(2)^2 + 12(2) + 4 = -12 + 24 + 4 = 16$.

23. $f(x) = 4x^5 - 3x^2 + 5x - 2$ is a polynomial of degree 5.

25. $f(x) = 3x^2 + 5x^{1/2} - 1$ is not a polynomial because the variable x is raised to the $\dfrac{1}{2}$ power,

which is not a nonnegative integer.

27. $f(x) = (x + 2)^3$

Using the graph of $y = x^3$, shift left 2 units.

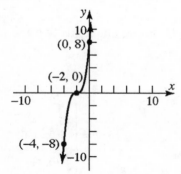

29. $f(x) = -(x - 1)^4$

Using the graph of $y = x^4$, shift right 1 unit, then reflect about the x-axis.

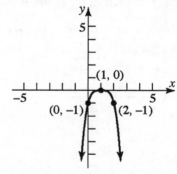

31. $f(x) = (x-1)^4 + 2$

Using the graph of $y = x^4$, shift right 1 unit, then shift up 2 units.

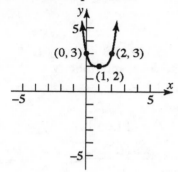

33. $f(x) = x(x+2)(x+4)$

(a) y-intercept: $f(0) = (0)(0+2)(0+4) = 0$

x-intercepts: solve $f(x) = 0$

$x(x+2)(x+4) = 0 \Rightarrow x = 0$ or $x = -2$ or $x = -4$

(b) The graph crosses the x-axis at $x = -4$, $x = -2$ and $x = 0$ since each zero has multiplicity 1.

(c) The function resembles $y = x^3$ for large values of $|x|$.

(d) The graph has at most 2 turning points.

(e) The x-intercepts yield the intervals $(-\infty,-4)$, $(-4,-2)$, $(-2,0)$ and $(0,\infty)$.

Interval	$(-\infty, -4)$	$(-4, -2)$	$(-2, 0)$	$(0, \infty)$
Number Chosen	-5	-3	-1	1
Value of f	$f(-5) = -15$	$f(-3) = 3$	$f(-1) = -3$	$f(1) = 15$
Location of Graph	Below x-axis	Above x-axis	Below x-axis	Above x-axis
Point on Graph	$(-5, -15)$	$(-3, 3)$	$(-1, -3)$	$(1, 15)$

f is below the x-axis on the intervals $(-\infty,-4)$ and $(-2,0)$

f is above the x-axis on the intervals $(-4,-2)$ and $(0,\infty)$

(f) Graphing:

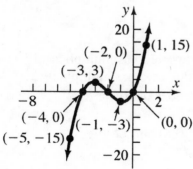

35. $f(x) = (x - 2)^2(x + 4)$

(a) y-intercept: $f(0) = (0 - 2)^2(0 + 4) = 16$
x-intercepts: solve $f(x) = 0$
$$(x - 2)^2(x + 4) = 0 \Rightarrow x = 2 \text{ or } x = -4$$

(b)   The graph crosses the x-axis at $x = -4$ since this zero has multiplicity 1.
The graph touches the x-axis at $x = 2$ since this zero has multiplicity 2.

(c)   The function resembles $y = x^3$ for large values of $|x|$.

(d)   The graph has at most 2 turning points.

(e)   The x-intercepts yield the intervals $(-\infty, -4)$, $(-4, 2)$ and $(2, \infty)$.

Interval	$(-\infty, -4)$	$(-4, 2)$	$(2, \infty)$
Number Chosen	-5	-2	3
Value of f	$f(-5) = -49$	$f(-2) = 32$	$f(3) = 7$
Location of Graph	Below x-axis	Above x-axis	Above x-axis
Point on Graph	$(-5, -49)$	$(-2, 32)$	$(3, 7)$

f is below the x-axis on the interval $(-\infty, -4)$
f is above the x-axis on the intervals $(-4, 2)$ and $(2, \infty)$

(f)   Graphing:

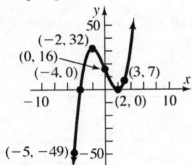

37. $f(x) = -2x^3 + 4x^2 = -2x^2(x - 2)$

(a) y-intercept: $f(0) = -2(0)^3 + 4(0)^2 = 0$
x-intercepts: solve $f(x) = 0$
$$-2x^2(x - 2) = 0 \Rightarrow x = 0 \text{ or } x = 2$$

(b)   The graph crosses the x-axis at $x = 2$ since this zero has multiplicity 1.
The graph touches the x-axis at $x = 0$ since this zero has multiplicity 2.

(c)   The function resembles $y = -2x^3$ for large values of $|x|$.

(d)   The graph has at most 2 turning points.

(e)   The x-intercepts yield the intervals $(-\infty, 0)$, $(0, 2)$ and $(2, \infty)$.

	0	2	
Interval	$(-\infty, 0)$	$(0, 2)$	$(2, \infty)$
Number Chosen	-1	1	3
Value of f	$f(-1) = 6$	$f(1) = 2$	$f(3) = -18$
Location of Graph	Above x-axis	Above x-axis	Below x-axis
Point on Graph	$(-1, 6)$	$(1, 2)$	$(3, -18)$

f is below the x-axis on the interval $(2, \infty)$

f is above the x-axis on the intervals $(-\infty, 0)$ and $(0, 2)$

(f)   Graphing:

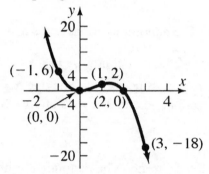

39. $f(x) = (x-1)^2(x+3)(x+1)$

(a) y-intercept: $f(0) = (0-1)^2(0+3)(0+1) = 3$

x-intercepts: solve $f(x) = 0$

$$(x-1)^2(x+3)(x+1) = 0 \Rightarrow x = 1 \text{ or } x = -3 \text{ or } x = -1$$

(b)   The graph crosses the x-axis at $x = -3$ and $x = -1$ since each zero has multiplicity 1. The graph touches the x-axis at $x = 1$ since this zero has multiplicity 2.

(c)   The function resembles $y = x^4$ for large values of $|x|$.

(d)   The graph has at most 3 turning points.

(e)   The x-intercepts yield the intervals $(-\infty, -3)$, $(-3, -1)$, $(-1, 1)$ and $(1, \infty)$.

	-3	-1	1	
Interval	$(-\infty, -3)$	$(-3, -1)$	$(-1, 1)$	$(1, \infty)$
Number Chosen	-4	-2	0	2
Value of f	$f(-4) = 75$	$f(-2) = -9$	$f(0) = 3$	$f(2) = 15$
Location of Graph	Above x-axis	Below x-axis	Above x-axis	Above x-axis
Point on Graph	$(-4, 75)$	$(-2, -9)$	$(0, 3)$	$(2, 15)$

f is below the x-axis on the interval $(-3, 1)$.

f is above the x-axis on the intervals $(-\infty, -3)$, $(-1, 1)$ and $(1, \infty)$.

(f)   Graphing:

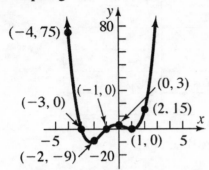

41. $R(x) = \dfrac{x+2}{x^2-9} = \dfrac{x+2}{(x+3)(x-3)}$ is in lowest terms. The denominator has zeros at –3 and 3.

Thus, the domain is $\{x \mid x \neq -3, x \neq 3\}$. The degree of the numerator, $p(x) = x+2$, is $n = 1$.
The degree of the denominator, $q(x) = x^2 - 9$, is $m = 2$. Since $n < m$, the line $y = 0$ is a
horizontal asymptote. Since the denominator is zero at –3 and 3, $x = -3$ and $x = 3$ are
vertical asymptotes.

43. $R(x) = \dfrac{x^2+3x+2}{(x+2)^2} = \dfrac{(x+2)(x+1)}{(x+2)^2} = \dfrac{x+1}{x+2}$ is in lowest terms. The denominator has a zero

at –2. Thus, the domain is $\{x \mid x \neq -2\}$. The degree of the numerator, $p(x) = x^2 + 3x + 2$,

is $n = 2$. The degree of the denominator, $q(x) = (x+2)^2 = x^2 + 4x + 4$, is $m = 2$. Since $n = m$,

the line $y = \dfrac{1}{1} = 1$ is a horizontal asymptote.

Since the denominator of $y = \dfrac{x+1}{x+2}$ is zero at 2, $x = 2$ is a vertical asymptote.

45. $R(x) = \dfrac{2x-6}{x}$ $p(x) = 2x - 6;\ q(x) = x;\ n = 1;\ m = 1$

Step 1:   Domain: $\{x \mid x \neq 0\}$

Step 2:   (a)   The x-intercept is the zero of $p(x)$: 3
          (b)   There is no y-intercept because 0 is not in the domain.

Step 3: $R(-x) = \dfrac{2(-x)-6}{-x} = \dfrac{-2x-6}{-x} = \dfrac{2x+6}{x}$; this is neither $R(x)$ nor $-R(x)$,

          so there is no symmetry.

Step 4: $R(x) = \dfrac{2x-6}{x}$ is in lowest terms.

          The vertical asymptote is the zero of $q(x)$: $x = 0$.

Step 5:   Since $n = m$, the line $y = \dfrac{2}{1} = 2$ is the horizontal asymptote.

Solve to find intersection points:

$$\frac{2x - 6}{x} = 2$$

$$2x - 6 = 2x$$

$$-6 \neq 0$$

$R(x)$ does not intersect $y = 2$.

Step 6:

Interval	$(-\infty, 0)$	$(0, 3)$	$(3, \infty)$
Number Chosen	-2	1	4
Value of R	$R(-2) = 5$	$R(1) = -4$	$R(4) = \frac{1}{2}$
Location of Graph	Above x-axis	Below x-axis	Above x-axis
Point on Graph	$(-2, 5)$	$(1, -4)$	$\left(4, \frac{1}{2}\right)$

Step 7:   Graphing:

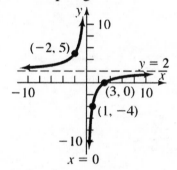

47. $H(x) = \dfrac{x + 2}{x(x - 2)}$ $p(x) = x + 2;\ q(x) = x(x - 2) = x^2 - 2x;\ n = 1;\ m = 2$

Step 1:   Domain: $\left\{x \mid x \neq 0,\ x \neq 2\right\}$.

Step 2:   (a)   The x-intercept is the zero of $p(x)$: -2

(b)   There is no y-intercept because 0 is not in the domain.

Step 3: $H(-x) = \dfrac{-x + 2}{-x(-x - 2)} = \dfrac{-x + 2}{x^2 + 2x}$; this is neither $H(x)$ nor $-H(x)$,

so there is no symmetry.

Step 4: $H(x) = \dfrac{x + 2}{x(x - 2)}$ is in lowest terms.

The vertical asymptotes are the zeros of $q(x)$: $x = 0$ and $x = 2$.

Step 5:   Since $n < m$, the line $y = 0$ is the horizontal asymptote.

Solve to find intersection points:

$$\frac{x + 2}{x(x - 2)} = 0$$

$$x + 2 = 0$$

$$x = -2$$

$H(x)$ intersects $y = 0$ at $(-2, 0)$.

Step 6:

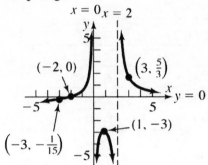

Interval	$(-\infty, -2)$	$(-2, 0)$	$(0, 2)$	$(2, \infty)$
Number Chosen	-3	-1	1	3
Value of H	$H(-3) = -\frac{1}{15}$	$H(-1) = \frac{1}{3}$	$H(1) = -3$	$H(3) = \frac{5}{3}$
Location of Graph	Below x-axis	Above x-axis	Below x-axis	Above x-axis
Point on Graph	$\left(-3, -\frac{1}{15}\right)$	$\left(-1, \frac{1}{3}\right)$	$(1, -3)$	$\left(3, \frac{5}{3}\right)$

Step 7:   Graphing:

49. $R(x) = \dfrac{x^2 + x - 6}{x^2 - x - 6} = \dfrac{(x+3)(x-2)}{(x-3)(x+2)}$ $p(x) = x^2 + x - 6$; $q(x) = x^2 - x - 6$;

Step 1:   Domain: $\left\{x \mid x \neq -2, \, x \neq 3\right\}$.

Step 2:   (a)   The x-intercepts are the zeros of $p(x)$: -3 and 2.

   (b)   The y-intercept is $R(0) = \dfrac{0^2 + 0 - 6}{0^2 - 0 - 6} = \dfrac{-6}{-6} = 1$.

Step 3: $R(-x) = \dfrac{(-x)^2 + (-x) - 6}{(-x)^2 - (-x) - 6} = \dfrac{x^2 - x - 6}{x^2 + x - 6}$; this is neither $R(x)$ nor $-R(x)$,

   so there is no symmetry.

Step 4: $R(x) = \dfrac{x^2 + x - 6}{x^2 - x - 6}$ is in lowest terms.

   The vertical asymptotes are the zeros of $q(x)$: $x = -2$ and $x = 3$.

Step 5:   Since $n = m$, the line $y = \dfrac{1}{1} = 1$ is the horizontal asymptote.

   Solve to find intersection points:

$$\frac{x^2 + x - 6}{x^2 - x - 6} = 1$$
$$x^2 + x - 6 = x^2 - x - 6$$
$$2x = 0$$
$$x = 0$$

   $R(x)$ intersects $y = 1$ at $(0, 1)$.

Step 6:

Interval	$(-\infty, -3)$	$(-3, -2)$	$(-2, 2)$	$(2, 3)$	$(3, \infty)$
Number Chosen	-4	-2.5	0	2.5	4
Value of R	$R(-4) \approx 0.43$	$R(-2.5) \approx -0.82$	$R(0) = 1$	$R(2.5) \approx -1.22$	$R(4) = \frac{7}{3}$
Location of Graph	Above x-axis	Below x-axis	Above x-axis	Below x-axis	Above x-axis
Point on Graph	$(-4, 0.43)$	$(-2.5, -0.82)$	$(0, 1)$	$(2.5, -1.22)$	$\left(4, \frac{7}{3}\right)$

Above the table, a number line with points at -3, -2, 2, 3.

Step 7: Graphing:

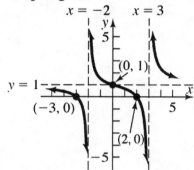

51. $F(x) = \dfrac{x^3}{x^2 - 4} = \dfrac{x^3}{(x+2)(x-2)}$ $p(x) = x^3$; $q(x) = x^2 - 4$; $n = 3$; $m = 2$

Step 1: Domain: $\left\{x \mid x \neq -2,\ x \neq 2\right\}$.

Step 2: (a) The x-intercept is the zero of $p(x)$: 0.

(b) The y-intercept is $F(0) = \dfrac{0^3}{0^2 - 4} = \dfrac{0}{-4} = 0$.

Step 3: $F(-x) = \dfrac{(-x)^3}{(-x)^2 - 4} = \dfrac{-x^3}{x^2 - 4} = -F(x)$; $F(x)$ is symmetric with respect to the origin.

Step 4: $F(x) = \dfrac{x^3}{x^2 - 4}$ is in lowest terms.

The vertical asymptotes are the zeros of $q(x)$: $x = -2$ and $x = 2$.

Step 5: Since $n = m + 1$, there is an oblique asymptote. Dividing:

$$x^2 - 4 \overline{\smash{\big)}\ \begin{aligned} & x \\[-2pt] & x^3 \end{aligned}}$$

$$\frac{x^3}{x^2 - 4} = x + \frac{4x}{x^2 - 4}$$

$$\underline{x^3 \qquad -4x}$$
$$4x$$

The oblique asymptote is $y = x$.

Solve to find intersection points:

$$\frac{x^3}{x^2 - 4} = x$$

$$x^3 = x^3 - 4x$$

$$4x = 0 \Rightarrow x = 0$$

$F(x)$ intersects $y = x$ at $(0, 0)$.

Step 6:

Interval	$(-\infty, -2)$	$(-2, 0)$	$(0, 2)$	$(2, \infty)$
Number Chosen	-3	-1	1	3
Value of F	$F(-3) = -\frac{27}{5}$	$F(-1) = \frac{1}{3}$	$F(1) = -\frac{1}{3}$	$F(3) = \frac{27}{5}$
Location of Graph	Below x-axis	Above x-axis	Below x-axis	Above x-axis
Point on Graph	$\left(-3, -\frac{27}{5}\right)$	$\left(-1, \frac{1}{3}\right)$	$\left(1, -\frac{1}{3}\right)$	$\left(3, \frac{27}{5}\right)$

Step 7:   Graphing:

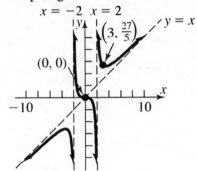

53. $R(x) = \dfrac{2x^4}{(x-1)^2}$ $p(x) = 2x^4;$ $q(x) = (x-1)^2;$ $n = 4;$ $m = 2$

Step 1:   Domain: $\{x \mid x \neq 1\}$.

Step 2:   (a)   The x-intercept is the zero of $p(x)$: 0.

   (b)   The y-intercept is $R(0) = \dfrac{2(0)^4}{(0-1)^2} = \dfrac{0}{1} = 0$.

Step 3: $R(-x) = \dfrac{2(-x)^4}{(-x-1)^2} = \dfrac{2x^4}{(x+1)^2}$; this is neither $R(x)$ nor $-R(x)$, so there

   is no symmetry.

Step 4: $R(x) = \dfrac{2x^4}{(x-1)^2}$ is in lowest terms.

   The vertical asymptote is the zero of $q(x)$: $x = 1$.

Step 5:   Since $n > m + 1$, there is no horizontal asymptote and no oblique asymptote.

Step 6:

Interval	$(-\infty, 0)$	$(0, 1)$	$(1, \infty)$
Number Chosen	-2	$\frac{1}{2}$	2
Value of R	$R(-2) \approx \frac{32}{9}$	$R\left(\frac{1}{2}\right) = \frac{1}{2}$	$R(2) = 32$
Location of Graph	Above x-axis	Above x-axis	Above x-axis
Point on Graph	$\left(-2, \frac{32}{9}\right)$	$\left(\frac{1}{2}, \frac{1}{2}\right)$	$(2, 32)$

Step 7: Graphing:

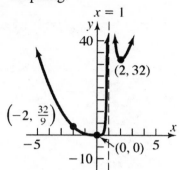

55. $G(x) = \dfrac{x^2-4}{x^2-x-2} = \dfrac{(x+2)(x-2)}{(x-2)(x+1)} = \dfrac{x+2}{x+1}$ $p(x) = x^2-4;\ q(x) = x^2-x-2;$

Step 1: Domain: $\{x \mid x \neq -1,\ x \neq 2\}$.

Step 2: (a) The x-intercept is the zero of $y = x+2$: -2; Note: 2 is not a zero because reduced form must be used to find the zeros.

(b) The y-intercept is $G(0) = \dfrac{0^2-4}{0^2-0-2} = \dfrac{-4}{-2} = 2$.

Step 3: $G(-x) = \dfrac{(-x)^2-4}{(-x)^2-(-x)-2} = \dfrac{x^2-4}{x^2+x-2}$; this is neither $G(x)$ nor $-G(x)$, so there is no symmetry.

Step 4: In lowest terms, $G(x) = \dfrac{x+2}{x+1},\ x \neq 2$.

The vertical asymptote is the zero of $f(x) = x+1$: $x = -1$; Note: $x = 2$ is not a vertical asymptote because reduced form must be used to find the asymptotes.

The graph has a hole at $\left(2, \dfrac{4}{3}\right)$.

Step 5: Since $n = m$, the line $y = \dfrac{1}{1} = 1$ is the horizontal asymptote.

Solve to find intersection points:
$$\frac{x^2-4}{x^2-x-2} = 1$$
$$x^2-4 = x^2-x-2$$
$$x = 2$$
$G(x)$ does not intersect $y = 1$ because $G(x)$ is not defined at $x = 2$.

Step 6:

Interval	$(-\infty, -2)$	$(-2, -1)$	$(-1, 2)$	$(2, \infty)$
Number Chosen	-3	-1.5	0	3
Value of G	$G(-3) = \frac{1}{2}$	$G(-1.5) = -1$	$G(0) = 2$	$G(3) = 1.25$
Location of Graph	Above x-axis	Below x-axis	Above x-axis	Above x-axis
Point on Graph	$\left(-3, \frac{1}{2}\right)$	$(-1.5, -1)$	$(0, 2)$	$(3, 1.25)$

Step 7:   Graphing:

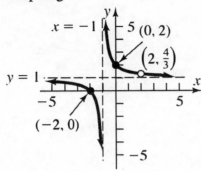

57. $2x^2 + 5x - 12 < 0$ $f(x) = 2x^2 + 5x - 12$

$(x+4)(2x-3) < 0$ $x = -4, x = \dfrac{3}{2}$ are the zeros of f.

Interval	$(-\infty, -4)$	$\left(-4, \dfrac{3}{2}\right)$	$\left(\dfrac{3}{2}, \infty\right)$
Number Chosen	-5	0	2
Value of f	13	-12	6
Conclusion	Positive	Negative	Positive

The solution set is $\left\{ x \,\middle|\, -4 < x < \dfrac{3}{2} \right\}$.

59. $\dfrac{6}{x+3} \geq 1$ $f(x) = \dfrac{3-x}{x+3}$

$\dfrac{6}{x+3} - 1 \geq 0$ The zeros and values where f is undefined are $x = -3$ and $x = 3$.

$\dfrac{6 - 1(x+3)}{x+3} \geq 0$

$\dfrac{3-x}{x+3} \geq 0$

Interval	$(-\infty, -3)$	$(-3, 3)$	$(3, \infty)$
Number Chosen	-4	0	4
Value of f	-7	1	$-\dfrac{1}{7}$
Conclusion	Negative	Positive	Negative

The solution set is $\left\{ x \,\middle|\, -3 < x \leq 3 \right\}$.

61.

$$\frac{2x-6}{1-x} < 2 \qquad f(x) = \frac{4x-8}{1-x}$$

The zeros and values where f is undefined are $x = 1$ and $x = 2$.

$$\frac{2x-6}{1-x} - 2 < 0$$

$$\frac{2x-6-2(1-x)}{1-x} < 0$$

$$\frac{4x-8}{1-x} < 0$$

Interval	$(-\infty,1)$	$(1,2)$	$(2,\infty)$
Number Chosen	0	1.5	3
Value of f	-8	4	-2
Conclusion	Negative	Positive	Negative

The solution set is $\left\{ x \mid x < 1 \text{ or } x > 2 \right\}$.

63.

$$\frac{(x-2)(x-1)}{x-3} \geq 0 \qquad f(x) = \frac{(x-2)(x-1)}{x-3}$$

The zeros and values where f is undefined are
$x = 1$, $x = 2$ and $x = 3$.

Interval	$(-\infty,1)$	$(1,2)$	$(2,3)$	$(3,\infty)$
Number Chosen	0	1.5	2.5	4
Value of f	$-\dfrac{2}{3}$	$\dfrac{1}{6}$	-1.5	6
Conclusion	Negative	Positive	Negative	Positive

The solution set is $\left\{ x \mid 1 \leq x \leq 2 \text{ or } x > 3 \right\}$.

65.

$$\frac{x^2-8x+12}{x^2-16} > 0 \qquad f(x) = \frac{x^2-8x+12}{x^2-16}$$

The zeros and values where f is undefined are

$$\frac{(x-2)(x-6)}{(x+4)(x-4)} > 0 \qquad x = -4, \ x = 2, \ x = 4 \text{ and } x = 6.$$

Interval	$(-\infty,-4)$	$(-4,2)$	$(2,4)$	$(4,6)$	$(6,\infty)$
Number Chosen	-5	0	3	5	7
Value of f	$\dfrac{77}{9}$	-0.75	$\dfrac{3}{7}$	$-\dfrac{1}{3}$	$\dfrac{5}{33}$
Conclusion	Positive	Negative	Positive	Negative	Positive

The solution set is $\left\{ x \mid x < -4 \text{ or } 2 < x < 4 \text{ or } x > 6 \right\}$.

67.

$$
\begin{array}{r|rrrr}
1) & 8 & -3 & 1 & 4 \\
& & 8 & 5 & 6 \\
\hline
& 8 & 5 & 6 & 10
\end{array}
$$

$$8x^3 - 3x^2 + x + 4 = (x-1)(8x^2+5x+6) + \frac{10}{x-1}$$

$$q(x) = 8x^2 + 5x + 6; \qquad R = 10$$

g is not a factor of f.

69.

$$\begin{array}{r} -2\overline{\smash{)}1 \quad -2 \quad 0 \quad 15 \quad -2} \\ -2 \quad 8 \quad -16 \quad 2 \\ \hline 1 \quad -4 \quad 8 \quad -1 \quad 0 \end{array}$$

$$x^4 - 2x^3 + 15x - 2 = (x+2)(x^3 - 4x^2 + 8x - 1) + \frac{0}{x+2}$$

$q(x) = x^3 - 4x^2 + 8x - 1; \qquad R = 0$

g is a factor of f.

71.

$$\begin{array}{r} 4\overline{\smash{)}12 \quad 0 \quad -8 \quad 0 \quad 0 \quad 0 \quad 1} \\ 48 \quad 192 \quad 736 \quad 2944 \quad 11{,}776 \quad 47{,}104 \\ \hline 12 \quad 48 \quad 184 \quad 736 \quad 2944 \quad 11{,}776 \quad 47{,}105 \end{array}$$

$f(4) = 47{,}105$

73.

Examining $f(x)$, there are 4 variations in sign; thus, there are four positive real zeros or two positive real zeros or no positive real zeros.

Examining $f(-x) = 12(-x)^8 - (-x)^7 + 8(-x)^4 - 2(-x)^3 + (-x) + 3$

$\qquad = 12x^8 + x^7 + 8x^4 + 2x^3 - x + 3,$

there are 2 variations in sign; thus, there are two negative real zeros or no negative real zeros.

75. $f(x) = 12x^8 - x^7 + 6x^4 - x^3 + x - 3$

p must be a factor of -3: $\ p = \pm 1, \pm 3$

q must be a factor of 12: $\ q = \pm 1, \pm 2, \pm 3, \pm 4, \pm 6, \pm 12$

The possible rational zeros are: $\ \dfrac{p}{q} = \pm 1, \pm 3, \pm \dfrac{1}{2}, \pm \dfrac{3}{2}, \pm \dfrac{1}{3}, \pm \dfrac{1}{4}, \pm \dfrac{3}{4}, \pm \dfrac{1}{6}, \pm \dfrac{1}{12}$

77. $f(x) = x^3 - 3x^2 - 6x + 8$

Step 1: $f(x)$ has at most 3 real zeros.

Step 2:   By Descartes' Rule of Signs, there are two positive real zeros or no positive real zeros.

$f(-x) = (-x)^3 - 3(-x)^2 - 6(-x) + 8 = -x^3 - 3x^2 + 6x + 8$, there is one negative real zero.

Step 3:   Possible rational zeros:

$p = \pm 1, \pm 2, \pm 4, \pm 8; \quad q = \pm 1;$

$\dfrac{p}{q} = \pm 1, \pm 2, \pm 4, \pm 8$

Step 4:   Using the Bounds on Zeros Theorem:

$a_2 = -3, \quad a_1 = -6, \quad a_0 = 8$

$\text{Max}\left\{1, |8| + |-6| + |-3|\right\} = \text{Max}\left\{1, 17\right\} = 17$

$1 + \text{Max}\left\{|8|, |-6|, |-3|\right\} = 1 + 8 = 9$

The smaller of the two numbers is 9. Thus, every real zero of f lies between -9 and 9.

Step 5:   Using synthetic division:
We try $x + 2$:

$$-2{\overline{)\begin{array}{rrrr} 1 & -3 & -6 & 8 \\ & -2 & 10 & -8 \end{array}}}$$
$$\begin{array}{rrrr} 1 & -5 & 4 & 0 \end{array}$$

$x + 2$ is a factor. The other factor is the quotient: $x^2 - 5x + 4$.

Thus, $f(x) = (x + 2)\left(x^2 - 5x + 4\right) = (x + 2)(x - 1)(x - 4)$.

The zeros are –2, 1, and 4, each of multiplicity 1.

79. $f(x) = 4x^3 + 4x^2 - 7x + 2$

Step 1: $f(x)$ has at most 3 real zeros.

Step 2:   By Descartes' Rule of Signs, there are two positive real zeros or no positive real zeros.

$f(-x) = 4(-x)^3 + 4(-x)^2 - 7(-x) + 2 = -4x^3 + 4x^2 + 7x + 2$; thus, there is one negative real zero.

Step 3:   Possible rational zeros:
$p = \pm 1, \pm 2; \quad q = \pm 1, \pm 2, \pm 4;$

$$\frac{p}{q} = \pm 1, \pm 2, \pm \frac{1}{2}, \pm \frac{1}{4}$$

Step 4:   Using the Bounds on Zeros Theorem:

$$f(x) = 4\left(x^3 + x^2 - \frac{7}{4}x + \frac{1}{2}\right) \Rightarrow a_2 = 1, \quad a_1 = -\frac{7}{4}, \quad a_0 = \frac{1}{2}$$

$$\text{Max}\left\{1, \left|\frac{1}{2}\right| + \left|-\frac{7}{4}\right| + |1|\right\} = \text{Max}\left\{1, \frac{13}{4}\right\} = \frac{13}{4} = 3.25$$

$$1 + \text{Max}\left\{\left|\frac{1}{2}\right|, \left|-\frac{7}{4}\right|, |1|\right\} = 1 + \frac{7}{4} = \frac{11}{4} = 2.75$$

The smaller of the two numbers is 2.75. Thus, every real zero of f lies between –2.75 and 2.75.

Step 5:   Using synthetic division:
We try $x + 2$:

$$-2{\overline{)\begin{array}{rrrr} 4 & 4 & -7 & 2 \\ & -8 & 8 & -2 \end{array}}}$$
$$\begin{array}{rrrr} 4 & -4 & 1 & 0 \end{array}$$

$x + 2$ is a factor. The other factor is the quotient: $4x^2 - 4x + 1$.

Thus, $f(x) = (x + 2)\left(4x^2 - 4x + 1\right) = (x + 2)(2x - 1)(2x - 1)$.

The zeros are –2, of multiplicity 1 and $\dfrac{1}{2}$, of multiplicity 2.

81. $f(x) = x^4 - 4x^3 + 9x^2 - 20x + 20$

    Step 1: $f(x)$ has at most 4 real zeros.

    Step 2:   By Descartes' Rule of Signs, there are four positive real zeros or two positive real zeros or no positive real zeros.

$$f(-x) = (-x)^4 - 4(-x)^3 + 9(-x)^2 - 20(-x) + 20$$
$$= x^4 + 4x^3 + 9x^2 + 20x + 20;$$

              thus, there are no negative real zeros.

    Step 3:   Possible rational zeros:

$$p = \pm 1, \pm 2, \pm 4, \pm 5, \pm 10, \pm 20; \quad q = \pm 1;$$

$$\frac{p}{q} = \pm 1, \pm 2, \pm 4, \pm 5, \pm 10, \pm 20$$

    Step 4:   Using the Bounds on Zeros Theorem:

$$a_3 = -4, \quad a_2 = 9, \quad a_1 = -20, \quad a_0 = 20$$

$$\text{Max}\left\{1, |20| + |-20| + |9| + |-4|\right\} = \text{Max}\{1, 53\} = 53$$

$$1 + \text{Max}\left\{|20|, |-20|, |9|, |-4|\right\} = 1 + 20 = 21$$

              The smaller of the two numbers is 21. Thus, every real zero of f lies between -21 and 21.

    Step 5:   Using synthetic division:

We try $x - 2$:

$$\begin{array}{r|rrrr} 2 & 1 & -4 & 9 & -20 & 20 \\ & & 2 & -4 & 10 & -20 \\ \hline & 1 & -2 & 5 & -10 & 0 \end{array}$$

$x - 2$ is a factor and the quotient is $x^3 - 2x^2 + 5x - 10$

We try $x - 2$ on $x^3 - 2x^2 + 5x - 10$

$$\begin{array}{r|rrrr} 2 & 1 & -2 & 5 & -10 \\ & & 2 & 0 & 10 \\ \hline & 1 & 0 & 5 & 0 \end{array}$$

$x - 2$ is a factor and the quotient is $x^2 + 5$

$x - 2$ is a factor twice. The other factor is the quotient: $x^2 + 5$.

Thus, $f(x) = (x - 2)(x - 2)(x^2 + 5) = (x - 2)^2(x^2 + 5)$.

Since $x^2 + 5 = 0$ has no real solutions, the only zero is 2, of multiplicity 2.

83. $2x^4 + 2x^3 - 11x^2 + x - 6 = 0$

    The solutions of the equation are the zeros of $f(x) = 2x^4 + 2x^3 - 11x^2 + x - 6$.

    Step 1: $f(x)$ has at most 4 real zeros.

    Step 2:   By Descartes' Rule of Signs, there are three positive real zeros or there is one positive real zero.

$$f(-x) = 2(-x)^4 + 2(-x)^3 - 11(-x)^2 + (-x) - 6 = 2x^4 - 2x^3 - 11x^2 - x - 6;$$

              thus, there is one negative real zero.

    Step 3:   Possible rational zeros:

$$p = \pm 1, \pm 2, \pm 3, \pm 6; \quad q = \pm 1, \pm 2;$$

$$\frac{p}{q} = \pm 1, \pm 2, \pm 3, \pm 6, \pm \frac{1}{2}, \pm \frac{3}{2}$$

Step 4:   Using the Bounds on Zeros Theorem:

$$f(x) = 2\left(x^4 + x^3 - \frac{11}{2}x^2 + \frac{1}{2}x - 3\right) \Rightarrow a_3 = 1, \ a_2 = -\frac{11}{2}, \ a_1 = \frac{1}{2}, \ a_0 = -3$$

$$\text{Max}\left\{1, |-3| + \left|\frac{1}{2}\right| + \left|-\frac{11}{2}\right| + |1|\right\} = \text{Max}\{1, 10\} = 10$$

$$1 + \text{Max}\left\{|-3|, \left|\frac{1}{2}\right|, \left|-\frac{11}{2}\right|, |1|\right\} = 1 + \frac{11}{2} = \frac{13}{2} = 6.5$$

The smaller of the two numbers is 6.5. Thus, every real zero of f lies between -6.5 and 6.5.

Step 5:   Using synthetic division:

We try $x + 3$:

$$\begin{array}{r|rrrr} -3) & 2 & 2 & -11 & 1 & -6 \\ & & -6 & 12 & -3 & 6 \\ \hline & 2 & -4 & 1 & -2 & 0 \end{array}$$

$x + 3$ is a factor and

the quotient is $2x^3 - 4x^2 + x - 2$

We try $x - 2$ on $2x^3 - 4x^2 + x - 2$

$$\begin{array}{r|rrrr} 2) & 2 & -4 & 1 & -2 \\ & & 4 & 0 & 2 \\ \hline & 2 & 0 & 1 & 0 \end{array}$$

$x - 2$ is a factor and

the quotient is $2x^2 + 1$

$x + 3$ and $x - 2$ are factors. The other factor is the quotient: $2x^2 + 1$.

Thus, $f(x) = (x + 3)(x - 2)\left(2x^2 + 1\right)$.

Since $2x^2 + 1 = 0$ has no real solutions, the solution set is $\{-3, 2\}$.

85. $2x^4 + 7x^3 + x^2 - 7x - 3 = 0$

The solutions of the equation are the zeros of $f(x) = 2x^4 + 7x^3 + x^2 - 7x - 3$.

Step 1: $f(x)$ has at most 4 real zeros.

Step 2:   By Descartes' Rule of Signs, there is one positive real zero.

$f(-x) = 2(-x)^4 + 7(-x)^3 + (-x)^2 - 7(-x) - 3 = 2x^4 - 7x^3 + x^2 + 7x - 3$;

thus, there are three negative real zeros or there is one negative real zero.

Step 3:   Possible rational zeros:

$p = \pm 1, \pm 3; \quad q = \pm 1, \pm 2;$

$$\frac{p}{q} = \pm 1, \pm 3, \pm \frac{1}{2}, \pm \frac{3}{2}$$

Step 4:   Using the Bounds on Zeros Theorem:

$$f(x) = 2\left(x^4 + \frac{7}{2}x^3 + \frac{1}{2}x^2 - \frac{7}{2}x - \frac{3}{2}\right) \Rightarrow a_3 = \frac{7}{2}, \ a_2 = \frac{1}{2}, \ a_1 = -\frac{7}{2}, \ a_0 = -\frac{3}{2}$$

$$\text{Max}\left\{1, \left|-\frac{3}{2}\right| + \left|-\frac{7}{2}\right| + \left|\frac{1}{2}\right| + \left|\frac{7}{2}\right|\right\} = \text{Max}\{1, 9\} = 9$$

$$1 + \text{Max}\left\{\left|-\frac{3}{2}\right|, \left|-\frac{7}{2}\right|, \left|\frac{1}{2}\right|, \left|\frac{7}{2}\right|\right\} = 1 + \frac{7}{2} = \frac{9}{2} = 4.5$$

The smaller of the two numbers is 4.5. Thus, every real zero of f lies between -4.5 and 4.5.

Step 5: Using synthetic division:

We try $x + 3$:

$$-3\overline{)\begin{array}{rrrrr} 2 & 7 & 1 & -7 & -3 \\ & -6 & -3 & 6 & 3 \\ \hline 2 & 1 & -2 & -1 & 0 \end{array}}$$

$x + 3$ is a factor and

the quotient is $2x^3 + x^2 - 2x - 1$

We try $x + 1$ on $2x^3 + x^2 - 2x - 1$

$$-1\overline{)\begin{array}{rrrr} 2 & 1 & -2 & -1 \\ & -2 & 1 & 1 \\ \hline 2 & -1 & -1 & 0 \end{array}}$$

$x + 1$ is a factor and

the quotient is $2x^2 - x - 1$

$x + 3$ and $x + 1$ are factors. The other factor is the quotient: $2x^2 - x - 1$.

Thus, $f(x) = (x + 3)(x + 1)(2x^2 - x - 1) = (x + 3)(x + 1)(2x + 1)(x - 1)$.

The solution set is $\left\{-3, \ -1, \ -\dfrac{1}{2}, \ 1\right\}$.

87.　$f(x) = x^3 - 3x^2 - 6x + 8$.

Step 1: $f(x)$ has at most 3 real zeros.

Step 2: By Descartes' Rule of Signs, there are two positive real zeros or no positive real zeros.

$$f(-x) = (-x)^3 - 3(-x)^2 - 6(-x) + 8 = -x^3 - 3x^2 + 6x + 8;$$

thus, there is one negative real zero.

Step 3: Possible rational zeros:

$p = \pm 1, \pm 2, \pm 4, \pm 8; \quad q = \pm 1;$

$\dfrac{p}{q} = \pm 1, \pm 2, \pm 4, \pm 8$

Step 4: Using synthetic division:

We try $x - 1$:

$$1\overline{)\begin{array}{rrrr} 1 & -3 & -6 & 8 \\ & 1 & -2 & -8 \\ \hline 1 & -2 & -8 & 0 \end{array}}$$

$x - 1$ is a factor and the quotient is $x^2 - 2x - 8$

Thus, $f(x) = (x - 1)(x^2 - 2x - 8) = (x - 1)(x - 4)(x + 2)$.

The complex zeros are 1, 4, and –2, each of multiplicity 1

89.　$f(x) = 4x^3 + 4x^2 - 7x + 2$.

Step 1: $f(x)$ has at most 3 real zeros.

Step 2: By Descartes' Rule of Signs, there are two positive real zeros or no positive real zeros.

$$f(-x) = 4(-x)^3 + 4(-x)^2 - 7(-x) + 2 = -4x^3 + 4x^2 + 7x + 2;$$

thus, there is one negative real zero.

Step 3: Possible rational zeros:

$p = \pm 1, \pm 2; \quad q = \pm 1, \pm 2, \pm 4;$

$\dfrac{p}{q} = \pm 1, \pm \dfrac{1}{2}, \pm \dfrac{1}{4}, \pm 2$

Step 4:   Using synthetic division:
We try $x + 2$:

$$
\begin{array}{r|rrrr}
-2) & 4 & 4 & -7 & 2 \\
    &   & -8 & 8 & -2 \\
\hline
    & 4 & -4 & 1 & 0
\end{array}
$$

$x + 2$ is a factor and the quotient is $4x^2 - 4x + 1$

Thus, $f(x) = (x+2)\left(4x^2 - 4x + 1\right) = (x+2)(2x-1)(2x-1) = (x+2)(2x-1)^2$.

The complex zeros are -2, of multiplicity 1, and $\dfrac{1}{2}$, of multiplicity 2.

91. $f(x) = x^4 - 4x^3 + 9x^2 - 20x + 20$.
Step 1: $f(x)$ has at most 4 real zeros.
Step 2:   By Descartes' Rule of Signs, there are four positive real zeros or two positive real zeros or no positive real zeros.
$f(-x) = (-x)^4 - 4(-x)^3 + 9(-x)^2 - 20(-x) + 20 = x^4 + 4x^3 + 9x^2 + 20x + 20;$
thus, there are no negative real zeros.
Step 3:   Possible rational zeros:
$p = \pm 1, \pm 2, \pm 4, \pm 5, \pm 10, \pm 20; \quad q = \pm 1;$

$\dfrac{p}{q} = \pm 1, \pm 2, \pm 4, \pm 5, \pm 10, \pm 20$

Step 4:   Using synthetic division:
We try $x - 2$:

$$
\begin{array}{r|rrrrr}
2) & 1 & -4 & 9 & -20 & 20 \\
   &   & 2 & -4 & 10 & -20 \\
\hline
   & 1 & -2 & 5 & -10 & 0
\end{array}
$$

$x - 2$ is a factor and the quotient is $x^3 - 2x^2 + 5x - 10$

Thus, $f(x) = (x-2)\left(x^3 - 2x^2 + 5x - 10\right)$.

We can factor $x^3 - 2x^2 + 5x - 10$ by grouping.
$$x^3 - 2x^2 + 5x - 10 = x^2(x-2) + 5(x-2)$$
$$= (x-2)(x^2 + 5)$$
$$= (x-2)\left(x + \sqrt{5}i\right)\left(x - \sqrt{5}i\right)$$

$f(x) = (x-2)^2\left(x + \sqrt{5}i\right)\left(x - \sqrt{5}i\right)$

The complex zeros are 2, of multiplicity 2, and $\sqrt{5}i$ and $-\sqrt{5}i$, each of multiplicity 1.

93. $f(x) = 2x^4 + 2x^3 - 11x^2 + x - 6$.
Step 1: $f(x)$ has at most 4 real zeros.
Step 2:   By Descartes' Rule of Signs, there are three positive real zeros or there is one positive real zero.
$f(-x) = 2(-x)^4 + 2(-x)^3 - 11(-x)^2 + (-x) - 6 = 2x^4 - 2x^3 - 11x^2 - x - 6;$
thus, there is one negative real zero.

Step 3:   Possible rational zeros:
$$p = \pm 1, \pm 2, \pm 3, \pm 6; \quad q = \pm 1, \pm 2;$$

$$\frac{p}{q} = \pm 1, \pm \frac{1}{2}, \pm 2, \pm 3, \pm \frac{3}{2} \pm 6$$

Step 4:   Using synthetic division:
We try $x - 2$:

$$
\begin{array}{r|rrrr}
2 & 2 & 2 & -11 & 1 & -6 \\
  &   & 4 & 12 & 2 & 6 \\
\hline
  & 2 & 6 & 1 & 3 & 0
\end{array}
$$

$x - 2$ is a factor and the quotient is $2x^3 + 6x^2 + x + 3$

Thus, $f(x) = (x - 2)(2x^3 + 6x^2 + x + 3)$.

We can factor $2x^3 + 6x^2 + x + 3$ by grouping.
$$2x^3 + 6x^2 + x + 3 = 2x^2(x + 3) + (x + 3)$$
$$= (x + 3)(2x^2 + 1)$$
$$= (x + 3)(\sqrt{2}x + i)(\sqrt{2}x - i)$$
$$f(x) = (x - 2)(x + 3)(\sqrt{2}x + i)(\sqrt{2}x - i)$$

The complex zeros are $2, -3, -\dfrac{\sqrt{2}}{2}i$, and $\dfrac{\sqrt{2}}{2}i$, each of multiplicity 1

95. $f(x) = x^3 - x^2 - 4x + 2$
$a_2 = -1, \quad a_1 = -4, \quad a_0 = 2$
$\text{Max}\left\{1, |2| + |-4| + |-1|\right\} = \text{Max}\left\{1, 7\right\} = 7$
$1 + \text{Max}\left\{|2|, |-4|, |-1|\right\} = 1 + 4 = 5$
The smaller of the two numbers is 5, so every real zero of f lies between -5 and 5.

97. $f(x) = 2x^3 - 7x^2 - 10x + 35 = 2\left(x^3 - \dfrac{7}{2}x^2 - 5x + \dfrac{35}{2}\right)$

$a_2 = -\dfrac{7}{2}, \quad a_1 = -5, \quad a_0 = \dfrac{35}{2}$

$\text{Max}\left\{1, \left|\dfrac{35}{2}\right| + |-5| + \left|-\dfrac{7}{2}\right|\right\} = \text{Max}\left\{1, 26\right\} = 26$

$1 + \text{Max}\left\{\left|\dfrac{35}{2}\right|, |-5|, \left|-\dfrac{7}{2}\right|\right\} = 1 + \dfrac{35}{2} = \dfrac{37}{2} = 18.5$

The smaller of the two numbers is 18.5, so every real zero of f lies between -18.5 and 18.5.

99. $f(x) = 3x^3 - x - 1;$ $[0, 1]$
$f(0) = -1 < 0$ and $f(1) = 1 > 0$
Since one value is positive and
one is negative, there is a zero
in the interval.

101. $f(x) = 8x^4 - 4x^3 - 2x - 1;$ $[0, 1]$
$f(0) = -1 < 0$ and $f(1) = 1 > 0$
Since one value is positive and
one is negative, there is a zero
in the interval.

103. $f(x) = x^3 - x - 2$
$f(1) = -2;$ $f(2) = 4$
So by the Intermediate Value Theorem, f has a zero on the interval [1,2].

Subdivide the interval [1,2] into 10 equal subintervals:
[1,1.1]; [1.1,1.2]; [1.2,1.3]; [1.3,1.4]; [1.4,1.5]; [1.5,1.6]; [1.6,1.7]; [1.7,1.8];
[1.8,1.9]; [1.9,2]
$f(1) = -2; f(1.1) = -1.769$
$f(1.1) = -1.769; f(1.2) = -1.472$
$f(1.2) = -1.472; f(1.3) = -1.103$
$f(1.3) = -1.103; f(1.4) = -0.656$
$f(1.4) = -0.656; f(1.5) = -0.125$
$f(1.5) = -0.125; f(1.6) = 0.496$
So f has a real zero on the interval [1.5,1.6].

Subdivide the interval [1.5,1.6] into 10 equal subintervals:
[1.5,1.51]; [1.51,1.52]; [1.52,1.53]; [1.53,1.54]; [1.54,1.55]; [1.55,1.56];[1.56,1.57];
[1.57,1.58]; [1.58,1.59]; [1.59,1.6]
$f(1.5) = -0.125; f(1.51) \approx -0.0670$
$f(1.51) \approx -0.0670; f(1.52) \approx -0.0082$
$f(1.52) \approx -0.0082; f(1.53) \approx 0.0516$
So f has a real zero on the interval [1.52,1.53], therefore the zero is 1.52, correct to
two decimal places.

105. $f(x) = 8x^4 - 4x^3 - 2x - 1$
$f(0) = -1;$ $f(1) = 1,$
So by the Intermediate Value Theorem, f has a zero on the interval [0,1].

Subdivide the interval [0,1] into 10 equal subintervals:
[0,0.1]; [0.1,0.2]; [0.2,0.3]; [0.3,0.4]; [0.4,0.5]; [0.5,0.6]; [0.6,0.7]; [0.7,0.8];
[0.8,0.9]; [0.9,1]
$f(0) = -1; f(0.1) = -1.2032$
$f(0.1) = -1.2032; f(0.2) = -1.4192$
$f(0.2) = -1.4192; f(0.3) = -1.6432$
$f(0.3) = -1.6432; f(0.4) = -1.8512$
$f(0.4) = -1.8512; f(0.5) = -2$
$f(0.5) = -2; f(0.6) = -2.0272$

$f(0.6) = -2.0272; f(0.7) = -1.8512$
$f(0.7) = -1.8512; f(0.8) = -1.3712$
$f(0.8) = -1.3712; f(0.9) = -0.4672$
$f(0.9) = -0.4672; f(1) = 1$
So f has a real zero on the interval $[0.9,1]$.

Subdivide the interval $[0.9,1]$ into 10 equal subintervals:
$[0.9,0.91]$; $[0.91,0.92]$; $[0.92,0.93]$; $[0.93,0.94]$; $[0.94,0.95]$; $[0.95,0.96]$;$[0.96,0.97]$;
$[0.97,0.98]$; $[0.98,0.99]$; $[0.99,1]$
$f(0.9) = -0.4672; f(0.91) \approx -0.3483$
$f(0.91) \approx -0.3483; f(0.92) \approx -0.2236$
$f(0.92) \approx -0.2236; f(0.93) \approx -0.0930$
$f(0.93) \approx -0.0930; f(0.94) \approx 0.0437$
So f has a real zero on the interval $[0.93,0.94]$, therefore the zero is 0.93, correct to two decimal places.

107. Since complex zeros appear in conjugate pairs, $4 - i$, the conjugate of $4 + i$, is the remaining zero of f.

109. Since complex zeros appear in conjugate pairs, $-i$, the conjugate of i, and $1 - i$, the conjugate of $1 + i$, are the remaining zeros of f.

111. The distance between the point $P(x,y)$ and $Q(3,1)$ is $d(P,Q) = \sqrt{(x-3)^2 + (y-1)^2}$. If P is on the line $y = x$, then the distance is

$$d(P,Q) = \sqrt{(x-3)^2 + (x-1)^2}$$
$$d^2(x) = (x-3)^2 + (x-1)^2 = x^2 - 6x + 9 + x^2 - 2x + 1$$
$$= 2x^2 - 8x + 10$$

Since $d^2(x) = 2x^2 - 8x + 10$ is a quadratic function with $a = 2 > 0$, the vertex corresponds to the minimum value for the function.

The vertex occurs at $x = -\dfrac{b}{2a} = -\dfrac{-8}{2(2)} = 2$. Therefore the point on the line $y = x$

closest to the point $(3,1)$ is $(2,2)$.

113. Consider the diagram

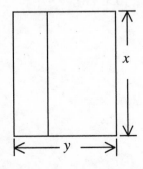

Total amount of fence $= 3x + 2y = 10{,}000$

$$y = \frac{10{,}000 - 3x}{2} = 5000 - \frac{3}{2}x$$

Total area enclosed $= (x)(y) = (x)\left(5000 - \frac{3}{2}x\right)$

$A(x) = 5000x - \dfrac{3}{2}x^2 = -\dfrac{3}{2}x^2 + 5000x$ is a quadratic function with $a = -\dfrac{3}{2} < 0$.

So the vertex corresponds to the maximum value for this function.

The vertex occurs when

$x = -\dfrac{b}{2a} = -\dfrac{5000}{2\left(-\dfrac{3}{2}\right)} = \dfrac{5000}{3}$

The maximum area is: $A\left(\dfrac{5000}{3}\right) = -\dfrac{3}{2}\left(\dfrac{5000}{3}\right)^2 + 5000\left(\dfrac{5000}{3}\right)$

$= -\dfrac{3}{2}\left(\dfrac{25,000,000}{9}\right) + \dfrac{25,000,000}{3}$

$= -\dfrac{12,500,000}{3} + \dfrac{25,000,000}{3}$

$= \dfrac{12,500,000}{3} \approx 4,166,666.67$ square meters

115. Consider the diagram

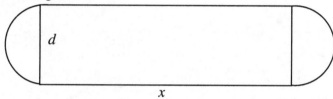

d = diameter of the semicircles = width of the rectangle
x = length of the rectangle
100 = outside dimension length

$100 = 2x + 2(\text{circumference of a semicircle})$

$100 = 2x + \text{circumference of a circle}$

$100 = 2x + \pi d$

$\Rightarrow x = \dfrac{100 - \pi d}{2} = 50 - \dfrac{1}{2}\pi d$

Total area enclosed = area of the rectangle + 2(area of a semicircle)

= area of the rectangle + area of a circle

$= (x)(d) + \pi r^2 = (x)(d) + \pi\left(\dfrac{d}{2}\right)^2$

$= \left(50 - \dfrac{1}{2}\pi d\right)(d) + \pi\left(\dfrac{d}{2}\right)^2 = 50d - \dfrac{1}{2}\pi d^2 + \dfrac{1}{4}\pi d^2$

$= 50d - \dfrac{1}{4}\pi d^2 = -\dfrac{1}{4}\pi d^2 + 50d$

$A(d) = -\dfrac{1}{4}\pi d^2 + 50d$ is a quadratic function with $a = -\dfrac{1}{4}\pi < 0$. Therefore the vertex corresponds to the maximum value for the function.

The vertex occurs when $x = -\dfrac{b}{2a} = -\dfrac{50}{2\left(-\dfrac{1}{4}\pi\right)} = \dfrac{100}{\pi}$.

The maximum area is $A\left(\dfrac{100}{\pi}\right) = -\dfrac{1}{4}\pi\left(\dfrac{100}{\pi}\right)^2 + 50\left(\dfrac{100}{\pi}\right) \approx 795.77$ square meters.

117. $C(x) = 4.9x^2 - 617.40x + 19{,}600$; $a = 4.9, b = -617.40, c = 19{,}600$. Since $a = 4.9 > 0$, the graph opens up, so the vertex is a minimum point.

(a)   The minimum marginal cost occurs at $x = -\dfrac{b}{2a} = -\dfrac{-617.40}{2(4.9)} = \dfrac{617.40}{9.8} = 63$.

Thus, 63 golf clubs should be manufactured in order to minimize the marginal cost.

(b)   The minimum marginal cost is

$$C\left(-\dfrac{b}{2a}\right) = C(63) = 4.9(63)^2 - (617.40)(63) + 19600 = \$151.90$$

119. (a)

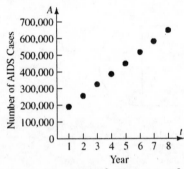

(b) $A(11) = -212(11)^3 + 2429(11)^2 + 59{,}569(11) + 130{,}003 = 796{,}999$ cases

(c) and (d)   Graphing the cubic function of best fit:

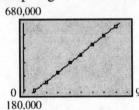

(e)   Answers will vary.

121. Answers will vary.

123. (a)   The degree is even.
(b)   The leading coefficient is positive.
(c)   The function is even, because it is symmetric with respect to the y-axis.
(d) x^2 is a factor because the curve touches the x-axis at the origin.
(e)   The minimum degree is 8.
(f)   Answers will vary.

Polynomial and Rational Functions

3.CR   Cumulative Review

1. $P = (1,3)$, $Q = (-4,2)$

$$d_{P,Q} = \sqrt{(-4-1)^2 + (2-3)^2}$$

$$= \sqrt{(-5)^2 + (-1)^2}$$

$$= \sqrt{25+1}$$

$$= \sqrt{26}$$

3. $x^2 - 3x < 4$ $f(x) = x^2 - 3x - 4$

$x^2 - 3x - 4 < 0$ $x = -1$, $x = 4$ are the zeros of f.

$(x-4)(x+1) < 0$

Interval	$(-\infty, -1)$	$(-1, 4)$	$(4, \infty)$
Number Chosen	-2	0	5
Value of f	6	-4	6
Conclusion	Positive	Negative	Positive

The solution set is
$\{x \mid -1 < x < 4\}$.

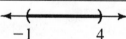

5. Parallel to $y = 2x + 1$;
    Slope 2, Containing the point (3, 5)
    Using the point-slope formula yields:

$$y - y_1 = m(x - x_1)$$

$$y - 5 = 2(x - 3)$$

$$y - 5 = 2x - 6$$

$$y = 2x - 1$$

7. This relation is not a function because the ordered pairs (3, 6) and (3, 8) have the same first element, but different second elements.

9. $3x + 2 \le 5x - 1$

$3 \le 2x$

$\dfrac{3}{2} \le x$

$x \ge \dfrac{3}{2}$

The solution set is $\left\{ x \middle| x \ge \dfrac{3}{2} \right\}$.

11. $y = x^3 - 9x$

x-intercepts: $0 = x^3 - 9x = x\left(x^2 - 9\right)$

$= x(x + 3)(x - 3) \Rightarrow x = 0, -3,$ and $3 \Rightarrow (0,0), (-3,0), (3,0)$

y-intercepts: $y = 0^3 - 9(0) = 0 \Rightarrow (0,0)$

Test for symmetry:

x-axis: Replace y by $-y$: $-y = x^3 - 9x$,

which is not equivalent to $y = x^3 - 9x$.

y-axis: Replace x by $-x$: $y = (-x)^3 - 9(-x) = -x^3 + 9x$,

which is not equivalent to $y = x^3 - 9x$.

Origin: Replace x by $-x$ and y by $-y$: $-y = (-x)^3 - 9(-x)$

$-y = -x^3 + 9x$,

which is equivalent to $y = x^3 - 9x$.

Therefore, the graph is symmetric with respect to origin.

13. Not a function, since the graph fails the Vertical Line Test, for example, when $x = 0$.

15. $f(x) = \dfrac{x + 5}{x - 1}$

(a) Domain $\{x | x \ne 1\}$.

(b) $f(2) = \dfrac{2 + 5}{2 - 1} = \dfrac{7}{1} = 7 \ne 6$; $(2,6)$ is not on the graph of f.

(c) $f(3) = \dfrac{3 + 5}{3 - 1} = \dfrac{8}{2} = 4$; $(3,4)$ is on the graph of f.

(d) Solve for x:

$\dfrac{x + 5}{x - 1} = 9$

$x + 5 = 9(x - 1)$

$x + 5 = 9x - 9$

$14 = 8x$

$x = \dfrac{14}{8} = 1.75$

Therefore, $(1.75, 9)$ is on the graph of f.

17. $f(x) = 2x^2 - 4x + 1$
       $a = 2$, $b = -4$, $c = 1$. Since $a = 2 > 0$, the graph opens up.

       The x-coordinate of the vertex is $x = -\dfrac{b}{2a} = -\dfrac{-4}{2(2)} = 1$.

       The y-coordinate of the vertex is $f\left(-\dfrac{b}{2a}\right) = f(1) = 2(1)^2 - 4(1) + 1 = -1$.

       Thus, the vertex is $(1, -1)$.
       The axis of symmetry is the line $x = 1$.
       The discriminant is:
       $$b^2 - 4ac = (-4)^2 - 4(2)(1) = 8 > 0,$$
       so the graph has two x-intercepts.
       The x-intercepts are found by solving:
       $$2x^2 - 4x + 1 = 0$$
       $$x = \frac{-(-4) \pm \sqrt{8}}{2(2)} = \frac{4 \pm 2\sqrt{2}}{4} = \frac{2 \pm \sqrt{2}}{2}$$
       The x-intercepts are $\dfrac{2 - \sqrt{2}}{2}$ and $\dfrac{2 + \sqrt{2}}{2}$.
       The y-intercept is $f(0) = 1$.

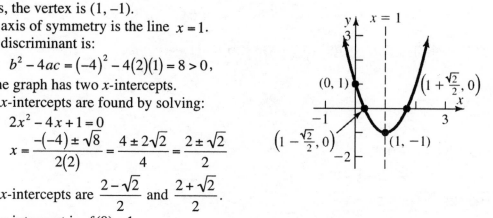

19.   (a) x-intercepts: $(-5,0)$, $(-1,0)$, $(5,0)$; y-intercept: $(0,-3)$
       (b)   The graph is not symmetric with respect to the origin, x-axis or y-axis.
       (c)   The function is neither even nor odd.
       (d) f is increasing on $(-\infty,-3)$ and $(2,\infty)$; f is decreasing on $(-3,2)$;
       (e) f has a local maximum at $x = -3$, and the local maximum is the point $(-3,5)$.
       (f) f has a local minimum at $x = 2$, and the local minimum is the point $(2,-6)$.

21. $f(x) = \begin{cases} 2x + 1 & \text{if } -3 < x < 2 \\ -3x + 4 & \text{if } x \geq 2 \end{cases}$

       (a)   Domain: $\{x | x > -3\}$

       (b) x-intercept: $\left(-\dfrac{1}{2},0\right)$
              y-intercept: $(0,1)$

       (c)

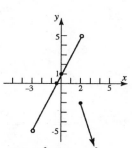

       (d)   Range: $\{y | y < 5\}$

23. $f(x) = x^2 - 5x + 1$ $g(x) = -4x - 7$
       (a) $(f + g)(x) = x^2 - 5x + 1 + (-4x - 7)$
              $= x^2 - 9x - 6$
              The domain is: $\{x | x \text{ is a real number}\}$.

       (b) $\left(\dfrac{f}{g}\right)(x) = \dfrac{f(x)}{g(x)} = \dfrac{x^2 - 5x + 1}{-4x - 7}$

              The domain is: $\left\{x \middle| x \neq -\dfrac{7}{4}\right\}$.

Exponential and Logarithmic Functions

4.1   Composite Functions

1. $f(3) = -4(3)^2 + 5(3) = -4(9) + 15 = -36 + 15 = -11$

3. $f(x) = \dfrac{x^2 - 1}{x^2 - 4}$

  $x^2 - 4 \neq 0$

  $(x + 2)(x - 2) \neq 0$

  $x \neq -2, \quad x \neq 2$

  Domain: $\{x \mid x \neq -2, x \neq 2\}$.

5.   False

7.   (a) $(f \circ g)(1) = f\big(g(1)\big) = f(0) = -1$
   (b) $(f \circ g)(-1) = f\big(g(-1)\big) = f(0) = -1$
   (c) $(g \circ f)(-1) = g\big(f(-1)\big) = g(-3) = 8$
   (d) $(g \circ f)(0) = g\big(f(0)\big) = g(-1) = 0$
   (e) $(g \circ g)(-2) = g\big(g(-2)\big) = g(3) = 8$
   (f) $(f \circ f)(-1) = f\big(f(-1)\big) = f(-3) = -7$

9. $f(x) = 2x \qquad g(x) = 3x^2 + 1$
   (a) $(f \circ g)(4) = f(g(4)) = f\big(3(4)^2 + 1\big) = f(49) = 2(49) = 98$
   (b) $(g \circ f)(2) = g(f(2)) = g(2 \cdot 2) = g(4) = 3(4)^2 + 1 = 48 + 1 = 49$
   (c) $(f \circ f)(1) = f(f(1)) = f(2(1)) = f(2) = 2(2) = 4$
   (d) $(g \circ g)(0) = g(g(0)) = g\big(3(0)^2 + 1\big) = g(1) = 3(1)^2 + 1 = 4$

11. $f(x) = 4x^2 - 3 \qquad g(x) = 3 - \dfrac{1}{2}x^2$

   (a) $(f \circ g)(4) = f(g(4)) = f\left(3 - \dfrac{1}{2}(4)^2\right) = f(-5) = 4(-5)^2 - 3 = 97$

   (b) $(g \circ f)(2) = g(f(2)) = g(4(2)^2 - 3) = g(13) = 3 - \dfrac{1}{2}(13)^2 = 3 - \dfrac{169}{2} = -\dfrac{163}{2}$

(c) $(f \circ f)(1) = f(f(1)) = f(4(1)^2 - 3) = f(1) = 4(1)^2 - 3 = 1$

(d) $(g \circ g)(0) = g(g(0)) = g\left(3 - \frac{1}{2}(0)^2\right) = g(3) = 3 - \frac{1}{2}(3)^2 = 3 - \frac{9}{2} = -\frac{3}{2}$

13. $f(x) = \sqrt{x}$ $g(x) = 2x$

(a) $(f \circ g)(4) = f(g(4)) = f(2(4)) = f(8) = \sqrt{8} = 2\sqrt{2}$

(b) $(g \circ f)(2) = g(f(2)) = g\left(\sqrt{2}\right) = 2\sqrt{2}$

(c) $(f \circ f)(1) = f(f(1)) = f\left(\sqrt{1}\right) = f(1) = \sqrt{1} = 1$

(d) $(g \circ g)(0) = g(g(0)) = g(2(0)) = g(0) = 2(0) = 0$

15. $f(x) = |x|$ $g(x) = \dfrac{1}{x^2 + 1}$

(a) $(f \circ g)(4) = f(g(4)) = f\left(\dfrac{1}{4^2 + 1}\right) = f\left(\dfrac{1}{17}\right) = \left|\dfrac{1}{17}\right| = \dfrac{1}{17}$

(b) $(g \circ f)(2) = g(f(2)) = g(|2|) = g(2) = \dfrac{1}{2^2 + 1} = \dfrac{1}{5}$

(c) $(f \circ f)(1) = f(f(1)) = f(|1|) = f(1) = |1| = 1$

(d) $(g \circ g)(0) = g(g(0)) = g\left(\dfrac{1}{0^2 + 1}\right) = g(1) = \dfrac{1}{1^2 + 1} = \dfrac{1}{2}$

17. $f(x) = \dfrac{3}{x + 1}$ $g(x) = \sqrt[3]{x}$

(a) $(f \circ g)(4) = f(g(4)) = f\left(\sqrt[3]{4}\right) = \dfrac{3}{\sqrt[3]{4} + 1}$

(b) $(g \circ f)(2) = g(f(2)) = g\left(\dfrac{3}{2 + 1}\right) = g\left(\dfrac{3}{3}\right) = \sqrt[3]{1} = 1$

(c) $(f \circ f)(1) = f(f(1)) = f\left(\dfrac{3}{1 + 1}\right) = f\left(\dfrac{3}{2}\right) = \dfrac{3}{\frac{3}{2} + 1} = \dfrac{3}{\frac{5}{2}} = \dfrac{6}{5}$

(d) $(g \circ g)(0) = g(g(0)) = g\left(\sqrt[3]{0}\right) = g(0) = \sqrt[3]{0} = 0$

19. The domain of g is $\{x \mid x \neq 0\}$. The domain of f is $\{x \mid x \neq 1\}$.
     Thus, $g(x) \neq 1$, so we solve:
$$g(x) = 1$$
$$\frac{2}{x} = 1$$
$$x = 2$$
     Thus, $x \neq 2$; so the domain of $f \circ g$ is $\{x \mid x \neq 0, \ x \neq 2\}$.

21. The domain of g is $\{x \mid x \neq 0\}$. The domain of f is $\{x \mid x \neq 1\}$.
     Thus, $g(x) \neq 1$, so we solve:
     $$g(x) = 1$$
     $$-\frac{4}{x} = 1$$
     $$x = -4$$
     Thus, $x \neq -4$; so the domain of $f \circ g$ is $\{x \mid x \neq -4, x \neq 0\}$.

23. The domain of g is $\{x \mid x \text{ is any real number}\}$. The domain of f is $\{x \mid x \geq 0\}$.
     Thus, $g(x) \geq 0$, so we solve:
     $$2x + 3 \geq 0$$
     $$x \geq -\frac{3}{2}$$
     Thus, the domain of $f \circ g$ is $\left\{x \mid x \geq -\frac{3}{2}\right\}$.

25. The domain of g is $\{x \mid x \geq 1\}$. The domain of f is $\{x \mid x \text{ is any real number}\}$.
     Thus, the domain of $f \circ g$ is $\{x \mid x \geq 1\}$.

27. $f(x) = 2x + 3 \qquad g(x) = 3x$
     The domain of f is $\{x \mid x \text{ is any real number}\}$. The domain of g is $\{x \mid x \text{ is any real number}\}$.
     (a) $(f \circ g)(x) = f(g(x)) = f(3x) = 2(3x) + 3 = 6x + 3$
          Domain: $\{x \mid x \text{ is any real number}\}$.
     (b) $(g \circ f)(x) = g(f(x)) = g(2x + 3) = 3(2x + 3) = 6x + 9$
          Domain: $\{x \mid x \text{ is any real number}\}$.
     (c) $(f \circ f)(x) = f(f(x)) = f(2x + 3) = 2(2x + 3) + 3 = 4x + 6 + 3 = 4x + 9$
          Domain: $\{x \mid x \text{ is any real number}\}$.
     (d) $(g \circ g)(x) = g(g(x)) = g(3x) = 3(3x) = 9x$
          Domain: $\{x \mid x \text{ is any real number}\}$.

29. $f(x) = 3x + 1 \qquad g(x) = x^2$
     The domain of f is $\{x \mid x \text{ is any real number}\}$. The domain of g is $\{x \mid x \text{ is any real number}\}$.
     (a) $(f \circ g)(x) = f(g(x)) = f(x^2) = 3x^2 + 1$
          Domain: $\{x \mid x \text{ is any real number}\}$.
     (b) $(g \circ f)(x) = g(f(x)) = g(3x + 1) = (3x + 1)^2 = 9x^2 + 6x + 1$
          Domain: $\{x \mid x \text{ is any real number}\}$.
     (c) $(f \circ f)(x) = f(f(x)) = f(3x + 1) = 3(3x + 1) + 1 = 9x + 3 + 1 = 9x + 4$
          Domain: $\{x \mid x \text{ is any real number}\}$.
     (d) $(g \circ g)(x) = g(g(x)) = g(x^2) = (x^2)^2 = x^4$
          Domain: $\{x \mid x \text{ is any real number}\}$.

31. $f(x) = x^2$ $g(x) = x^2 + 4$

The domain of f is $\{x \mid x \text{ is any real number}\}$. The domain of g is $\{x \mid x \text{ is any real number}\}$.

(a) $(f \circ g)(x) = f(g(x)) = f(x^2 + 4) = (x^2 + 4)^2 = x^4 + 8x^2 + 16$

Domain: $\{x \mid x \text{ is any real number}\}$.

(b) $(g \circ f)(x) = g(f(x)) = g(x^2) = (x^2)^2 + 4 = x^4 + 4$

Domain: $\{x \mid x \text{ is any real number}\}$.

(c) $(f \circ f)(x) = f(f(x)) = f(x^2) = (x^2)^2 = x^4$

Domain: $\{x \mid x \text{ is any real number}\}$.

(d) $(g \circ g)(x) = g(g(x)) = g(x^2 + 4) = (x^2 + 4)^2 + 4$

$$= x^4 + 8x^2 + 16 + 4 = x^4 + 8x^2 + 20$$

Domain: $\{x \mid x \text{ is any real number}\}$.

33. $f(x) = \dfrac{3}{x - 1}$ $g(x) = \dfrac{2}{x}$

The domain of f is $\{x \mid x \neq 1\}$. The domain of g is $\{x \mid x \neq 0\}$.

(a) $(f \circ g)(x) = f(g(x)) = f\left(\dfrac{2}{x}\right) = \dfrac{3}{\dfrac{2}{x} - 1} = \dfrac{3}{\dfrac{2 - x}{x}} = \dfrac{3x}{2 - x}$

Domain $\{x \mid x \neq 0, x \neq 2\}$.

(b) $(g \circ f)(x) = g(f(x)) = g\left(\dfrac{3}{x - 1}\right) = \dfrac{2}{\dfrac{3}{x - 1}} = \dfrac{2(x - 1)}{3}$

Domain $\{x \mid x \neq 1\}$

(c) $(f \circ f)(x) = f(f(x)) = f\left(\dfrac{3}{x - 1}\right) = \dfrac{3}{\dfrac{3}{x - 1} - 1} = \dfrac{3}{\dfrac{3 - (x - 1)}{x - 1}} = \dfrac{3(x - 1)}{4 - x}$

Domain $\{x \mid x \neq 1, x \neq 4\}$.

(d) $(g \circ g)(x) = g(g(x)) = g\left(\dfrac{2}{x}\right) = \dfrac{2}{\dfrac{2}{x}} = \dfrac{2x}{2} = x$

Domain $\{x \mid x \neq 0\}$.

35. $f(x) = \dfrac{x}{x - 1}$ $g(x) = -\dfrac{4}{x}$

The domain of f is $\{x \mid x \neq 1\}$. The domain of g is $\{x \mid x \neq 0\}$.

(a) $(f \circ g)(x) = f(g(x)) = f\left(-\dfrac{4}{x}\right) = \dfrac{-\dfrac{4}{x}}{-\dfrac{4}{x} - 1} = \dfrac{-\dfrac{4}{x}}{\dfrac{-4 - x}{x}} = \dfrac{-4}{-4 - x} = \dfrac{4}{4 + x}$

Domain $\{x \mid x \neq -4, x \neq 0\}$.

(b) $(g \circ f)(x) = g(f(x)) = g\left(\dfrac{x}{x-1}\right) = -\dfrac{4}{\dfrac{x}{x-1}} = \dfrac{-4(x-1)}{x}$

Domain $\{x \mid x \neq 0, x \neq 1\}$.

(c) $(f \circ f)(x) = f(f(x)) = f\left(\dfrac{x}{x-1}\right) = \dfrac{\dfrac{x}{x-1}}{\dfrac{x}{x-1}-1} = \dfrac{\dfrac{x}{x-1}}{\dfrac{x-(x-1)}{x-1}} = \dfrac{\dfrac{x}{x-1}}{\dfrac{1}{x-1}} = \dfrac{x}{x-1} \cdot \dfrac{x-1}{1} = \dfrac{x}{1} = x$

Domain $\{x \mid x \neq 1\}$.

(d) $(g \circ g)(x) = g(g(x)) = g\left(\dfrac{-4}{x}\right) = -\dfrac{4}{-\dfrac{4}{x}} = \dfrac{-4x}{-4} = x$

Domain $\{x \mid x \neq 0\}$.

37. $f(x) = \sqrt{x}$ $g(x) = 2x + 3$

The domain of f is $\{x \mid x \geq 0\}$. The domain of g is $\{x \mid x \text{ is any real number}\}$.

(a) $(f \circ g)(x) = f(g(x)) = f(2x+3) = \sqrt{2x+3}$

Domain $\left\{x \mid x \geq -\dfrac{3}{2}\right\}$.

(b) $(g \circ f)(x) = g(f(x)) = g\left(\sqrt{x}\right) = 2\sqrt{x} + 3$

Domain $\{x \mid x \geq 0\}$.

(c) $(f \circ f)(x) = f(f(x)) = f\left(\sqrt{x}\right) = \sqrt{\sqrt{x}} = x^{1/4} = \sqrt[4]{x}$

Domain $\{x \mid x \geq 0\}$.

(d) $(g \circ g)(x) = g(g(x)) = g(2x+3) = 2(2x+3) + 3 = 4x + 6 + 3 = 4x + 9$

Domain $\{x \mid x \text{ is any real number}\}$.

39. $f(x) = x^2 + 1$ $g(x) = \sqrt{x-1}$

The domain of f is $\{x \mid x \text{ is any real number}\}$. The domain of g is $\{x \mid x \geq 1\}$.

(a) $(f \circ g)(x) = f(g(x)) = f\left(\sqrt{x-1}\right) = \left(\sqrt{x-1}\right)^2 + 1 = x - 1 + 1 = x$

Domain $\{x \mid x \geq 1\}$.

(b) $(g \circ f)(x) = g(f(x)) = g\left(x^2 + 1\right) = \sqrt{x^2 + 1 - 1} = \sqrt{x^2} = |x|$

Domain $\{x \mid x \text{ is any real number}\}$.

(c) $(f \circ f)(x) = f(f(x)) = f\left(x^2 + 1\right) = \left(x^2 + 1\right)^2 + 1 = x^4 + 2x^2 + 1 + 1 = x^4 + 2x^2 + 2$

Domain $\{x \mid x \text{ is any real number}\}$.

(d) $(g \circ g)(x) = g(g(x)) = g\left(\sqrt{x-1}\right) = \sqrt{\sqrt{x-1} - 1}$

$\sqrt{x-1} - 1 \geq 0 \Rightarrow \sqrt{x-1} \geq 1 \Rightarrow x - 1 \geq 1 \Rightarrow x \geq 2$

Domain $\{x \mid x \geq 2\}$.

41. $f(x) = ax + b \qquad g(x) = cx + d$
     The domain of f is $\{x \mid x \text{ is any real number}\}$. The domain of g is $\{x \mid x \text{ is any real number}\}$.
     (a) $(f \circ g)(x) = f(g(x)) = f(cx + d) = a(cx + d) + b = acx + ad + b$
          Domain $\{x \mid x \text{ is any real number}\}$.
     (b) $(g \circ f)(x) = g(f(x)) = g(ax + b) = c(ax + b) + d = acx + bc + d$
          Domain $\{x \mid x \text{ is any real number}\}$.
     (c) $(f \circ f)(x) = f(f(x)) = f(ax + b) = a(ax + b) + b = a^2 x + ab + b$
          Domain $\{x \mid x \text{ is any real number}\}$.
     (d) $(g \circ g)(x) = g(g(x)) = g(cx + d) = c(cx + d) + d = c^2 x + cd + d$
          Domain $\{x \mid x \text{ is any real number}\}$.

43. $(f \circ g)(x) = f(g(x)) = f\left(\dfrac{1}{2}x\right) = 2\left(\dfrac{1}{2}x\right) = x$

     $(g \circ f)(x) = g(f(x)) = g(2x) = \dfrac{1}{2}(2x) = x$

45. $(f \circ g)(x) = f(g(x)) = f\left(\sqrt[3]{x}\right) = \left(\sqrt[3]{x}\right)^3 = x$

     $(g \circ f)(x) = g(f(x)) = g\left(x^3\right) = \sqrt[3]{x^3} = x$

47. $(f \circ g)(x) = f(g(x)) = f\left(\dfrac{1}{2}(x + 6)\right) = 2\left(\dfrac{1}{2}(x + 6)\right) - 6 = x + 6 - 6 = x$

     $(g \circ f)(x) = g(f(x)) = g(2x - 6) = \dfrac{1}{2}((2x - 6) + 6) = \dfrac{1}{2}(2x) = x$

49. $(f \circ g)(x) = f(g(x)) = f\left(\dfrac{1}{a}(x - b)\right) = a\left(\dfrac{1}{a}(x - b)\right) + b = x - b + b = x$

     $(g \circ f)(x) = g(f(x)) = g(ax + b) = \dfrac{1}{a}((ax + b) - b) = \dfrac{1}{a}(ax) = x$

51. $H(x) = (2x + 3)^4 \qquad\qquad f(x) = x^4, \quad g(x) = 2x + 3$

53. $H(x) = \sqrt{x^2 + 1} \qquad\qquad f(x) = \sqrt{x}, \quad g(x) = x^2 + 1$

55. $H(x) = |2x + 1| \qquad\qquad f(x) = |x|, \quad g(x) = 2x + 1$

57. $f(x) = 2x^3 - 3x^2 + 4x - 1 \qquad g(x) = 2$
     $(f \circ g)(x) = f(g(x)) = f(2) = 2(2)^3 - 3(2)^2 + 4(2) - 1 = 16 - 12 + 8 - 1 = 11$
     $(g \circ f)(x) = g(f(x)) = g\left(2x^3 - 3x^2 + 4x - 1\right) = 2$

59. $f(x) = 2x^2 + 5$ $g(x) = 3x + a$
$(f \circ g)(x) = f(g(x)) = f(3x + a) = 2(3x + a)^2 + 5$
When $x = 0$, $(f \circ g)(0) = 23$
Solving:
$$2(3 \cdot 0 + a)^2 + 5 = 23$$
$$2a^2 + 5 = 23$$
$$2a^2 = 18$$
$$a^2 = 9$$
$$a = -3 \text{ or } a = 3$$

61. $S(r) = 4\pi r^2$ $r(t) = \dfrac{2}{3}t^3, \ t \geq 0$

$$S(r(t)) = S\left(\frac{2}{3}t^3\right) = 4\pi\left(\frac{2}{3}t^3\right)^2 = 4\pi\left(\frac{4}{9}t^6\right) = \frac{16}{9}\pi t^6$$

63. $N(t) = 100t - 5t^2, \ 0 \leq t \leq 10$ $C(N) = 15{,}000 + 8000N$
$C(N(t)) = C\left(100t - 5t^2\right)$

$$= 15{,}000 + 8000\left(100t - 5t^2\right)$$
$$= 15{,}000 + 800{,}000t - 40{,}000t^2$$

65. $p = -\dfrac{1}{4}x + 100$ $0 \leq x \leq 400$

$$\frac{1}{4}x = 100 - p \Rightarrow x = 4(100 - p)$$

$$C = \frac{\sqrt{x}}{25} + 600$$

$$= \frac{\sqrt{4(100 - p)}}{25} + 600$$

$$= \frac{2\sqrt{100 - p}}{25} + 600, \quad 0 \leq p \leq 100$$

67. $V = \pi r^2 h$ $h = 2r \Rightarrow V(r) = \pi r^2(2r) = 2\pi r^3$

69. $f(x) =$ number of Euros bought for x dollars; $g(x) =$ number of yen bought for x Euros
   (a) $f(x) = 0.857118x$
   (b) $g(x) = 128.6054x$
   (c) $g(f(x)) = 128.6054(0.857118x) = 110.2300032372x$
   (d) $g(f(1000)) = 110.2300032372(1000) = 110{,}230.0032372$ yen

71.   Given that f is odd and g is even, we know that
$f(-x) = -f(x)$ and $g(-x) = g(x)$ for all x in the domain of f and g, respectively.
The composite function $f \circ g = f(g(x))$ has the following property:
$f(g(-x)) = f(g(x))$ since g is even, therefore, $f \circ g$ is even
The composite function $g \circ f = g(f(x))$ has the following property:
$g(f(-x)) = g(-f(x))$ since f is odd

$\qquad = g(f(x))$ since g is even, therefore $g \circ f$ is even

Chapter 4

Exponential and Logarithmic Functions

4.2   Inverse Functions

1. The set of ordered pairs is a function because there are no ordered pairs with the same first element and different second elements.

3. Increasing on the interval $(-\infty, \infty)$.

5. $y = x$

7. False

9. (a)

Domain	Range
$200	20 hours
$300	25 hours
$350	30 hours
$425	40 hours

(b) Inverse is a function since each element in the domain corresponds to one and only one element in the range.

11. (a)

Domain	Range
	20 hours
$200	25 hours
$350	30 hours
$425	40 hours

(b) Inverse is not a function since domain element $200 corresponds to two different elements in the range.

13. (a) $\{(6, 2), (6, -3), (9, 4), (10, 1)\}$

(b) Inverse is not a function since domain element 6 corresponds to two different elements in the range.

15. (a) $\{(0, 0), (1, 1), (16, 2), (81, 3)\}$

(b) Inverse is a function since each element in the domain corresponds to one and only one element in the range.

17. Every horizontal line intersects the graph of f at exactly one point. One-to-one.

19. There are horizontal lines that intersect the graph of f at more than one point. Not one-to-one.

21. Every horizontal line intersects the graph of f at exactly one point. One-to-one.

23. Graphing the inverse:

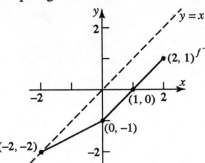

25. Graphing the inverse:

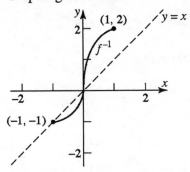

27. Graphing the inverse:

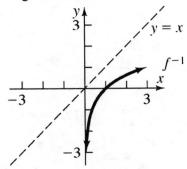

29. $f(x) = 3x + 4, \qquad g(x) = \dfrac{1}{3}(x - 4)$

$f(g(x)) = f\left(\dfrac{1}{3}(x - 4)\right)$

$\qquad = 3\left(\dfrac{1}{3}(x - 4)\right) + 4$

$\qquad = (x - 4) + 4 = x$

$g(f(x)) = g(3x + 4)$

$\qquad = \dfrac{1}{3}\big((3x + 4) - 4\big)$

$\qquad = \dfrac{1}{3}(3x) = x$

31. $f(x) = 4x - 8, \qquad g(x) = \dfrac{x}{4} + 2$

$f(g(x)) = f\left(\dfrac{x}{4} + 2\right)$

$\qquad = 4\left(\dfrac{x}{4} + 2\right) - 8$

$\qquad = (x + 8) - 8 = x$

$g(f(x)) = g(4x - 8)$

$\qquad = \dfrac{4x - 8}{4} + 2$

$\qquad = x - 2 + 2 = x$

33. $f(x) = x^3 - 8, \qquad g(x) = \sqrt[3]{x + 8}$

$f(g(x)) = f\left(\sqrt[3]{x + 8}\right)$

$\qquad = \left(\sqrt[3]{x + 8}\right)^3 - 8$

$\qquad = (x + 8) - 8 = x$

$g(f(x)) = g(x^3 - 8)$

$\qquad = \sqrt[3]{(x^3 - 8) + 8}$

$\qquad = \sqrt[3]{x^3} = x$

35. $f(x) = \dfrac{1}{x}, \qquad g(x) = \dfrac{1}{x}$

$f(g(x)) = f\left(\dfrac{1}{x}\right)$

$= \dfrac{1}{\dfrac{1}{x}} = x$

$g(f(x)) = g\left(\dfrac{1}{x}\right)$

$= \dfrac{1}{\dfrac{1}{x}} = x$

37. $f(x) = \dfrac{2x+3}{x+4}, \qquad g(x) = \dfrac{4x-3}{2-x}$

$f(g(x)) = f\left(\dfrac{4x-3}{2-x}\right)$

$= \dfrac{2\left(\dfrac{4x-3}{2-x}\right) + 3}{\dfrac{4x-3}{2-x} + 4}$

$= \dfrac{\dfrac{8x-6+6-3x}{2-x}}{\dfrac{4x-3+8-4x}{2-x}}$

$= \dfrac{\dfrac{5x}{2-x}}{\dfrac{5}{2-x}}$

$= \dfrac{5x}{2-x} \cdot \dfrac{2-x}{5} = x$

$g(f(x)) = g\left(\dfrac{2x+3}{x+4}\right)$

$= \dfrac{4\left(\dfrac{2x+3}{x+4}\right) - 3}{2 - \dfrac{2x+3}{x+4}}$

$= \dfrac{\dfrac{8x+12-3x-12}{x+4}}{\dfrac{2x+8-2x-3}{x+4}}$

$= \dfrac{\dfrac{5x}{x+4}}{\dfrac{5}{x+4}}$

$= \dfrac{5x}{x+4} \cdot \dfrac{x+4}{5} = x$

39. $f(x) = 3x$

$y = 3x$

$x = 3y \quad$ Inverse

$y = \dfrac{x}{3}$

$f^{-1}(x) = \dfrac{x}{3}$

Verify: $f\left(f^{-1}(x)\right) = f\left(\dfrac{x}{3}\right) = 3\left(\dfrac{x}{3}\right) = x$

$f^{-1}\left(f(x)\right) = f^{-1}(3x) = \dfrac{3x}{3} = x$

Domain of f = range of $f^{-1} = (-\infty, \infty)$

Range of f = domain of $f^{-1} = (-\infty, \infty)$

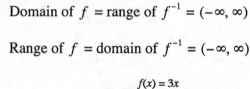

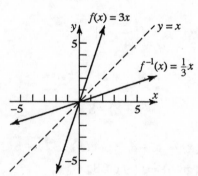

41. $f(x) = 4x + 2$

$y = 4x + 2$

$x = 4y + 2$ Inverse

$4y = x - 2$

$y = \dfrac{x-2}{4}$

$f^{-1}(x) = \dfrac{x-2}{4}$

Domain of f = range of $f^{-1} = (-\infty, \infty)$

Range of f = domain of $f^{-1} = (-\infty, \infty)$

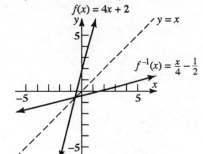

Verify: $f\left(f^{-1}(x)\right) = f\left(\dfrac{x-2}{4}\right) = 4\left(\dfrac{x-2}{4}\right) + 2 = x - 2 + 2 = x$

$f^{-1}\left(f(x)\right) = f^{-1}(4x + 2) = \dfrac{(4x+2)-2}{4} = \dfrac{4x}{4} = x$

43. $f(x) = x^3 - 1$

$y = x^3 - 1$

$x = y^3 - 1$ Inverse

$y^3 = x + 1$

$y = \sqrt[3]{x+1}$

$f^{-1}(x) = \sqrt[3]{x+1}$

Domain of f = range of $f^{-1} = (-\infty, \infty)$

Range of f = domain of $f^{-1} = (-\infty, \infty)$

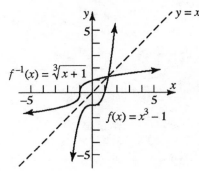

Verify: $f\left(f^{-1}(x)\right) = f\left(\sqrt[3]{x+1}\right) = \left(\sqrt[3]{x+1}\right)^3 - 1 = x + 1 - 1 = x$

$f^{-1}\left(f(x)\right) = f^{-1}\left(x^3 - 1\right) = \sqrt[3]{\left(x^3 - 1\right)+1} = \sqrt[3]{x^3} = x$

45. $f(x) = x^2 + 4, \quad x \geq 0$

$\qquad y = x^2 + 4 \quad x \geq 0$

$\qquad x = y^2 + 4 \quad y \geq 0 \quad$ Inverse

$\qquad y^2 = x - 4 \quad y \geq 0$

$\qquad y = \sqrt{x - 4}$

$f^{-1}(x) = \sqrt{x - 4}$

Verify:

$f\left(f^{-1}(x)\right) = f\left(\sqrt{x - 4}\right)$

$\qquad = \left(\sqrt{x - 4}\right)^2 + 4$

$\qquad = x - 4 + 4$

$\qquad = x$

$f^{-1}\left(f(x)\right) = f^{-1}\left(x^2 + 4\right)$

$\qquad = \sqrt{\left(x^2 + 4\right) - 4}$

$\qquad = \sqrt{x^2}$

$\qquad = |x|$

$\qquad = x, x \geq 0$

Domain of f = range of $f^{-1} = [0, \infty)$

Range of f = domain of $f^{-1} = [4, \infty)$

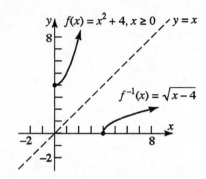

47. $f(x) = \dfrac{4}{x}$

$\qquad y = \dfrac{4}{x}$

$\qquad x = \dfrac{4}{y} \quad$ Inverse

$\qquad xy = 4$

$\qquad y = \dfrac{4}{x}$

$f^{-1}(x) = \dfrac{4}{x}$

Domain of f = range of f^{-1}
$\qquad\qquad$ = all real numbers except 0

Range of f = domain of f^{-1}
$\qquad\qquad$ = all real numbers except 0

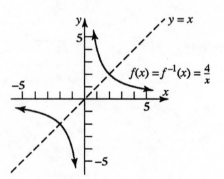

Verify: $f\left(f^{-1}(x)\right) = f\left(\dfrac{4}{x}\right) = \dfrac{4}{\dfrac{4}{x}} = 4 \cdot \left(\dfrac{x}{4}\right) = x$

$\qquad\qquad f^{-1}\left(f(x)\right) = f^{-1}\left(\dfrac{4}{x}\right) = \dfrac{4}{\dfrac{4}{x}} = 4 \cdot \left(\dfrac{x}{4}\right) = x$

49. $f(x) = \dfrac{1}{x-2}$

$y = \dfrac{1}{x-2}$

$x = \dfrac{1}{y-2}$ Inverse

$x(y-2) = 1$

$xy - 2x = 1$

$xy = 2x + 1$

$y = \dfrac{2x+1}{x}$

$f^{-1}(x) = \dfrac{2x+1}{x}$

Domain of f = range of f^{-1}
    = all real numbers except 2

Range of f = domain of f^{-1}
    = all real numbers except 0

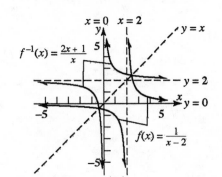

Verify: $f\left(f^{-1}(x)\right) = f\left(\dfrac{2x+1}{x}\right) = \dfrac{1}{\dfrac{2x+1}{x}-2} = \dfrac{1}{\dfrac{2x+1-2x}{x}} = \dfrac{1}{\dfrac{1}{x}} = 1 \cdot \left(\dfrac{x}{1}\right) = x$

$f^{-1}(f(x)) = f^{-1}\left(\dfrac{1}{x-2}\right) = \dfrac{2\left(\dfrac{1}{x-2}\right)+1}{\dfrac{1}{x-2}} = \dfrac{\dfrac{2+x-2}{x-2}}{\dfrac{1}{x-2}} = \left(\dfrac{x}{x-2}\right) \cdot \left(\dfrac{x-2}{1}\right) = x$

51. $f(x) = \dfrac{2}{3+x}$

$y = \dfrac{2}{3+x}$

$x = \dfrac{2}{3+y}$ Inverse

$x(3+y) = 2$

$3x + xy = 2$

$xy = 2 - 3x$

$y = \dfrac{2-3x}{x}$

$f^{-1}(x) = \dfrac{2-3x}{x}$

Domain of f = range of f^{-1}
    = all real numbers except -3

Range of f = domain of f^{-1}
    = all real numbers except 0

Verify: $f\left(f^{-1}(x)\right) = f\left(\dfrac{2-3x}{x}\right) = \dfrac{2}{3+\dfrac{2-3x}{x}} = \dfrac{2}{\dfrac{3x+2-3x}{x}} = \dfrac{2}{\dfrac{2}{x}} = 2 \cdot \left(\dfrac{x}{2}\right) = x$

$f^{-1}(f(x)) = f^{-1}\left(\dfrac{2}{3+x}\right) = \dfrac{2-3\left(\dfrac{2}{3+x}\right)}{\dfrac{2}{3+x}} = \dfrac{\dfrac{6+2x-6}{3+x}}{\dfrac{2}{3+x}} = \left(\dfrac{2x}{3+x}\right) \cdot \left(\dfrac{3+x}{2}\right) = x$

53. $f(x) = \dfrac{3x}{x+2}$

$y = \dfrac{3x}{x+2}$

$x = \dfrac{3y}{y+2}$ Inverse

$x(y+2) = 3y$

$xy + 2x = 3y$

$xy - 3y = -2x$

$y(x-3) = -2x$

$y = \dfrac{-2x}{x-3}$

$f^{-1}(x) = \dfrac{-2x}{x-3}$

Domain of f = range of f^{-1}
  = all real numbers except –2
Range of f = domain of f^{-1}
  = all real numbers except 3

Verify:

$$f\left(f^{-1}(x)\right) = f\left(\dfrac{-2x}{x-3}\right) = \dfrac{3\left(\dfrac{-2x}{x-3}\right)}{\dfrac{-2x}{x-3}+2}$$

$$= \dfrac{\dfrac{-6x}{x-3}}{\dfrac{-2x+2x-6}{x-3}} = \left(\dfrac{-6x}{x-3}\right)\cdot\left(\dfrac{x-3}{-6}\right) = x$$

$$f^{-1}\left(f(x)\right) = f^{-1}\left(\dfrac{3x}{x+2}\right) = \dfrac{-2\left(\dfrac{3x}{x+2}\right)}{\dfrac{3x}{x+2}-3}$$

$$= \dfrac{\dfrac{-6x}{x+2}}{\dfrac{3x-3x-6}{x+2}} = \left(\dfrac{-6x}{x+2}\right)\cdot\left(\dfrac{x+2}{-6}\right) = x$$

55. $f(x) = \dfrac{2x}{3x-1}$

$y = \dfrac{2x}{3x-1}$

$x = \dfrac{2y}{3y-1}$ Inverse

$x(3y-1) = 2y$

$3xy - x = 2y$

$3xy - 2y = x$

$y(3x-2) = x$

$y = \dfrac{x}{3x-2}$

$f^{-1}(x) = \dfrac{x}{3x-2}$

Domain of f = range of f^{-1} = all real numbers except $\dfrac{1}{3}$

Range of f = domain of f^{-1} = all real numbers except $\dfrac{2}{3}$

Verify:

$$f\left(f^{-1}(x)\right) = f\left(\dfrac{x}{3x-2}\right) = \dfrac{2\left(\dfrac{x}{3x-2}\right)}{3\left(\dfrac{x}{3x-2}\right)-1}$$

$$= \dfrac{\dfrac{2x}{3x-2}}{\dfrac{3x-3x+2}{3x-2}} = \dfrac{\dfrac{2x}{3x-2}}{\dfrac{2}{3x-2}} = \left(\dfrac{2x}{3x-2}\right)\cdot\left(\dfrac{3x-2}{2}\right) = x$$

$$f^{-1}\left(f(x)\right) = f\left(\dfrac{2x}{3x-1}\right) = \dfrac{\dfrac{2x}{3x-1}}{3\left(\dfrac{2x}{3x-1}\right)-2} = \dfrac{\dfrac{2x}{3x-1}}{\dfrac{6x-6x+2}{3x-1}} = \dfrac{\dfrac{2x}{3x-1}}{\dfrac{2}{3x-1}} = \left(\dfrac{2x}{3x-1}\right)\cdot\left(\dfrac{3x-1}{2}\right) = x$$

57. $f(x) = \dfrac{3x+4}{2x-3}$

$y = \dfrac{3x+4}{2x-3}$

$x = \dfrac{3y+4}{2y-3}$ Inverse

$x(2y-3) = 3y+4$

$2xy - 3x = 3y+4$

$2xy - 3y = 3x+4$

$y(2x-3) = 3x+4$

$y = \dfrac{3x+4}{2x-3}$

$f^{-1}(x) = \dfrac{3x+4}{2x-3}$

Domain of f = range of f^{-1}

= all real numbers except $\dfrac{3}{2}$

Range of f = domain of f^{-1}

= all real numbers except $\dfrac{3}{2}$

Verify: $f\left(f^{-1}(x)\right) = f\left(\dfrac{3x+4}{2x-3}\right) = \dfrac{3\left(\dfrac{3x+4}{2x-3}\right)+4}{2\left(\dfrac{3x+4}{2x-3}\right)-3} = \dfrac{\dfrac{9x+12+8x-12}{2x-3}}{\dfrac{6x+8-6x+9}{2x-3}} = \dfrac{\dfrac{17x}{2x-3}}{\dfrac{17}{2x-3}}$

$= \dfrac{17x}{2x-3} \cdot \dfrac{2x-3}{17} = x$

$f^{-1}\left(f(x)\right) = f^{-1}\left(\dfrac{3x+4}{2x-3}\right) = \dfrac{3\left(\dfrac{3x+4}{2x-3}\right)+4}{2\left(\dfrac{3x+4}{2x-3}\right)-3} = \dfrac{\dfrac{9x+12+8x-12}{2x-3}}{\dfrac{6x+8-6x+9}{2x-3}} = \dfrac{\dfrac{17x}{2x-3}}{\dfrac{17}{2x-3}}$

$= \dfrac{17x}{2x-3} \cdot \dfrac{2x-3}{17} = x$

59. $f(x) = \dfrac{2x+3}{x+2}$

$y = \dfrac{2x+3}{x+2}$

$x = \dfrac{2y+3}{y+2}$ Inverse

$x(y+2) = 2y+3$

$xy + 2x = 2y+3$

$xy - 2y = -2x+3$

$y(x-2) = -2x+3$

$y = \dfrac{-2x+3}{x-2}$

$f^{-1}(x) = \dfrac{-2x+3}{x-2}$

Domain of f = range of f^{-1}

= all real numbers except -2

Range of f = domain of f^{-1}

= all real numbers except 2

Verify: $f\left(f^{-1}(x)\right) = f\left(\dfrac{-2x+3}{x-2}\right) = \dfrac{2\left(\dfrac{-2x+3}{x-2}\right)+3}{\dfrac{-2x+3}{x-2}+2} = \dfrac{\dfrac{-4x+6+3x-6}{x-2}}{\dfrac{-2x+3+2x-4}{x-2}} = \dfrac{\dfrac{-x}{x-2}}{\dfrac{-1}{x-2}}$

$$= \left(\dfrac{-x}{x-2}\right)\cdot\left(\dfrac{x-2}{-1}\right) = x$$

$f^{-1}\left(f(x)\right) = f^{-1}\left(\dfrac{2x+3}{x+2}\right) = \dfrac{-2\left(\dfrac{2x+3}{x+2}\right)+3}{\dfrac{2x+3}{x+2}-2} = \dfrac{\dfrac{-4x-6+3x+6}{x+2}}{\dfrac{2x+3-2x-4}{x+2}} = \dfrac{\dfrac{-x}{x+2}}{\dfrac{-1}{x+2}}$

$$= \left(\dfrac{-x}{x+2}\right)\cdot\left(\dfrac{x+2}{-1}\right) = x$$

61. $f(x) = \dfrac{x^2-4}{2x^2},\ x>0$

$y = \dfrac{x^2-4}{2x^2},\ x>0$

$x = \dfrac{y^2-4}{2y^2},\ y>0$ Inverse

$2xy^2 = y^2-4,\quad x<\dfrac{1}{2}$

$2xy^2-y^2 = -4,\qquad x<\dfrac{1}{2}$

$y^2(2x-1) = -4,\qquad x<\dfrac{1}{2}$

$y^2 = \dfrac{-4}{2x-1},\qquad x<\dfrac{1}{2}$

$y = \sqrt{\dfrac{-4}{2x-1}},\qquad x<\dfrac{1}{2}$

$f^{-1}(x) = \sqrt{\dfrac{-4}{2x-1}},\qquad x<\dfrac{1}{2}$

Domain of f = range of $f^{-1} = (0,\infty)$

Range of f = domain of $f^{-1} = \left(-\infty,\dfrac{1}{2}\right)$

Verify:

$f\left(f^{-1}(x)\right) = f\left(\sqrt{\dfrac{-4}{2x-1}}\right) = \dfrac{\left(\sqrt{\dfrac{-4}{2x-1}}\right)^2-4}{2\left(\sqrt{\dfrac{-4}{2x-1}}\right)^2}$

$= \dfrac{\dfrac{-4}{2x-1}-4}{2\left(\dfrac{-4}{2x-1}\right)} = \dfrac{\dfrac{-4-8x+4}{2x-1}}{\dfrac{-8}{2x-1}}$

$= \left(\dfrac{-8x}{2x-1}\right)\cdot\left(\dfrac{2x-1}{-8}\right) = x$

$f^{-1}\left(f(x)\right) = f^{-1}\left(\dfrac{x^2-4}{2x^2}\right) = \sqrt{\dfrac{-4}{2\left(\dfrac{x^2-4}{2x^2}\right)-1}} = \sqrt{\dfrac{-4}{\dfrac{x^2-4}{x^2}-1}}$

$= \sqrt{\dfrac{-4}{\dfrac{x^2-4-x^2}{x^2}}} = \sqrt{\dfrac{-4}{\dfrac{-4}{x^2}}} = \sqrt{(-4)\left(\dfrac{x^2}{-4}\right)}$

$= \sqrt{x^2} = x,\ x>0$

63.   (a) $f(-1)=0$ (b) $f(1)=2$ (c) $f^{-1}(1)=0$ (d) $f^{-1}(2)=1$

65. $f(x) = mx + b, \quad m \neq 0$

$y = mx + b$

$x = my + b \quad$ Inverse

$x - b = my$

$y = \dfrac{x - b}{m}$

$f^{-1}(x) = \dfrac{x - b}{m}, \quad m \neq 0$

67. The graph of f^{-1} lies in quadrant I. Whenever (a,b) is on the graph of f, then (b,a) is on the graph of f^{-1}. Since both coordinates of (a, b) are positive, both coordinates of (b, a) are positive and it is in quadrant I.

69. $f(x) = |x|, x \geq 0$ is one-to-one. Thus, $f(x) = x, x \geq 0$ and $f^{-1}(x) = x, x \geq 0$.

71. (a) $H(C) = 2.15C - 10.53$

$H = 2.15C - 10.53$

$H + 10.53 = 2.15C$

$\dfrac{H + 10.53}{2.15} = C$

$\dfrac{H}{2.15} + \dfrac{10.53}{2.15} = C$

$\dfrac{20}{43}H + \dfrac{1053}{215} = C$

$C(H) = \dfrac{20}{43}H + \dfrac{1053}{215}$

(b) $C(26) = \dfrac{20}{43}(26) + \dfrac{1053}{215} \approx 16.99$ inches

73. $p(x) = 300 - 50x, \quad x \geq 0$

$p = 300 - 50x$

$50x = 300 - p$

$x = \dfrac{300 - p}{50}$

$x(p) = \dfrac{300 - p}{50}, \quad p \leq 300$

75. $f(x) = \dfrac{ax + b}{cx + d}$

$y = \dfrac{ax + b}{cx + d}$

$x = \dfrac{ay + b}{cy + d} \quad$ Inverse

$x(cy + d) = ay + b$

$cxy + dx = ay + b$

$cxy - ay = b - dx$

$y(cx - a) = b - dx$

$y = \dfrac{b - dx}{cx - a}$

$f^{-1}(x) = \dfrac{-dx + b}{cx - a}$

Therefore, $f = f^{-1}$ provided $\dfrac{ax + b}{cx + d} = \dfrac{-dx + b}{cx - a}$, this is true if $a = -d$.

77.   Answers will vary.

79.   The only way the function $y = f(x)$ can be both even and one-to-one is if the domain of $y = f(x)$ is $\{x | x = 0\}$. Otherwise, its graph will fail the Horizontal Line Test.

81.   If the graph of a function and its inverse intersect, they must intersect at a point on the line $y = x$. However, the graphs do not have to intersect.

Exponential and Logarithmic Functions

4.3 Exponential Functions

1. $64; \left(\sqrt[3]{8}\right)^2 = 2^2 = 4; \dfrac{1}{3^2} = \dfrac{1}{9}$

3. $\dfrac{f(x) - f(c)}{x - c} = \dfrac{3x - 5 - (3c - 5)}{x - c}$

   $= \dfrac{3x - 5 - 3c + 5}{x - c}$

   $= \dfrac{3x - 3c}{x - c}$

   $= \dfrac{3(x - c)}{x - c}$

   $= 3$

5. False

7. 1

9. False

11. (a) $3^{2.2} \approx 11.212$
    (b) $3^{2.23} \approx 11.587$
    (c) $3^{2.236} \approx 11.664$
    (d) $3^{\sqrt{5}} \approx 11.665$

13. (a) $2^{3.14} \approx 8.815$
    (b) $2^{3.141} \approx 8.821$
    (c) $2^{3.1415} \approx 8.824$
    (d) $2^{\pi} \approx 8.825$

15. (a) $3.1^{2.7} \approx 21.217$
    (b) $3.14^{2.71} \approx 22.217$
    (c) $3.141^{2.718} \approx 22.440$
    (d) $\pi^{e} \approx 22.459$

17. $e^{1.2} \approx 3.320$

19. $e^{-0.85} \approx 0.427$

21.

x	$y = f(x)$	$\dfrac{f(x+1)}{f(x)}$
-1	3	$\dfrac{6}{3} = 2$
0	6	$\dfrac{12}{6} = 2$
1	12	$\dfrac{18}{12} = \dfrac{3}{2}$
2	18	
3	30	

Not an exponential function since the ratio of consecutive terms is not constant.

23.

x	$y = H(x)$	$\dfrac{H(x+1)}{H(x)}$
-1	$\dfrac{1}{4}$	$\dfrac{1}{(1/4)} = 4$
0	1	$\dfrac{4}{1} = 4$
1	4	$\dfrac{16}{4} = 4$
2	16	$\dfrac{64}{16} = 4$
3	64	

Yes, an exponential function since the ratio of consecutive terms is constant with $a = 4$. So the base is 4.

25.

x	$y = f(x)$	$\dfrac{f(x+1)}{f(x)}$
-1	$\dfrac{3}{2}$	$\dfrac{3}{(3/2)} = 2$
0	3	$\dfrac{6}{3} = 2$
1	6	$\dfrac{12}{6} = 2$
2	12	$\dfrac{24}{12} = 2$
3	24	

Yes, an exponential function since the ratio of consecutive terms is constant with $a = 2$. So the base is 2.

27.

x	$y = H(x)$	$\dfrac{H(x+1)}{H(x)}$
-1	2	$\dfrac{4}{2} = 2$
0	4	$\dfrac{6}{4} = \dfrac{3}{2}$
1	6	
2	8	
3	10	

Not an exponential function since the ratio of consecutive terms is not constant.

29. *B*   31. *D*   33. *A*   35. *E*

37. $f(x) = 2^x + 1$
Using the graph of $y = 2^x$, shift the graph
up 1 unit.
Domain: $(-\infty, \infty)$
Range: $(1, \infty)$
Horizontal Asymptote: $y = 1$

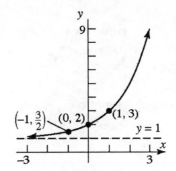

39. $f(x) = 3^{-x} - 2$
Using the graph of $y = 3^x$, reflect the
graph about the y-axis, and shift down 2
units.
Domain: $(-\infty, \infty)$
Range: $(-2, \infty)$
Horizontal Asymptote: $y = -2$

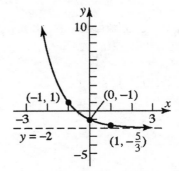

41. $f(x) = 2 + 3\left(4^x\right)$
Using the graph of $y = 4^x$, stretch the
graph vertically by a factor of 3, and shift
up 2 units.
Domain: $(-\infty, \infty)$
Range: $(2, \infty)$
Horizontal Asymptote: $y = 2$

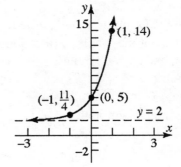

43. $f(x) = 2 + 3^{x/2}$
Using the graph of $y = 3^x$, stretch the
graph horizontally by a factor of 2, and
shift up 2 units.
Domain: $(-\infty, \infty)$
Range: $(2, \infty)$
Horizontal Asymptote: $y = 2$

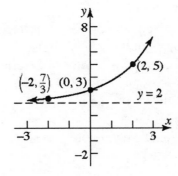

45. $f(x) = e^{-x}$
Using the graph of $y = e^x$, reflect the
graph about the y-axis.
Domain: $(-\infty, \infty)$
Range: $(0, \infty)$
Horizontal Asymptote: $y = 0$

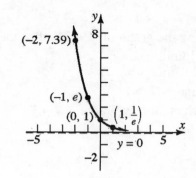

47. $f(x) = e^{x+2}$
Using the graph of $y = e^x$, shift the graph
2 units to the left.
Domain: $(-\infty, \infty)$
Range: $(0, \infty)$
Horizontal Asymptote: $y = 0$

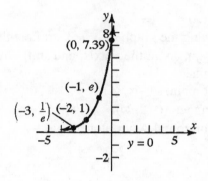

49. $f(x) = 5 - e^{-x}$
Using the graph of $y = e^x$, reflect the
graph about the y-axis, reflect about the
x-axis, and shift up 5 units.
Domain: $(-\infty, \infty)$
Range: $(-\infty, 5)$
Horizontal Asymptote: $y = 5$

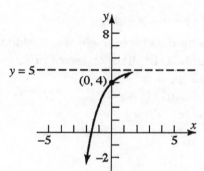

51. $f(x) = 2 - e^{-x/2}$
Using the graph of $y = e^x$, reflect the
graph about the y-axis, stretch
horizontally by a factor of 2, reflect about
the x-axis, and shift up 2 units.
Domain: $(-\infty, \infty)$
Range: $(-\infty, 2)$
Horizontal Asymptote: $y = 2$

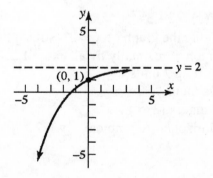

53.
$$2^{2x+1} = 4$$
$$2^{2x+1} = 2^2$$
$$2x + 1 = 2$$
$$2x = 1$$
$$x = \frac{1}{2}$$

The solution set is $\left\{\frac{1}{2}\right\}$.

55.
$$3^{x^3} = 9^x$$
$$3^{x^3} = \left(3^2\right)^x$$
$$3^{x^3} = 3^{2x}$$
$$x^3 = 2x$$
$$x^3 - 2x = 0$$
$$x\left(x^2 - 2\right) = 0$$
$$x = 0$$

or

$$x^2 - 2 = 0 \Rightarrow x^2 = 2 \Rightarrow x = \pm\sqrt{2}$$

The solution set is $\left\{-\sqrt{2},\ 0,\ \sqrt{2}\right\}$.

57.
$$8^{x^2-2x} = \frac{1}{2}$$
$$\left(2^3\right)^{x^2-2x} = 2^{-1}$$
$$2^{3x^2-6x} = 2^{-1}$$
$$3x^2 - 6x = -1$$
$$3x^2 - 6x + 1 = 0$$

$$x = \frac{-(-6) \pm \sqrt{(-6)^2 - 4(3)(1)}}{2(3)}$$

$$= \frac{6 \pm \sqrt{24}}{6}$$

$$= \frac{6 \pm 2\sqrt{6}}{6} = \frac{3 \pm \sqrt{6}}{3}$$

The solution set is $\left\{1 - \dfrac{\sqrt{6}}{3},\ 1 + \dfrac{\sqrt{6}}{3}\right\}$.

59.
$$2^x \cdot 8^{-x} = 4^x$$
$$2^x \cdot \left(2^3\right)^{-x} = \left(2^2\right)^x$$
$$2^x \cdot 2^{-3x} = 2^{2x}$$
$$2^{-2x} = 2^{2x}$$
$$-2x = 2x$$
$$-4x = 0 \Rightarrow x = 0$$

The solution set is $\{0\}$.

61.
$$\left(\frac{1}{5}\right)^{2-x} = 25$$
$$\left(5^{-1}\right)^{2-x} = 5^2$$
$$5^{x-2} = 5^2$$
$$x - 2 = 2 \Rightarrow x = 4$$

The solution set is $\{4\}$.

63.
$$4^x = 8$$
$$\left(2^2\right)^x = 2^3$$
$$2^{2x} = 2^3$$
$$2x = 3 \Rightarrow x = \frac{3}{2}$$

The solution set is $\left\{\dfrac{3}{2}\right\}$.

65.
$$e^{x^2} = e^{3x} \cdot \frac{1}{e^2}$$
$$e^{x^2} = e^{3x-2}$$
$$x^2 = 3x - 2$$
$$x^2 - 3x + 2 = 0$$
$$(x-1)(x-2) = 0 \Rightarrow x = 1 \text{ or } x = 2$$
The solution set is $\{1, 2\}$.

67. Given $4^x = 7$,
$$4^{-2x} = \left(4^x\right)^{-2}$$
$$= 7^{-2}$$
$$= \frac{1}{7^2}$$
$$= \frac{1}{49}$$

69. Given $3^{-x} = 2$,
$$3^{2x} = \left(3^{-x}\right)^{-2}$$
$$= 2^{-2}$$
$$= \frac{1}{2^2}$$
$$= \frac{1}{4}$$

71. We need a function of the form $f(x) = k \cdot a^{p \cdot x}$, with $a > 0$, $a \neq 1$.

The graph contains the points $\left(-1, \frac{1}{3}\right)$, $(0, 1)$, $(1, 3)$ and $(2, 9)$.

In other words, $f(-1) = \frac{1}{3}$, $f(0) = 1$, $f(1) = 3$ and $f(2) = 9$.

Therefore, $f(0) = k \cdot a^{p \cdot (0)} = k \cdot a^0 = k \cdot 1 = k \Rightarrow k = 1$.

and $f(1) = a^{p \cdot (1)} = a^p \Rightarrow a^p = 3$.

Let's use $a = 3$, $p = 1$. Then $f(x) = 3^x$.

Now we need to verify that this function yields the other known points on the graph.
$$f(-1) = 3^{-1} = \frac{1}{3}; \qquad f(2) = 3^2 = 9$$

So we have the function $f(x) = 3^x$.

73. We need a function of the form $f(x) = k \cdot a^{p \cdot x}$, with $a > 0$, $a \neq 1$.

The graph contains the points $\left(-1, -\frac{1}{6}\right)$, $(0, -1)$, $(1, -6)$ and $(2, -36)$.

In other words, $f(-1) = -\frac{1}{6}$, $f(0) = -1$, $f(1) = -6$ and $f(2) = -36$.

Therefore, $f(0) = k \cdot a^{p \cdot (0)} = k \cdot a^0 = k \cdot 1 = k \Rightarrow k = -1$.

and $f(1) = -a^{p \cdot (1)} = -a^p \Rightarrow -a^p = -6 \Rightarrow a^p = 6$.

Let's use $a = 6$, $p = 1$. Then $f(x) = -6^x$.

Now we need to verify that this function yields the other known points on the graph.
$$f(-1) = -6^{-1} = -\frac{1}{6}; \qquad f(2) = -6^2 = -36$$

So we have the function $f(x) = -6^x$.

75. $p = 100e^{-0.03n}$

  (a) $p = 100e^{-0.03(10)} = 100e^{-0.3} \approx 100(0.741) = 74.1\%$ of light

  (b) $p = 100e^{-0.03(25)} = 100e^{-0.75} \approx 100(0.472) = 47.2\%$ of light

77. $w(d) = 50e^{-0.004d}$

  (a) $w(30) = 50e^{-0.004(30)} = 50e^{-0.12} \approx 50(0.887) = 44.35$ watts

  (b) $w(365) = 50e^{-0.004(365)} = 50e^{-1.46} \approx 50(0.232) = 11.61$ watts

79. $D(h) = 5e^{-0.4h}$

  $D(1) = 5e^{-0.4(1)} = 5e^{-0.4} \approx 5(0.670) = 3.35$ milligrams

  $D(6) = 5e^{-0.4(6)} = 5e^{-2.4} \approx 5(0.091) = 0.45$ milligram

81. $F(t) = 1 - e^{-0.1t}$

  (a) $F(10) = 1 - e^{-0.1(10)} = 1 - e^{-1} \approx 1 - 0.368 = 0.632 = 63.2\%$

   The probability that a car will arrive within 10 minutes of 12:00 PM is 63.2%.

  (b) $F(40) = 1 - e^{-0.1(40)} = 1 - e^{-4} \approx 1 - 0.018 = 0.982 = 98.2\%$

   The probability that a car will arrive within 40 minutes of 12:00 PM is 98.2%

  (c) As $t \to \infty$, $F(t) = 1 - e^{-0.1t} \to 1 - 0 = 1$

  (d) Graphing the function:

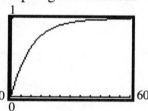

  (e) $F(7) \approx 50$, so 7 minutes are needed for the probability to reach 50%.

83. $P(x) = \dfrac{20^x e^{-20}}{x!}$

  (a) $P(15) = \dfrac{20^{15} e^{-20}}{15!} \approx 0.0516 = 5.16\%$

   The probability that 15 cars will arrive between 5:00 PM and 6:00 PM is 5.16%.

  (b) $P(20) = \dfrac{20^{20} e^{-20}}{20!} \approx 0.0888 = 8.88\%$

   The probability that 20 cars will arrive between 5:00 PM and 6:00 PM is 8.88%.

85. $p(x) = 16{,}630(0.90)^x$

  (a) $p(3) = 16{,}630(0.90)^3 \approx 16{,}630(0.729) \approx \$12{,}123$

   A 3-year-old Civic DX Sedan costs $12,123.

  (b) $p(9) = 16{,}630(0.90)^9 \approx 16{,}630(0.387) = \$6442.80$

   A 9-year-old Civic DX Sedan costs $6443.

87. $I = \dfrac{E}{R}\left[1 - e^{-\left(\frac{R}{L}\right)t}\right]$

(a) $I_1 = \dfrac{120}{10}\left[1 - e^{-\left(\frac{10}{5}\right)0.3}\right] = 12\left[1 - e^{-0.6}\right] \approx 5.414$ amperes after 0.3 second

$I_1 = \dfrac{120}{10}\left[1 - e^{-\left(\frac{10}{5}\right)0.5}\right] = 12\left[1 - e^{-1}\right] \approx 7.585$ amperes after 0.5 second

$I_1 = \dfrac{120}{10}\left[1 - e^{-\left(\frac{10}{5}\right)1}\right] = 12\left[1 - e^{-2}\right] \approx 10.376$ amperes after 1 second

(b) As $t \to \infty$, $e^{-\left(\frac{10}{5}\right)t} \to 0$. Therefore, the maximum current is 12 amperes.

(c), (f) Graphing the function:

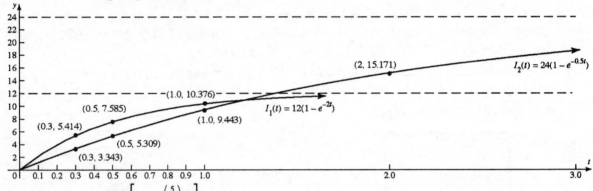

(d) $I_2 = \dfrac{120}{5}\left[1 - e^{-\left(\frac{5}{10}\right)0.3}\right] = 24\left[1 - e^{-0.15}\right] \approx 3.343$ amperes after 0.3 second

$I_2 = \dfrac{120}{5}\left[1 - e^{-\left(\frac{5}{10}\right)0.5}\right] = 24\left[1 - e^{-0.25}\right] \approx 5.309$ amperes after 0.5 second

$I_2 = \dfrac{120}{5}\left[1 - e^{-\left(\frac{5}{10}\right)1}\right] = 24\left[1 - e^{-0.5}\right] \approx 9.443$ amperes after 1 second

(e) As $t \to \infty$, $e^{-\left(\frac{5}{10}\right)t} \to 0$. Therefore, the maximum current is 24 amperes.

89. $2 + \dfrac{1}{2!} + \dfrac{1}{3!} + \dfrac{1}{4!} + \ldots + \dfrac{1}{n!}$

$n = 4; \quad 2 + \dfrac{1}{2!} + \dfrac{1}{3!} + \dfrac{1}{4!} \approx 2.7083$

$n = 6; \quad 2 + \dfrac{1}{2!} + \dfrac{1}{3!} + \dfrac{1}{4!} + \dfrac{1}{5!} + \dfrac{1}{6!} \approx 2.7181$

$n = 8; \quad 2 + \dfrac{1}{2!} + \dfrac{1}{3!} + \dfrac{1}{4!} + \dfrac{1}{5!} + \dfrac{1}{6!} + \dfrac{1}{7!} + \dfrac{1}{8!} \approx 2.7182788$

$n = 10; \quad 2 + \dfrac{1}{2!} + \dfrac{1}{3!} + \dfrac{1}{4!} + \dfrac{1}{5!} + \dfrac{1}{6!} + \dfrac{1}{7!} + \dfrac{1}{8!} + \dfrac{1}{9!} + \dfrac{1}{10!} \approx 2.7182818$

$e \approx 2.718281828$

91. $f(x) = a^x$

$$\frac{f(x+h) - f(x)}{h} = \frac{a^{x+h} - a^x}{h} = \frac{a^x a^h - a^x}{h} = \frac{a^x(a^h - 1)}{h} = a^x\left(\frac{a^h - 1}{h}\right)$$

93. $f(x) = a^x$

$$f(-x) = a^{-x} = \frac{1}{a^x} = \frac{1}{f(x)}$$

95. $R = 10^{\left(\frac{4221}{T+459.4} - \frac{4221}{D+459.4} + 2\right)}$

(a) $R = 10^{\left(\frac{4221}{50+459.4} - \frac{4221}{41+459.4} + 2\right)} \approx 70.95\%$

(b) $R = 10^{\left(\frac{4221}{68+459.4} - \frac{4221}{59+459.4} + 2\right)} \approx 72.62\%$

(c) $R = 10^{\left(\frac{4221}{T+459.4} - \frac{4221}{T+459.4} + 2\right)} = 10^2 = 100\%$

97. $\sinh x = \frac{1}{2}\left(e^x - e^{-x}\right)$

(a) $f(-x) = \sinh(-x) = \frac{1}{2}\left(e^{-x} - e^x\right) = -\frac{1}{2}\left(e^x - e^{-x}\right) = -\sinh x = -f(x)$
       Therefore, $f(x) = \sinh x$ is an odd function.

(b)   Graphing:

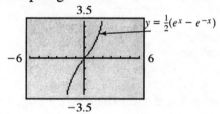

99. Since the number of bacteria doubles every minute, half of the container is full one minute before it is full. Thus, it takes 59 minutes to fill the container.

101. Answers will vary.

103. Using the laws of exponents, we have: $a^{-x} = \frac{1}{a^x} = \left(\frac{1}{a}\right)^x$. So $y = a^{-x}$ and $y = \left(\frac{1}{a}\right)^x$ will have the same graph.

Chapter 4

Exponential and Logarithmic Functions

4.4 Logarithmic Functions

1. $3x - 7 \le 8 - 2x$

   $5x \le 15$

   $x \le 3$

   The solution set is $\{x \mid x \le 3\}$.

3. $\dfrac{x-1}{x+4} > 0$ $f(x) = \dfrac{x-1}{x+4}$

   f is zero or undefined when $x = 1$ or $x = -4$.

Interval	$(-\infty, -4)$	$(-4, 1)$	$(1, \infty)$
Number Chosen	-5	0	2
Value of f	6	-0.25	$\dfrac{1}{6}$
Conclusion	Positive	Negative	Positive

   The solution set is $\{x \mid x < -4 \text{ or } x > 1\}$.

5. $(1, 0)$, $(a, 1)$, $\left(\dfrac{1}{a}, -1\right)$

7. False

9. $9 = 3^2$ is equivalent to $2 = \log_3 9$

11. $a^2 = 1.6$ is equivalent to $2 = \log_a 1.6$

13. $1.1^2 = M$ is equivalent to $2 = \log_{1.1} M$

15. $2^x = 7.2$ is equivalent to $x = \log_2 7.2$

17. $x^{\sqrt{2}} = \pi$ is equivalent to $\sqrt{2} = \log_x \pi$

19. $e^x = 8$ is equivalent to $x = \ln 8$

21. $\log_2 8 = 3$ is equivalent to $2^3 = 8$

23. $\log_a 3 = 6$ is equivalent to $a^6 = 3$

25. $\log_3 2 = x$ is equivalent to $3^x = 2$

27. $\log_2 M = 1.3$ is equivalent to $2^{1.3} = M$

29. $\log_{\sqrt{2}} \pi = x$ is equivalent to $\left(\sqrt{2}\right)^x = \pi$

31. $\ln 4 = x$ is equivalent to $e^x = 4$

33. $\log_2 1 = 0$ since $2^0 = 1$

35. $\log_5 25 = 2$ since $5^2 = 25$

37. $\log_{1/2} 16 = -4$ since $\left(\dfrac{1}{2}\right)^{-4} = 2^4 = 16$

39. $\log_{10} \sqrt{10} = \dfrac{1}{2}$ since $10^{1/2} = \sqrt{10}$

238

41. $\log_{\sqrt{2}} 4 = 4$ since $\left(\sqrt{2}\right)^4 = 4$

43. $\ln\sqrt{e} = \dfrac{1}{2}$ since $e^{1/2} = \sqrt{e}$

45. $f(x) = \ln(x - 3)$ requires
$x - 3 > 0$

$x > 3$
The domain of f is $\{x \mid x > 3\}$.

47. $F(x) = \log_2 x^2$ requires $x^2 > 0$.
$x^2 > 0$ for all $x \neq 0$.
The domain of F is $\{x \mid x \neq 0\}$.

49. $f(x) = 3 - 2\log_4 \dfrac{x}{2}$ requires $\dfrac{x}{2} > 0$.

$\dfrac{x}{2} > 0$

$x > 0$
The domain of f is $\{x \mid x > 0\}$.

51. $f(x) = \ln\left(\dfrac{1}{x+1}\right)$ requires

$\dfrac{1}{x+1} > 0$

$p(x) = \dfrac{1}{x+1}$

p is undefined when $x = -1$.

Interval	$(-\infty, -1)$	$(-1, \infty)$
Number Chosen	-2	0
Value of p	-1	1
Conclusion	Negative	Positive

The domain of f is $\{x \mid x > -1\}$.

53. $g(x) = \log_5\left(\dfrac{x+1}{x}\right)$ requires

$\dfrac{x+1}{x} > 0$

$p(x) = \dfrac{x+1}{x}$

$p(x)$ is zero or undefined when $x = -1$ or $x = 0$.

Interval	$(-\infty, -1)$	$(-1, 0)$	$(0, \infty)$
Number Chosen	-2	-0.5	1
Value of p	0.5	-1	2
Conclusion	Positive	Negative	Positive

The domain of g is $\{x \mid x < -1 \text{ or } x > 0\}$.

55. $f(x) = \sqrt{\ln x}$ requires $\ln x \geq 0$ and $x > 0$
$\ln x \geq 0$

$\Rightarrow x \geq e^0$

$x \geq 1$
The domain of h is $\{x \mid x \geq 1\}$.

57. $\ln\left(\dfrac{5}{3}\right) \approx 0.511$

59. $\dfrac{\ln(10/3)}{0.04} \approx 30.099$

61. For $f(x) = \log_a x$, find a so that $f(2) = \log_a 2 = 2$ or $a^2 = 2$ or $a = \sqrt{2}$.
(The base a must be positive by definition.)

63. $y = \log_3 x$

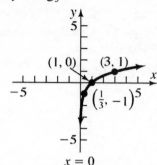

$x = 0$

65. $y = \log_{1/4} x$

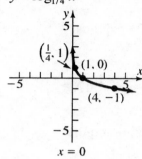

$x = 0$

67.   B   69.   D   71.   A   73.   E

75. $f(x) = \ln(x + 4)$
Using the graph of $y = \ln x$, shift the
graph 4 units to the left.
Domain: $(-4, \infty)$
Range: $(-\infty, \infty)$
Vertical Asymptote: $x = -4$

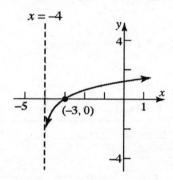

77. $f(x) = 2 + \ln x = \ln x + 2$
Using the graph of $y = \ln x$, shift up 2
units.
Domain: $(0, \infty)$
Range: $(-\infty, \infty)$
Vertical Asymptote: $x = 0$

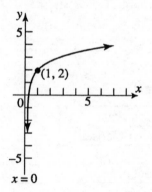

79. $g(x) = \ln(2x)$
Using the graph of $y = \ln x$, compress the
graph horizontally by a factor of $\dfrac{1}{2}$.
Domain: $(0, \infty)$
Range: $(-\infty, \infty)$
Vertical Asymptote: $x = 0$

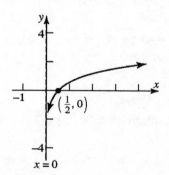

81. $f(x) = 3\ln x$
Using the graph of $y = \ln x$, stretch the
graph vertically by a factor of 3.
Domain: $(0, \infty)$
Range: $(-\infty, \infty)$
Vertical Asymptote: $x = 0$

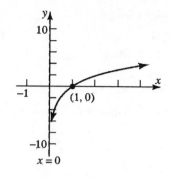

83. $f(x) = \log(x - 4)$
Using the graph of $y = \log x$, shift 4 units
to the right.
Domain: $(4, \infty)$
Range: $(-\infty, \infty)$
Vertical Asymptote: $x = 4$

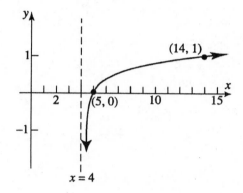

85. $h(x) = 4\log x$
Using the graph of $y = \log x$, stretch the
graph vertically by a factor of 4.
Domain: $(0, \infty)$
Range: $(-\infty, \infty)$
Vertical Asymptote: $x = 0$

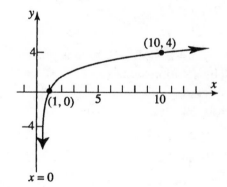

87. $F(x) = \log(2x)$
Using the graph of $y = \log x$, compress
the graph horizontally by a factor of $\dfrac{1}{2}$.
Domain: $(0, \infty)$
Range: $(-\infty, \infty)$
Vertical Asymptote: $x = 0$

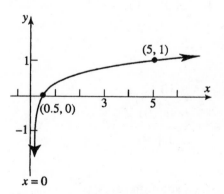

89. $f(x) = 3 + \log(x+2) = \log(x+2) + 3$
Using the graph of $y = \log x$, shift 2 units
to the left, and shift up 3 units.
Domain: $(-2, \infty)$
Range: $(-\infty, \infty)$
Vertical Asymptote: $x = -2$

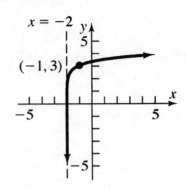

91. $\log_3 x = 2$

    $x = 3^2$

    $x = 9$
The solution set is $\{9\}$.

93. $\log_2(2x+1) = 3$

    $2x+1 = 2^3$

    $2x+1 = 8$

    $2x = 7$

    $x = \dfrac{7}{2}$
The solution set is $\left\{\dfrac{7}{2}\right\}$.

95. $\log_x 4 = 2$

    $x^2 = 4$

    $x = 2$ ($x \neq -2$, base is positive)
The solution set is $\{2\}$.

97. $\ln e^x = 5$

    $e^x = e^5$

    $x = 5$
The solution set is $\{5\}$.

99. $\log_4 64 = x$

    $4^x = 64$

    $4^x = 4^3$

    $x = 3$
The solution set is $\{3\}$.

101. $\log_3 243 = 2x+1$

    $3^{2x+1} = 243$

    $3^{2x+1} = 3^5$

    $2x+1 = 5$

    $2x = 4$

    $x = 2$
The solution set is $\{2\}$.

103. $e^{3x} = 10$

    $3x = \ln(10)$

    $x = \dfrac{\ln(10)}{3}$
The solution set is $\left\{\dfrac{\ln(10)}{3}\right\}$.

105. $e^{2x+5} = 8$

    $2x+5 = \ln(8)$

    $2x = \ln(8) - 5$

    $x = \dfrac{\ln(8)-5}{2}$
The solution set is $\left\{\dfrac{\ln(8)-5}{2}\right\}$.

107. $\log_3\left(x^2+1\right)=2$

$\qquad x^2+1=3^2$

$\qquad x^2+1=9$

$\qquad x^2=8$

$\qquad\qquad x=-\sqrt{8}\quad\text{or}\quad x=\sqrt{8}$

$\qquad\qquad x=-2\sqrt{2}\quad\text{or}\quad x=2\sqrt{2}$

The solution set is $\left\{-2\sqrt{2},\ 2\sqrt{2}\right\}$.

109. $\log_2 8^x=-3$

$\qquad 8^x=2^{-3}$

$\qquad \left(2^3\right)^x=2^{-3}$

$\qquad 2^{3x}=2^{-3}$

$\qquad 3x=-3\Rightarrow x=-1$

The solution set is $\{-1\}$.

111. $\mathrm{pH}=-\log_{10}\left[\mathrm{H}^+\right]$

  (a) $\mathrm{pH}=-\log_{10}[0.1]=-(-1)=1$

  (b) $\mathrm{pH}=-\log_{10}[0.01]=-(-2)=2$

  (c) $\mathrm{pH}=-\log_{10}[0.001]=-(-3)=3$

  (d)   As the H^+ decreases, the pH increases.

  (e) $3.5=-\log_{10}\left[\mathrm{H}^+\right]\Rightarrow -3.5=\log_{10}\left[\mathrm{H}^+\right]\Rightarrow \left[\mathrm{H}^+\right]=10^{-3.5}\approx 3.16\times 10^{-4}=0.000316$

  (f) $7.4=-\log_{10}\left[\mathrm{H}^+\right]\Rightarrow -7.4=\log_{10}\left[\mathrm{H}^+\right]\Rightarrow \left[\mathrm{H}^+\right]=10^{-7.4}\approx 3.981\times 10^{-8}=0.00000003981$

113. $p=760e^{-0.145h}$

  (a) $320=760e^{-0.145h}$

$\qquad \dfrac{320}{760}=e^{-0.145h}$

$\qquad \ln\!\left(\dfrac{320}{760}\right)=-0.145h$

$\qquad h=\dfrac{\ln\!\left(\dfrac{320}{760}\right)}{-0.145}\approx 5.97\text{ km}$

  (b) $667=760e^{-0.145h}$

$\qquad \dfrac{667}{760}=e^{-0.145h}$

$\qquad \ln\!\left(\dfrac{667}{760}\right)=-0.145h$

$\qquad h=\dfrac{\ln\!\left(\dfrac{667}{760}\right)}{-0.145}\approx 0.90\text{ km}$

115. $F(t)=1-e^{-0.1t}$

  (a) $0.5=1-e^{-0.1t}$

$\qquad -0.5=-e^{-0.1t}$

$\qquad 0.5=e^{-0.1t}\Rightarrow \ln(0.5)=-0.1t$

$\qquad t=\dfrac{\ln(0.5)}{-0.1}\approx 6.93$

Approximately 7 minutes.

  (b) $0.8=1-e^{-0.1t}$

$\qquad -0.2=-e^{-0.1t}$

$\qquad 0.2=e^{-0.1t}\Rightarrow \ln(0.2)=-0.1t$

$\qquad t=\dfrac{\ln(0.2)}{-0.1}\approx 16.09$

Approximately 16 minutes.

  (c)   It is impossible for the probability to reach 100% because $e^{-0.1t}$ will never equal zero; thus, $F(t)=1-e^{-0.1t}$ will never equal 1.

117. $D = 5e^{-0.4h}$

$2 = 5e^{-0.4h}$

$0.4 = e^{-0.4h}$

$\ln(0.4) = -0.4h$

$h = \dfrac{\ln(0.4)}{-0.4} \approx 2.29$

Approximately 2.3 hours.

119. $I = \dfrac{E}{R}\left[1 - e^{-(R/L)t}\right]$

0.5 ampere:

$0.5 = \dfrac{12}{10}\left[1 - e^{-(10/5)t}\right]$

$\dfrac{5}{12} = 1 - e^{-2t}$

$e^{-2t} = \dfrac{7}{12}$

$-2t = \ln(7/12)$

$t = \dfrac{\ln(7/12)}{-2} \approx 0.2695$ seconds

1.0 ampere:

$1.0 = \dfrac{12}{10}\left[1 - e^{-(10/5)t}\right]$

$\dfrac{10}{12} = 1 - e^{-2t}$

$e^{-2t} = \dfrac{1}{6}$

$-2t = \ln(1/6)$

$t = \dfrac{\ln(1/6)}{-2} \approx 0.8959$ seconds

Graphing:

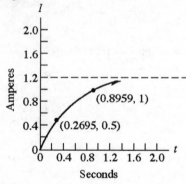

121. $L(10^{-7}) = 10\log\left(\dfrac{10^{-7}}{10^{-12}}\right) = 10\log\left(10^{5}\right) = 10 \cdot 5 = 50$ decibels

123. $L(10^{-1}) = 10\log\left(\dfrac{10^{-1}}{10^{-12}}\right) = 10\log\left(10^{11}\right) = 10 \cdot 11 = 110$ decibels

125. $M(125{,}892) = \log\left(\dfrac{125{,}892}{10^{-3}}\right) \approx 8.1$

127. $R = 3e^{kx}$

(a)
$$10 = 3e^{k(0.06)}$$
$$3.3333 = e^{0.06k}$$
$$\ln(3.3333) = 0.06\,k$$
$$k = \frac{\ln(3.3333)}{0.06}$$
$$k \approx 20.07$$

(b)
$$R = 3e^{20.066(0.17)}$$
$$R = 3e^{3.41122}$$
$$R \approx 91\%$$

(c)
$$100 = 3e^{20.066x}$$
$$33.3333 = e^{20.066x}$$
$$\ln(33.3333) = 20.066x$$
$$x = \frac{\ln(33.3333)}{20.07}$$
$$x \approx 0.175$$

(d)
$$15 = 3e^{20.066x}$$
$$5 = e^{20.066x}$$
$$\ln(5) = 20.066x$$
$$x = \frac{\ln(5)}{20.066}$$
$$x \approx 0.08$$

129. If the base of a logarithmic function equals 1, we would have the following:
$$f(x) = \log_1(x) \Rightarrow f^{-1}(x) = 1^x = 1 \text{ for every real number } x.$$
In other words, f^{-1} would be a constant function and, therefore, f^{-1} would not be one-to-one.

Exponential and Logarithmic Functions

4.5   Properties of Logarithms

1. sum

3. $r\log_a M$

5. False

7. $\log_3 3^{71} = 71$

9. $\ln e^{-4} = -4$

11. $2^{\log_2 7} = 7$

13. $\log_8 2 + \log_8 4 = \log_8(4 \cdot 2) = \log_8 8 = 1$

15. $\log_6 18 - \log_6 3 = \log_6 \dfrac{18}{3} = \log_6 6 = 1$

17. $\begin{aligned} \log_2 6 \cdot \log_6 4 &= \log_6 4^{\log_2 6} \\ &= \log_6(2^2)^{\log_2 6} \\ &= \log_6(2)^{2\log_2 6} \\ &= \log_6(2)^{\log_2 6^2} \\ &= \log_6 6^2 \\ &= 2 \end{aligned}$

19. $3^{\log_3 5 - \log_3 4} = 3^{\log_3 \frac{5}{4}} = \dfrac{5}{4}$

21. $e^{\log_{e^2} 16}$
Simplify the exponent.
Let $a = \log_{e^2} 16$, then $(e^2)^a = 16$.
$$e^{2a} = 16$$
$$e^{2a} = 4^2$$
$$(e^{2a})^{1/2} = (4^2)^{1/2}$$
$$e^a = 4$$
$$a = \ln 4$$
Thus, $e^{\log_{e^2} 16} = e^{\ln 4} = 4$.

23. $\ln 6 = \ln(3 \cdot 2) = \ln 3 + \ln 2 = b + a$

25. $\ln 1.5 = \ln \dfrac{3}{2} = \ln 3 - \ln 2 = b - a$

27. $\ln 8 = \ln 2^3 = 3 \cdot \ln 2 = 3a$

29. $\ln \sqrt[5]{6} = \ln 6^{1/5} = \dfrac{1}{5} \cdot \ln 6 = \dfrac{1}{5} \cdot \ln(2 \cdot 3) = \dfrac{1}{5} \cdot (\ln 2 + \ln 3) = \dfrac{1}{5} \cdot (a + b)$

31. $\log_5(25x) = \log_5 25 + \log_5 x = 2 + \log_5 x$

33. $\log_2 z^3 = 3\log_2 z$ 35. $\ln(ex) = \ln e + \ln x = 1 + \ln x$

37. $\ln(xe^x) = \ln x + \ln e^x = \ln x + x$

39. $\log_a(u^2 v^3) = \log_a u^2 + \log_a v^3 = 2\log_a u + 3\log_a v$

41. $\ln(x^2 \sqrt{1-x}) = \ln x^2 + \ln \sqrt{1-x} = \ln x^2 + \ln(1-x)^{1/2} = 2\ln x + \dfrac{1}{2}\ln(1-x)$

43. $\log_2 \left(\dfrac{x^3}{x-3} \right) = \log_2 x^3 - \log_2(x-3) = 3\log_2 x - \log_2(x-3)$

45. $\log \left[\dfrac{x(x+2)}{(x+3)^2} \right] = \log(x(x+2)) - \log(x+3)^2 = \log x + \log(x+2) - 2\log(x+3)$

47. $\ln \left[\dfrac{x^2 - x - 2}{(x+4)^2} \right]^{1/3} = \dfrac{1}{3}\ln \left[\dfrac{(x-2)(x+1)}{(x+4)^2} \right]$

$\qquad\qquad\qquad = \dfrac{1}{3}\left[\ln(x-2)(x+1) - \ln(x+4)^2 \right]$

$\qquad\qquad\qquad = \dfrac{1}{3}\left[\ln(x-2) + \ln(x+1) - 2\ln(x+4) \right]$

$\qquad\qquad\qquad = \dfrac{1}{3}\ln(x-2) + \dfrac{1}{3}\ln(x+1) - \dfrac{2}{3}\ln(x+4)$

49. $\ln \dfrac{5x\sqrt{1+3x}}{(x-4)^3} = \ln(5x\sqrt{1+3x}) - \ln(x-4)^3$

$\qquad\qquad\qquad = \ln 5 + \ln x + \ln \sqrt{1+3x} - 3\ln(x-4)$

$\qquad\qquad\qquad = \ln 5 + \ln x + \ln(1+3x)^{1/2} - 3\ln(x-4)$

$\qquad\qquad\qquad = \ln 5 + \ln x + \dfrac{1}{2}\ln(1+3x) - 3\ln(x-4)$

51. $3\log_5 u + 4\log_5 v = \log_5 u^3 + \log_5 v^4 = \log_5(u^3 v^4)$

53. $\log_3 \sqrt{x} - \log_3 x^3 = \log_3 \left(\dfrac{\sqrt{x}}{x^3} \right) = \log_3 \left(\dfrac{x^{1/2}}{x^3} \right) = \log_3 x^{-5/2} = -\dfrac{5}{2}\log_3 x$

55. $\log_4(x^2-1) - 5\log_4(x+1) = \log_4(x^2-1) - \log_4(x+1)^5$

$$= \log_4\left(\frac{x^2-1}{(x+1)^5}\right)$$

$$= \log_4\left(\frac{(x+1)(x-1)}{(x+1)^5}\right)$$

$$= \log_4\left(\frac{x-1}{(x+1)^4}\right)$$

57. $\ln\left(\frac{x}{x-1}\right) + \ln\left(\frac{x+1}{x}\right) - \ln(x^2-1) = \ln\left[\frac{x}{x-1} \cdot \frac{x+1}{x}\right] - \ln(x^2-1)$

$$= \ln\left[\frac{x+1}{x-1} \div (x^2-1)\right]$$

$$= \ln\left[\frac{x+1}{(x-1)(x-1)(x+1)}\right]$$

$$= \ln\left(\frac{1}{(x-1)^2}\right)$$

$$= \ln(x-1)^{-2}$$

$$= -2\ln(x-1)$$

59. $8\log_2\sqrt{3x-2} - \log_2\left(\frac{4}{x}\right) + \log_2 4 = \log_2\left(\sqrt{3x-2}\right)^8 - (\log_2 4 - \log_2 x) + \log_2 4$

$$= \log_2(3x-2)^4 - \log_2 4 + \log_2 x + \log_2 4$$

$$= \log_2\left[x(3x-2)^4\right]$$

61. $2\log_a(5x^3) - \frac{1}{2}\log_a(2x+3) = \log_a(5x^3)^2 - \log_a(2x-3)^{1/2} = \log_a\left[\frac{25x^6}{(2x-3)^{1/2}}\right]$

63. $2\log_2(x+1) - \log_2(x+3) - \log_2(x-1) = \log_2(x+1)^2 - \log_2(x+3) - \log_2(x-1)$

$$= \log_2\frac{(x+1)^2}{(x+3)} - \log_2(x-1)$$

$$= \log_2\frac{\dfrac{(x+1)^2}{(x+3)}}{(x-1)}$$

$$= \log_2\frac{(x+1)^2}{(x+3)(x-1)}$$

65. $\log_3 21 = \dfrac{\log 21}{\log 3} \approx \dfrac{1.32222}{0.47712} \approx 2.771$

67. $\log_{1/3} 71 = \dfrac{\log 71}{\log(1/3)} = \dfrac{\log 71}{-\log 3} \approx \dfrac{1.85126}{-0.47712} \approx -3.880$

69. $\log_{\sqrt{2}} 7 = \dfrac{\log 7}{\log \sqrt{2}} \approx \dfrac{0.84510}{0.15051} \approx 5.615$

71. $\log_\pi e = \dfrac{\ln e}{\ln \pi} \approx \dfrac{1}{1.14473} \approx 0.874$

73. $y = \log_4 x = \dfrac{\ln x}{\ln 4}$ or $y = \dfrac{\log x}{\log 4}$

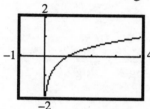

75. $y = \log_2(x+2) = \dfrac{\ln(x+2)}{\ln 2}$

or $y = \dfrac{\log(x+2)}{\log 2}$

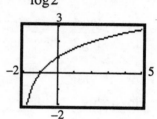

77. $y = \log_{x-1}(x+1) = \dfrac{\ln(x+1)}{\ln(x-1)}$

or $y = \dfrac{\log(x+1)}{\log(x-1)}$

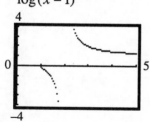

79. $\ln y = \ln x + \ln C$
     $\ln y = \ln(xC)$
     $y = Cx$

81. $\ln y = \ln x + \ln(x+1) + \ln C$
     $\ln y = \ln\big(x(x+1)C\big)$
     $y = Cx(x+1)$

83. $\ln y = 3x + \ln C$
     $\ln y = \ln e^{3x} + \ln C$
     $\ln y = \ln\big(Ce^{3x}\big)$
     $y = Ce^{3x}$

85. $\ln(y-3) = -4x + \ln C$
     $\ln(y-3) = \ln e^{-4x} + \ln C$
     $\ln(y-3) = \ln\big(Ce^{-4x}\big)$
     $y - 3 = Ce^{-4x}$
     $y = Ce^{-4x} + 3$

87. $3\ln y = \dfrac{1}{2}\ln(2x+1) - \dfrac{1}{3}\ln(x+4) + \ln C$

$\ln y^3 = \ln(2x+1)^{1/2} - \ln(x+4)^{1/3} + \ln C$

$\ln y^3 = \ln\left[\dfrac{C(2x+1)^{1/2}}{(x+4)^{1/3}}\right]$

$y^3 = \dfrac{C(2x+1)^{1/2}}{(x+4)^{1/3}}$

$y = \left[\dfrac{C(2x+1)^{1/2}}{(x+4)^{1/3}}\right]^{1/3}$

$y = \dfrac{\sqrt[3]{C}(2x+1)^{1/6}}{(x+4)^{1/9}}$

89. $\log_2 3 \cdot \log_3 4 \cdot \log_4 5 \cdot \log_5 6 \cdot \log_6 7 \cdot \log_7 8$

$= \dfrac{\log 3}{\log 2} \cdot \dfrac{\log 4}{\log 3} \cdot \dfrac{\log 5}{\log 4} \cdot \dfrac{\log 6}{\log 5} \cdot \dfrac{\log 7}{\log 6} \cdot \dfrac{\log 8}{\log 7} = \dfrac{\log 8}{\log 2} = \dfrac{\log 2^3}{\log 2} = \dfrac{3\log 2}{\log 2} = 3$

91. $\log_2 3 \cdot \log_3 4 \cdots \log_n(n+1) \cdot \log_{n+1} 2$

$= \dfrac{\log 3}{\log 2} \cdot \dfrac{\log 4}{\log 3} \cdots \dfrac{\log(n+1)}{\log n} \cdot \dfrac{\log 2}{\log(n+1)} = \dfrac{\log 2}{\log 2} = 1$

93. $\log_a\left(x + \sqrt{x^2 - 1}\right) + \log_a\left(x - \sqrt{x^2 - 1}\right) = \log_a\left[\left(x + \sqrt{x^2 - 1}\right)\left(x - \sqrt{x^2 - 1}\right)\right]$:

$= \log_a\left[x^2 - \left(x^2 - 1\right)\right]$

$= \log_a\left[x^2 - x^2 + 1\right]$

$= \log_a 1$

$= 0$

95. $2x + \ln\left(1 + e^{-2x}\right) = \ln e^{2x} + \ln\left(1 + e^{-2x}\right)$

$= \ln\left(e^{2x}\left(1 + e^{-2x}\right)\right)$

$= \ln\left(e^{2x} + e^0\right)$

$= \ln\left(e^{2x} + 1\right)$

97. $f(x) = \log_a x$

$x = a^{f(x)}$

$x^{-1} = a^{-f(x)} = \left(a^{-1}\right)^{f(x)} = \left(\dfrac{1}{a}\right)^{f(x)}$

Therefore, $\log_{1/a} x^{-1} = f(x) \Rightarrow -\log_{1/a} x = f(x) \Rightarrow -f(x) = \log_{1/a} x$.

99. $f(x) = \log_a x$

$a^{f(x)} = x$

$\dfrac{1}{a^{f(x)}} = \dfrac{1}{x}$

$a^{-f(x)} = \dfrac{1}{x}$

Therefore, $-f(x) = \log_a \dfrac{1}{x} = f\left(\dfrac{1}{x}\right)$.

101.   If $A = \log_a M$ and $B = \log_a N$, then $a^A = M$ and $a^B = N$.

$\log_a\left(\dfrac{M}{N}\right) = \log_a\left(\dfrac{a^A}{a^B}\right)$

$= \log_a a^{A-B}$

$= A - B$

$= \log_a M - \log_a N$

103.   The domain of $f(x) = \log_a x^2$ is $\{x \mid x \neq 0\}$. The domain of $g(x) = 2\log_a x$ is $\{x \mid x > 0\}$.
These two domains are different because the logarithm property $\log_a x^n = n \cdot \log_a x$ holds only when $\log_a x$ exists.

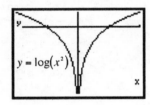   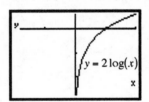

Chapter 4

Exponential and Logarithmic Functions

4.6   Logarithmic and Exponential Equations

1. $\log_4(x+2) = \log_4 8$
    $$x+2 = 8$$
    $$x = 6$$

    The solution set is $\{6\}$.

3. $\dfrac{1}{2}\log_3 x = 2\log_3 2$
    $$\log_3 x^{1/2} = \log_3 2^2$$
    $$x^{1/2} = 4$$
    $$x = 16$$

    The solution set is $\{16\}$.

5. $2\log_5 x = 3\log_5 4$
    $$\log_5 x^2 = \log_5 4^3$$
    $$x^2 = 64$$
    $$x = \pm 8$$
    Since $\log_5(-8)$ is undefined, the solution set is $\{8\}$.

7. $3\log_2(x-1) + \log_2 4 = 5$
    $$\log_2(x-1)^3 + \log_2 4 = 5$$
    $$\log_2\left(4(x-1)^3\right) = 5$$
    $$4(x-1)^3 = 2^5$$
    $$(x-1)^3 = \frac{32}{4}$$
    $$(x-1)^3 = 8$$
    $$x-1 = 2$$
    $$x = 3$$
    The solution set is $\{3\}$.

9. $\log x + \log(x+15) = 2$
    $$\log\left(x(x+15)\right) = 2$$
    $$x(x+15) = 10^2$$
    $$x^2 + 15x - 100 = 0$$
    $$(x+20)(x-5) = 0$$
    $$x = -20 \text{ or } x = 5$$
    Since $\log(-20)$ is undefined, the solution set is $\{5\}$.

11. $\ln x + \ln(x+2) = 4$

$\ln(x(x+2)) = 4$

$x(x+2) = e^4$

$x^2 + 2x - e^4 = 0$

$x = \dfrac{-2 \pm \sqrt{2^2 - 4(1)(-e^4)}}{2(1)}$

$= \dfrac{-2 \pm \sqrt{4 + 4e^4}}{2}$

$= \dfrac{-2 \pm 2\sqrt{1 + e^4}}{2}$

$= -1 \pm \sqrt{1 + e^4}$

$x = -1 - \sqrt{1 + e^4}$ or $x = -1 + \sqrt{1 + e^4}$

≈ -8.456 ≈ 6.456

Since $\ln(-8.456)$ is undefined, the solution set is $\left\{-1 + \sqrt{1 + e^4} \approx 6.456\right\}$.

13. $2^{2x} + 2^x - 12 = 0$

$\left(2^x\right)^2 + 2^x - 12 = 0$

$\left(2^x - 3\right)\left(2^x + 4\right) = 0$

$2^x - 3 = 0$ or $2^x + 4 = 0$

$2^x = 3$ or $2^x = -4$

$x = \log_2 3$ No solution

$= \dfrac{\log 3}{\log 2} \approx 1.585$

The solution set is $\left\{\log_2 3 \approx 1.585\right\}$.

15. $3^{2x} + 3^{x+1} - 4 = 0$

$\left(3^x\right)^2 + 3 \cdot 3^x - 4 = 0$

$\left(3^x - 1\right)\left(3^x + 4\right) = 0$

$3^x - 1 = 0$ or $3^x + 4 = 0$

$3^x = 1$ or $3^x = -4$

$x = 0$ No solution

The solution set is $\{0\}$.

17.
$$2^x = 10$$
$$\log(2^x) = \log 10$$
$$x \log 2 = 1$$
$$x = \frac{1}{\log 2} \approx 3.322$$

The solution set is $\left\{ \dfrac{1}{\log 2} \approx 3.322 \right\}$

19.
$$8^{-x} = 1.2$$
$$\log(8^{-x}) = \log(1.2)$$
$$-x \log 8 = \log(1.2)$$
$$x = \frac{\log(1.2)}{-\log 8} \approx -0.088$$

The solution set is $\left\{ \dfrac{\log(1.2)}{-\log 8} \approx -0.088 \right\}$.

21.
$$3^{1-2x} = 4^x$$
$$\log(3^{1-2x}) = \log(4^x)$$
$$(1-2x)\log 3 = x \log 4$$
$$\log 3 - 2x \log 3 = x \log 4$$
$$\log 3 = x \log 4 + 2x \log 3$$
$$\log 3 = x(\log 4 + 2 \log 3)$$
$$x = \frac{\log 3}{\log 4 + 2 \log 3} \approx 0.307$$

The solution set is $\left\{ \dfrac{\log 3}{\log 4 + 2 \log 3} \approx 0.307 \right\}$.

23.
$$\left(\frac{3}{5}\right)^x = 7^{1-x}$$
$$\log\left(\left(\frac{3}{5}\right)^x\right) = \log(7^{1-x})$$
$$x \log\left(\frac{3}{5}\right) = (1-x)\log 7$$
$$x(\log 3 - \log 5) = \log 7 - x \log 7$$
$$x \log 3 - x \log 5 + x \log 7 = \log 7$$
$$x(\log 3 - \log 5 + \log 7) = \log 7$$
$$x = \frac{\log 7}{\log 3 - \log 5 + \log 7} \approx 1.356$$

The solution set is $\left\{ \dfrac{\log 7}{\log 3 - \log 5 + \log 7} \approx 1.356 \right\}$.

25.
$$1.2^x = (0.5)^{-x}$$
$$\log 1.2^x = \log(0.5)^{-x}$$
$$x\log(1.2) = -x\log(0.5)$$
$$x\log(1.2) + x\log(0.5) = 0$$
$$x\left(\log(1.2) + \log(0.5)\right) = 0$$
$$x = 0$$

The solution set is $\{0\}$.

27.
$$\pi^{1-x} = e^x$$
$$\ln \pi^{1-x} = \ln e^x$$
$$(1-x)\ln \pi = x$$
$$\ln \pi - x\ln \pi = x$$
$$\ln \pi = x + x\ln \pi$$
$$\ln \pi = x(1 + \ln \pi)$$
$$x = \frac{\ln \pi}{1 + \ln \pi} \approx 0.534$$

The solution set is $\left\{\dfrac{\ln \pi}{1 + \ln \pi} \approx 0.534\right\}$.

29.
$$5\left(2^{3x}\right) = 8$$
$$2^{3x} = \frac{8}{5}$$
$$\log 2^{3x} = \log\left(\frac{8}{5}\right)$$
$$3x\log 2 = \log 8 - \log 5$$
$$x = \frac{\log 8 - \log 5}{3\log 2} \approx 0.226$$

The solution set is $\left\{\dfrac{\log 8 - \log 5}{3\log 2} \approx 0.226\right\}$.

31. $\log_a(x-1) - \log_a(x+6) = \log_a(x-2) - \log_a(x+3)$
$$\log_a\left(\frac{x-1}{x+6}\right) = \log_a\left(\frac{x-2}{x+3}\right)$$
$$\frac{x-1}{x+6} = \frac{x-2}{x+3}$$
$$(x-1)(x+3) = (x-2)(x+6)$$
$$x^2 + 2x - 3 = x^2 + 4x - 12$$
$$2x - 3 = 4x - 12$$
$$9 = 2x$$
$$x = \frac{9}{2}$$

Since each of the original logarithms is defined for $x = \dfrac{9}{2}$, the solution set is $\left\{\dfrac{9}{2}\right\}$.

33. $\log_{1/3}(x^2 + x) - \log_{1/3}(x^2 - x) = -1$

$$\log_{1/3}\left(\frac{x^2 + x}{x^2 - x}\right) = -1$$

$$\frac{x^2 + x}{x^2 - x} = \left(\frac{1}{3}\right)^{-1}$$

$$\frac{x(x+1)}{x(x-1)} = 3$$

$$x + 1 = 3(x - 1)$$

$$x + 1 = 3x - 3$$

$$-2x = -4$$

$$x = 2$$

Since each of the original logarithms is defined for $x = 2$, the solution set is $\{2\}$.

35. $\log_2(x + 1) - \log_4 x = 1$

$$\log_2(x + 1) - \frac{\log_2 x}{\log_2 4} = 1$$

$$\log_2(x + 1) - \frac{\log_2 x}{2} = 1$$

$$2\log_2(x + 1) - \log_2 x = 2$$

$$\log_2(x + 1)^2 - \log_2 x = 2$$

$$\log_2\left(\frac{(x+1)^2}{x}\right) = 2$$

$$\frac{(x+1)^2}{x} = 2^2$$

$$x^2 + 2x + 1 = 4x$$

$$x^2 - 2x + 1 = 0$$

$$(x - 1)^2 = 0$$

$$x - 1 = 0$$

$$x = 1$$

Since each of the original logarithms is defined for $x = 1$, the solution set is $\{1\}$.

37. $\log_{16} x + \log_4 x + \log_2 x = 7$

$$\frac{\log_2 x}{\log_2 16} + \frac{\log_2 x}{\log_2 4} + \log_2 x = 7$$

$$\frac{\log_2 x}{4} + \frac{\log_2 x}{2} + \log_2 x = 7$$

$$\log_2 x + 2\log_2 x + 4\log_2 x = 28$$

$$7\log_2 x = 28$$

$$\log_2 x = 4$$

$$x = 2^4 = 16$$

Since each of the original logarithms is defined for $x = 16$, the solution set is $\{16\}$.

39.

$$\left(\sqrt[3]{2}\right)^{2-x} = 2^{x^2}$$

$$\left(2^{1/3}\right)^{2-x} = 2^{x^2}$$

$$2^{\frac{1}{3}(2-x)} = 2^{x^2}$$

$$\frac{1}{3}(2-x) = x^2$$

$$2 - x = 3x^2$$

$$3x^2 + x - 2 = 0$$

$$(3x - 2)(x + 1) = 0$$

$$x = \frac{2}{3} \text{ or } x = -1$$

The solution set is $\left\{-1, \dfrac{2}{3}\right\}$.

41.

$$\frac{e^x + e^{-x}}{2} = 1$$

$$e^x + e^{-x} = 2$$

$$e^x\left(e^x + e^{-x}\right) = 2e^x$$

$$e^{2x} + 1 = 2e^x$$

$$(e^x)^2 - 2e^x + 1 = 0$$

$$\left(e^x - 1\right)^2 = 0$$

$$e^x - 1 = 0$$

$$e^x = 1$$

$$x = 0$$

The solution set is $\{0\}$.

43.

$$\frac{e^x - e^{-x}}{2} = 2$$

$$e^x - e^{-x} = 4$$

$$e^x\left(e^x - e^{-x}\right) = 4e^x$$

$$e^{2x} - 1 = 4e^x$$

$$(e^x)^2 - 4e^x - 1 = 0$$

$$e^x = \frac{-(-4) \pm \sqrt{(-4)^2 - 4(1)(-1)}}{2(1)}$$

$$= \frac{4 \pm \sqrt{20}}{2}$$

$$= \frac{4 \pm 2\sqrt{5}}{2} = 2 \pm \sqrt{5}$$

$$x = \ln\left(2 - \sqrt{5}\right) \text{ or } x = \ln\left(2 + \sqrt{5}\right)$$

$$x \approx \ln(-0.236) \text{ or } x \approx 1.444$$

Since $\ln(-0.236)$ is undefined, the solution set is $\left\{\ln\left(2 + \sqrt{5}\right) \approx 1.444\right\}$.

45. Using INTERSECT to solve:
$$y_1 = \ln(x)\,/\,\ln(5) + \ln(x)\,/\,\ln(3)$$
$$y_2 = 1$$

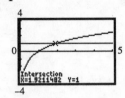

The solution is $x \approx 1.92$.

47. Using INTERSECT to solve:
$$y_1 = \ln(x+1)\,/\,\ln(5) - \ln(x-2)\,/\,\ln(4)$$
$$y_2 = 1$$

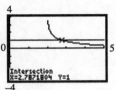

The solution is $x \approx 2.79$.

49. Using INTERSECT to solve:
$$y_1 = e^x;\ y_2 = -x$$

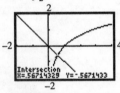

The solution is $x \approx -0.57$.

51. Using INTERSECT to solve:
$$y_1 = e^x;\ y_2 = x^2$$

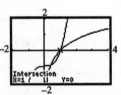

The solution is $x \approx -0.70$.

53. Using INTERSECT to solve:
$$y_1 = \ln x;\ y_2 = -x$$

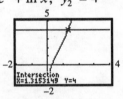

The solution is $x \approx 0.57$.

55. Using INTERSECT to solve:
$$y_1 = \ln x;\ y_2 = x^3 - 1$$

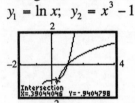

The solutions are $x \approx 0.39$ and $x = 1$.

57. Using INTERSECT to solve:
$$y_1 = e^x + \ln x;\ y_2 = 4$$

The solution is $x \approx 1.32$.

59. Using INTERSECT to solve:
$$y_1 = e^{-x};\ y_2 = \ln x$$

The solution is $x \approx 1.31$.

Exponential and Logarithmic Functions

4.7   Compound Interest

1. $P = \$500$, $r = 0.06$, $t = 6$ months $= 0.5$ year
    $I = Prt = (500)(0.06)(0.5) = \$15.00$

3. $P = \$100$, $r = 0.04$, $n = 4$, $t = 2$
    $A = P\left(1 + \dfrac{r}{n}\right)^{nt} = 100\left(1 + \dfrac{0.04}{4}\right)^{(4)(2)} = \$108.29$

5. $P = \$500$, $r = 0.08$, $n = 4$, $t = 2.5$
    $A = P\left(1 + \dfrac{r}{n}\right)^{nt} = 500\left(1 + \dfrac{0.08}{4}\right)^{(4)(2.5)} = \$609.50$

7. $P = \$600$, $r = 0.05$, $n = 365$, $t = 3$
    $A = P\left(1 + \dfrac{r}{n}\right)^{nt} = 600\left(1 + \dfrac{0.05}{365}\right)^{(365)(3)} = \$697.09$

9. $P = \$10$, $r = 0.11$, $t = 2$
    $A = Pe^{rt} = 40e^{(0.07)(3)} = \$49.35$

11. $P = \$100$, $r = 0.10$, $t = 2.25$
    $A = Pe^{rt} = 100e^{(0.10)(2.25)} = \$125.23$

13. $A = \$100$, $r = 0.06$, $n = 12$, $t = 2$
    $P = A\left(1 + \dfrac{r}{n}\right)^{-nt} = 100\left(1 + \dfrac{0.06}{12}\right)^{(-12)(2)} = \$88.72$

15. $A = \$1000$, $r = 0.06$, $n = 365$, $t = 2.5$
    $P = A\left(1 + \dfrac{r}{n}\right)^{-nt} = 1000\left(1 + \dfrac{0.06}{365}\right)^{(-365)(2.5)} = \$860.72$

17. $A = \$600$, $r = 0.04$, $n = 4$, $t = 2$
    $P = A\left(1 + \dfrac{r}{n}\right)^{-nt} = 600\left(1 + \dfrac{0.04}{4}\right)^{(-4)(2)} = \$554.09$

19. $A = \$80$, $r = 0.09$, $t = 3.25$
    $P = Ae^{-rt} = 80e^{(-0.09)(3.25)} = \$59.71$

21. $A = \$400$, $r = 0.10$, $t = 1$
    $P = Ae^{-rt} = 400e^{(-0.10)(1)} = \$361.93$

23. $1000 invested for 1 year at $5\frac{1}{4}\%$ compounded quarterly yields

$$100\left(1+\frac{0.0525}{4}\right)^{(4)(1)} = \$1053.54.$$

Since \$53.54 is 5.354% of \$1000, the effective interest rate is 5.354%.

25. $2P = P(1+r)^3$

$2 = (1+r)^3$

$\sqrt[3]{2} = 1+r$

$r = \sqrt[3]{2} - 1$

$\approx 1.26 - 1$

$= 0.26$

$r \approx 26\%$

27. 6% compounded quarterly:

$$A = 10,000\left(1+\frac{0.06}{4}\right)^{(4)(1)} = \$10,613.64$$

$6\frac{1}{4}\%$ compounded annually:

$$A = 10,000(1+0.0625)^1 = \$10,625$$

$6\frac{1}{4}\%$ compounded annually yields the larger amount.

29. 9% compounded monthly:

$$A = 10,000\left(1+\frac{0.09}{12}\right)^{(12)(1)} = \$10,938.07$$

8.8% compounded daily:

$$A = 10,000\left(1+\frac{0.088}{365}\right)^{365} = \$10,919.77$$

9% compounded monthly yields the larger amount.

31. Compounded monthly:

$$2P = P\left(1+\frac{0.08}{12}\right)^{12t}$$

$2 \approx (1.006667)^{12t}$

$\ln 2 \approx 12t\,\ln(1.006667)$

$t \approx \dfrac{\ln 2}{12\,\ln(1.006667)} \approx 8.69$ years

Compounded continuously:

$$2P = Pe^{0.08t}$$

$2 = e^{0.08t}$

$\ln 2 = 0.08t$

$t = \dfrac{\ln 2}{0.08} \approx 8.66$ years

33. Compounded monthly:

$$150 = 100\left(1+\frac{0.08}{12}\right)^{12t}$$

$1.5 \approx (1.006667)^{12t}$

$\ln 1.5 \approx 12t\,\ln(1.006667)$

$t \approx \dfrac{\ln 1.5}{12\,\ln(1.006667)} \approx 5.085$ years

Compounded continuously:

$$150 = 100e^{0.08t}$$

$1.5 = e^{0.08\,t}$

$\ln 1.5 = 0.08t$

$t = \dfrac{\ln 1.5}{0.08} \approx 5.068$ years

35. $25,000 = 10,000e^{0.06t}$

$\qquad 2.5 = e^{0.06t}$

$\qquad \ln 2.5 = 0.06t$

$\qquad t = \dfrac{\ln 2.5}{0.06} \approx 15.27$ years

37. $A = 90,000(1 + 0.03)^5 = \$104,335$

39. $P = 15,000e^{(-0.05)(3)} \approx \$12,910.62$

41. $A = 1500(1 + 0.15)^5 = 1500(1.15)^5 \approx \$3017$

43. $850,000 = 650,000(1 + r)^3$

$\qquad \dfrac{85}{65} = (1 + r)^3$

$\qquad \sqrt[3]{\dfrac{85}{65}} = 1 + r$

$\qquad r \approx \sqrt[3]{1.3077} - 1 \approx 0.0935$

$\qquad r \approx 9.35\%$

45. 5.6% compounded continuously:
$A = 1000e^{(0.056)(1)} = \$1057.60$
Jim will not have enough money to buy the computer.
5.9% compounded monthly:
$A = 1000\left(1 + \dfrac{0.059}{12}\right)^{12} = \$1060.62$
The second bank offers the better deal.

47. Will: 9% compounded semiannually:
$A = 2000\left(1 + \dfrac{0.09}{2}\right)^{(2)(20)} = \$11,632.73$

Henry: 8.5% compounded continuously:
$A = 2000e^{(0.085)(20)} = \$10,947.89$
Will has more money after 20 years.

49. $P = 50,000;\ t = 5$
(a) Simple interest at 12% per annum:
$A = 50,000 + 50,000(0.12)(5) = \$80,000$
(b) 11.5% compounded monthly:
$A = 50,000\left(1 + \dfrac{0.115}{12}\right)^{(12)(5)} = \$88,613.59$
(c) 11.25% compounded continuously:
$A = 50,000e^{(0.1125)(5)} = \$87,752.73$
Subtract \$50,000 from each to get the amount of interest:
(a) \$30,000   (b) \$38,613.59   (c) \$37.752.73
Option (a) results in the least interest.

51. (a) $A = \$10{,}000, \ r = 0.10, \ n = 12, \ t = 20$

$$P = 10{,}000\left(1 + \frac{0.10}{12}\right)^{(-12)(20)} = \$1364.62$$

(b) $A = \$10{,}000, \ r = 0.10, \ t = 20$

$$P = 10{,}000e^{(-0.10)(20)} = \$1353.35$$

53. $A = \$10{,}000, \ r = 0.08, \ n = 1, \ t = 10$

$$P = 10{,}000\left(1 + \frac{0.08}{1}\right)^{(-1)(10)} = \$4631.93$$

55. (a) $t = \dfrac{\ln 2}{1 \cdot \ln\left(1 + \dfrac{0.12}{1}\right)} = \dfrac{\ln 2}{\ln(1.12)} \approx 6.12 \text{ years}$

(b) $t = \dfrac{\ln 3}{4 \cdot \ln\left(1 + \dfrac{0.06}{4}\right)} = \dfrac{\ln 3}{4\ln(1.015)} \approx 18.45 \text{ years}$

(c) $mP = P\left(1 + \dfrac{r}{n}\right)^{nt}$

$$m = \left(1 + \frac{r}{n}\right)^{nt}$$

$$\ln m = nt \cdot \ln\left(1 + \frac{r}{n}\right)$$

$$t = \frac{\ln m}{n \cdot \ln\left(1 + \dfrac{r}{n}\right)}$$

57–59. Answers will vary.

Chapter 4

Exponential and Logarithmic Functions

4.8 Exponential Growth and Decay; Newton's Law; Logistic Models

1. $P(t) = 500e^{0.02t}$
    (a) $P(0) = 500e^{(0.02)\cdot(0)} = 500$ flies
    (b) growth rate $= 2\%$
    (c) $P(10) = 500e^{(0.02)\cdot(10)} = 611$ flies
    (d) Find t when $P = 800$:
    $$800 = 500e^{0.02t}$$
    $$1.6 = e^{0.02t}$$
    $$\ln 1.6 = 0.02t$$
    $$t = \frac{\ln 1.6}{0.02} \approx 23.5 \text{ days}$$

    (e) Find t when $P = 1000$:
    $$1000 = 500e^{0.02t}$$
    $$2 = e^{0.02t}$$
    $$\ln 2 = 0.02t$$
    $$t = \frac{\ln 2}{0.02} \approx 34.7 \text{ days}$$

3. $A(t) = A_0 e^{-0.0244t} = 500e^{-0.0244t}$
    (a) decay rate $= 2.44\%$
    (b) $A(10) = 500e^{(-0.0244)(10)} \approx 391.74$ grams
    (c) Find t when $A = 400$:
    $$400 = 500e^{-0.0244t}$$
    $$0.8 = e^{-0.0244t}$$
    $$\ln 0.8 = -0.0244t$$
    $$t = \frac{\ln 0.8}{-0.0244} \approx 9.15 \text{ years}$$

    (d) Find t when $A = 250$:
    $$250 = 500e^{-0.0244t}$$
    $$0.5 = e^{-0.0244t}$$
    $$\ln 0.5 = -0.0244t$$
    $$t = \frac{\ln 0.5}{-0.0244} \approx 28.4 \text{ years}$$

5. Use $N(t) = N_0 e^{kt}$ and solve for k:
    $$1800 = 1000e^{k(1)}$$
    $$1.8 = e^{k}$$
    $$k = \ln 1.8$$
    When $t = 3$:
    $$N(3) = 1000e^{(\ln 1.8)(3)} = 5832 \text{ mosquitos}$$

    Find t when $N(t) = 10{,}000$:
    $$10{,}000 = 1000e^{(\ln 1.8)t}$$
    $$10 = e^{(\ln 1.8)t}$$
    $$\ln 10 = (\ln 1.8)t$$
    $$t = \frac{\ln 10}{\ln 1.8} \approx 3.9 \text{ days}$$

7. Use $P(t) = P_0 e^{kt}$ and solve for k:

$$2P_0 = P_0 e^{k(1.5)}$$

$$2 = e^{1.5k}$$

$$\ln 2 = 1.5k$$

$$k = \frac{\ln 2}{1.5}$$

When $t = 2$: $P(2) = 10,000 e^{\left(\frac{\ln 2}{1.5}\right)(2)} = 25,198$ is the population 2 years from now.

9. Use $A = A_0 e^{kt}$ and solve for k:

$$0.5 A_0 = A_0 e^{k(1590)}$$

$$0.5 = e^{1590k}$$

$$\ln 0.5 = 1590k$$

$$k = \frac{\ln 0.5}{1590}$$

When $A_0 = 10$ and $t = 50$: $A = 5 e^{\left(\frac{\ln 0.5}{1590}\right)(50)} \approx 10.220$ grams

11. Use $A = A_0 e^{kt}$ and solve for k:

half-life $= 5600$ years

$$0.5 A_0 = A_0 e^{k(5600)}$$

$$0.5 = e^{5600k}$$

$$\ln 0.5 = 5600k$$

$$k = \frac{\ln 0.5}{5600}$$

Solve for t when $A = 0.3 A_0$:

$$0.3 A_0 = A_0 e^{\left(\frac{\ln 0.5}{5600}\right)t}$$

$$0.3 = e^{\left(\frac{\ln 0.5}{5600}\right)t}$$

$$\ln 0.3 = \left(\frac{\ln 0.5}{5600}\right)t$$

$$t = \frac{\ln 0.3}{\frac{\ln 0.5}{5600}}$$

$$t \approx 9710$$

The tree died approximately 9710 years ago.

13. (a) Using $u = T + (u_0 - T)e^{kt}$ where $t = 5$, $T = 70$, $u_0 = 450$, $u = 300$:

$$300 = 70 + (450 - 70)e^{k(5)}$$

$$230 = 380 e^{5k}$$

$$\frac{230}{380} = e^{5k}$$

$$\ln\left(\frac{230}{380}\right) = 5k$$

$$k = \frac{\ln\left(\frac{230}{380}\right)}{5} \approx -0.1004$$

$T = 70$, $u_0 = 450$, $u = 135$:

$$135 = 70 + (450 - 70)e^{-0.1004\,t}$$

$$65 = 380 e^{-0.1004\,t}$$

$$\frac{65}{380} = e^{-0.1004\,t}$$

$$\ln\left(\frac{65}{380}\right) = -0.1004t$$

$$t = \frac{\ln\left(\frac{65}{380}\right)}{-0.1004} \approx 17.6 \text{ minutes}$$

The pizza will be cool enough to eat at 5:18 PM.

(b) $T = 70,\ u_0 = 450,\ u = 160$:

$$160 = 70 + (450 - 70)e^{-0.1004\,t}$$

$$90 = 380e^{-0.1004\,t}$$

$$\frac{90}{380} = e^{-0.1004\,t}$$

$$\ln\left(\frac{90}{380}\right) = -0.1004\,t \Rightarrow t = \frac{\ln(90/380)}{-0.1004} \approx 14.3$$

The pizza will be 160°F after about 14.3 minutes.

(c)   As time passes, the temperature gets closer to 70°F.

15. Using $u = T + (u_0 - T)e^{kt}$ where $t = 3$,
$T = 35,\ u_0 = 8,\ u = 15$:

$$15 = 35 + (8 - 35)e^{k(3)}$$

$$-20 = -27e^{3k}$$

$$\frac{20}{27} = e^{3k}$$

$$\ln\left(\frac{20}{27}\right) = 3k \Rightarrow k = \frac{\ln(20/27)}{3}$$

At $t = 5$: $u = 35 + (8 - 35)e^{\left(\frac{\ln(20/27)}{3}\right)(5)} \approx 18.63°C$

At $t = 10$: $u = 35 + (8 - 35)e^{\left(\frac{\ln(20/27)}{3}\right)(10)} \approx 25.1°C$

17. Use $A = A_0 e^{kt}$ and solve for k :

$$15 = 25e^{k(10)}$$

$$0.6 = e^{10k}$$

$$\ln 0.6 = 10k$$

$$k = \frac{\ln 0.6}{10}$$

When $A_0 = 25$ and $t = 24$:

$$A = 25e^{\left(\frac{\ln 0.6}{10}\right)(24)} \approx 7.33 \text{ kilograms}$$

Find t when $A = 0.5A_0$:

$$0.5 = 25\,e^{\left(\frac{\ln 0.6}{10}\right)t}$$

$$0.02 = e^{\left(\frac{\ln 0.6}{10}\right)t}$$

$$\ln 0.02 = \left(\frac{\ln 0.6}{10}\right)t \Rightarrow t = \frac{\ln 0.02}{\left(\frac{\ln 0.6}{10}\right)} \approx 76.6 \text{ hours}$$

19. Use $A = A_0 e^{kt}$ and solve for k :

$$0.5A_0 = A_0 e^{k(8)}$$

$$0.5 = e^{8k}$$

$$\ln 0.5 = 8k$$

$$k = \frac{\ln 0.5}{8}$$

Find t when $A = 0.1A_0$:

$$0.1A_0 = A_0 e^{\left(\frac{\ln 0.5}{8}\right)t}$$

$$0.1 = e^{\left(\frac{\ln 0.5}{8}\right)t}$$

$$\ln 0.1 = \left(\frac{\ln 0.5}{8}\right)t \Rightarrow t = \frac{\ln 0.1}{\frac{\ln 0.5}{8}} \approx 26.6 \text{ days}$$

The farmers need to wait about 27 days before
using the hay.

21. (a) $P(0) = \dfrac{0.9}{1 + 6e^{-0.32(0)}} = \dfrac{0.9}{1 + 6 \cdot 1} = \dfrac{0.9}{7} = 0.1286$

In 2004, 12.86% of U.S. households owned a DVD.

(b) The maximum proportion is the carrying capacity, $c = 0.9 = 90\%$.

(c) Find t such that $P = 0.8$:

$$0.8 = \frac{0.9}{1 + 6e^{-0.32t}}$$

$$0.8\left(1 + 6e^{-0.32t}\right) = 0.9$$

$$1 + 6e^{-0.32t} = 1.125$$

$$6e^{-0.32t} = 0.125$$

$$e^{-0.32t} = \frac{0.125}{6}$$

$$-0.32t = \ln\!\left(\frac{0.125}{6}\right)$$

$$t = \frac{\ln\!\left(\dfrac{0.125}{6}\right)}{-0.32} \approx 12.1$$

80% of households will own a DVD in 2016 ($t = 12$).

23. (a) As $t \to \infty$, $e^{-0.439t} \to 0$. Thus, $P(t) \to 1000$.

The carrying capacity is 1000 bacteria. Growth rate = 43.9%.

(b) $P(0) = \dfrac{1000}{1 + 32.33e^{-0.439(0)}} = \dfrac{1000}{33.33} = 30$ bacteria

(c) Find t such that $P = 800$:

$$800 = \frac{1000}{1 + 32.33e^{-0.439t}}$$

$$800\left(1 + 32.33e^{-0.439t}\right) = 1000$$

$$1 + 32.33e^{-0.439t} = 1.25$$

$$32.33e^{-0.439t} = 0.25$$

$$e^{-0.439t} = \frac{0.25}{32.33}$$

$$-0.439t = \ln\!\left(\frac{0.25}{32.33}\right)$$

$$t = \frac{\ln\!\left(\dfrac{0.25}{32.33}\right)}{-0.439} \approx 11.08$$

The amount of bacteria will be 800 after approximately 11.08 hours.

Chapter 4

Exponential and Logarithmic Functions

4.9 Fitting Data to Exponential, Logarithmic, and Logistic Functions

1. (a) Scatter diagram

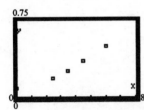

(b) Using EXPonential REGression on the data yields: $y = 0.0903(1.3384)^x$

(c) $y = 0.0903(1.3384)^x = 0.0903\left(e^{\ln(1.3384)}\right)^x = 0.0903e^{\ln(1.3384)x}$

$N(t) = 0.0903e^{\ln(1.3384)t} = 0.0903e^{0.2915t}$

(d) Graphing: $y_1 = 0.0903e^{0.2915x}$

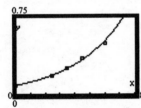

(e) $N(7) = 0.0903e^{(0.2915)\cdot 7} \approx 0.695$ bacteria

(f) Find t when $N(t) = 0.75$

$0.0903e^{(0.2915)\cdot t} = 0.75$

$e^{(0.2915)\cdot t} = \dfrac{0.75}{0.0903}$

$0.2915t = \ln\left(\dfrac{0.75}{0.0903}\right)$

$t \approx \dfrac{\ln\left(\dfrac{0.75}{0.0903}\right)}{0.2915} \approx 7.26$ hours

3.   (a)   Scatter diagram

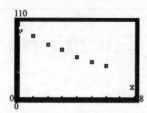

      (b)   Using EXPonential REGression on the data yields: $y = 100.3263(0.8769)^x$

      (c) $y = 100.3263(0.8769)^x = 100.3263\left(e^{\ln(0.8769)}\right)^x = 100.3263e^{\ln(0.8769)x}$

             $A(t) = 100.3263e^{(-0.1314)t}$

      (d)   Graphing: $y_1 = 100.3263e^{(-0.1314)x}$

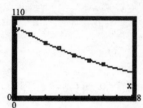

      (e)   Find t when $A(t) = 0.5 \cdot A_0$

             $100.3263e^{(-0.1314)t} = (0.5)(100.3263)$

             $$e^{(-0.1314)t} = 0.5$$

             $$-0.1314t = \ln 0.5$$

             $$t = \frac{\ln 0.5}{-0.1314} \approx 5.28 \text{ weeks}$$

      (f) $A(50) = 100.3263e^{(-0.1314)\cdot 50} \approx 0.141$ grams

      (g)   Find t when $A(t) = 20$

             $100.3263e^{(-0.1314)t} = 20$

             $$e^{(-0.1314)t} = \frac{20}{100.3263}$$

             $$-0.1314t = \ln\left(\frac{20}{100.3263}\right)$$

             $$t = \frac{\ln\left(\dfrac{20}{100.3263}\right)}{-0.1314} \approx 12.27 \text{ weeks}$$

5.   (a)   Let $x = 1$ correspond to 1994, $x = 2$ correspond to 1995, etc.

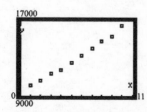

(b)   Using EXPonential REGression on the data yields: $y = 9478.4453(1.0566)^x$

(c)   The average annual rate of return over the 10 years is $0.0566 = 5.66\%$.

(d)   In the year 2021, $x = 28$, so $y = 9478.4453(1.0566)^{28} = \$44{,}282.70$.

(e)   Find x when $y = 50,000$

$$9478.4453(1.0566)^x = 50{,}000$$

$$(1.0566)^x = \frac{50{,}000}{9478.4453}$$

$$x\ln 1.0566 = \ln\left(\frac{50{,}000}{9478.4453}\right)$$

$$x = \frac{\ln\left(\dfrac{50{,}000}{9478.4453}\right)}{\ln 1.0566} \approx 30.21 \text{ years, that is, in the year 2023.}$$

7.   (a)   Scatter diagram

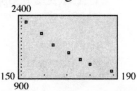

(b)   Using LnREGression on the data yields: $y = 32{,}741.02369 - 6070.956754\ln x$

(c)   Graphing $y_1 = 32{,}741.02369 - 6070.956754\ln x$

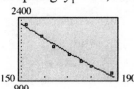

(d)   Find x when $y = 1650$:

$$1650 = 32{,}741.02369 - 6070.956754\ln x$$

$$-31{,}091.02369 = -6070.956754\ln x$$

$$\frac{-31{,}091.02369}{-6070.956754} = \ln x$$

$$5.1213 \approx \ln x$$

$$e^{5.1213} \approx x$$

$$x \approx 168 \text{ computers}$$

9.   (a)   Let $x = 0$ correspond to 1900, $x = 1$ correspond to 1910, etc.

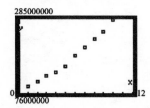

(b)   Using LOGISTIC REGression on the data yields:

$$y = \frac{799{,}475{,}916.5}{1 + 9.1968e^{-0.1603x}}$$

(c)   Graphing $y_1 = \dfrac{799{,}475{,}916.5}{1 + 9.1968e^{-0.1603x}}$:

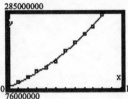

(d)   As $x \to \infty$, $y = \dfrac{799{,}475{,}916.5}{1 + 9.1968e^{-0.1603x}} \to \dfrac{799{,}475{,}916.5}{1 + 0} = 799{,}475{,}916.5$

Therefore, the carrying capacity of the United States is approximately 799,475,916 people.

(e)   In the year 2004, $x = 10.4$, so $y = \dfrac{799{,}475{,}916.5}{1 + 9.1968e^{-0.1603(10.4)}} \approx 292{,}177{,}932$ people

(f)   Find x when $y = 300{,}000{,}000$

$$\frac{799{,}475{,}916.5}{1 + 9.1968e^{-0.1603x}} = 300{,}000{,}000$$

$$799{,}475{,}916.5 = 300{,}000{,}000\left(1 + 9.1968e^{-0.1603x}\right)$$

$$\frac{799{,}475{,}916.5}{300{,}000{,}000} = 1 + 9.1968e^{-0.1603x}$$

$$\frac{799{,}475{,}916.5}{300{,}000{,}000} - 1 = 9.1968e^{-0.1603x}$$

$$1.6649 \approx 9.1968e^{-0.1603x}$$

$$\frac{1.6649}{9.1968} \approx e^{-0.1603x}$$

$$\ln\left(\frac{1.6649}{9.1968}\right) \approx -0.1603x$$

$$\frac{\ln\left(\dfrac{1.6649}{9.1968}\right)}{-0.1603} \approx x$$

$$x \approx 10.7$$

Therefore, the United States population will be 300,000,000 in the year 2007.

11.   (a)   Let $x = 0$ correspond to 1900, $x = 1$ correspond to 1910, etc.

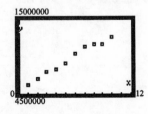

(b)   Using LOGISTIC REGression on the data yields:

$$y = \frac{14{,}471{,}245.24}{1 + 2.01527e^{-0.2458x}}$$

(c)   Graphing $y_1 = \dfrac{14{,}471{,}245.24}{1 + 2.01527e^{-0.2458x}}$:

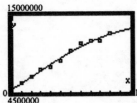

(d)   As $x \to \infty$, $y = \dfrac{14{,}471{,}245.24}{1 + 2.01527e^{-0.2458x}} \to \dfrac{14{,}471{,}245.24}{1 + 0} = 14{,}471{,}245.24$

Therefore, the carrying capacity of Illinois is approximately 14,471,245 people.

(e)   In the year 2010, $x = 11$, so $y = \dfrac{14{,}471{,}245.24}{1 + 2.01527e^{-0.2458(11)}} \approx 12{,}750{,}854$ people.

Exponential and Logarithmic Functions

4.R   Chapter Review

1. $f(x) = 3x - 5 \qquad g(x) = 1 - 2x^2$

   (a) $(f \circ g)(2) = f(g(2)) = f\left(1 - 2(2)^2\right) = f(-7) = 3(-7) - 5 = -26$

   (b) $(g \circ f)(-2) = g(f(-2)) = g\left(3(-2) - 5\right) = g(-11) = 1 - 2(-11)^2 = -241$

   (c) $(f \circ f)(4) = f(f(4)) = f\left(3(4) - 5\right) = f(7) = 3(7) - 5 = 16$

   (d) $(g \circ g)(-1) = g(g(-1)) = g\left(1 - 2(-1)^2\right) = g(-1) = 1 - 2(-1)^2 = -1$

3. $f(x) = \sqrt{x + 2} \qquad g(x) = 2x^2 + 1$

   (a) $(f \circ g)(2) = f(g(2)) = f\left(2(2)^2 + 1\right) = f(9) = \sqrt{9 + 2} = \sqrt{11}$

   (b) $(g \circ f)(-2) = g(f(-2)) = g\left(\sqrt{-2 + 2}\right) = g(0) = 2(0)^2 + 1 = 1$

   (c) $(f \circ f)(4) = f(f(4)) = f\left(\sqrt{4 + 2}\right) = f\left(\sqrt{6}\right) = \sqrt{\sqrt{6} + 2}$

   (d) $(g \circ g)(-1) = g(g(-1)) = g\left(2(-1)^2 + 1\right) = g(3) = 2(3)^2 + 1 = 19$

5. $f(x) = e^x \qquad g(x) = 3x - 2$

   (a) $(f \circ g)(2) = f(g(2)) = f\left(3(2) - 2\right) = f(4) = e^4$

   (b) $(g \circ f)(-2) = g(f(-2)) = g\left(e^{-2}\right) = g\left(\dfrac{1}{e^2}\right) = 3\left(\dfrac{1}{e^2}\right) - 2 = \dfrac{3}{e^2} - 2$

   (c) $(f \circ f)(4) = f(f(4)) = f\left(e^4\right) = e^{e^4}$

   (d) $(g \circ g)(-1) = g(g(-1)) = g\left(3(-1) - 2\right) = g(-5) = 3(-5) - 2 = -17$

7. $f(x) = 2 - x \qquad g(x) = 3x + 1$

   The domain of f is $\{x \mid x \text{ is any real number}\}$. The domain of g is $\{x \mid x \text{ is any real number}\}$.

   $(f \circ g)(x) = f(g(x)) = f(3x + 1) = 2 - (3x + 1) = 2 - 3x - 1 = 1 - 3x$

   Domain: $\{x \mid x \text{ is any real number}\}$.

   $(g \circ f)(x) = g(f(x)) = g(2 - x) = 3(2 - x) + 1 = 6 - 3x + 1 = 7 - 3x$

   Domain: $\{x \mid x \text{ is any real number}\}$.

   $(f \circ f)(x) = f(f(x)) = f(2 - x) = 2 - (2 - x) = 2 - 2 + x = x$

   Domain: $\{x \mid x \text{ is any real number}\}$.

   $(g \circ g)(x) = g(g(x)) = g(3x + 1) = 3(3x + 1) + 1 = 9x + 3 + 1 = 9x + 4$

   Domain: $\{x \mid x \text{ is any real number}\}$.

9. $f(x) = 3x^2 + x + 1$ $g(x) = |3x|$

The domain of f is $\{x \mid x \text{ is any real number}\}$. The domain of g is $\{x \mid x \text{ is any real number}\}$.

$(f \circ g)(x) = f(g(x)) = f(|3x|) = 3(|3x|)^2 + (|3x|) + 1 = 27x^2 + 3|x| + 1$

Domain: $\{x \mid x \text{ is any real number}\}$.

$(g \circ f)(x) = g(f(x)) = g(3x^2 + x + 1) = |3(3x^2 + x + 1)| = |9x^2 + 3x + 3|$

Domain: $\{x \mid x \text{ is any real number}\}$.

$(f \circ f)(x) = f(f(x)) = f(3x^2 + x + 1) = 3(3x^2 + x + 1)^2 + (3x^2 + x + 1) + 1$

$\qquad = 3(9x^4 + 6x^3 + 7x^2 + 2x + 1) + 3x^2 + x + 1 + 1$

$\qquad = 27x^4 + 18x^3 + 24x^2 + 7x + 5$

Domain: $\{x \mid x \text{ is any real number}\}$.

$(g \circ g)(x) = g(g(x)) = g(|3x|) = |3|3x|| = 9|x|$

Domain: $\{x \mid x \text{ is any real number}\}$.

11. $f(x) = \dfrac{x+1}{x-1}$ $g(x) = \dfrac{1}{x}$

The domain of f is $\{x \mid x \neq 1\}$. The domain of g is $\{x \mid x \neq 0\}$.

$(f \circ g)(x) = f(g(x)) = f\left(\dfrac{1}{x}\right) = \dfrac{\dfrac{1}{x}+1}{\dfrac{1}{x}-1} = \dfrac{\dfrac{1+x}{x}}{\dfrac{1-x}{x}} = \left(\dfrac{1+x}{x}\right)\left(\dfrac{x}{1-x}\right) = \dfrac{1+x}{1-x}$

Domain $\{x \mid x \neq 0, x \neq 1\}$.

$(g \circ f)(x) = g(f(x)) = g\left(\dfrac{x+1}{x-1}\right) = \dfrac{1}{\dfrac{x+1}{x-1}} = \dfrac{x-1}{x+1}$

Domain $\{x \mid x \neq -1, x \neq 1\}$.

$(f \circ f)(x) = f(f(x)) = f\left(\dfrac{x+1}{x-1}\right) = \dfrac{\dfrac{x+1}{x-1}+1}{\dfrac{x+1}{x-1}-1} = \dfrac{\dfrac{x+1+1(x-1)}{x-1}}{\dfrac{x+1-1(x-1)}{x-1}} = \dfrac{\dfrac{x+1+x-1}{x-1}}{\dfrac{x+1-x+1}{x-1}}$

$\qquad = \dfrac{\dfrac{2x}{x-1}}{\dfrac{2}{x-1}} = \left(\dfrac{2x}{x-1}\right)\left(\dfrac{x-1}{2}\right) = x$

Domain $\{x \mid x \neq 1\}$.

$(g \circ g)(x) = g(g(x)) = g\left(\dfrac{1}{x}\right) = \dfrac{1}{\dfrac{1}{x}} = x$

Domain $\{x \mid x \neq 0\}$.

13.   (a)   The inverse is $\{(2,1),(5,3),(8,5),(10,6)\}$.   (b)   The inverse is a function.

15.

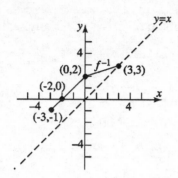

17. $f(x) = \dfrac{2x+3}{5x-2}$

$\qquad y = \dfrac{2x+3}{5x-2}$

$\qquad x = \dfrac{2y+3}{5y-2}$ Inverse

$x(5y-2) = 2y+3$

$5xy - 2x = 2y + 3$

$5xy - 2y = 2x + 3$

$y(5x-2) = 2x + 3$

$\qquad y = \dfrac{2x+3}{5x-2}$

$f^{-1}(x) = \dfrac{2x+3}{5x-2}$

Domain of f = range of f^{-1}

$\qquad$ = all real numbers except $\dfrac{2}{5}$.

Range of f = domain of f^{-1}

$\qquad$ = all real numbers except $\dfrac{2}{5}$.

19. $f(x) = \dfrac{1}{x-1}$

$\qquad y = \dfrac{1}{x-1}$

$\qquad x = \dfrac{1}{y-1}$ Inverse

$x(y-1) = 1$

$xy - x = 1$

$xy = x + 1$

$\qquad y = \dfrac{x+1}{x}$

$f^{-1}(x) = \dfrac{x+1}{x}$

Domain of f = range of f^{-1}
$\qquad$ = all real numbers except 1

Range of f = domain of f^{-1}
$\qquad$ = all real numbers except 0

21. $f(x) = \dfrac{3}{x^{1/3}}$

$y = \dfrac{3}{x^{1/3}}$

$x = \dfrac{3}{y^{1/3}}$ Inverse

$xy^{1/3} = 3$

$y^{1/3} = \dfrac{3}{x}$

$y = \dfrac{27}{x^3}$

$f^{-1}(x) = \dfrac{27}{x^3}$

Domain of f = range of f^{-1}
= all real numbers except 0
Range of f = domain of f^{-1}
= all real numbers except 0

23. (a) $f(4) = 3^4 = 81$

(c) $f(-2) = 3^{-2} = \dfrac{1}{9}$

(b) $g(9) = \log_3(9) = \log_3(3^2) = 2$

(d) $g\left(\dfrac{1}{27}\right) = \log_3\left(\dfrac{1}{27}\right) = \log_3(3^{-3}) = -3$

25. $5^2 = z$ is equivalent to $2 = \log_5 z$

27. $\log_5 u = 13$ is equivalent to $5^{13} = u$

29. $f(x) = \log(3x - 2)$ requires:
$3x - 2 > 0$

$x > \dfrac{2}{3}$

Domain: $\left\{ x \,\middle|\, x > \dfrac{2}{3} \right\}$

31. $H(x) = \log_2(x^2 - 3x + 2)$ requires $p(x) = (x - 2)(x - 1)$

$x^2 - 3x + 2 > 0$ $x = 2$ and $x = 1$ are the zeros of p.

$(x - 2)(x - 1) > 0$

Interval	$(-\infty, 1)$	$(1, 2)$	$(2, \infty)$
Number Chosen	0	1.5	3
Value of p	2	−0.25	2
Conclusion	Positive	Negative	Positive

The domain of is $\{ x \mid x < 1 \text{ or } x > 2 \}$.

33. $\log_2\left(\dfrac{1}{8}\right) = \log_2 2^{-3} = -3\log_2 2 = -3$

35. $\ln e^{\sqrt{2}} = \sqrt{2}$

37. $2^{\log_2 0.4} = 0.4$

39. $\log_3\left(\dfrac{uv^2}{w}\right) = \log_3 uv^2 - \log_3 w = \log_3 u + \log_3 v^2 - \log_3 w = \log_3 u + 2\log_3 v - \log_3 w$

41. $\log\left(x^2\sqrt{x^3+1}\right) = \log x^2 + \log\left(x^3+1\right)^{1/2} = 2\log x + \dfrac{1}{2}\log\left(x^3+1\right)$

43. $\ln\left(\dfrac{x\sqrt[3]{x^2+1}}{x-3}\right) = \ln\left(x\sqrt[3]{x^2+1}\right) - \ln(x-3)$

$$= \ln x + \ln\left(x^2+1\right)^{1/3} - \ln(x-3)$$

$$= \ln x + \dfrac{1}{3}\ln\left(x^2+1\right) - \ln(x-3)$$

45. $3\log_4 x^2 + \dfrac{1}{2}\log_4 \sqrt{x} = \log_4\left(x^2\right)^3 + \log_4\left(x^{1/2}\right)^{1/2}$

$$= \log_4 x^6 + \log_4 x^{1/4}$$

$$= \log_4\left(x^6 \cdot x^{1/4}\right)$$

$$= \log_4 x^{25/4}$$

$$= \dfrac{25}{4}\log_4 x$$

47. $\ln\left(\dfrac{x-1}{x}\right) + \ln\left(\dfrac{x}{x+1}\right) - \ln\left(x^2-1\right) = \ln\left(\dfrac{x-1}{x}\cdot\dfrac{x}{x+1}\right) - \ln\left(x^2-1\right)$

$$= \ln\left[\dfrac{\dfrac{x-1}{x+1}}{x^2-1}\right]$$

$$= \ln\left(\dfrac{x-1}{x+1}\cdot\dfrac{1}{(x-1)(x+1)}\right)$$

$$= \ln\dfrac{1}{(x+1)^2}$$

$$= \ln(x+1)^{-2}$$

$$= -2\ln(x+1)$$

49. $2\log 2 + 3\log x - \dfrac{1}{2}\left[\log(x+3) + \log(x-2)\right] = \log 2^2 + \log x^3 - \dfrac{1}{2}\log\left[(x+3)(x-2)\right]$

$$= \log\left(4x^3\right) - \log\left((x+3)(x-2)\right)^{1/2}$$

$$= \log\left[\dfrac{4x^3}{\left((x+3)(x-2)\right)^{1/2}}\right]$$

51. $\log_4 19 = \dfrac{\log 19}{\log 4} \approx 2.124$

53. $y = \log_3 x = \dfrac{\ln x}{\ln 3}$

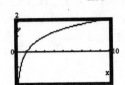

55. $f(x) = 2^{x-3}$

Using the graph of $y = 2^x$, shift the graph 3 units to the right.

Domain: $(-\infty, \infty)$

Range: $(0, \infty)$

Horizontal Asymptote: $y = 0$

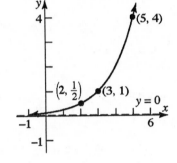

57. $f(x) = \dfrac{1}{2}\left(3^{-x}\right)$

Using the graph of $y = 3^x$, reflect the graph about the y-axis, and shrink vertically by a factor of $\dfrac{1}{2}$.

Domain: $(-\infty, \infty)$

Range: $(0, \infty)$

Horizontal Asymptote: $y = 0$

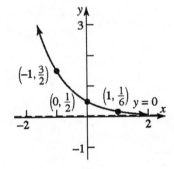

59. $f(x) = 1 - e^x$

Using the graph of $y = e^x$, reflect about the x-axis, and shift up 1 unit.

Domain: $(-\infty, \infty)$

Range: $(-\infty, 1)$

Horizontal Asymptote: $y = 1$

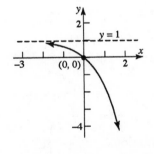

61. $f(x) = \dfrac{1}{2}\ln x$

Using the graph of $y = \ln x$, shrink vertically by a factor of $\dfrac{1}{2}$.

Domain: $(0, \infty)$

Range: $(-\infty, \infty)$

Vertical Asymptote: $x = 0$

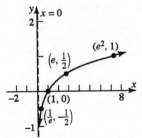

63. $f(x) = 3 - e^{-x}$
Using the graph of $y = e^x$, reflect the
graph about the y-axis, reflect about the
x-axis, and shift up 3 units.
Domain: $(-\infty, \infty)$
Range: $(-\infty, 3)$
Horizontal Asymptote: $y = 3$

65. $4^{1-2x} = 2$

$\left(2^2\right)^{1-2x} = 2$

$2^{2-4x} = 2^1$

$2 - 4x = 1$

$-4x = -1$

$x = \dfrac{1}{4}$

The solution set is $\left\{\dfrac{1}{4}\right\}$.

67. $3^{x^2+x} = \sqrt{3}$

$3^{x^2+x} = 3^{1/2}$

$x^2 + x = \dfrac{1}{2}$

$2x^2 + 2x - 1 = 0$

$x = \dfrac{-2 \pm \sqrt{2^2 - 4(2)(-1)}}{2(2)}$

$= \dfrac{-2 \pm \sqrt{12}}{4}$

$= \dfrac{-2 \pm 2\sqrt{3}}{4} = \dfrac{-1 \pm \sqrt{3}}{2}$

$x = \dfrac{-1 - \sqrt{3}}{2}$ or $x = \dfrac{-1 + \sqrt{3}}{2}$

The solution set is $\left\{\dfrac{-1-\sqrt{3}}{2}, \dfrac{-1+\sqrt{3}}{2}\right\}$.

69. $\log_x 64 = -3$

$x^{-3} = 64$

$\left(x^{-3}\right)^{-1/3} = 64^{-1/3}$

$x = \dfrac{1}{\sqrt[3]{64}} = \dfrac{1}{4}$

The solution set is $\left\{\dfrac{1}{4}\right\}$.

71. $5^x = 3^{x+2}$

$\log\left(5^x\right) = \log\left(3^{x+2}\right)$

$x \log 5 = (x + 2)\log 3$

$x \log 5 = x \log 3 + 2 \log 3$

$x \log 5 - x \log 3 = 2 \log 3$

$x(\log 5 - \log 3) = 2 \log 3$

$x = \dfrac{2 \log 3}{\log 5 - \log 3}$

The solution set is $\left\{\dfrac{2 \log 3}{\log 5 - \log 3}\right\}$.

73. $9^{2x} = 27^{3x-4}$

$\left(3^2\right)^{2x} = \left(3^3\right)^{3x-4}$

$3^{4x} = 3^{9x-12}$

$4x = 9x - 12$

$-5x = -12$

$x = \dfrac{12}{5}$

The solution set is $\left\{\dfrac{12}{5}\right\}$.

75. $\log_3 \sqrt{x-2} = 2$

$\sqrt{x-2} = 3^2$

$\sqrt{x-2} = 9$

$x - 2 = 9^2$

$x - 2 = 81$

$x = 83$

Check: $x = 83$:

$\log_3 \sqrt{83-2} = \log_3 \sqrt{81}$

$= \log_3 \sqrt{3^4}$

$= \log_3 3^2$

$= 2$

The solution set is $\{83\}$.

77. $8 = 4^{x^2} \cdot 2^{5x}$

$2^3 = \left(2^2\right)^{x^2} \cdot 2^{5x}$

$2^3 = 2^{2x^2+5x}$

$3 = 2x^2 + 5x$

$0 = 2x^2 + 5x - 3$

$0 = (2x-1)(x+3)$

$x = \dfrac{1}{2}$ or $x = -3$

The solution set is $\left\{-3, \dfrac{1}{2}\right\}$.

79. $\log_6(x+3) + \log_6(x+4) = 1$

$\log_6\big((x+3)(x+4)\big) = 1$

$(x+3)(x+4) = 6^1$

$x^2 + 7x + 12 = 6$

$x^2 + 7x + 6 = 0$

$(x+6)(x+1) = 0$

$x = -6$ or $x = -1$

Since $\log_6(-6+3) = \log_6(-3)$ is undefined, the solution set is $\{-1\}$.

81. $e^{1-x} = 5$

$1 - x = \ln 5$

$-x = -1 + \ln 5$

$x = 1 - \ln 5$

The solution set is $\{1 - \ln 5\}$.

83. $2^{3x} = 3^{2x+1}$

$\ln 2^{3x} = \ln 3^{2x+1}$

$3x \ln 2 = (2x+1)\ln 3$

$3x \ln 2 = 2x \ln 3 + \ln 3$

$3x \ln 2 - 2x \ln 3 = \ln 3$

$x(3\ln 2 - 2\ln 3) = \ln 3$

$x = \dfrac{\ln 3}{3\ln 2 - 2\ln 3}$

The solution set is $\left\{\dfrac{\ln 3}{3\ln 2 - 2\ln 3}\right\}$.

85. $h(300) = \big(30(0) + 8000\big)\log\left(\dfrac{760}{300}\right) \approx 8000\log(2.53333) \approx 3229.5$ meters

87. $P = 25e^{0.1d}$
(a) $P = 25e^{0.1(4)}$
$= 25e^{0.4}$
≈ 37.3 watts

(b) $50 = 25e^{0.1d}$
$2 = e^{0.1d}$
$\ln 2 = 0.1d$
$d = \dfrac{\ln 2}{0.1} \approx 6.9$ decibels

89.   (a) $n = \dfrac{\log 10{,}000 - \log 90{,}000}{\log(1 - 0.20)} \approx 9.85$ years

(b) $n = \dfrac{\log(0.5i) - \log(i)}{\log(1 - 0.15)} = \dfrac{\log\left(\dfrac{0.5i}{i}\right)}{\log 0.85} = \dfrac{\log 0.5}{\log 0.85} \approx 4.27$ years

91. $P = A\left(1 + \dfrac{r}{n}\right)^{-nt} = 85{,}000\left(1 + \dfrac{0.04}{2}\right)^{-2(18)} = \$41{,}668.97$

93. $A = A_0 e^{kt}$
$0.5A_0 = A_0 e^{k(5600)}$
$0.5 = e^{5600k}$
$\ln 0.5 = 5600k$
$k = \dfrac{\ln 0.5}{5600}$

$0.05A_0 = A_0 e^{\left(\frac{\ln 0.5}{5600}\right)t}$
$0.05 = e^{\left(\frac{\ln 0.5}{5600}\right)t}$
$\ln 0.05 = \left(\dfrac{\ln 0.5}{5600}\right)t \Rightarrow t = \dfrac{\ln 0.05}{\dfrac{\ln 0.5}{5600}} \approx 24{,}200$

Therefore, the man died approximately 24,200 years ago.

95. $P = P_0 e^{kt} = 6{,}302{,}486{,}693 e^{0.0167(7)} \approx 6{,}835{,}600{,}129$ people

97.   (a) $P(0) = \dfrac{0.8}{1 + 1.67e^{-0.16(0)}} = \dfrac{0.8}{1 + 1.67} = 0.2996$
In 2003, 30% of cars had a GPS.
(b)   The maximum proportion is the carrying capacity, $c = 0.80 = 80\%$
(c)   Graphing:

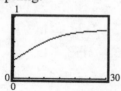

(d)   Find t such that $P(t) = 0.75$.

$$\frac{0.8}{1+1.67e^{-0.16t}} = 0.75$$

$$0.8 = 0.75\left(1+1.67e^{-0.16t}\right)$$

$$\frac{0.8}{0.75} = 1+1.67e^{-0.16t}$$

$$\frac{0.8}{0.75}-1 = 1.67e^{-0.16t}$$

$$\frac{\dfrac{0.8}{0.75}-1}{1.67} = e^{-0.16t}$$

$$\ln\left(\frac{\dfrac{0.8}{0.75}-1}{1.67}\right) = -0.16t$$

$$t = \frac{\ln\left(\dfrac{\dfrac{0.8}{0.75}-1}{1.67}\right)}{-0.16} \approx 20.13$$

So 75% of new cars will have GPS in 2023.

99. (a)   Scatter diagram

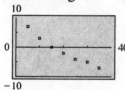

(b)   Using LnREGression on the data yields: $y = 18.9028 - 7.0963\ln x$

(c)   Graphing $y_1 = 18.9028 - 7.0963\ln x$

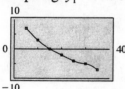

(d)   If $x = 23$, $y = 18.9028 - 7.0963\ln 23 \approx -3.33°F$.

Exponential and Logarithmic Functions

4.CR Cumulative Review

1. The graph represents a function since it passes the Vertical Line Test. The function is not a one-to-one function since the graph fails the Horizontal Line Test.

3. $x^2 + y^2 = 1$

    (a) $\left(\dfrac{1}{2}\right)^2 + \left(\dfrac{1}{2}\right)^2 = \dfrac{1}{4} + \dfrac{1}{4} = \dfrac{1}{2} \neq 1$; $\left(\dfrac{1}{2}, \dfrac{1}{2}\right)$ is not on the graph.

    (b) $\left(\dfrac{1}{2}\right)^2 + \left(\dfrac{\sqrt{3}}{2}\right)^2 = \dfrac{1}{4} + \dfrac{3}{4} = 1$; $\left(\dfrac{1}{2}, \dfrac{\sqrt{3}}{2}\right)$ is on the graph.

5. $2x - 4y = 16$

    x-intercept:
    $$2x - 4(0) = 16 \Rightarrow 2x = 16 \Rightarrow x = 8$$

    y-intercept:
    $$2(0) - 4y = 16 \Rightarrow -4y = 16 \Rightarrow y = -4$$

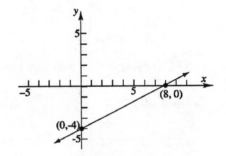

7. Given that the graph of $f(x) = ax^2 + bx + c$ has vertex $(4, -8)$ and passes through the point $(0, 24)$, we can conclude

    $$-\dfrac{b}{2a} = 4, \qquad f(4) = -8, \qquad \text{and} \qquad f(0) = 24$$

    Notice that $f(0) = 24 \Rightarrow a(0)^2 + b(0) + c = 24 \Rightarrow c = 24$

    Therefore $f(x) = ax^2 + bx + c = ax^2 + bx + 24$.

    Furthermore, $-\dfrac{b}{2a} = 4 \Rightarrow b = -8a$ and $f(4) = -8 \Rightarrow a(4)^2 + b(4) + 24 = -8$

    $$16a + 4b + 24 = -8$$
    $$16a + 4b = -32$$
    $$4a + b = -8$$

    Replacing b with $-8a$ in this equation yields
    $4a - 8a = -8 \Rightarrow -4a = -8 \Rightarrow a = 2$
    So $b = -8a = -8(2) = -16$.
    Therefore, we have the function $f(x) = 2x^2 - 16x + 24$.

9. $f(x) = x^2 + 2 \qquad g(x) = \dfrac{2}{x-3}$

$f(g(x)) = f\left(\dfrac{2}{x-3}\right) = \left(\dfrac{2}{x-3}\right)^2 + 2 = \dfrac{4}{(x-3)^2} + 2$

The domain of g is $\{x \mid x \neq 3\}$. The domain of $\{x \mid x$ is any real number$\}$.

So the domain of $f(g(x))$ is $\{x \mid x \neq 3\}$. $f(g(5)) = \dfrac{4}{(5-3)^2} + 2 = \dfrac{4}{2^2} + 2 = \dfrac{4}{4} + 2 = 3$

11.   (a) $g(x) = 3^x + 2$
Using the graph of $y = 3^x$, shift up 2 units.
Domain: $(-\infty, \infty)$
Range: $(2, \infty)$
Horizontal Asymptote: $y = 2$

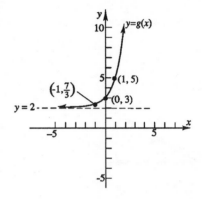

(b) $g(x) = 3^x + 2$

$\qquad y = 3^x + 2$

$\qquad x = 3^y + 2 \qquad$ Inverse

$\qquad x - 2 = 3^y$

$\ln(x - 2) = \ln(3^y)$

$\ln(x - 2) = y \cdot \ln(3)$

$\dfrac{\ln(x - 2)}{\ln(3)} = y$

$\qquad g^{-1}(x) = \dfrac{\ln(x - 2)}{\ln(3)} = \log_3(x - 2)$

Domain: $(2, \infty)$
Range: $(-\infty, \infty)$
Vertical Asymptote: $x = 2$

(c)

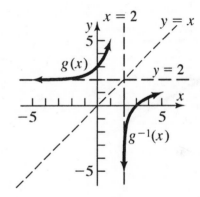

13. $\log_3(x + 1) + \log_3(2x - 3) = \log_9 9$

$\qquad \log_3((x + 1)(2x - 3)) = 1$

$\qquad\qquad (x + 1)(2x - 3) = 3^1$

$\qquad\qquad\qquad 2x^2 - x - 3 = 3$

$\qquad\qquad\qquad 2x^2 - x - 6 = 0$

$\qquad\qquad (2x + 3)(x - 2) = 0$

$\qquad\qquad x = -\dfrac{3}{2} \ \text{ or } \ x = 2$

Since $\log_3\left(-\dfrac{3}{2} + 1\right) = \log_3\left(-\dfrac{1}{2}\right)$ is undefined the solution set is $\{2\}$.

15.   (a)   Scatter diagram:

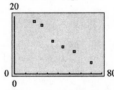

(b)   Answers will vary.
(c)   Answers will vary.

Trigonometric Functions

5.1 Angles and Their Measure

1. $C = 2\pi r$

3. standard position

5. $\dfrac{s}{t}; \dfrac{\theta}{t}$

7. True

9. False

11.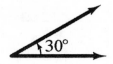

13.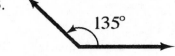

15.

17.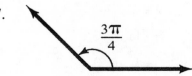

19.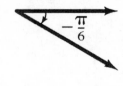

21.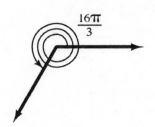

23. $40°10'25" = \left(40 + 10 \cdot \dfrac{1}{60} + 25 \cdot \dfrac{1}{60} \cdot \dfrac{1}{60}\right)° \approx (40 + 0.1667 + 0.00694)° \approx 40.17°$

25. $1°2'3" = \left(1 + 2 \cdot \dfrac{1}{60} + 3 \cdot \dfrac{1}{60} \cdot \dfrac{1}{60}\right)° \approx (1 + 0.0333 + 0.00083)° \approx 1.03°$

27. $9°9'9" = \left(9 + 9 \cdot \dfrac{1}{60} + 9 \cdot \dfrac{1}{60} \cdot \dfrac{1}{60}\right)° = (9 + 0.15 + 0.0025)° \approx 9.15°$

29. $40.32° = ?$

$0.32° = 0.32(1°) = 0.32(60') = 19.2'$

$0.2' = 0.2(1') = 0.2(60") = 12"$

$40.32° = 40° + 0.32° = 40° + 19.2' = 40° + 19' + 0.2' = 40° + 19' + 12" = 40°19'12"$

31. 18.255° = ?
$$0.255° = 0.255(1°) = 0.255(60') = 15.3'$$
$$0.3' = 0.3(1') = 0.3(60'') = 18''$$
$$18.255° = 18° + 0.255° = 18° + 15.3' = 18° + 15' + 0.3' = 18° + 15' + 18'' = 18°15'18''$$

33. 19.99° = ?
$$0.99° = 0.99(1°) = 0.99(60') = 59.4'$$
$$0.4' = 0.4(1') = 0.4(60'') = 24''$$
$$19.99° = 19° + 0.99° = 19° + 59.4' = 19° + 59' + 0.4' = 19° + 59' + 24'' = 19°59'24''$$

35. $30° = 30 \cdot \dfrac{\pi}{180}$ radian $= \dfrac{\pi}{6}$ radian

37. $240° = 240 \cdot \dfrac{\pi}{180}$ radian $= \dfrac{4\pi}{3}$ radians

39. $-60° = -60 \cdot \dfrac{\pi}{180}$ radian $= -\dfrac{\pi}{3}$ radian

41. $180° = 180 \cdot \dfrac{\pi}{180}$ radian $= \pi$ radians

43. $-135° = -135 \cdot \dfrac{\pi}{180}$ radian $= -\dfrac{3\pi}{4}$ radians

45. $-90° = -90 \cdot \dfrac{\pi}{180}$ radian $= -\dfrac{\pi}{2}$ radians

47. $\dfrac{\pi}{3} = \dfrac{\pi}{3} \cdot \dfrac{180}{\pi}$ degrees $= 60°$

49. $-\dfrac{5\pi}{4} = -\dfrac{5\pi}{4} \cdot \dfrac{180}{\pi}$ degrees $= -225°$

51. $\dfrac{\pi}{2} = \dfrac{\pi}{2} \cdot \dfrac{180}{\pi}$ degrees $= 90°$

53. $\dfrac{\pi}{12} = \dfrac{\pi}{12} \cdot \dfrac{180}{\pi}$ degrees $= 15°$

55. $-\dfrac{\pi}{2} = -\dfrac{\pi}{2} \cdot \dfrac{180}{\pi}$ degrees $= -90°$

57. $-\dfrac{\pi}{6} = -\dfrac{\pi}{6} \cdot \dfrac{180}{\pi}$ degrees $= -30°$

59. $17° = 17 \cdot \dfrac{\pi}{180}$ radian $= \dfrac{17\pi}{180}$ radian ≈ 0.30 radian

61. $-40° = -40 \cdot \dfrac{\pi}{180}$ radian $= -\dfrac{2\pi}{9}$ radian ≈ -0.70 radian

63. $125° = 125 \cdot \dfrac{\pi}{180}$ radian $= \dfrac{25\pi}{36}$ radians ≈ 2.18 radians

65. 3.14 radians $= 3.14 \cdot \dfrac{180}{\pi}$ degrees $\approx 179.91°$

67. 2 radians $= 2 \cdot \dfrac{180}{\pi}$ degrees $\approx 114.59°$

69. 6.32 radians $= 6.32 \cdot \dfrac{180}{\pi}$ degrees $\approx 362.11°$

71. $r = 10$ meters; $\theta = \dfrac{1}{2}$ radian; $s = r\theta = 10 \cdot \dfrac{1}{2} = 5$ meters

73. $\theta = \dfrac{1}{3}$ radian; $s = 2$ feet; $s = r\theta \Rightarrow r = \dfrac{s}{\theta} = \dfrac{2}{(1/3)} = 6$ feet

75. $r = 5$ miles; $s = 3$ miles; $s = r\theta \Rightarrow \theta = \dfrac{s}{r} = \dfrac{3}{5} = 0.6$ radian

77. $r = 2$ inches; $\theta = 30°$;   Convert to radians: $30° = 30 \cdot \dfrac{\pi}{180} = \dfrac{\pi}{6}$ radian

     $s = r\theta = 2 \cdot \dfrac{\pi}{6} = \dfrac{\pi}{3}$ inches

79. $r = 10$ meters; $\theta = \dfrac{1}{2}$ radian

     $A = \dfrac{1}{2} r^2 \theta = \dfrac{1}{2}(10)^2 \left(\dfrac{1}{2}\right) = \dfrac{100}{4} = 25$ square meters

81. $\theta = \dfrac{1}{3}$ radian; $A = 2$ square feet

         $A = \dfrac{1}{2} r^2 \theta \Rightarrow 2 = \dfrac{1}{2} r^2 \left(\dfrac{1}{3}\right) = \dfrac{1}{6} r^2$

         $2 = \dfrac{1}{6} r^2$

         $12 = r^2$

         $r = \sqrt{12} \approx 3.464$ feet

83. $r = 5$ miles; $A = 3$ square miles

         $A = \dfrac{1}{2} r^2 \theta \Rightarrow 3 = \dfrac{1}{2}(5)^2 \theta = \dfrac{25}{2} \theta$

         $3 = \dfrac{25}{2} \theta$

         $\dfrac{6}{25} = \theta$

         $\theta = 0.24$ radian

85. $r = 2$ inches; $\theta = 30°$;   Convert to radians: $30° = 30 \cdot \dfrac{\pi}{180} = \dfrac{\pi}{6}$ radian

     $A = \dfrac{1}{2} r^2 \theta = \dfrac{1}{2}(2)^2 \left(\dfrac{\pi}{6}\right) = \dfrac{1}{2} \cdot 4 \left(\dfrac{\pi}{6}\right) = \dfrac{\pi}{3} \approx 1.047$ square inches

87. $r = 2$ feet; $\theta = \dfrac{\pi}{3}$ radians

     $s = r\theta = 2 \cdot \dfrac{\pi}{3} = \dfrac{2\pi}{3} \approx 2.094$ feet

     $A = \dfrac{1}{2} r^2 \theta = \dfrac{1}{2}(2)^2 \left(\dfrac{\pi}{3}\right) = \dfrac{1}{2} \cdot 4 \left(\dfrac{\pi}{3}\right) = \dfrac{2\pi}{3} \approx 2.094$ square feet

89. $r = 12$ yards; $\theta = 70°$;   Convert to radians: $70° = 70 \cdot \dfrac{\pi}{180} = \dfrac{7\pi}{18}$ radians

     $s = r\theta = 12 \cdot \dfrac{7\pi}{18} \approx 14.661$ yards

     $A = \dfrac{1}{2} r^2 \theta = \dfrac{1}{2}(12)^2 \left(\dfrac{7\pi}{18}\right) = \dfrac{1}{2} \cdot 144 \left(\dfrac{7\pi}{18}\right) = 72 \left(\dfrac{7\pi}{18}\right) \approx 87.965$ square yards

91.   In 15 minutes, $r = 6$ inches; $\theta = \dfrac{15}{60}$ rev $= \dfrac{1}{4} \cdot 360° = 90° = \dfrac{\pi}{2}$ radians

$s = r\theta = 6 \cdot \dfrac{\pi}{2} = 3\pi$ inches ≈ 9.42 inches

In 25 minutes, $r = 6$ inches; $\theta = \dfrac{25}{60}$ rev $= \dfrac{5}{12} \cdot 360° = 150° = \dfrac{5\pi}{6}$ radians

$s = r\theta = 6 \cdot \dfrac{5\pi}{6} = 5\pi$ inches ≈ 15.71 inches

93. $r = 4$ m; $\theta = 45°$;   Convert to radians : $45° = 45 \cdot \dfrac{\pi}{180} = \dfrac{\pi}{4}$ radian

$A = \dfrac{1}{2} r^2 \theta = \dfrac{1}{2} (4)^2 \left(\dfrac{\pi}{4} \right) = \dfrac{1}{2} \cdot 16 \left(\dfrac{\pi}{4} \right) = 2\pi \approx 6.283$ square meters

95. $r = 30$ feet; $\theta = 135°$;   Convert to radians : $135° = 135 \cdot \dfrac{\pi}{180} = \dfrac{3\pi}{4}$ radians

$A = \dfrac{1}{2} r^2 \theta = \dfrac{1}{2} (30)^2 \left(\dfrac{3\pi}{4} \right) = \dfrac{1}{2} \cdot (900) \left(\dfrac{3\pi}{4} \right) = \dfrac{2700\pi}{8} \approx 1060.29$ square feet

97. $r = 5$ cm; $t = 20$ seconds; $\theta = \dfrac{1}{3}$ radian

$\omega = \dfrac{\theta}{t} = \dfrac{(1/3)}{20} = \dfrac{1}{3} \cdot \dfrac{1}{20} = \dfrac{1}{60}$ radian/sec

$v = \dfrac{s}{t} = \dfrac{r\theta}{t} = \dfrac{5 \cdot (1/3)}{20} = \dfrac{5}{3} \cdot \dfrac{1}{20} = \dfrac{1}{12}$ cm/sec

99. $d = 26$ inches; $r = 13$ inches; $v = 35$ mi / hr

$v = \dfrac{35 \text{ mi}}{\text{hr}} \cdot \dfrac{5280 \text{ ft}}{\text{mi}} \cdot \dfrac{12 \text{ in.}}{\text{ft}} \cdot \dfrac{1 \text{ hr}}{60 \text{ min}} = 36{,}960$ in./min

$\omega = \dfrac{v}{r} = \dfrac{36{,}960 \text{ in./min}}{13 \text{ in.}} \approx 2843.08$ radians/min $\approx \dfrac{2843.08 \text{ rad}}{\text{min}} \cdot \dfrac{1 \text{ rev}}{2\pi \text{ rad}} \approx 452.5$ rev/min

101. $r = 3960$ miles; $\theta = 35°9' - 29°57' = 5°12' = 5.2° = 5.2 \cdot \dfrac{\pi}{180} \approx 0.09076$ radian

$s = r\theta = 3960 \cdot 0.09076 \approx 359.4$ miles

103. $r = 3429.5$ miles; $\omega = 1$ rev / day $= 2\pi$ radians / day $= \dfrac{\pi}{12}$ radians / hr

$v = r\omega = 3429.5 \cdot \dfrac{\pi}{12} \approx 897.8$ miles/hr

105. $r = 2.39 \times 10^5$ miles;

$\omega = 1$ rev/27.3 days $= 2\pi$ radians/27.3 days $= \dfrac{\pi}{12 \cdot 27.3}$ radians/hr

$v = r\omega = \left(2.39 \times 10^5 \right) \cdot \dfrac{\pi}{327.6} \approx 2292$ miles/hr

107. $r_1 = 2$ inches; $r_2 = 8$ inches; $\omega_1 = 3$ rev / min $= 6\pi$ radians / min

Find ω_2:

$v_1 = v_2$

$r_1\omega_1 = r_2\omega_2 \Rightarrow 2(6\pi) = 8\omega_2$

$\omega_2 = \dfrac{12\pi}{8} = 1.5\pi$ radians/min $= \dfrac{1.5\pi}{2\pi}$ rev/min $= \dfrac{3}{4}$ rev/min

109. $r = 4$ feet; $\omega = 10$ rev / min $= 20\pi$ radians / min

$v = r\omega = 4 \cdot 20\pi = 80\pi \dfrac{\text{ft}}{\text{min}} = \dfrac{80\pi \text{ ft}}{\text{min}} \cdot \dfrac{1 \text{ mi}}{5280 \text{ ft}} \cdot \dfrac{60 \text{ min}}{\text{hr}} \approx 2.86$ mi/hr

111. $d = 8.5$ feet; $r = 4.25$ feet; $v = 9.55$ mi/hr

$\omega = \dfrac{v}{r} = \dfrac{9.55 \text{ mi/hr}}{4.25 \text{ ft}} = \dfrac{9.55 \text{ mi}}{\text{hr}} \cdot \dfrac{1}{4.25 \text{ ft}} \cdot \dfrac{5280 \text{ ft}}{\text{mi}} \cdot \dfrac{1 \text{ hr}}{60 \text{ min}} \cdot \dfrac{1 \text{ rev}}{2\pi} \approx 31.47$ rev/min

113. The earth makes one full rotation in 24 hours. The distance traveled in 24 hours is the circumference of the earth. At the equator the circumference is $2\pi(3960)$ miles. Therefore, the linear velocity a person must travel to keep up with the sun is:

$v = \dfrac{s}{t} = \dfrac{2\pi(3960)}{24} \approx 1037$ miles / hr

115. r_1 rotates at ω_1 rev/min; r_2 rotates at ω_2 rev/min

Since the linear speed of the belt connecting the pulleys is the same,

$v = r_1\omega_1 = r_2\omega_2 \Rightarrow \dfrac{r_1}{r_2} = \dfrac{\omega_2}{\omega_1}$

117. Answers will vary.

119. Linear speed measures the distance traveled per unit time, and angular speed measures the change in a central angle per unit time. In other words, linear speed describes distance traveled by a point located on the edge of a circle, and angular speed describes the turning rate of the circle itself.

121. Answers will vary.

Trigonometric Functions

5.2 Trigonometric Functions: Unit Circle Approach

1. $a^2 + b^2 = c^2$

3 True

5. $\left(-\dfrac{1}{2}, \dfrac{\sqrt{3}}{2}\right)$

7. $1 + \dfrac{1}{2} = \dfrac{3}{2}$

9. True

11. $P = \left(\dfrac{\sqrt{3}}{2}, \dfrac{1}{2}\right) \Rightarrow x = \dfrac{\sqrt{3}}{2}, y = \dfrac{1}{2}$

$\sin t = y = \dfrac{1}{2}$

$\csc t = \dfrac{1}{y} = \dfrac{1}{\dfrac{1}{2}} = 2$

$\cos t = x = \dfrac{\sqrt{3}}{2}$

$\sec t = \dfrac{1}{x} = \dfrac{1}{\dfrac{\sqrt{3}}{2}} = \dfrac{2}{\sqrt{3}} = \dfrac{2}{\sqrt{3}} \cdot \dfrac{\sqrt{3}}{\sqrt{3}} = \dfrac{2\sqrt{3}}{3}$

$\tan t = \dfrac{y}{x} = \dfrac{\dfrac{1}{2}}{\dfrac{\sqrt{3}}{2}} = \left(\dfrac{1}{2}\right) \cdot \left(\dfrac{2}{\sqrt{3}}\right) \cdot \dfrac{\sqrt{3}}{\sqrt{3}} = \dfrac{\sqrt{3}}{3}$

$\cot t = \dfrac{x}{y} = \dfrac{\dfrac{\sqrt{3}}{2}}{\dfrac{1}{2}} = \left(\dfrac{\sqrt{3}}{2}\right) \cdot \left(\dfrac{2}{1}\right) = \sqrt{3}$

13. $\left(-\dfrac{2}{5}, \dfrac{\sqrt{21}}{5}\right) \Rightarrow x = -\dfrac{2}{5}, y = \dfrac{\sqrt{21}}{5}$

$\sin t = y = \dfrac{\sqrt{21}}{5}$

$\csc t = \dfrac{1}{y} = \dfrac{1}{\dfrac{\sqrt{21}}{5}} = \dfrac{5}{\sqrt{21}} \cdot \dfrac{\sqrt{21}}{\sqrt{21}} = \dfrac{5\sqrt{21}}{21}$

$\cos t = x = -\dfrac{2}{5}$

$\sec t = \dfrac{1}{x} = \dfrac{1}{-\dfrac{2}{5}} = -\dfrac{5}{2}$

$\tan t = \dfrac{y}{x} = \dfrac{\dfrac{\sqrt{21}}{5}}{-\dfrac{2}{5}} = \dfrac{\sqrt{21}}{5} \cdot \left(-\dfrac{5}{2}\right) = -\dfrac{\sqrt{21}}{2}$

$\cot t = \dfrac{x}{y} = \dfrac{-\dfrac{2}{5}}{\dfrac{\sqrt{21}}{5}} = -\dfrac{2}{5} \cdot \dfrac{5}{\sqrt{21}} \cdot \dfrac{\sqrt{21}}{\sqrt{21}} = -\dfrac{2\sqrt{21}}{21}$

15. $P = \left(-\dfrac{\sqrt{2}}{2}, \dfrac{\sqrt{2}}{2}\right) \Rightarrow x = -\dfrac{\sqrt{2}}{2}, y = \dfrac{\sqrt{2}}{2}$

$\sin t = y = \dfrac{\sqrt{2}}{2}$

$\csc t = \dfrac{1}{y} = \dfrac{1}{\dfrac{\sqrt{2}}{2}} = \dfrac{2}{\sqrt{2}} \cdot \dfrac{\sqrt{2}}{\sqrt{2}} = \sqrt{2}$

$\cos t = x = -\dfrac{\sqrt{2}}{2}$

$\sec t = \dfrac{1}{x} = \dfrac{1}{-\dfrac{\sqrt{2}}{2}} = -\dfrac{2}{\sqrt{2}} \cdot \dfrac{\sqrt{2}}{\sqrt{2}} = -\sqrt{2}$

$\tan t = \dfrac{y}{x} = \dfrac{\dfrac{\sqrt{2}}{2}}{-\dfrac{\sqrt{2}}{2}} = \left(\dfrac{\sqrt{2}}{2}\right) \cdot \left(-\dfrac{2}{\sqrt{2}}\right) = -1$ $\cot t = \dfrac{x}{y} = \dfrac{-\dfrac{\sqrt{2}}{2}}{\dfrac{\sqrt{2}}{2}} = \left(-\dfrac{\sqrt{2}}{2}\right) \cdot \left(\dfrac{2}{\sqrt{2}}\right) = -1$

17. $\left(\dfrac{2\sqrt{2}}{3}, -\dfrac{1}{3}\right) \Rightarrow x = \dfrac{2\sqrt{2}}{3}, y = -\dfrac{1}{3}$

$\sin t = y = -\dfrac{1}{3}$

$\csc t = \dfrac{1}{y} = \dfrac{1}{-\dfrac{1}{3}} = -3$

$\cos t = x = \dfrac{2\sqrt{2}}{3}$

$\sec t = \dfrac{1}{x} = \dfrac{1}{\dfrac{2\sqrt{2}}{3}} = \dfrac{3}{2\sqrt{2}} \cdot \dfrac{\sqrt{2}}{\sqrt{2}} = \dfrac{3\sqrt{2}}{4}$

$\tan t = \dfrac{y}{x} = \dfrac{-\dfrac{1}{3}}{\dfrac{2\sqrt{2}}{3}} = -\dfrac{1}{3} \cdot \dfrac{3}{2\sqrt{2}} \cdot \dfrac{\sqrt{2}}{\sqrt{2}} = -\dfrac{\sqrt{2}}{4}$ $\cot t = \dfrac{x}{y} = \dfrac{\dfrac{2\sqrt{2}}{3}}{-\dfrac{1}{3}} = \dfrac{2\sqrt{2}}{3} \cdot \left(-\dfrac{3}{1}\right) = -2\sqrt{2}$

19.   The point on the unit circle that corresponds to $\dfrac{11\pi}{2} = 990°$ is $(0, -1)$, thus $\sin\dfrac{11\pi}{2} = -1$.

21.   The point on the unit circle that corresponds to $6\pi = 1080°$ is $(1, 0)$, thus $\tan(6\pi) = \dfrac{0}{1} = 0$.

23.   The point on the unit circle that corresponds to $\dfrac{11\pi}{2} = 990°$ is $(0, -1)$, thus $\csc\dfrac{11\pi}{2} = \dfrac{1}{-1} = -1$.

25.   The point on the unit circle that corresponds to $-\dfrac{3\pi}{2} = -270°$ is $(0, 1)$, thus $\cos\left(-\dfrac{3\pi}{2}\right) = 0$.

27.   The point on the unit circle that corresponds to $-\pi = -180°$ is $(-1, 0)$, thus $\sec(-\pi) = \dfrac{-1}{1} = -1$.

29. $\sin 45° + \cos 60° = \dfrac{\sqrt{2}}{2} + \dfrac{1}{2} = \dfrac{1 + \sqrt{2}}{2}$ 30. $\sin 30° - \cos 45° = \dfrac{1}{2} - \dfrac{\sqrt{2}}{2} = \dfrac{1 - \sqrt{2}}{2}$

31. $\sin 90° + \tan 45° = 1 + 1 = 2$

33. $\sin 45° \cdot \cos 45° = \dfrac{\sqrt{2}}{2} \cdot \dfrac{\sqrt{2}}{2} = \dfrac{2}{4} = \dfrac{1}{2}$

35. $\csc 45° \cdot \tan 60° = \sqrt{2} \cdot \sqrt{3} = \sqrt{6}$

37. $4 \sin 90° - 3 \tan 180° = 4 \cdot 1 - 3 \cdot 0 = 4$

39. $2\sin\dfrac{\pi}{3} - 3\tan\dfrac{\pi}{6} = 2 \cdot \dfrac{\sqrt{3}}{2} - 3 \cdot \dfrac{\sqrt{3}}{3} = \sqrt{3} - \sqrt{3} = 0$

41. $\sin\dfrac{\pi}{4} - \cos\dfrac{\pi}{4} = \dfrac{\sqrt{2}}{2} - \dfrac{\sqrt{2}}{2} = 0$

43. $2\sec\dfrac{\pi}{4} + 4\cot\dfrac{\pi}{3} = 2 \cdot \sqrt{2} + 4 \cdot \dfrac{\sqrt{3}}{3} = 2\sqrt{2} + \dfrac{4\sqrt{3}}{3}$

45. $\tan\pi - \cos 0 = 0 - 1 = -1$

47. $\csc\dfrac{\pi}{2} + \cot\dfrac{\pi}{2} = 1 + 0 = 1$

49. The point on the unit circle that corresponds to $\theta = \dfrac{2\pi}{3} = 120°$ is $\left(-\dfrac{1}{2}, \dfrac{\sqrt{3}}{2}\right)$.

$\sin\theta = \dfrac{\sqrt{3}}{2}$
$\qquad\qquad\qquad$
$\csc\theta = \dfrac{1}{\dfrac{\sqrt{3}}{2}} = \dfrac{2}{\sqrt{3}} \cdot \dfrac{\sqrt{3}}{\sqrt{3}} = \dfrac{2\sqrt{3}}{3}$

$\cos\theta = -\dfrac{1}{2}$
$\qquad\qquad\qquad$
$\sec\theta = \dfrac{1}{-\dfrac{1}{2}} = -2$

$\tan\theta = \dfrac{\dfrac{\sqrt{3}}{2}}{-\dfrac{1}{2}} = \dfrac{\sqrt{3}}{2} \cdot \left(-\dfrac{2}{1}\right) = -\sqrt{3}$
$\qquad$
$\cot\theta = \dfrac{-\dfrac{1}{2}}{\dfrac{\sqrt{3}}{2}} = -\dfrac{1}{2}\left(\dfrac{2}{\sqrt{3}}\right) \cdot \dfrac{\sqrt{3}}{\sqrt{3}} = -\dfrac{\sqrt{3}}{3}$

51. The point on the unit circle that corresponds to $\theta = 210° = \dfrac{7\pi}{6}$ is $\left(-\dfrac{\sqrt{3}}{2}, -\dfrac{1}{2}\right)$.

$\sin\theta = -\dfrac{1}{2}$
$\qquad\qquad\qquad$
$\csc\theta = \dfrac{1}{-\dfrac{1}{2}} = -2$

$\cos\theta = -\dfrac{\sqrt{3}}{2}$
$\qquad\qquad\qquad$
$\sec\theta = \dfrac{1}{-\dfrac{\sqrt{3}}{2}} = -\dfrac{2}{\sqrt{3}} \cdot \dfrac{\sqrt{3}}{\sqrt{3}} = -\dfrac{2\sqrt{3}}{3}$

$\tan\theta = \dfrac{-\dfrac{1}{2}}{-\dfrac{\sqrt{3}}{2}} = -\dfrac{1}{2}\left(-\dfrac{2}{\sqrt{3}}\right) \cdot \dfrac{\sqrt{3}}{\sqrt{3}} = \dfrac{\sqrt{3}}{3}$
$\qquad$
$\cot\theta = \dfrac{-\dfrac{\sqrt{3}}{2}}{-\dfrac{1}{2}} = -\dfrac{\sqrt{3}}{2} \cdot \left(-\dfrac{2}{1}\right) = \sqrt{3}$

53. The point on the unit circle that corresponds to $\theta = \dfrac{3\pi}{4} = 135°$ is $\left(-\dfrac{\sqrt{2}}{2}, \dfrac{\sqrt{2}}{2}\right)$.

$\sin\theta = \dfrac{\sqrt{2}}{2}$

$\csc\theta = \dfrac{1}{\dfrac{\sqrt{2}}{2}} = \dfrac{2}{\sqrt{2}} \cdot \dfrac{\sqrt{2}}{\sqrt{2}} = \sqrt{2}$

$\cos\theta = -\dfrac{\sqrt{2}}{2}$

$\sec\theta = \dfrac{1}{-\dfrac{\sqrt{2}}{2}} = \left(-\dfrac{2}{\sqrt{2}}\right) \cdot \dfrac{\sqrt{2}}{\sqrt{2}} = -\sqrt{2}$

$\tan\theta = \dfrac{\dfrac{\sqrt{2}}{2}}{-\dfrac{\sqrt{2}}{2}} = \dfrac{\sqrt{2}}{2} \cdot \left(-\dfrac{2}{\sqrt{2}}\right) = -1$

$\cot\theta = \dfrac{-\dfrac{\sqrt{2}}{2}}{\dfrac{\sqrt{2}}{2}} = -\dfrac{\sqrt{2}}{2} \cdot \dfrac{2}{\sqrt{2}} = -1$

55. The point on the unit circle that corresponds to $\theta = \dfrac{8\pi}{3} = 480°$ is $\left(-\dfrac{1}{2}, \dfrac{\sqrt{3}}{2}\right)$.

$\sin\theta = \dfrac{\sqrt{3}}{2}$

$\csc\theta = \dfrac{1}{\dfrac{\sqrt{3}}{2}} = \dfrac{2}{\sqrt{3}} \cdot \dfrac{\sqrt{3}}{\sqrt{3}} = \dfrac{2\sqrt{3}}{3}$

$\cos\theta = -\dfrac{1}{2}$

$\sec\theta = \dfrac{1}{-\dfrac{1}{2}} = -2$

$\tan\theta = \dfrac{\dfrac{\sqrt{3}}{2}}{-\dfrac{1}{2}} = \dfrac{\sqrt{3}}{2} \cdot \left(-\dfrac{2}{1}\right) = -\sqrt{3}$

$\cot\theta = \dfrac{-\dfrac{1}{2}}{\dfrac{\sqrt{3}}{2}} = -\dfrac{1}{2} \cdot \left(\dfrac{2}{\sqrt{3}}\right) \cdot \dfrac{\sqrt{3}}{\sqrt{3}} = -\dfrac{\sqrt{3}}{3}$

57. The point on the unit circle that corresponds to $\theta = 405° = \dfrac{9\pi}{4}$ is $\left(\dfrac{\sqrt{2}}{2}, \dfrac{\sqrt{2}}{2}\right)$.

$\sin\theta = \dfrac{\sqrt{2}}{2}$

$\csc\theta = \dfrac{1}{\dfrac{\sqrt{2}}{2}} = \dfrac{2}{\sqrt{2}} \cdot \dfrac{\sqrt{2}}{\sqrt{2}} = \sqrt{2}$

$\cos\theta = \dfrac{\sqrt{2}}{2}$

$\sec\theta = \dfrac{1}{\dfrac{\sqrt{2}}{2}} = \dfrac{2}{\sqrt{2}} \cdot \dfrac{\sqrt{2}}{\sqrt{2}} = \sqrt{2}$

$\tan\theta = \dfrac{\dfrac{\sqrt{2}}{2}}{\dfrac{\sqrt{2}}{2}} = \dfrac{\sqrt{2}}{2} \cdot \dfrac{2}{\sqrt{2}} = 1$

$\cot\theta = \dfrac{\dfrac{\sqrt{2}}{2}}{\dfrac{\sqrt{2}}{2}} = \dfrac{\sqrt{2}}{2} \cdot \dfrac{2}{\sqrt{2}} = 1$

59. The point on the unit circle that corresponds to $\theta = -\dfrac{\pi}{6} = -30°$ is $\left(\dfrac{\sqrt{3}}{2}, -\dfrac{1}{2}\right)$.

$\sin\theta = -\dfrac{1}{2}$
$\qquad\qquad\qquad\qquad$
$\csc\theta = \dfrac{1}{-\dfrac{1}{2}} = -2$

$\cos\theta = \dfrac{\sqrt{3}}{2}$
$\qquad\qquad\qquad\qquad$
$\sec\theta = \dfrac{1}{\dfrac{\sqrt{3}}{2}} = \dfrac{2}{\sqrt{3}} \cdot \dfrac{\sqrt{3}}{\sqrt{3}} = \dfrac{2\sqrt{3}}{3}$

$\tan\theta = \dfrac{-\dfrac{1}{2}}{\dfrac{\sqrt{3}}{2}} = -\dfrac{1}{2} \cdot \dfrac{2}{\sqrt{3}} \cdot \dfrac{\sqrt{3}}{\sqrt{3}} = -\dfrac{\sqrt{3}}{3}$
$\qquad$
$\cot\theta = \dfrac{\dfrac{\sqrt{3}}{2}}{-\dfrac{1}{2}} = \dfrac{\sqrt{3}}{2} \cdot \left(-\dfrac{2}{1}\right) = -\sqrt{3}$

61. The point on the unit circle that corresponds to $\theta = -45° = -\dfrac{\pi}{4}$ is $\left(\dfrac{\sqrt{2}}{2}, -\dfrac{\sqrt{2}}{2}\right)$.

$\sin\theta = -\dfrac{\sqrt{2}}{2}$
$\qquad\qquad\qquad\qquad$
$\csc\theta = \dfrac{1}{-\dfrac{\sqrt{2}}{2}} = -\dfrac{2}{\sqrt{2}} \cdot \dfrac{\sqrt{2}}{\sqrt{2}} = -\sqrt{2}$

$\cos\theta = \dfrac{\sqrt{2}}{2}$
$\qquad\qquad\qquad\qquad$
$\sec\theta = \dfrac{1}{\dfrac{\sqrt{2}}{2}} = \dfrac{2}{\sqrt{2}} \cdot \dfrac{\sqrt{2}}{\sqrt{2}} = \sqrt{2}$

$\tan\theta = \dfrac{-\dfrac{\sqrt{2}}{2}}{\dfrac{\sqrt{2}}{2}} = -\dfrac{\sqrt{2}}{2} \cdot \dfrac{2}{\sqrt{2}} = -1$
$\qquad$
$\cot\theta = \dfrac{\dfrac{\sqrt{2}}{2}}{-\dfrac{\sqrt{2}}{2}} = \dfrac{\sqrt{2}}{2} \cdot \left(-\dfrac{2}{\sqrt{2}}\right) = -1$

63. The point on the unit circle that corresponds to $\theta = \dfrac{5\pi}{2} = 450°$ is $(0, 1)$.

$\sin\theta = 1$
$\qquad\qquad\qquad\qquad$
$\csc\theta = \dfrac{1}{1} = 1$

$\cos\theta = 0$
$\qquad\qquad\qquad\qquad$
$\sec\theta = \dfrac{1}{0}$, not defined

$\tan\theta = \dfrac{1}{0}$, not defined
$\qquad\qquad$
$\cot\theta = \dfrac{0}{1} = 0$

65. The point on the unit circle that corresponds to $\theta = 720° = 4\pi$ is $(1, 0)$.

$\sin\theta = 0$
$\qquad\qquad\qquad\qquad$
$\csc\theta = \dfrac{1}{0}$, not defined

$\cos\theta = 1$
$\qquad\qquad\qquad\qquad$
$\sec\theta = \dfrac{1}{1} = 1$

$\tan\theta = \dfrac{0}{1} = 0$
$\qquad\qquad\qquad$
$\cot\theta = \dfrac{1}{0}$, not defined

67. Set the calculator to degree mode: $\sin 28° \approx 0.47$.

69. Set the calculator to degree mode: $\tan 21° \approx 0.38$.

71. Set the calculator to degree mode: $\sec 41° = \dfrac{1}{\cos 41°} \approx 1.33$.

73. Set the calculator to radian mode: $\sin \dfrac{\pi}{10} \approx 0.31$.

75. Set the calculator to radian mode: $\tan \dfrac{5\pi}{12} \approx 3.73$.

77. Set the calculator to radian mode: $\sec \dfrac{\pi}{12} = \dfrac{1}{\cos \dfrac{\pi}{12}} \approx 1.04$.

79. Set the calculator to radian mode: $\sin 1 \approx 0.84$.

81. Set the calculator to degree mode: $\sin 1° \approx 0.02$.

83. For the point $(-3, 4)$, $x = -3$, $y = 4$, $r = \sqrt{x^2 + y^2} = \sqrt{9 + 16} = \sqrt{25} = 5$

$\sin \theta = \dfrac{y}{r} = \dfrac{4}{5}$ $\cos \theta = \dfrac{x}{r} = \dfrac{-3}{5} = -\dfrac{3}{5}$ $\tan \theta = \dfrac{y}{x} = \dfrac{4}{-3} = -\dfrac{4}{3}$

$\csc \theta = \dfrac{r}{y} = \dfrac{5}{4}$ $\sec \theta = \dfrac{r}{x} = \dfrac{5}{-3} = -\dfrac{5}{3}$ $\cot \theta = \dfrac{x}{y} = \dfrac{-3}{4} = -\dfrac{3}{4}$

85. For the point $(2, -3)$, $x = 2$, $y = -3$, $r = \sqrt{x^2 + y^2} = \sqrt{4 + 9} = \sqrt{13}$

$\sin \theta = \dfrac{y}{r} = \dfrac{-3}{\sqrt{13}} \cdot \dfrac{\sqrt{13}}{\sqrt{13}} = -\dfrac{3\sqrt{13}}{13}$ $\cos \theta = \dfrac{x}{r} = \dfrac{2}{\sqrt{13}} \cdot \dfrac{\sqrt{13}}{\sqrt{13}} = \dfrac{2\sqrt{13}}{13}$ $\tan \theta = \dfrac{y}{x} = \dfrac{-3}{2} = -\dfrac{3}{2}$

$\csc \theta = \dfrac{r}{y} = \dfrac{\sqrt{13}}{-3} = -\dfrac{\sqrt{13}}{3}$ $\sec \theta = \dfrac{r}{x} = \dfrac{\sqrt{13}}{2}$ $\cot \theta = \dfrac{x}{y} = \dfrac{2}{-3} = -\dfrac{2}{3}$

87. For the point $(-2, -2)$, $x = -2$, $y = -2$, $r = \sqrt{x^2 + y^2} = \sqrt{4 + 4} = \sqrt{8} = 2\sqrt{2}$

$\sin \theta = \dfrac{y}{r} = \dfrac{-2}{2\sqrt{2}} \cdot \dfrac{\sqrt{2}}{\sqrt{2}} = -\dfrac{\sqrt{2}}{2}$ $\cos \theta = \dfrac{x}{r} = \dfrac{-2}{2\sqrt{2}} \cdot \dfrac{\sqrt{2}}{\sqrt{2}} = -\dfrac{\sqrt{2}}{2}$ $\tan \theta = \dfrac{y}{x} = \dfrac{-2}{-2} = 1$

$\csc \theta = \dfrac{r}{y} = \dfrac{2\sqrt{2}}{-2} = -\sqrt{2}$ $\sec \theta = \dfrac{r}{x} = \dfrac{2\sqrt{2}}{-2} = -\sqrt{2}$ $\cot \theta = \dfrac{x}{y} = \dfrac{-2}{-2} = 1$

89. For the point $(-3, -2)$, $x = -3$, $y = -2$, $r = \sqrt{x^2 + y^2} = \sqrt{9 + 4} = \sqrt{13}$

$\sin \theta = \dfrac{y}{r} = \dfrac{-2}{\sqrt{13}} \cdot \dfrac{\sqrt{13}}{\sqrt{13}} = -\dfrac{2\sqrt{13}}{13}$ $\cos \theta = \dfrac{x}{r} = \dfrac{-3}{\sqrt{13}} \cdot \dfrac{\sqrt{13}}{\sqrt{13}} = -\dfrac{3\sqrt{13}}{13}$ $\tan \theta = \dfrac{y}{x} = \dfrac{-2}{-3} = \dfrac{2}{3}$

$\csc \theta = \dfrac{r}{y} = \dfrac{\sqrt{13}}{-2} = -\dfrac{\sqrt{13}}{2}$ $\sec \theta = \dfrac{r}{x} = \dfrac{\sqrt{13}}{-3} = -\dfrac{\sqrt{13}}{3}$ $\cot \theta = \dfrac{x}{y} = \dfrac{-3}{-2} = \dfrac{3}{2}$

91. For the point $\left(\dfrac{1}{3}, -\dfrac{1}{4}\right)$, $x = \dfrac{1}{3}$, $y = -\dfrac{1}{4}$, $r = \sqrt{x^2 + y^2} = \sqrt{\dfrac{1}{9} + \dfrac{1}{16}} = \sqrt{\dfrac{25}{144}} = \dfrac{5}{12}$

$$\sin\theta = \dfrac{y}{r} = \dfrac{-\dfrac{1}{4}}{\dfrac{5}{12}} = -\dfrac{1}{4}\cdot\dfrac{12}{5} = -\dfrac{3}{5} \qquad \cos\theta = \dfrac{x}{r} = \dfrac{\dfrac{1}{3}}{\dfrac{5}{12}} = \dfrac{1}{3}\cdot\dfrac{12}{5} = \dfrac{4}{5} \qquad \tan\theta = \dfrac{y}{x} = \dfrac{-\dfrac{1}{4}}{\dfrac{1}{3}} = -\dfrac{1}{4}\cdot\dfrac{3}{1} = -\dfrac{3}{4}$$

$$\csc\theta = \dfrac{r}{y} = \dfrac{\dfrac{5}{12}}{-\dfrac{1}{4}} = \dfrac{5}{12}\cdot\left(-\dfrac{4}{1}\right) = -\dfrac{5}{3} \qquad \sec\theta = \dfrac{r}{x} = \dfrac{\dfrac{5}{12}}{\dfrac{1}{3}} = \dfrac{5}{12}\cdot\dfrac{3}{1} = \dfrac{5}{4} \qquad \cot\theta = \dfrac{x}{y} = \dfrac{\dfrac{1}{3}}{-\dfrac{1}{4}} = \dfrac{1}{3}\cdot\left(-\dfrac{4}{1}\right) = -\dfrac{4}{3}$$

93. $\sin 45° + \sin 135° + \sin 225° + \sin 315°$

$= \sin\dfrac{\pi}{4} + \sin\dfrac{3\pi}{4} + \sin\dfrac{5\pi}{4} + \sin\dfrac{7\pi}{4}$

$= \dfrac{\sqrt{2}}{2} + \dfrac{\sqrt{2}}{2} + \left(-\dfrac{\sqrt{2}}{2}\right) + \left(-\dfrac{\sqrt{2}}{2}\right) = 0$

95. Given: $\sin\theta = 0.1 \Rightarrow \theta$ in quadrant I or II
Therefore, $\theta + \pi$ is in quadrant III or IV $\Rightarrow \sin(\theta + \pi) = -0.1$

97. Given: $\tan\theta = 3 \Rightarrow \theta$ in quadrant I or III
Therefore, $\theta + \pi$ is in quadrant III or I $\Rightarrow \tan(\theta + \pi) = 3$

99. Given $\sin\theta = \dfrac{1}{5}$, then $\csc\theta = \dfrac{1}{\sin\theta} = \dfrac{1}{\dfrac{1}{5}} = 5$

101. $f(\theta) = \sin 60° = \dfrac{\sqrt{3}}{2}$

103. $f\left(\dfrac{\theta}{2}\right) = \sin\left(\dfrac{60°}{2}\right) = \sin 30° = \dfrac{1}{2}$

105. $[f(\theta)]^2 = [\sin 60°]^2 = \left(\dfrac{\sqrt{3}}{2}\right)^2 = \dfrac{3}{4}$

107. $f(2\theta) = \sin(2\cdot 60°) = \sin 120° = \dfrac{\sqrt{3}}{2}$

109. $2f(\theta) = 2\sin 60° = 2\cdot\dfrac{\sqrt{3}}{2} = \sqrt{3}$

111. $f(-\theta) = \sin(-60°) = -\dfrac{\sqrt{3}}{2}$

113. Complete the table:

θ	0.5	0.4	0.2	0.1	0.01	0.001	0.0001	0.00001
$\sin\theta$	0.4794	0.3894	0.1987	0.0998	0.0100	0.0010	0.0001	0.00001
$\dfrac{\sin\theta}{\theta}$	0.9589	0.9735	0.9933	0.9983	1.0000	1.0000	1.0000	1.0000

The ratio $\dfrac{\sin\theta}{\theta}$ approaches 1 as θ approaches 0.

115. Use the formula $R = \dfrac{v_0{}^2 \sin(2\theta)}{g}$ with $g = 32.2\text{ft / sec}^2$; $\theta = 45°$; $v_0 = 100\text{ ft / sec}$:

$$R = \frac{100^2 \sin(2(45°))}{32.2} \approx 310.56 \text{ feet}$$

Use the formula $H = \dfrac{v_0{}^2 \sin^2\theta}{2g}$ with $g = 32.2\text{ft / sec}^2$; $\theta = 45°$; $v_0 = 100\text{ ft / sec}$:

$$H = \frac{100^2 \sin^2(45°)}{2(32.2)} \approx 77.64 \text{ feet}$$

117. Use the formula $R = \dfrac{v_0{}^2 \sin(2\theta)}{g}$ with $g = 9.8\text{ m / sec}^2$; $\theta = 25°$; $v_0 = 500\text{ m / sec}$:

$$R = \frac{500^2 \sin(2(25°))}{9.8} \approx 19{,}542 \text{ meters}$$

Use the formula $H = \dfrac{v_0{}^2 \sin^2\theta}{2g}$ with $g = 9.8\text{ m / sec}^2$; $\theta = 25°$; $v_0 = 500\text{ m / sec}$:

$$H = \frac{500^2 \sin^2(25°)}{2(9.8)} \approx 2278 \text{ meters}$$

119. Use the formula $t = \sqrt{\dfrac{2a}{g \sin\theta \cos\theta}}$ with $g = 32\text{ ft / sec}^2$ and $a = 10$ feet :

(a) $t = \sqrt{\dfrac{2(10)}{32\sin(30°)\cos(30°)}} = \sqrt{\dfrac{20}{\left(32 \cdot \dfrac{1}{2} \cdot \dfrac{\sqrt{3}}{2}\right)}} = \sqrt{\dfrac{20}{8\sqrt{3}}} = \sqrt{\dfrac{5}{2\sqrt{3}}} \approx 1.20$ seconds

(b) $t = \sqrt{\dfrac{2(10)}{32\sin(45°)\cos(45°)}} = \sqrt{\dfrac{20}{\left(32 \cdot \dfrac{\sqrt{2}}{2} \cdot \dfrac{\sqrt{2}}{2}\right)}} = \sqrt{\dfrac{20}{16}} = \sqrt{\dfrac{5}{4}} \approx 1.12$ seconds

(c) $t = \sqrt{\dfrac{2(10)}{32\sin(60°)\cos(60°)}} = \sqrt{\dfrac{20}{\left(32 \cdot \dfrac{\sqrt{3}}{2} \cdot \dfrac{1}{2}\right)}} = \sqrt{\dfrac{20}{8\sqrt{3}}} = \sqrt{\dfrac{5}{2\sqrt{3}}} \approx 1.20$ seconds

121. (a) $T(30°) = 1 + \dfrac{2}{3\sin 30°} - \dfrac{1}{4\tan 30°} = 1 + \dfrac{2}{3 \cdot \dfrac{1}{2}} - \dfrac{1}{4 \cdot \dfrac{1}{\sqrt{3}}} = 1 + \dfrac{4}{3} - \dfrac{\sqrt{3}}{4} \approx 1.9$ hr

So Sally is on the paved road for $1 - \dfrac{1}{4\tan 30°} \approx 0.57$ hr.

(b) $T(45°) = 1 + \dfrac{2}{3\sin 45°} - \dfrac{1}{4\tan 45°} = 1 + \dfrac{2}{3 \cdot \dfrac{1}{\sqrt{2}}} - \dfrac{1}{4 \cdot 1} = 1 + \dfrac{2\sqrt{2}}{3} - \dfrac{1}{4} \approx 1.69$ hr

So Sally is on the paved road for $1 - \dfrac{1}{4\tan 45°} = 0.75$ hr.

(c) $T(60°) = 1 + \dfrac{2}{3\sin 60°} - \dfrac{1}{4\tan 60°} = 1 + \dfrac{2}{3 \cdot \dfrac{\sqrt{3}}{2}} - \dfrac{1}{4 \cdot \sqrt{3}}$

$= 1 + \dfrac{4}{3\sqrt{3}} - \dfrac{1}{4\sqrt{3}} \approx 1.63 \text{ hr}$

So Sally is on the paved road for $1 - \dfrac{1}{4\tan 60°} \approx 0.86 \text{ hr}$.

(d) $T(90°) = 1 + \dfrac{2}{3\sin 90°} - \dfrac{1}{4\tan 90°}$.

But $\tan 90°$ is undefined, so we can't use the function formula for this path.
The distance would be 2 miles in the sand and 8 miles on the road. The total time
would be: $\dfrac{2}{3} + 1 = \dfrac{5}{3} \approx 1.67$ hours.

123. (a) $R = \dfrac{\left(32^2\right)\sqrt{2}}{32} \cdot \left[\sin(2(60°)) - \cos(2(60°)) - 1\right] \approx 16.6 \text{ ft}$

(b)   Graph:

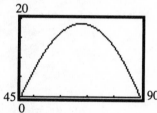

(c)   Using MAXIMUM, R is largest when $\theta \approx 67.5°$.

125–127.   Answers will vary.

Trigonometric Functions

5.3   Properties of the Trigonometric Functions

1. the set of all real numbers

3   the set of nonnegative real numbers

5. 2π, π

7. $[-1,1]$

9. False

11. $\sin 405° = \sin(360° + 45°) = \sin 45° = \dfrac{\sqrt{2}}{2}$

13. $\tan 405° = \tan(180° + 180° + 45°) = \tan 45° = 1$

15. $\csc 450° = \csc(360° + 90°) = \csc 90° = 1$

17. $\cot 390° = \cot(180° + 180° + 30°) = \cot 30° = \sqrt{3}$

19. $\cos\dfrac{33\pi}{4} = \cos\left(\dfrac{\pi}{4} + \dfrac{32\pi}{4}\right) = \cos\left(\dfrac{\pi}{4} + 8\pi\right) = \cos\left(\dfrac{\pi}{4} + 4\cdot 2\pi\right) = \cos\dfrac{\pi}{4} = \dfrac{\sqrt{2}}{2}$

21. $\tan(21\pi) = \tan(0 + 21\pi) = \tan 0 = 0$

23. $\sec\dfrac{17\pi}{4} = \sec\left(\dfrac{\pi}{4} + \dfrac{16\pi}{4}\right) = \sec\left(\dfrac{\pi}{4} + 4\pi\right) = \sec\left(\dfrac{\pi}{4} + 2\cdot 2\pi\right) = \sec\dfrac{\pi}{4} = \sqrt{2}$

25. $\tan\dfrac{19\pi}{6} = \tan\left(\dfrac{\pi}{6} + \dfrac{18\pi}{6}\right) = \tan\left(\dfrac{\pi}{6} + 3\pi\right) = \tan\dfrac{\pi}{6} = \dfrac{\sqrt{3}}{3}$

27. Since $\sin\theta > 0$ for points in quadrants I and II, and $\cos\theta < 0$ for points in quadrants II and III, the angle θ lies in quadrant II.

29. Since $\sin\theta < 0$ for points in quadrants III and IV, and $\tan\theta < 0$ for points in quadrants II and IV, the angle θ lies in quadrant IV.

31. Since $\cos\theta > 0$ for points in quadrants I and IV, and $\tan\theta < 0$ for points in quadrants II and IV, the angle θ lies in quadrant IV.

33.   Since $\sec\theta < 0$ for points in quadrants II and III, and $\sin\theta > 0$ for points in quadrants I and II, the angle θ lies in quadrant II.

35. $\sin\theta = -\dfrac{3}{5}, \quad \cos\theta = \dfrac{4}{5}$

$\tan\theta = \dfrac{\sin\theta}{\cos\theta} = \dfrac{-\dfrac{3}{5}}{\dfrac{4}{5}} = -\dfrac{3}{5}\cdot\dfrac{5}{4} = -\dfrac{3}{4}$

$\cot\theta = \dfrac{1}{\tan\theta} = -\dfrac{4}{3}$

$\sec\theta = \dfrac{1}{\cos\theta} = \dfrac{1}{\dfrac{4}{5}} = \dfrac{5}{4}$

$\csc\theta = \dfrac{1}{\sin\theta} = \dfrac{1}{-\dfrac{3}{5}} = -\dfrac{5}{3}$

37. $\sin\theta = \dfrac{2\sqrt{5}}{5}, \quad \cos\theta = \dfrac{\sqrt{5}}{5}$

$\tan\theta = \dfrac{\sin\theta}{\cos\theta} = \dfrac{\dfrac{2\sqrt{5}}{5}}{\dfrac{\sqrt{5}}{5}} = \dfrac{2\sqrt{5}}{5}\cdot\dfrac{5}{\sqrt{5}} = 2$

$\cot\theta = \dfrac{1}{\tan\theta} = \dfrac{1}{2}$

$\sec\theta = \dfrac{1}{\cos\theta} = \dfrac{1}{\dfrac{\sqrt{5}}{5}} = \dfrac{5}{\sqrt{5}}\cdot\dfrac{\sqrt{5}}{\sqrt{5}} = \sqrt{5}$

$\csc\theta = \dfrac{1}{\sin\theta} = \dfrac{1}{\dfrac{2\sqrt{5}}{5}} = \dfrac{5}{2\sqrt{5}}\cdot\dfrac{\sqrt{5}}{\sqrt{5}} = \dfrac{\sqrt{5}}{2}$

39. $\sin\theta = \dfrac{1}{2}, \quad \cos\theta = \dfrac{\sqrt{3}}{2}$

$\tan\theta = \dfrac{\sin\theta}{\cos\theta} = \dfrac{\dfrac{1}{2}}{\dfrac{\sqrt{3}}{2}} = \dfrac{1}{2}\cdot\dfrac{2}{\sqrt{3}}\cdot\dfrac{\sqrt{3}}{\sqrt{3}} = \dfrac{\sqrt{3}}{3}$

$\cot\theta = \dfrac{1}{\tan\theta} = \dfrac{1}{\dfrac{\sqrt{3}}{3}} = \dfrac{3}{\sqrt{3}}\cdot\dfrac{\sqrt{3}}{\sqrt{3}} = \sqrt{3}$

$\sec\theta = \dfrac{1}{\cos\theta} = \dfrac{1}{\dfrac{\sqrt{3}}{2}} = \dfrac{2}{\sqrt{3}}\cdot\dfrac{\sqrt{3}}{\sqrt{3}} = \dfrac{2\sqrt{3}}{3}$

$\csc\theta = \dfrac{1}{\sin\theta} = \dfrac{1}{\dfrac{1}{2}} = 2$

41. $\sin\theta = -\dfrac{1}{3}, \quad \cos\theta = \dfrac{2\sqrt{2}}{3}$

$\tan\theta = \dfrac{\sin\theta}{\cos\theta} = \dfrac{-\dfrac{1}{3}}{\dfrac{2\sqrt{2}}{3}} = -\dfrac{1}{3}\cdot\dfrac{3}{2\sqrt{2}}\cdot\dfrac{\sqrt{2}}{\sqrt{2}} = -\dfrac{\sqrt{2}}{4}$

$\cot\theta = \dfrac{1}{\tan\theta} = \dfrac{1}{-\dfrac{\sqrt{2}}{4}} = -\dfrac{4}{\sqrt{2}}\cdot\dfrac{\sqrt{2}}{\sqrt{2}} = -2\sqrt{2}$

$\sec\theta = \dfrac{1}{\cos\theta} = \dfrac{1}{\dfrac{2\sqrt{2}}{3}} = \dfrac{3}{2\sqrt{2}}\cdot\dfrac{\sqrt{2}}{\sqrt{2}} = \dfrac{3\sqrt{2}}{4}$

$\csc\theta = \dfrac{1}{\sin\theta} = \dfrac{1}{-\dfrac{1}{3}} = -3$

43.　$\sin\theta = \dfrac{12}{13}, \quad \theta$ in quadrant II

Solve for $\cos\theta$:

$$\sin^2\theta + \cos^2\theta = 1$$
$$\cos^2\theta = 1 - \sin^2\theta$$
$$\cos\theta = \pm\sqrt{1 - \sin^2\theta}$$

Since θ is in quadrant II, $\cos\theta < 0$.

$$\cos\theta = -\sqrt{1 - \sin^2\theta} = -\sqrt{1 - \left(\dfrac{12}{13}\right)^2} = -\sqrt{1 - \dfrac{144}{169}} = -\sqrt{\dfrac{25}{169}} = -\dfrac{5}{13}$$

$$\tan\theta = \dfrac{\sin\theta}{\cos\theta} = \dfrac{\dfrac{12}{13}}{-\dfrac{5}{13}} = \dfrac{12}{13}\cdot\left(-\dfrac{13}{5}\right) = -\dfrac{12}{5}$$

$$\cot\theta = \dfrac{1}{\tan\theta} = \dfrac{1}{-\dfrac{12}{5}} = -\dfrac{5}{12}$$

$$\sec\theta = \dfrac{1}{\cos\theta} = \dfrac{1}{-\dfrac{5}{13}} = -\dfrac{13}{5}$$

$$\csc\theta = \dfrac{1}{\sin\theta} = \dfrac{1}{\dfrac{12}{13}} = \dfrac{13}{12}$$

45.　$\cos\theta = -\dfrac{4}{5}, \quad \theta$ in quadrant III

Solve for $\sin\theta$:

$$\sin^2\theta + \cos^2\theta = 1$$
$$\sin^2\theta = 1 - \cos^2\theta$$
$$\sin\theta = \pm\sqrt{1 - \cos^2\theta}$$

Since θ is in quadrant III, $\sin\theta < 0$.

$$\sin\theta = -\sqrt{1 - \cos^2\theta} = -\sqrt{1 - \left(-\dfrac{4}{5}\right)^2} = -\sqrt{1 - \dfrac{16}{25}} = -\sqrt{\dfrac{9}{25}} = -\dfrac{3}{5}$$

$$\tan\theta = \dfrac{\sin\theta}{\cos\theta} = \dfrac{-\dfrac{3}{5}}{-\dfrac{4}{5}} = -\dfrac{3}{5}\cdot\left(-\dfrac{5}{4}\right) = \dfrac{3}{4}$$

$$\cot\theta = \dfrac{1}{\tan\theta} = \dfrac{1}{\dfrac{3}{4}} = \dfrac{4}{3}$$

$$\sec\theta = \dfrac{1}{\cos\theta} = \dfrac{1}{-\dfrac{4}{5}} = -\dfrac{5}{4}$$

$$\csc\theta = \dfrac{1}{\sin\theta} = \dfrac{1}{-\dfrac{3}{5}} = -\dfrac{5}{3}$$

47.　$\sin\theta = \dfrac{5}{13}, \quad 90° < \theta < 180° \Rightarrow \theta$ in quadrant II

Solve for $\cos\theta$:

$$\sin^2\theta + \cos^2\theta = 1$$
$$\cos^2\theta = 1 - \sin^2\theta$$
$$\cos\theta = \pm\sqrt{1 - \sin^2\theta}$$

Since θ is in quadrant II, $\cos\theta < 0$.

$$\cos\theta = -\sqrt{1 - \sin^2\theta} = -\sqrt{1 - \left(\dfrac{5}{13}\right)^2} = -\sqrt{1 - \dfrac{25}{169}} = -\sqrt{\dfrac{144}{169}} = -\dfrac{12}{13}$$

$$\tan\theta = \frac{\sin\theta}{\cos\theta} = \frac{\frac{5}{13}}{-\frac{12}{13}} = \frac{5}{13}\cdot\left(-\frac{13}{12}\right) = -\frac{5}{12}$$

$$\cot\theta = \frac{1}{\tan\theta} = \frac{1}{-\frac{5}{12}} = -\frac{12}{5}$$

$$\sec\theta = \frac{1}{\cos\theta} = \frac{1}{-\frac{12}{13}} = -\frac{13}{12}$$

$$\csc\theta = \frac{1}{\sin\theta} = \frac{1}{\frac{5}{13}} = \frac{13}{5}$$

49. $\cos\theta = -\dfrac{1}{3},\ \dfrac{\pi}{2} < \theta < \pi \Rightarrow \theta$ in quadrant II

Solve for $\sin\theta$:
$$\sin^2\theta + \cos^2\theta = 1$$
$$\sin^2\theta = 1 - \cos^2\theta$$
$$\sin\theta = \pm\sqrt{1 - \cos^2\theta}$$

Since θ is in quadrant II, $\sin\theta > 0$.

$$\sin\theta = \sqrt{1 - \cos^2\theta} = \sqrt{1 - \left(-\frac{1}{3}\right)^2} = \sqrt{1 - \frac{1}{9}} = \sqrt{\frac{8}{9}} = \frac{2\sqrt{2}}{3}$$

$$\tan\theta = \frac{\sin\theta}{\cos\theta} = \frac{\frac{2\sqrt{2}}{3}}{-\frac{1}{3}} = \frac{2\sqrt{2}}{3}\cdot\left(-\frac{3}{1}\right) = -2\sqrt{2}$$

$$\cot\theta = \frac{1}{\tan\theta} = \frac{1}{-2\sqrt{2}}\cdot\frac{\sqrt{2}}{\sqrt{2}} = -\frac{\sqrt{2}}{4}$$

$$\sec\theta = \frac{1}{\cos\theta} = \frac{1}{-\frac{1}{3}} = -3$$

$$\csc\theta = \frac{1}{\sin\theta} = \frac{1}{\frac{2\sqrt{2}}{3}} = \frac{3}{2\sqrt{2}}\cdot\frac{\sqrt{2}}{\sqrt{2}} = \frac{3\sqrt{2}}{4}$$

51. $\sin\theta = \dfrac{2}{3},\ \tan\theta < 0 \Rightarrow \theta$ in quadrant II

Solve for $\cos\theta$:
$$\sin^2\theta + \cos^2\theta = 1$$
$$\cos^2\theta = 1 - \sin^2\theta$$
$$\cos\theta = \pm\sqrt{1 - \sin^2\theta}$$

Since θ is in quadrant II, $\cos\theta < 0$.

$$\cos\theta = -\sqrt{1 - \sin^2\theta} = -\sqrt{1 - \left(\frac{2}{3}\right)^2} = -\sqrt{1 - \frac{4}{9}} = -\sqrt{\frac{5}{9}} = -\frac{\sqrt{5}}{3}$$

$$\tan\theta = \frac{\sin\theta}{\cos\theta} = \frac{\frac{2}{3}}{-\frac{\sqrt{5}}{3}} = \frac{2}{3}\cdot\left(-\frac{3}{\sqrt{5}}\right)\cdot\frac{\sqrt{5}}{\sqrt{5}} = -\frac{2\sqrt{5}}{5}$$

$$\cot\theta = \frac{1}{\tan\theta} = \frac{1}{-\frac{2\sqrt{5}}{5}} = -\frac{5}{2\sqrt{5}}\cdot\frac{\sqrt{5}}{\sqrt{5}} = -\frac{\sqrt{5}}{2}$$

$$\sec\theta = \frac{1}{\cos\theta} = \frac{1}{-\frac{\sqrt{5}}{3}} = -\frac{3}{\sqrt{5}}\cdot\frac{\sqrt{5}}{\sqrt{5}} = -\frac{3\sqrt{5}}{5}$$

$$\csc\theta = \frac{1}{\sin\theta} = \frac{1}{\frac{2}{3}} = \frac{3}{2}$$

53. $\sec\theta = 2$, $\sin\theta < 0 \Rightarrow \theta$ in quadrant IV

Solve for $\cos\theta$:
$$\cos\theta = \frac{1}{\sec\theta} = \frac{1}{2}$$

Solve for $\sin\theta$:
$$\sin^2\theta + \cos^2\theta = 1$$
$$\sin^2\theta = 1 - \cos^2\theta$$
$$\sin\theta = \pm\sqrt{1 - \cos^2\theta}$$

$$\sin\theta = -\sqrt{1 - \cos^2\theta} = -\sqrt{1 - \left(\frac{1}{2}\right)^2} = -\sqrt{1 - \frac{1}{4}} = -\sqrt{\frac{3}{4}} = -\frac{\sqrt{3}}{2}$$

$$\tan\theta = \frac{\sin\theta}{\cos\theta} = \frac{-\dfrac{\sqrt{3}}{2}}{\dfrac{1}{2}} = -\frac{\sqrt{3}}{2} \cdot \frac{2}{1} = -\sqrt{3}$$

$$\cot\theta = \frac{1}{\tan\theta} = \frac{1}{-\sqrt{3}} \cdot \frac{\sqrt{3}}{\sqrt{3}} = -\frac{\sqrt{3}}{3}$$

$$\csc\theta = \frac{1}{\sin\theta} = \frac{1}{-\dfrac{\sqrt{3}}{2}} = -\frac{2}{\sqrt{3}} \cdot \frac{\sqrt{3}}{\sqrt{3}} = -\frac{2\sqrt{3}}{3}$$

55. $\tan\theta = \frac{3}{4}$, $\sin\theta < 0 \Rightarrow \theta$ in quadrant III

Solve for $\sec\theta$:
$$\sec^2\theta = 1 + \tan^2\theta$$
$$\sec\theta = \pm\sqrt{1 + \tan^2\theta}$$

Since θ is in quadrant III, $\sec\theta < 0$.

$$\sec\theta = -\sqrt{1 + \tan^2\theta} = -\sqrt{1 + \left(\frac{3}{4}\right)^2} = -\sqrt{1 + \frac{9}{16}} = -\sqrt{\frac{25}{16}} = -\frac{5}{4}$$

$$\cos\theta = \frac{1}{\sec\theta} = \frac{1}{-\dfrac{5}{4}} = -\frac{4}{5}$$

$$\sin\theta = -\sqrt{1 - \cos^2\theta} = -\sqrt{1 - \left(-\frac{4}{5}\right)^2} = -\sqrt{1 - \frac{16}{25}} = -\sqrt{\frac{9}{25}} = -\frac{3}{5}$$

$$\csc\theta = \frac{1}{\sin\theta} = \frac{1}{-\dfrac{3}{5}} = -\frac{5}{3}$$

$$\cot\theta = \frac{1}{\tan\theta} = \frac{1}{\dfrac{3}{4}} = \frac{4}{3}$$

57. $\tan\theta = -\frac{1}{3}$, $\sin\theta > 0 \Rightarrow \theta$ in quadrant II

Solve for $\sec\theta$:
$$\sec^2\theta = 1 + \tan^2\theta$$
$$\sec\theta = \pm\sqrt{1 + \tan^2\theta}$$

Since θ is in quadrant II, $\sec\theta < 0$.

$$\sec\theta = -\sqrt{1+\tan^2\theta} = -\sqrt{1+\left(-\frac{1}{3}\right)^2} = -\sqrt{1+\frac{1}{9}} = -\sqrt{\frac{10}{9}} = -\frac{\sqrt{10}}{3}$$

$$\cos\theta = \frac{1}{\sec\theta} = \frac{1}{-\frac{\sqrt{10}}{3}} = -\frac{3}{\sqrt{10}} \cdot \frac{\sqrt{10}}{\sqrt{10}} = -\frac{3\sqrt{10}}{10}$$

$\sin\theta > 0$ was given.

$$\sin\theta = \sqrt{1-\cos^2\theta} = \sqrt{1-\left(-\frac{3\sqrt{10}}{10}\right)^2} = \sqrt{1-\frac{90}{100}} = \sqrt{\frac{10}{100}} = \frac{\sqrt{10}}{10}$$

$$\csc\theta = \frac{1}{\sin\theta} = \frac{1}{\frac{\sqrt{10}}{10}} = \frac{10}{\sqrt{10}} \cdot \frac{\sqrt{10}}{\sqrt{10}} = \sqrt{10} \qquad\qquad \cot\theta = \frac{1}{\tan\theta} = \frac{1}{-\frac{1}{3}} = -3$$

59. $\sin(-60°) = -\sin 60° = -\dfrac{\sqrt{3}}{2}$

61. $\tan(-30°) = -\tan 30° = -\dfrac{\sqrt{3}}{3}$

63. $\sec(-60°) = \sec 60° = 2$

65. $\sin(-90°) = -\sin 90° = -1$

67. $\tan\left(-\dfrac{\pi}{4}\right) = -\tan\dfrac{\pi}{4} = -1$

69. $\cos\left(-\dfrac{\pi}{4}\right) = \cos\dfrac{\pi}{4} = \dfrac{\sqrt{2}}{2}$

71. $\tan(-\pi) = -\tan\pi = 0$

73. $\csc\left(-\dfrac{\pi}{4}\right) = -\csc\dfrac{\pi}{4} = -\sqrt{2}$

75. $\sec\left(-\dfrac{\pi}{6}\right) = \sec\dfrac{\pi}{6} = \dfrac{2\sqrt{3}}{3}$

77. $\sin^2 40° + \cos^2 40° = 1$

79. $\sin 80° \csc 80° = \sin 80° \cdot \dfrac{1}{\sin 80°} = 1$

81. $\tan 40° - \dfrac{\sin 40°}{\cos 40°} = \tan 40° - \tan 40° = 0$

83. $\cos 400° \cdot \sec 40° = \cos(40°+360°) \cdot \sec 40° = \cos 40° \cdot \sec 40° = \cos 40° \cdot \dfrac{1}{\cos 40°} = 1$

85. $\sin\left(-\dfrac{\pi}{12}\right)\csc\dfrac{25\pi}{12} = -\sin\dfrac{\pi}{12}\csc\dfrac{25\pi}{12} = -\sin\dfrac{\pi}{12}\csc\left(\dfrac{\pi}{12}+\dfrac{24\pi}{12}\right)$

$$= -\sin\dfrac{\pi}{12}\csc\left(\dfrac{\pi}{12}+2\pi\right) = -\sin\dfrac{\pi}{12}\csc\dfrac{\pi}{12} = -\sin\dfrac{\pi}{12} \cdot \dfrac{1}{\sin\dfrac{\pi}{12}} = -1$$

87. $\dfrac{\sin(-20°)}{\cos 380°} + \tan 200° = \dfrac{-\sin 20°}{\cos(20°+360°)} + \tan(20°+180°)$

$$= \dfrac{-\sin 20°}{\cos 20°} + \tan 20° = -\tan 20° + \tan 20° = 0$$

89. If $\sin\theta = 0.3$, then $\sin\theta + \sin(\theta + 2\pi) + \sin(\theta + 4\pi) = 0.3 + 0.3 + 0.3 = 0.9$

91. If $\tan\theta = 3$, then $\tan\theta + \tan(\theta + \pi) + \tan(\theta + 2\pi) = 3 + 3 + 3 = 9$

93. $\sin 1° + \sin 2° + \sin 3° + \cdots + \sin 357° + \sin 358° + \sin 359°$
$= \sin 1° + \sin 2° + \sin 3° + \cdots + \sin(360° - 3°) + \sin(360° - 2°) + \sin(360° - 1°)$
$= \sin 1° + \sin 2° + \sin 3° + \cdots + \sin(-3°) + \sin(-2°) + \sin(-1°)$
$= \sin 1° + \sin 2° + \sin 3° + \cdots - \sin 3° - \sin 2° - \sin 2° = \sin 180° = 0$

95. The domain of the sine function is the set of all real numbers.

97. $f(\theta) = \tan\theta$ is not defined for numbers that are odd multiples of $\frac{\pi}{2}$.

99. $f(\theta) = \sec\theta$ is not defined for numbers that are odd multiples of $\frac{\pi}{2}$.

101. The range of the sine function is the set of all real numbers between -1 and 1, inclusive.

103. The range of the tangent function is the set of all real numbers.

105. The range of the secant function is the set of all real number greater than or equal to 1 and all real numbers less than or equal to -1.

107. The sine function is odd because $\sin(-\theta) = -\sin\theta$. Its graph is symmetric with respect to the origin.

109. The tangent function is odd because $\tan(-\theta) = -\tan\theta$. Its graph is symmetric with respect to the origin.

111. The secant function is even because $\sec(-\theta) = \sec\theta$. Its graph is symmetric with respect to the y-axis.

113. (a) $f(-a) = -f(a) = -\frac{1}{3}$
      (b) $f(a) + f(a + 2\pi) + f(a + 4\pi) = f(a) + f(a) + f(a) = \frac{1}{3} + \frac{1}{3} + \frac{1}{3} = 1$

115. (a) $f(-a) = -f(a) = -2$
      (b) $f(a) + f(a + \pi) + f(a + 2\pi) = f(a) + f(a) + f(a) = 2 + 2 + 2 = 6$

117. (a) $f(-a) = f(a) = -4$
      (b) $f(a) + f(a + 2\pi) + f(a + 4\pi) = f(a) + f(a) + f(a) = -4 + (-4) + (-4) = -12$

119. Since $\tan\theta = \dfrac{500}{1500} = \dfrac{1}{3} = \dfrac{y}{x}$, for $0 < \theta < \dfrac{\pi}{2}$. $r^2 = x^2 + y^2 = 9 + 1 = 10 \Rightarrow r = \sqrt{10}$.

Thus, $\sin\theta = \dfrac{1}{\sqrt{10}}$.

$$T = 5 - \dfrac{5}{3 \cdot \dfrac{1}{3}} + \dfrac{5}{\dfrac{1}{\sqrt{10}}} = 5 - 5 + 5\sqrt{10} = 5\sqrt{10} \approx 15.8 \text{ minutes}$$

121. Let $P = (x, y)$ be the point on the unit circle that corresponds to an angle θ.

Consider the equation $\tan\theta = \dfrac{y}{x} = a$. Then $y = ax$. Now $x^2 + y^2 = 1$, so $x^2 + a^2 x^2 = 1$.

Thus, $x = \pm\dfrac{1}{\sqrt{1 + a^2}}$ and $y = \pm\dfrac{a}{\sqrt{1 + a^2}}$; that is, for any real number a, there is a point $P = (x, y)$ on the unit circle for which $\tan\theta = a$. In other words, $-\infty < \tan\theta < \infty$, and the range of the tangent function is the set of all real numbers.

123. Suppose there is a number p, $0 < p < 2\pi$, for which $\sin(\theta + p) = \sin\theta$ for all θ. If $\theta = 0$, then $\sin(0 + p) = \sin p = \sin 0 = 0$; so that $p = \pi$. If $\theta = \dfrac{\pi}{2}$, then $\sin\left(\dfrac{\pi}{2} + p\right) = \sin\left(\dfrac{\pi}{2}\right)$.

But $p = \pi$. Thus, $\sin\left(\dfrac{3\pi}{2}\right) = -1 = \sin\left(\dfrac{\pi}{2}\right) = 1$, or $-1 = 1$. This is impossible. The smallest positive number p for which $\sin(\theta + p) = \sin\theta$ for all θ is therefore $p = 2\pi$.

125. $\sec\theta = \dfrac{1}{\cos\theta}$: since $\cos\theta$ has period 2π, so does $\sec\theta$.

127. If $P = (a, b)$ is the point on the unit circle corresponding to θ, then $Q = (-a, -b)$ is the point on the unit circle corresponding to $\theta + \pi$.

Thus, $\tan(\theta + \pi) = \dfrac{-b}{-a} = \dfrac{b}{a} = \tan\theta$. If there exists a number p, $0 < p < \pi$, for which $\tan(\theta + p) = \tan\theta$ for all θ, then $\theta = 0 \Rightarrow \tan(p) = \tan(0) = 0$.

But this means that p is a multiple of π. Since no multiple of π exists in the interval $(0, \pi)$, this is impossible. Therefore, the period of $f(\theta) = \tan\theta$ is π.

129. Let $P = (a, b)$ be the point on the unit circle corresponding to θ.

Then $\csc\theta = \dfrac{1}{b} = \dfrac{1}{\sin\theta}$; $\sec\theta = \dfrac{1}{a} = \dfrac{1}{\cos\theta}$; $\cot\theta = \dfrac{a}{b} = \dfrac{1}{\dfrac{b}{a}} = \dfrac{1}{\tan\theta}$.

131. $(\sin\theta\cos\phi)^2 + (\sin\theta\sin\phi)^2 + \cos^2\theta$

$= \sin^2\theta\cos^2\phi + \sin^2\theta\sin^2\phi + \cos^2\theta$

$= \sin^2\theta(\cos^2\phi + \sin^2\phi) + \cos^2\theta = \sin^2\theta + \cos^2\theta$

$= 1$

133–135.   Answers will vary.

Trigonometric Functions

5.4   Graphs of the Sine and Cosine Functions

1. $y = 3x^2$

Using the graph of $y = x^2$, vertically stretch the graph by a factor of 3.

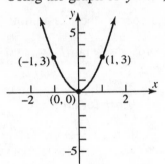

3. $1; ..., -\dfrac{3\pi}{2}, \dfrac{\pi}{2}, \dfrac{5\pi}{2}, \dfrac{9\pi}{2}, ...$

5. $3; \dfrac{2\pi}{6} = \dfrac{\pi}{3}$

7.   False

9.   0

11.   The graph of $y = \sin x$ is increasing for $-\dfrac{\pi}{2} < x < \dfrac{\pi}{2}$.

13.   The largest value of $y = \sin x$ is 1.

15. $\sin x = 0$ when $x = 0, \pi, 2\pi$.

17. $\sin x = 1$ when $x = -\dfrac{3\pi}{2}, \dfrac{\pi}{2}$; $\sin x = -1$ when $x = -\dfrac{\pi}{2}, \dfrac{3\pi}{2}$.

19.   B, C, F

21. $y = 3\sin x$; The graph of $y = \sin x$ is stretched vertically by a factor of 3.

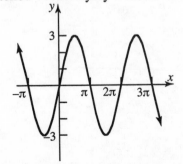

23. $y = -\cos x$; The graph of $y = \cos x$ is reflected across the x-axis.

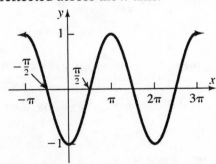

25. $y = \sin x - 1$; The graph of $y = \sin x$ is shifted down 1 unit.

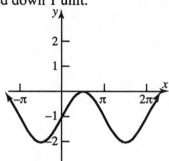

27. $y = \sin(x - \pi)$; The graph of $y = \sin x$ is shifted right π units.

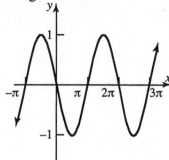

29. $y = \sin(\pi x)$; The graph of $y = \sin x$ is compressed horizontally by a factor of $\dfrac{1}{\pi}$.

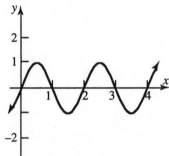

31. $y = 2\sin x + 2$; The graph of $y = \sin x$ is stretched vertically by a factor of 2 and shifted up 2 units.

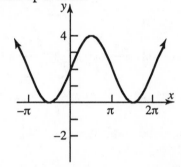

33. $y = 4\cos(2x)$; The graph of $y = \cos x$ is compressed horizontally by a factor of $\dfrac{1}{2}$, then stretched vertically by a factor of 4.

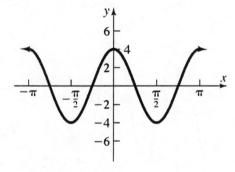

35. $y = -2\sin x + 2$; The graph of $y = \sin x$ is stretched vertically by a factor of 2, reflected across the x-axis, then shifted up 2 units.

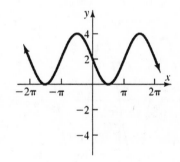

37. $y = 2\sin x$

This is in the form $y = A\sin(\omega x)$ where $A = 2$ and $\omega = 1$.

Thus, the amplitude is $|A| = |2| = 2$ and the period is $T = \dfrac{2\pi}{\omega} = \dfrac{2\pi}{1} = 2\pi$.

39. $y = -4\cos(2x)$

This is in the form $y = A\cos(\omega x)$ where $A = -4$ and $\omega = 2$.

Thus, the amplitude is $|A| = |-4| = 4$ and the period is $T = \dfrac{2\pi}{\omega} = \dfrac{2\pi}{2} = \pi$.

41. $y = 6\sin(\pi x)$

This is in the form $y = A\sin(\omega x)$ where $A = 6$ and $\omega = \pi$.

Thus, the amplitude is $|A| = |6| = 6$ and the period is $T = \dfrac{2\pi}{\omega} = \dfrac{2\pi}{\pi} = 2$.

43. $y = -\dfrac{1}{2}\cos\left(\dfrac{3}{2}x\right)$

This is in the form $y = A\cos(\omega x)$ where $A = -\dfrac{1}{2}$ and $\omega = \dfrac{3}{2}$.

Thus, the amplitude is $|A| = \left|-\dfrac{1}{2}\right| = \dfrac{1}{2}$ and the period is $T = \dfrac{2\pi}{\omega} = \dfrac{2\pi}{\dfrac{3}{2}} = \dfrac{4\pi}{3}$.

45. $y = \dfrac{5}{3}\sin\left(-\dfrac{2\pi}{3}x\right) = -\dfrac{5}{3}\sin\left(\dfrac{2\pi}{3}x\right)$

This is in the form $y = A\sin(\omega x)$ where $A = -\dfrac{5}{3}$ and $\omega = \dfrac{2\pi}{3}$.

Thus, the amplitude is $|A| = \left|-\dfrac{5}{3}\right| = \dfrac{5}{3}$ and the period is $T = \dfrac{2\pi}{\omega} = \dfrac{2\pi}{\dfrac{2\pi}{3}} = 3$.

47. F	49. A	50. I	51. H
53. C	55. J	57. A	59. B

61. $y = 5\sin(4x) \quad A = 5; \quad T = \dfrac{\pi}{2}$

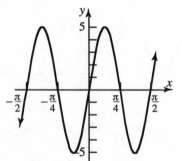

63. $y = 5\cos(\pi x) \quad A = 5; \quad T = 2$

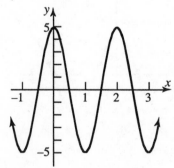

65. $y = -2\cos(2\pi x)$ $A = -2;\ \ T = 1$

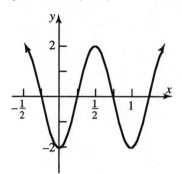

67. $y = -4\sin\left(\dfrac{1}{2}x\right)$ $A = -4;\ \ T = 4\pi$

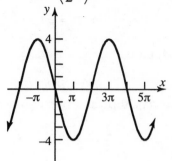

69. $y = \dfrac{3}{2}\sin\left(-\dfrac{2}{3}x\right) = -\dfrac{3}{2}\sin\left(\dfrac{2}{3}x\right)$

$A = -\dfrac{3}{2};\ \ T = 3\pi$

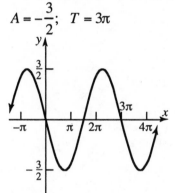

71. $|A| = 3;\ \ T = \pi;\ \ \omega = \dfrac{2\pi}{T} = \dfrac{2\pi}{\pi} = 2;\qquad y = \pm 3\sin(2x)$

73. $|A| = 3;\ \ T = 2;\ \ \omega = \dfrac{2\pi}{T} = \dfrac{2\pi}{2} = \pi;\qquad y = \pm 3\sin(\pi x)$

75.   The graph is a cosine graph with an amplitude of 5 and a period of 8.

Find ω: $8 = \dfrac{2\pi}{\omega}\ \Rightarrow\ 8\omega = 2\pi\ \Rightarrow\ \omega = \dfrac{2\pi}{8} = \dfrac{\pi}{4}$

The equation is: $y = 5\cos\left(\dfrac{\pi}{4}x\right)$.

77.   The graph is a reflected cosine graph with an amplitude of 3 and a period of 4π.

Find ω: $4\pi = \dfrac{2\pi}{\omega}\ \Rightarrow\ 4\pi\omega = 2\pi\ \Rightarrow\ \omega = \dfrac{2\pi}{4\pi} = \dfrac{1}{2}$

The equation is: $y = -3\cos\left(\dfrac{1}{2}x\right)$.

79.   The graph is a sine graph with an amplitude of $\frac{3}{4}$ and a period of 1.

Find ω: $1 = \frac{2\pi}{\omega} \Rightarrow \omega = 2\pi$

The equation is: $y = \frac{3}{4}\sin(2\pi x)$.

81.   The graph is a reflected sine graph with an amplitude of 1 and a period of $\frac{4\pi}{3}$.

Find ω: $\frac{4\pi}{3} = \frac{2\pi}{\omega} \Rightarrow 4\pi\omega = 6\pi \Rightarrow \omega = \frac{6\pi}{4\pi} = \frac{3}{2}$

The equation is: $y = -\sin\left(\frac{3}{2}x\right)$.

83.   The graph is a reflected cosine graph, shifted up 1 unit, with an amplitude of 1 and a period of $\frac{3}{2}$.

Find ω: $\frac{3}{2} = \frac{2\pi}{\omega} \Rightarrow 3\omega = 4\pi \Rightarrow \omega = \frac{4\pi}{3}$

The equation is: $y = -\cos\left(\frac{4\pi}{2}x\right) + 1$.

85.   The graph is a sine graph with an amplitude of 3 and a period of 4.

Find ω: $4 = \frac{2\pi}{\omega} \Rightarrow 4\omega = 2\pi \Rightarrow \omega = \frac{2\pi}{4} = \frac{\pi}{2}$

The equation is: $y = 3\sin\left(\frac{\pi}{2}x\right)$.

87.   The graph is a reflected cosine graph with an amplitude of 4 and a period of $\frac{2\pi}{3}$.

Find ω: $\frac{2\pi}{3} = \frac{2\pi}{\omega} \Rightarrow 2\pi\omega = 6\pi \Rightarrow \omega = \frac{6\pi}{2\pi} = 3$

The equation is: $y = -4\cos(3x)$.

89. $I = 220\sin(60\pi t),\ t \geq 0$

Period: $T = \frac{2\pi}{\omega} = \frac{2\pi}{60\pi} = \frac{1}{30}$

Amplitude: $|A| = |220| = 220$

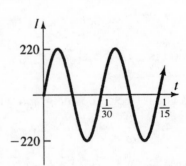

91. $V = 220\sin(120\pi t)$

  (a) Amplitude : $|A| = |220| = 220$

      Period: $T = \dfrac{2\pi}{\omega} = \dfrac{2\pi}{120\pi} = \dfrac{1}{60}$

  (b), (e)

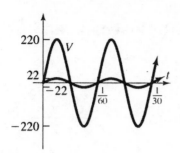

  (c) $V = IR$

       $220\sin(120\pi t) = 10I$

       $22\sin(120\pi t) = I$

  (d) Amplitude : $|A| = |22| = 22$

      Period: $T = \dfrac{2\pi}{\omega} = \dfrac{2\pi}{120\pi} = \dfrac{1}{60}$

93. (a) $P = \dfrac{V^2}{R} = \dfrac{\left(V_0\sin(2\pi f)t\right)^2}{R} = \dfrac{V_0^{\,2}\sin^2(2\pi f)t}{R} = \dfrac{V_0^{\,2}}{R}\sin^2(2\pi ft)$

  (b) The graph is the reflected cosine graph translated up a distance equivalent to the amplitude. The period is $\dfrac{1}{2f}$, so $\omega = 4\pi f$. The amplitude is $\dfrac{1}{2}\cdot\dfrac{V_0^{\,2}}{R} = \dfrac{V_0^{\,2}}{2R}$.

      The equation is: $P = -\dfrac{V_0^{\,2}}{2R}\cos(4\pi f)t + \dfrac{V_0^{\,2}}{2R} = \dfrac{V_0^{\,2}}{R}\dfrac{1}{2}\left(1-\cos(4\pi f)t\right)$

  (c) Comparing the formulas:

       $\sin^2(2\pi ft) = \dfrac{1}{2}\left(1-\cos(4\pi ft)\right)$

95. $y = |\cos x|, \quad -2\pi \le x \le 2\pi$

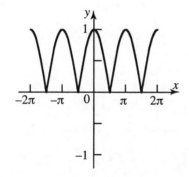

97–101. Answers will vary.

Trigonometric Functions

5.5 Graphs of the Tangent, Cotangent, Cosecant, and Secant Functions

1. $x = 4$

3. origin; $x = \ldots, -\dfrac{3\pi}{2}, -\dfrac{\pi}{2}, \dfrac{\pi}{2}, \dfrac{3\pi}{2}, \ldots$

5. $y = \cos x$

7. 0

9. 1

11. $\sec x = 1$ for $x = -2\pi,\ 0,\ 2\pi;$ $\sec x = -1$ for $x = -\pi,\ \pi$

13. $y = \sec x$ has vertical asymptotes for $x = -\dfrac{3\pi}{2}, -\dfrac{\pi}{2}, \dfrac{\pi}{2}, \dfrac{3\pi}{2}$

15. $y = \tan x$ has vertical asymptotes for $x = -\dfrac{3\pi}{2}, -\dfrac{\pi}{2}, \dfrac{\pi}{2}, \dfrac{3\pi}{2}$

17. D

19. B

21. $y = -\sec x$; The graph of $y = \sec x$ is reflected across the x-axis.

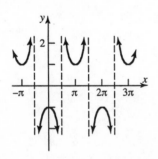

23. $y = \sec\!\left(x - \dfrac{\pi}{2}\right)$; The graph of $y = \sec x$ is

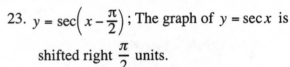

shifted right $\dfrac{\pi}{2}$ units.

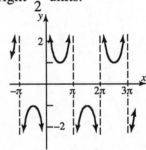

25. $y = \tan(x - \pi)$; The graph of $y = \tan x$ is shifted right π units.

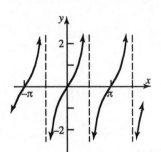

27. $y = 3\tan(2x)$; The graph of $y = \tan x$ is compressed horizontally by a factor of $\dfrac{1}{2}$ and stretched vertically by a factor of 3.

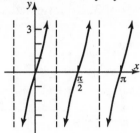

29. $y = \sec(2x)$; The graph of $y = \sec x$ is compressed horizontally by a factor of $\dfrac{1}{2}$.

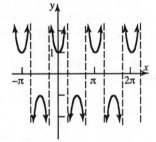

31. $y = \cot(\pi x)$; The graph of $y = \cot x$ is compressed horizontally by a factor of $\dfrac{1}{\pi}$.

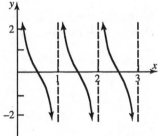

33. $y = -3\tan(4x)$; The graph of $y = \tan x$ is compressed horizontally by a factor of $\dfrac{1}{4}$, stretched vertically by a factor of 3 and reflected across the x-axis.

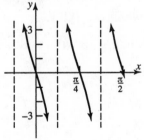

35. $y = 2\sec\left(\dfrac{1}{2}x\right)$; The graph of $y = \sec x$ is stretched horizontally by a factor of 2, and stretched vertically by a factor of 2.

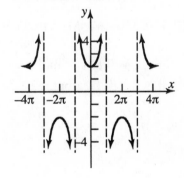

37. $y = -3\csc\left(x + \dfrac{\pi}{4}\right)$; The graph of $y = \csc x$ is shifted left $\dfrac{\pi}{4}$ units, stretched vertically by a factor of 3 and reflected across the x-axis.

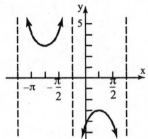

39. $y = \dfrac{1}{2}\cot\left(x - \dfrac{\pi}{4}\right)$; The graph of $y = \cot x$ is shifted right $\dfrac{\pi}{4}$ units and compressed vertically by a factor of $\dfrac{1}{2}$.

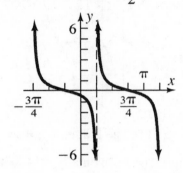

41. (a) Consider the length of the line segment in two sections, x, the portion across the hall that is 3 feet wide and y, the portion across that hall that is 4 feet wide. Then,

$$\cos\theta = \frac{3}{x} \implies x = \frac{3}{\cos\theta} \quad \text{and} \quad \sin\theta = \frac{4}{y} \implies y = \frac{4}{\sin\theta}$$

$$L = x + y = \frac{3}{\cos\theta} + \frac{4}{\sin\theta}$$

(b) Graph:

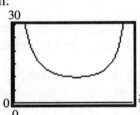

(c) Use MINIMUM to find the least value: L is least when $\theta = 0.83$.

(d) $L \approx \dfrac{3}{\cos(0.83)} + \dfrac{4}{\sin(0.83)} \approx 9.86$ feet. Note that rounding up will result in a ladder that won't fit around the corner. Answers will vary.

Trigonometric Functions

5.6 Phase Shift; Sinusoidal Curve Fitting

1. phase shift

3. $y = 4\sin(2x - \pi)$

Amplitude: $|A| = |4| = 4$

Period: $T = \dfrac{2\pi}{\omega} = \dfrac{2\pi}{2} = \pi$

Phase Shift: $\dfrac{\phi}{\omega} = \dfrac{\pi}{2}$

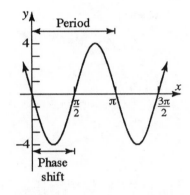

5. $y = 2\cos\left(3x + \dfrac{\pi}{2}\right)$

Amplitude: $|A| = |2| = 2$

Period: $T = \dfrac{2\pi}{\omega} = \dfrac{2\pi}{3}$

Phase Shift: $\dfrac{\phi}{\omega} = \dfrac{-\dfrac{\pi}{2}}{3} = -\dfrac{\pi}{6}$

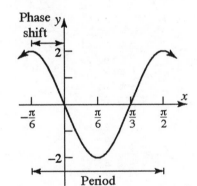

7. $y = -3\sin\left(2x + \dfrac{\pi}{2}\right)$

Amplitude: $|A| = |-3| = 3$

Period: $T = \dfrac{2\pi}{\omega} = \dfrac{2\pi}{2} = \pi$

Phase Shift: $\dfrac{\phi}{\omega} = \dfrac{-\dfrac{\pi}{2}}{2} = -\dfrac{\pi}{4}$

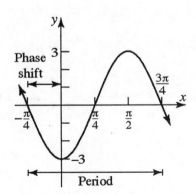

9. $y = 4\sin(\pi x + 2)$

Amplitude : $\left|A\right| = \left|4\right| = 4$

Period : $T = \dfrac{2\pi}{\omega} = \dfrac{2\pi}{\pi} = 2$

Phase Shift : $\dfrac{\phi}{\omega} = \dfrac{-2}{\pi} = -\dfrac{2}{\pi}$

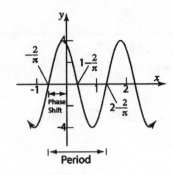

11. $y = 3\cos(\pi x - 2)$

Amplitude : $\left|A\right| = \left|3\right| = 3$

Period : $T = \dfrac{2\pi}{\omega} = \dfrac{2\pi}{\pi} = 2$

Phase Shift : $\dfrac{\phi}{\omega} = \dfrac{2}{\pi}$

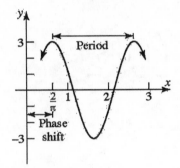

13. $y = 3\sin\left(-2x + \dfrac{\pi}{2}\right) = 3\sin\left(-\left(2x - \dfrac{\pi}{2}\right)\right)$

$\qquad\qquad\qquad = -3\sin\left(2x - \dfrac{\pi}{2}\right)$

Amplitude : $\left|A\right| = \left|-3\right| = 3$

Period : $T = \dfrac{2\pi}{\omega} = \dfrac{2\pi}{2} = \pi$

Phase Shift : $\dfrac{\phi}{\omega} = \dfrac{\dfrac{\pi}{2}}{2} = \dfrac{\pi}{4}$

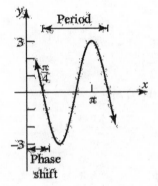

15. $\left|A\right| = 2;\ \ T = \pi;\ \ \dfrac{\phi}{\omega} = \dfrac{1}{2};\ \ \omega = \dfrac{2\pi}{T} = \dfrac{2\pi}{\pi} = 2;\ \ \dfrac{\phi}{\omega} = \dfrac{\phi}{2} = \dfrac{1}{2}\ \Rightarrow\ \phi = 1$

$y = \pm 2\sin(2x - 1) = \pm 2\sin\left[2\left(x - \dfrac{1}{2}\right)\right]$

17. $\left|A\right| = 3;\ \ T = 3\pi;\ \ \dfrac{\phi}{\omega} = -\dfrac{1}{3};\ \ \omega = \dfrac{2\pi}{T} = \dfrac{2\pi}{3\pi} = \dfrac{2}{3};$

$\dfrac{\phi}{\omega} = \dfrac{\phi}{\dfrac{2}{3}} = -\dfrac{1}{3}\ \Rightarrow\ \phi = -\dfrac{1}{3}\cdot\dfrac{2}{3} = -\dfrac{2}{9}$

$y = \pm 3\sin\left(\dfrac{2}{3}x + \dfrac{2}{9}\right) = \pm 3\sin\left[\dfrac{2}{3}\left(x + \dfrac{1}{3}\right)\right]$

19. $I = 120\sin\left(30\pi t - \dfrac{\pi}{3}\right), \quad t \geq 0$

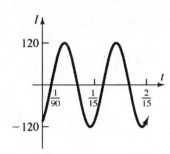

Period: $T = \dfrac{2\pi}{\omega} = \dfrac{2\pi}{30\pi} = \dfrac{1}{15}$

Amplitude: $|A| = |120| = 120$

Phase Shift: $\dfrac{\phi}{\omega} = \dfrac{\frac{\pi}{3}}{30\pi} = \dfrac{1}{90}$

21. (a) Draw a scatter diagram:

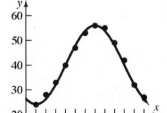

(b) Amplitude: $A = \dfrac{56.0 - 24.2}{2} = \dfrac{31.8}{2} = 15.9$

Vertical Shift: $\dfrac{56.0 + 24.2}{2} = \dfrac{80.2}{2} = 40.1$

$\omega = \dfrac{2\pi}{12} = \dfrac{\pi}{6}$

Phase shift (use $y = 24.2, \; x = 1$):

$24.2 = 15.9\sin\left(\dfrac{\pi}{6} \cdot 1 - \phi\right) + 40.1$

$-15.9 = 15.9\sin\left(\dfrac{\pi}{6} - \phi\right)$

$-1 = \sin\left(\dfrac{\pi}{6} - \phi\right)$

$-\dfrac{\pi}{2} = \dfrac{\pi}{6} - \phi$

$\phi = \dfrac{2\pi}{3}$

Thus, $y = 15.9\sin\left(\dfrac{\pi}{6}x - \dfrac{2\pi}{3}\right) + 40.1.$

(c)

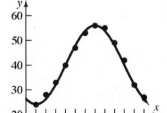

(d) $y = 15.62\sin(0.517x - 2.096) + 40.377$

(e)

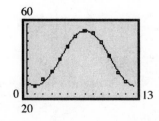

23. (a) Draw a scatter diagram:

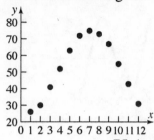

(b) Amplitude: $A = \dfrac{75.4 - 25.5}{2} = \dfrac{49.9}{2} = 24.95$

Vertical Shift: $\dfrac{75.4 + 25.5}{2} = \dfrac{100.9}{2} = 50.45$

$\omega = \dfrac{2\pi}{12} = \dfrac{\pi}{6}$

Phase shift (use $y = 25.5$, $x = 1$):

$$25.5 = 24.95 \sin\left(\dfrac{\pi}{6} \cdot 1 - \phi\right) + 50.45$$

$$-24.95 = 24.95 \sin\left(\dfrac{\pi}{6} - \phi\right)$$

$$-1 = \sin\left(\dfrac{\pi}{6} - \phi\right)$$

$$-\dfrac{\pi}{2} = \dfrac{\pi}{6} - \phi$$

$$\phi = \dfrac{2\pi}{3}$$

Thus, $y = 24.95 \sin\left(\dfrac{\pi}{6}x - \dfrac{2\pi}{3}\right) + 50.45$.

(c)

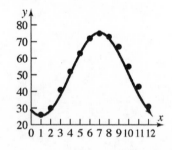

(d) $y = 25.693 \sin(0.476x - 1.814) + 49.854$

(e)

25. (a) $3.6333 + 12.5 = 16.1333$ hours which is at 4:08 PM.

(b) Amplitude: $A = \dfrac{8.2 - (-0.6)}{2} = \dfrac{8.8}{2} = 4.4$

Vertical Shift: $\dfrac{8.2 + (-0.6)}{2} = \dfrac{7.6}{2} = 3.8$

$\omega = \dfrac{2\pi}{12.5} = \dfrac{\pi}{6.25}$

Phase shift (use $y = -0.6$, $x = 10.1333$):

$$-0.6 = 4.4\sin\left(\frac{\pi}{6.25}\cdot 10.1333 - \phi\right) + 3.8$$

$$-4.4 = 4.4\sin\left(\frac{\pi}{6.25}\cdot 10.1333 - \phi\right)$$

$$-1 = \sin\left(\frac{10.1333\pi}{6.25} - \phi\right)$$

$$-\frac{\pi}{2} = \frac{10.1333\pi}{6.25} - \phi$$

$$\phi \approx 6.6643$$

Thus, $y = 4.4\sin\left(\frac{\pi}{6.25}x - 6.6643\right) + 3.8$

(c)

(d) $y = 4.4\sin\left(\frac{\pi}{6.25}(16.1333) - 6.6643\right) + 3.8 \approx 8.2$ feet

27. (a)   Amplitude : $A = \dfrac{12.75 - 10.583}{2} = \dfrac{2.167}{2} = 1.0835$

Vertical Shift : $\dfrac{12.75 + 10.583}{2} = \dfrac{23.333}{2} = 11.6665$

$\omega = \dfrac{2\pi}{365}$

Phase shift (use $y = 10.583$, $x = 355$):

$$10.583 = 1.0835\sin\left(\frac{2\pi}{365}\cdot 355 - \phi\right) + 11.6665$$

$$-1.0835 = 1.0835\sin\left(\frac{2\pi}{365}\cdot 355 - \phi\right)$$

$$-1 = \sin\left(\frac{710\pi}{365} - \phi\right)$$

$$-\frac{\pi}{2} = \frac{710\pi}{365} - \phi$$

$$\phi \approx 7.6818$$

Thus, $y = 1.0835\sin\left(\frac{2\pi}{365}x - 7.6818\right) + 11.6665$.

(b) $y = 1.0835\sin\left(\frac{2\pi}{365}(91) - 7.6818\right) + 11.6665 \approx 11.85$ hours

(c)

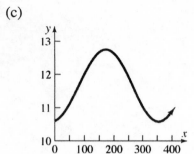

(d)   Answers will vary.

29.   (a)   Amplitude: $A = \dfrac{16.233 - 5.45}{2} = \dfrac{10.783}{2} = 5.3915$

Vertical Shift: $\dfrac{16.233 + 5.45}{2} = \dfrac{21.683}{2} = 10.8415$

$\omega = \dfrac{2\pi}{365}$

Phase shift (use $y = 5.45$, $x = 355$):

$$5.45 = 5.3915 \sin\left(\dfrac{2\pi}{365} \cdot 355 - \phi\right) + 10.8415$$

$$-5.3915 = 5.3915 \sin\left(\dfrac{2\pi}{365} \cdot 355 - \phi\right)$$

$$-1 = \sin\left(\dfrac{710\pi}{365} - \phi\right)$$

$$-\dfrac{\pi}{2} = \dfrac{710\pi}{365} - \phi$$

$$\phi \approx 7.6818$$

Thus, $y = 5.3915 \sin\left(\dfrac{2\pi}{365} x - 7.6818\right) + 10.8415$.

(b) $y = 5.3915 \sin\left(\dfrac{2\pi}{365}(91) - 7.6818\right) + 10.8415 \approx 11.74$ hours

(c)

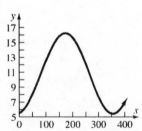

(d)   Answers will vary.

31.   Answers will vary.

Trigonometric Functions

5.R   Chapter Review

1. $135° = 135 \cdot \dfrac{\pi}{180}$ radian $= \dfrac{3\pi}{4}$ radians

3. $18° = 18 \cdot \dfrac{\pi}{180}$ radian $= \dfrac{\pi}{10}$ radian

5. $\dfrac{3\pi}{4} = \dfrac{3\pi}{4} \cdot \dfrac{180}{\pi}$ degrees $= 135°$

7. $-\dfrac{5\pi}{2} = -\dfrac{5\pi}{2} \cdot \dfrac{180}{\pi}$ degrees $= -450°$

9. $\tan\dfrac{\pi}{4} - \sin\dfrac{\pi}{6} = 1 - \dfrac{1}{2} = \dfrac{1}{2}$

11. $3\sin 45° - 4\tan\dfrac{\pi}{6} = 3 \cdot \dfrac{\sqrt{2}}{2} - 4 \cdot \dfrac{\sqrt{3}}{3} = \dfrac{3\sqrt{2}}{2} - \dfrac{4\sqrt{3}}{3}$

13. $6\cos\dfrac{3\pi}{4} + 2\tan\left(-\dfrac{\pi}{3}\right) = 6\left(-\dfrac{\sqrt{2}}{2}\right) + 2\left(-\sqrt{3}\right) = -3\sqrt{2} - 2\sqrt{3}$

15. $\sec\left(-\dfrac{\pi}{3}\right) - \cot\left(-\dfrac{5\pi}{4}\right) = \sec\dfrac{\pi}{3} + \cot\dfrac{5\pi}{4} = 2 + 1 = 3$

17. $\tan\pi + \sin\pi = 0 + 0 = 0$

19. $\cos 540° - \tan(-45°) = -1 - (-1) = -1 + 1 = 0$

21. $\sin^2 20° + \dfrac{1}{\sec^2 20°} = \sin^2 20° + \cos^2 20° = 1$

23. $\sec 50° \cdot \cos 50° = \dfrac{1}{\cos 50°} \cdot \cos 50° = 1$

25. $\sec^2 20° - \tan^2 20° = \tan^2 20° + 1 - \tan^2 20° = 1$

27. $\sin(-40°) \cdot \csc(40°) = -\sin(40°) \cdot \dfrac{1}{\sin(40°)} = -1$

29. $\cos(410°) \cdot \sec(-50°) = \cos(360° + 50°) \cdot \sec(50°) = \cos(50°) \cdot \dfrac{1}{\cos(50°)} = 1$

31. $\sin\theta = \dfrac{4}{5}, \quad \theta \text{ acute} \Rightarrow \theta \text{ in quadrant I}$

Solve for $\cos\theta$:
$$\sin^2\theta + \cos^2\theta = 1$$
$$\cos^2\theta = 1 - \sin^2\theta$$
$$\cos\theta = \pm\sqrt{1 - \sin^2\theta}$$

Since θ is in quadrant I, $\cos\theta > 0$.

$$\cos\theta = \sqrt{1 - \sin^2\theta} = \sqrt{1 - \left(\dfrac{4}{5}\right)^2} = \sqrt{1 - \dfrac{16}{25}} = \sqrt{\dfrac{9}{25}} = \dfrac{3}{5}$$

$\tan\theta = \dfrac{\sin\theta}{\cos\theta} = \dfrac{\frac{4}{5}}{\frac{3}{5}} = \dfrac{4}{5} \cdot \dfrac{5}{3} = \dfrac{4}{3}$

$\cot\theta = \dfrac{1}{\tan\theta} = \dfrac{1}{\frac{4}{3}} = \dfrac{3}{4}$

$\sec\theta = \dfrac{1}{\cos\theta} = \dfrac{1}{\frac{3}{5}} = \dfrac{5}{3}$

$\csc\theta = \dfrac{1}{\sin\theta} = \dfrac{1}{\frac{4}{5}} = \dfrac{5}{4}$

33. $\tan\theta = \dfrac{12}{5}, \quad \sin\theta < 0 \Rightarrow \theta \text{ in quadrant III}$

Solve for $\sec\theta$:
$$\sec^2\theta = \tan^2\theta + 1$$
$$\sec\theta = \pm\sqrt{\tan^2\theta + 1}$$

Since θ is in quadrant III, $\sec\theta < 0$.

$$\sec\theta = -\sqrt{\tan^2\theta + 1} = -\sqrt{\left(\dfrac{12}{5}\right)^2 + 1} = -\sqrt{\dfrac{144}{25} + 1} = -\sqrt{\dfrac{169}{25}} = -\dfrac{13}{5}$$

$$\cos\theta = \dfrac{1}{\sec\theta} = \dfrac{1}{-\frac{13}{5}} = -\dfrac{5}{13}$$

Solve for $\sin\theta$:
$$\sin^2\theta + \cos^2\theta = 1$$
$$\sin^2\theta = 1 - \cos^2\theta$$
$$\sin\theta = \pm\sqrt{1 - \cos^2\theta}$$

Since θ is in quadrant III, $\sin\theta < 0$.

$$\sin\theta = -\sqrt{1-\cos^2\theta} = -\sqrt{1-\left(-\frac{5}{13}\right)^2} = -\sqrt{1-\frac{25}{169}} = -\sqrt{\frac{144}{169}} = -\frac{12}{13}$$

$$\cot\theta = \frac{1}{\tan\theta} = \frac{1}{\frac{12}{5}} = \frac{5}{12} \qquad\qquad \csc\theta = \frac{1}{\sin\theta} = \frac{1}{-\frac{12}{13}} = -\frac{13}{12}$$

35. $\sec\theta = -\dfrac{5}{4}, \quad \tan\theta < 0 \Rightarrow \theta$ in quadrant II

Solve for $\tan\theta$:
$$\tan^2\theta + 1 = \sec^2\theta$$
$$\tan\theta = \pm\sqrt{\sec^2\theta - 1}$$
Since θ is in quadrant II, $\tan\theta < 0$.

$$\tan\theta = -\sqrt{\sec^2\theta - 1} = -\sqrt{\left(-\frac{5}{4}\right)^2 - 1} = -\sqrt{\frac{25}{16} - 1} = -\sqrt{\frac{9}{16}} = -\frac{3}{4}$$

$$\cos\theta = \frac{1}{\sec\theta} = \frac{1}{-\frac{5}{4}} = -\frac{4}{5}$$

Solve for $\sin\theta$:
$$\sin^2\theta + \cos^2\theta = 1$$
$$\sin^2\theta = 1 - \cos^2\theta$$
$$\sin\theta = \pm\sqrt{1-\cos^2\theta}$$
Since θ is in quadrant II, $\sin\theta > 0$.

$$\sin\theta = \sqrt{1-\cos^2\theta} = \sqrt{1-\left(-\frac{4}{5}\right)^2} = \sqrt{1-\frac{16}{25}} = \sqrt{\frac{9}{25}} = \frac{3}{5}$$

$$\cot\theta = \frac{1}{\tan\theta} = \frac{1}{-\frac{1}{4}} = -\frac{4}{3} \qquad\qquad \csc\theta = \frac{1}{\sin\theta} = \frac{1}{\frac{3}{5}} = \frac{5}{3}$$

37. $\sin\theta = \dfrac{12}{13}, \quad \theta$ in quadrant II

Solve for $\cos\theta$:
$$\sin^2\theta + \cos^2\theta = 1$$
$$\cos^2\theta = 1 - \sin^2\theta$$
$$\cos\theta = \pm\sqrt{1-\sin^2\theta}$$
Since θ is in quadrant II, $\cos\theta < 0$.

$$\cos\theta = -\sqrt{1-\sin^2\theta} = -\sqrt{1-\left(\frac{12}{13}\right)^2} = -\sqrt{1-\frac{144}{169}} = -\sqrt{\frac{25}{169}} = -\frac{5}{13}$$

$$\tan\theta = \frac{\sin\theta}{\cos\theta} = \frac{\frac{12}{13}}{-\frac{5}{13}} = \frac{12}{13}\cdot\left(-\frac{13}{5}\right) = -\frac{12}{5} \qquad \cot\theta = \frac{1}{\tan\theta} = \frac{1}{-\frac{12}{5}} = -\frac{5}{12}$$

$$\sec\theta = \frac{1}{\cos\theta} = \frac{1}{-\dfrac{5}{13}} = -\frac{13}{5} \qquad\qquad \csc\theta = \frac{1}{\sin\theta} = \frac{1}{\dfrac{12}{13}} = \frac{13}{12}$$

39. $\sin\theta = -\dfrac{5}{13}, \quad \dfrac{3\pi}{2} < \theta < 2\pi \Rightarrow \theta$ in quadrant IV

Solve for $\cos\theta$:
$$\sin^2\theta + \cos^2\theta = 1$$
$$\cos^2\theta = 1 - \sin^2\theta$$
$$\cos\theta = \pm\sqrt{1 - \sin^2\theta}$$
Since θ is in quadrant IV, $\cos\theta > 0$.

$$\cos\theta = \sqrt{1 - \sin^2\theta} = \sqrt{1 - \left(-\frac{5}{13}\right)^2} = \sqrt{1 - \frac{25}{169}} = \sqrt{\frac{144}{169}} = \frac{12}{13}$$

$$\tan\theta = \frac{\sin\theta}{\cos\theta} = \frac{-\dfrac{5}{13}}{\dfrac{12}{13}} = -\frac{5}{13}\cdot\left(\frac{13}{12}\right) = -\frac{5}{12} \qquad \cot\theta = \frac{1}{\tan\theta} = \frac{1}{-\dfrac{5}{12}} = -\frac{12}{5}$$

$$\sec\theta = \frac{1}{\cos\theta} = \frac{1}{\dfrac{12}{13}} = \frac{13}{12} \qquad\qquad \csc\theta = \frac{1}{\sin\theta} = \frac{1}{-\dfrac{5}{13}} = -\frac{13}{5}$$

41. $\tan\theta = \dfrac{1}{3}, \quad 180° < \theta < 270° \Rightarrow \theta$ in quadrant III

Solve for $\sec\theta$:
$$\sec^2\theta = \tan^2\theta + 1$$
$$\sec\theta = \pm\sqrt{\tan^2\theta + 1}$$
Since θ is in quadrant III, $\sec\theta < 0$.

$$\sec\theta = -\sqrt{\tan^2\theta + 1} = -\sqrt{\left(\frac{1}{3}\right)^2 + 1} = -\sqrt{\frac{1}{9} + 1} = -\sqrt{\frac{10}{9}} = -\frac{\sqrt{10}}{3}$$

$$\cos\theta = \frac{1}{\sec\theta} = \frac{1}{-\dfrac{\sqrt{10}}{3}} = -\frac{3}{\sqrt{10}}\cdot\frac{\sqrt{10}}{\sqrt{10}} = -\frac{3\sqrt{10}}{10}$$

Solve for $\sin\theta$:
$$\sin^2\theta + \cos^2\theta = 1$$
$$\sin^2\theta = 1 - \cos^2\theta$$
$$\sin\theta = \pm\sqrt{1 - \cos^2\theta}$$
Since θ is in quadrant III, $\sin\theta < 0$.

$$\sin\theta = -\sqrt{1 - \cos^2\theta} = -\sqrt{1 - \left(-\frac{3}{\sqrt{10}}\right)^2} = -\sqrt{1 - \frac{9}{10}} = -\sqrt{\frac{1}{10}} = -\frac{1}{\sqrt{10}}\cdot\frac{\sqrt{10}}{\sqrt{10}} = -\frac{\sqrt{10}}{10}$$

$$\cot\theta = \frac{1}{\tan\theta} = \frac{1}{\dfrac{1}{3}} = 3 \qquad\qquad \csc\theta = \frac{1}{\sin\theta} = \frac{1}{-\dfrac{1}{\sqrt{10}}} = -\sqrt{10}$$

43. $\sec\theta = 3,\ \dfrac{3\pi}{2} < \theta < 2\pi \Rightarrow \theta$ in quadrant IV

Solve for $\tan\theta$:

$$\tan^2\theta + 1 = \sec^2\theta$$

$$\tan\theta = \pm\sqrt{\sec^2\theta - 1}$$

Since θ is in quadrant IV, $\tan\theta < 0$.

$$\tan\theta = -\sqrt{\sec^2\theta - 1} = -\sqrt{3^2 - 1} = -\sqrt{9 - 1} = -\sqrt{8} = -2\sqrt{2}$$

$$\cos\theta = \frac{1}{\sec\theta} = \frac{1}{3}$$

Solve for $\sin\theta$:

$$\sin^2\theta + \cos^2\theta = 1$$

$$\sin^2\theta = 1 - \cos^2\theta$$

$$\sin\theta = \pm\sqrt{1 - \cos^2\theta}$$

Since θ is in quadrant IV, $\sin\theta < 0$.

$$\sin\theta = -\sqrt{1 - \cos^2\theta} = -\sqrt{1 - \left(\frac{1}{3}\right)^2} = -\sqrt{1 - \frac{1}{9}} = -\sqrt{\frac{8}{9}} = -\frac{\sqrt{8}}{3} = -\frac{2\sqrt{2}}{3}$$

$$\cot\theta = \frac{1}{\tan\theta} = \frac{1}{-2\sqrt{2}} \cdot \frac{\sqrt{2}}{\sqrt{2}} = -\frac{\sqrt{2}}{4} \qquad \csc\theta = \frac{1}{\sin\theta} = \frac{1}{-\dfrac{2\sqrt{2}}{3}} = -\frac{3}{2\sqrt{2}} \cdot \frac{\sqrt{2}}{\sqrt{2}} = -\frac{3\sqrt{2}}{4}$$

45. $\cot\theta = -2,\ \dfrac{\pi}{2} < \theta < \pi \Rightarrow \theta$ in quadrant II

Solve for $\csc\theta$:

$$\csc^2\theta = \cot^2\theta + 1$$

$$\csc\theta = \pm\sqrt{\cot^2\theta + 1}$$

Since θ is in quadrant II, $\csc\theta > 0$.

$$\csc\theta = \sqrt{\cot^2\theta + 1} = \sqrt{(-2)^2 + 1} = \sqrt{4 + 1} = \sqrt{5}$$

$$\sin\theta = \frac{1}{\csc\theta} = \frac{1}{\sqrt{5}} \cdot \frac{\sqrt{5}}{\sqrt{5}} = \frac{\sqrt{5}}{5}$$

Solve for $\cos\theta$:

$$\sin^2\theta + \cos^2\theta = 1$$

$$\cos^2\theta = 1 - \sin^2\theta$$

$$\cos\theta = \pm\sqrt{1 - \sin^2\theta}$$

Since θ is in quadrant II, $\cos\theta < 0$.

$$\cos\theta = -\sqrt{1 - \sin^2\theta} = -\sqrt{1 - \left(\frac{1}{\sqrt{5}}\right)^2} = -\sqrt{1 - \frac{1}{5}} = -\sqrt{\frac{4}{5}} = -\frac{2}{\sqrt{5}} \cdot \frac{\sqrt{5}}{\sqrt{5}} = -\frac{2\sqrt{5}}{5}$$

$$\tan\theta = \frac{1}{\cot\theta} = \frac{1}{-2} = -\frac{1}{2} \qquad \sec\theta = \frac{1}{\cos\theta} = \frac{1}{-\dfrac{2}{\sqrt{5}}} = -\frac{\sqrt{5}}{2}$$

47. $y = 2\sin(4x)$ The graph of $y = \sin x$ is stretched vertically by a factor of 2 and compressed horizontally by a factor of $\frac{1}{4}$.

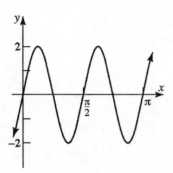

49. $y = -2\cos\left(x + \frac{\pi}{2}\right)$ The graph of $y = \cos x$ is shifted $\frac{\pi}{2}$ units to the left, stretched vertically by a factor of 2, and reflected across the x-axis.

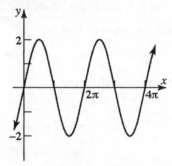

51. $y = \tan(x + \pi)$ The graph of $y = \tan x$ is shifted π units to the left.

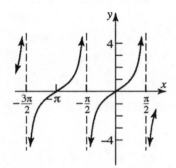

53. $y = -2\tan(3x)$ The graph of $y = \tan x$ is stretched vertically by a factor of 2, reflected across the x-axis, and compressed horizontally by a factor of $\frac{1}{3}$.

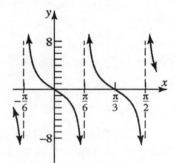

55. $y = \cot\left(x + \frac{\pi}{4}\right)$ The graph of $y = \cot x$ is shifted $\frac{\pi}{4}$ units to the left.

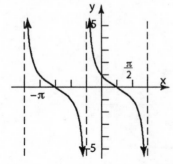

57. $y = \sec\left(x - \frac{\pi}{4}\right)$ The graph of $y = \sec x$ is shifted $\frac{\pi}{4}$ units to the right.

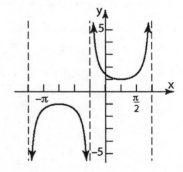

59. $y = 4\cos x$

Amplitude $= |4| = 4$

Period $= 2\pi$

61. $y = -8\sin\left(\dfrac{\pi}{2}x\right)$

Amplitude $= |-8| = 8$

Period $= \dfrac{2\pi}{\dfrac{\pi}{2}} = 4$

63. $y = 4\sin(3x)$

Amplitude : $|A| = |4| = 4$

Period : $T = \dfrac{2\pi}{\omega} = \dfrac{2\pi}{3}$

Phase Shift : $\dfrac{\phi}{\omega} = \dfrac{0}{3} = 0$

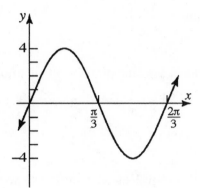

65. $y = 2\sin(2x - \pi)$

Amplitude : $|A| = |2| = 2$

Period : $T = \dfrac{2\pi}{\omega} = \dfrac{2\pi}{2} = \pi$

Phase Shift : $\dfrac{\phi}{\omega} = \dfrac{\pi}{2} = \dfrac{\pi}{2}$

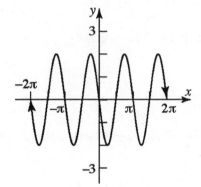

67. $y = \dfrac{1}{2}\sin\left(\dfrac{3}{2}x - \pi\right)$

Amplitude : $|A| = \left|\dfrac{1}{2}\right| = \dfrac{1}{2}$

Period : $T = \dfrac{2\pi}{\omega} = \dfrac{2\pi}{\dfrac{3}{2}} = \dfrac{4\pi}{3}$

Phase Shift : $\dfrac{\phi}{\omega} = \dfrac{\pi}{\dfrac{3}{2}} = \dfrac{2\pi}{3}$

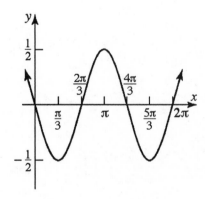

69. $y = -\dfrac{2}{3}\cos(\pi x - 6)$

Amplitude: $|A| = \left|-\dfrac{2}{3}\right| = \dfrac{2}{3}$

Period: $T = \dfrac{2\pi}{\omega} = \dfrac{2\pi}{\pi} = 2$

Phase Shift: $\dfrac{\phi}{\omega} = \dfrac{6}{\pi}$

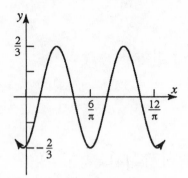

71.   The graph is a cosine graph with an amplitude of 5 and a period of 8π.

Find ω: $8\pi = \dfrac{2\pi}{\omega} \Rightarrow 8\pi\omega = 2\pi \Rightarrow \omega = \dfrac{2\pi}{8\pi} = \dfrac{1}{4}$

The equation is: $y = 5\cos\left(\dfrac{1}{4}x\right)$.

73.   The graph is a reflected cosine graph with an amplitude of 6 and a period of 8.

Find ω: $8 = \dfrac{2\pi}{\omega} \Rightarrow 8\omega = 2\pi \Rightarrow \omega = \dfrac{2\pi}{8} = \dfrac{\pi}{4}$

The equation is: $y = -6\cos\left(\dfrac{\pi}{4}x\right)$.

75.   Use calculator in radian mode: $\sin\dfrac{\pi}{8} \approx 0.38$

77. θ in quadrant III $\Rightarrow \sin\theta < 0,\ \cos\theta < 0,\ \sec\theta < 0,\ \csc\theta < 0,\ \tan\theta > 0,$ and $\cot\theta > 0$.

79. $P = \left(-\dfrac{1}{3}, \dfrac{2\sqrt{2}}{3}\right) \Rightarrow x = -\dfrac{1}{3},\ y = \dfrac{2\sqrt{2}}{3}$

$\sin t = y = \dfrac{2\sqrt{2}}{3}$

$\csc t = \dfrac{1}{y} = \dfrac{1}{\dfrac{2\sqrt{2}}{3}} = \dfrac{3}{2\sqrt{2}} \cdot \dfrac{\sqrt{2}}{\sqrt{2}} = \dfrac{3\sqrt{2}}{4}$

$\cos t = x = -\dfrac{1}{3}$

$\sec t = \dfrac{1}{x} = \dfrac{1}{-\dfrac{1}{3}} = -3$

$\tan t = \dfrac{y}{x} = \dfrac{\dfrac{2\sqrt{2}}{3}}{-\dfrac{1}{3}} = \left(\dfrac{2\sqrt{2}}{3}\right)\cdot\left(-\dfrac{3}{1}\right) = -2\sqrt{2}$

$\cot t = \dfrac{x}{y} = \dfrac{-\dfrac{1}{3}}{\dfrac{2\sqrt{2}}{3}} = \left(-\dfrac{1}{3}\right)\cdot\dfrac{3}{2\sqrt{2}}\cdot\dfrac{\sqrt{2}}{\sqrt{2}} = -\dfrac{\sqrt{2}}{4}$

81.   The domain of $y = \sec x$ is $\left\{x\,\middle|\,x \text{ is any real number, except odd multiples of } \dfrac{\pi}{2}\right\}$.

The range of $y = \sec x$ is $\{y\,|\,y < -1 \text{ or } y > 1\}$.

83. $r = 2$ feet, $\theta = 30° \Rightarrow \theta = \dfrac{\pi}{6}$

   $s = r\theta = 2 \cdot \dfrac{\pi}{6} = \dfrac{\pi}{3}$ feet

   $A = \dfrac{1}{2} \cdot r^2\theta = \dfrac{1}{2} \cdot (2)^2 \cdot \dfrac{\pi}{6} = \dfrac{\pi}{3}$ square feet

85. $v = 180$ mi/hr, $d = \dfrac{1}{2}$ mile $\Rightarrow r = \dfrac{1}{4} = 0.25$ mile

   $\omega = \dfrac{v}{r} = \dfrac{180 \text{ mi/hr}}{0.25 \text{ mi}} = 720$ rad/hr $= \dfrac{720 \text{ rad}}{\text{hr}} \cdot \dfrac{1 \text{ rev}}{2\pi \text{ rad}} = \dfrac{360 \text{ rev}}{\pi \text{ hr}} \approx 114.6$ rev/hr

87.   Since there are two lights on opposite sides and the light is seen every 5 seconds, the beacon makes 1 revolution every 10 seconds.

   $\omega = \dfrac{1 \text{ rev}}{10 \text{ sec}} \cdot \dfrac{2\pi \text{ radians}}{1 \text{ rev}} = \dfrac{\pi}{5}$ radian/second

89. $E(t) = 120\sin(120\pi t), \quad t \geq 0$

   (a)   The maximum value of E is the amplitude, which is 120.

   (b)   Period $= \dfrac{2\pi}{120\pi} = \dfrac{1}{60}$ second

   (c)   Graphing:

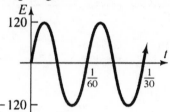

91.   (a)   Draw a scatter diagram:

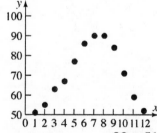

   (b)   Amplitude : $A = \dfrac{90 - 51}{2} = \dfrac{39}{2} = 19.5$

       Vertical Shift : $\dfrac{90 + 51}{2} = \dfrac{141}{2} = 70.5$

       $\omega = \dfrac{2\pi}{12} = \dfrac{\pi}{6}$

       Phase shift (use $y = 51$, $x = 1$):

$$51 = 19.5 \sin\left(\frac{\pi}{6} \cdot 1 - \phi\right) + 70.5$$

$$-19.5 = 19.5 \sin\left(\frac{\pi}{6} - \phi\right)$$

$$-1 = \sin\left(\frac{\pi}{6} - \phi\right)$$

$$-\frac{\pi}{2} = \frac{\pi}{6} - \phi$$

$$\phi = \frac{2\pi}{3}$$

Thus, $y = 19.5 \sin\left(\frac{\pi}{6} x - \frac{2\pi}{3}\right) + 70.5$.

(c)

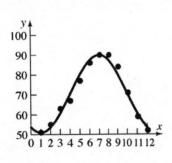

(d) $y = 19.518 \sin(0.541x - 2.283) + 71.01$

(e)

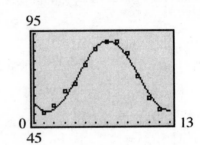

93.   (a)   Amplitude: $A = \dfrac{13.367 - 9.667}{2} = \dfrac{3.7}{2} = 1.85$

Vertical Shift: $\dfrac{13.367 + 9.667}{2} = \dfrac{23.034}{2} = 11.517$

$\omega = \dfrac{2\pi}{365}$

Phase shift (use $y = 9.667, x = 355$):

$$9.667 = 1.85 \sin\left(\frac{2\pi}{365} \cdot 355 - \phi\right) + 11.517$$

$$-1.85 = 1.85 \sin\left(\frac{2\pi}{365} \cdot 355 - \phi\right)$$

$$-1 = \sin\left(\frac{710\pi}{365} - \phi\right)$$

$$-\frac{\pi}{2} = \frac{710\pi}{365} - \phi$$

$$\phi \approx 7.6818$$

Thus, $y = 1.85 \sin\left(\dfrac{2\pi}{365} x - 7.6818\right) + 11.517$.

(b) $y = 1.85 \sin\left(\dfrac{2\pi}{365}(91) - 7.6818\right) + 11.517 \approx 11.83$ hours

(c)

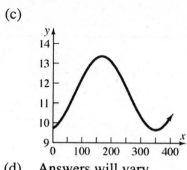

(d)   Answers will vary.

Trigonometric Functions

5.CR   Cumulative Review

1. $2x^2 + x - 1 = 0$
    $(2x - 1)(x + 1) = 0$

    $x = \dfrac{1}{2}$ or $x = -1$

3. radius $= 4$, center $(0, -2)$
    Using $(x - h)^2 + (y - k)^2 = r^2$
    $(x - 0)^2 + (y - (-2))^2 = 4^2$
    $x^2 + (y + 2)^2 = 16$
    $x^2 + y^2 + 4y + 4 = 16$
    $x^2 + y^2 + 4y - 12 = 0$

5. $x^2 + y^2 - 2x + 4y - 4 = 0$
    $x^2 - 2x + 1 + y^2 + 4y + 4 = 4 + 1 + 4$

    $(x - 1)^2 + (y + 2)^2 = 9$

    $(x - 1)^2 + (y + 2)^2 = 3^2$

    This equation yields a circle with
    radius 3 and center $(1, -2)$.

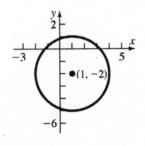

7. (a) $y = x^2$

(b) $y = x^3$

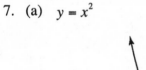

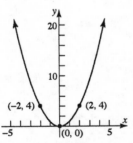

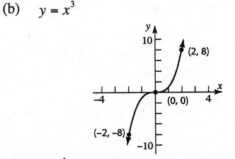

(c) $y = e^x$

(d) $y = \ln x$

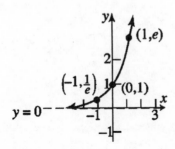

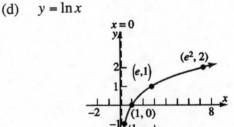

(e) $y = \sin x$

(f) $y = \tan x$

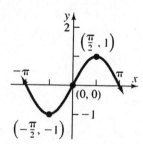

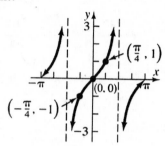

9. $\left(\sin 14°\right)^2 + \left(\cos 14°\right)^2 - 3 = 1 - 3 = -2$

11. $\tan\dfrac{\pi}{4} - 3\cos\dfrac{\pi}{6} + \csc\dfrac{\pi}{6} = 1 - 3\left(\dfrac{\sqrt{3}}{2}\right) + 2 = 3 - \dfrac{3\sqrt{3}}{2} = \dfrac{6 - 3\sqrt{3}}{2}$

13.   The graph is a cosine graph with an amplitude 3 and a period 12.

Find ω: $12 = \dfrac{2\pi}{\omega} \;\Rightarrow\; 12\omega = 2\pi \;\Rightarrow\; \omega = \dfrac{2\pi}{12} = \dfrac{\pi}{6}$

The equation is: $y = 3\cos\left(\dfrac{\pi}{6}x\right)$.

15.   (a)   A polynomial function of degree 3 whose x-intercepts are -2, 3, and 5 will have the form $f(x) = a(x+2)(x-3)(x-5)$, since the x-intercepts correspond to the zeros of the function. Given a y-intercept of 5, we have

$$f(0) = 5$$

$$a(0+2)(0-3)(0-5) = 5$$

$$30a = 5 \Rightarrow a = \frac{1}{6}$$

Therefore, we have the function $f(x) = \dfrac{1}{6}(x+2)(x-3)(x-5)$.

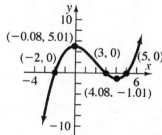

(b)   A rational function whose x-intercepts are -2, 3, and 5 and that has the line $x = 2$ as a vertical asymptote will have the form $f(x) = \dfrac{a(x+2)(x-3)(x-5)}{x-2}$, since the x-intercepts correspond to the zeros of the numerator, and the vertical asymptote corresponds to the zero of the denominator.

Given a y-intercept of 5, we have

$$f(0) = 5$$

$$\frac{a(0+2)(0-3)(0-5)}{0-2} = 5$$

$$30a = -10$$

$$a = -\frac{1}{3}$$

Therefore, we have the function

$$f(x) = \frac{-\frac{1}{3}(x+2)(x-3)(x-5)}{x-2}$$

$$= \frac{-\frac{1}{3}(x+2)(x^2-8x+15)}{(x-2)}$$

$$= \frac{-\frac{1}{3}(x^3-6x^2-x+30)}{(x-2)}$$

$$= \frac{x^3-6x^2-x+30}{-3(x-2)} = \frac{x^3-6x^2-x+30}{-3x+6}$$

$$f(x) = \frac{x^3-6x^2-x+30}{6-3x}$$

Analytic Trigonometry

6.1 The Inverse Sine, Cosine, and Tangent Functions

1. Domain $\{x \mid x \text{ is any real number}\}$, Range $\{y \mid -1 \le y \le 1\}$

3. False

5. $1; \dfrac{\sqrt{3}}{2}$

7. $x = \sin y$

9. $\dfrac{\pi}{5}$

11. True

13. $\sin^{-1} 0$

We are finding the angle θ, $-\dfrac{\pi}{2} \le \theta \le \dfrac{\pi}{2}$, whose sine equals 0.

$$\sin\theta = 0 \qquad -\dfrac{\pi}{2} \le \theta \le \dfrac{\pi}{2}$$

$$\theta = 0$$

$$\sin^{-1} 0 = 0$$

15. $\sin^{-1}(-1)$

We are finding the angle θ, $-\dfrac{\pi}{2} \le \theta \le \dfrac{\pi}{2}$, whose sine equals -1.

$$\sin\theta = -1 \qquad -\dfrac{\pi}{2} \le \theta \le \dfrac{\pi}{2}$$

$$\theta = -\dfrac{\pi}{2}$$

$$\sin^{-1}(-1) = -\dfrac{\pi}{2}$$

17. $\tan^{-1} 0$

We are finding the angle θ, $-\dfrac{\pi}{2} < \theta < \dfrac{\pi}{2}$, whose tangent equals 0.

$$\tan\theta = 0 \qquad -\dfrac{\pi}{2} < \theta < \dfrac{\pi}{2}$$

$$\theta = 0$$

$$\tan^{-1} 0 = 0$$

19. $\sin^{-1} \dfrac{\sqrt{2}}{2}$

We are finding the angle θ, $-\dfrac{\pi}{2} \le \theta \le \dfrac{\pi}{2}$, whose sine equals $\dfrac{\sqrt{2}}{2}$.

$$\sin\theta = \dfrac{\sqrt{2}}{2} \qquad -\dfrac{\pi}{2} \le \theta \le \dfrac{\pi}{2}$$

$$\theta = \dfrac{\pi}{4}$$

$$\sin^{-1} \dfrac{\sqrt{2}}{2} = \dfrac{\pi}{4}$$

21. $\tan^{-1} \sqrt{3}$

We are finding the angle θ, $-\dfrac{\pi}{2} < \theta < \dfrac{\pi}{2}$, whose tangent equals $\sqrt{3}$.

$$\tan\theta = \sqrt{3} \qquad -\dfrac{\pi}{2} < \theta < \dfrac{\pi}{2}$$

$$\theta = \dfrac{\pi}{3}$$

$$\tan^{-1} \sqrt{3} = \dfrac{\pi}{3}$$

23. $\cos^{-1}\left(-\dfrac{\sqrt{3}}{2}\right)$

We are finding the angle θ, $0 \le \theta \le \pi$, whose cosine equals $-\dfrac{\sqrt{3}}{2}$.

$$\cos\theta = -\dfrac{\sqrt{3}}{2} \qquad 0 \le \theta \le \pi$$

$$\theta = \dfrac{5\pi}{6}$$

$$\cos^{-1}\left(-\dfrac{\sqrt{3}}{2}\right) = \dfrac{5\pi}{6}$$

25. $\sin^{-1} 0.1 \approx 0.10$

27. $\tan^{-1} 5 \approx 1.37$

29. $\cos^{-1} \dfrac{7}{8} \approx 0.51$

31. $\tan^{-1}(-0.4) \approx -0.38$

33. $\sin^{-1}(-0.12) \approx -0.12$

35. $\cos^{-1} \dfrac{\sqrt{2}}{3} \approx 1.08$

37. $\sin\left[\sin^{-1}(0.54)\right] = 0.54$

39. $\cos^{-1}\left[\cos\left(\dfrac{4\pi}{5}\right)\right] = \dfrac{4\pi}{5}$

41. $\tan\left[\tan^{-1}(-3.5)\right] = -3.5$

43. $\sin^{-1}\left[\sin\left(-\dfrac{3\pi}{7}\right)\right] = -\dfrac{3\pi}{7}$

45. Yes, $\sin^{-1}\left[\sin\left(-\dfrac{\pi}{6}\right)\right] = -\dfrac{\pi}{6}$, since $\sin^{-1}\left[\sin(x)\right] = x$ where $-\dfrac{\pi}{2} \le x \le \dfrac{\pi}{2}$

and $-\dfrac{\pi}{6}$ is in the restricted domain of $f(x) = \sin(x)$.

47. No, $\sin\left[\sin^{-1}(2)\right] \ne 2$. since $\sin\left[\sin^{-1}(x)\right] = x$ where $-1 \le x \le 1$
     and 2 is not in the domain of $f(x) = \sin^{-1}(x)$.

49. No, $\cos^{-1}\left[\cos\left(-\dfrac{\pi}{6}\right)\right] \ne -\dfrac{\pi}{6}$, since $\cos^{-1}\left[\cos(x)\right] = x$ where $0 \le x \le \pi$

and $-\dfrac{\pi}{6}$ is not in the restricted domain of $f(x) = \cos(x)$.

51. Yes, $\cos\left[\cos^{-1}\left(-\dfrac{1}{2}\right)\right] = -\dfrac{1}{2}$, since $\cos\left[\cos^{-1}(x)\right] = x$ where $-1 \le x \le 1$

and $-\dfrac{1}{2}$ is in the domain of $f(x) = \cos^{-1}(x)$.

53. Yes, $\tan^{-1}\left[\tan\left(-\dfrac{\pi}{3}\right)\right] = -\dfrac{\pi}{3}$, since $\tan^{-1}\left[\tan(x)\right] = x$ where $-\dfrac{\pi}{2} < x < \dfrac{\pi}{2}$

and $-\dfrac{\pi}{3}$ is in the restricted domain of $f(x) = \tan(x)$.

55. Yes, $\tan\left[\tan^{-1}(2)\right] = 2$, since $\tan\left[\tan^{-1}(x)\right] = x$ where $-\infty < x < \infty$.

57. Note that $\theta = 29°45' = 29.75$ degrees.

(a) $D = 24 \cdot \left[1 - \dfrac{\cos^{-1}\left(\tan\left(23.5 \cdot \dfrac{\pi}{180}\right)\tan\left(29.75 \cdot \dfrac{\pi}{180}\right)\right)}{\pi}\right] \approx 13.92$ hours

(b) $D = 24 \cdot \left[1 - \dfrac{\cos^{-1}\left(\tan\left(0 \cdot \dfrac{\pi}{180}\right)\tan\left(29.75 \cdot \dfrac{\pi}{180}\right)\right)}{\pi}\right] \approx 12$ hours

(c) $D = 24 \cdot \left[1 - \dfrac{\cos^{-1}\left(\tan\left(22.8 \cdot \dfrac{\pi}{180}\right)\tan\left(29.75 \cdot \dfrac{\pi}{180}\right)\right)}{\pi}\right] \approx 13.85$ hours

59.   Note that $\theta = 21°18' = 21.3$ degrees.

(a) $D = 24 \cdot \left(1 - \dfrac{\cos^{-1}\left(\tan\left(23.5 \cdot \dfrac{\pi}{180}\right)\tan\left(21.3 \cdot \dfrac{\pi}{180}\right)\right)}{\pi}\right) \approx 13.30$ hours

(b) $D = 24 \cdot \left(1 - \dfrac{\cos^{-1}\left(\tan\left(0 \cdot \dfrac{\pi}{180}\right)\tan\left(21.3 \cdot \dfrac{\pi}{180}\right)\right)}{\pi}\right) \approx 12$ hours

(c) $D = 24 \cdot \left(1 - \dfrac{\cos^{-1}\left(\tan\left(22.8 \cdot \dfrac{\pi}{180}\right)\tan\left(21.3 \cdot \dfrac{\pi}{180}\right)\right)}{\pi}\right) \approx 13.26$ hours

61.   (a) $D = 24 \cdot \left(1 - \dfrac{\cos^{-1}\left(\tan\left(23.5 \cdot \dfrac{\pi}{180}\right)\tan\left(0 \cdot \dfrac{\pi}{180}\right)\right)}{\pi}\right) \approx 12$ hours

(b) $D = 24 \cdot \left(1 - \dfrac{\cos^{-1}\left(\tan\left(0 \cdot \dfrac{\pi}{180}\right)\tan\left(0 \cdot \dfrac{\pi}{180}\right)\right)}{\pi}\right) \approx 12$ hours

(c) $D = 24 \cdot \left(1 - \dfrac{\cos^{-1}\left(\tan\left(22.8 \cdot \dfrac{\pi}{180}\right)\tan\left(0 \cdot \dfrac{\pi}{180}\right)\right)}{\pi}\right) \approx 12$ hours

(d)   There are approximately 12 hours of daylight every day at the equator.

63.   At the latitude of Cadillac Mountain, the effective radius of the earth is 2710 miles.

$1530 \text{ ft} \cdot \dfrac{1 \text{ mile}}{5280 \text{ feet}} \approx 0.29 \text{ mile}$

$\cos\theta = \dfrac{2710}{2710.29}$

$\theta = \cos^{-1}\left(\dfrac{2710}{2710.29}\right)$

$\approx 0.01463 \text{ radians}$

$s = r\theta$

$= 2710(0.01463) = 39.65 \text{ miles}$

$\dfrac{2\pi(2710)}{24} = \dfrac{39.65}{t}$

$t \approx 0.05589 \text{ hour} \approx 3.35 \text{ minutes}$

Chapter 6

Analytic Trigonometry

6.2   The Inverse Trigonometric Functions (Continued)

1.   Domain $\left\{x \middle| x \neq (2n+1) \cdot \dfrac{\pi}{2}, n \text{ is any integer}\right\}$, Range $\left\{y \middle| |y| \geq 1\right\}$

3. $-\dfrac{2}{\sqrt{5}}$

5. $\dfrac{\sqrt{2}}{2}$

7.   True

9. $\cos\left(\sin^{-1}\dfrac{\sqrt{2}}{2}\right)$

Find the angle θ, $-\dfrac{\pi}{2} \leq \theta \leq \dfrac{\pi}{2}$, whose sine equals $\dfrac{\sqrt{2}}{2}$.

$$\sin\theta = \dfrac{\sqrt{2}}{2} \qquad -\dfrac{\pi}{2} \leq \theta \leq \dfrac{\pi}{2}$$

$$\theta = \dfrac{\pi}{4}$$

$$\cos\left(\sin^{-1}\dfrac{\sqrt{2}}{2}\right) = \cos\dfrac{\pi}{4} = \dfrac{\sqrt{2}}{2}$$

11. $\tan\left(\cos^{-1}\left(-\dfrac{\sqrt{3}}{2}\right)\right)$

Find the angle θ, $0 \leq \theta \leq \pi$, whose cosine equals $-\dfrac{\sqrt{3}}{2}$.

$$\cos\theta = -\dfrac{\sqrt{3}}{2} \qquad 0 \leq \theta \leq \pi$$

$$\theta = \dfrac{5\pi}{6}$$

$$\tan\left(\cos^{-1}\left(-\dfrac{\sqrt{3}}{2}\right)\right) = \tan\dfrac{5\pi}{6} = -\dfrac{\sqrt{3}}{3}$$

13. $\sec\left(\cos^{-1}\dfrac{1}{2}\right)$

Find the angle θ, $0 \le \theta \le \pi$, whose cosine equals $\dfrac{1}{2}$.

$$\cos\theta = \dfrac{1}{2} \qquad 0 \le \theta \le \pi$$

$$\theta = \dfrac{\pi}{3}$$

$$\sec\left(\cos^{-1}\dfrac{1}{2}\right) = \sec\dfrac{\pi}{3} = 2$$

15. $\csc\left(\tan^{-1}1\right)$

Find the angle θ, $-\dfrac{\pi}{2} < \theta < \dfrac{\pi}{2}$, whose tangent equals 1.

$$\tan\theta = 1 \qquad -\dfrac{\pi}{2} < \theta < \dfrac{\pi}{2}$$

$$\theta = \dfrac{\pi}{4}$$

$$\csc\left(\tan^{-1}1\right) = \csc\dfrac{\pi}{4} = \sqrt{2}$$

17. $\sin\left(\tan^{-1}(-1)\right)$

Find the angle θ, $-\dfrac{\pi}{2} < \theta < \dfrac{\pi}{2}$, whose tangent equals -1.

$$\tan\theta = -1 \qquad -\dfrac{\pi}{2} < \theta < \dfrac{\pi}{2}$$

$$\theta = -\dfrac{\pi}{4}$$

$$\sin\left(\tan^{-1}(-1)\right) = \sin\left(-\dfrac{\pi}{4}\right) = -\dfrac{\sqrt{2}}{2}$$

19. $\sec\left(\sin^{-1}\left(-\dfrac{1}{2}\right)\right)$

Find the angle θ, $-\dfrac{\pi}{2} \le \theta \le \dfrac{\pi}{2}$, whose sine equals $-\dfrac{1}{2}$.

$$\sin\theta = -\dfrac{1}{2} \qquad -\dfrac{\pi}{2} \le \theta \le \dfrac{\pi}{2}$$

$$\theta = -\dfrac{\pi}{6}$$

$$\sec\left(\sin^{-1}\left(-\dfrac{1}{2}\right)\right) = \sec\left(-\dfrac{\pi}{6}\right) = \dfrac{2\sqrt{3}}{3}$$

21. $\cos^{-1}\left(\cos\dfrac{5\pi}{4}\right) = \cos^{-1}\left(-\dfrac{\sqrt{2}}{2}\right)$

Find the angle θ, $0 \le \theta \le \pi$, whose cosine equals $-\dfrac{\sqrt{2}}{2}$.

$$\cos\theta = -\dfrac{\sqrt{2}}{2} \qquad 0 \le \theta \le \pi$$

$$\theta = \dfrac{3\pi}{4}$$

$$\cos^{-1}\left(\cos\dfrac{5\pi}{4}\right) = \dfrac{3\pi}{4}$$

23. $\sin^{-1}\left(\sin\left(-\dfrac{7\pi}{6}\right)\right) = \sin^{-1}\dfrac{1}{2}$

Find the angle θ, $-\dfrac{\pi}{2} \le \theta \le \dfrac{\pi}{2}$, whose sine equals $\dfrac{1}{2}$.

$$\sin\theta = \dfrac{1}{2} \qquad -\dfrac{\pi}{2} \le \theta \le \dfrac{\pi}{2}$$

$$\theta = \dfrac{\pi}{6}$$

$$\sin^{-1}\left(\sin\left(-\dfrac{7\pi}{6}\right)\right) = \dfrac{\pi}{6}$$

25. $\tan\left(\sin^{-1}\dfrac{1}{3}\right)$

Let $\theta = \sin^{-1}\dfrac{1}{3}$. Since $\sin\theta = \dfrac{1}{3}$ and $-\dfrac{\pi}{2} \le \theta \le \dfrac{\pi}{2}$, θ is in quadrant I and we can let $y = 1$ and $r = 3$.
Solve for x:

$$x^2 + 1 = 9$$

$$x^2 = 8$$

$$x = \pm\sqrt{8} = \pm 2\sqrt{2}$$

Since θ is in quadrant I, $x = 2\sqrt{2}$.

$$\tan\left(\sin^{-1}\dfrac{1}{2}\right) = \tan\theta = \dfrac{y}{x} = \dfrac{1}{2\sqrt{2}}\dfrac{\sqrt{2}}{\sqrt{2}} = \dfrac{\sqrt{2}}{4}$$

27. $\sec\left(\tan^{-1}\dfrac{1}{2}\right)$

Let $\theta = \tan^{-1}\dfrac{1}{2}$. Since $\tan\theta = \dfrac{1}{2}$ and $-\dfrac{\pi}{2} < \theta < \dfrac{\pi}{2}$, θ is in quadrant I and we can let $x = 2$ and $y = 1$.

Solve for r:

$2^2 + 1 = r^2$

$r^2 = 5$

$r = \sqrt{5}$

θ is in quadrant I.

$\sec\left(\tan^{-1}\dfrac{1}{2}\right) = \sec\theta = \dfrac{r}{x} = \dfrac{\sqrt{5}}{2}$

29. $\cot\left(\sin^{-1}\left(-\dfrac{\sqrt{2}}{3}\right)\right)$

Let $\theta = \sin^{-1}\left(-\dfrac{\sqrt{2}}{3}\right)$. Since $\sin\theta = -\dfrac{\sqrt{2}}{3}$ and $-\dfrac{\pi}{2} \le \theta \le \dfrac{\pi}{2}$, θ is in quadrant IV and we can

let $y = -\sqrt{2}$ and $r = 3$.

Solve for x:

$x^2 + 2 = 9$

$x^2 = 7$

$x = \pm\sqrt{7}$

Since θ is in quadrant IV, $x = \sqrt{7}$.

$\cot\left(\sin^{-1}\left(-\dfrac{\sqrt{2}}{3}\right)\right) = \cot\theta = \dfrac{x}{y} = \dfrac{\sqrt{7}}{-\sqrt{2}}\dfrac{\sqrt{2}}{\sqrt{2}} = -\dfrac{\sqrt{14}}{2}$

31. $\sin\left(\tan^{-1}(-3)\right)$

Let $\theta = \tan^{-1}(-3)$. Since $\tan\theta = -3$ and $-\dfrac{\pi}{2} < \theta < \dfrac{\pi}{2}$, θ is in quadrant IV and we can

let $x = 1$ and $y = -3$.

Solve for r:

$1 + 9 = r^2$

$r^2 = 10$

$r = \sqrt{10}$

θ is in quadrant IV.

$\sin\left(\tan^{-1}(-3)\right) = \sin\theta = \dfrac{y}{r} = \dfrac{-3}{\sqrt{10}}\dfrac{\sqrt{10}}{\sqrt{10}} = -\dfrac{3\sqrt{10}}{10}$

33. $\sec\left(\sin^{-1}\dfrac{2\sqrt{5}}{5}\right)$

Let $\theta = \sin^{-1}\dfrac{2\sqrt{5}}{5}$. Since $\sin\theta = \dfrac{2\sqrt{5}}{5}$ and $-\dfrac{\pi}{2} \le \theta \le \dfrac{\pi}{2}$, θ is in quadrant I and we can

let $y = 2\sqrt{5}$ and $r = 5$.

Solve for x:

$x^2 + 20 = 25$

$x^2 = 5$

$x = \pm\sqrt{5}$

Since θ is in quadrant I, $x = \sqrt{5}$.

$\sec\left(\sin^{-1}\left(\dfrac{2\sqrt{5}}{5}\right)\right) = \sec\theta = \dfrac{r}{x} = \dfrac{5}{\sqrt{5}}\dfrac{\sqrt{5}}{\sqrt{5}} = \sqrt{5}$

35. $\sin^{-1}\left(\cos\dfrac{3\pi}{4}\right) = \sin^{-1}\left(-\dfrac{\sqrt{2}}{2}\right) = -\dfrac{\pi}{4}$

37. $\cot^{-1}\sqrt{3}$

We are finding the angle θ, $0 < \theta < \pi$, whose cotangent equals $\sqrt{3}$.

$\cot\theta = \sqrt{3} \quad 0 < \theta < \pi$

$\theta = \dfrac{\pi}{6}$

$\cot^{-1}\sqrt{3} = \dfrac{\pi}{6}$

39. $\csc^{-1}(-1)$

We are finding the angle θ, $-\dfrac{\pi}{2} \le \theta \le \dfrac{\pi}{2}$, $\theta \ne 0$, whose cosecant equals -1.

$\csc\theta = -1 \quad -\dfrac{\pi}{2} \le \theta \le \dfrac{\pi}{2}, \ \theta \ne 0$

$\theta = -\dfrac{\pi}{2}$

$\csc^{-1}(-1) = -\dfrac{\pi}{2}$

41. $\sec^{-1}\dfrac{2\sqrt{3}}{3}$

We are finding the angle θ, $0 \le \theta \le \pi$, $\theta \ne \dfrac{\pi}{2}$, whose secant equals $\dfrac{2\sqrt{3}}{3}$.

$\sec\theta = \dfrac{2\sqrt{3}}{3} \quad 0 \le \theta \le \pi, \ \theta \ne \dfrac{\pi}{2}$

$\theta = \dfrac{\pi}{6}$

$\sec^{-1}\dfrac{2\sqrt{3}}{3} = \dfrac{\pi}{6}$

43. $\cot^{-1}\left(-\dfrac{\sqrt{3}}{3}\right)$

We are finding the angle θ, $0 < \theta < \pi$, whose cotangent equals $-\dfrac{\sqrt{3}}{3}$.

$$\cot\theta = -\dfrac{\sqrt{3}}{3} \quad 0 < \theta < \pi$$

$$\theta = \dfrac{2\pi}{3}$$

$$\cot^{-1}\left(-\dfrac{\sqrt{3}}{3}\right) = \dfrac{2\pi}{3}$$

45. $\sec^{-1}4 = \cos^{-1}\dfrac{1}{4}$

We are finding the angle θ, $0 \le \theta \le \pi$, whose cosine equals $\dfrac{1}{4}$.

$$\cos\theta = \dfrac{1}{4} \Rightarrow \theta \text{ in quadrant I}$$

The calculator yields $\theta = \cos^{-1}\dfrac{1}{4} \approx 1.32$, which is an angle in quadrant I.
Therefore, $\sec^{-1}(4) \approx 1.32$.

47. $\cot^{-1}2 = \tan^{-1}\dfrac{1}{2}$

We are finding the angle θ, $0 \le \theta \le \pi$, whose tangent equals $\dfrac{1}{2}$.

$$\tan\theta = \dfrac{1}{2} \Rightarrow \theta \text{ in quadrant I}$$

The calculator yields $\theta = \tan^{-1}\dfrac{1}{2} \approx 0.46$, which is an angle in quadrant I.
Therefore, $\cot^{-1}(2) \approx 0.46$.

49. $\csc^{-1}(-3) = \sin^{-1}\left(-\dfrac{1}{3}\right)$

We are finding the angle θ, $-\dfrac{\pi}{2} \le \theta \le \dfrac{\pi}{2}$, whose sine equals $-\dfrac{1}{3}$.

$$\sin\theta = -\dfrac{1}{3} \Rightarrow \theta \text{ in quadrant IV}$$

The calculator yields $\theta = \sin^{-1}\left(-\dfrac{1}{3}\right) \approx -0.34$, which is an angle in quadrant IV.
Therefore $\csc^{-1}(-3) \approx -0.34$.

51. $\cot^{-1}\left(-\sqrt{5}\right) = \tan^{-1}\left(-\dfrac{1}{\sqrt{5}}\right)$

We are finding the angle θ, $0 \le \theta \le \pi$, whose tangent equals $-\dfrac{1}{\sqrt{5}}$.

$\tan\theta = -\dfrac{1}{\sqrt{5}} \Rightarrow \theta$ in quadrant II

The calculator yields $\tan^{-1}\left(-\dfrac{1}{\sqrt{5}}\right) \approx -0.42$, which is an angle in quadrant IV.

Since θ is in quadrant II, $\theta \approx -0.42 + \pi \approx 2.72$.

Therefore, $\cot^{-1}\left(-\sqrt{5}\right) \approx 2.72$.

53. $\csc^{-1}\left(-\dfrac{3}{2}\right) = \sin^{-1}\left(-\dfrac{2}{3}\right)$

We are finding the angle θ, $-\dfrac{\pi}{2} \le \theta \le \dfrac{\pi}{2}$, $\theta \ne 0$, whose sine equals $-\dfrac{2}{3}$.

$\sin\theta = -\dfrac{2}{3} \Rightarrow \theta$ in quadrant IV

The calculator yields $\sin^{-1}\left(-\dfrac{2}{3}\right) \approx -0.73$, which is an angle in quadrant IV.

Therefore, $\csc^{-1}\left(-\dfrac{3}{2}\right) \approx -0.73$

55. $\cot^{-1}\left(-\dfrac{3}{2}\right) = \tan^{-1}\left(-\dfrac{2}{3}\right)$

We are finding the angle θ, $0 \le \theta \le \pi$, whose tangent equals $-\dfrac{2}{3}$.

$\tan\theta = -\dfrac{2}{3} \Rightarrow \theta$ in quadrant II

The calculator yields $\tan^{-1}\left(-\dfrac{2}{3}\right) \approx -0.59$, which is an angle in quadrant IV.

Since θ is in quadrant II, $\theta \approx -0.59 + \pi \approx 2.55$.

Therefore, $\cot^{-1}\left(-\dfrac{3}{2}\right) \approx 2.55$

57. $y = \cot^{-1} x$

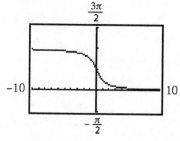

59. $y = \csc^{-1} x$

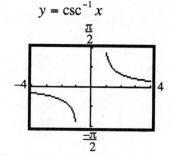

61. Answers will vary.

Analytic Trigonometry

6.3 Trigonometric Identities

1. True

3. identity, conditional

5. 0

7. True

9. $\tan\theta\cdot\csc\theta = \dfrac{\sin\theta}{\cos\theta}\cdot\dfrac{1}{\sin\theta} = \dfrac{1}{\cos\theta}$

11. $\dfrac{\cos\theta}{1-\sin\theta}\cdot\dfrac{1+\sin\theta}{1+\sin\theta} = \dfrac{\cos\theta(1+\sin\theta)}{1-\sin^2\theta} = \dfrac{\cos\theta(1+\sin\theta)}{\cos^2\theta} = \dfrac{1+\sin\theta}{\cos\theta}$

13. $\dfrac{\sin\theta+\cos\theta}{\cos\theta} + \dfrac{\cos\theta-\sin\theta}{\sin\theta} = \dfrac{\sin^2\theta+\sin\theta\cos\theta+\cos\theta(\cos\theta-\sin\theta)}{\sin\theta\cos\theta}$

   $= \dfrac{\sin^2\theta+\sin\theta\cos\theta+\cos^2\theta-\cos\theta\sin\theta}{\sin\theta\cos\theta} = \dfrac{\sin^2\theta+\cos^2\theta+\sin\theta\cos\theta-\cos\theta\sin\theta}{\sin\theta\cos\theta}$

   $= \dfrac{1}{\sin\theta\cos\theta}$

15. $\dfrac{(\sin\theta+\cos\theta)(\sin\theta+\cos\theta)-1}{\sin\theta\cos\theta} = \dfrac{\sin^2\theta+2\sin\theta\cos\theta+\cos^2\theta-1}{\sin\theta\cos\theta}$

   $= \dfrac{\sin^2\theta+\cos^2\theta+2\sin\theta\cos\theta-1}{\sin\theta\cos\theta} = \dfrac{1+2\sin\theta\cos\theta-1}{\sin\theta\cos\theta} = \dfrac{2\sin\theta\cos\theta}{\sin\theta\cos\theta} = 2$

17. $\dfrac{3\sin^2\theta+4\sin\theta+1}{\sin^2\theta+2\sin\theta+1} = \dfrac{(3\sin\theta+1)(\sin\theta+1)}{(\sin\theta+1)(\sin\theta+1)} = \dfrac{3\sin\theta+1}{\sin\theta+1}$

19. $\csc\theta\cdot\cos\theta = \dfrac{1}{\sin\theta}\cdot\cos\theta = \dfrac{\cos\theta}{\sin\theta} = \cot\theta$

21. $1+\tan^2(-\theta) = 1+(-\tan\theta)^2 = 1+\tan^2\theta = \sec^2\theta$

23. $\cos\theta(\tan\theta+\cot\theta) = \cos\theta\left(\dfrac{\sin\theta}{\cos\theta}+\dfrac{\cos\theta}{\sin\theta}\right) = \cos\theta\left(\dfrac{\sin^2\theta+\cos^2\theta}{\cos\theta\sin\theta}\right) = \dfrac{1}{\sin\theta} = \csc\theta$

25. $\tan\theta\cot\theta-\cos^2\theta = \tan\theta\cdot\dfrac{1}{\tan\theta}-\cos^2\theta = 1-\cos^2\theta = \sin^2\theta$

27. $(\sec\theta-1)(\sec\theta+1) = \sec^2\theta-1 = \tan^2\theta$

29. $(\sec\theta + \tan\theta)(\sec\theta - \tan\theta) = \sec^2\theta - \tan^2\theta = 1$

31. $\cos^2\theta(1 + \tan^2\theta) = \cos^2\theta \cdot \sec^2\theta = \cos^2\theta \cdot \dfrac{1}{\cos^2\theta} = 1$

33. $(\sin\theta + \cos\theta)^2 + (\sin\theta - \cos\theta)^2$
$$= \sin^2\theta + 2\sin\theta\cos\theta + \cos^2\theta + \sin^2\theta - 2\sin\theta\cos\theta + \cos^2\theta$$
$$= 2\sin^2\theta + 2\cos^2\theta = 2(\sin^2\theta + \cos^2\theta) = 2\cdot 1 = 2$$

35. $\sec^4\theta - \sec^2\theta = \sec^2\theta(\sec^2\theta - 1) = (\tan^2\theta + 1)\tan^2\theta = \tan^4\theta + \tan^2\theta$

37. $\sec\theta - \tan\theta = \dfrac{1}{\cos\theta} - \dfrac{\sin\theta}{\cos\theta} = \left(\dfrac{1 - \sin\theta}{\cos\theta}\right)\cdot\left(\dfrac{1 + \sin\theta}{1 + \sin\theta}\right) = \dfrac{1 - \sin^2\theta}{\cos\theta(1 + \sin\theta)}$
$$= \dfrac{\cos^2\theta}{\cos\theta(1 + \sin\theta)} = \dfrac{\cos\theta}{1 + \sin\theta}$$

39. $3\sin^2\theta + 4\cos^2\theta = 3\sin^2\theta + 3\cos^2\theta + \cos^2\theta = 3(\sin^2\theta + \cos^2\theta) + \cos^2\theta$
$$= 3\cdot 1 + \cos^2\theta = 3 + \cos^2\theta$$

41. $1 - \dfrac{\cos^2\theta}{1 + \sin\theta} = 1 - \dfrac{1 - \sin^2\theta}{1 + \sin\theta} = 1 - \dfrac{(1 - \sin\theta)(1 + \sin\theta)}{1 + \sin\theta} = 1 - 1 + \sin\theta = \sin\theta$

43. $\dfrac{1 + \tan\theta}{1 - \tan\theta} = \dfrac{1 + \dfrac{1}{\cot\theta}}{1 - \dfrac{1}{\cot\theta}} = \dfrac{\dfrac{\cot\theta + 1}{\cot\theta}}{\dfrac{\cot\theta - 1}{\cot\theta}} = \dfrac{\cot\theta + 1}{\cot\theta}\cdot\dfrac{\cot\theta}{\cot\theta - 1} = \dfrac{\cot\theta + 1}{\cot\theta - 1}$

45. $\dfrac{\sec\theta}{\csc\theta} + \dfrac{\sin\theta}{\cos\theta} = \dfrac{\dfrac{1}{\cos\theta}}{\dfrac{1}{\sin\theta}} + \dfrac{\sin\theta}{\cos\theta} = \dfrac{\sin\theta}{\cos\theta} + \dfrac{\sin\theta}{\cos\theta} = \tan\theta + \tan\theta = 2\tan\theta$

47. $\dfrac{1 + \sin\theta}{1 - \sin\theta} = \dfrac{1 + \dfrac{1}{\csc\theta}}{1 - \dfrac{1}{\csc\theta}} = \dfrac{\dfrac{\csc\theta + 1}{\csc\theta}}{\dfrac{\csc\theta - 1}{\csc\theta}} = \dfrac{\csc\theta + 1}{\csc\theta}\cdot\dfrac{\csc\theta}{\csc\theta - 1} = \dfrac{\csc\theta + 1}{\csc\theta - 1}$

49. $\dfrac{1 - \sin\theta}{\cos\theta} + \dfrac{\cos\theta}{1 - \sin\theta} = \dfrac{(1 - \sin\theta)^2 + \cos^2\theta}{\cos\theta(1 - \sin\theta)} = \dfrac{1 - 2\sin\theta + \sin^2\theta + \cos^2\theta}{\cos\theta(1 - \sin\theta)}$
$$= \dfrac{1 - 2\sin\theta + 1}{\cos\theta(1 - \sin\theta)} = \dfrac{2 - 2\sin\theta}{\cos\theta(1 - \sin\theta)} = \dfrac{2(1 - \sin\theta)}{\cos\theta(1 - \sin\theta)} = \dfrac{2}{\cos\theta} = 2\sec\theta$

51. $\dfrac{\sin\theta}{\sin\theta - \cos\theta} = \dfrac{\sin\theta}{\sin\theta - \cos\theta}\cdot\dfrac{\dfrac{1}{\sin\theta}}{\dfrac{1}{\sin\theta}} = \dfrac{1}{1 - \dfrac{\cos\theta}{\sin\theta}} = \dfrac{1}{1 - \cot\theta}$

53. $(\sec\theta - \tan\theta)^2 = \sec^2\theta - 2\sec\theta\tan\theta + \tan^2\theta = \dfrac{1}{\cos^2\theta} - 2\cdot\dfrac{1}{\cos\theta}\cdot\dfrac{\sin\theta}{\cos\theta} + \dfrac{\sin^2\theta}{\cos^2\theta}$

$= \dfrac{1 - 2\sin\theta + \sin^2\theta}{\cos^2\theta} = \dfrac{(1-\sin\theta)(1-\sin\theta)}{1 - \sin^2\theta} = \dfrac{(1-\sin\theta)(1-\sin\theta)}{(1-\sin\theta)(1+\sin\theta)} = \dfrac{1-\sin\theta}{1+\sin\theta}$

55. $\dfrac{\cos\theta}{1 - \tan\theta} + \dfrac{\sin\theta}{1 - \cot\theta} = \dfrac{\cos\theta}{1 - \dfrac{\sin\theta}{\cos\theta}} + \dfrac{\sin\theta}{1 - \dfrac{\cos\theta}{\sin\theta}} = \dfrac{\cos\theta}{\dfrac{\cos\theta - \sin\theta}{\cos\theta}} + \dfrac{\sin\theta}{\dfrac{\sin\theta - \cos\theta}{\sin\theta}}$

$= \dfrac{\cos^2\theta}{\cos\theta - \sin\theta} + \dfrac{\sin^2\theta}{\sin\theta - \cos\theta} = \dfrac{\cos^2\theta - \sin^2\theta}{\cos\theta - \sin\theta}$

$= \dfrac{(\cos\theta - \sin\theta)(\cos\theta + \sin\theta)}{\cos\theta - \sin\theta} = \cos\theta + \sin\theta = \sin\theta + \cos\theta$

57. $\tan\theta + \dfrac{\cos\theta}{1+\sin\theta} = \dfrac{\sin\theta}{\cos\theta} + \dfrac{\cos\theta}{1+\sin\theta} = \dfrac{\sin\theta(1+\sin\theta) + \cos^2\theta}{\cos\theta(1+\sin\theta)}$

$= \dfrac{\sin\theta + \sin^2\theta + \cos^2\theta}{\cos\theta(1+\sin\theta)} = \dfrac{\sin\theta + 1}{\cos\theta(1+\sin\theta)} = \dfrac{1}{\cos\theta} = \sec\theta$

59. $\dfrac{\tan\theta + \sec\theta - 1}{\tan\theta - \sec\theta + 1} = \dfrac{\tan\theta + (\sec\theta - 1)}{\tan\theta - (\sec\theta - 1)} \cdot \dfrac{\tan\theta + (\sec\theta - 1)}{\tan\theta + (\sec\theta - 1)}$

$= \dfrac{\tan^2\theta + 2\tan\theta(\sec\theta - 1) + \sec^2\theta - 2\sec\theta + 1}{\tan^2\theta - (\sec^2\theta - 2\sec\theta + 1)}$

$= \dfrac{\sec^2\theta - 1 + 2\tan\theta(\sec\theta - 1) + \sec^2\theta - 2\sec\theta + 1}{\sec^2\theta - 1 - \sec^2\theta + 2\sec\theta - 1}$

$= \dfrac{2\sec^2\theta - 2\sec\theta + 2\tan\theta(\sec\theta - 1)}{2\sec\theta - 2}$

$= \dfrac{2\sec\theta(\sec\theta - 1) + 2\tan\theta(\sec\theta - 1)}{2\sec\theta - 2}$

$= \dfrac{2(\sec\theta - 1)(\sec\theta + \tan\theta)}{2(\sec\theta - 1)} = \sec\theta + \tan\theta = \tan\theta + \sec\theta$

61. $\dfrac{\tan\theta - \cot\theta}{\tan\theta + \cot\theta} = \dfrac{\dfrac{\sin\theta}{\cos\theta} - \dfrac{\cos\theta}{\sin\theta}}{\dfrac{\sin\theta}{\cos\theta} + \dfrac{\cos\theta}{\sin\theta}} = \dfrac{\dfrac{\sin^2\theta - \cos^2\theta}{\cos\theta\sin\theta}}{\dfrac{\sin^2\theta + \cos^2\theta}{\cos\theta\sin\theta}} = \dfrac{\sin^2\theta - \cos^2\theta}{1} = \sin^2\theta - \cos^2\theta$

63. $\dfrac{\tan\theta - \cot\theta}{\tan\theta + \cot\theta} + 1 = \dfrac{\dfrac{\sin\theta}{\cos\theta} - \dfrac{\cos\theta}{\sin\theta}}{\dfrac{\sin\theta}{\cos\theta} + \dfrac{\cos\theta}{\sin\theta}} + 1 = \dfrac{\dfrac{\sin^2\theta - \cos^2\theta}{\cos\theta\sin\theta}}{\dfrac{\sin^2\theta + \cos^2\theta}{\cos\theta\sin\theta}} + 1 = \dfrac{\sin^2\theta - \cos^2\theta}{1} + 1$

$= \sin^2\theta - \cos^2\theta + 1 = \sin^2\theta + (1 - \cos^2\theta) = \sin^2\theta + \sin^2\theta = 2\sin^2\theta$

65. $\dfrac{\sec\theta + \tan\theta}{\cot\theta + \cos\theta} = \dfrac{\dfrac{1}{\cos\theta} + \dfrac{\sin\theta}{\cos\theta}}{\dfrac{\cos\theta}{\sin\theta} + \cos\theta} = \dfrac{\dfrac{1+\sin\theta}{\cos\theta}}{\dfrac{\cos\theta + \cos\theta\sin\theta}{\sin\theta}} = \dfrac{1+\sin\theta}{\cos\theta} \cdot \dfrac{\sin\theta}{\cos\theta(1+\sin\theta)}$

$= \dfrac{\sin\theta}{\cos\theta} \cdot \dfrac{1}{\cos\theta} = \tan\theta\sec\theta$

67. $\dfrac{1-\tan^2\theta}{1+\tan^2\theta} + 1 = \dfrac{1-\tan^2\theta + 1 + \tan^2\theta}{1+\tan^2\theta} = \dfrac{2}{\sec^2\theta} = 2 \cdot \dfrac{1}{\sec^2\theta} = 2\cos^2\theta$

69. $\dfrac{\sec\theta - \csc\theta}{\sec\theta\csc\theta} = \dfrac{\dfrac{1}{\cos\theta} - \dfrac{1}{\sin\theta}}{\dfrac{1}{\cos\theta} \cdot \dfrac{1}{\sin\theta}} = \dfrac{\dfrac{\sin\theta - \cos\theta}{\cos\theta\sin\theta}}{\dfrac{1}{\cos\theta\sin\theta}} = \sin\theta - \cos\theta$

71. $\sec\theta - \cos\theta - \sin\theta\tan\theta = \dfrac{1}{\cos\theta} - \cos\theta - \sin\theta \cdot \dfrac{\sin\theta}{\cos\theta} = \dfrac{1-\cos^2\theta - \sin^2\theta}{\cos\theta}$

$= \dfrac{\sin^2\theta - \sin^2\theta}{\cos\theta} = 0$

73. $\dfrac{1}{1-\sin\theta} + \dfrac{1}{1+\sin\theta} = \dfrac{1+\sin\theta + 1 - \sin\theta}{(1-\sin\theta)(1+\sin\theta)} = \dfrac{2}{1-\sin^2\theta} = \dfrac{2}{\cos^2\theta} = 2\sec^2\theta$

75. $\dfrac{\sec\theta}{1-\sin\theta} = \left(\dfrac{\sec\theta}{1-\sin\theta}\right) \cdot \left(\dfrac{1+\sin\theta}{1+\sin\theta}\right) = \dfrac{\sec\theta(1+\sin\theta)}{1-\sin^2\theta} = \dfrac{\sec\theta(1+\sin\theta)}{\cos^2\theta}$

$= \dfrac{1}{\cos\theta} \cdot \dfrac{1+\sin\theta}{\cos^2\theta} = \dfrac{1+\sin\theta}{\cos^3\theta}$

77. $\dfrac{(\sec\theta - \tan\theta)^2 + 1}{\csc\theta(\sec\theta - \tan\theta)} = \dfrac{\sec^2\theta - 2\sec\theta\tan\theta + \tan^2\theta + 1}{\csc\theta(\sec\theta - \tan\theta)} = \dfrac{2\sec^2\theta - 2\sec\theta\tan\theta}{\csc\theta(\sec\theta - \tan\theta)}$

$= \dfrac{2\sec\theta(\sec\theta - \tan\theta)}{\csc\theta(\sec\theta - \tan\theta)} = \dfrac{2\sec\theta}{\csc\theta} = \dfrac{2 \cdot \dfrac{1}{\cos\theta}}{\dfrac{1}{\sin\theta}} = 2 \cdot \dfrac{1}{\cos\theta} \cdot \dfrac{\sin\theta}{1} = 2\tan\theta$

79. $\dfrac{\sin\theta + \cos\theta}{\cos\theta} - \dfrac{\sin\theta - \cos\theta}{\sin\theta} = \dfrac{\sin\theta}{\cos\theta} + \dfrac{\cos\theta}{\cos\theta} - \dfrac{\sin\theta}{\sin\theta} + \dfrac{\cos\theta}{\sin\theta} = \dfrac{\sin\theta}{\cos\theta} + 1 - 1 + \dfrac{\cos\theta}{\sin\theta}$

$= \dfrac{\sin^2\theta + \cos^2\theta}{\cos\theta\sin\theta} = \dfrac{1}{\cos\theta\sin\theta} = \sec\theta\csc\theta$

81. $\dfrac{\sin^3\theta + \cos^3\theta}{\sin\theta + \cos\theta} = \dfrac{(\sin\theta + \cos\theta)(\sin^2\theta - \sin\theta\cos\theta + \cos^2\theta)}{\sin\theta + \cos\theta} = 1 - \sin\theta\cos\theta$

83. $\dfrac{\cos^2\theta - \sin^2\theta}{1-\tan^2\theta} = \dfrac{\cos^2\theta - \sin^2\theta}{1 - \dfrac{\sin^2\theta}{\cos^2\theta}} = \dfrac{\cos^2\theta - \sin^2\theta}{\dfrac{\cos^2\theta - \sin^2\theta}{\cos^2\theta}} = \cos^2\theta$

85.
$$\frac{(2\cos^2\theta-1)^2}{\cos^4\theta-\sin^4\theta}=\frac{\left[2\cos^2\theta-(\sin^2\theta+\cos^2\theta)\right]^2}{(\cos^2\theta-\sin^2\theta)(\cos^2\theta+\sin^2\theta)}$$

$$=\frac{(\cos^2\theta-\sin^2\theta)^2}{(\cos^2\theta-\sin^2\theta)(\cos^2\theta+\sin^2\theta)}=\frac{\cos^2\theta-\sin^2\theta}{\cos^2\theta+\sin^2\theta}$$

$$=\cos^2\theta-\sin^2\theta=1-\sin^2\theta-\sin^2\theta=1-2\sin^2\theta$$

87.
$$\frac{1+\sin\theta+\cos\theta}{1+\sin\theta-\cos\theta}=\frac{(1+\sin\theta)+\cos\theta}{(1+\sin\theta)-\cos\theta}\cdot\frac{(1+\sin\theta)+\cos\theta}{(1+\sin\theta)+\cos\theta}$$

$$=\frac{1+2\sin\theta+\sin^2\theta+2\cos\theta(1+\sin\theta)+\cos^2\theta}{1+2\sin\theta+\sin^2\theta-\cos^2\theta}$$

$$=\frac{1+2\sin\theta+\sin^2\theta+2\cos\theta(1+\sin\theta)+(1-\sin^2\theta)}{1+2\sin\theta+\sin^2\theta-(1-\sin^2\theta)}$$

$$=\frac{2+2\sin\theta+2\cos\theta(1+\sin\theta)}{2\sin\theta+2\sin^2\theta}=\frac{2(1+\sin\theta)+2\cos\theta(1+\sin\theta)}{2\sin\theta(1+\sin\theta)}$$

$$=\frac{2(1+\sin\theta)(1+\cos\theta)}{2\sin\theta(1+\sin\theta)}=\frac{1+\cos\theta}{\sin\theta}$$

89.
$$(a\sin\theta+b\cos\theta)^2+(a\cos\theta-b\sin\theta)^2$$
$$=a^2\sin^2\theta+2ab\sin\theta\cos\theta+b^2\cos^2\theta+a^2\cos^2\theta-2ab\sin\theta\cos\theta+b^2\sin^2\theta$$
$$=a^2(\sin^2\theta+\cos^2\theta)+b^2(\sin^2\theta+\cos^2\theta)=a^2+b^2$$

91.
$$\frac{\tan\alpha+\tan\beta}{\cot\alpha+\cot\beta}=\frac{\tan\alpha+\tan\beta}{\dfrac{1}{\tan\alpha}+\dfrac{1}{\tan\beta}}=\frac{\tan\alpha+\tan\beta}{\dfrac{\tan\beta+\tan\alpha}{\tan\alpha\tan\beta}}$$

$$=(\tan\alpha+\tan\beta)\cdot\left(\frac{\tan\alpha\tan\beta}{\tan\alpha+\tan\beta}\right)=\tan\alpha\tan\beta$$

93.
$$(\sin\alpha+\cos\beta)^2+(\cos\beta+\sin\alpha)(\cos\beta-\sin\alpha)$$
$$=\sin^2\alpha+2\sin\alpha\cos\beta+\cos^2\beta+\cos^2\beta-\sin^2\alpha$$
$$=2\sin\alpha\cos\beta+2\cos^2\beta=2\cos\beta(\sin\alpha+\cos\beta)$$

95.
$$\ln|\sec\theta|=\ln\left|\frac{1}{\cos\theta}\right|=\ln\left|\cos\theta\right|^{-1}=-\ln|\cos\theta|$$

97.
$$\ln|1+\cos\theta|+\ln|1-\cos\theta|=\ln\left(|1+\cos\theta|\cdot|1-\cos\theta|\right)=\ln\left|1-\cos^2\theta\right|$$
$$=\ln\left|\sin^2\theta\right|=2\ln|\sin\theta|$$

99. Show that $\sec\left(\tan^{-1}v\right)=\sqrt{1+v^2}$.

Let $\alpha=\tan^{-1}v$. Then $\tan\alpha=v,\ -\dfrac{\pi}{2}<\alpha<\dfrac{\pi}{2}$.

$$\sec\left(\tan^{-1}v\right)=\sec\alpha=\sqrt{1+\tan^2\alpha}=\sqrt{1+v^2}$$

101. Show that $\tan\left(\cos^{-1}v\right) = \dfrac{\sqrt{1-v^2}}{v}$.

Let $\alpha = \cos^{-1}v$. Then $\cos\alpha = v,\ 0 \le \alpha \le \pi$.

$\tan\left(\cos^{-1}v\right) = \tan\alpha = \dfrac{\sin\alpha}{\cos\alpha} = \dfrac{\sqrt{1-\cos^2\alpha}}{\cos\alpha} = \dfrac{\sqrt{1-v^2}}{v}$

103. Show that $\cos\left(\sin^{-1}v\right) = \sqrt{1-v^2}$.

Let $\alpha = \sin^{-1}v$. Then $\sin\alpha = v,\ -\dfrac{\pi}{2} \le \alpha \le \dfrac{\pi}{2}$.

$\cos\left(\sin^{-1}v\right) = \cos\alpha = \sqrt{1-\sin^2\alpha} = \sqrt{1-v^2}$

105–107.   Answers will vary.

Analytic Trigonometry

6.4 Sum and Difference Formulas

1. $\sqrt{(5-2)^2 + (1-(-3))^2} = \sqrt{3^2 + 4^2}$
    $= \sqrt{9+16} = \sqrt{25} = 5$

3. (a) $\dfrac{\sqrt{2}}{2} \cdot \dfrac{1}{2} = \dfrac{\sqrt{2}}{4}$

    (b) $1 - \dfrac{1}{2} = \dfrac{1}{2}$

5. –   7. False

9. $\sin\dfrac{5\pi}{12} = \sin\left(\dfrac{3\pi}{12} + \dfrac{2\pi}{12}\right) = \sin\dfrac{\pi}{4} \cdot \cos\dfrac{\pi}{6} + \cos\dfrac{\pi}{4} \cdot \sin\dfrac{\pi}{6} = \dfrac{\sqrt{2}}{2} \cdot \dfrac{\sqrt{3}}{2} + \dfrac{\sqrt{2}}{2} \cdot \dfrac{1}{2}$
    $= \dfrac{1}{4}\left(\sqrt{6} + \sqrt{2}\right)$

11. $\cos\dfrac{7\pi}{12} = \cos\left(\dfrac{4\pi}{12} + \dfrac{3\pi}{12}\right) = \cos\dfrac{\pi}{3} \cdot \cos\dfrac{\pi}{4} - \sin\dfrac{\pi}{3} \cdot \sin\dfrac{\pi}{4} = \dfrac{1}{2} \cdot \dfrac{\sqrt{2}}{2} - \dfrac{\sqrt{3}}{2} \cdot \dfrac{\sqrt{2}}{2}$
    $= \dfrac{1}{4}\left(\sqrt{2} - \sqrt{6}\right)$

13. $\cos 165° = \cos(120° + 45°) = \cos 120° \cdot \cos 45° - \sin 120° \cdot \sin 45°$
    $= -\dfrac{1}{2} \cdot \dfrac{\sqrt{2}}{2} - \dfrac{\sqrt{3}}{2} \cdot \dfrac{\sqrt{2}}{2} = -\dfrac{1}{4}\left(\sqrt{2} + \sqrt{6}\right)$

15. $\tan 15° = \tan(45° - 30°) = \dfrac{\tan 45° - \tan 30°}{1 + \tan 45° \cdot \tan 30°} = \dfrac{1 - \dfrac{\sqrt{3}}{3}}{1 + 1 \cdot \dfrac{\sqrt{3}}{3}} = \dfrac{\dfrac{3 - \sqrt{3}}{3}}{\dfrac{3 + \sqrt{3}}{3}}$
    $= \left(\dfrac{3 - \sqrt{3}}{3 + \sqrt{3}}\right) \cdot \left(\dfrac{3 - \sqrt{3}}{3 - \sqrt{3}}\right) = \dfrac{9 - 6\sqrt{3} + 3}{9 - 3} = \dfrac{12 - 6\sqrt{3}}{6} = \dfrac{6\left(2 - \sqrt{3}\right)}{6} = 2 - \sqrt{3}$

17. $\sin\dfrac{17\pi}{12} = \sin\left(\dfrac{15\pi}{12} + \dfrac{2\pi}{12}\right) = \sin\dfrac{5\pi}{4} \cdot \cos\dfrac{\pi}{6} + \cos\dfrac{5\pi}{4} \cdot \sin\dfrac{\pi}{6} = -\dfrac{\sqrt{2}}{2} \cdot \dfrac{\sqrt{3}}{2} + \left(-\dfrac{\sqrt{2}}{2}\right) \cdot \dfrac{1}{2}$
    $= -\dfrac{1}{4}\left(\sqrt{6} + \sqrt{2}\right)$

19. $\sec\left(-\dfrac{\pi}{12}\right) = \dfrac{1}{\cos\left(-\dfrac{\pi}{12}\right)} = \dfrac{1}{\cos\left(\dfrac{3\pi}{12} - \dfrac{4\pi}{12}\right)} = \dfrac{1}{\cos\dfrac{\pi}{4}\cdot\cos\dfrac{\pi}{3} + \sin\dfrac{\pi}{4}\cdot\sin\dfrac{\pi}{3}}$

$= \dfrac{1}{\dfrac{\sqrt{2}}{2}\cdot\dfrac{1}{2} + \dfrac{\sqrt{2}}{2}\cdot\dfrac{\sqrt{3}}{2}} = \dfrac{1}{\dfrac{\sqrt{2} + \sqrt{6}}{4}} = \left(\dfrac{4}{\sqrt{2} + \sqrt{6}}\right)\cdot\left(\dfrac{\sqrt{2} - \sqrt{6}}{\sqrt{2} - \sqrt{6}}\right)$

$= \dfrac{4\left(\sqrt{2} - \sqrt{6}\right)}{2 - 6} = \dfrac{4\left(\sqrt{2} - \sqrt{6}\right)}{-4} = -\left(\sqrt{2} - \sqrt{6}\right) = \sqrt{6} - \sqrt{2}$

21. $\sin 20°\cdot\cos 10° + \cos 20°\cdot\sin 10° = \sin(20° + 10°) = \sin 30° = \dfrac{1}{2}$

23. $\cos 70°\cdot\cos 20° - \sin 70°\cdot\sin 20° = \cos(70° + 20°) = \cos 90° = 0$

25. $\dfrac{\tan 20° + \tan 25°}{1 - \left(\tan 20°\right)\left(\tan 25°\right)} = \tan(20° + 25°) = \tan 45° = 1$

27. $\sin\dfrac{\pi}{12}\cdot\cos\dfrac{7\pi}{12} - \cos\dfrac{\pi}{12}\cdot\sin\dfrac{7\pi}{12} = \sin\left(\dfrac{\pi}{12} - \dfrac{7\pi}{12}\right) = \sin\left(-\dfrac{6\pi}{12}\right) = \sin\left(-\dfrac{\pi}{2}\right) = -1$

29. $\cos\dfrac{\pi}{12}\cdot\cos\dfrac{5\pi}{12} + \sin\dfrac{5\pi}{12}\cdot\sin\dfrac{\pi}{12} = \cos\left(\dfrac{\pi}{12} - \dfrac{5\pi}{12}\right) = \cos\left(-\dfrac{4\pi}{12}\right) = \cos\left(-\dfrac{\pi}{3}\right) = \cos\dfrac{\pi}{3} = \dfrac{1}{2}$

31. $\sin\alpha = \dfrac{3}{5},\ 0 < \alpha < \dfrac{\pi}{2};$ $\cos\beta = \dfrac{2\sqrt{5}}{5},\ -\dfrac{\pi}{2} < \beta < 0$

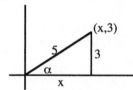

$x^2 + 3^2 = 5^2,\ x > 0$

$\quad x^2 = 25 - 9 = 16,\ x > 0$

$\quad\quad x = 4$

$\cos\alpha = \dfrac{4}{5},\ \tan\alpha = \dfrac{3}{4}$

$\left(2\sqrt{5}\right)^2 + y^2 = 5^2,\ y < 0$

$\quad y^2 = 25 - 20 = 5,\ y < 0$

$\quad\quad y = -\sqrt{5}$

$\sin\beta = -\dfrac{\sqrt{5}}{5},\ \tan\beta = \dfrac{-\sqrt{5}}{2\sqrt{5}} = -\dfrac{1}{2}$

(a) $\sin(\alpha + \beta) = \sin\alpha\cos\beta + \cos\alpha\sin\beta = \dfrac{3}{5}\cdot\dfrac{2\sqrt{5}}{5} + \dfrac{4}{5}\cdot\left(-\dfrac{\sqrt{5}}{5}\right) = \dfrac{6\sqrt{5} - 4\sqrt{5}}{25} = \dfrac{2\sqrt{5}}{25}$

(b) $\cos(\alpha + \beta) = \cos\alpha\cos\beta - \sin\alpha\sin\beta = \dfrac{4}{5}\cdot\dfrac{2\sqrt{5}}{5} - \dfrac{3}{5}\cdot\left(-\dfrac{\sqrt{5}}{5}\right) = \dfrac{8\sqrt{5} + 3\sqrt{5}}{25} = \dfrac{11\sqrt{5}}{25}$

(c) $\sin(\alpha - \beta) = \sin\alpha\cos\beta - \cos\alpha\sin\beta = \dfrac{3}{5} \cdot \dfrac{2\sqrt{5}}{5} - \dfrac{4}{5} \cdot \left(-\dfrac{\sqrt{5}}{5}\right) = \dfrac{6\sqrt{5} + 4\sqrt{5}}{25}$

$= \dfrac{10\sqrt{5}}{25} = \dfrac{2\sqrt{5}}{5}$

(d) $\tan(\alpha - \beta) = \dfrac{\tan\alpha - \tan\beta}{1 + \tan\alpha \cdot \tan\beta} = \dfrac{\dfrac{3}{4} - \left(-\dfrac{1}{2}\right)}{1 + \left(\dfrac{3}{4}\right)\left(-\dfrac{1}{2}\right)} = \dfrac{\dfrac{5}{4}}{\dfrac{5}{8}} = 2$

33. $\tan\alpha = -\dfrac{4}{3}, \ \dfrac{\pi}{2} < \alpha < \pi;$ $\cos\beta = \dfrac{1}{2}, \ 0 < \beta < \dfrac{\pi}{2}$

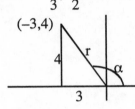

$r^2 = (-3)^2 + 4^2 = 25$

$r = 5$

$\sin\alpha = \dfrac{4}{5}, \ \cos\alpha = \dfrac{-3}{5} = -\dfrac{3}{5}$

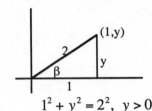

$1^2 + y^2 = 2^2, \ y > 0$

$y^2 = 4 - 1 = 3, \ y > 0$

$y = \sqrt{3}$

$\sin\beta = \dfrac{\sqrt{3}}{2}, \quad \tan\beta = \dfrac{\sqrt{3}}{1} = \sqrt{3}$

(a) $\sin(\alpha + \beta) = \sin\alpha\cos\beta + \cos\alpha\sin\beta = \left(\dfrac{4}{5}\right)\cdot\left(\dfrac{1}{2}\right) + \left(-\dfrac{3}{5}\right)\cdot\left(\dfrac{\sqrt{3}}{2}\right) = \dfrac{4 - 3\sqrt{3}}{10}$

(b) $\cos(\alpha + \beta) = \cos\alpha\cos\beta - \sin\alpha\sin\beta = \left(-\dfrac{3}{5}\right)\cdot\left(\dfrac{1}{2}\right) - \left(\dfrac{4}{5}\right)\cdot\left(\dfrac{\sqrt{3}}{2}\right) = \dfrac{-3 - 4\sqrt{3}}{10}$

(c) $\sin(\alpha - \beta) = \sin\alpha\cos\beta - \cos\alpha\sin\beta = \left(\dfrac{4}{5}\right)\cdot\left(\dfrac{1}{2}\right) - \left(-\dfrac{3}{5}\right)\cdot\left(\dfrac{\sqrt{3}}{2}\right) = \dfrac{4 + 3\sqrt{3}}{10}$

(d) $\tan(\alpha - \beta) = \dfrac{\tan\alpha - \tan\beta}{1 + \tan\alpha\tan\beta} = \dfrac{-\dfrac{4}{3} - \sqrt{3}}{1 + \left(-\dfrac{4}{3}\right)\cdot\sqrt{3}} = \dfrac{\dfrac{-4 - 3\sqrt{3}}{3}}{\dfrac{3 - 4\sqrt{3}}{3}} = \left(\dfrac{-4 - 3\sqrt{3}}{3 - 4\sqrt{3}}\right)\cdot\left(\dfrac{3 + 4\sqrt{3}}{3 + 4\sqrt{3}}\right)$

$= \dfrac{-48 - 25\sqrt{3}}{-39} = \dfrac{48 + 25\sqrt{3}}{39}$

35. $\sin\alpha = \dfrac{5}{13}, \ -\dfrac{3\pi}{2} < \alpha < -\pi;$ $\tan\beta = -\sqrt{3}, \ \dfrac{\pi}{2} < \beta < \pi$

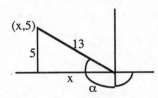

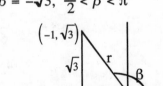

$$x^2 + 5^2 = 13^2, \quad x < 0 \qquad\qquad r^2 = (-1)^2 + \sqrt{3}^2 = 4$$

$$x^2 = 169 - 25 = 144, \quad x < 0 \qquad\qquad r = 2$$

$$x = -12$$

$$\cos\alpha = \frac{-12}{13} = -\frac{12}{13}, \quad \tan\alpha = -\frac{5}{12} \qquad\qquad \sin\beta = \frac{\sqrt{3}}{2}, \quad \cos\beta = \frac{-1}{2} = -\frac{1}{2}$$

(a) $\sin(\alpha + \beta) = \sin\alpha\cos\beta + \cos\alpha\sin\beta = \left(\frac{5}{13}\right)\cdot\left(-\frac{1}{2}\right) + \left(-\frac{12}{13}\right)\cdot\left(\frac{\sqrt{3}}{2}\right) = \dfrac{-5 - 12\sqrt{3}}{26}$

(b) $\cos(\alpha + \beta) = \cos\alpha\cos\beta - \sin\alpha\sin\beta = \left(-\frac{12}{13}\right)\cdot\left(-\frac{1}{2}\right) - \left(\frac{5}{13}\right)\cdot\left(\frac{\sqrt{3}}{2}\right) = \dfrac{12 - 5\sqrt{3}}{26}$

(c) $\sin(\alpha - \beta) = \sin\alpha\cos\beta - \cos\alpha\sin\beta = \left(\frac{5}{13}\right)\cdot\left(-\frac{1}{2}\right) - \left(-\frac{12}{13}\right)\cdot\left(\frac{\sqrt{3}}{2}\right) = \dfrac{-5 + 12\sqrt{3}}{26}$

(d) $\tan(\alpha - \beta) = \dfrac{\tan\alpha - \tan\beta}{1 + \tan\alpha\tan\beta} = \dfrac{-\dfrac{5}{12} - \left(-\sqrt{3}\right)}{1 + \left(-\dfrac{5}{12}\right)\cdot\left(-\sqrt{3}\right)} = \dfrac{\dfrac{-5 + 12\sqrt{3}}{12}}{\dfrac{12 + 5\sqrt{3}}{12}}$

$$= \left(\dfrac{-5 + 12\sqrt{3}}{12 + 5\sqrt{3}}\right)\cdot\left(\dfrac{12 - 5\sqrt{3}}{12 - 5\sqrt{3}}\right) = \dfrac{-240 + 169\sqrt{3}}{69}$$

37. $\sin\theta = \dfrac{1}{3}, \quad \theta$ in quadrant II

(a) $\cos\theta = -\sqrt{1 - \sin^2\theta} = -\sqrt{1 - \left(\frac{1}{3}\right)^2} = -\sqrt{1 - \frac{1}{9}} = -\sqrt{\frac{8}{9}} = -\dfrac{2\sqrt{2}}{3}$

(b) $\sin\left(\theta + \dfrac{\pi}{6}\right) = \sin\theta\cdot\cos\dfrac{\pi}{6} + \cos\theta\cdot\sin\dfrac{\pi}{6} = \left(\frac{1}{3}\right)\left(\frac{\sqrt{3}}{2}\right) + \left(-\frac{2\sqrt{2}}{3}\right)\left(\frac{1}{2}\right) = \dfrac{\sqrt{3} - 2\sqrt{2}}{6}$

(c) $\cos\left(\theta - \dfrac{\pi}{3}\right) = \cos\theta\cdot\cos\dfrac{\pi}{3} + \sin\theta\cdot\sin\dfrac{\pi}{3} = \left(-\frac{2\sqrt{2}}{3}\right)\left(\frac{1}{2}\right) + \left(\frac{1}{3}\right)\left(\frac{\sqrt{3}}{2}\right) = \dfrac{-2\sqrt{2} + \sqrt{3}}{6}$

(d) $\tan\left(\theta + \dfrac{\pi}{4}\right) = \dfrac{\tan\theta + \tan\dfrac{\pi}{4}}{1 - \tan\theta\cdot\tan\dfrac{\pi}{4}} = \dfrac{-\dfrac{1}{2\sqrt{2}} + 1}{1 - \left(-\dfrac{1}{2\sqrt{2}}\right)\cdot 1} = \dfrac{\dfrac{-1 + 2\sqrt{2}}{2\sqrt{2}}}{\dfrac{2\sqrt{2} + 1}{2\sqrt{2}}}$

$$= \left(\dfrac{2\sqrt{2} - 1}{2\sqrt{2} + 1}\right)\cdot\left(\dfrac{2\sqrt{2} - 1}{2\sqrt{2} - 1}\right) = \dfrac{8 - 4\sqrt{2} + 1}{8 - 1} = \dfrac{9 - 4\sqrt{2}}{7}$$

39. $\sin\left(\dfrac{\pi}{2} + \theta\right) = \sin\dfrac{\pi}{2}\cdot\cos\theta + \cos\dfrac{\pi}{2}\cdot\sin\theta = 1\cdot\cos\theta + 0\cdot\sin\theta = \cos\theta$

41. $\sin(\pi - \theta) = \sin\pi\cdot\cos\theta - \cos\pi\cdot\sin\theta = 0\cdot\cos\theta - (-1)\sin\theta = \sin\theta$

43. $\sin(\pi + \theta) = \sin\pi\cdot\cos\theta + \cos\pi\cdot\sin\theta = 0\cdot\cos\theta + (-1)\sin\theta = -\sin\theta$

45. $\tan(\pi - \theta) = \dfrac{\tan\pi - \tan\theta}{1 + \tan\pi \cdot \tan\theta} = \dfrac{0 - \tan\theta}{1 + 0 \cdot \tan\theta} = \dfrac{-\tan\theta}{1} = -\tan\theta$

47. $\sin\left(\dfrac{3\pi}{2} + \theta\right) = \sin\dfrac{3\pi}{2} \cdot \cos\theta + \cos\dfrac{3\pi}{2} \cdot \sin\theta = -1 \cdot \cos\theta + 0 \cdot \sin\theta = -\cos\theta$

49. $\sin(\alpha + \beta) + \sin(\alpha - \beta) = \sin\alpha\cos\beta + \cos\alpha\sin\beta + \sin\alpha\cos\beta - \cos\alpha\sin\beta$
        $= 2\sin\alpha\cos\beta$

51. $\dfrac{\sin(\alpha + \beta)}{\sin\alpha\cos\beta} = \dfrac{\sin\alpha\cos\beta + \cos\alpha\sin\beta}{\sin\alpha\cos\beta} = \dfrac{\sin\alpha\cos\beta}{\sin\alpha\cos\beta} + \dfrac{\cos\alpha\sin\beta}{\sin\alpha\cos\beta} = 1 + \cot\alpha\tan\beta$

53. $\dfrac{\cos(\alpha + \beta)}{\cos\alpha\cos\beta} = \dfrac{\cos\alpha\cos\beta - \sin\alpha\sin\beta}{\cos\alpha\cos\beta} = \dfrac{\cos\alpha\cos\beta}{\cos\alpha\cos\beta} - \dfrac{\sin\alpha\sin\beta}{\cos\alpha\cos\beta} = 1 - \tan\alpha\tan\beta$

55. $\dfrac{\sin(\alpha + \beta)}{\sin(\alpha - \beta)} = \dfrac{\sin\alpha\cos\beta + \cos\alpha\sin\beta}{\sin\alpha\cos\beta - \cos\alpha\sin\beta} = \dfrac{\dfrac{\sin\alpha\cos\beta}{\cos\alpha\cos\beta} + \dfrac{\cos\alpha\sin\beta}{\cos\alpha\cos\beta}}{\dfrac{\sin\alpha\cos\beta}{\cos\alpha\cos\beta} - \dfrac{\cos\alpha\sin\beta}{\cos\alpha\cos\beta}} = \dfrac{\tan\alpha + \tan\beta}{\tan\alpha - \tan\beta}$

57. $\cot(\alpha + \beta) = \dfrac{\cos(\alpha + \beta)}{\sin(\alpha + \beta)} = \dfrac{\cos\alpha\cos\beta - \sin\alpha\sin\beta}{\sin\alpha\cos\beta + \cos\alpha\sin\beta}$

        $= \dfrac{\dfrac{\cos\alpha\cos\beta}{\sin\alpha\sin\beta} - \dfrac{\sin\alpha\sin\beta}{\sin\alpha\sin\beta}}{\dfrac{\sin\alpha\cos\beta}{\sin\alpha\sin\beta} + \dfrac{\cos\alpha\sin\beta}{\sin\alpha\sin\beta}} = \dfrac{\cot\alpha\cot\beta - 1}{\cot\beta + \cot\alpha}$

59. $\sec(\alpha + \beta) = \dfrac{1}{\cos(\alpha + \beta)} = \dfrac{1}{\cos\alpha\cos\beta - \sin\alpha\sin\beta}$

        $= \dfrac{\dfrac{1}{\sin\alpha\sin\beta}}{\dfrac{\cos\alpha\cos\beta}{\sin\alpha\sin\beta} - \dfrac{\sin\alpha\sin\beta}{\sin\alpha\sin\beta}} = \dfrac{\csc\alpha\csc\beta}{\cot\alpha\cot\beta - 1}$

61. $\sin(\alpha - \beta)\sin(\alpha + \beta) = \left(\sin\alpha\cos\beta - \cos\alpha\sin\beta\right)\left(\sin\alpha\cos\beta + \cos\alpha\sin\beta\right)$
        $= \sin^2\alpha\cos^2\beta - \cos^2\alpha\sin^2\beta = \sin^2\alpha(1 - \sin^2\beta) - (1 - \sin^2\alpha)\sin^2\beta$
        $= \sin^2\alpha - \sin^2\alpha\sin^2\beta - \sin^2\beta + \sin^2\alpha\sin^2\beta = \sin^2\alpha - \sin^2\beta$

63. $\sin(\theta + k\pi) = \sin\theta \cdot \cos(k\pi) + \cos\theta \cdot \sin(k\pi) = \sin\theta(-1)^k + \cos\theta \cdot 0 = (-1)^k\sin\theta$, k any integer

65. $\sin\left(\sin^{-1}\dfrac{1}{2} + \cos^{-1}0\right) = \sin\left(\dfrac{\pi}{6} + \dfrac{\pi}{2}\right) = \sin\dfrac{2\pi}{3} = \dfrac{\sqrt{3}}{2}$

67. $\sin\left[\sin^{-1}\dfrac{3}{5} - \cos^{-1}\left(-\dfrac{4}{5}\right)\right]$

Let $\alpha = \sin^{-1}\dfrac{3}{5}$ and $\beta = \cos^{-1}\left(-\dfrac{4}{5}\right)$. α is in quadrant I; β is in quadrant II.

Then $\sin\alpha = \dfrac{3}{5}$, $0 \le \alpha \le \dfrac{\pi}{2}$, $\cos\beta = -\dfrac{4}{5}$, $\dfrac{\pi}{2} \le \beta \le \pi$.

$\cos\alpha = \sqrt{1-\sin^2\alpha} = \sqrt{1-\left(\dfrac{3}{5}\right)^2} = \sqrt{1-\dfrac{9}{25}} = \sqrt{\dfrac{16}{25}} = \dfrac{4}{5}$

$\sin\beta = \sqrt{1-\cos^2\beta} = \sqrt{1-\left(-\dfrac{4}{5}\right)^2} = \sqrt{1-\dfrac{16}{25}} = \sqrt{\dfrac{9}{25}} = \dfrac{3}{5}$

$\sin\left[\sin^{-1}\dfrac{3}{5} - \cos^{-1}\left(-\dfrac{4}{5}\right)\right] = \sin(\alpha-\beta) = \sin\alpha\cos\beta - \cos\alpha\sin\beta$

$= \left(\dfrac{3}{5}\right)\cdot\left(-\dfrac{4}{5}\right) - \left(\dfrac{4}{5}\right)\cdot\left(\dfrac{3}{5}\right) = -\dfrac{12}{25} - \dfrac{12}{25} = -\dfrac{24}{25}$

69. $\cos\left(\tan^{-1}\dfrac{4}{3} + \cos^{-1}\dfrac{5}{13}\right)$

Let $\alpha = \tan^{-1}\dfrac{4}{3}$ and $\beta = \cos^{-1}\dfrac{5}{13}$. α is in quadrant I; β is in quadrant I.

Then $\tan\alpha = \dfrac{4}{3}$, $0 < \alpha < \dfrac{\pi}{2}$, $\cos\beta = \dfrac{5}{13}$, $0 \le \beta \le \dfrac{\pi}{2}$.

$\sec\alpha = \sqrt{1+\tan^2\alpha} = \sqrt{1+\left(\dfrac{4}{3}\right)^2} = \sqrt{1+\dfrac{16}{9}} = \sqrt{\dfrac{25}{9}} = \dfrac{5}{3}$; $\cos\alpha = \dfrac{3}{5}$

$\sin\alpha = \sqrt{1-\cos^2\alpha} = \sqrt{1-\left(\dfrac{3}{5}\right)^2} = \sqrt{1-\dfrac{9}{25}} = \sqrt{\dfrac{16}{25}} = \dfrac{4}{5}$

$\sin\beta = \sqrt{1-\cos^2\beta} = \sqrt{1-\left(\dfrac{5}{13}\right)^2} = \sqrt{1-\dfrac{25}{169}} = \sqrt{\dfrac{144}{169}} = \dfrac{12}{13}$

$\cos\left(\tan^{-1}\dfrac{4}{3} + \cos^{-1}\dfrac{5}{13}\right) = \cos(\alpha+\beta) = \cos\alpha\cos\beta - \sin\alpha\sin\beta$

$= \left(\dfrac{3}{5}\right)\cdot\left(\dfrac{5}{13}\right) - \left(\dfrac{4}{5}\right)\cdot\left(\dfrac{12}{13}\right) = \dfrac{15}{65} - \dfrac{48}{65} = -\dfrac{33}{65}$

71. $\cos\left(\sin^{-1}\dfrac{5}{13} - \tan^{-1}\dfrac{3}{4}\right)$

Let $\alpha = \sin^{-1}\dfrac{5}{13}$ and $\beta = \tan^{-1}\dfrac{3}{4}$. α is in quadrant I; β is in quadrant I.

Then $\sin\alpha = \dfrac{5}{13}$, $0 \le \alpha \le \dfrac{\pi}{2}$, and $\tan\beta = \dfrac{3}{4}$, $0 < \beta < \dfrac{\pi}{2}$.

$$\cos\alpha = \sqrt{1-\sin^2\alpha} = \sqrt{1-\left(\frac{5}{13}\right)^2} = \sqrt{1-\frac{25}{169}} = \sqrt{\frac{144}{169}} = \frac{12}{13}$$

$$\sec\beta = \sqrt{1+\tan^2\beta} = \sqrt{1+\left(\frac{3}{4}\right)^2} = \sqrt{1+\frac{9}{16}} = \sqrt{\frac{25}{16}} = \frac{5}{4}; \quad \cos\beta = \frac{4}{5}$$

$$\sin\beta = \sqrt{1-\cos^2\beta} = \sqrt{1-\left(\frac{4}{5}\right)^2} = \sqrt{1-\frac{16}{25}} = \sqrt{\frac{9}{25}} = \frac{3}{5}$$

$$\cos\left[\sin^{-1}\frac{5}{13} - \tan^{-1}\frac{3}{4}\right] = \cos(\alpha-\beta)$$

$$= \cos\alpha\cos\beta + \sin\alpha\sin\beta = \frac{12}{13}\cdot\frac{4}{5} + \frac{5}{13}\cdot\frac{3}{5} = \frac{48}{65} + \frac{15}{65} = \frac{63}{65}$$

73. $\tan\left(\sin^{-1}\frac{3}{5} + \frac{\pi}{6}\right)$

Let $\alpha = \sin^{-1}\frac{3}{5}$. α is in quadrant I.

Then $\sin\alpha = \frac{3}{5}, \ 0\le\alpha\le\frac{\pi}{2}$.

$$\cos\alpha = \sqrt{1-\sin^2\alpha} = \sqrt{1-\left(\frac{3}{5}\right)^2} = \sqrt{1-\frac{9}{25}} = \sqrt{\frac{16}{25}} = \frac{4}{5}$$

$$\tan\alpha = \frac{\sin\alpha}{\cos\alpha} = \frac{\frac{3}{5}}{\frac{4}{5}} = \frac{3}{5}\cdot\frac{5}{4} = \frac{3}{4}$$

$$\tan\left(\sin^{-1}\frac{3}{5} + \frac{\pi}{6}\right) = \frac{\tan\left(\sin^{-1}\frac{3}{5}\right) + \tan\frac{\pi}{6}}{1-\tan\left(\sin^{-1}\frac{3}{5}\right)\cdot\tan\frac{\pi}{6}} = \frac{\frac{3}{4} + \frac{\sqrt{3}}{3}}{1 - \frac{3}{4}\cdot\frac{\sqrt{3}}{3}}$$

$$= \frac{\frac{9+4\sqrt{3}}{12}}{\frac{12-3\sqrt{3}}{12}} = \left(\frac{9+4\sqrt{3}}{12}\right)\left(\frac{12}{12-3\sqrt{3}}\right) = \left(\frac{9+4\sqrt{3}}{12-3\sqrt{3}}\right)\left(\frac{12+3\sqrt{3}}{12+3\sqrt{3}}\right)$$

$$= \frac{108+75\sqrt{3}+36}{144-27} = \frac{144+75\sqrt{3}}{117} = \frac{48+25\sqrt{3}}{39}$$

75. $\tan\left(\sin^{-1}\frac{4}{5} + \cos^{-1}1\right)$

Let $\alpha = \sin^{-1}\frac{4}{5}$ and $\beta = \cos^{-1}1$; α is in quadrant I.

Then $\sin\alpha = \frac{4}{5}, \ 0\le\alpha\le\frac{\pi}{2}$ and $\cos\beta = 1, \ 0\le\beta\le\pi$.

$$\cos\beta = 1,\ 0 \le \beta \le \pi \Rightarrow \beta = 0 \Rightarrow \cos^{-1}1 = 0$$

$$\cos\alpha = \sqrt{1 - \sin^2\alpha} = \sqrt{1 - \left(\frac{4}{5}\right)^2} = \sqrt{1 - \frac{16}{25}} = \sqrt{\frac{9}{25}} = \frac{3}{5}$$

$$\tan\alpha = \frac{\sin\alpha}{\cos\alpha} = \frac{\frac{4}{5}}{\frac{3}{5}} = \frac{4}{5} \cdot \frac{5}{3} = \frac{4}{3}$$

$$\tan\left(\sin^{-1}\frac{4}{5} + \cos^{-1}1\right) = \frac{\tan\left(\sin^{-1}\frac{4}{5}\right) + \tan\left(\cos^{-1}1\right)}{1 - \tan\left(\sin^{-1}\frac{4}{5}\right)\tan\left(\cos^{-1}1\right)} = \frac{\frac{4}{3} + 0}{1 - \frac{4}{3}\cdot 0} = \frac{\frac{4}{3}}{1} = \frac{4}{3}$$

77. $\cos\left(\cos^{-1}u + \sin^{-1}v\right)$

Let $\alpha = \cos^{-1}u$ and $\beta = \sin^{-1}v$.

Then $\cos\alpha = u,\ 0 \le \alpha \le \pi$, and $\sin\beta = v, -\frac{\pi}{2} \le \beta \le \frac{\pi}{2}$

$\sin\alpha = \sqrt{1 - \cos^2\alpha} = \sqrt{1 - u^2}$

$\cos\beta = \sqrt{1 - \sin^2\beta} = \sqrt{1 - v^2}$

$\cos\left(\cos^{-1}u + \sin^{-1}v\right) = \cos(\alpha + \beta) = \cos\alpha\cos\beta - \sin\alpha\sin\beta = u\sqrt{1 - v^2} - v\sqrt{1 - u^2}$

79. $\sin\left(\tan^{-1}u - \sin^{-1}v\right)$

Let $\alpha = \tan^{-1}u$ and $\beta = \sin^{-1}v$.

Then $\tan\alpha = u,\ -\frac{\pi}{2} < \alpha < \frac{\pi}{2}$, and $\sin\beta = v, -\frac{\pi}{2} \le \beta \le \frac{\pi}{2}$

$\sec\alpha = \sqrt{\tan^2\alpha + 1} = \sqrt{u^2 + 1}; \quad \cos\alpha = \frac{1}{\sqrt{u^2 + 1}}$

$\sin\alpha = \sqrt{1 - \cos^2\alpha} = \sqrt{1 - \frac{1.}{u^2 + 1}} = \sqrt{\frac{u^2 + 1 - 1}{u^2 + 1}} = \sqrt{\frac{u^2}{u^2 + 1}} = \frac{u}{\sqrt{u^2 + 1}}$

$\cos\beta = \sqrt{1 - \sin^2\beta} = \sqrt{1 - v^2}$

$\sin\left(\tan^{-1}u - \sin^{-1}v\right) = \sin(\alpha - \beta) = \sin\alpha\cos\beta - \cos\alpha\sin\beta$

$$= \frac{u}{\sqrt{u^2 + 1}} \cdot \sqrt{1 - v^2} - \frac{1}{\sqrt{u^2 + 1}} \cdot v = \frac{u\sqrt{1 - v^2} - v}{\sqrt{u^2 + 1}}$$

81. $\tan\left(\sin^{-1}u - \cos^{-1}v\right)$

Let $\alpha = \sin^{-1}u$ and $\beta = \cos^{-1}v$.

Then $\sin\alpha = u,\ -\frac{\pi}{2} \le \alpha \le \frac{\pi}{2}$, and $\cos\beta = v, 0 \le \beta \le \pi$

$\cos\alpha = \sqrt{1 - \sin^2\alpha} = \sqrt{1 - u^2}; \quad \tan\alpha = \frac{\sin\alpha}{\cos\alpha} = \frac{u}{\sqrt{1 - u^2}}$

$\sin\beta = \sqrt{1 - \cos^2\beta} = \sqrt{1 - v^2}; \quad \tan\beta = \frac{\sin\beta}{\cos\beta} = \frac{\sqrt{1 - v^2}}{v}$

$$\tan\left(\sin^{-1}u - \cos^{-1}v\right) = \tan(\alpha - \beta) = \frac{\tan\alpha - \tan\beta}{1 + \tan\alpha\tan\beta} = \frac{\dfrac{u}{\sqrt{1-u^2}} - \dfrac{\sqrt{1-v^2}}{v}}{1 + \dfrac{u}{\sqrt{1-u^2}}\cdot\dfrac{\sqrt{1-v^2}}{v}}$$

$$= \frac{\dfrac{uv - \sqrt{1-u^2}\,\sqrt{1-v^2}}{v\sqrt{1-u^2}}}{\dfrac{v\sqrt{1-u^2} + u\sqrt{1-v^2}}{v\sqrt{1-u^2}}} = \frac{uv - \sqrt{1-u^2}\,\sqrt{1-v^2}}{v\sqrt{1-u^2} + u\sqrt{1-v^2}}$$

83. Show that $\sin^{-1}v + \cos^{-1}v = \dfrac{\pi}{2}$.

Let $\alpha = \sin^{-1}v$ and $\beta = \cos^{-1}v$.

Then $\sin\alpha = v = \cos\beta$, and since $\sin\alpha = \cos\left(\dfrac{\pi}{2} - \alpha\right)$, $\cos\left(\dfrac{\pi}{2} - \alpha\right) = \cos\beta$.

If $v \geq 0$, then $0 \leq \alpha \leq \dfrac{\pi}{2}$, so that $\left(\dfrac{\pi}{2} - \alpha\right)$ and β both lie in the interval $\left[0, \dfrac{\pi}{2}\right]$.

If $v < 0$, then $-\dfrac{\pi}{2} \leq \alpha < 0$, so that $\left(\dfrac{\pi}{2} - \alpha\right)$ and β both lie in the interval $\left(\dfrac{\pi}{2}, \pi\right]$. Either

way, $\cos\left(\dfrac{\pi}{2} - \alpha\right) = \cos\beta$ implies $\dfrac{\pi}{2} - \alpha = \beta$, or $\alpha + \beta = \dfrac{\pi}{2}$. Thus, $\sin^{-1}v + \cos^{-1}v = \dfrac{\pi}{2}$.

85. Show that $\tan^{-1}\left(\dfrac{1}{v}\right) = \dfrac{\pi}{2} - \tan^{-1}v$, if $v > 0$.

Let $\alpha = \tan^{-1}\left(\dfrac{1}{v}\right)$ and $\beta = \tan^{-1}v$. Because $\dfrac{1}{v}$ must be defined, $v \neq 0$ and so $\alpha, \beta \neq 0$.

Then $\tan\alpha = \dfrac{1}{v} = \dfrac{1}{\tan\beta} = \cot\beta$, and since $\tan\alpha = \cot\left(\dfrac{\pi}{2} - \alpha\right)$, $\cot\left(\dfrac{\pi}{2} - \alpha\right) = \cot\beta$.

Because $v > 0$, $0 < \alpha < \dfrac{\pi}{2}$ and so $\dfrac{\pi}{2} - \alpha$ and β both lie in the interval $\left(0, \dfrac{\pi}{2}\right)$.

Then, $\cot\left(\dfrac{\pi}{2} - \alpha\right) = \cot\beta$ implies $\dfrac{\pi}{2} - \alpha = \beta$ or $\alpha = \dfrac{\pi}{2} - \beta$.

Thus, $\tan^{-1}\left(\dfrac{1}{v}\right) = \dfrac{\pi}{2} - \tan^{-1}v$, if $v > 0$.

87. $\sin\left(\sin^{-1}v + \cos^{-1}v\right) = \sin\left(\sin^{-1}v\right)\cos\left(\cos^{-1}v\right) + \cos\left(\sin^{-1}v\right)\sin\left(\cos^{-1}v\right)$

$= v \cdot v + \sqrt{1-v^2}\,\sqrt{1-v^2} = v^2 + 1 - v^2 = 1$

89. $\dfrac{\sin(x+h) - \sin x}{h} = \dfrac{\sin x\cos h + \cos x\sin h - \sin x}{h} = \dfrac{\cos x\sin h - \sin x + \sin x\cos h}{h}$

$= \dfrac{\cos x\sin h - \sin x(1 - \cos h)}{h} = \cos x \cdot \left(\dfrac{\sin h}{h}\right) - \sin x \cdot \left(\dfrac{1 - \cos h}{h}\right)$

91. $\tan\left(\dfrac{\pi}{2} - \theta\right) = \dfrac{\tan\dfrac{\pi}{2} - \tan\theta}{1 + \tan\dfrac{\pi}{2} \cdot \tan\theta}$. This is impossible because $\tan\dfrac{\pi}{2}$ is undefined.

$\tan\left(\dfrac{\pi}{2} - \theta\right) = \dfrac{\sin\left(\dfrac{\pi}{2} - \theta\right)}{\cos\left(\dfrac{\pi}{2} - \theta\right)} = \dfrac{\cos\theta}{\sin\theta} = \cot\theta$

93. $\tan\theta = \tan\left(\theta_2 - \theta_1\right) = \dfrac{\tan\theta_2 - \tan\theta_1}{1 + \tan\theta_2\tan\theta_1} = \dfrac{m_2 - m_1}{1 + m_2 m_1}$

95. The first step in the derivation, $\tan\left(\theta + \dfrac{\pi}{2}\right) = \dfrac{\tan\theta + \tan\dfrac{\pi}{2}}{1 - \tan\theta \cdot \tan\dfrac{\pi}{2}}$, is impossible because

$\tan\dfrac{\pi}{2}$ is undefined.

Chapter **6**

Analytic Trigonometry

6.5 Double-angle and Half-angle Formulas

1. $\sin^2\theta,\ 2\cos^2\theta,\ 2\sin^2\theta$ 3. $\sin\theta$ 5. False

7. $\sin\theta=\dfrac{3}{5},\ \ 0<\theta<\dfrac{\pi}{2};\quad$ thus, $0<\dfrac{\theta}{2}<\dfrac{\pi}{4}\ \Rightarrow\ \dfrac{\theta}{2}$ is in quadrant I.

$y=3,\ r=5$

$x^2+3^2=5^2,\ x>0\ \Rightarrow\ x^2=25-9=16,\ x>0\ \Rightarrow\ x=4$

$\cos\theta=\dfrac{4}{5}$

(a) $\sin(2\theta)=2\sin\theta\cos\theta=2\cdot\dfrac{3}{5}\cdot\dfrac{4}{5}=\dfrac{24}{25}$

(b) $\cos(2\theta)=\cos^2\theta-\sin^2\theta=\left(\dfrac{4}{5}\right)^2-\left(\dfrac{3}{5}\right)^2=\dfrac{16}{25}-\dfrac{9}{25}=\dfrac{7}{25}$

(c) $\sin\dfrac{\theta}{2}=\sqrt{\dfrac{1-\cos\theta}{2}}=\sqrt{\dfrac{1-\frac{4}{5}}{2}}=\sqrt{\dfrac{\frac{1}{5}}{2}}=\sqrt{\dfrac{1}{10}}=\dfrac{1}{\sqrt{10}}\dfrac{\sqrt{10}}{\sqrt{10}}=\dfrac{\sqrt{10}}{10}$

(d) $\cos\dfrac{\theta}{2}=\sqrt{\dfrac{1+\cos\theta}{2}}=\sqrt{\dfrac{1+\frac{4}{5}}{2}}=\sqrt{\dfrac{\frac{9}{5}}{2}}=\sqrt{\dfrac{9}{10}}=\dfrac{3}{\sqrt{10}}\dfrac{\sqrt{10}}{\sqrt{10}}=\dfrac{3\sqrt{10}}{10}$

9. $\tan\theta=\dfrac{4}{3},\ \ \pi<\theta<\dfrac{3\pi}{2};\quad$ thus, $\dfrac{\pi}{2}<\dfrac{\theta}{2}<\dfrac{3\pi}{4}\ \Rightarrow\ \dfrac{\theta}{2}$ is in quadrant II.

$x=-3,\ y=-4$

$r^2=(-3)^2+(-4)^2=9+16=25\ \Rightarrow\ r=5$

$\sin\theta=-\dfrac{4}{5},\ \ \cos\theta=-\dfrac{3}{5}$

(a) $\sin(2\theta)=2\sin\theta\cos\theta=2\cdot\left(-\dfrac{4}{5}\right)\cdot\left(-\dfrac{3}{5}\right)=\dfrac{24}{25}$

(b) $\cos(2\theta)=\cos^2\theta-\sin^2\theta=\left(-\dfrac{3}{5}\right)^2-\left(-\dfrac{4}{5}\right)^2=\dfrac{9}{25}-\dfrac{16}{25}=-\dfrac{7}{25}$

(c) $\sin\dfrac{\theta}{2} = \sqrt{\dfrac{1-\cos\theta}{2}} = \sqrt{\dfrac{1-\left(-\dfrac{3}{5}\right)}{2}} = \sqrt{\dfrac{\dfrac{8}{5}}{2}} = \sqrt{\dfrac{4}{5}} = \dfrac{2}{\sqrt{5}} \dfrac{\sqrt{5}}{\sqrt{5}} = \dfrac{2\sqrt{5}}{5}$

(d) $\cos\dfrac{\theta}{2} = -\sqrt{\dfrac{1+\cos\theta}{2}} = -\sqrt{\dfrac{1+\left(-\dfrac{3}{5}\right)}{2}} = -\sqrt{\dfrac{\dfrac{2}{5}}{2}} = -\sqrt{\dfrac{1}{5}} = -\dfrac{1}{\sqrt{5}} \dfrac{\sqrt{5}}{\sqrt{5}} = -\dfrac{\sqrt{5}}{5}$

11. $\cos\theta = -\dfrac{\sqrt{6}}{3}, \ \dfrac{\pi}{2} < \theta < \pi;$ thus, $\dfrac{\pi}{4} < \dfrac{\theta}{2} < \dfrac{\pi}{2} \Rightarrow \ \dfrac{\theta}{2}$ is in quadrant I.

$x = -\sqrt{6}, \ r = 3$

$(-\sqrt{6})^2 + y^2 = 3^2 \ \Rightarrow \ y^2 = 9 - 6 = 3 \ \Rightarrow \ y = \sqrt{3}$

$\sin\theta = \dfrac{\sqrt{3}}{3}$

(a) $\sin(2\theta) = 2\sin\theta\cos\theta = 2\cdot\left(\dfrac{\sqrt{3}}{3}\right)\cdot\left(-\dfrac{\sqrt{6}}{3}\right) = -\dfrac{2\sqrt{18}}{9} = -\dfrac{6\sqrt{2}}{9} = -\dfrac{2\sqrt{2}}{3}$

(b) $\cos(2\theta) = \cos^2\theta - \sin^2\theta = \left(-\dfrac{\sqrt{6}}{3}\right)^2 - \left(\dfrac{\sqrt{3}}{3}\right)^2 = \dfrac{6}{9} - \dfrac{3}{9} = \dfrac{3}{9} = \dfrac{1}{3}$

(c) $\sin\dfrac{\theta}{2} = \sqrt{\dfrac{1-\cos\theta}{2}} = \sqrt{\dfrac{1-\left(-\dfrac{\sqrt{6}}{3}\right)}{2}} = \sqrt{\dfrac{\dfrac{3+\sqrt{6}}{3}}{2}} = \sqrt{\dfrac{3+\sqrt{6}}{6}}$

(d) $\cos\dfrac{\theta}{2} = \sqrt{\dfrac{1+\cos\theta}{2}} = \sqrt{\dfrac{1+\left(-\dfrac{\sqrt{6}}{3}\right)}{2}} = \sqrt{\dfrac{\dfrac{3-\sqrt{6}}{3}}{2}} = \sqrt{\dfrac{3-\sqrt{6}}{6}}$

13. $\sec\theta = 3, \ \sin\theta > 0 \Rightarrow \ 0 < \theta < \dfrac{\pi}{2};$ thus, $0 < \dfrac{\theta}{2} < \dfrac{\pi}{4} \Rightarrow \dfrac{\theta}{2}$ is in quadrant I.

$\cos\theta = \dfrac{1}{3}, \ x = 1, \ r = 3$

$1^2 + y^2 = 3^2 \ \Rightarrow \ y^2 = 9 - 1 = 8 \ \Rightarrow \ y = 2\sqrt{2}$

$\sin\theta = \dfrac{2\sqrt{2}}{3}$

(a) $\sin(2\theta) = 2\sin\theta\cos\theta = 2\cdot\dfrac{2\sqrt{2}}{3}\cdot\dfrac{1}{3} = \dfrac{4\sqrt{2}}{9}$

(b) $\cos(2\theta) = \cos^2\theta - \sin^2\theta = \left(\dfrac{1}{3}\right)^2 - \left(\dfrac{2\sqrt{2}}{3}\right)^2 = \dfrac{1}{9} - \dfrac{8}{9} = -\dfrac{7}{9}$

(c) $\sin\dfrac{\theta}{2} = \sqrt{\dfrac{1-\cos\theta}{2}} = \sqrt{\dfrac{1-\dfrac{1}{3}}{2}} = \sqrt{\dfrac{\dfrac{2}{3}}{2}} = \sqrt{\dfrac{1}{3}} = \dfrac{1}{\sqrt{3}}\dfrac{\sqrt{3}}{\sqrt{3}} = \dfrac{\sqrt{3}}{3}$

(d) $\cos\dfrac{\theta}{2} = \sqrt{\dfrac{1+\cos\theta}{2}} = \sqrt{\dfrac{1+\dfrac{1}{3}}{2}} = \sqrt{\dfrac{\dfrac{4}{3}}{2}} = \sqrt{\dfrac{2}{3}} = \dfrac{\sqrt{2}}{\sqrt{3}}\dfrac{\sqrt{3}}{\sqrt{3}} = \dfrac{\sqrt{6}}{3}$

15. $\cot\theta = -2$, $\sec\theta < 0 \Rightarrow \dfrac{\pi}{2} < \theta < \pi$; thus, $\dfrac{\pi}{4} < \dfrac{\theta}{2} < \dfrac{\pi}{2} \Rightarrow \dfrac{\theta}{2}$ is in quadrant I.

   $x = -2$, $y = 1$

   $r^2 = (-2)^2 + 1^2 = 4+1 = 5 \ \Rightarrow \ r = \sqrt{5}$

   $\sin\theta = \dfrac{1}{\sqrt{5}} = \dfrac{\sqrt{5}}{5}$, $\cos\theta = -\dfrac{2}{\sqrt{5}} = -\dfrac{2\sqrt{5}}{5}$

(a) $\sin(2\theta) = 2\sin\theta\cos\theta = 2\cdot\left(\dfrac{\sqrt{5}}{5}\right)\cdot\left(-\dfrac{2\sqrt{5}}{5}\right) = -\dfrac{20}{25} = -\dfrac{4}{5}$

(b) $\cos(2\theta) = \cos^2\theta - \sin^2\theta = \left(-\dfrac{2\sqrt{5}}{5}\right)^2 - \left(\dfrac{\sqrt{5}}{5}\right)^2 = \dfrac{20}{25} - \dfrac{5}{25} = \dfrac{15}{25} = \dfrac{3}{5}$

(c) $\sin\dfrac{\theta}{2} = \sqrt{\dfrac{1-\cos\theta}{2}} = \sqrt{\dfrac{1-\left(-\dfrac{2\sqrt{5}}{5}\right)}{2}} = \sqrt{\dfrac{\dfrac{5+2\sqrt{5}}{5}}{2}} = \sqrt{\dfrac{5+2\sqrt{5}}{10}}$

(d) $\cos\dfrac{\theta}{2} = \sqrt{\dfrac{1+\cos\theta}{2}} = \sqrt{\dfrac{1+\left(-\dfrac{2\sqrt{5}}{5}\right)}{2}} = \sqrt{\dfrac{\dfrac{5-2\sqrt{5}}{5}}{2}} = \sqrt{\dfrac{5-2\sqrt{5}}{10}}$

17. $\tan\theta = -3$, $\sin\theta < 0 \Rightarrow \dfrac{3\pi}{2} < \theta < 2\pi$; thus, $\dfrac{3\pi}{4} < \dfrac{\theta}{2} < \pi \Rightarrow \dfrac{\theta}{2}$ is in quadrant II.

   $x = 1$, $y = -3$

   $r^2 = 1^2 + (-3)^2 = 1+9 = 10 \ \Rightarrow \ r = \sqrt{10}$

   $\sin\theta = \dfrac{-3}{\sqrt{10}} = -\dfrac{3\sqrt{10}}{10}$, $\cos\theta = \dfrac{1}{\sqrt{10}} = \dfrac{\sqrt{10}}{10}$

(a) $\sin(2\theta) = 2\sin\theta\cos\theta = 2\cdot\left(-\dfrac{3\sqrt{10}}{10}\right)\cdot\left(\dfrac{\sqrt{10}}{10}\right) = -\dfrac{6}{10} = -\dfrac{3}{5}$

(b) $\cos(2\theta) = \cos^2\theta - \sin^2\theta = \left(\dfrac{\sqrt{10}}{10}\right)^2 - \left(-\dfrac{3\sqrt{10}}{10}\right)^2 = \dfrac{10}{100} - \dfrac{90}{100} = -\dfrac{80}{100} = -\dfrac{4}{5}$

(c) $\sin\dfrac{\theta}{2} = \sqrt{\dfrac{1-\cos\theta}{2}} = \sqrt{\dfrac{1-\dfrac{\sqrt{10}}{10}}{2}} = \sqrt{\dfrac{\dfrac{10-\sqrt{10}}{10}}{2}} = \sqrt{\dfrac{10-\sqrt{10}}{20}} = \dfrac{1}{2}\sqrt{\dfrac{10-\sqrt{10}}{5}}$

(d) $\cos\dfrac{\theta}{2} = -\sqrt{\dfrac{1+\cos\theta}{2}} = -\sqrt{\dfrac{1+\dfrac{\sqrt{10}}{10}}{2}} = -\sqrt{\dfrac{\dfrac{10+\sqrt{10}}{10}}{2}} = -\sqrt{\dfrac{10+\sqrt{10}}{20}} = -\dfrac{1}{2}\sqrt{\dfrac{10+\sqrt{10}}{5}}$

19. $\sin 22.5° = \sin\left(\dfrac{45°}{2}\right) = \sqrt{\dfrac{1-\cos 45°}{2}} = \sqrt{\dfrac{1-\dfrac{\sqrt{2}}{2}}{2}} = \sqrt{\dfrac{2-\sqrt{2}}{4}} = \dfrac{\sqrt{2-\sqrt{2}}}{2}$

21. $\tan\dfrac{7\pi}{8} = \tan\left(\dfrac{\dfrac{7\pi}{4}}{2}\right) = -\sqrt{\dfrac{1-\cos\dfrac{7\pi}{4}}{1+\cos\dfrac{7\pi}{4}}} = -\sqrt{\dfrac{1-\dfrac{\sqrt{2}}{2}}{1+\dfrac{\sqrt{2}}{2}}} = -\sqrt{\left(\dfrac{2-\sqrt{2}}{2+\sqrt{2}}\right)\cdot\left(\dfrac{2-\sqrt{2}}{2-\sqrt{2}}\right)}$

$= -\sqrt{\dfrac{\left(2-\sqrt{2}\right)^2}{2}} = -\left(\dfrac{2-\sqrt{2}}{\sqrt{2}}\right) = -\left(\sqrt{2}-1\right) = 1-\sqrt{2}$

23. $\cos 165° = \cos\left(\dfrac{330°}{2}\right) = -\sqrt{\dfrac{1+\cos 330°}{2}} = -\sqrt{\dfrac{1+\dfrac{\sqrt{3}}{2}}{2}} = -\sqrt{\dfrac{2+\sqrt{3}}{4}} = -\dfrac{\sqrt{2+\sqrt{3}}}{2}$

25. $\sec\dfrac{15\pi}{8} = \dfrac{1}{\cos\dfrac{15\pi}{8}} = \dfrac{1}{\cos\left(\dfrac{\dfrac{15\pi}{4}}{2}\right)} = \dfrac{1}{\sqrt{\dfrac{1+\cos\dfrac{15\pi}{4}}{2}}} = \dfrac{1}{\sqrt{\dfrac{1+\dfrac{\sqrt{2}}{2}}{2}}} = \dfrac{1}{\sqrt{\dfrac{2+\sqrt{2}}{4}}}$

$= \left(\dfrac{2}{\sqrt{2+\sqrt{2}}}\right)\cdot\left(\dfrac{\sqrt{2+\sqrt{2}}}{\sqrt{2+\sqrt{2}}}\right) = \left(\dfrac{2\sqrt{2+\sqrt{2}}}{2+\sqrt{2}}\right)\cdot\left(\dfrac{2-\sqrt{2}}{2-\sqrt{2}}\right)$

$= \dfrac{2\left(2-\sqrt{2}\right)\sqrt{2+\sqrt{2}}}{2} = \left(2-\sqrt{2}\right)\sqrt{2+\sqrt{2}}$

27. $\sin\left(-\dfrac{\pi}{8}\right) = \sin\left(\dfrac{\left(-\dfrac{\pi}{4}\right)}{2}\right) = -\sqrt{\dfrac{1-\cos\left(-\dfrac{\pi}{4}\right)}{2}} = -\sqrt{\dfrac{1-\dfrac{\sqrt{2}}{2}}{2}} = -\sqrt{\dfrac{2-\sqrt{2}}{4}} = -\dfrac{\sqrt{2-\sqrt{2}}}{2}$

29. $\sin^4\theta = \left(\sin^2\theta\right)^2 = \left(\dfrac{1-\cos(2\theta)}{2}\right)^2 = \dfrac{1}{4}\left(1-2\cos(2\theta)+\cos^2(2\theta)\right)$

$= \dfrac{1}{4} - \dfrac{1}{2}\cos(2\theta) + \dfrac{1}{4}\cos^2(2\theta) = \dfrac{1}{4} - \dfrac{1}{2}\cos(2\theta) + \dfrac{1}{4}\left(\dfrac{1+\cos(4\theta)}{2}\right)$

$= \dfrac{1}{4} - \dfrac{1}{2}\cos(2\theta) + \dfrac{1}{8} + \dfrac{1}{8}\cos(4\theta) = \dfrac{3}{8} - \dfrac{1}{2}\cos(2\theta) + \dfrac{1}{8}\cos(4\theta)$

31. $\sin(4\theta) = \sin(2(2\theta)) = 2\sin(2\theta)\cos(2\theta) = 2(2\sin\theta\cos\theta)\left(1-2\sin^2\theta\right)$

$= \cos\theta\left(4\sin\theta - 8\sin^3\theta\right)$

33. Use the results of problem 31 to help solve the problem:
$\sin(5\theta) = \sin(4\theta + \theta) = \sin(4\theta)\cos\theta + \cos(4\theta)\sin\theta$

$= \cos\theta\left(4\sin\theta - 8\sin^3\theta\right)\cos\theta + \cos(2(2\theta))\sin\theta$

$= \cos^2\theta\left(4\sin\theta - 8\sin^3\theta\right) + \left(1 - 2\sin^2(2\theta)\right)\sin\theta$

$= \left(1 - \sin^2\theta\right)\left(4\sin\theta - 8\sin^3\theta\right) + \sin\theta\left(1 - 2(2\sin\theta\cos\theta)^2\right)$

$= 4\sin\theta - 12\sin^3\theta + 8\sin^5\theta + \sin\theta\left(1 - 8\sin^2\theta\cos^2\theta\right)$

$= 4\sin\theta - 12\sin^3\theta + 8\sin^5\theta + \sin\theta - 8\sin^3\theta\left(1 - \sin^2\theta\right)$

$= 5\sin\theta - 12\sin^3\theta + 8\sin^5\theta - 8\sin^3\theta + 8\sin^5\theta$

$= 5\sin\theta - 20\sin^3\theta + 16\sin^5\theta$

35. $\cos^4\theta - \sin^4\theta = \left(\cos^2\theta + \sin^2\theta\right)\left(\cos^2\theta - \sin^2\theta\right) = 1\cdot\cos(2\theta) = \cos(2\theta)$

37. $\cot(2\theta) = \dfrac{1}{\tan(2\theta)} = \dfrac{1}{\dfrac{2\tan\theta}{1-\tan^2\theta}} = \dfrac{1-\tan^2\theta}{2\tan\theta} = \dfrac{1-\dfrac{1}{\cot^2\theta}}{\dfrac{2}{\cot\theta}} = \dfrac{\dfrac{\cot^2\theta-1}{\cot^2\theta}}{\dfrac{2}{\cot\theta}} = \dfrac{\cot^2\theta-1}{2\cot\theta}$

39. $\sec(2\theta) = \dfrac{1}{\cos(2\theta)} = \dfrac{1}{2\cos^2\theta-1} = \dfrac{1}{\dfrac{2}{\sec^2\theta}-1} = \dfrac{1}{\dfrac{2-\sec^2\theta}{\sec^2\theta}} = \dfrac{\sec^2\theta}{2-\sec^2\theta}$

41. $\cos^2(2\theta) - \sin^2(2\theta) = \cos(2(2\theta)) = \cos(4\theta)$

43. $\dfrac{\cos(2\theta)}{1+\sin(2\theta)} = \dfrac{\cos^2\theta-\sin^2\theta}{1+2\sin\theta\cos\theta} = \dfrac{(\cos\theta-\sin\theta)(\cos\theta+\sin\theta)}{\cos^2\theta+\sin^2\theta+2\sin\theta\cos\theta}$

$= \dfrac{(\cos\theta-\sin\theta)(\cos\theta+\sin\theta)}{(\cos\theta+\sin\theta)(\cos\theta+\sin\theta)} = \dfrac{\cos\theta-\sin\theta}{\cos\theta+\sin\theta} = \dfrac{\dfrac{\cos\theta}{\sin\theta}-\dfrac{\sin\theta}{\sin\theta}}{\dfrac{\cos\theta}{\sin\theta}+\dfrac{\sin\theta}{\sin\theta}} = \dfrac{\cot\theta-1}{\cot\theta+1}$

45. $\sec^2\left(\dfrac{\theta}{2}\right) = \dfrac{1}{\cos^2\left(\dfrac{\theta}{2}\right)} = \dfrac{1}{\dfrac{1+\cos\theta}{2}} = \dfrac{2}{1+\cos\theta}$

47. $\cot^2\left(\dfrac{\theta}{2}\right) = \dfrac{1}{\tan^2\left(\dfrac{\theta}{2}\right)} = \dfrac{1}{\dfrac{1-\cos\theta}{1+\cos\theta}} = \dfrac{1+\cos\theta}{1-\cos\theta} = \dfrac{1+\dfrac{1}{\sec\theta}}{1-\dfrac{1}{\sec\theta}} = \dfrac{\dfrac{\sec\theta+1}{\sec\theta}}{\dfrac{\sec\theta-1}{\sec\theta}} = \dfrac{\sec\theta+1}{\sec\theta-1}$

49. $\dfrac{1-\tan^2\dfrac{\theta}{2}}{1+\tan^2\dfrac{\theta}{2}} = \dfrac{1-\dfrac{1-\cos\theta}{1+\cos\theta}}{1+\dfrac{1-\cos\theta}{1+\cos\theta}} = \dfrac{\dfrac{1+\cos\theta-(1-\cos\theta)}{1+\cos\theta}}{\dfrac{1+\cos\theta+1-\cos\theta}{1+\cos\theta}} = \dfrac{2\cos\theta}{2} = \cos\theta$

51. $\dfrac{\sin(3\theta)}{\sin\theta} - \dfrac{\cos(3\theta)}{\cos\theta} = \dfrac{\sin(3\theta)\cos\theta - \cos(3\theta)\sin\theta}{\sin\theta\cos\theta} = \dfrac{\sin(3\theta-\theta)}{\sin\theta\cos\theta} = \dfrac{\sin 2\theta}{\sin\theta\cos\theta} = \dfrac{2\sin\theta\cos\theta}{\sin\theta\cos\theta} = 2$

53. $\tan(3\theta) = \tan(2\theta+\theta) = \dfrac{\tan(2\theta)+\tan\theta}{1-\tan(2\theta)\tan\theta} = \dfrac{\dfrac{2\tan\theta}{1-\tan^2\theta}+\tan\theta}{1-\dfrac{2\tan\theta}{1-\tan^2\theta}\cdot\tan\theta}$

$= \dfrac{\dfrac{2\tan\theta+\tan\theta-\tan^3\theta}{1-\tan^2\theta}}{\dfrac{1-\tan^2\theta-2\tan^2\theta}{1-\tan^2\theta}} = \dfrac{3\tan\theta-\tan^3\theta}{1-3\tan^2\theta}$

55. $\dfrac{1}{2}\cdot\left(\ln\left|1-\cos(2\theta)\right|-\ln 2\right) = \dfrac{1}{2}\cdot\ln\left|\dfrac{1-\cos 2\theta}{2}\right| = \ln\left(\left|\dfrac{1-\cos(2\theta)}{2}\right|^{1/2}\right) = \ln\left(\left|\sin^2\theta\right|^{1/2}\right) = \ln\left|\sin\theta\right|$

57. $\sin\left(2\sin^{-1}\dfrac{1}{2}\right) = \sin\left(2\left(\dfrac{\pi}{6}\right)\right) = \sin\dfrac{\pi}{3} = \dfrac{\sqrt{3}}{2}$

59. $\cos\left(2\sin^{-1}\dfrac{3}{5}\right) = 1-2\sin^2\left(\sin^{-1}\dfrac{3}{5}\right) = 1-2\left(\dfrac{3}{5}\right)^2 = 1-2\left(\dfrac{9}{25}\right) = 1-\dfrac{18}{25} = \dfrac{7}{25}$

61. $\tan\left[2\cos^{-1}\left(-\dfrac{3}{5}\right)\right]$

Let $\alpha = \cos^{-1}\left(-\dfrac{3}{5}\right)$. α is in quadrant II.

Then $\cos\alpha = -\dfrac{3}{5}$, $\dfrac{\pi}{2} \le \alpha \le \pi$.

$$\sec\alpha = -\frac{5}{3}; \quad \tan\alpha = -\sqrt{\sec^2\alpha - 1} = -\sqrt{\left(-\frac{5}{3}\right)^2 - 1} = -\sqrt{\frac{25}{9} - 1} = -\sqrt{\frac{16}{9}} = -\frac{4}{3}$$

$$\tan\left[2\cos^{-1}\left(-\frac{3}{5}\right)\right] = \tan 2\alpha = \frac{2\tan\alpha}{1 - \tan^2\alpha} = \frac{2\left(-\frac{4}{3}\right)}{1 - \left(-\frac{4}{3}\right)^2} = \frac{-\frac{8}{3}}{1 - \frac{16}{9}} = \frac{-\frac{8}{3}}{-\frac{7}{9}} = \left(-\frac{8}{3}\right)\cdot\left(-\frac{9}{7}\right) = \frac{24}{7}$$

63. $\sin\left(2\cos^{-1}\frac{4}{5}\right)$

Let $\alpha = \cos^{-1}\frac{4}{5}$. α is in quadrant I.

Then $\cos\alpha = \frac{4}{5}, \ 0 \le \alpha \le \frac{\pi}{2}$.

$$\sin\alpha = \sqrt{1 - \cos^2\alpha} = \sqrt{1 - \left(\frac{4}{5}\right)^2} = \sqrt{1 - \frac{16}{25}} = \sqrt{\frac{9}{25}} = \frac{3}{5}$$

$$\sin\left[2\cos^{-1}\frac{4}{5}\right] = \sin 2\alpha = 2\sin\alpha\cos\alpha = 2\cdot\frac{3}{5}\cdot\frac{4}{5} = \frac{24}{25}$$

65. $\sin^2\left[\frac{1}{2}\cdot\cos^{-1}\frac{3}{5}\right] = \dfrac{1 - \cos\left(\cos^{-1}\frac{3}{5}\right)}{2} = \dfrac{1 - \frac{3}{5}}{2} = \dfrac{\frac{2}{5}}{2} = \frac{1}{5}$

67. $\sec\left(2\tan^{-1}\frac{3}{4}\right)$

Let $\alpha = \tan^{-1}\left(\frac{3}{4}\right)$. α is in quadrant I.

Then $\tan\alpha = \frac{3}{4}, \ 0 < \alpha < \frac{\pi}{2}$.

$$\sec\alpha = \sqrt{\tan^2\alpha + 1} = \sqrt{\left(\frac{3}{4}\right)^2 + 1} = \sqrt{\frac{9}{16} + 1} = \sqrt{\frac{25}{16}} = \frac{5}{4}; \quad \cos\alpha = \frac{4}{5}$$

$$\sec\left[2\tan^{-1}\frac{3}{4}\right] = \sec(2\alpha) = \frac{1}{\cos(2\alpha)} = \frac{1}{2\cos^2\alpha - 1} = \frac{1}{2\left(\frac{4}{5}\right)^2 - 1} = \frac{1}{2\cdot\left(\frac{16}{25}\right) - 1} = \frac{1}{\frac{7}{25}} = \frac{25}{7}$$

69. $\sin(2\theta) = 2\sin\theta\cos\theta = \dfrac{2\sin\theta}{\cos\theta}\cdot\dfrac{\cos^2\theta}{1} = \dfrac{2\cdot\dfrac{\sin\theta}{\cos\theta}}{\dfrac{1}{\cos^2\theta}} = \dfrac{2\tan\theta}{\sec^2\theta} = \dfrac{2\tan\theta}{1 + \tan^2\theta}\cdot\dfrac{4}{4}$

$$= \frac{4(2\tan\theta)}{4 + (2\tan\theta)^2} = \frac{4x}{4 + x^2}$$

71. $\dfrac{1}{2} \cdot \sin^2 x + C = -\dfrac{1}{4} \cdot \cos(2x)$

$C = -\dfrac{1}{4} \cdot \cos(2x) - \dfrac{1}{2} \cdot \sin^2 x = -\dfrac{1}{4} \cdot \left(\cos(2x) + 2\sin^2 x\right)$

$= -\dfrac{1}{4} \cdot \left(1 - 2\sin^2 x + 2\sin^2 x\right) = -\dfrac{1}{4} \cdot (1)$

$= -\dfrac{1}{4}$

73.

$$z = \tan\left(\dfrac{\alpha}{2}\right)$$

$$z = \dfrac{1 - \cos\alpha}{\sin\alpha}$$

$$z\sin\alpha = 1 - \cos\alpha$$

$$z\sin\alpha = 1 - \sqrt{1 - \sin^2\alpha}$$

$$z\sin\alpha - 1 = -\sqrt{1 - \sin^2\alpha}$$

$$z^2\sin^2\alpha - 2z\sin\alpha + 1 = 1 - \sin^2\alpha$$

$$z^2\sin^2\alpha + \sin^2\alpha = 2z\sin\alpha$$

$$\sin^2\alpha(z^2 + 1) = 2z\sin\alpha$$

$$\sin\alpha(z^2 + 1) = 2z$$

$$\sin\alpha = \dfrac{2z}{z^2 + 1}$$

75. Let b represent the base of the triangle.

$\cos\dfrac{\theta}{2} = \dfrac{h}{s} \ \Rightarrow \ h = s\cos\dfrac{\theta}{2}$ and $\sin\dfrac{\theta}{2} = \dfrac{(b/2)}{s} \ \Rightarrow \ b = 2s\sin\dfrac{\theta}{2}$

$A = \dfrac{1}{2}b \cdot h = \dfrac{1}{2} \cdot \left(2s\sin\dfrac{\theta}{2}\right)\left(s\cos\dfrac{\theta}{2}\right) = s^2\sin\dfrac{\theta}{2}\cos\dfrac{\theta}{2} = \dfrac{1}{2} \cdot s^2\sin\theta$

77. $f(x) = \sin^2 x = \dfrac{1 - \cos(2x)}{2}$

Starting with the graph of $y = \cos x$,
compress horizontally by a factor of 2,
reflect across the x-axis, shift 1 unit up,
and shrink vertically by a factor of 2.

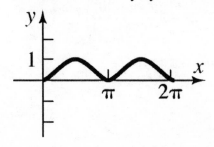

79. $\sin \dfrac{\pi}{24} = \sin\!\left(\dfrac{\dfrac{\pi}{12}}{2}\right) = \sqrt{\dfrac{1-\cos\dfrac{\pi}{12}}{2}} = \sqrt{\dfrac{1-\left(\dfrac{1}{4}\left(\sqrt{6}+\sqrt{2}\right)\right)}{2}} = \sqrt{\dfrac{1}{2} - \dfrac{1}{8}\left(\sqrt{6}+\sqrt{2}\right)}$

$= \sqrt{\dfrac{8-2\left(\sqrt{6}+\sqrt{2}\right)}{16}} = \dfrac{\sqrt{8-2\left(\sqrt{6}+\sqrt{2}\right)}}{4} = \dfrac{\sqrt{2\left(4-\left(\sqrt{6}+\sqrt{2}\right)\right)}}{4} = \dfrac{\sqrt{2}}{4}\sqrt{4-\sqrt{6}-\sqrt{2}}$

$\cos \dfrac{\pi}{24} = \cos\!\left(\dfrac{\dfrac{\pi}{12}}{2}\right) = \sqrt{\dfrac{1+\cos\dfrac{\pi}{12}}{2}} = \sqrt{\dfrac{1+\left(\dfrac{1}{4}\left(\sqrt{6}+\sqrt{2}\right)\right)}{2}} = \sqrt{\dfrac{1}{2} + \dfrac{1}{8}\left(\sqrt{6}+\sqrt{2}\right)}$

$= \sqrt{\dfrac{8+2\left(\sqrt{6}+\sqrt{2}\right)}{16}} = \dfrac{\sqrt{8+2\left(\sqrt{6}+\sqrt{2}\right)}}{4} = \dfrac{\sqrt{2\left(4+\sqrt{6}+\sqrt{2}\right)}}{4} = \dfrac{\sqrt{2}}{4}\sqrt{4+\sqrt{6}+\sqrt{2}}$

81. $\sin^3\theta + \sin^3(\theta+120°) + \sin^3(\theta+240°)$

$= \sin^3\theta + \left(\sin\theta\cos(120°) + \cos\theta\sin(120°)\right)^3 + \left(\sin\theta\cos(240°) + \cos\theta\sin(240°)\right)^3$

$= \sin^3\theta + \left(-\dfrac{1}{2}\cdot\sin\theta + \dfrac{\sqrt{3}}{2}\cdot\cos\theta\right)^3 + \left(-\dfrac{1}{2}\cdot\sin\theta - \dfrac{\sqrt{3}}{2}\cdot\cos\theta\right)^3$

$= \sin^3\theta + \dfrac{1}{8}\cdot\left(-\sin^3\theta + 3\sqrt{3}\sin^2\theta\cos\theta - 9\sin\theta\cos^2\theta + 3\sqrt{3}\cos^3\theta\right)$

$\qquad - \dfrac{1}{8}\left(\sin^3\theta + 3\sqrt{3}\sin^2\theta\cos\theta + 9\sin\theta\cos^2\theta + 3\sqrt{3}\cos^3\theta\right)$

$= \sin^3\theta - \dfrac{1}{8}\cdot\sin^3\theta + \dfrac{3\sqrt{3}}{8}\cdot\sin^2\theta\cos\theta - \dfrac{9}{8}\cdot\sin\theta\cos^2\theta + \dfrac{3\sqrt{3}}{8}\cdot\cos^3\theta$

$\qquad - \dfrac{1}{8}\cdot\sin^3\theta - \dfrac{3\sqrt{3}}{8}\cdot\sin^2\theta\cos\theta - \dfrac{9}{8}\cdot\sin\theta\cos^2\theta - \dfrac{3\sqrt{3}}{8}\cdot\cos^3\theta$

$= \dfrac{3}{4}\cdot\sin^3\theta - \dfrac{9}{4}\cdot\sin\theta\cos^2\theta = \dfrac{3}{4}\cdot\left(\sin^3\theta - 3\sin\theta\left(1-\sin^2\theta\right)\right)$

$= \dfrac{3}{4}\cdot\left(\sin^3\theta - 3\sin\theta + 3\sin^3\theta\right) = \dfrac{3}{4}\cdot\left(4\sin^3\theta - 3\sin\theta\right) = -\dfrac{3}{4}\cdot\sin(3\theta)$

(See the formula for $\sin(3\theta)$ on page 628 of the text.)

83. (a) $R(\theta) = \dfrac{v_0^{\,2}\sqrt{2}}{16}\cos\theta(\sin\theta - \cos\theta) = \dfrac{v_0^{\,2}\sqrt{2}}{16}(\cos\theta\sin\theta - \cos^2\theta)$

$= \dfrac{v_0^{\,2}\sqrt{2}}{16}\cdot\dfrac{1}{2}(2\cos\theta\sin\theta - 2\cos^2\theta) = \dfrac{v_0^{\,2}\sqrt{2}}{32}\left(\sin 2\theta - 2\left(\dfrac{1+\cos 2\theta}{2}\right)\right)$

$= \dfrac{v_0^{\,2}\sqrt{2}}{32}\left(\sin(2\theta) - 1 - \cos(2\theta)\right) = \dfrac{v_0^{\,2}\sqrt{2}}{32}\left(\sin(2\theta) - \cos(2\theta) - 1\right)$

(b)

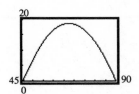

(c)   Using the MAXIMUM feature on the calculator:
       R has the largest value when $\theta \approx 67.5°$.

85.   Answers will vary.

Analytic Trigonometry

6.6 Product-to-Sum and Sum-to-Product Formulas

For Problems 1–9, use the formulas:

$$\sin\alpha\sin\beta = \frac{1}{2}\cdot\left[\cos(\alpha-\beta)-\cos(\alpha+\beta)\right] \qquad \cos\alpha\cos\beta = \frac{1}{2}\cdot\left[\cos(\alpha-\beta)+\cos(\alpha+\beta)\right]$$

$$\sin\alpha\cos\beta = \frac{1}{2}\cdot\left[\sin(\alpha+\beta)+\sin(\alpha-\beta)\right]$$

1. $\sin(4\theta)\sin(2\theta) = \frac{1}{2}\cdot\left[\cos(4\theta-2\theta)-\cos(4\theta+2\theta)\right] = \frac{1}{2}\cdot\left[\cos(2\theta)-\cos(6\theta)\right]$

3. $\sin(4\theta)\cos(2\theta) = \frac{1}{2}\cdot\left[\sin(4\theta+2\theta)+\sin(4\theta-2\theta)\right] = \frac{1}{2}\cdot\left[\sin(6\theta)+\sin(2\theta)\right]$

5. $\cos(3\theta)\cos(5\theta) = \frac{1}{2}\cdot\left[\cos(3\theta-5\theta)+\cos(3\theta+5\theta)\right] = \frac{1}{2}\cdot\left[\cos(-2\theta)+\cos(8\theta)\right]$

   $= \frac{1}{2}\cdot\left[\cos(2\theta)+\cos(8\theta)\right]$

7. $\sin\theta\sin(2\theta) = \frac{1}{2}\cdot\left[\cos(\theta-2\theta)-\cos(\theta+2\theta)\right] = \frac{1}{2}\cdot\left[\cos(-\theta)-\cos(3\theta)\right]$

   $= \frac{1}{2}\cdot\left[\cos\theta-\cos(3\theta)\right]$

9. $\sin\dfrac{3\theta}{2}\cos\dfrac{\theta}{2} = \dfrac{1}{2}\cdot\left[\sin\!\left(\dfrac{3\theta}{2}+\dfrac{\theta}{2}\right)+\sin\!\left(\dfrac{3\theta}{2}-\dfrac{\theta}{2}\right)\right] = \dfrac{1}{2}\cdot\left[\sin(2\theta)+\sin\theta\right]$

For Problems 11–17, use the formulas:

$$\sin\alpha+\sin\beta = 2\sin\!\left(\frac{\alpha+\beta}{2}\right)\cos\!\left(\frac{\alpha-\beta}{2}\right) \qquad \sin\alpha-\sin\beta = 2\sin\!\left(\frac{\alpha-\beta}{2}\right)\cos\!\left(\frac{\alpha+\beta}{2}\right)$$

$$\cos\alpha+\cos\beta = 2\cos\!\left(\frac{\alpha+\beta}{2}\right)\cos\!\left(\frac{\alpha-\beta}{2}\right) \qquad \cos\alpha-\cos\beta = -2\sin\!\left(\frac{\alpha+\beta}{2}\right)\sin\!\left(\frac{\alpha-\beta}{2}\right)$$

11. $\sin(4\theta)-\sin(2\theta) = 2\sin\!\left(\dfrac{4\theta-2\theta}{2}\right)\cos\!\left(\dfrac{4\theta+2\theta}{2}\right) = 2\sin\theta\cos(3\theta)$

13. $\cos(2\theta) + \cos(4\theta) = 2\cos\left(\dfrac{2\theta + 4\theta}{2}\right)\cos\left(\dfrac{2\theta - 4\theta}{2}\right) = 2\cos(3\theta)\cos(-\theta) = 2\cos(3\theta)\cos\theta$

15. $\sin\theta + \sin(3\theta) = 2\sin\left(\dfrac{\theta + 3\theta}{2}\right)\cos\left(\dfrac{\theta - 3\theta}{2}\right) = 2\sin(2\theta)\cos(-\theta) = 2\sin(2\theta)\cos\theta$

17. $\cos\dfrac{\theta}{2} - \cos\dfrac{3\theta}{2} = -2\sin\left(\dfrac{\dfrac{\theta}{2} + \dfrac{3\theta}{2}}{2}\right)\sin\left(\dfrac{\dfrac{\theta}{2} - \dfrac{3\theta}{2}}{2}\right) = -2\sin\theta\sin\left(-\dfrac{\theta}{2}\right) = -2\sin\theta\left(-\sin\dfrac{\theta}{2}\right) = 2\sin\theta\sin\dfrac{\theta}{2}$

19. $\dfrac{\sin\theta + \sin(3\theta)}{2\sin(2\theta)} = \dfrac{2\sin(2\theta)\cos(-\theta)}{2\sin(2\theta)} = \cos(-\theta) = \cos\theta$

21. $\dfrac{\sin(4\theta) + \sin(2\theta)}{\cos(4\theta) + \cos(2\theta)} = \dfrac{2\sin(3\theta)\cos\theta}{2\cos(3\theta)\cos\theta} = \dfrac{\sin(3\theta)}{\cos(3\theta)} = \tan(3\theta)$

23. $\dfrac{\cos\theta - \cos(3\theta)}{\sin\theta + \sin(3\theta)} = \dfrac{-2\sin(2\theta)\sin(-\theta)}{2\sin(2\theta)\cos(-\theta)} = \dfrac{-(-\sin\theta)}{\cos\theta} = \tan\theta$

25. $\sin\theta\big[\sin\theta + \sin(3\theta)\big] = \sin\theta\big[2\sin(2\theta)\cos(-\theta)\big] = \cos\theta\big[2\sin(2\theta)\sin\theta\big]$

   $= \cos\theta\left[2 \cdot \dfrac{1}{2}(\cos\theta - \cos(3\theta))\right] = \cos\theta(\cos\theta - \cos(3\theta))$

27. $\dfrac{\sin(4\theta) + \sin(8\theta)}{\cos(4\theta) + \cos(8\theta)} = \dfrac{2\sin(6\theta)\cos(-2\theta)}{2\cos(6\theta)\cos(-2\theta)} = \dfrac{\sin(6\theta)}{\cos(6\theta)} = \tan(6\theta)$

29. $\dfrac{\sin(4\theta) + \sin(8\theta)}{\sin(4\theta) - \sin(8\theta)} = \dfrac{2\sin(6\theta)\cos(-2\theta)}{2\sin(-2\theta)\cos(6\theta)} = \dfrac{\sin(6\theta)\cos(2\theta)}{-\sin(2\theta)\cos(6\theta)}$

   $= -\tan(6\theta)\cot(2\theta) = -\dfrac{\tan(6\theta)}{\tan(2\theta)}$

31. $\dfrac{\sin\alpha + \sin\beta}{\sin\alpha - \sin\beta} = \dfrac{2\sin\left(\dfrac{\alpha + \beta}{2}\right)\cos\left(\dfrac{\alpha - \beta}{2}\right)}{2\sin\left(\dfrac{\alpha - \beta}{2}\right)\cos\left(\dfrac{\alpha + \beta}{2}\right)} = \tan\left(\dfrac{\alpha + \beta}{2}\right)\cot\left(\dfrac{\alpha - \beta}{2}\right)$

33. $\dfrac{\sin\alpha + \sin\beta}{\cos\alpha + \cos\beta} = \dfrac{2\sin\left(\dfrac{\alpha + \beta}{2}\right)\cos\left(\dfrac{\alpha - \beta}{2}\right)}{2\cos\left(\dfrac{\alpha + \beta}{2}\right)\cos\left(\dfrac{\alpha - \beta}{2}\right)} = \tan\left(\dfrac{\alpha + \beta}{2}\right)$

35. $1 + \cos(2\theta) + \cos(4\theta) + \cos(6\theta) = \cos 0 + \cos(6\theta) + \cos(2\theta) + \cos(4\theta)$

   $= 2\cos(3\theta)\cos(-3\theta) + 2\cos(3\theta)\cos(-\theta) = 2\cos^2(3\theta) + 2\cos(3\theta)\cos\theta$

   $= 2\cos(3\theta)\big(\cos(3\theta) + \cos\theta\big) = 2\cos(3\theta)2\cos(2\theta)\cos\theta$

   $= 4\cos\theta\cos(2\theta)\cos(3\theta)$

37. (a) $y = \sin[2\pi(852)t] + \sin[2\pi(1209)t]$

$$= 2\sin\left(\frac{2\pi(852)t + 2\pi(1209)t}{2}\right)\cos\left(\frac{2\pi(852)t - 2\pi(1209)t}{2}\right)$$

$$= 2\sin(2061\pi t)\cos(-357\pi t)$$

$$= 2\sin(2061\pi t)\cos(357\pi t)$$

(b) The maximum value of y is 2.

(c)

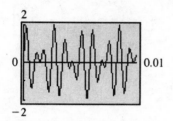

39. $\sin(2\alpha) + \sin(2\beta) + \sin(2\gamma)$

$$= 2\sin\left(\frac{2\alpha + 2\beta}{2}\right)\cos\left(\frac{2\alpha - 2\beta}{2}\right) + \sin(2\gamma)$$

$$= 2\sin(\alpha + \beta)\cos(\alpha - \beta) + 2\sin\gamma\cos\gamma$$

$$= 2\sin(\pi - \gamma)\cos(\alpha - \beta) + 2\sin\gamma\cos\gamma$$

$$= 2\sin\gamma\cos(\alpha - \beta) + 2\sin\gamma\cos\gamma = 2\sin\gamma[\cos(\alpha - \beta) + \cos\gamma]$$

$$= 2\sin\gamma\left(2\cos\left(\frac{\alpha - \beta + \gamma}{2}\right)\cos\left(\frac{\alpha - \beta - \gamma}{2}\right)\right)$$

$$= 4\sin\gamma\cos\left(\frac{\pi}{2} - \beta\right)\cos\left(\alpha - \frac{\pi}{2}\right) = 4\sin\gamma\sin\beta\sin\alpha$$

$$= 4\sin\alpha\sin\beta\sin\gamma$$

41. Add the sum formulas for $\sin(\alpha + \beta)$ and $\sin(\alpha - \beta)$ and solve for $\sin\alpha\cos\beta$:

$$\sin(\alpha + \beta) = \sin\alpha\cos\beta + \cos\alpha\sin\beta$$

$$\sin(\alpha - \beta) = \sin\alpha\cos\beta - \cos\alpha\sin\beta$$

$$\sin(\alpha + \beta) + \sin(\alpha - \beta) = 2\sin\alpha\cos\beta$$

$$\sin\alpha\cos\beta = \frac{1}{2} \cdot [\sin(\alpha + \beta) + \sin(\alpha - \beta)]$$

43. $2\cos\left(\frac{\alpha + \beta}{2}\right)\cos\left(\frac{\alpha - \beta}{2}\right) = 2 \cdot \frac{1}{2}\left[\cos\left(\frac{\alpha + \beta}{2} - \frac{\alpha - \beta}{2}\right) + \cos\left(\frac{\alpha + \beta}{2} + \frac{\alpha - \beta}{2}\right)\right]$

$$= \cos\left(\frac{2\beta}{2}\right) + \cos\left(\frac{2\alpha}{2}\right) = \cos\beta + \cos\alpha$$

Therefore, $\cos\alpha + \cos\beta = 2\cos\left(\frac{\alpha + \beta}{2}\right)\cos\left(\frac{\alpha - \beta}{2}\right)$

Chapter 6

Analytic Trigonometry

6.7 Trigonometric Equations (I)

1. $3x - 5 = -x + 1$
   $4x = 6$
   $x = \dfrac{6}{4} = \dfrac{3}{2}$
   The solution set is $\left\{ \dfrac{3}{2} \right\}$.

3. $\dfrac{\pi}{6}, \dfrac{5\pi}{6}$

5. False

7. $2\sin\theta + 3 = 2$
   $2\sin\theta = -1 \Rightarrow \sin\theta = -\dfrac{1}{2}$
   $\theta = \dfrac{7\pi}{6} + 2k\pi$ or $\theta = \dfrac{11\pi}{6} + 2k\pi$, k is any integer
   The solutions on the interval $0 \le \theta < 2\pi$ are $\theta = \dfrac{7\pi}{6}, \dfrac{11\pi}{6}$.

9. $4\cos^2\theta = 1$
   $\cos^2\theta = \dfrac{1}{4} \Rightarrow \cos\theta = \pm\dfrac{1}{2}$
   $\theta = \dfrac{\pi}{3} + k\pi$ or $\theta = \dfrac{2\pi}{3} + k\pi$, k is any integer
   The solutions on the interval $0 \le \theta < 2\pi$ are $\theta = \dfrac{\pi}{3}, \dfrac{2\pi}{3}, \dfrac{4\pi}{3}, \dfrac{5\pi}{3}$.

11. $2\sin^2\theta - 1 = 0$
    $2\sin^2\theta = 1 \Rightarrow \sin^2\theta = \dfrac{1}{2} \Rightarrow \sin\theta = \pm\sqrt{\dfrac{1}{2}} = \pm\dfrac{\sqrt{2}}{2}$
    $\theta = \dfrac{\pi}{4} + k\pi$ or $\theta = \dfrac{3\pi}{4} + k\pi$, k is any integer
    The solutions on the interval $0 \le \theta < 2\pi$ are $\theta = \dfrac{\pi}{4}, \dfrac{3\pi}{4}, \dfrac{5\pi}{4}, \dfrac{7\pi}{4}$.

13. $\sin(3\theta) = -1$

$$3\theta = \frac{3\pi}{2} + 2k\pi \implies \theta = \frac{\pi}{2} + \frac{2k\pi}{3}, \ k \text{ is any integer}$$

The solutions on the interval $0 \le \theta < 2\pi$ are $\theta = \frac{\pi}{2}, \frac{7\pi}{6}, \frac{11\pi}{6}$.

15. $\cos(2\theta) = -\frac{1}{2}$

$$2\theta = \frac{2\pi}{3} + 2k\pi \implies \theta = \frac{\pi}{3} + k\pi, k \text{ is any integer}$$

$$2\theta = \frac{4\pi}{3} + 2k\pi \implies \theta = \frac{2\pi}{3} + k\pi, k \text{ is any integer}$$

The solutions on the interval $0 \le \theta < 2\pi$ are $\theta = \frac{\pi}{3}, \frac{2\pi}{3}, \frac{4\pi}{3}, \frac{5\pi}{3}$.

17. $\sec\left(\frac{3\theta}{2}\right) = -2$

$$\frac{3\theta}{2} = \frac{2\pi}{3} + 2k\pi \implies \theta = \frac{4\pi}{9} + \frac{4k\pi}{3}, k \text{ is any integer}$$

$$\frac{3\theta}{2} = \frac{4\pi}{3} + 2k\pi \implies \theta = \frac{8\pi}{9} + \frac{4k\pi}{3}, k \text{ is any integer}$$

The solutions on the interval $0 \le \theta < 2\pi$ are $\theta = \frac{4\pi}{9}, \frac{8\pi}{9}, \frac{16\pi}{9}$

19. $2\sin\theta + 1 = 0 \implies 2\sin\theta = -1 \implies \sin\theta = -\frac{1}{2}$

$$\theta = \frac{7\pi}{6} + 2k\pi \quad \text{or} \quad \theta = \frac{11\pi}{6} + 2k\pi, \ k \text{ is any integer}$$

The solutions on the interval $0 \le \theta < 2\pi$ are $\theta = \frac{7\pi}{6}, \frac{11\pi}{6}$.

21. $\tan\theta + 1 = 0 \implies \tan\theta = -1$

$$\theta = \frac{3\pi}{4} + k\pi, \ k \text{ is any integer}$$

The solutions on the interval $0 \le \theta < 2\pi$ are $\theta = \frac{3\pi}{4}, \frac{7\pi}{4}$.

23. $4\sec\theta + 6 = -2 \implies 4\sec\theta = -8 \implies \sec\theta = -2$

$$\theta = \frac{2\pi}{3} + 2k\pi \quad \text{or} \quad \theta = \frac{4\pi}{3} + 2k\pi, \ k \text{ is any integer}$$

The solutions on the interval $0 \le \theta < 2\pi$ are $\theta = \frac{2\pi}{3}, \frac{4\pi}{3}$.

25. $3\sqrt{2}\cos\theta + 2 = -1 \implies 3\sqrt{2}\cos\theta = -3 \implies \cos\theta = -\frac{1}{\sqrt{2}} = -\frac{\sqrt{2}}{2}$

$$\theta = \frac{3\pi}{4} + 2k\pi \quad \text{or} \quad \theta = \frac{5\pi}{4} + 2k\pi, \ k \text{ is any integer}$$

The solutions on the interval $0 \le \theta < 2\pi$ are $\theta = \frac{3\pi}{4}, \frac{5\pi}{4}$.

27. $\cos\left(2\theta - \dfrac{\pi}{2}\right) = -1$

$2\theta - \dfrac{\pi}{2} = \pi + 2k\pi \implies 2\theta = \dfrac{3\pi}{2} + 2k\pi \implies \theta = \dfrac{3\pi}{4} + k\pi,\ k$ is any integer

The solutions on the interval $0 \le \theta < 2\pi$ are $\theta = \dfrac{3\pi}{4},\ \dfrac{7\pi}{4}$.

29. $\tan\left(\dfrac{\theta}{2} + \dfrac{\pi}{3}\right) = 1$

$\dfrac{\theta}{2} + \dfrac{\pi}{3} = \dfrac{\pi}{4} + k\pi \implies \dfrac{\theta}{2} = -\dfrac{\pi}{12} + k\pi \implies \theta = -\dfrac{\pi}{6} + 2k\pi,\ k$ is any integer

The solution on the interval $0 \le \theta < 2\pi$ is $\theta = \dfrac{11\pi}{6}$.

31. $\sin\theta = \dfrac{1}{2}$

$\theta = \dfrac{\pi}{6} + 2k\pi$ or $\theta = \dfrac{5\pi}{6} + 2k\pi,\ k$ is any integer

Six solutions are $\theta = \dfrac{\pi}{6}, \dfrac{5\pi}{6}, \dfrac{13\pi}{6}, \dfrac{17\pi}{6}, \dfrac{25\pi}{6}, \dfrac{29\pi}{6}$.

33. $\tan\theta = -\dfrac{\sqrt{3}}{3}$

$\theta = \dfrac{5\pi}{6} + k\pi,\ k$ is any integer

Six solutions are $\theta = \dfrac{5\pi}{6}, \dfrac{11\pi}{6}, \dfrac{17\pi}{6}, \dfrac{23\pi}{6}, \dfrac{29\pi}{6}, \dfrac{35\pi}{6}$.

35. $\cos\theta = 0$

$\theta = \dfrac{\pi}{2} + 2k\pi$ or $\theta = \dfrac{3\pi}{2} + 2k\pi,\ k$ is any integer

Six solutions are $\theta = \dfrac{\pi}{2}, \dfrac{3\pi}{2}, \dfrac{5\pi}{2}, \dfrac{7\pi}{2}, \dfrac{9\pi}{2}, \dfrac{11\pi}{2}$.

37. $\cos(2\theta) = -\dfrac{1}{2}$

$2\theta = \dfrac{2\pi}{3} + 2k\pi \implies \theta = \dfrac{\pi}{3} + k\pi,\ k$ is any integer

$2\theta = \dfrac{4\pi}{3} + 2k\pi \implies \theta = \dfrac{2\pi}{3} + k\pi,\ k$ is any integer

Six solutions are $\theta = \dfrac{\pi}{3}, \dfrac{2\pi}{3}, \dfrac{4\pi}{3}, \dfrac{5\pi}{3}, \dfrac{7\pi}{3}, \dfrac{8\pi}{3}$.

39. $\sin\dfrac{\theta}{2} = -\dfrac{\sqrt{3}}{2}$

$\dfrac{\theta}{2} = \dfrac{4\pi}{3} + 2k\pi \implies \theta = \dfrac{8\pi}{3} + 4k\pi$, k is any integer

$\dfrac{\theta}{2} = \dfrac{5\pi}{3} + 2k\pi \implies \theta = \dfrac{10\pi}{3} + 4k\pi$, k is any integer

Six solutions are $\theta = \dfrac{8\pi}{3}, \dfrac{10\pi}{3}, \dfrac{20\pi}{3}, \dfrac{22\pi}{3}, \dfrac{32\pi}{3}, \dfrac{34\pi}{3}$.

41.

Thus, $\theta \approx 0.41$ or $\theta \approx \pi - 0.41 \approx 2.73$.
The solution set is $\{0.41,\ 2.73\}$.

43. $\tan\theta = 5$

$\theta = \tan^{-1}(5) \approx 1.37$

Thus, $\theta \approx 1.37$ or $\theta \approx \pi + 1.37 \approx 4.51$.
The solution set is $\{1.37,\ 4.51\}$.

45. $\cos\theta = -0.9$

$\theta = \cos^{-1}(-0.9) \approx 2.69$

Thus, $\theta \approx 2.69$ or $\theta \approx 2\pi - 2.69 \approx 3.59$.
The solution set is $\{2.69,\ 3.69\}$.

47. $\sec\theta = -4$

$\cos\theta = -\dfrac{1}{4}$

$\theta = \cos^{-1}\left(-\dfrac{1}{4}\right) \approx 1.82$

Thus, $\theta \approx 1.82$ or $\theta \approx 2\pi - 1.82 \approx 4.46$.
The solution set is $\{1.82,\ 4.46\}$.

49. $5\tan\theta + 9 = 0$

$5\tan\theta = -9$

$\tan\theta = -\dfrac{9}{5}$

$\theta = \tan^{-1}\left(-\dfrac{9}{5}\right) \approx -1.064$

Thus, $\theta \approx -1.064 + \pi \approx 2.08$ or
$\theta \approx -1.064 + 2\pi \approx 5.22$.
The solution set is $\{2.08,\ 5.22\}$.

51. $3\sin\theta - 2 = 0$

$3\sin\theta = 2$

$\sin\theta = \dfrac{2}{3}$

$\theta = \sin^{-1}\left(\dfrac{2}{3}\right) \approx 0.73$

Thus, $\theta \approx 0.73$ or $\theta \approx \pi - 0.73 \approx 2.41$.
The solution set is $\{0.73,\ 2.41\}$

53. $f(x) = 3\sin x$

(a) $f(x) = \dfrac{3}{2} \implies 3\sin x = \dfrac{3}{2}$

$3\sin x = \dfrac{3}{2}$

$\sin x = \dfrac{1}{2}$

$x = \dfrac{\pi}{6} + 2k\pi$ or $x = \dfrac{5\pi}{6} + 2k\pi$, k any integer

(b) $f(x) > \dfrac{3}{2} \Rightarrow 3\sin x > \dfrac{3}{2}$

$3\sin x > \dfrac{3}{2} \Rightarrow \sin x > \dfrac{1}{2}$

Graphing $y_1 = \sin x$ and $y_2 = \dfrac{1}{2}$ on the interval $[0, 2\pi)$ and using INTERSECT, we can

see that $y_1 > y_2$ for $\dfrac{\pi}{6} < x < \dfrac{5\pi}{6}$.

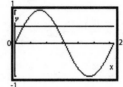

55. $f(x) = 4\tan x$

(a) $f(x) = -4 \Rightarrow 4\tan x = -4$
$4\tan x = -4$

$\tan x = -1 \Rightarrow x = -\dfrac{\pi}{4} + k\pi,\ k$ any integer

(b) $f(x) < -4 \Rightarrow 4\tan x < -4$
$4\tan x < -4 \Rightarrow \tan x < -1$

Graphing $y_1 = \tan x$ and $y_2 = -1$ on the interval $\left(-\dfrac{\pi}{2}, \dfrac{\pi}{2}\right)$ and using INTERSECT,

we can see that $y_1 < y_2$ for $-\dfrac{\pi}{2} < x < -\dfrac{\pi}{4}$.

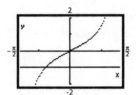

57. $h(t) = 125\sin\left(0.157t - \dfrac{\pi}{2}\right) + 125$

(a) Solve $h(t) = 125\sin\left(0.157t - \dfrac{\pi}{2}\right) + 125 = 125$ on the interval $[0, 40]$.

$1125\sin\left(0.157t - \dfrac{\pi}{2}\right) + 125 = 125$

$125\sin\left(0.157t - \dfrac{\pi}{2}\right) + 125 = 125$

$125\sin\left(0.157t - \dfrac{\pi}{2}\right) = 0 \Rightarrow \sin\left(0.157t - \dfrac{\pi}{2}\right) = 0$

$0.157t - \dfrac{\pi}{2} = k\pi,\ k$ any integer

$0.157t = k\pi + \dfrac{\pi}{2},\ k$ any integer $\Rightarrow t = \dfrac{(k\pi + \pi/2)}{0.157},\ k$ any integer

For $k = 0$, $t = \dfrac{0 + \pi/2}{0.157} \approx 10.01$ seconds. For $k = 1$, $t = \dfrac{\pi + \pi/2}{0.157} \approx 30.02$ seconds.

For $k = 2$, $t = \dfrac{2\pi + \pi/2}{0.157} \approx 50.03$ seconds.

So during the first 40 seconds, an individual on the Ferris Wheel is exactly 125 feet above the ground when $t \approx 10.01$ seconds and again when $t \approx 30.02$ seconds.

(b)   Solve $h(t) = 125\sin\left(0.157t - \dfrac{\pi}{2}\right) + 125 = 250$ on the interval $[0,80]$.

$$125\sin\left(0.157t - \frac{\pi}{2}\right) + 125 = 250$$

$$125\sin\left(0.157t - \frac{\pi}{2}\right) = 125$$

$$\sin\left(0.157t - \frac{\pi}{2}\right) = 1$$

$$0.157t - \frac{\pi}{2} = \frac{\pi}{2} + 2k\pi, \ k \text{ any integer}$$

$$0.157t = \pi + 2k\pi, \ k \text{ any integer}$$

$$t = \frac{\pi + 2k\pi}{0.157}, \ k \text{ any integer}$$

For $k = 0$, $t = \dfrac{\pi}{0.157} \approx 20.01$ seconds. For $k = 1$, $t = \dfrac{3\pi}{0.157} \approx 60.03$ seconds.

For $k = 5$, $t = \dfrac{5\pi}{0.157} \approx 100.01$ seconds.

So during the first 80 seconds, an individual on the Ferris Wheel is exactly 250 feet above the ground when $t \approx 20.01$ seconds and again when $t \approx 60.030$ seconds.

(c)   Solve $h(t) = 125\sin\left(0.157t - \dfrac{\pi}{2}\right) + 125 > 125$ on the interval $[0,40]$.

$$125\sin\left(0.157t - \frac{\pi}{2}\right) + 125 > 125$$

$$125\sin\left(0.157t - \frac{\pi}{2}\right) > 0 \Rightarrow \sin\left(0.157t - \frac{\pi}{2}\right) > 0$$

Graphing $y_1 = \sin\left(0.157x - \dfrac{\pi}{2}\right)$ and $y_2 = 0$ on the interval $[0,40]$ and using

INTERSECT, we can see that $y_1 > y_2$ for $10.01 < x < 30.02$.

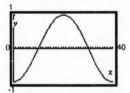

So during the first 40 seconds, an individual on the Ferris Wheel is more than 125 feet above the ground between 10.01 seconds and 30.02 seconds.

59. $d(x) = 70\sin(0.65x) + 150$

(a) $d(0) = 70\sin(0.65(0)) + 150 = 70\sin(0) + 150 = 150$ miles.

(b)   Solve $d(x) = 70\sin(0.65x) + 150 = 100$ on the interval $[0, 20]$.

$$70\sin(0.65x) + 150 = 100$$

$$70\sin(0.65x) = -50$$

$$\sin(0.65x) = -\frac{5}{7}$$

$$0.65x = \sin^{-1}\left(-\frac{5}{7}\right)$$

$$x = \frac{\sin^{-1}(-5/7)}{0.65} \approx \frac{3.94 + 2k\pi}{0.65} \quad \text{or} \quad \frac{5.49 + 2k\pi}{0.65}, \ k \text{ any integer}$$

For $k = 0$, $x \approx \dfrac{3.94 + 0}{0.65} \approx 6.06$ minutes or $x \approx \dfrac{5.49 + 0}{0.65} \approx 8.44$ minutes.

For $k = 1$, $x \approx \dfrac{3.94 + 2\pi}{0.65} \approx 15.72$ minutes or $x \approx \dfrac{5.49 + 2\pi}{0.65} \approx 18.11$ minutes.

For $k = 2$, $x \approx \dfrac{3.94 + 4\pi}{0.65} \approx 25.39$ minutes or $x \approx \dfrac{5.49 + 4\pi}{0.65} \approx 27.78$ minutes.

So during the first 20 minutes in the holding pattern, the plane is exactly 100 miles from the airport when $x \approx 6.06$ minutes, $x \approx 8.44$ minutes, $x \approx 15.72$ minutes, and $x \approx 18.11$ minutes.

(c)   Solve $d(x) = 70\sin(0.65x) + 150 > 100$ on the interval $[0, 20]$.

$$70\sin(0.65x) + 150 > 100$$

$$70\sin(0.65x) > -50 \Rightarrow \sin(0.65x) > -\frac{5}{7}$$

Graphing $y_1 = \sin(0.65x)$ and $y_2 = -\dfrac{5}{7}$ on the interval $[0, 20]$ and using INTERSECT, we can see that $y_1 > y_2$ for $0 < x < 6.06$, $8.44 < x < 15.72$ and $18.11 < x < 20$

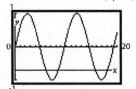

So during the first 20 minutes in the holding pattern, the plane is more than 100 miles from the airport when $0 < x < 6.06$ minutes, when $8.44 < x < 15.72$ minutes and when $18.11 < x < 20$ minutes.

(d)   The minimum value of $\sin(0.65x)$ is -1. Thus, the least distance that the plane is from the airport is $70(-1) + 150 = 80$ miles. The plane is never within 70 miles of the airport while in the holding pattern.

61. $\dfrac{\sin 40°}{\sin \theta_2} = 1.33$

$\sin 40° = 1.33 \sin \theta_2$

$\sin \theta_2 = \dfrac{\sin 40°}{1.33} \approx 0.4833$

$\theta_2 = \sin^{-1}(0.4833) \approx 28.90°$

63. Calculate the index of refraction for each:

$\theta_1 = 10°, \ \theta_2 = 7°45' = 7.75°$ $\qquad \dfrac{\sin \theta_1}{\sin \theta_2} = \dfrac{\sin 10°}{\sin 7.75°} \approx 1.2877$

$\theta_1 = 20°, \ \theta_2 = 15°30' = 15.5°$ $\qquad \dfrac{\sin \theta_1}{\sin \theta_2} = \dfrac{\sin 20°}{\sin 15.5°} \approx 1.2798$

$\theta_1 = 30°, \ \theta_2 = 22°30' = 22.5°$ $\qquad \dfrac{\sin \theta_1}{\sin \theta_2} = \dfrac{\sin 30°}{\sin 22.5°} \approx 1.3066$

$\theta_1 = 40°, \ \theta_2 = 29°0' = 29°$ $\qquad \dfrac{\sin \theta_1}{\sin \theta_2} = \dfrac{\sin 40°}{\sin 29°} \approx 1.3259$

$\theta_1 = 50°, \ \theta_2 = 35°0' = 35°$ $\qquad \dfrac{\sin \theta_1}{\sin \theta_2} = \dfrac{\sin 50°}{\sin 35°} \approx 1.3356$

$\theta_1 = 60°, \ \theta_2 = 40°30' = 40.5°$ $\qquad \dfrac{\sin \theta_1}{\sin \theta_2} = \dfrac{\sin 60°}{\sin 40.5°} \approx 1.3335$

$\theta_1 = 70°, \ \theta_2 = 45°30' = 45.5°$ $\qquad \dfrac{\sin \theta_1}{\sin \theta_2} = \dfrac{\sin 70°}{\sin 45.5°} \approx 1.3175$

$\theta_1 = 80°, \ \theta_2 = 50°0' = 50°$ $\qquad \dfrac{\sin \theta_1}{\sin \theta_2} = \dfrac{\sin 80°}{\sin 50°} \approx 1.2856$

The results range from 1.28 to 1.34 and are surprisingly close to Snell's Law.

65. Calculate the index of refraction:

$\theta_1 = 40°, \ \theta_2 = 26°$ $\qquad \dfrac{\sin \theta_1}{\sin \theta_2} = \dfrac{\sin 40°}{\sin 26°} \approx 1.47$

67. If θ is the original angle of incidence and ϕ is the angle of refraction, then $\dfrac{\sin \theta}{\sin \phi} = n_2$. The angle of incidence of the emerging beam is also ϕ, and the index of refraction is $\dfrac{1}{n_2}$. Thus, θ is the angle of refraction of the emerging beam. The two beams are parallel since the original angle of incidence and the angle of refraction of the emerging beam are equal.

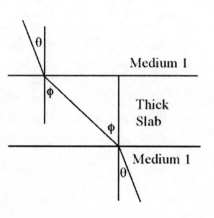

382

Chapter 6

Analytic Trigonometry

6.8   Trigonometric Equations (II)

1.
$$4x^2 - x - 5 = 0$$
$$(4x-5)(x+1) = 0$$

$$4x-5 = 0 \Rightarrow x = \frac{5}{4}$$
$$x+1 = 0 \Rightarrow x = -1$$

The solution set is $\left\{-1, \dfrac{4}{5}\right\}$.

3.
$$2\cos^2\theta + \cos\theta = 0$$
$$\cos\theta(2\cos\theta + 1) = 0$$

$$\cos\theta = 0 \Rightarrow \theta = \frac{\pi}{2}, \frac{3\pi}{2}$$
or $2\cos\theta + 1 = 0$

$$\cos\theta = -\frac{1}{2} \Rightarrow \theta = \frac{2\pi}{3}, \frac{4\pi}{3}$$

The solution set is $\left\{\dfrac{\pi}{2}, \dfrac{2\pi}{3}, \dfrac{4\pi}{3}, \dfrac{3\pi}{2}\right\}$.

5.
$$2\sin^2\theta - \sin\theta - 1 = 0$$
$$(2\sin\theta + 1)(\sin\theta - 1) = 0$$
$$2\sin\theta + 1 = 0$$

$$\sin\theta = -\frac{1}{2} \Rightarrow \theta = \frac{7\pi}{6}, \frac{11\pi}{6}$$
or $\sin\theta - 1 = 0$

$$\sin\theta = 1 \Rightarrow \theta = \frac{\pi}{2}$$

The solution set is $\left\{\dfrac{\pi}{2}, \dfrac{7\pi}{6}, \dfrac{11\pi}{6}\right\}$.

7.
$$(\tan\theta - 1)(\sec\theta - 1) = 0$$
$$\tan\theta - 1 = 0$$

$$\tan\theta = 1 \Rightarrow \theta = \frac{\pi}{4}, \frac{5\pi}{4}$$
or $\sec\theta - 1 = 0$

$$\sec\theta = 1 \Rightarrow \theta = 0$$

The solution set is $\left\{0, \dfrac{\pi}{4}, \dfrac{5\pi}{4}\right\}$.

9.
$$\sin^2\theta - \cos^2\theta = 1 + \cos\theta$$
$$(1 - \cos^2\theta) - \cos^2\theta = 1 + \cos\theta$$
$$1 - 2\cos^2\theta = 1 + \cos\theta$$
$$2\cos^2\theta + \cos\theta = 0 \Rightarrow \cos\theta(2\cos\theta + 1) = 0$$

$$\cos\theta = 0 \Rightarrow \theta = \frac{\pi}{2}, \frac{3\pi}{2}$$
or $2\cos\theta + 1 = 0$

$$\cos\theta = -\frac{1}{2} \Rightarrow \theta = \frac{2\pi}{3}, \frac{4\pi}{3}$$

The solution set is $\left\{\dfrac{\pi}{2}, \dfrac{2\pi}{3}, \dfrac{4\pi}{3}, \dfrac{3\pi}{2}\right\}$.

11.
$$\sin^2\theta = 6(\cos\theta + 1)$$
$$1 - \cos^2\theta = 6\cos\theta + 6$$
$$\cos^2\theta + 6\cos\theta + 5 = 0$$
$$(\cos\theta + 5)(\cos\theta + 1) = 0$$
$$\cos\theta + 5 = 0$$
$$\cos\theta = -5, \text{ which is impossible}$$
$$\text{or } \cos\theta + 1 = 0$$
$$\cos\theta = -1 \Rightarrow \theta = \pi$$
The solution set is $\{\pi\}$.

13.
$$\cos(2\theta) + 6\sin^2\theta = 4$$
$$1 - 2\sin^2\theta + 6\sin^2\theta = 4$$
$$4\sin^2\theta = 3$$
$$\sin^2\theta = \frac{3}{4}$$
$$\sin\theta = \pm\sqrt{\frac{3}{4}}$$
$$\sin\theta = \pm\frac{\sqrt{3}}{2} \Rightarrow \theta = \frac{\pi}{3}, \frac{2\pi}{3}, \frac{4\pi}{3}, \frac{5\pi}{3}$$
The solution set is $\left\{0, \frac{\pi}{3}, \frac{2\pi}{3}, \frac{4\pi}{3}, \frac{5\pi}{3}\right\}$.

15.
$$\cos\theta = \sin\theta$$
$$\frac{\sin\theta}{\cos\theta} = 1$$
$$\tan\theta = 1 \Rightarrow \theta = \frac{\pi}{4}, \frac{5\pi}{4}$$
The solution set is $\left\{\frac{\pi}{4}, \frac{5\pi}{4}\right\}$.

17.
$$\tan\theta = 2\sin\theta$$
$$\frac{\sin\theta}{\cos\theta} = 2\sin\theta$$
$$\sin\theta = 2\sin\theta\cos\theta$$
$$0 = 2\sin\theta\cos\theta - \sin\theta$$
$$0 = \sin\theta(2\cos\theta - 1)$$
$$2\cos\theta - 1 = 0$$
$$\cos\theta = \frac{1}{2} \Rightarrow \theta = \frac{\pi}{3}, \frac{5\pi}{3}$$
$$\text{or } \sin\theta = 0 \Rightarrow \theta = 0, \pi$$
The solution set is $\left\{0, \frac{\pi}{3}, \pi, \frac{5\pi}{3}\right\}$.

19. $\sin\theta = \csc\theta$

$$\sin\theta = \frac{1}{\sin\theta}$$

$$\sin^2\theta = 1$$

$$\sin\theta = \pm 1 \Rightarrow \theta = \frac{\pi}{2}, \frac{3\pi}{2}$$

The solution set is $\left\{\dfrac{\pi}{2}, \dfrac{3\pi}{2}\right\}$.

21. $\cos(2\theta) = \cos\theta$

$$2\cos^2\theta - 1 = \cos\theta$$

$$2\cos^2\theta - \cos\theta - 1 = 0$$

$$(2\cos\theta + 1)(\cos\theta - 1) = 0$$

$$2\cos\theta + 1 = 0$$

$$\cos\theta = -\frac{1}{2} \Rightarrow \theta = \frac{2\pi}{3}, \frac{4\pi}{3}$$

or $\cos\theta - 1 = 0$

$$\cos\theta = 1 \Rightarrow \theta = 0$$

The solution set is $\left\{0, \dfrac{2\pi}{3}, \dfrac{4\pi}{3}\right\}$.

23. $\sin(2\theta) + \sin(4\theta) = 0$

$$\sin(2\theta) + 2\sin(2\theta)\cos(2\theta) = 0$$

$$\sin(2\theta)(1 + 2\cos(2\theta)) = 0$$

$$1 + 2\cos(2\theta) = 0$$

$$\cos(2\theta) = -\frac{1}{2} \Rightarrow 2\theta = \frac{2\pi}{3} + 2k\pi \Rightarrow \theta = \frac{\pi}{3} + k\pi$$

$$2\theta = \frac{4\pi}{3} + 2k\pi \Rightarrow \theta = \frac{2\pi}{3} + k\pi$$

or $\sin(2\theta) = 0 \Rightarrow 2\theta = 0 + k\pi \Rightarrow \theta = \dfrac{k\pi}{2}$

The solution set is $\left\{0, \dfrac{\pi}{3}, \dfrac{\pi}{2}, \dfrac{2\pi}{3}, \pi, \dfrac{4\pi}{3}, \dfrac{3\pi}{2}, \dfrac{5\pi}{3}\right\}$.

25. $\cos(4\theta) - \cos(6\theta) = 0$

$$-2\sin(5\theta)\sin(-\theta) = 0$$

$$2\sin(5\theta)\sin\theta = 0$$

$$\sin(5\theta) = 0 \Rightarrow 5\theta = 0 + k\pi \Rightarrow \theta = \frac{k\pi}{5}$$

or $\sin\theta = 0 \Rightarrow \theta = 0 + k\pi$

The solution set is $\left\{0, \dfrac{\pi}{5}, \dfrac{2\pi}{5}, \dfrac{3\pi}{5}, \dfrac{4\pi}{5}, \pi, \dfrac{6\pi}{5}, \dfrac{7\pi}{5}, \dfrac{8\pi}{5}, \dfrac{9\pi}{5}\right\}$.

27.
$$1 + \sin\theta = 2\cos^2\theta$$
$$1 + \sin\theta = 2(1 - \sin^2\theta)$$
$$2\sin^2\theta + \sin\theta - 1 = 0$$
$$(2\sin\theta - 1)(\sin\theta + 1) = 0$$
$$2\sin\theta - 1 = 0$$
$$\sin\theta = \frac{1}{2} \Rightarrow \theta = \frac{\pi}{6}, \frac{5\pi}{6}$$
$$\text{or } \sin\theta + 1 = 0$$
$$\sin\theta = -1 \Rightarrow \theta = \frac{3\pi}{2}$$

The solution set is $\left\{ \dfrac{\pi}{6}, \dfrac{5\pi}{6}, \dfrac{3\pi}{2} \right\}$.

29.
$$2\sin^2\theta - 5\sin\theta + 3 = 0$$
$$(\sin\theta - 1)(2\sin\theta - 3) = 0$$
$$\sin\theta - 1 = 0$$
$$\sin\theta = 1 \Rightarrow \theta = \frac{\pi}{2}$$
$$\text{or } 2\sin\theta - 3 = 0$$
$$\sin\theta = \frac{3}{2}, \text{ which is impossible}$$

The solution set is $\left\{ \dfrac{\pi}{2} \right\}$.

31.
$$3(1 - \cos\theta) = \sin^2\theta$$
$$3 - 3\cos\theta = 1 - \cos^2\theta$$
$$\cos^2\theta - 3\cos\theta + 2 = 0$$
$$(\cos\theta - 1)(\cos\theta - 2) = 0$$
$$\cos\theta - 1 = 0$$
$$\cos\theta = 1 \Rightarrow \theta = 0$$
$$\text{or } \cos\theta - 2 = 0$$
$$\cos\theta = 2, \text{ which is impossible}$$

The solution set is $\{0\}$.

33.

$$\tan^2\theta = \frac{3}{2}\sec\theta$$

$$\sec^2\theta - 1 = \frac{3}{2}\sec\theta$$

$$2\sec^2\theta - 2 = 3\sec\theta$$

$$2\sec^2\theta - 3\sec\theta - 2 = 0$$

$$(2\sec\theta + 1)(\sec\theta - 2) = 0$$

$$2\sec\theta + 1 = 0$$

$$\sec\theta = -\frac{1}{2} \Rightarrow \cos\theta = -2, \text{ which is impossible}$$

$$\text{or } \sec\theta - 2 = 0$$

$$\sec\theta = 2 \Rightarrow \theta = \frac{\pi}{3}, \frac{5\pi}{3}$$

The solution set is $\left\{\dfrac{\pi}{3}, \dfrac{5\pi}{3}\right\}$.

35. $3 - \sin\theta = \cos(2\theta)$

$3 - \sin\theta = 1 - 2\sin^2\theta \Rightarrow 2\sin^2\theta - \sin\theta + 2 = 0$
This is a quadratic equation in $\sin\theta$. The discriminant is
$b^2 - 4ac = (-1)^2 - 4(2)(2) = 1 - 16 = -15 < 0$.
The equation has no real solutions.

37. $\sec^2\theta + \tan\theta = 0$

$\tan^2\theta + 1 + \tan\theta = 0 \Rightarrow \tan^2\theta + \tan\theta + 1 = 0$
This is a quadratic equation in $\tan\theta$. The discriminant is $b^2 - 4ac = 1^2 - 4(1)(1) = 1 - 4 = -3 < 0$.
The equation has no real solutions.

39. $\sin\theta - \sqrt{3}\cos\theta = 1$

Divide each side by 2: $\dfrac{1}{2}\sin\theta - \dfrac{\sqrt{3}}{2}\cos\theta = \dfrac{1}{2}$

Rewrite using the difference of two angles formula where $\phi = \dfrac{\pi}{3}$, so $\cos\phi = \dfrac{1}{2}$ and $\sin\phi = \dfrac{\sqrt{3}}{2}$.

$$\sin\theta\cos\phi - \cos\theta\sin\phi = \frac{1}{2} \Rightarrow \sin(\theta - \phi) = \frac{1}{2}$$

$$\theta - \phi = \frac{\pi}{6} \quad \text{or} \quad \theta - \phi = \frac{5\pi}{6}$$

$$\theta - \frac{\pi}{3} = \frac{\pi}{6} \quad \text{or} \quad \theta - \frac{\pi}{3} = \frac{5\pi}{6}$$

$$\theta = \frac{\pi}{2} \quad \text{or} \quad \theta = \frac{7\pi}{6}$$

The solution set is $\left\{\dfrac{\pi}{2}, \dfrac{7\pi}{6}\right\}$.

41.

$$\tan(2\theta) + 2\sin\theta = 0$$

$$\frac{\sin(2\theta)}{\cos(2\theta)} + 2\sin\theta = 0$$

$$\frac{\sin(2\theta) + 2\sin\theta\cos(2\theta)}{\cos(2\theta)} = 0$$

$$2\sin\theta\cos\theta + 2\sin\theta(2\cos^2\theta - 1) = 0$$

$$2\sin\theta(\cos\theta + 2\cos^2\theta - 1) = 0$$

$$2\sin\theta(2\cos^2\theta + \cos\theta - 1) = 0$$

$$2\sin\theta(2\cos\theta - 1)(\cos\theta + 1) = 0$$

$$2\cos\theta - 1 = 0$$

$$\cos\theta = \frac{1}{2} \Rightarrow \theta = \frac{\pi}{3}, \frac{5\pi}{3}$$

or $2\sin\theta = 0$

$$\sin\theta = 0 \Rightarrow \theta = 0, \pi$$

or $\cos\theta + 1 = 0$

$$\cos\theta = -1 \Rightarrow \theta = \pi$$

The solution set is $\left\{0, \dfrac{\pi}{3}, \pi, \dfrac{5\pi}{3}\right\}$.

43. $\sin\theta + \cos\theta = \sqrt{2}$

Divide each side by $\sqrt{2}$: $\dfrac{1}{\sqrt{2}}\sin\theta + \dfrac{1}{\sqrt{2}}\cos\theta = 1$

Rewrite in the sum of two angles form where $\phi = \dfrac{\pi}{4}$, so $\cos\phi = \dfrac{1}{\sqrt{2}}$ and $\sin\phi = \dfrac{1}{\sqrt{2}}$.

$$\sin\theta\cos\phi + \cos\theta\sin\phi = 1$$

$$\sin(\theta + \phi) = 1$$

$$\theta + \phi = \frac{\pi}{2}$$

$$\theta + \frac{\pi}{4} = \frac{\pi}{2} \Rightarrow \theta = \frac{\pi}{4}$$

The solution set is $\left\{\dfrac{\pi}{4}\right\}$.

45.   Use INTERSECT to solve by graphing $y_1 = \cos x,\ \ y_2 = e^x$.

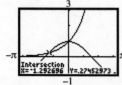

The solution set is $\{-1.29, 0\}$.

47.   Use INTERSECT to solve by graphing $y_1 = 2\sin x,\ y_2 = 0.7x$.

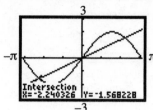 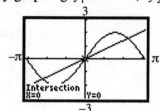 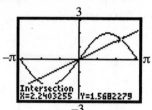

The solution set is $\{-2.24, 0, 2.24\}$.

49.   Use INTERSECT to solve by graphing $y_1 = \cos x,\ y_2 = x^2$.

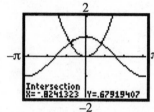 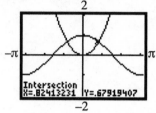

The solution set is $\{-0.82, 0.82\}$.

51. $x + 5\cos x = 0$
Find the intersection of
$y_1 = x + 5\cos x$ and $y_2 = 0$:

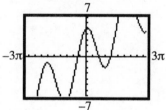

The solution set is $\{-1.31, 1.98, 3.84\}$.

53. $22x - 17\sin x = 3$
Find the intersection of
$y_1 = 22x - 17\sin x$ and $y_2 = 3$:

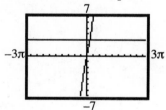

The solution set is $\{0.52\}$.

55. $\sin x + \cos x = x$
Find the intersection of
$y_1 = \sin x + \cos x$ and $y_2 = x$:

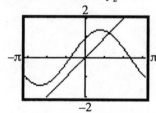

The solution set is $\{1.26\}$.

57. $x^2 - 2\cos x = 0$
Find the intersection of
$y_1 = x^2 - 2\cos x$ and $y_2 = 0$:

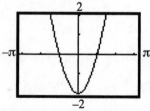

The solution set is $\{-1.02, 1.02\}$.

59.　$x^2 - 2\sin(2x) = 3x$

Find the intersection of

$y_1 = x^2 - 2\sin 2x$ and $y_2 = 3x$:

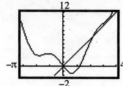

The solution set is $\{0, 2.15\}$.

61.　$6\sin x - e^x = 2, \ x > 0$

Find the intersection of

$y_1 = 6\sin x - e^x$ and $y_2 = 2$:

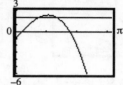

The solution set is $\{0.76, 1.35\}$.

63.　(a)　　　　$\cos(2\theta) + \cos\theta = 0, \ 0° < \theta < 90°$

$$2\cos^2\theta - 1 + \cos\theta = 0$$

$$2\cos^2\theta + \cos\theta - 1 = 0$$

$$(2\cos\theta - 1)(\cos\theta + 1) = 0$$

$$2\cos\theta - 1 = 0$$

$$\cos\theta = \frac{1}{2} \Rightarrow \theta = 60°, \ 300°$$

or　$\cos\theta + 1 = 0$

$$\cos\theta = -1 \Rightarrow \theta = 180°$$

The solution is 60°.

(b)　　$\cos(2\theta) + \cos\theta = 0, \ 0° < \theta < 90°$

$$2\cos\left(\frac{3\theta}{2}\right)\cos\left(\frac{\theta}{2}\right) = 0$$

$$\cos\left(\frac{3\theta}{2}\right) = 0 \Rightarrow \frac{3\theta}{2} = 90° \Rightarrow \theta = 60°$$

or　$\dfrac{3\theta}{2} = 270° \Rightarrow \theta = 180°$

or　$\cos\left(\frac{\theta}{2}\right) = 0 \Rightarrow \frac{\theta}{2} = 90° \Rightarrow \theta = 180°$

or　$\dfrac{\theta}{2} = 270° \Rightarrow \theta = 540°$

The solution is 60°.

(c)　$A(60°) = 16\sin(60°)(\cos(60°) + 1) = 16 \cdot \dfrac{\sqrt{3}}{2}\left(\dfrac{1}{2} + 1\right) = 12\sqrt{3} \text{ in}^2 \approx 20.78 \text{ in}^2$

(d)　Graph and use the MAXIMUM feature:

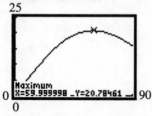

The maximum area is approximately 20.78 in² when the angle is 60°.

65. Graph:

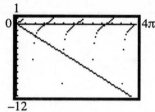

The first two positive solutions are 2.03 and 4.91.

67. (a) $R(\theta) = 107 \Rightarrow \dfrac{(34.8)^2 \sin(2\theta)}{9.8} = 107$

$\sin(2\theta) = \dfrac{107(9.8)}{(34.8)^2} \approx 0.8659$

$2\theta = \sin^{-1}(0.8659) \approx 59.98^\circ$ or 120.02°

$\theta \approx 29.99^\circ$ or 60.01°

(b) The maximum distance occurs when $\sin(2\theta)$ is greatest. That is, when $\sin(2\theta) = 1$.
     But $\sin(2\theta) = 1$ when $2\theta = 90^\circ \Rightarrow \theta = 45^\circ$.
     The maximum distance is

$R(45^\circ) = \dfrac{(34.8)^2 \sin(2 \cdot 45^\circ)}{9.8} = \dfrac{(34.8)^2 \sin(90^\circ)}{9.8} \approx 123.58$ meters

(c) Graph: $y_1 = \dfrac{(34.8)^2 \sin(2x)}{9.8}$

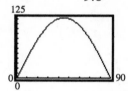

(d) Verifying (a): $107 = \dfrac{(34.8)^2 \sin(2\theta)}{9.8} \Rightarrow \dfrac{(34.8)^2 \sin(2\theta)}{9.8} - 107 = 0$.

Using ROOT or ZERO on $y_1 = \dfrac{(34.8)^2 \sin(2x)}{9.8} - 107$ yields $x \approx 29.99^\circ$ or 60.01°.

Verifying (b): Using MAXIMUM on $y_1 = \dfrac{(34.8)^2 \sin(2x)}{9.8}$ yields a maximum

of $y \approx 123.58$ meters.

Analytic Trigonometry

6.R   Chapter Review

1. $\sin^{-1} 1$

Find the angle θ, $-\dfrac{\pi}{2} \le \theta \le \dfrac{\pi}{2}$, whose sine equals 1.

$$\sin\theta = 1 \qquad -\frac{\pi}{2} \le \theta \le \frac{\pi}{2}$$

$$\theta = \frac{\pi}{2} \Rightarrow \sin^{-1} 1 = \frac{\pi}{2}$$

3. $\tan^{-1} 1$

Find the angle θ, $-\dfrac{\pi}{2} < \theta < \dfrac{\pi}{2}$, whose tangent equals 1.

$$\tan\theta = 1 \qquad -\frac{\pi}{2} < \theta < \frac{\pi}{2}$$

$$\theta = \frac{\pi}{4} \Rightarrow \tan^{-1} 1 = \frac{\pi}{4}$$

5. $\cos^{-1}\left(-\dfrac{\sqrt{3}}{2}\right)$

Find the angle θ, $0 \le \theta \le \pi$, whose cosine equals $-\dfrac{\sqrt{3}}{2}$.

$$\cos\theta = -\frac{\sqrt{3}}{2} \qquad 0 \le \theta \le \pi$$

$$\theta = \frac{5\pi}{6} \Rightarrow \cos^{-1}\left(-\frac{\sqrt{3}}{2}\right) = \frac{5\pi}{6}$$

7. $\sec^{-1}\sqrt{2}$

Find the angle θ, $0 \le \theta \le \pi$, whose secant equals $\sqrt{2}$.

$$\sec\theta = \sqrt{2} \qquad 0 \le \theta \le \pi$$

$$\theta = \frac{\pi}{4} \Rightarrow \sec^{-1}\sqrt{2} = \frac{\pi}{4}$$

9. $\tan\left(\sin^{-1}\left(-\dfrac{\sqrt{3}}{2}\right)\right)$

Find the angle θ, $-\dfrac{\pi}{2} \le \theta \le \dfrac{\pi}{2}$, whose sine equals $-\dfrac{\sqrt{3}}{2}$.

$$\sin\theta = -\dfrac{\sqrt{3}}{2} \qquad -\dfrac{\pi}{2} \le \theta \le \dfrac{\pi}{2}$$

$$\theta = -\dfrac{\pi}{3} \Rightarrow \tan\left(\sin^{-1}\left(-\dfrac{\sqrt{3}}{2}\right)\right) = \tan\left(-\dfrac{\pi}{3}\right) = -\sqrt{3}$$

11. $\sec\left(\tan^{-1}\dfrac{\sqrt{3}}{3}\right)$

Find the angle θ, $-\dfrac{\pi}{2} < \theta < \dfrac{\pi}{2}$, whose tangent is $\dfrac{\sqrt{3}}{3}$

$$\tan\theta = \dfrac{\sqrt{3}}{3}, \qquad -\dfrac{\pi}{2} < \theta < \dfrac{\pi}{2}$$

$$\theta = \dfrac{\pi}{6} \Rightarrow \sec\left(\tan^{-1}\dfrac{\sqrt{3}}{3}\right) = \sec\dfrac{\pi}{6} = \dfrac{2\sqrt{3}}{3}$$

13. $\sin\left(\tan^{-1}\dfrac{3}{4}\right)$

Since $\tan\theta = \dfrac{3}{4}$, $-\dfrac{\pi}{2} < \theta < \dfrac{\pi}{2}$, let $x = 4$ and $y = 3$.

Solve for r:

$16 + 9 = r^2 \Rightarrow r^2 = 25 \Rightarrow r = 5$

θ is in quadrant I.

$$\sin\left(\tan^{-1}\dfrac{3}{4}\right) = \sin\theta = \dfrac{y}{r} = \dfrac{3}{5}$$

15. $\tan\left(\sin^{-1}\left(-\dfrac{4}{5}\right)\right)$

Since $\sin\theta = -\dfrac{4}{5}$, $-\dfrac{\pi}{2} \le \theta \le \dfrac{\pi}{2}$, let $y = -4$ and $r = 5$.

Solve for x:

$x^2 + 16 = 25 \Rightarrow x^2 = 9 \Rightarrow x = \pm 3$

Since θ is in quadrant IV, $x = 3$.

$$\tan\left(\sin^{-1}\left(-\dfrac{4}{5}\right)\right) = \tan\theta = \dfrac{y}{x} = \dfrac{-4}{3} = -\dfrac{4}{3}$$

17. $\sin^{-1}\left(\cos\dfrac{2\pi}{3}\right) = \sin^{-1}\left(-\dfrac{1}{2}\right) = -\dfrac{\pi}{6}$

19. $\tan^{-1}\left(\tan\dfrac{7\pi}{4}\right) = \tan^{-1}(-1) = -\dfrac{\pi}{4}$

21. $\tan\theta\cot\theta - \sin^2\theta = \tan\theta\cdot\dfrac{1}{\tan\theta} - \sin^2\theta = 1 - \sin^2\theta = \cos^2\theta$

23. $\cos^2\theta(1+\tan^2\theta) = \cos^2\theta\cdot\sec^2\theta = \cos^2\theta\cdot\dfrac{1}{\cos^2\theta} = 1$

25. $4\cos^2\theta + 3\sin^2\theta = \cos^2\theta + 3\cos^2\theta + 3\sin^2\theta = \cos^2\theta + 3(\cos^2\theta + \sin^2\theta)$
$= \cos^2\theta + 3\cdot 1 = \cos^2\theta + 3 = 3 + \cos^2\theta$

27. $\dfrac{1-\cos\theta}{\sin\theta} + \dfrac{\sin\theta}{1-\cos\theta} = \dfrac{(1-\cos\theta)^2 + \sin^2\theta}{\sin\theta(1-\cos\theta)} = \dfrac{1 - 2\cos\theta + \cos^2\theta + \sin^2\theta}{\sin\theta(1-\cos\theta)}$
$= \dfrac{1-2\cos\theta+1}{\sin\theta(1-\cos\theta)} = \dfrac{2-2\cos\theta}{\sin\theta(1-\cos\theta)} = \dfrac{2(1-\cos\theta)}{\sin\theta(1-\cos\theta)} = \dfrac{2}{\sin\theta} = 2\csc\theta$

29. $\dfrac{\cos\theta}{\cos\theta-\sin\theta} = \dfrac{\cos\theta}{\cos\theta-\sin\theta}\cdot\left(\dfrac{\frac{1}{\cos\theta}}{\frac{1}{\cos\theta}}\right) = \dfrac{1}{1 - \frac{\sin\theta}{\cos\theta}} = \dfrac{1}{1-\tan\theta}$

31. $\dfrac{\csc\theta}{1+\csc\theta} = \dfrac{\frac{1}{\sin\theta}}{1+\frac{1}{\sin\theta}} = \dfrac{\frac{1}{\sin\theta}}{\frac{\sin\theta+1}{\sin\theta}} = \left(\dfrac{1}{1+\sin\theta}\right)\cdot\left(\dfrac{1-\sin\theta}{1-\sin\theta}\right) = \dfrac{1-\sin\theta}{1-\sin^2\theta} = \dfrac{1-\sin\theta}{\cos^2\theta}$

33. $\csc\theta - \sin\theta = \dfrac{1}{\sin\theta} - \sin\theta = \dfrac{1-\sin^2\theta}{\sin\theta} = \dfrac{\cos^2\theta}{\sin\theta} = \cos\theta\cdot\dfrac{\cos\theta}{\sin\theta} = \cos\theta\cot\theta$

35. $\dfrac{1-\sin\theta}{\sec\theta} = \cos\theta(1-\sin\theta) = \cos\theta(1-\sin\theta)\cdot\dfrac{1+\sin\theta}{1+\sin\theta} = \dfrac{\cos\theta(1-\sin^2\theta)}{1+\sin\theta} = \dfrac{\cos\theta(\cos^2\theta)}{1+\sin\theta} = \dfrac{\cos^3\theta}{1+\sin\theta}$

37. $\cot\theta - \tan\theta = \dfrac{\cos\theta}{\sin\theta} - \dfrac{\sin\theta}{\cos\theta} = \dfrac{\cos^2\theta - \sin^2\theta}{\sin\theta\cos\theta} = \dfrac{1-\sin^2\theta - \sin^2\theta}{\sin\theta\cos\theta} = \dfrac{1-2\sin^2\theta}{\sin\theta\cos\theta}$

39. $\dfrac{\cos(\alpha+\beta)}{\cos\alpha\sin\beta} = \dfrac{\cos\alpha\cos\beta - \sin\alpha\sin\beta}{\cos\alpha\sin\beta} = \dfrac{\cos\alpha\cos\beta}{\cos\alpha\sin\beta} - \dfrac{\sin\alpha\sin\beta}{\cos\alpha\sin\beta} = \cot\beta - \tan\alpha$

41. $\dfrac{\cos(\alpha-\beta)}{\cos\alpha\cos\beta} = \dfrac{\cos\alpha\cos\beta + \sin\alpha\sin\beta}{\cos\alpha\cos\beta} = \dfrac{\cos\alpha\cos\beta}{\cos\alpha\cos\beta} + \dfrac{\sin\alpha\sin\beta}{\cos\alpha\cos\beta} = 1 + \tan\alpha\tan\beta$

43. $(1+\cos\theta)\left(\tan\left(\dfrac{\theta}{2}\right)\right) = (1+\cos\theta)\cdot\dfrac{\sin\theta}{1+\cos\theta} = \sin\theta$

45. $2\cot\theta\cot(2\theta) = 2\cdot\dfrac{\cos\theta}{\sin\theta}\cdot\dfrac{\cos(2\theta)}{\sin(2\theta)} = \dfrac{2\cos\theta(\cos^2\theta - \sin^2\theta)}{\sin\theta(2\sin\theta\cos\theta)} = \dfrac{\cos^2\theta - \sin^2\theta}{\sin^2\theta} = \cot^2\theta - 1$

47. $1 - 8\sin^2\theta\cos^2\theta = 1 - 2(2\sin\theta\cos\theta)^2 = 1 - 2\sin^2(2\theta) = \cos(4\theta)$

49. $\dfrac{\sin(2\theta)+\sin(4\theta)}{\cos(2\theta)+\cos(4\theta)} = \dfrac{2\sin(3\theta)\cos(-\theta)}{2\cos(3\theta)\cos(-\theta)} = \dfrac{\sin(3\theta)}{\cos(3\theta)} = \tan(3\theta)$

51. $\dfrac{\cos(2\theta)-\cos(4\theta)}{\cos(2\theta)+\cos(4\theta)} - \tan\theta\tan(3\theta) = \dfrac{-2\sin(3\theta)\sin(-\theta)}{2\cos(3\theta)\cos(-\theta)} - \tan\theta\tan(3\theta)$

$= \dfrac{2\sin(3\theta)\sin\theta}{2\cos(3\theta)\cos\theta} - \tan\theta\tan(3\theta) = \tan(3\theta)\tan\theta - \tan\theta\tan(3\theta) = 0$

53. $\sin165° = \sin(120°+45°) = \sin120°\cdot\cos45° + \cos120°\cdot\sin45°$

$= \left(\dfrac{\sqrt{3}}{2}\right)\cdot\left(\dfrac{\sqrt{2}}{2}\right) + \left(-\dfrac{1}{2}\right)\cdot\left(\dfrac{\sqrt{2}}{2}\right) = \dfrac{1}{4}\left(\sqrt{6}-\sqrt{2}\right)$

55. $\cos\dfrac{5\pi}{12} = \cos\left(\dfrac{3\pi}{12}+\dfrac{2\pi}{12}\right) = \cos\dfrac{\pi}{4}\cdot\cos\dfrac{\pi}{6} - \sin\dfrac{\pi}{4}\cdot\sin\dfrac{\pi}{6} = \dfrac{\sqrt{2}}{2}\cdot\dfrac{\sqrt{3}}{2} - \dfrac{\sqrt{2}}{2}\cdot\dfrac{1}{2} = \dfrac{1}{4}\left(\sqrt{6}-\sqrt{2}\right)$

57. $\cos80°\cdot\cos20° + \sin80°\cdot\sin20° = \cos(80°-20°) = \cos60° = \dfrac{1}{2}$

59. $\tan\dfrac{\pi}{8} = \tan\left(\dfrac{\frac{\pi}{4}}{2}\right) = \sqrt{\dfrac{1-\cos\frac{\pi}{4}}{1+\cos\frac{\pi}{4}}} = \sqrt{\dfrac{1-\frac{\sqrt{2}}{2}}{1+\frac{\sqrt{2}}{2}}} = \sqrt{\left(\dfrac{2-\sqrt{2}}{2+\sqrt{2}}\right)\cdot\left(\dfrac{2-\sqrt{2}}{2-\sqrt{2}}\right)}$

$= \sqrt{\dfrac{\left(2-\sqrt{2}\right)^2}{2}} = \left(\dfrac{2-\sqrt{2}}{\sqrt{2}}\right)\cdot\dfrac{\sqrt{2}}{\sqrt{2}} = \dfrac{2\sqrt{2}-2}{2} = \sqrt{2}-1$

61. $\sin\alpha = \dfrac{4}{5},\ 0 < \alpha < \dfrac{\pi}{2}; \qquad \sin\beta = \dfrac{5}{13},\ \dfrac{\pi}{2} < \beta < \pi$

$\cos\alpha = \dfrac{3}{5},\ \tan\alpha = \dfrac{4}{3},\ \cos\beta = -\dfrac{12}{13},\ \tan\beta = -\dfrac{5}{12},\quad 0 < \dfrac{\alpha}{2} < \dfrac{\pi}{4},\quad \dfrac{\pi}{4} < \dfrac{\beta}{2} < \dfrac{\pi}{2}$

(a) $\sin(\alpha+\beta) = \sin\alpha\cos\beta + \cos\alpha\sin\beta = \left(\dfrac{4}{5}\right)\cdot\left(-\dfrac{12}{13}\right) + \left(\dfrac{3}{5}\right)\cdot\left(\dfrac{5}{13}\right) = \dfrac{-48+15}{65} = -\dfrac{33}{65}$

(b) $\cos(\alpha+\beta) = \cos\alpha\cos\beta - \sin\alpha\sin\beta = \left(\dfrac{3}{5}\right)\cdot\left(-\dfrac{12}{13}\right) - \left(\dfrac{4}{5}\right)\cdot\left(\dfrac{5}{13}\right) = \dfrac{-36-20}{65} = -\dfrac{56}{65}$

(c) $\sin(\alpha-\beta) = \sin\alpha\cos\beta - \cos\alpha\sin\beta = \left(\dfrac{4}{5}\right)\cdot\left(-\dfrac{12}{13}\right) - \left(\dfrac{3}{5}\right)\cdot\left(\dfrac{5}{13}\right) = \dfrac{-48-15}{65} = -\dfrac{63}{65}$

(d) $\tan(\alpha+\beta) = \dfrac{\tan\alpha+\tan\beta}{1-\tan\alpha\tan\beta} = \dfrac{\frac{4}{3}+\left(-\frac{5}{12}\right)}{1-\left(\frac{4}{3}\right)\cdot\left(-\frac{5}{12}\right)} = \dfrac{\frac{11}{12}}{\frac{14}{9}} = \dfrac{11}{12}\cdot\dfrac{9}{14} = \dfrac{33}{56}$

(e) $\sin(2\alpha) = 2\sin\alpha\cos\alpha = 2\cdot\dfrac{4}{5}\cdot\dfrac{3}{5} = \dfrac{24}{25}$

(f) $\cos(2\beta) = \cos^2\beta - \sin^2\beta = \left(-\dfrac{12}{13}\right)^2 - \left(\dfrac{5}{13}\right)^2 = \dfrac{144}{169} - \dfrac{25}{169} = \dfrac{119}{169}$

(g)　$\sin\dfrac{\beta}{2}=\sqrt{\dfrac{1-\cos\beta}{2}}=\sqrt{\dfrac{1-\left(-\dfrac{12}{13}\right)}{2}}=\sqrt{\dfrac{\dfrac{25}{13}}{2}}=\sqrt{\dfrac{25}{26}}=\dfrac{5}{\sqrt{26}}=\dfrac{5\sqrt{26}}{26}$

(h)　$\cos\dfrac{\alpha}{2}=\sqrt{\dfrac{1+\cos\alpha}{2}}=\sqrt{\dfrac{1+\dfrac{3}{5}}{2}}=\sqrt{\dfrac{\dfrac{8}{5}}{2}}=\sqrt{\dfrac{4}{5}}=\dfrac{2}{\sqrt{5}}=\dfrac{2\sqrt{5}}{5}$

63.　$\sin\alpha=-\dfrac{3}{5},\ \pi<\alpha<\dfrac{3\pi}{2};\qquad \cos\beta=\dfrac{12}{13},\ \dfrac{3\pi}{2}<\beta<2\pi$

$\cos\alpha=-\dfrac{4}{5},\ \tan\alpha=\dfrac{3}{4},\ \sin\beta=-\dfrac{5}{13},\ \tan\beta=-\dfrac{5}{12},\ \dfrac{\pi}{2}<\dfrac{\alpha}{2}<\dfrac{3\pi}{4},\ \dfrac{3\pi}{4}<\dfrac{\beta}{2}<\pi$

(a)　$\sin(\alpha+\beta)=\sin\alpha\cos\beta+\cos\alpha\sin\beta=\left(-\dfrac{3}{5}\right)\cdot\left(\dfrac{12}{13}\right)+\left(-\dfrac{4}{5}\right)\cdot\left(-\dfrac{5}{13}\right)=\dfrac{-36+20}{65}=-\dfrac{16}{65}$

(b)　$\cos(\alpha+\beta)=\cos\alpha\cos\beta-\sin\alpha\sin\beta=\left(-\dfrac{4}{5}\right)\cdot\left(\dfrac{12}{13}\right)-\left(-\dfrac{3}{5}\right)\cdot\left(-\dfrac{5}{13}\right)=\dfrac{-48-15}{65}=-\dfrac{63}{65}$

(c)　$\sin(\alpha-\beta)=\sin\alpha\cos\beta-\cos\alpha\sin\beta=\left(-\dfrac{3}{5}\right)\cdot\left(\dfrac{12}{13}\right)-\left(-\dfrac{4}{5}\right)\cdot\left(-\dfrac{5}{13}\right)=\dfrac{-36-20}{65}=-\dfrac{56}{65}$

(d)　$\tan(\alpha+\beta)=\dfrac{\tan\alpha+\tan\beta}{1-\tan\alpha\tan\beta}=\dfrac{\dfrac{3}{4}+\left(-\dfrac{5}{12}\right)}{1-\left(\dfrac{3}{4}\right)\cdot\left(-\dfrac{5}{12}\right)}=\dfrac{\dfrac{1}{3}}{\dfrac{21}{16}}=\dfrac{1}{3}\cdot\dfrac{16}{21}=\dfrac{16}{63}$

(e)　$\sin(2\alpha)=2\sin\alpha\cos\alpha=2\cdot\left(-\dfrac{3}{5}\right)\cdot\left(-\dfrac{4}{5}\right)=\dfrac{24}{25}$

(f)　$\cos(2\beta)=\cos^2\beta-\sin^2\beta=\left(\dfrac{12}{13}\right)^2-\left(-\dfrac{5}{13}\right)^2=\dfrac{144}{169}-\dfrac{25}{169}=\dfrac{119}{169}$

(g)　$\sin\dfrac{\beta}{2}=\sqrt{\dfrac{1-\cos\beta}{2}}=\sqrt{\dfrac{1-\dfrac{12}{13}}{2}}=\sqrt{\dfrac{\dfrac{1}{13}}{2}}=\sqrt{\dfrac{1}{26}}=\dfrac{1}{\sqrt{26}}=\dfrac{\sqrt{26}}{26}$

(h)　$\cos\dfrac{\alpha}{2}=-\sqrt{\dfrac{1+\cos\alpha}{2}}=-\sqrt{\dfrac{1+\left(-\dfrac{4}{5}\right)}{2}}=-\sqrt{\dfrac{\dfrac{1}{5}}{2}}=-\sqrt{\dfrac{1}{10}}=-\dfrac{1}{\sqrt{10}}=-\dfrac{\sqrt{10}}{10}$

65.　$\tan\alpha=\dfrac{3}{4},\ \pi<\alpha<\dfrac{3\pi}{2};\qquad \tan\beta=\dfrac{12}{5},\ 0<\beta<\dfrac{\pi}{2}$

$\sin\alpha=-\dfrac{3}{5},\ \cos\alpha=-\dfrac{4}{5},\ \sin\beta=\dfrac{12}{13},\ \cos\beta=\dfrac{5}{13},\ \dfrac{\pi}{2}<\dfrac{\alpha}{2}<\dfrac{3\pi}{4},\ 0<\dfrac{\beta}{2}<\dfrac{\pi}{4}$

(a)　$\sin(\alpha+\beta)=\sin\alpha\cos\beta+\cos\alpha\sin\beta=\left(-\dfrac{3}{5}\right)\cdot\left(\dfrac{5}{13}\right)+\left(-\dfrac{4}{5}\right)\cdot\left(\dfrac{12}{13}\right)=\dfrac{-15-48}{65}=-\dfrac{63}{65}$

(b)　$\cos(\alpha+\beta)=\cos\alpha\cos\beta-\sin\alpha\sin\beta=\left(-\dfrac{4}{5}\right)\cdot\left(\dfrac{5}{13}\right)-\left(-\dfrac{3}{5}\right)\cdot\left(\dfrac{12}{13}\right)=\dfrac{-20+36}{65}=\dfrac{16}{65}$

(c)　$\sin(\alpha-\beta)=\sin\alpha\cos\beta-\cos\alpha\sin\beta=\left(-\dfrac{3}{5}\right)\cdot\left(\dfrac{5}{13}\right)-\left(-\dfrac{4}{5}\right)\cdot\left(\dfrac{12}{13}\right)=\dfrac{-15+48}{65}=\dfrac{33}{65}$

(d) $\tan(\alpha + \beta) = \dfrac{\tan\alpha + \tan\beta}{1 - \tan\alpha\tan\beta} = \dfrac{\dfrac{3}{4} + \dfrac{12}{5}}{1 - \left(\dfrac{3}{4}\right)\cdot\left(\dfrac{12}{5}\right)} = \dfrac{\dfrac{15 + 48}{20}}{-\dfrac{4}{5}} = \left(\dfrac{63}{20}\right)\cdot\left(-\dfrac{5}{4}\right) = -\dfrac{63}{16}$

(e) $\sin(2\alpha) = 2\sin\alpha\cos\alpha = 2\cdot\left(-\dfrac{3}{5}\right)\cdot\left(-\dfrac{4}{5}\right) = \dfrac{24}{25}$

(f) $\cos(2\beta) = \cos^2\beta - \sin^2\beta = \left(\dfrac{5}{13}\right)^2 - \left(\dfrac{12}{13}\right)^2 = \dfrac{25}{169} - \dfrac{144}{169} = -\dfrac{119}{169}$

(g) $\sin\dfrac{\beta}{2} = \sqrt{\dfrac{1 - \cos\beta}{2}} = \sqrt{\dfrac{1 - \dfrac{5}{13}}{2}} = \sqrt{\dfrac{\dfrac{8}{13}}{2}} = \sqrt{\dfrac{4}{13}} = \dfrac{2}{\sqrt{13}} = \dfrac{2\sqrt{13}}{13}$

(h) $\cos\dfrac{\alpha}{2} = -\sqrt{\dfrac{1 + \cos\alpha}{2}} = -\sqrt{\dfrac{1 + \left(-\dfrac{4}{5}\right)}{2}} = -\sqrt{\dfrac{\dfrac{1}{5}}{2}} = -\sqrt{\dfrac{1}{10}} = -\dfrac{1}{\sqrt{10}} = -\dfrac{\sqrt{10}}{10}$

67. $\sec\alpha = 2, \ -\dfrac{\pi}{2} < \alpha < 0; \qquad \sec\beta = 3, \ \dfrac{3\pi}{2} < \beta < 2\pi$

$\sin\alpha = -\dfrac{\sqrt{3}}{2}, \cos\alpha = \dfrac{1}{2}, \tan\alpha = -\sqrt{3}, \sin\beta = -\dfrac{2\sqrt{2}}{3}, \cos\beta = \dfrac{1}{3}, \tan\beta = -2\sqrt{2},$

$-\dfrac{\pi}{4} < \dfrac{\alpha}{2} < 0, \quad \dfrac{3\pi}{4} < \dfrac{\beta}{2} < \pi$

(a) $\sin(\alpha + \beta) = \sin\alpha\cos\beta + \cos\alpha\sin\beta = \left(-\dfrac{\sqrt{3}}{2}\right)\cdot\left(\dfrac{1}{3}\right) + \left(\dfrac{1}{2}\right)\cdot\left(-\dfrac{2\sqrt{2}}{3}\right) = \dfrac{-\sqrt{3} - 2\sqrt{2}}{6}$

(b) $\cos(\alpha + \beta) = \cos\alpha\cos\beta - \sin\alpha\sin\beta = \left(\dfrac{1}{2}\right)\cdot\left(\dfrac{1}{3}\right) - \left(-\dfrac{\sqrt{3}}{2}\right)\cdot\left(-\dfrac{2\sqrt{2}}{3}\right) = \dfrac{1 - 2\sqrt{6}}{6}$

(c) $\sin(\alpha - \beta) = \sin\alpha\cos\beta - \cos\alpha\sin\beta = \left(-\dfrac{\sqrt{3}}{2}\right)\cdot\left(\dfrac{1}{3}\right) - \left(\dfrac{1}{2}\right)\cdot\left(-\dfrac{2\sqrt{2}}{3}\right) = \dfrac{-\sqrt{3} + 2\sqrt{2}}{6}$

(d) $\tan(\alpha + \beta) = \dfrac{\tan\alpha + \tan\beta}{1 - \tan\alpha\tan\beta} = \dfrac{-\sqrt{3} + \left(-2\sqrt{2}\right)}{1 - \left(-\sqrt{3}\right)\left(-2\sqrt{2}\right)} = \left(\dfrac{-\sqrt{3} - 2\sqrt{2}}{1 - 2\sqrt{6}}\right)\cdot\left(\dfrac{1 + 2\sqrt{6}}{1 + 2\sqrt{6}}\right)$

$= \dfrac{-9\sqrt{3} - 8\sqrt{2}}{-23} = \dfrac{9\sqrt{3} + 8\sqrt{2}}{23}$

(e) $\sin(2\alpha) = 2\sin\alpha\cos\alpha = 2\cdot\left(-\dfrac{\sqrt{3}}{2}\right)\cdot\left(\dfrac{1}{2}\right) = -\dfrac{\sqrt{3}}{2}$

(f) $\cos(2\beta) = \cos^2\beta - \sin^2\beta = \left(\dfrac{1}{3}\right)^2 - \left(-\dfrac{2\sqrt{2}}{3}\right)^2 = \dfrac{1}{9} - \dfrac{8}{9} = -\dfrac{7}{9}$

(g) $\sin\dfrac{\beta}{2} = \sqrt{\dfrac{1 - \cos\beta}{2}} = \sqrt{\dfrac{1 - \dfrac{1}{3}}{2}} = \sqrt{\dfrac{\dfrac{2}{3}}{2}} = \sqrt{\dfrac{1}{3}} = \dfrac{1}{\sqrt{3}} = \dfrac{\sqrt{3}}{3}$

(h) $\cos\dfrac{\alpha}{2} = \sqrt{\dfrac{1 + \cos\alpha}{2}} = \sqrt{\dfrac{1 + \dfrac{1}{2}}{2}} = \sqrt{\dfrac{\dfrac{3}{2}}{2}} = \sqrt{\dfrac{3}{4}} = \dfrac{\sqrt{3}}{2}$

69. $\sin\alpha = -\dfrac{2}{3},\ \pi < \alpha < \dfrac{3\pi}{2};\qquad \cos\beta = -\dfrac{2}{3},\ \pi < \beta < \dfrac{3\pi}{2}$

$\cos\alpha = -\dfrac{\sqrt{5}}{3},\ \tan\alpha = \dfrac{2\sqrt{5}}{5},\ \sin\beta = -\dfrac{\sqrt{5}}{3},\ \tan\beta = \dfrac{\sqrt{5}}{2},\ \dfrac{\pi}{2} < \dfrac{\alpha}{2} < \dfrac{3\pi}{4},\ \dfrac{\pi}{2} < \dfrac{\beta}{2} < \dfrac{3\pi}{4}$

(a) $\sin(\alpha+\beta) = \sin\alpha\cos\beta + \cos\alpha\sin\beta = \left(-\dfrac{2}{3}\right)\cdot\left(-\dfrac{2}{3}\right) + \left(-\dfrac{\sqrt{5}}{3}\right)\cdot\left(-\dfrac{\sqrt{5}}{3}\right) = \dfrac{4+5}{9} = 1$

(b) $\cos(\alpha+\beta) = \cos\alpha\cos\beta - \sin\alpha\sin\beta = \left(-\dfrac{\sqrt{5}}{3}\right)\cdot\left(-\dfrac{2}{3}\right) - \left(-\dfrac{2}{3}\right)\cdot\left(-\dfrac{\sqrt{5}}{3}\right) = \dfrac{2\sqrt{5} - 2\sqrt{5}}{9} = 0$

(c) $\sin(\alpha-\beta) = \sin\alpha\cos\beta - \cos\alpha\sin\beta = \left(-\dfrac{2}{3}\right)\cdot\left(-\dfrac{2}{3}\right) - \left(-\dfrac{\sqrt{5}}{3}\right)\cdot\left(-\dfrac{\sqrt{5}}{3}\right) = \dfrac{4-5}{9} = -\dfrac{1}{9}$

(d) $\tan(\alpha+\beta) = \dfrac{\tan\alpha + \tan\beta}{1 - \tan\alpha\tan\beta} = \dfrac{\dfrac{2\sqrt{5}}{5} + \dfrac{\sqrt{5}}{2}}{1 - \left(\dfrac{2\sqrt{5}}{5}\right)\cdot\left(\dfrac{\sqrt{5}}{2}\right)} = \dfrac{\dfrac{4\sqrt{5} + 5\sqrt{5}}{10}}{\dfrac{10-10}{10}} = \dfrac{\dfrac{9\sqrt{5}}{10}}{\dfrac{0}{}};\ \text{undefined}$

(e) $\sin(2\alpha) = 2\sin\alpha\cos\alpha = 2\cdot\left(-\dfrac{2}{3}\right)\cdot\left(-\dfrac{\sqrt{5}}{3}\right) = \dfrac{4\sqrt{5}}{9}$

(f) $\cos(2\beta) = \cos^2\beta - \sin^2\beta = \left(-\dfrac{2}{3}\right)^2 - \left(-\dfrac{\sqrt{5}}{3}\right)^2 = \dfrac{4}{9} - \dfrac{5}{9} = -\dfrac{1}{9}$

(g) $\sin\dfrac{\beta}{2} = \sqrt{\dfrac{1-\cos\beta}{2}} = \sqrt{\dfrac{1-\left(-\dfrac{2}{3}\right)}{2}} = \sqrt{\dfrac{\dfrac{5}{3}}{2}} = \sqrt{\dfrac{5}{6}} = \dfrac{\sqrt{30}}{6}$

(h) $\cos\dfrac{\alpha}{2} = -\sqrt{\dfrac{1+\cos\alpha}{2}} = -\sqrt{\dfrac{1+\left(-\dfrac{\sqrt{5}}{3}\right)}{2}} = -\sqrt{\dfrac{\dfrac{3-\sqrt{5}}{3}}{2}} = -\sqrt{\dfrac{3-\sqrt{5}}{6}}$

$= -\dfrac{\sqrt{6\left(3-\sqrt{5}\right)}}{6} = -\dfrac{\sqrt{6}\sqrt{3-\sqrt{5}}}{6}$

71. $\cos\left(\sin^{-1}\dfrac{3}{5} - \cos^{-1}\dfrac{1}{2}\right)$

Let $\alpha = \sin^{-1}\dfrac{3}{5}$ and $\beta = \cos^{-1}\dfrac{1}{2}$. α is in quadrant I; β is in quadrant I.

Then $\sin\alpha = \dfrac{3}{5},\ 0 \le \alpha \le \dfrac{\pi}{2}$, and $\cos\beta = \dfrac{1}{2}, 0 \le \beta \le \dfrac{\pi}{2}$.

$\cos\alpha = \sqrt{1-\sin^2\alpha} = \sqrt{1-\left(\dfrac{3}{5}\right)^2} = \sqrt{1-\dfrac{9}{25}} = \sqrt{\dfrac{16}{25}} = \dfrac{4}{5}$

$\sin\beta = \sqrt{1-\cos^2\beta} = \sqrt{1-\left(\dfrac{1}{2}\right)^2} = \sqrt{1-\dfrac{1}{4}} = \sqrt{\dfrac{3}{4}} = \dfrac{\sqrt{3}}{2}$

$\cos\left(\sin^{-1}\dfrac{3}{5} - \cos^{-1}\dfrac{1}{2}\right) = \cos(\alpha-\beta) = \cos\alpha\cos\beta + \sin\alpha\sin\beta = \left(\dfrac{4}{5}\right)\cdot\left(\dfrac{1}{2}\right) + \left(\dfrac{3}{5}\right)\cdot\left(\dfrac{\sqrt{3}}{2}\right) = \dfrac{4+3\sqrt{3}}{10}$

73. $\tan\left[\sin^{-1}\left(-\dfrac{1}{2}\right) - \tan^{-1}\dfrac{3}{4}\right]$

Let $\alpha = \sin^{-1}\left(-\dfrac{1}{2}\right)$ and $\beta = \tan^{-1}\dfrac{3}{4}$. α is in quadrant IV; β is in quadrant I.

Then $\sin\alpha = -\dfrac{1}{2}$, $0 \le \alpha \le \dfrac{\pi}{2}$, and $\tan\beta = \dfrac{3}{4}$, $0 < \beta < \dfrac{\pi}{2}$.

$\cos\alpha = \sqrt{1 - \sin^2\alpha} = \sqrt{1 - \left(-\dfrac{1}{2}\right)^2} = \sqrt{1 - \dfrac{1}{4}} = \sqrt{\dfrac{3}{4}} = \dfrac{\sqrt{3}}{2}$; $\quad \tan\alpha = -\dfrac{1}{\sqrt{3}} = -\dfrac{\sqrt{3}}{3}$

$\tan\left[\sin^{-1}\left(-\dfrac{1}{2}\right) - \tan^{-1}\dfrac{3}{4}\right] = \tan(\alpha - \beta) = \dfrac{\tan\alpha - \tan\beta}{1 + \tan\alpha\tan\beta} = \dfrac{-\dfrac{\sqrt{3}}{3} - \dfrac{3}{4}}{1 + \left(-\dfrac{\sqrt{3}}{3}\right)\cdot\left(\dfrac{3}{4}\right)}$

$= \dfrac{\dfrac{-4\sqrt{3} - 9}{12}}{1 - \dfrac{3\sqrt{3}}{12}} = \left(\dfrac{-9 - 4\sqrt{3}}{12 - 3\sqrt{3}}\right)\cdot\left(\dfrac{12 + 3\sqrt{3}}{12 + 3\sqrt{3}}\right) = \dfrac{-144 - 75\sqrt{3}}{117} = \dfrac{-48 - 25\sqrt{3}}{39} = -\dfrac{48 + 25\sqrt{3}}{39}$

75. $\sin\left[2\cos^{-1}\left(-\dfrac{3}{5}\right)\right]$

Let $\alpha = \cos^{-1}\left(-\dfrac{3}{5}\right)$. α is in quadrant II.

Then $\cos\alpha = -\dfrac{3}{5}$, $\dfrac{\pi}{2} \le \alpha \le \pi$.

$\sin\alpha = \sqrt{1 - \cos^2\alpha} = \sqrt{1 - \left(-\dfrac{3}{5}\right)^2} = \sqrt{1 - \dfrac{9}{25}} = \sqrt{\dfrac{16}{25}} = \dfrac{4}{5}$

$\sin\left[2\cos^{-1}\left(-\dfrac{3}{5}\right)\right] = \sin 2\alpha = 2\sin\alpha\cos\alpha = 2\cdot\left(\dfrac{4}{5}\right)\cdot\left(-\dfrac{3}{5}\right) = -\dfrac{24}{25}$

77. $\cos\theta = \dfrac{1}{2}$

$\theta = \dfrac{\pi}{3} + 2k\pi$ or $\theta = \dfrac{5\pi}{3} + 2k\pi$, k is any integer

The solution set is $\left\{\dfrac{\pi}{3}, \dfrac{5\pi}{3}\right\}$.

79. $2\cos\theta + \sqrt{2} = 0$

$2\cos\theta = -\sqrt{2} \Rightarrow \cos\theta = -\dfrac{\sqrt{2}}{2}$

$\theta = \dfrac{3\pi}{4} + 2k\pi$ or $\theta = \dfrac{5\pi}{4} + 2k\pi$, k is any integer

The solution set is $\left\{\dfrac{3\pi}{4}, \dfrac{5\pi}{4}\right\}$.

81. $\sin(2\theta) + 1 = 0$

$\sin(2\theta) = -1$

$2\theta = \dfrac{3\pi}{2} + 2k\pi \Rightarrow \quad \theta = \dfrac{3\pi}{4} + k\pi$, k is any integer

The solution set is $\left\{\dfrac{3\pi}{4}, \dfrac{7\pi}{4}\right\}$.

83. $\tan(2\theta) = 0$

$2\theta = 0 + k\pi \Rightarrow \theta = \dfrac{k\pi}{2}$, k is any integer

The solution set is $\left\{0, \dfrac{\pi}{2}, \pi, \dfrac{3\pi}{2}\right\}$.

85. $\sec^2\theta = 4$

$\sec\theta = \pm 2 \Rightarrow \cos\theta = \pm\dfrac{1}{2}$

$\theta = \dfrac{\pi}{3} + k\pi$, k is any integer

$\theta = \dfrac{2\pi}{3} + k\pi$, k is any integer

The solution set is $\left\{\dfrac{\pi}{3}, \dfrac{2\pi}{3}, \dfrac{4\pi}{3}, \dfrac{5\pi}{3}\right\}$.

87. $\sin\theta = \tan\theta$

$\sin\theta = \dfrac{\sin\theta}{\cos\theta}$

$\sin\theta\cos\theta = \sin\theta$

$\sin\theta\cos\theta - \sin\theta = 0$

$\sin\theta(\cos\theta - 1) = 0$

$\cos\theta - 1 = 0$

$\cos\theta = 1 \Rightarrow \theta = 0$

or $\sin\theta = 0 \Rightarrow \theta = 0, \pi$

The solution set is $\{0, \pi\}$.

89. $\sin\theta + \sin(2\theta) = 0$

$\sin\theta + 2\sin\theta\cos\theta = 0$

$\sin\theta(1 + 2\cos\theta) = 0$

$1 + 2\cos\theta = 0$

$\cos\theta = -\dfrac{1}{2} \Rightarrow \theta = \dfrac{2\pi}{3}, \dfrac{4\pi}{3}$

or $\sin\theta = 0 \Rightarrow \theta = 0, \pi$

The solution set is $\left\{0, \dfrac{2\pi}{3}, \pi, \dfrac{4\pi}{3}\right\}$.

91.
$$\sin(2\theta) - \cos\theta - 2\sin\theta + 1 = 0$$
$$2\sin\theta\cos\theta - \cos\theta - 2\sin\theta + 1 = 0$$
$$\cos\theta(2\sin\theta - 1) - 1(2\sin\theta - 1) = 0$$
$$(2\sin\theta - 1)(\cos\theta - 1) = 0$$
$$2\sin\theta - 1 = 0$$
$$\sin\theta = \frac{1}{2} \Rightarrow \theta = \frac{\pi}{6}, \frac{5\pi}{6}$$
$$\text{or}\ \ \cos\theta - 1 = 0$$
$$\cos\theta = 1 \Rightarrow \theta = 0$$

The solution set is $\left\{ 0, \dfrac{\pi}{6}, \dfrac{5\pi}{6} \right\}$.

93.
$$2\sin^2\theta - 3\sin\theta + 1 = 0$$
$$(2\sin\theta - 1)(\sin\theta - 1) = 0$$
$$2\sin\theta - 1 = 0$$
$$\sin\theta = \frac{1}{2} \Rightarrow \theta = \frac{\pi}{6}, \frac{5\pi}{6}$$
$$\text{or}\ \ \sin\theta - 1 = 0$$
$$\sin\theta = 1 \Rightarrow \theta = \frac{\pi}{2}$$

The solution set is $\left\{ \dfrac{\pi}{6}, \dfrac{\pi}{2}, \dfrac{5\pi}{6} \right\}$.

95.
$$4\sin^2\theta = 1 + 4\cos\theta$$
$$4(1 - \cos^2\theta) = 1 + 4\cos\theta$$
$$4 - 4\cos^2\theta = 1 + 4\cos\theta$$
$$4\cos^2\theta + 4\cos\theta - 3 = 0$$
$$(2\cos\theta - 1)(2\cos\theta + 3) = 0$$
$$2\cos\theta - 1 = 0$$
$$\cos\theta = \frac{1}{2} \Rightarrow \theta = \frac{\pi}{3}, \frac{5\pi}{3}$$
$$\text{or}\ \ 2\cos\theta + 3 = 0$$
$$\cos\theta = -\frac{3}{2}, \text{ which is impossible}$$

The solution set is $\left\{ \dfrac{\pi}{3}, \dfrac{5\pi}{3} \right\}$.

97.
$$\sin(2\theta) = \sqrt{2}\cos\theta$$
$$2\sin\theta\cos\theta = \sqrt{2}\cos\theta$$
$$2\sin\theta\cos\theta - \sqrt{2}\cos\theta = 0$$
$$\cos\theta\left(2\sin\theta - \sqrt{2}\right) = 0$$
$$\cos\theta = 0 \Rightarrow \theta = \frac{\pi}{2}, \frac{3\pi}{2}$$
or $2\sin\theta - \sqrt{2} = 0$
$$\sin\theta = \frac{\sqrt{2}}{2} \Rightarrow \theta = \frac{\pi}{4}, \frac{3\pi}{4}$$

The solution set is $\left\{\dfrac{\pi}{4}, \dfrac{\pi}{2}, \dfrac{3\pi}{4}, \dfrac{3\pi}{2}\right\}$.

99. $\sin\theta - \cos\theta = 1$

Divide each side by $\sqrt{2}$: $\dfrac{1}{\sqrt{2}}\sin\theta - \dfrac{1}{\sqrt{2}}\cos\theta = \dfrac{1}{\sqrt{2}}$

Rewrite using the difference of two angles formula where $\phi = \dfrac{\pi}{4}$, so $\cos\phi = \dfrac{1}{\sqrt{2}}$ and $\sin\phi = \dfrac{1}{\sqrt{2}}$.

$$\sin\theta\cos\phi - \cos\theta\sin\phi = \frac{1}{\sqrt{2}}$$
$$\sin(\theta - \phi) = \frac{\sqrt{2}}{2}$$
$$\theta - \phi = \frac{\pi}{4} \quad \text{or} \quad \theta - \phi = \frac{3\pi}{4}$$
$$\theta - \frac{\pi}{4} = \frac{\pi}{4} \quad \text{or} \quad \theta - \frac{\pi}{4} = \frac{3\pi}{4}$$
$$\theta = \frac{\pi}{2} \quad \text{or} \quad \theta = \pi$$

The solution set is $\left\{\dfrac{\pi}{2}, \pi\right\}$.

101. $\sin^{-1}0.7 \approx 0.78$

103. $\tan^{-1}(-2) \approx -1.11$

105. $\sec^{-1}3 = \cos^{-1}\dfrac{1}{3} \approx 1.23$

107. $2x = 5\cos x$
    Find the intersection of
    $y_1 = 2x$ and $y_2 = 5\cos x$:

    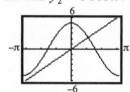

    The solution set is $\{1.11\}$.

109. $2\sin x + 3\cos x = 4x$
    Find the intersection of
    $y_1 = 2\sin x + 3\cos x$ and $y_2 = 4x$:

    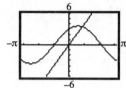

    The solution set is $\{0.87\}$.

111. $\sin x = \ln x$
    Find the intersection of
    $y_1 = \sin x$ and $y_2 = \ln x$:

    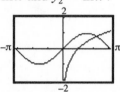

    The solution set is $\{2.22\}$.

113. Using a half-angle formula:

$$\sin 15° = \sin\left(\frac{30°}{2}\right) = \sqrt{\frac{1 - \cos 30°}{2}} = \sqrt{\frac{1 - \dfrac{\sqrt{3}}{2}}{2}} = \sqrt{\frac{2 - \sqrt{3}}{4}} = \frac{\sqrt{2 - \sqrt{3}}}{2}$$

Note: since $15°$ lies in quadrant I, we have $\sin 15° > 0$.
Using a difference formula:
$$\sin 15° = \sin(45° - 30°) = \sin 45° \cdot \cos 30° - \cos 45° \cdot \sin 30°$$

$$= \frac{\sqrt{2}}{2} \cdot \frac{\sqrt{3}}{2} - \frac{\sqrt{2}}{2} \cdot \frac{1}{2} = \frac{1}{4}\left(\sqrt{6} - \sqrt{2}\right)$$

$$= \frac{\sqrt{2}\sqrt{3} - \sqrt{2}}{4} = \frac{\sqrt{2}\left(\sqrt{3} - 1\right)}{4} = \sqrt{\left(\frac{\sqrt{2}\left(\sqrt{3} - 1\right)}{4}\right)^2}$$

$$= \sqrt{\frac{2\left(\sqrt{3} - 1\right)^2}{16}} = \sqrt{\frac{2\left(3 - 2\sqrt{3} + 1\right)}{16}} = \sqrt{\frac{2\left(4 - 2\sqrt{3}\right)}{16}}$$

$$= \sqrt{\frac{4\left(2 - \sqrt{3}\right)}{16}} = \sqrt{\frac{2 - \sqrt{3}}{4}}$$

$$= \frac{\sqrt{2 - \sqrt{3}}}{2}$$

Analytic Trigonometry

6.CR   Cumulative Review

1. $3x^2 + x - 1 = 0$

$$x = \frac{-b \pm \sqrt{b^2 - 4ac}}{2a} = \frac{-1 \pm \sqrt{1^2 - 4(3)(-1)}}{2(3)}$$

$$= \frac{-1 \pm \sqrt{1 + 12}}{6} = \frac{-1 \pm \sqrt{13}}{6}$$

The solution set is $\left\{ \dfrac{-1 - \sqrt{13}}{6}, \dfrac{-1 + \sqrt{13}}{6} \right\}$.

3. $3x + y^2 = 9$

   x-intercept: $3x + 0^2 = 9 \Rightarrow 3x = 9 \Rightarrow x = 3;\ (3,0)$

   y-intercepts: $3(0) + y^2 = 9 \Rightarrow y^2 = 9 \Rightarrow y = \pm 3;\ (0,-3),(0,3)$

   Test for symmetry:

   x-axis: Replace y by $-y$: $3x + (-y)^2 = 9 \Rightarrow 3x + y^2 = 9$

   which is equivalent to $3x + y^2 = 9$.

   y-axis: Replace x by $-x$: $3(-x) + y^2 = 9 \Rightarrow -3x + y^2 = 9$,

   which is not equivalent to $3x + y^2 = 9$.

   Origin :   Replace x by $-x$ and y by $-y$: $3(-x) + (-y)^2 = 9 \Rightarrow -3x + y^2 = 9$,

   which is not equivalent to $3x + y^2 = 9$.

   Therefore, the graph is symmetric with respect to the x-axis.

5. $y = 3e^x - 2$

   Using the graph of $y = e^x$, stretch
   vertically by a factor of 3, and vertically
   shift down 2 units.

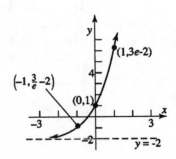

7. (a) $y = x^3$ Inverse function: $y = \sqrt[3]{x}$

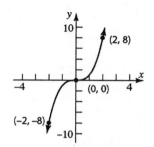

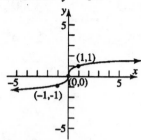

(b) $y = e^x$ Inverse function: $y = \ln x$

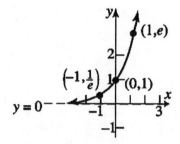

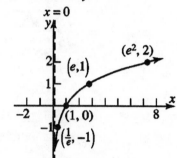

(c) $y = \sin x$, $-\dfrac{\pi}{2} \le x \le \dfrac{\pi}{2}$ Inverse function: $y = \sin^{-1} x$

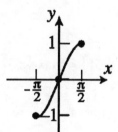

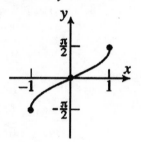

(d) $y = \cos x$, $0 \le x \le \pi$ Inverse function: $y = \cos^{-1} x$

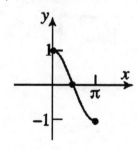

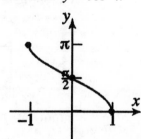

9. $\cos\left(\tan^{-1}2\right)$

Find the angle θ, $-\dfrac{\pi}{2} < \theta < \dfrac{\pi}{2}$, whose tangent equals 2.

$\tan\theta = 2$, $-\dfrac{\pi}{2} < \theta < \dfrac{\pi}{2} \Rightarrow \theta$ is in Quadrant I

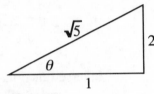

$\cos\theta = \dfrac{1}{\sqrt{5}} \cdot \dfrac{\sqrt{5}}{\sqrt{5}} = \dfrac{\sqrt{5}}{5}$

Therefore, $\cos\left(\tan^{-1}2\right) = \dfrac{\sqrt{5}}{5}$.

11. $f(x) = 2x^5 - x^4 - 4x^3 + 2x^2 + 2x - 1$

(a) Step 1: $f(x)$ has at most 5 real zeros.

Step 2:   Possible rational zeros: $p = \pm 1$; $q = \pm 1, \pm 2$; $\dfrac{p}{q} = \pm 1, \pm\dfrac{1}{2}$

Step 3:   Using the Bounds on Zeros Theorem:

$f(x) = 2\left(x^5 - 0.5x^4 - 2x^3 + x^2 + x - 0.5\right)$

$a_4 = -0.5$, $a_3 = -2$, $a_2 = 1$, $a_1 = 1$, $a_0 = -0.5$

Max $\left\{1, |-0.5| + |\, 1\,| + |\, 1\, | + |-2| + |-0.5|\right\} = $ Max $\left\{1, 5\right\} = 5$

$1 + $ Max $\left\{|-0.5|, |\, 1\,|, |\, 1\, |, |-2|, |-0.5|\right\} = 1 + 2 = 3$

The smaller of the two numbers is 3. Thus, every zero of f lies between -3 and 3.

Graphing using the bounds and ZOOM-FIT: (Second graph has a better window.)

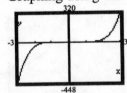   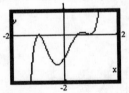

Step 4:   From the graph it appears that there are x-intercepts at -1, $\dfrac{1}{2}$, and 1.

Using synthetic division with -1:

$$
\begin{array}{r|rrrrrr}
-1) & 2 & -1 & -4 & 2 & 2 & -1 \\
    &   & -2 & 3 & 1 & -3 & 1 \\
\hline
    & 2 & -3 & -1 & 3 & -1 & 0
\end{array}
$$

Since the remainder is 0, $x - (-1) = x + 1$ is a factor.

The other factor is the quotient: $2x^4 - 3x^3 - x^2 + 3x - 1$.

Using synthetic division with 1 on the quotient:

$$\begin{array}{r|rrrrr} 1) & 2 & -3 & -1 & 3 & -1 \\ & & 2 & -1 & -2 & 1 \\ \hline & 2 & -1 & -2 & 1 & 0 \end{array}$$

Since the remainder is 0, $x - 1$ is a factor.

The other factor is the quotient: $2x^3 - x^2 - 2x + 1$.

Using synthetic division with $\dfrac{1}{2}$ on the quotient:

$$\begin{array}{r|rrrr} \frac{1}{2}) & 2 & -1 & -2 & 1 \\ & & 1 & 0 & -1 \\ \hline & 2 & 0 & -2 & 0 \end{array}$$

Since the remainder is 0, $x - \dfrac{1}{2}$ is a factor.

The other factor is the quotient: $2x^2 - 2 = 2(x^2 - 1) = 2(x + 1)(x - 1)$.

Factoring,

$$f(x) = (x + 1)(x - 1)\left(x - \frac{1}{2}\right)2(x + 1)(x - 1) = (2x - 1)(x + 1)^2(x - 1)^2$$

The real zeros are -1 and 1 (multiplicity 2), and $\dfrac{1}{2}$ (multiplicity 1).

(b) x-intercepts: $(1,0)$, $\left(\dfrac{1}{2},0\right)$, $(-1,0)$; y-intercept: $(0,-1)$

(c) f resembles the graph of $y = 2x^5$ for large $|x|$.

(d)   graphing utility:

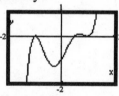

(e)   4 turning points; use MAXIMUM and MINIMUM to locate local maxima at $(-1,0)$, $(0.69,0.1)$ and local minima at $(1,0)$, $(-0.29,-1.33)$

(f)   Graphing by hand

Interval	$(-\infty, -1)$	$(-1, 0.5)$	$(0.5, 1)$	$(1, \infty)$
Number Chosen	-2	0	0.6	2
Value of f	$f(-2) = -45$	$f(0) = -1$	$f(0.6) \approx 0.8$	$f(2) = 27$
Location of Graph	Below x-axis	Below x-axis	Above x-axis	Above x-axis
Point on Graph	$(-2, -45)$	$(0,-1)$	$(0.6, 0.8)$	$(2, 27)$

f is above the x-axis for $\left(\dfrac{1}{2},1\right)$, and $(1,\infty)$

f is below the x-axis for $(-\infty,-1)$ and $\left(-1,\dfrac{1}{2}\right)$

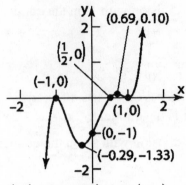

(g) f is increasing on $(-\infty,-1)$, $(-0.29,0.69)$, and $(1,\infty)$
        f is decreasing on $(-1,-0.29)$, and $(0.69,1)$

Applications of Trigonometric Functions
7.1 Right Triangle Trigonometry; Applications

1. $a = \sqrt{5^2 - 3^2} = \sqrt{25 - 9} = \sqrt{16} = 4$

3. False

3. False

5. angle of elevation

7. True

9. opposite = 5; adjacent = 12
Find the hypotenuse:
$5^2 + 12^2 = (\text{hypotenuse})^2 \Rightarrow (\text{hypotenuse})^2 = 25 + 144 = 169 \Rightarrow \text{hypotenuse} = 13$

$\sin\theta = \dfrac{\text{opp}}{\text{hyp}} = \dfrac{5}{13}$ $\qquad \cos\theta = \dfrac{\text{adj}}{\text{hyp}} = \dfrac{12}{13}$ $\qquad \tan\theta = \dfrac{\text{opp}}{\text{adj}} = \dfrac{5}{12}$

$\csc\theta = \dfrac{\text{hyp}}{\text{opp}} = \dfrac{13}{5}$ $\qquad \sec\theta = \dfrac{\text{hyp}}{\text{adj}} = \dfrac{13}{12}$ $\qquad \cot\theta = \dfrac{\text{adj}}{\text{opp}} = \dfrac{12}{5}$

11. opposite = 2; adjacent = 3
Find the hypotenuse:
$2^2 + 3^2 = (\text{hypotenuse})^2 \Rightarrow (\text{hypotenuse})^2 = 4 + 9 = 13 \Rightarrow \text{hypotenuse} = \sqrt{13}$

$\sin\theta = \dfrac{\text{opp}}{\text{hyp}} = \dfrac{2}{\sqrt{13}} \cdot \dfrac{\sqrt{13}}{\sqrt{13}} = \dfrac{2\sqrt{13}}{13}$ $\qquad \cos\theta = \dfrac{\text{adj}}{\text{hyp}} = \dfrac{3}{\sqrt{13}} \cdot \dfrac{\sqrt{13}}{\sqrt{13}} = \dfrac{3\sqrt{13}}{13}$ $\qquad \tan\theta = \dfrac{\text{opp}}{\text{adj}} = \dfrac{2}{3}$

$\csc\theta = \dfrac{\text{hyp}}{\text{opp}} = \dfrac{\sqrt{13}}{2}$ $\qquad \sec\theta = \dfrac{\text{hyp}}{\text{adj}} = \dfrac{\sqrt{13}}{3}$ $\qquad \cot\theta = \dfrac{\text{adj}}{\text{opp}} = \dfrac{3}{2}$

13. adjacent = 2; hypotenuse = 4
Find the opposite side:
$(\text{opposite})^2 + 2^2 = 4^2 \Rightarrow (\text{opposite})^2 = 16 - 4 = 12 \Rightarrow \text{opposite} = \sqrt{12} = 2\sqrt{3}$

$\sin\theta = \dfrac{\text{opp}}{\text{hyp}} = \dfrac{2\sqrt{3}}{4} = \dfrac{\sqrt{3}}{2}$ $\qquad \cos\theta = \dfrac{\text{adj}}{\text{hyp}} = \dfrac{2}{4} = \dfrac{1}{2}$ $\qquad \tan\theta = \dfrac{\text{opp}}{\text{adj}} = \dfrac{2\sqrt{3}}{2} = \sqrt{3}$

$\csc\theta = \dfrac{\text{hyp}}{\text{opp}} = \dfrac{4}{2\sqrt{3}} \cdot \dfrac{\sqrt{3}}{\sqrt{3}} = \dfrac{2\sqrt{3}}{3}$ $\qquad \sec\theta = \dfrac{\text{hyp}}{\text{adj}} = \dfrac{4}{2} = 2$ $\qquad \cot\theta = \dfrac{\text{adj}}{\text{opp}} = \dfrac{2}{2\sqrt{3}} \cdot \dfrac{\sqrt{3}}{\sqrt{3}} = \dfrac{\sqrt{3}}{3}$

15. opposite = $\sqrt{2}$; adjacent = 1
Find the hypotenuse:
$\left(\sqrt{2}\right)^2 + 1^2 = (\text{hypotenuse})^2 \Rightarrow (\text{hypotenuse})^2 = 2 + 1 = 3 \Rightarrow \text{hypotenuse} = \sqrt{3}$

$$\sin\theta = \frac{\text{opp}}{\text{hyp}} = \frac{\sqrt{2}}{\sqrt{3}}\frac{\sqrt{3}}{\sqrt{3}} = \frac{\sqrt{6}}{3} \qquad \cos\theta = \frac{\text{adj}}{\text{hyp}} = \frac{1}{\sqrt{3}}\frac{\sqrt{3}}{\sqrt{3}} = \frac{\sqrt{3}}{3} \qquad \tan\theta = \frac{\text{opp}}{\text{adj}} = \frac{\sqrt{2}}{1} = \sqrt{2}$$

$$\csc\theta = \frac{\text{hyp}}{\text{opp}} = \frac{\sqrt{3}}{\sqrt{2}}\frac{\sqrt{2}}{\sqrt{2}} = \frac{\sqrt{6}}{2} \qquad \sec\theta = \frac{\text{hyp}}{\text{adj}} = \frac{\sqrt{3}}{1} = \sqrt{3} \qquad \cot\theta = \frac{\text{adj}}{\text{opp}} = \frac{1}{\sqrt{2}}\frac{\sqrt{2}}{\sqrt{2}} = \frac{\sqrt{2}}{2}$$

17. opposite = 1; hypotenuse = $\sqrt{5}$
Find the adjacent side:

$$1^2 + (\text{adjacent})^2 = \left(\sqrt{5}\right)^2 \Rightarrow (\text{adjacent})^2 = 5 - 1 = 4 \Rightarrow \text{adjacent} = 2 .$$

$$\sin\theta = \frac{\text{opp}}{\text{hyp}} = \frac{1}{\sqrt{5}}\frac{\sqrt{5}}{\sqrt{5}} = \frac{\sqrt{5}}{5} \qquad \cos\theta = \frac{\text{adj}}{\text{hyp}} = \frac{2}{\sqrt{5}}\frac{\sqrt{5}}{\sqrt{5}} = \frac{2\sqrt{5}}{5} \qquad \tan\theta = \frac{\text{opp}}{\text{adj}} = \frac{1}{2}$$

$$\csc\theta = \frac{\text{hyp}}{\text{opp}} = \frac{\sqrt{5}}{1} = \sqrt{5} \qquad \sec\theta = \frac{\text{hyp}}{\text{adj}} = \frac{\sqrt{5}}{2} \qquad \cot\theta = \frac{\text{adj}}{\text{opp}} = \frac{2}{1} = 2$$

19. $\sin 38° - \cos 52° = \sin 38° - \sin(90° - 52°) = \sin 38° - \sin 38° = 0$

21. $\dfrac{\cos 10°}{\sin 80°} = \dfrac{\sin(90° - 10°)}{\sin 80°} = \dfrac{\sin 80°}{\sin 80°} = 1$

23. $1 - \cos^2 20° - \cos^2 70° = \sin^2 20° - \sin^2(90° - 70°) = \sin^2 20° - \sin^2 20° = 0$

25. $\tan 20° - \dfrac{\cos 70°}{\cos 20°} = \tan 20° - \dfrac{\sin(90° - 70°)}{\cos 20°} = \tan 20° - \dfrac{\sin 20°}{\cos 20°} = \tan 20° - \tan 20° = 0$

27. $\cos 35° \cdot \sin 55° + \cos 55° \cdot \sin 35° = \cos 35° \cdot \cos 35° + \sin 35° \cdot \sin 35° = \cos^2 35° + \sin^2 35° = 1$

29. $b = 5, \ \beta = 20°$

$$\sin\beta = \frac{b}{c} \Rightarrow \sin 20° = \frac{5}{c} \Rightarrow c = \frac{5}{\sin 20°} \approx 14.62$$

$$\tan\beta = \frac{b}{a} \Rightarrow \tan 20° = \frac{5}{a} \Rightarrow a = \frac{5}{\tan 20°} \approx 13.74$$

$$\alpha = 90° - \beta = 90° - 20° = 70°$$

31. $a = 6, \ \beta = 40°$

$$\cos\beta = \frac{a}{c} \Rightarrow \cos 40° = \frac{6}{c} \Rightarrow c = \frac{6}{\cos 40°} \approx 7.83$$

$$\tan\beta = \frac{b}{a} \Rightarrow \tan 40° = \frac{b}{6} \Rightarrow b = 6\tan 40° \approx 5.03$$

$$\alpha = 90° - \beta = 90° - 40° = 50°$$

33. $b = 4, \ \alpha = 10°$

$$\tan\alpha = \frac{a}{b} \Rightarrow \tan 10° = \frac{a}{4} \Rightarrow a = 4\tan 10° \approx 0.71$$

$$\cos\alpha = \frac{b}{c} \Rightarrow \cos 10° = \frac{4}{c} \Rightarrow c = \frac{4}{\cos 10°} \approx 4.06$$

$$\beta = 90° - \alpha = 90° - 10° = 80°$$

35. $a = 5, \ \alpha = 25°$

$$\tan\alpha = \frac{a}{b} \Rightarrow \tan 25° = \frac{5}{b} \Rightarrow b = \frac{5}{\tan 25°} \approx 10.72$$

$$\sin\alpha = \frac{a}{c} \Rightarrow \sin 25° = \frac{5}{c} \Rightarrow c = \frac{5}{\sin 25°} \approx 11.83$$

$$\beta = 90° - \alpha = 90° - 25° = 65°$$

37. $c = 9, \ \beta = 20°$

$$\sin\beta = \frac{b}{c} \Rightarrow \sin 20° = \frac{b}{9} \Rightarrow b = 9\sin 20° \approx 3.08$$

$$\cos\beta = \frac{a}{c} \Rightarrow \cos 20° = \frac{a}{9} \Rightarrow a = 9\cos 20° \approx 8.46$$

$$\beta = 90° - \alpha = 90° - 20° = 70°$$

39. $a = 5, \ b = 3$

$$c^2 = a^2 + b^2 = 5^2 + 3^2 = 25 + 9 = 34 \Rightarrow c = \sqrt{34} \approx 5.83$$

$$\tan\alpha = \frac{a}{b} = \frac{5}{3} \Rightarrow \alpha \approx 59.0°$$

$$\beta = 90° - \alpha = 90° - 59.0° = 31.0°$$

41. $a = 2, \ c = 5$

$$c^2 = a^2 + b^2 \Rightarrow b^2 = c^2 - a^2 = 5^2 - 2^2 = 25 - 4 = 21 \Rightarrow b = \sqrt{21} \approx 4.58$$

$$\sin\alpha = \frac{a}{c} = \frac{2}{5} \Rightarrow \alpha \approx 23.6°$$

$$\beta = 90° - \alpha = 90° - 23.6° = 66.4°$$

43. $c = 8, \ \alpha = 35°$

$$\sin 35° = \frac{a}{8} \Rightarrow a = 8\sin 35° \approx 4.59 \text{ in.}$$

$$\cos 35° = \frac{b}{8} \Rightarrow b = 8\cos 35° \approx 6.55 \text{ in.}$$

45. $\alpha = 25°, \ a = 5$

$$\sin 25° = \frac{5}{c} \Rightarrow c = \frac{5}{\sin 25°} \approx 11.83 \text{ in.}$$

$\alpha = 25°, \ b = 5$

$$\cos 25° = \frac{5}{c} \Rightarrow c = \frac{5}{\cos 25°} \approx 5.52 \text{ in.}$$

47. $c = 5, \ a = 2$

$$\sin\alpha = \frac{2}{5} = 0.4000 \Rightarrow \alpha \approx 23.6 \Rightarrow \beta = 90° - \alpha = 90° - 23.6° \approx 66.4°$$

49. $$\tan 35° = \frac{b}{100} \Rightarrow b = 100\tan 35° \approx 70.02 \text{ feet}$$

51. $$\tan 85.361° = \frac{a}{80} \Rightarrow a = 80\tan 85.361° \approx 985.91 \text{ feet}$$

53. 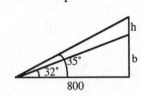 $\tan 20° = \dfrac{50}{x} \Rightarrow x = \dfrac{50}{\tan 20°} \approx 137.37$ meters

55. $\sin 70° = \dfrac{x}{22} \Rightarrow x = 22 \sin 70° \approx 20.67$ feet

57. $\tan 32° = \dfrac{500}{x} \Rightarrow x = \dfrac{500}{\tan 32°}$

$\tan 23° = \dfrac{500}{y} \Rightarrow y = \dfrac{500}{\tan 23°}$

Distance $= x + y = \dfrac{500}{\tan 32°} + \dfrac{500}{\tan 23°} \approx 1978.09$ feet

59. Let h represent the height of Lincoln's face.

$\tan 32° = \dfrac{b}{800} \Rightarrow b = 800 \tan 32°$

$\tan 35° = \dfrac{b + h}{800} \Rightarrow b + h = 800 \tan 35°$

$h = (b + h) - b = 800 \tan 35° - 800 \tan 32° \approx 60.27$ feet

61. $\sin 21° = \dfrac{190}{x} \Rightarrow x = \dfrac{190}{\sin 21°} \approx 530.18$ ft

63. $\tan 35.1° = \dfrac{x}{789} \Rightarrow x = 789 \tan 35.1° \approx 554.52$ ft

65. (a) $\tan 15° = \dfrac{30}{x} \Rightarrow x = \dfrac{30}{\tan 15°} \approx 111.96$ feet

The truck is traveling at 111.96 ft/sec.

$\dfrac{111.96 \text{ ft}}{\text{sec}} \cdot \dfrac{1 \text{ mile}}{5280 \text{ ft}} \cdot \dfrac{3600 \text{ sec}}{\text{hr}} \approx 76.3$ mph

(b) $\tan 20° = \dfrac{30}{x} \Rightarrow x = \dfrac{30}{\tan 20°} \approx 82.42$ feet

The truck is traveling at 82.42 ft/sec.

$\dfrac{82.42 \text{ ft}}{\text{sec}} \cdot \dfrac{1 \text{ mile}}{5280 \text{ ft}} \cdot \dfrac{3600 \text{ sec}}{\text{hr}} \approx 56.20$ mph

(c) A ticket is issued for traveling at a speed of 60 mi/hr or more.

$\dfrac{60 \text{ mi}}{\text{hr}} \cdot \dfrac{5280 \text{ ft}}{\text{mi}} \cdot \dfrac{1 \text{hr}}{3600 \text{ sec}} = 88$ ft / sec.

If $\tan \theta \le \dfrac{30}{88}$, the trooper should issue a ticket.

$\tan \theta \le \dfrac{30}{88} \Rightarrow \theta \le \tan^{-1}\left(\dfrac{30}{88}\right) \approx 18.8°$

A ticket is issued if $\theta \le 18.8°$.

67. Find angle θ: (see the figure)

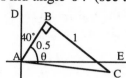

$$\tan\theta = \frac{1}{0.5} = 2 \Rightarrow \theta = \tan^{-1} 2 \approx 63.4°$$
$$\angle DAC = 40° + 63.4° = 103.4°$$
$$\angle EAC = 103.4° - 90° = 13.4°$$

The bearing the control tower should use is S76.6°E.

69. The height of the beam above the wall is $46 - 20 = 26$ feet.
$$\tan\theta = \frac{26}{10} = 2.6 \Rightarrow \theta = \tan^{-1} 2.6 \approx 69.0°$$

71. The length of the highway $= x + y + z$

$$\sin 40° = \frac{1}{x} \Rightarrow x = \frac{1}{\sin 40°}$$

$$\sin 50° = \frac{1}{z} \Rightarrow z = \frac{1}{\sin 50°}$$

$$\tan 40° = \frac{1}{a} \Rightarrow a = \frac{1}{\tan 40°}$$

$$\tan 50° = \frac{1}{b} \Rightarrow b = \frac{1}{\tan 50°}$$

$$a + y + b = 3 \Rightarrow y = 3 - a - b$$

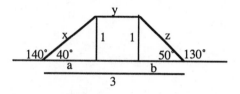

The length of the highway is: $\dfrac{1}{\sin 40°} + \dfrac{1}{\sin 50°} + 3 - \dfrac{1}{\tan 40°} - \dfrac{1}{\tan 50°} \approx$ miles.

73. With the camera 10 feet from George, the amount that will be seen by the lens above or below the 4-foot level is a, where $\tan 20° = \dfrac{a}{10}$. Thus, $a = 10\tan 20° \approx 3.64$ feet.

So George's head will be seen by the lens, but his feet will not.

In order to see George's head and feet, the camera must be x feet from George.

Solve: $\tan 20° = \dfrac{4}{x} \Rightarrow x = \dfrac{4}{\tan 20°} \approx 10.99$ feet

The camera will need to be moved back 1 foot to see George's feet and head.

75. Let $\theta =$ the central angle formed by the top of the lighthouse, the center of the Earth and the point P on the Earth's surface where the line of sight from the top of the lighthouse is tangent to the Earth. Note also that 362 feet $= \dfrac{362}{5280}$ miles.

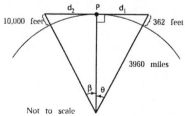

$$\theta = \cos^{-1}\left(\frac{3960}{3960 + 362/5280}\right) \approx 0.33715°$$

413

Verify the airplane information:
    Let β = the central angle formed by the plane, the center of the Earth and the point P.

$$\beta = \cos^{-1}\left(\frac{3960}{3960 + 10,000/5280}\right) \approx 1.77169°$$

Note that $\tan\theta = \dfrac{d_1}{3690} \Rightarrow d_1 = 3960\tan\theta$ and $\tan\beta = \dfrac{d_2}{3690} \Rightarrow d_2 = 3960\tan\beta$

So

$$d_1 + d_2 = 3960\tan\theta + 3960\tan\beta \approx 3960\tan\left(0.33715°\right) + 3960\tan\left(1.77169°\right) \approx 146 \text{ miles}$$

    To express this distance in nautical miles, we express the total angle $\theta + \beta$ in minutes. That is, $\theta + \beta \approx \left(0.33715° + 1.77169°\right) \cdot 60 \approx 126.5$ nautical miles. Therefore, a plane flying at an altitude of 10,000 feet can see the lighthouse 120 miles away.

Verify the ship information:

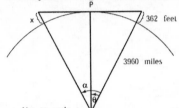

Let α = the central angle formed by 40 nautical miles

then. $\alpha = \dfrac{40}{60} = \dfrac{2°}{3}$.

$$\cos(\alpha - \theta) = \frac{3960}{3960 + x} \Rightarrow \cos\left(\frac{2°}{3} - 0.33715°\right) = \frac{3960}{3960 + x}$$

$$(3960 + x)\cos\left(\frac{2°}{3} - 0.33715°\right) = 3960 \Rightarrow 3960 + x = \frac{3960}{\cos\left(\dfrac{2°}{3} - 0.33715°\right)}$$

$$x = \frac{3960}{\cos\left(\dfrac{2°}{3} - 0.33715°\right)} - 3960$$

$$\approx 0.06549 \text{ mile} \approx (0.06549 \text{ mi})\left(5280\frac{\text{ft}}{\text{mi}}\right) \approx 346 \text{ feet}$$

    Therefore, a ship that is 346 feet above sea level can see the lighthouse from a distance of 40 nautical miles.

77.   Answers will vary.

Applications of Trigonometric Functions

7.2 The Law of Sines

1. $\sin\alpha\cos\beta - \cos\alpha\sin\beta$

3. $\sin\theta = 2$
   There is no angle θ with $\sin\theta > 1$. Therefore, the equation has no solution.

5. $\dfrac{\sin\alpha}{a} = \dfrac{\sin\beta}{b} = \dfrac{\sin\gamma}{c}$

7. True

9. $c = 5$, $\beta = 45°$, $\gamma = 95°$
   $\alpha = 180° - \beta - \gamma = 180° - 45° - 95° = 40°$
   $$\frac{\sin\alpha}{a} = \frac{\sin\gamma}{c} \Rightarrow \frac{\sin40°}{a} = \frac{\sin95°}{5} \Rightarrow a = \frac{5\sin40°}{\sin95°} \approx 3.23$$
   $$\frac{\sin\beta}{b} = \frac{\sin\gamma}{c} \Rightarrow \frac{\sin45°}{b} = \frac{\sin95°}{5} \Rightarrow b = \frac{5\sin45°}{\sin95°} \approx 3.55$$

11. $b = 3$, $\alpha = 50°$, $\gamma = 85°$
    $\beta = 180° - \alpha - \gamma = 180° - 50° - 85° = 45°$
    $$\frac{\sin\alpha}{a} = \frac{\sin\beta}{b} \Rightarrow \frac{\sin50°}{a} = \frac{\sin45°}{3} \Rightarrow a = \frac{3\sin50°}{\sin45°} \approx 3.25$$
    $$\frac{\sin\gamma}{c} = \frac{\sin\beta}{b} \Rightarrow \frac{\sin85°}{c} = \frac{\sin45°}{3} \Rightarrow c = \frac{3\sin85°}{\sin45°} \approx 4.23$$

13. $b = 7$, $\alpha = 40°$, $\beta = 45°$
    $\gamma = 180° - \alpha - \beta = 180° - 40° - 45° = 95°$
    $$\frac{\sin\alpha}{a} = \frac{\sin\beta}{b} \Rightarrow \frac{\sin40°}{a} = \frac{\sin45°}{7} \Rightarrow a = \frac{7\sin40°}{\sin45°} \approx 6.36$$
    $$\frac{\sin\gamma}{c} = \frac{\sin\beta}{b} \Rightarrow \frac{\sin95°}{c} = \frac{\sin45°}{7} \Rightarrow c = \frac{7\sin95°}{\sin45°} \approx 9.86$$

15. $b = 2$, $\beta = 40°$, $\gamma = 100°$
    $\alpha = 180° - \beta - \gamma = 180° - 40° - 100° = 40°$
    $$\frac{\sin\alpha}{a} = \frac{\sin\beta}{b} \Rightarrow \frac{\sin40°}{a} = \frac{\sin40°}{2} \Rightarrow a = \frac{2\sin40°}{\sin40°} = 2$$
    $$\frac{\sin\gamma}{c} = \frac{\sin\beta}{b} \Rightarrow \frac{\sin100°}{c} = \frac{\sin40°}{2} \Rightarrow c = \frac{2\sin100°}{\sin40°} \approx 3.06$$

17. $\alpha = 40°,\ \beta = 20°,\ a = 2$
$\gamma = 180° - \alpha - \beta = 180° - 40° - 20° = 120°$

$$\frac{\sin\alpha}{a} = \frac{\sin\beta}{b} \Rightarrow \frac{\sin40°}{2} = \frac{\sin20°}{b} \Rightarrow b = \frac{2\sin20°}{\sin40°} \approx 1.06$$

$$\frac{\sin\gamma}{c} = \frac{\sin\alpha}{a} \Rightarrow \frac{\sin120°}{c} = \frac{\sin40°}{2} \Rightarrow c = \frac{2\sin120°}{\sin40°} \approx 2.69$$

19. $\beta = 70°,\ \gamma = 10°,\ b = 5$
$\alpha = 180° - \beta - \gamma = 180° - 70° - 10° = 100°$

$$\frac{\sin\alpha}{a} = \frac{\sin\beta}{b} \Rightarrow \frac{\sin100°}{a} = \frac{\sin70°}{5} \Rightarrow a = \frac{5\sin100°}{\sin70°} \approx 5.24$$

$$\frac{\sin\gamma}{c} = \frac{\sin\beta}{b} \Rightarrow \frac{\sin10°}{c} = \frac{\sin70°}{5} \Rightarrow c = \frac{5\sin10°}{\sin70°} \approx 0.92$$

21. $\alpha = 110°,\ \gamma = 30°,\ c = 3$
$\beta = 180° - \alpha - \gamma = 180° - 110° - 30° = 40°$

$$\frac{\sin\alpha}{a} = \frac{\sin\gamma}{c} \Rightarrow \frac{\sin110°}{a} = \frac{\sin30°}{3} \Rightarrow a = \frac{3\sin110°}{\sin30°} \approx 5.64$$

$$\frac{\sin\gamma}{c} = \frac{\sin\beta}{b} \Rightarrow \frac{\sin30°}{3} = \frac{\sin40°}{b} \Rightarrow b = \frac{3\sin40°}{\sin30°} \approx 3.86$$

23. $\alpha = 40°,\ \beta = 40°,\ c = 2$
$\gamma = 180° - \alpha - \beta = 180° - 40° - 40° = 100°$

$$\frac{\sin\alpha}{a} = \frac{\sin\gamma}{c} \Rightarrow \frac{\sin40°}{a} = \frac{\sin100°}{2} \Rightarrow a = \frac{2\sin40°}{\sin100°} \approx 1.31$$

$$\frac{\sin\beta}{b} = \frac{\sin\gamma}{c} \Rightarrow \frac{\sin40°}{b} = \frac{\sin100°}{2} \Rightarrow b = \frac{2\sin40°}{\sin100°} \approx 1.31$$

25. $a = 3,\ b = 2,\ \alpha = 50°$

$$\frac{\sin\beta}{b} = \frac{\sin\alpha}{a} \Rightarrow \frac{\sin\beta}{2} = \frac{\sin(50°)}{3} \Rightarrow \sin\beta = \frac{2\sin(50°)}{3} \approx 0.5107$$

$\beta = \sin^{-1}(0.5107)$

$\beta = 30.7°\ \text{or}\ \beta = 149.3°$
The second value is discarded because $\alpha + \beta > 180°$.
$\gamma = 180° - \alpha - \beta = 180° - 50° - 30.7° = 99.3°$

$$\frac{\sin\gamma}{c} = \frac{\sin\alpha}{a} \Rightarrow \frac{\sin99.3°}{c} = \frac{\sin50°}{3} \Rightarrow c = \frac{3\sin99.3°}{\sin50°} \approx 3.86$$

One triangle: $\beta \approx 30.7°,\ \gamma \approx 99.3°,\ c \approx 3.86$

27. $b = 5$, $c = 3$, $\beta = 100°$

$$\frac{\sin\beta}{b} = \frac{\sin\gamma}{c} \Rightarrow \frac{\sin 100°}{5} = \frac{\sin\gamma}{3} \Rightarrow \sin\gamma = \frac{3\sin 100°}{5} \approx 0.5909$$

$\gamma = \sin^{-1}(0.5909)$

$\gamma = 36.2°$ or $\gamma = 143.8°$

The second value is discarded because $\beta + \gamma > 180°$.

$\alpha = 180° - \beta - \gamma = 180° - 100° - 36.2° = 43.8°$

$$\frac{\sin\beta}{b} = \frac{\sin\alpha}{a} \Rightarrow \frac{\sin 100°}{5} = \frac{\sin 43.8°}{a} \Rightarrow a = \frac{5\sin 43.8°}{\sin 100°} \approx 3.51$$

One triangle: $\alpha \approx 43.8°$, $\gamma \approx 36.2°$, $a \approx 3.51$

29. $a = 4$, $b = 5$, $\alpha = 60°$

$$\frac{\sin\beta}{b} = \frac{\sin\alpha}{a} \Rightarrow \frac{\sin\beta}{5} = \frac{\sin 60°}{4} \Rightarrow \sin\beta = \frac{5\sin 60°}{4} \approx 1.0825$$

There is no angle β for which $\sin\beta > 1$. Therefore, there is no triangle with the given measurements.

31. $b = 4$, $c = 6$, $\beta = 20°$

$$\frac{\sin\beta}{b} = \frac{\sin\gamma}{c} \Rightarrow \frac{\sin 20°}{4} = \frac{\sin\gamma}{6} \Rightarrow \sin\gamma = \frac{6\sin 20°}{4} \approx 0.5130$$

$\gamma = \sin^{-1}(0.5130)$

$\gamma_1 = 30.9°$ or $\gamma_2 = 149.1°$

For both values, $\beta + \gamma < 180°$. Therefore, there are two triangles.

$\alpha_1 = 180° - \beta - \gamma_1 = 180° - 20° - 30.9° = 129.1°$

$$\frac{\sin\beta}{b} = \frac{\sin\alpha_1}{a_1} \Rightarrow \frac{\sin 20°}{4} = \frac{\sin 129.1°}{a_1} \Rightarrow a_1 = \frac{4\sin 129.1°}{\sin 20°} \approx 9.08$$

$\alpha_2 = 180° - \beta - \gamma_2 = 180° - 20° - 149.1° = 10.9°$

$$\frac{\sin\beta}{b} = \frac{\sin\alpha_2}{a_2} \Rightarrow \frac{\sin 20°}{4} = \frac{\sin 10.9°}{a_2} \Rightarrow a_2 = \frac{4\sin 10.9°}{\sin 20°} \approx 2.21$$

Two triangles: $\alpha_1 \approx 129.1°$, $\gamma_1 \approx 30.9°$, $a_1 \approx 9.08$

        or $\alpha_2 \approx 10.9°$, $\gamma_2 \approx 149.1°$, $a_2 \approx 2.21$

33. $a = 2$, $c = 1$, $\gamma = 100°$

$$\frac{\sin\gamma}{c} = \frac{\sin\alpha}{a} \Rightarrow \frac{\sin 100°}{1} = \frac{\sin\alpha}{2} \Rightarrow \sin\alpha = \frac{2\sin 100°}{1} \approx 1.9696$$

There is no angle α for which $\sin\alpha > 1$. Therefore, there is no triangle with the given measurements.

35. $a = 2$, $c = 1$, $\gamma = 25°$

$$\frac{\sin\alpha}{a} = \frac{\sin\gamma}{c} \Rightarrow \frac{\sin\alpha}{2} = \frac{\sin 25°}{1} \Rightarrow \sin\alpha = \frac{2\sin 25°}{1} \approx 0.8452$$

$\alpha = \sin^{-1}(0.8452)$

$\alpha_1 = 57.7°$ or $\alpha_2 = 122.3°$

For both values, $\alpha + \gamma < 180°$. Therefore, there are two triangles.

$\beta_1 = 180° - \alpha_1 - \gamma = 180° - 57.7° - 25° = 97.3°$

$\dfrac{\sin\beta_1}{b_1} = \dfrac{\sin\gamma}{c} \Rightarrow \dfrac{\sin 97.3°}{b_1} = \dfrac{\sin 25°}{1} \Rightarrow b_1 = \dfrac{1\sin 97.3°}{\sin 25°} \approx 2.35$

$\beta_2 = 180° - \alpha_2 - \gamma = 180° - 122.3° - 25° = 32.7°$

$\dfrac{\sin\beta_2}{b_2} = \dfrac{\sin\gamma}{c} \Rightarrow \dfrac{\sin 32.7°}{b_2} = \dfrac{\sin 25°}{1} \Rightarrow b_2 = \dfrac{1\sin 32.7°}{\sin 25°} \approx 1.28$

Two triangles: $\alpha_1 \approx 57.7°$, $\beta_1 \approx 97.3°$, $b_1 \approx 2.35$
            or $\alpha_2 \approx 122.3°$, $\beta_2 \approx 32.7°$, $b_2 \approx 1.28$

37. (a) Find γ ; then use the Law of Sines:

$\gamma = 180° - 60° - 55° = 65°$

$\dfrac{\sin 55°}{a} = \dfrac{\sin 65°}{150} \Rightarrow a = \dfrac{150\sin 55°}{\sin 65°} \approx 135.58$ miles

$\dfrac{\sin 60°}{b} = \dfrac{\sin 65°}{150} \Rightarrow b = \dfrac{150\sin 60°}{\sin 65°} \approx 143.33$ miles

   (b) $t = \dfrac{a}{r} = \dfrac{135.6}{200} \approx 0.68$ hours or ≈ 41 minutes

39. $\angle CAB = 180° - 25° = 155°$ $\angle ABC = 180° - 155° - 15° = 10°$

Let c represent the distance from A to B.

$\dfrac{\sin 15°}{c} = \dfrac{\sin 10°}{1000} \Rightarrow c = \dfrac{1000\sin 15°}{\sin 10°} \approx 1490.48$

The length of the proposed ski lift is approximately 1490 feet.

41. Find the distance from B to the plane:

$\gamma = 180° - 40° - 35° = 105°$ $(\gamma = \angle APB)$

$\dfrac{\sin 40°}{x} = \dfrac{\sin 105°}{1000} \Rightarrow x = \dfrac{1000\sin 40°}{\sin 105°} \approx 665.46$ feet

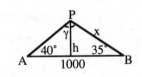

Find the height:

$\sin 35° = \dfrac{h}{x} = \dfrac{h}{665.46} \Rightarrow h = (665.46)\sin 35° \approx 381.69$ feet

The plane is about 381.69 feet high.

43. (a) $\angle ABC = 180° - 40° = 140°$

     Find the angle at city C:

$\dfrac{\sin C}{150} = \dfrac{\sin(140°)}{300} \Rightarrow \sin C = \dfrac{150\sin(140°)}{300} \approx 0.3214$

$C = \sin^{-1}(0.3214) \approx 18.7°$

Let y be the distance from city A to city C.

Find the angle at city A:
$$A = 180° - 140° - 18.7° = 21.3°$$

$$\frac{\sin 21.3°}{y} = \frac{\sin 140°}{300} \Rightarrow y = \frac{300 \sin 21.3°}{\sin 140°} \approx 169.54 \text{ miles}$$

The distance from city B to city C is approximately 169.54 miles.

(b)   To find the angle to turn, subtract angle C from 180°:
$$180° - 18.7° = 161.3°$$

The pilot needs to turn through an angle of 161.3° to return to city A.

45.   Find angle β ($\angle ACB$):
$$\frac{\sin \beta}{123} = \frac{\sin 60°}{184.5} \Rightarrow \sin \beta = \frac{123 \sin 60°}{184.5} \approx 0.5774$$

$$\beta = \sin^{-1}(0.5774) \approx 35.3°$$

$$\angle CAB = 180° - 60° - 35.3° \approx 84.7°$$

Find the perpendicular distance:
$$\sin 84.7° = \frac{h}{184.5} \Rightarrow h = 184.5 \sin 84.7° = 183.71 \text{ feet}$$

47. $\alpha = 180° - 140° = 40°$ $\beta = 180° - 135° = 45°$
$$\gamma = 180° - 40° - 45° = 95°$$

$$\frac{\sin 40°}{a} = \frac{\sin 95°}{2} \Rightarrow a = \frac{2 \sin 40°}{\sin 95°} \approx 1.290 \text{ mi}$$

$$\frac{\sin 45°}{b} = \frac{\sin 95°}{2} \Rightarrow b = \frac{2 \sin 45°}{\sin 95°} \approx 1.420 \text{ mi}$$

$$\overline{BE} = 1.290 - 0.125 = 1.165 \text{ mi}$$

$$\overline{AD} = 1.420 - 0.125 = 1.295 \text{ mi}$$

For the isosceles triangle,
$$\angle CDE = \angle CED = \frac{180° - 95°}{2} = 42.5°$$

$$\frac{\sin 95°}{DE} = \frac{\sin 42.5°}{0.125} \Rightarrow DE = \frac{0.125 \sin 95°}{\sin 42.5°} \approx 0.184 \text{ miles}$$

The approximate length of the highway is $1.165 + 1.295 + 0.184 = 2.64$ miles.

49.   Using the Law of Sines:
$$\frac{\sin 105°}{88} = \frac{\sin 25°}{L}$$

$$L = \frac{88 \sin 25°}{\sin 105°} \approx 38.50 \text{ inches}$$

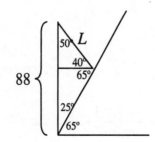

51.

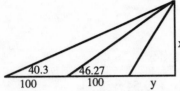

Using the Law of Sines twice yields two equations relating x and y.
Equation 1:

$$\frac{\sin 46.27°}{x} = \frac{\sin(90° - 46.27°)}{y + 100}$$

$$(y + 100)\sin 46.27° = x \sin 43.73°$$

$$y \sin 46.27° + 100 \sin 46.27° = x \sin 43.73°$$

$$y = \frac{x \sin 43.73° - 100 \sin 46.27°}{\sin 46.27°}$$

Equation 2:

$$\frac{\sin 40.3°}{x} = \frac{\sin(90° - 40.3°)}{y + 200}$$

$$(y + 200)\sin 40.3° = x \sin 49.7°$$

$$y \sin 40.3° + 200 \sin 40.3° = x \sin 49.7°$$

$$y = \frac{x \sin 49.7° - 200 \sin 40.3°}{\sin 40.3°}$$

Set the two equations equal to each other and solve for x.

$$\frac{x \sin 43.73° - 100 \sin 46.27°}{\sin 46.27°} = \frac{x \sin 49.7° - 200 \sin 40.3°}{\sin 40.3°}$$

$$x \sin 43.73° \cdot \sin 40.3° - 100 \sin 46.27° \cdot \sin 40.3° = x \sin 49.7° \cdot \sin 46.27° - 200 \sin 40.3° \cdot \sin 46.27°$$

$$x \sin 43.73° \cdot \sin 40.3° - x \sin 49.7° \cdot \sin 46.27° = 100 \sin 46.27° \cdot \sin 40.3° - 200 \sin 40.3° \cdot \sin 46.27°$$

$$x = \frac{100 \sin 46.27° \cdot \sin 40.3° - 200 \sin 40.3° \cdot \sin 46.27°}{\sin 43.73° \cdot \sin 40.3° - \sin 49.7° \cdot \sin 46.27°}$$

$$\approx 449.36 \text{ feet}$$

The current height of the pyramid is about 449.36 feet.

53.   Using the Law of Sines:

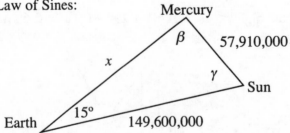

$$\frac{\sin 15°}{57,910,000} = \frac{\sin \beta}{149,600,000} \Rightarrow \sin \beta = \frac{149,600,000 \cdot \sin 15°}{57,910,000} = \frac{14,960 \cdot \sin 15°}{5791}$$

$$\beta = \sin^{-1}\left(\frac{14,960 \cdot \sin 15°}{5791}\right) \approx 41.96° \text{ or } \beta \approx 138.04°$$

$$\gamma \approx 180° - 41.96° - 15° = 123.04° \text{ or } \gamma \approx 180° - 138.04° - 15° = 26.96°$$

$$\frac{\sin 15°}{57,910,000} = \frac{\sin \gamma}{x} \Rightarrow x = \frac{57,910,000 \cdot \sin \gamma}{\sin 15°}$$

$$x = \frac{57,910,000 \cdot \sin 123.04°}{\sin 15°} \approx 187,564,951.5 \text{ km}$$

or $\quad x = \dfrac{57,910,000 \cdot \sin 26.96°}{\sin 15°} \approx 101,439,834.5 \text{ km}$

So the approximate possible distances between Earth and Mercury are 101,439,834.5 km and 187,564,951.5 km.

55.

Using the Law of Sines twice yields two equations relating x and h.

Equation 1:

$$\frac{\sin 30°}{h} = \frac{\sin 60°}{x}$$

$$x = \frac{h \sin 60°}{\sin 30°}$$

Equation 2:

$$\frac{\sin 20°}{h} = \frac{\sin 70°}{x + 40}$$

$$x = \frac{h \sin 70°}{\sin 20°} - 40$$

Set the two equations equal to each other and solve for h.

$$\frac{h \sin 60°}{\sin 30°} = \frac{h \sin 70°}{\sin 20°} - 40$$

$$h\left(\frac{\sin 60°}{\sin 30°} - \frac{\sin 70°}{\sin 20°}\right) = -40$$

$$h = \frac{-40}{\dfrac{\sin 60°}{\sin 30°} - \dfrac{\sin 70°}{\sin 20°}}$$

$$\approx 39.39 \text{ feet}$$

The height of the tree is about 39.39 feet.

57. Find the distance from B to the helicopter:

$$\gamma = 180° - 40° - 25° = 115° \qquad (\gamma = \angle APB)$$

$$\frac{\sin 40°}{x} = \frac{\sin 115°}{100} \Rightarrow x = \frac{100 \sin 40°}{\sin 115°} \approx 70.9 \text{ feet}$$

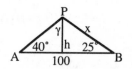

Find the height:

$$\sin 25° = \frac{h}{x} \approx \frac{h}{70.9} \Rightarrow h \approx (70.9)\sin 25° \approx 29.96 \text{ feet}$$

The helicopter is about 30 feet high.

59. $\dfrac{a-b}{c} = \dfrac{a}{c} - \dfrac{b}{c} = \dfrac{\sin\alpha}{\sin\gamma} - \dfrac{\sin\beta}{\sin\gamma} = \dfrac{\sin\alpha - \sin\beta}{\sin\gamma} = \dfrac{2\sin\left(\dfrac{\alpha-\beta}{2}\right)\cos\left(\dfrac{\alpha+\beta}{2}\right)}{\sin\left(2\cdot\dfrac{\gamma}{2}\right)}$

$= \dfrac{2\sin\left(\dfrac{\alpha-\beta}{2}\right)\cos\left(\dfrac{\alpha+\beta}{2}\right)}{2\sin\left(\dfrac{\gamma}{2}\right)\cos\left(\dfrac{\gamma}{2}\right)} = \dfrac{\sin\left(\dfrac{\alpha-\beta}{2}\right)\cos\left(\dfrac{\pi}{2}-\dfrac{\gamma}{2}\right)}{\sin\left(\dfrac{\gamma}{2}\right)\cos\left(\dfrac{\gamma}{2}\right)} = \dfrac{\sin\left(\dfrac{\alpha-\beta}{2}\right)\sin\left(\dfrac{\gamma}{2}\right)}{\sin\left(\dfrac{\gamma}{2}\right)\cos\left(\dfrac{\gamma}{2}\right)}$

$= \dfrac{\sin\left(\dfrac{\alpha-\beta}{2}\right)}{\cos\left(\dfrac{\gamma}{2}\right)}$

$= \dfrac{\sin\left[\dfrac{1}{2}(\alpha-\beta)\right]}{\cos\left(\dfrac{1}{2}\gamma\right)}$

61. Derive the Law of Tangents:

$\dfrac{a-b}{a+b} = \dfrac{\dfrac{a-b}{c}}{\dfrac{a+b}{c}} = \dfrac{\dfrac{\sin\left(\dfrac{1}{2}(\alpha-\beta)\right)}{\cos\left(\dfrac{1}{2}\gamma\right)}}{\dfrac{\cos\left(\dfrac{1}{2}(\alpha-\beta)\right)}{\sin\left(\dfrac{1}{2}\gamma\right)}} = \dfrac{\sin\left(\dfrac{1}{2}(\alpha-\beta)\right)}{\cos\left(\dfrac{1}{2}\gamma\right)} \cdot \dfrac{\sin\left(\dfrac{1}{2}\gamma\right)}{\cos\left(\dfrac{1}{2}(\alpha-\beta)\right)}$

$= \tan\left(\dfrac{1}{2}(\alpha-\beta)\right)\tan\left(\dfrac{1}{2}\gamma\right) = \tan\left(\dfrac{1}{2}(\alpha-\beta)\right)\tan\left(\dfrac{1}{2}(\pi-(\alpha+\beta))\right)$

$= \tan\left(\dfrac{1}{2}(\alpha-\beta)\right)\tan\left(\dfrac{\pi}{2}-\left(\dfrac{\alpha+\beta}{2}\right)\right) = \tan\left(\dfrac{1}{2}(\alpha-\beta)\right)\cot\left(\dfrac{\alpha+\beta}{2}\right)$

$= \dfrac{\tan\left(\dfrac{1}{2}(\alpha-\beta)\right)}{\tan\left(\dfrac{1}{2}(\alpha+\beta)\right)}$

63–65.   Answers will vary.

Applications of Trigonometric Functions

7.3 The Law of Cosines

1. $d = \sqrt{(x_2 - x_1)^2 + (y_2 - y_1)^2}$

3. Cosines

5. Cosines

7. False

9. $a = 2, \ c = 4, \ \beta = 45° \qquad b^2 = a^2 + c^2 - 2ac\cos\beta$

$b^2 = 2^2 + 4^2 - 2\cdot 2 \cdot 4\cos 45° = 20 - 16\cdot\dfrac{\sqrt{2}}{2} = 20 - 8\sqrt{2}$

$b \approx 2.95$

$a^2 = b^2 + c^2 - 2bc\cos\alpha \Rightarrow 2bc\cos\alpha = b^2 + c^2 - a^2 \Rightarrow \cos\alpha = \dfrac{b^2 + c^2 - a^2}{2bc}$

$\cos\alpha = \dfrac{2.95^2 + 4^2 - 2^2}{2(2.95)(4)} = \dfrac{20.7025}{23.6} \Rightarrow \alpha \approx 28.7°$

$\gamma = 180° - \alpha - \beta \approx 180° - 28.7° - 45° \approx 106.3°$

11. $a = 2, \ b = 3, \ \gamma = 95° \qquad c^2 = a^2 + b^2 - 2ab\cos\gamma$

$c^2 = 2^2 + 3^2 - 2\cdot 2 \cdot 3\cos 95° = 13 - 12\cos 95°$

$c \approx 3.75$

$a^2 = b^2 + c^2 - 2bc\cos\alpha \Rightarrow \cos\alpha = \dfrac{b^2 + c^2 - a^2}{2bc} = \dfrac{3^2 + 3.75^2 - 2^2}{2(3)(3.75)} = \dfrac{19.0625}{22.5} \Rightarrow \alpha \approx 32.1°$

$\beta = 180° - \alpha - \gamma \approx 180° - 32.1° - 95° = 52.9°$

13. $a = 6, \ b = 5, \ c = 8$

$a^2 = b^2 + c^2 - 2bc\cos\alpha \Rightarrow \cos\alpha = \dfrac{b^2 + c^2 - a^2}{2bc} = \dfrac{5^2 + 8^2 - 6^2}{2(5)(8)} = \dfrac{53}{80} \Rightarrow \alpha \approx 48.5°$

$b^2 = a^2 + c^2 - 2ac\cos\beta \Rightarrow \cos\beta = \dfrac{a^2 + c^2 - b^2}{2ac} = \dfrac{6^2 + 8^2 - 5^2}{2(6)(8)} = \dfrac{75}{96} \Rightarrow \beta \approx 38.6°$

$\gamma = 180° - \alpha - \beta \approx 180° - 48.5° - 38.6° \approx 92.9°$

15. $a = 9$, $b = 6$, $c = 4$

$$a^2 = b^2 + c^2 - 2bc\cos\alpha \Rightarrow \cos\alpha = \frac{b^2 + c^2 - a^2}{2bc} = \frac{6^2 + 4^2 - 9^2}{2(6)(4)} = -\frac{29}{48} \Rightarrow \alpha \approx 127.2°$$

$$b^2 = a^2 + c^2 - 2ac\cos\beta \Rightarrow \cos\beta = \frac{a^2 + c^2 - b^2}{2ac} = \frac{9^2 + 4^2 - 6^2}{2(9)(4)} = \frac{61}{72} \Rightarrow \beta \approx 32.1°$$

$$\gamma = 180° - \alpha - \beta \approx 180° - 127.2° - 32.1° = 20.7°$$

17. $a = 3$, $b = 4$, $\gamma = 40°$
$$c^2 = a^2 + b^2 - 2ab\cos\gamma$$
$$c^2 = 3^2 + 4^2 - 2\cdot 3\cdot 4\cos 40° = 25 - 24\cos 40°$$

$$c \approx 2.57$$

$$a^2 = b^2 + c^2 - 2bc\cos\alpha \Rightarrow \cos\alpha = \frac{b^2 + c^2 - a^2}{2bc} = \frac{4^2 + 2.57^2 - 3^2}{2(4)(2.57)} = \frac{13.6049}{20.56} \Rightarrow \alpha \approx 48.6°$$

$$\beta = 180° - \alpha - \gamma \approx 180° - 48.6° - 40° = 91.4°$$

19. $b = 1$, $c = 3$, $\alpha = 80°$
$$a^2 = b^2 + c^2 - 2bc\cos\alpha$$

$$a^2 = 1^2 + 3^2 - 2\cdot 1\cdot 3\cos 80° = 10 - 6\cos 80°$$

$$a \approx 2.99$$

$$b^2 = a^2 + c^2 - 2ac\cos\beta \Rightarrow \cos\beta = \frac{a^2 + c^2 - b^2}{2ac} = \frac{2.99^2 + 3^2 - 1^2}{2(2.99)(3)} = \frac{16.9401}{17.94} \Rightarrow \beta \approx 19.2°$$

$$\gamma = 180° - \alpha - \beta \approx 180° - 80° - 19.2° = 80.8°$$

21. $a = 3$, $c = 2$, $\beta = 110°$
$$b^2 = a^2 + c^2 - 2ac\cos\beta$$

$$b^2 = 3^2 + 2^2 - 2\cdot 3\cdot 2\cos 110° = 13 - 12\cos 110°$$

$$b \approx 4.14$$

$$c^2 = a^2 + b^2 - 2ab\cos\gamma \Rightarrow \cos\gamma = \frac{a^2 + b^2 - c^2}{2ab} = \frac{3^2 + 4.14^2 - 2^2}{2(3)(4.14)} = \frac{22.1396}{24.84} \Rightarrow \gamma \approx 27.0°$$

$$\alpha = 180° - \beta - \gamma \approx 180° - 110° - 27.0° = 43.0°$$

23. $a = 2$, $b = 2$, $\gamma = 50°$
$$c^2 = a^2 + b^2 - 2ab\cos\gamma$$

$$c^2 = 2^2 + 2^2 - 2\cdot 2\cdot 2\cos 50° = 8 - 8\cos 50°$$

$$c \approx 1.69$$

$$a^2 = b^2 + c^2 - 2bc\cos\alpha \Rightarrow \cos\alpha = \frac{b^2 + c^2 - a^2}{2bc} = \frac{2^2 + 1.69^2 - 2^2}{2(2)(1.69)} = \frac{2.8561}{6.76} \Rightarrow \alpha \approx 65.0°$$

$$\beta = 180° - \alpha - \gamma \approx 180° - 65.0° - 50° = 65.0°$$

25. $a = 12$, $b = 13$, $c = 5$

$$a^2 = b^2 + c^2 - 2bc\cos\alpha \Rightarrow \cos\alpha = \frac{b^2 + c^2 - a^2}{2bc} = \frac{13^2 + 5^2 - 12^2}{2(13)(5)} = \frac{50}{130} \Rightarrow \alpha \approx 67.4°$$

$$b^2 = a^2 + c^2 - 2ac\cos\beta \Rightarrow \cos\beta = \frac{a^2 + c^2 - b^2}{2ac} = \frac{12^2 + 5^2 - 13^2}{2(12)(5)} = 0 \Rightarrow \beta = 90°$$

$$\gamma = 180° - \alpha - \beta \approx 180° - 67.4° - 90° = 22.6°$$

27. $a = 2$, $b = 2$, $c = 2$

$$a^2 = b^2 + c^2 - 2bc\cos\alpha \Rightarrow \cos\alpha = \frac{b^2 + c^2 - a^2}{2bc} = \frac{2^2 + 2^2 - 2^2}{2(2)(2)} = 0.5 \Rightarrow \alpha = 60°$$

$$b^2 = a^2 + c^2 - 2ac\cos\beta \Rightarrow \cos\beta = \frac{a^2 + c^2 - b^2}{2ac} = \frac{2^2 + 2^2 - 2^2}{2(2)(2)} = 0.5 \Rightarrow \beta = 60°$$

$$\gamma = 180° - \alpha - \beta \approx 180° - 60° - 60° = 60°$$

29. $a = 5$, $b = 8$, $c = 9$

$$a^2 = b^2 + c^2 - 2bc\cos\alpha \Rightarrow \cos\alpha = \frac{b^2 + c^2 - a^2}{2bc} = \frac{8^2 + 9^2 - 5^2}{2(8)(9)} = \frac{120}{144} \Rightarrow \alpha \approx 33.6°$$

$$b^2 = a^2 + c^2 - 2ac\cos\beta \Rightarrow \cos\beta = \frac{a^2 + c^2 - b^2}{2ac} = \frac{5^2 + 9^2 - 8^2}{2(5)(9)} = \frac{42}{90} \Rightarrow \beta \approx 62.2°$$

$$\gamma = 180° - \alpha - \beta \approx 180° - 33.6° - 62.2° = 84.2°$$

31. $a = 10$, $b = 8$, $c = 5$

$$a^2 = b^2 + c^2 - 2bc\cos\alpha \Rightarrow \cos\alpha = \frac{b^2 + c^2 - a^2}{2bc} = \frac{8^2 + 5^2 - 10^2}{2(8)(5)} \approx -\frac{11}{80} \Rightarrow \alpha \approx 97.9°$$

$$b^2 = a^2 + c^2 - 2ac\cos\beta \Rightarrow \cos\beta = \frac{a^2 + c^2 - b^2}{2ac} = \frac{10^2 + 5^2 - 8^2}{2(10)(5)} = \frac{61}{100} \Rightarrow \beta \approx 52.4°$$

$$\gamma = 180° - \alpha - \beta \approx 180° - 97.9° - 52.4° = 29.7°$$

33. Find the third side of the triangle using the Law of Cosines:
$a = 50$, $b = 70$, $\gamma = 70°$
$c^2 = a^2 + b^2 - 2ab\cos\gamma = 50^2 + 70^2 - 2 \cdot 50 \cdot 70\cos 70° = 7400 - 7000\cos 70° \Rightarrow c \approx 70.75$
The houses are approximately 70.75 feet apart.

35. After 10 hours the ship will have traveled 150 nautical miles along its altered course.
Use the Law of Cosines to find the distance from Barbados on the new course.
$a = 600$, $b = 150$, $\gamma = 20°$
$c^2 = a^2 + b^2 - 2ab\cos\gamma = 600^2 + 150^2 - 2 \cdot 600 \cdot 150\cos 20° = 382{,}500 - 180{,}000\cos 20°$

$c \approx 461.9$ nautical miles

(a)　Use the Law of Cosines to find the angle opposite the side of 600:

$$\cos\alpha = \frac{b^2 + c^2 - a^2}{2bc}$$

$$\cos\alpha = \frac{150^2 + 461.9^2 - 600^2}{2(150)(461.9)} = -\frac{124{,}148.39}{138{,}570} \Rightarrow \alpha \approx 153.6°$$

The captain needs to turn the ship through an angle of $180° - 153.6° = 26.4°$.

(b)　$t = \dfrac{461.9}{15} \approx 30.8$ hours are required for the second leg of the trip. The total time for the trip will be 40.8 hours.

37.　(a)　Find x in the figure:

$$x^2 = 60.5^2 + 90^2 - 2(60.5)90\cos 45°$$

$$x \approx 63.7 \text{ feet}$$

It is about 63.7 feet from the pitching rubber to first base.

(b)　Use the Pythagorean Theorem to find y in the figure:

$$90^2 + 90^2 = (60.5 + y)^2 \Rightarrow 8100 + 8100 = (60.5 + y)^2$$

$$16{,}200 = (60.5 + y)^2 \Rightarrow 60.5 + y \approx 127.3 \Rightarrow y \approx 66.8 \text{ feet}$$

It is about 66.8 feet from the pitching rubber to second base.

(c)　Find β in the figure by using the Law of Cosines:

$$\cos\beta = \frac{60.5^2 + 63.7^2 - 90^2}{2(60.5)(63.7)} = -\frac{382.06}{7707.7} \Rightarrow \beta \approx 92.8°$$

The pitcher needs to turn through an angle of 92.8° to face first base.

39.　(a)　Find x by using the Law of Cosines:

$$x^2 = 500^2 + 100^2 - 2(500)100\cos 80°$$

$$= 260{,}000 - 100{,}000\cos 80°$$

$$x \approx 492.6 \text{ feet}$$

The guy wire needs to be about 492.6 feet long.

(b)　Use the Pythagorean Theorem to find the value of y:

$$y^2 = 100^2 + 250^2 = 72{,}500$$

$$y = 269.3 \text{ feet}$$

The guy wire needs to be about 269.3 feet long.

41.　Find x by using the Law of Cosines:

$$x^2 = 400^2 + 90^2 - 2(400)90\cos(45°)$$

$$= 168{,}100 - 36{,}000\sqrt{2}$$

$$x \approx 342.33 \text{ feet}$$

It is approximately 342.33 feet from dead center to third base.

43. Use the Law of Cosines:

$$L^2 = x^2 + r^2 - 2xr\cos\theta$$

$$x^2 - 2xr\cos\theta + r^2 - L^2 = 0$$

$$x = \frac{2r\cos\theta + \sqrt{(2r\cos\theta)^2 - 4(1)(r^2 - L^2)}}{2(1)}$$

$$x = \frac{2r\cos\theta + \sqrt{4r^2\cos^2\theta - 4(r^2 - L^2)}}{2}$$

$$x = r\cos\theta + \sqrt{r^2\cos^2\theta + L^2 - r^2}$$

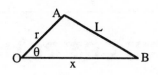

45. $$\cos\frac{\gamma}{2} = \sqrt{\frac{1 + \cos\gamma}{2}} = \sqrt{\frac{1 + \dfrac{a^2 + b^2 - c^2}{2ab}}{2}} = \sqrt{\frac{2ab + a^2 + b^2 - c^2}{4ab}} = \sqrt{\frac{(a + b)^2 - c^2}{4ab}}$$

$$= \sqrt{\frac{(a + b + c)(a + b - c)}{4ab}} = \sqrt{\frac{2s(2s - c - c)}{4ab}} = \sqrt{\frac{4s(s - c)}{4ab}} = \sqrt{\frac{s(s - c)}{ab}}$$

47. $$\frac{\cos\alpha}{a} + \frac{\cos\beta}{b} + \frac{\cos\gamma}{c} = \frac{b^2 + c^2 - a^2}{2bca} + \frac{a^2 + c^2 - b^2}{2acb} + \frac{a^2 + b^2 - c^2}{2abc}$$

$$= \frac{b^2 + c^2 - a^2 + a^2 + c^2 - b^2 + a^2 + b^2 - c^2}{2abc} = \frac{a^2 + b^2 + c^2}{2abc}$$

49–51. Answers will vary.

Applications of Trigonometric Functions

7.4   The Area of a Triangle

1. $\dfrac{1}{2}bh$

3. False

5. $a = 2,\ c = 4,\ \beta = 45°$

$A = \dfrac{1}{2}ac\sin\beta = \dfrac{1}{2}(2)(4)\sin 45° \approx 2.83$

7. $a = 2,\ b = 3,\ \gamma = 95°$

$A = \dfrac{1}{2}ab\sin\gamma = \dfrac{1}{2}(2)(3)\sin 95° \approx 2.99$

9. $a = 6,\ b = 5,\ c = 8$

$s = \dfrac{1}{2}(a+b+c) = \dfrac{1}{2}(6+5+8) = \dfrac{19}{2}$

$A = \sqrt{s(s-a)(s-b)(s-c)} = \sqrt{\left(\dfrac{19}{2}\right)\left(\dfrac{7}{2}\right)\left(\dfrac{9}{2}\right)\left(\dfrac{3}{2}\right)} = \sqrt{\dfrac{3591}{16}} \approx 14.98$

11. $a = 9,\ b = 6,\ c = 4$

$s = \dfrac{1}{2}(a+b+c) = \dfrac{1}{2}(9+6+4) = \dfrac{19}{2}$

$A = \sqrt{s(s-a)(s-b)(s-c)} = \sqrt{\left(\dfrac{19}{2}\right)\left(\dfrac{1}{2}\right)\left(\dfrac{7}{2}\right)\left(\dfrac{11}{2}\right)} = \sqrt{\dfrac{1463}{16}} \approx 9.56$

13. $a = 3,\ b = 4,\ \gamma = 40°$

$A = \dfrac{1}{2}ab\sin\gamma = \dfrac{1}{2}(3)(4)\sin 40° \approx 3.86$

15. $b = 1,\ c = 3,\ \alpha = 80°$

$A = \dfrac{1}{2}bc\sin\alpha = \dfrac{1}{2}(1)(3)\sin 80° \approx 1.48$

17. $a = 3,\ c = 2,\ \beta = 110°$

$A = \dfrac{1}{2}ac\sin\beta = \dfrac{1}{2}(3)(2)\sin 110° \approx 2.82$

19. $a = 12$, $b = 13$, $c = 5$

$s = \dfrac{1}{2}(a+b+c) = \dfrac{1}{2}(12+13+5) = 15$

$A = \sqrt{s(s-a)(s-b)(s-c)} = \sqrt{(15)(3)(2)(10)} = \sqrt{900} = 30$

21. $a = 2$, $b = 2$, $c = 2$

$s = \dfrac{1}{2}(a+b+c) = \dfrac{1}{2}(2+2+2) = 3$

$A = \sqrt{s(s-a)(s-b)(s-c)} = \sqrt{(3)(1)(1)(1)} = \sqrt{3} \approx 1.73$

23. $a = 5$, $b = 8$, $c = 9$

$s = \dfrac{1}{2}(a+b+c) = \dfrac{1}{2}(5+8+9) = 11$

$A = \sqrt{s(s-a)(s-b)(s-c)} = \sqrt{(11)(6)(3)(2)} = \sqrt{396} \approx 19.90$

25. Use the Law of Sines in the area of the triangle formula:

$A = \dfrac{1}{2}ab\sin\gamma = \dfrac{1}{2}a\sin\gamma\left(\dfrac{a\sin\beta}{\sin\alpha}\right) = \dfrac{a^2\sin\beta\sin\gamma}{2\sin\alpha}$

27. $\alpha = 40°$, $\beta = 20°$, $a = 2$ $\gamma = 180° - \alpha - \beta = 180° - 40° - 20° = 120°$

$A = \dfrac{a^2\sin\beta\sin\gamma}{2\sin\alpha} = \dfrac{2^2\sin 20° \cdot \sin 120°}{2\sin 40°} \approx 0.92$

29. $\beta = 70°$, $\gamma = 10°$, $b = 5$ $\alpha = 180° - \beta - \gamma = 180° - 70° - 10° = 100°$

$A = \dfrac{b^2\sin\alpha\sin\gamma}{2\sin\beta} = \dfrac{5^2\sin 100° \cdot \sin 10°}{2\sin 70°} \approx 2.27$

31. $\alpha = 110°$, $\gamma = 30°$, $c = 3$ $\beta = 180° - \alpha - \gamma = 180° - 110° - 30° = 40°$

$A = \dfrac{c^2\sin\alpha\sin\beta}{2\sin\gamma} = \dfrac{3^2\sin 110° \cdot \sin 40°}{2\sin 30°} \approx 5.44$

33. Area of a sector $= \dfrac{1}{2}r^2\theta$ where θ is in radians.

$\theta = 70 \cdot \dfrac{\pi}{180} = \dfrac{7\pi}{18}$

Area of the sector $= \dfrac{1}{2} \cdot 8^2 \cdot \dfrac{7\pi}{18} = \dfrac{112\pi}{9}$ square feet

Area of the triangle $= \dfrac{1}{2} \cdot 8 \cdot 8\sin 70° = 32\sin 70°$ square feet

Area of the segment $= \dfrac{112\pi}{9} - 32\sin 70° \approx 9.03$ square feet

35. Find the area of the lot using Heron's Formula:
     $a = 100, \ b = 50, \ c = 75$

$$s = \frac{1}{2}(a+b+c) = \frac{1}{2}(100+50+75) = \frac{225}{2}$$

$$A = \sqrt{s(s-a)(s-b)(s-c)} = \sqrt{\left(\frac{225}{2}\right)\left(\frac{25}{2}\right)\left(\frac{125}{2}\right)\left(\frac{75}{2}\right)} = \sqrt{\frac{52,734,375}{16}} \approx 1815.46$$

     Cost $= (\$3)(1815.46) = \$5446.38$

37. The area of the shaded region = the area of the semicircle – the area of the triangle.

Area of the semicircle $= \frac{1}{2}\pi r^2 = \frac{1}{2}\pi(4)^2 = 8\pi$ square centimeters

The triangle is a right triangle. Find the other leg:

$6^2 + b^2 = 8^2 \Rightarrow b^2 = 64 - 36 = 28 \Rightarrow b = \sqrt{28} = 2\sqrt{7}$

Area of the triangle $= \frac{1}{2}\cdot 6\cdot 2\sqrt{7} = 6\sqrt{7}$ square centimeters

Area of the shaded region $= 8\pi - 6\sqrt{7} \approx 9.26$ square centimeters

39. The area is the sum of the area of a triangle and a sector.

Area of the triangle $= \frac{1}{2}r\cdot r\sin(\pi - \theta) = \frac{1}{2}r^2 \sin(\pi - \theta)$

Area of the sector $= \frac{1}{2}r^2\theta$

$$A = \frac{1}{2}r^2 \sin(\pi - \theta) + \frac{1}{2}r^2\theta = \frac{1}{2}r^2(\sin(\pi - \theta) + \theta)$$

$$= \frac{1}{2}r^2(\sin\pi\cos\theta - \cos\pi\sin\theta + \theta)$$

$$= \frac{1}{2}r^2(0 + \sin\theta + \theta)$$

$$= \frac{1}{2}r^2(\theta + \sin\theta)$$

41. $\sin\theta = \frac{y}{1} = y; \qquad \cos\theta = \frac{x}{1} = x$

     (a) $A = 2xy = 2\cos\theta\sin\theta$

     (b) $2\cos\theta\sin\theta = 2\sin\theta\cos\theta = \sin(2\theta)$

     (c)   The largest value of the sine function is 1. Solve:
           $\sin(2\theta) = 1$

           $2\theta = \frac{\pi}{2} \Rightarrow \theta = \frac{\pi}{4} = 45°$

     (d) $x = \cos\frac{\pi}{4} = \frac{\sqrt{2}}{2}; \qquad y = \sin\frac{\pi}{4} = \frac{\sqrt{2}}{2}$

           The dimensions are $\frac{\sqrt{2}}{2}$ by $\sqrt{2}$.

43. (a) Area $\Delta OAC = \frac{1}{2}|OC| \cdot |AC| = \frac{1}{2} \cdot \frac{|OC|}{1} \cdot \frac{|AC|}{1} = \frac{1}{2}\cos\alpha\sin\alpha = \frac{1}{2}\sin\alpha\cos\alpha$

(b) Area $\Delta OCB = \frac{1}{2}|OC| \cdot |BC| = \frac{1}{2} \cdot |OB|^2 \cdot \frac{|OC|}{|OB|} \cdot \frac{|BC|}{|OB|} = \frac{1}{2}|OB|^2\cos\beta\sin\beta$

$\qquad\qquad = \frac{1}{2}|OB|^2\sin\beta\cos\beta$

(c) Area $\Delta OAB = \frac{1}{2}|BD| \cdot |OA| = \frac{1}{2}|BD| \cdot 1 = \frac{1}{2} \cdot |OB| \cdot \frac{|BD|}{|OB|} = \frac{1}{2}|OB|\sin(\alpha + \beta)$

(d) $\dfrac{\cos\alpha}{\cos\beta} = \dfrac{\frac{|OC|}{|OA|}}{\frac{|OC|}{|OB|}} = \dfrac{|OC|}{1} \cdot \dfrac{|OB|}{|OC|} = |OB|$

(e) Area ΔOAB = Area ΔOAC + Area ΔOCB

$\frac{1}{2}|OB|\sin(\alpha + \beta) = \frac{1}{2}\sin\alpha\cos\alpha + \frac{1}{2}|OB|^2\sin\beta\cos\beta$

$\dfrac{\cos\alpha}{\cos\beta}\sin(\alpha + \beta) = \sin\alpha\cos\alpha + \dfrac{\cos^2\alpha}{\cos^2\beta}\sin\beta\cos\beta$

$\sin(\alpha + \beta) = \dfrac{\cos\beta}{\cos\alpha}\sin\alpha\cos\alpha + \dfrac{\cos\alpha}{\cos\beta}\sin\beta\cos\beta$

$\sin(\alpha + \beta) = \sin\alpha\cos\beta + \cos\alpha\sin\beta$

45. The grazing area must be considered in sections. A_1 represents $\frac{3}{4}$ of a circle:

$A_1 = \frac{3}{4}\pi(100)^2 = 7500\pi \approx 23{,}562$ square feet

Angles are needed to find A_2 and A_3: (see the figure)
In $\Delta ABC, \angle CBA = 45°, AB = 10, AC = 90$

Find $\angle BCA$:

$\dfrac{\sin\angle CBA}{90} = \dfrac{\sin\angle BCA}{10} \Rightarrow \dfrac{\sin 45°}{90} = \dfrac{\sin\angle BCA}{10}$

$\sin\angle BCA = \dfrac{10\sin 45°}{90} \approx 0.0786$

$\quad \angle BCA \approx 4.5°$

$m\angle BAC = 180° - 45° - 4.5° = 130.5°$

$m\angle DAC = 130.5° - 90° = 40.5°$

Area of $A_3 = \frac{1}{2}(10)(90)\sin 40.5° \approx 292$ square feet

The angle for sector A_2 is $90° - 40.5° = 49.5°$.

Area of sector $A_2 = \frac{1}{2}(90)^2\left((49.5)\cdot\frac{\pi}{180}\right) \approx 3499$ square feet

Since the cow can go in either direction around the barn, A_2 must be doubled. Total grazing area is: $23,562 + 2(3499) + 292 = 30,852$ square feet.

47. $K = \frac{1}{2}h_1 a$, so $h_1 = \frac{2K}{a}$, similarly $h_2 = \frac{2K}{b}$ and $h_3 = \frac{2K}{c}$.

$$\frac{1}{h_1} + \frac{1}{h_2} + \frac{1}{h_3} = \frac{a}{2K} + \frac{b}{2K} + \frac{c}{2K} = \frac{a+b+c}{2K} = \frac{2s}{2K} = \frac{s}{K}$$

49. $h = \frac{a\sin\beta\sin\gamma}{\sin\alpha}$ where h is the altitude to side a.

In $\triangle OAB$, c is opposite angle AOB. The two adjacent angles are $\frac{\alpha}{2}$ and $\frac{\beta}{2}$.

Then $r = \dfrac{c\sin\frac{\alpha}{2}\sin\frac{\beta}{2}}{\sin(\angle AOB)}$.

$$\angle AOB = \pi - \left(\frac{\alpha}{2} + \frac{\beta}{2}\right)$$

$$\sin(\angle AOB) = \sin\left(\pi - \left(\frac{\alpha}{2} + \frac{\beta}{2}\right)\right) = \sin\left(\frac{\alpha}{2} + \frac{\beta}{2}\right) = \sin\left(\frac{\alpha+\beta}{2}\right) = \cos\left(\frac{\pi}{2} - \left(\frac{\alpha+\beta}{2}\right)\right)$$

$$= \cos\left(\frac{\pi - (\alpha+\beta)}{2}\right) = \cos\frac{\gamma}{2}$$

Thus, $r = \dfrac{c\sin\frac{\alpha}{2}\sin\frac{\beta}{2}}{\cos\frac{\gamma}{2}}$.

51. Use the result of Problem 50:
$$\cot\frac{\alpha}{2} + \cot\frac{\beta}{2} + \cot\frac{\gamma}{2} = \frac{s-a}{r} + \frac{s-b}{r} + \frac{s-c}{r} = \frac{s-a+s-b+s-c}{r}$$
$$= \frac{3s-(a+b+c)}{r} = \frac{3s-2s}{r} = \frac{s}{r}$$

53. Answers will vary.

Chapter 7

Applications of Trigonometric Functions

7.5   Simple Harmonic Motion; Damped Motion; Combining Waves

1. $|5| = 5$; $\dfrac{2\pi}{4} = \dfrac{\pi}{2}$

3.   simple harmonic motion; damped motion

5. $d = -5\cos(\pi t)$

7. $d = -6\cos(2t)$

9. $d = -5\sin(\pi t)$

11. $d = -6\sin(2t)$

13. $d = 5\sin(3t)$
    (a)   Simple harmonic
    (b)   5 meters
    (c) $\dfrac{2\pi}{3}$ seconds
    (d) $\dfrac{3}{2\pi}$ oscillation/second

15. $d = 6\cos(\pi t)$
    (a)   Simple harmonic
    (b)   6 meters
    (c)   2 seconds
    (d) $\dfrac{1}{2}$ oscillation/second

17. $d = -3\sin\left(\dfrac{1}{2}t\right)$

    (a)   Simple harmonic
    (b)   3 meters
    (c) 4π seconds
    (d) $\dfrac{1}{4\pi}$ oscillation/second

19. $d = 6 + 2\cos(2\pi t)$
    (a)   Simple harmonic
    (b)   2 meters
    (c)   1 second
    (d)   1 oscillation/second

21. $d(t) = e^{-t/\pi}\cos(2t)$, $0 \le t \le 2\pi$

23. $d(t) = e^{-t/2\pi}\cos(t)$, $0 \le t \le 2\pi$

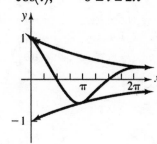

25. $f(x) = x + \cos x$

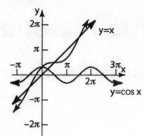

27. $f(x) = x - \sin x$

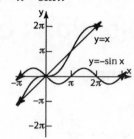

29. $f(x) = \sin x + \cos x$

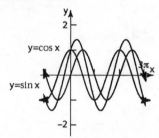

31. $f(x) = \sin x + \sin(2x)$

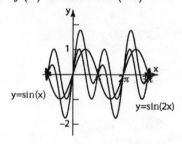

33.   (a)   Graph:

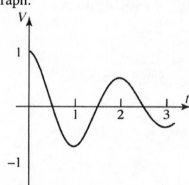

(b)   The graph of V touches the graph of $y = e^{-t/3}$ when $t = 0, 2$.
The graph of V touches the graph of $y = -e^{-t/3}$ when $t = 1, 3$.

(c)   To solve the inequality $-0.4 < e^{-t/3} \cdot \cos(\pi t) < 0.4$ on the interval $0 \le t \le 3$, we consider the graphs of $y = -0.4$; $y = e^{-t/3} \cdot \cos(\pi t)$; and $y = 0.4$.

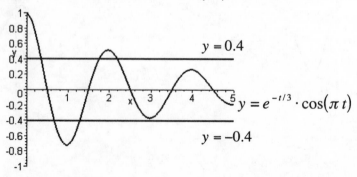

On the interval $0 \leq t \leq 3$, we can use the INTERSECT feature on a calculator to determine that $y = e^{-t/3} \cdot \cos(\pi t)$ intersects $y = 0.4$ when $t \approx 0.35$, $t \approx 1.75$, and $t \approx 2.19$, $y = e^{-t/3} \cdot \cos(\pi t)$ intersects $y = -0.4$ when $t \approx 0.67$ and $t \approx 1.29$ and the graph shows that $-0.4 < e^{-t/3} \cdot \cos(\pi t) < 0.4$ when $t = 3$.

Therefore, the voltage V is between -0.4 and 0.4 on the intervals $0.35 < t < 0.67$, $1.29 < t < 1.75$ and $2.19 < t \leq 3$.

35. $y = \sin(2\pi(852)t) + \sin(2\pi(1209)t)$

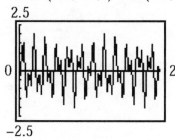

37.   Graph $f(x) = \dfrac{\sin x}{x}$:

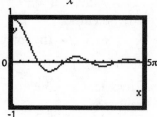

As x approaches 0, $\dfrac{\sin x}{x}$ approaches 1.

39.   Graphing:

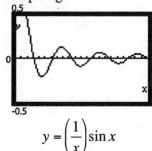

$$y = \left(\frac{1}{x}\right)\sin x$$

$$y = \left(\frac{1}{x^2}\right)\sin x$$

$$y = \left(\frac{1}{x^3}\right)\sin x$$

As x gets larger, the graph of $y = \left(\dfrac{1}{x^n}\right)\sin x$ gets closer to $y = 0$.

41.   Answers will vary.

435

Applications of Trigonometric Functions

7.R Chapter Review

1. opposite = 4; adjacent = 3
    Find the hypotenuse:
    $$4^2 + 3^2 = (\text{hypotenuse})^2$$
    $$(\text{hypotenuse})^2 = 16 + 9 = 25$$
    $$\text{hypotenuse} = 5$$

    $\sin\theta = \dfrac{\text{opp}}{\text{hyp}} = \dfrac{4}{5}$ $\qquad \cos\theta = \dfrac{\text{adj}}{\text{hyp}} = \dfrac{3}{5}$ $\qquad \tan\theta = \dfrac{\text{opp}}{\text{adj}} = \dfrac{4}{3}$

    $\csc\theta = \dfrac{\text{hyp}}{\text{opp}} = \dfrac{5}{4}$ $\qquad \sec\theta = \dfrac{\text{hyp}}{\text{adj}} = \dfrac{5}{3}$ $\qquad \cot\theta = \dfrac{\text{adj}}{\text{opp}} = \dfrac{3}{4}$

3. adjacent = 2; hypotenuse = 4
    Find the opposite side:
    $$(\text{opposite})^2 + 2^2 = 4^2$$
    $$(\text{opposite})^2 = 16 - 4 = 12$$
    $$\text{opposite} = \sqrt{12} = 2\sqrt{3}$$

    $\sin\theta = \dfrac{\text{opp}}{\text{hyp}} = \dfrac{2\sqrt{3}}{4} = \dfrac{\sqrt{3}}{2}$ $\qquad \cos\theta = \dfrac{\text{adj}}{\text{hyp}} = \dfrac{2}{4} = \dfrac{1}{2}$ $\qquad \tan\theta = \dfrac{\text{opp}}{\text{adj}} = \dfrac{2\sqrt{3}}{2} = \sqrt{3}$

    $\csc\theta = \dfrac{\text{hyp}}{\text{opp}} = \dfrac{4}{2\sqrt{3}} \dfrac{\sqrt{3}}{\sqrt{3}} = \dfrac{2\sqrt{3}}{3}$ $\qquad \sec\theta = \dfrac{\text{hyp}}{\text{adj}} = \dfrac{4}{2} = 2$ $\qquad \cot\theta = \dfrac{\text{adj}}{\text{opp}} = \dfrac{2}{2\sqrt{3}} \dfrac{\sqrt{3}}{\sqrt{3}} = \dfrac{\sqrt{3}}{3}$

5. $\cos 62° - \sin 28° = \cos(90° - 28°) - \sin 28° = \sin 28° - \sin 28° = 0$

7. $\dfrac{\sec 55°}{\csc 35°} = \dfrac{\sec 55°}{\csc(90° - 55°)} = \dfrac{\sec 55°}{\sec 55°} = 1$

9. $\cos^2 40° + \cos^2 50° = \cos^2 40° + \cos^2(90° - 40°) = \cos^2 40° + \sin^2 40° = 1$

11. $c = 10, \quad \beta = 20°$

    $$\sin\beta = \frac{b}{c} \Rightarrow \sin 20° = \frac{b}{10} \Rightarrow b = 10\sin 20° \approx 3.42$$

    $$\cos\beta = \frac{a}{c} \Rightarrow \cos 20° = \frac{a}{10} \Rightarrow a = 10\cos 20° \approx 9.40$$

    $$\alpha = 90° - \beta = 90° - 20° = 70°$$

13. $b = 2, \ c = 5$

$$c^2 = a^2 + b^2$$

$$a^2 = c^2 - b^2 = 5^2 - 2^2 = 25 - 4 = 21 \Rightarrow a = \sqrt{21} \approx 4.58$$

$$\sin\beta = \frac{b}{c} = \frac{2}{5} \Rightarrow \beta \approx 23.6°$$

$$\alpha = 90° - \beta \approx 90° - 23.6° = 66.4°$$

15. $\alpha = 50°, \ \beta = 30°, \ a = 1$

$$\gamma = 180° - \alpha - \beta = 180° - 50° - 30° = 100°$$

$$\frac{\sin\alpha}{a} = \frac{\sin\beta}{b} \Rightarrow \frac{\sin 50°}{1} = \frac{\sin 30°}{b} \Rightarrow b = \frac{1\sin 30°}{\sin 50°} \approx 0.65$$

$$\frac{\sin\gamma}{c} = \frac{\sin\alpha}{a} \Rightarrow \frac{\sin 100°}{c} = \frac{\sin 50°}{1} \Rightarrow c = \frac{1\sin 100°}{\sin 50°} \approx 1.29$$

17. $\alpha = 100°, \ c = 2, \ a = 5$

$$\frac{\sin\gamma}{c} = \frac{\sin\alpha}{a} \Rightarrow \frac{\sin\gamma}{2} = \frac{\sin 100°}{5} \Rightarrow \sin\gamma = \frac{2\sin 100°}{5} \approx 0.3939$$

$$\gamma \approx 23.2° \ \text{ or } \ \gamma \approx 156.8°$$

The second value is discarded because $\alpha + \gamma > 180°$.

$$\beta = 180° - \alpha - \gamma \approx 180° - 100° - 23.2° = 56.8°$$

$$\frac{\sin\beta}{b} = \frac{\sin\alpha}{a} \Rightarrow \frac{\sin 56.8°}{b} = \frac{\sin 100°}{5}$$

$$b = \frac{5\sin 56.8°}{\sin 100°} \approx 4.25$$

19. $a = 3, \ c = 1, \ \gamma = 110°$

$$\frac{\sin\gamma}{c} = \frac{\sin\alpha}{a} \Rightarrow \frac{\sin 110°}{1} = \frac{\sin\alpha}{3} \Rightarrow \sin\alpha = \frac{3\sin 110°}{1} \approx 2.8191$$

There is no angle α for which $\sin\alpha > 1$. Therefore, there is no triangle with the given measurements.

21. $a = 3, \ c = 1, \ \beta = 100°$

$$b^2 = a^2 + c^2 - 2ac\cos\beta = 3^2 + 1^2 - 2 \cdot 3 \cdot 1\cos 100° = 10 - 6\cos 100° \Rightarrow b \approx 3.32$$

$$a^2 = b^2 + c^2 - 2bc\cos\alpha$$

$$\cos\alpha = \frac{b^2 + c^2 - a^2}{2bc} = \frac{3.32^2 + 1^2 - 3^2}{2(3.32)(1)} = \frac{3.0224}{6.64} \Rightarrow \alpha \approx 62.9°$$

$$\gamma = 180° - \alpha - \beta \approx 180° - 62.9° - 100° = 17.1°$$

23. $a = 2, \ b = 3, \ c = 1$

$$a^2 = b^2 + c^2 - 2bc\cos\alpha \Rightarrow \cos\alpha = \frac{b^2 + c^2 - a^2}{2bc} = \frac{3^2 + 1^2 - 2^2}{2(3)(1)} = 1 \Rightarrow \alpha \approx 0°$$

No triangle exists with an angle of 0°.

25. $a = 1,\ b = 3,\ \gamma = 40°$

$c^2 = a^2 + b^2 - 2ab\cos\gamma$

$c^2 = 1^2 + 3^2 - 2\cdot 1\cdot 3\cos 40° = 10 - 6\cos 40° \Rightarrow c \approx 2.32$

$a^2 = b^2 + c^2 - 2bc\cos\alpha \Rightarrow \cos\alpha = \dfrac{b^2 + c^2 - a^2}{2bc} = \dfrac{3^2 + 2.32^2 - 1^2}{2(3)(2.32)} = \dfrac{13.3824}{13.92} \Rightarrow \alpha \approx 16.0°$

$\beta = 180° - \alpha - \gamma \approx 180° - 16.0° - 40° = 124.0°$

27. $a = 5,\ b = 3,\ \alpha = 80°$

$\dfrac{\sin\beta}{b} = \dfrac{\sin\alpha}{a} \Rightarrow \dfrac{\sin\beta}{3} = \dfrac{\sin 80°}{5}$

$\sin\beta = \dfrac{3\sin 80°}{5} \approx 0.5909$

$\beta \approx 36.2°$ or $\beta \approx 143.8°$

The second value is discarded because $\alpha + \beta > 180°$.

$\gamma = 180° - \alpha - \beta \approx 180° - 80° - 36.2° = 63.8°$

$\dfrac{\sin\gamma}{c} = \dfrac{\sin\alpha}{a} \Rightarrow \dfrac{\sin 63.8°}{c} = \dfrac{\sin 80°}{5}$

$c = \dfrac{5\sin 63.8°}{\sin 80°} \approx 4.56$

29. $a = 1,\ b = \dfrac{1}{2},\ c = \dfrac{4}{3}$

$a^2 = b^2 + c^2 - 2bc\cos\alpha$

$\cos\alpha = \dfrac{b^2 + c^2 - a^2}{2bc} = \dfrac{(1/2)^2 + (4/3)^2 - 1^2}{2(1/2)(4/3)} = \dfrac{\frac{36}{37}}{\frac{4}{3}} \Rightarrow \alpha \approx 39.6°$

$b^2 = a^2 + c^2 - 2ac\cos\beta$

$\cos\beta = \dfrac{a^2 + c^2 - b^2}{2ac} = \dfrac{1^2 + (4/3)^2 - (1/2)^2}{2(1)(4/3)} = \dfrac{\frac{91}{36}}{\frac{8}{3}} \Rightarrow \beta \approx 18.6°$

$\gamma = 180° - \alpha - \beta \approx 180° - 39.6° - 18.6° \approx 121.8°$

31. $a = 3,\ \alpha = 10°,\ b = 4$

$\dfrac{\sin\beta}{b} = \dfrac{\sin\alpha}{a} \Rightarrow \dfrac{\sin\beta}{4} = \dfrac{\sin 10°}{3}$

$\sin\beta = \dfrac{4\sin 10°}{3} \approx 0.2315$

$\beta_1 \approx 13.4°$ or $\beta_2 \approx 166.6°$

For both values, $\alpha + \beta < 180°$.

Therefore, there are two triangles.

$$\gamma_1 = 180° - \alpha - \beta_1 \approx 180° - 10° - 13.4° \approx 156.6°$$

$$\frac{\sin\alpha}{a} = \frac{\sin\gamma_1}{c_1} \Rightarrow \frac{\sin10°}{3} = \frac{\sin156.6°}{c_1} \Rightarrow c_1 = \frac{3\sin156.6°}{\sin10°} \approx 6.86$$

$$\gamma_2 = 180° - \alpha - \beta_2 = 180° - 10° - 166.6° \approx 3.4°$$

$$\frac{\sin\alpha}{a} = \frac{\sin\gamma_2}{c_2} \Rightarrow \frac{\sin10°}{3} = \frac{\sin3.4°}{c_2} \Rightarrow c_2 = \frac{3\sin3.4°}{\sin10°} \approx 1.02$$

Two triangles: $\beta_1 \approx 13.4°$, $\gamma_1 \approx 156.6°$, $c_1 \approx 6.86$
         or $\beta_2 \approx 166.6°$, $\gamma_2 \approx 3.4°$, $c_2 \approx 1.02$

33. $c = 5$, $b = 4$, $\alpha = 70°$

$$a^2 = b^2 + c^2 - 2bc\cos\alpha$$

$$a^2 = 4^2 + 5^2 - 2\cdot4\cdot5\cos70° = 41 - 40\cos70° \Rightarrow a \approx 5.23$$

$$c^2 = a^2 + b^2 - 2ab\cos\gamma$$

$$\cos\gamma = \frac{a^2 + b^2 - c^2}{2ab} = \frac{5.23^2 + 4^2 - 5^2}{2(5.23)(4)} = \frac{18.3529}{41.48} \Rightarrow \gamma \approx 64.0°$$

$$\beta = 180° - \alpha - \gamma \approx 180° - 70° - 64.0° \approx 46.0°$$

35. $a = 2$, $b = 3$, $\gamma = 40°$

$$A = \frac{1}{2}ab\sin\gamma = \frac{1}{2}(2)(3)\sin40° \approx 1.93$$

37. $b = 4$, $c = 10$, $\alpha = 70°$

$$A = \frac{1}{2}bc\sin\alpha = \frac{1}{2}(4)(10)\sin70° \approx 18.79$$

39. $a = 4$, $b = 3$, $c = 5$

$$s = \frac{1}{2}(a+b+c) = \frac{1}{2}(4+3+5) = 6$$

$$A = \sqrt{s(s-a)(s-b)(s-c)} = \sqrt{(6)(2)(3)(1)} = \sqrt{36} = 6$$

41. $a = 4$, $b = 2$, $c = 5$

$$s = \frac{1}{2}(a+b+c) = \frac{1}{2}(4+2+5) = \frac{11}{2}$$

$$A = \sqrt{s(s-a)(s-b)(s-c)} = \sqrt{(11/2)(3/2)(7/2)(1/2)} = \sqrt{231/16} \approx 3.80$$

43. $\alpha = 50°$, $\beta = 30°$, $a = 1$ $\gamma = 180° - \alpha - \beta = 180° - 50° - 30° = 100°$

$$A = \frac{a^2\sin\beta\sin\gamma}{2\sin\alpha} = \frac{1^2\sin30°\cdot\sin100°}{2\sin50°} \approx 0.32$$

45.   Use right triangle methods:

$$\tan65° = \frac{500}{b} \Rightarrow b = \frac{500}{\tan65°} \approx 233.15$$

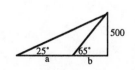

$$\tan 25° = \frac{500}{a+b} \Rightarrow a+b = \frac{500}{\tan 25°} \approx 1072.25$$

$a \approx 1072.25 - 233.15 = 839.1$ feet
The lake is approximately 839 feet long.

47. $\tan 25° = \dfrac{b}{50} \Rightarrow b = 50\tan 25° \approx 23.32$ feet

49. 1454 ft ≈ 0.2754 mile

$$\tan 5° \approx \frac{0.2754}{a+1}$$

$$a+1 = \frac{0.2754}{\tan 5°} \approx 3.15 \Rightarrow a \approx 2.15 \text{ miles}$$

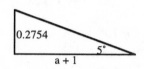

The boat is about 2.15 miles offshore.

51. $\angle ABC = 180° - 20° = 160°$

Find the angle at city C: $\dfrac{\sin C}{100} = \dfrac{\sin 160°}{300} \Rightarrow \sin C = \dfrac{100\sin 160°}{300} \Rightarrow C \approx 6.55°$

Find the angle at city A: $A \approx 180° - 160° - 6.55° = 13.45°$

$$\frac{\sin 13.45°}{y} = \frac{\sin 160°}{300} \Rightarrow y = \frac{300\sin 13.45°}{\sin 160°} \approx 204.07 \text{ miles}$$

The distance from city B to city C is approximately 204 miles.

53. Draw a line perpendicular to the shore:
  (a) $\angle ACB = 12° + 30° = 42°$

$\angle ABC = 90° - 30° = 60°$

$\angle CAB = 90° - 12° = 78°$

$$\frac{\sin 60°}{b} = \frac{\sin 42°}{2} \Rightarrow b = \frac{2\sin 60°}{\sin 42°} \approx 2.59 \text{ miles}$$

  (b) $\dfrac{\sin 78°}{a} = \dfrac{\sin 42°}{2} \Rightarrow a = \dfrac{2\sin 78°}{\sin 42°} \approx 2.92$ miles

  (c) $\cos 12° = \dfrac{x}{2.59} \Rightarrow x = 2.59\cos 12° \approx 2.53$ miles

55. (a) After 4 hours, the sailboat would have sailed $18(4) = 72$ miles.
    Find the third side of the triangle to determine the distance from the island:
      $a = 72, \ b = 200, \ \gamma = 15°$

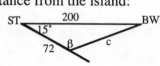

$$c^2 = a^2 + b^2 - 2ab\cos\gamma$$

$$c^2 = 72^2 + 200^2 - 2\cdot 72\cdot 200\cos 15°$$

$$= 45{,}184 - 28{,}800\cos 15° \Rightarrow c \approx 131.8 \text{ miles}$$

The sailboat is about 131.8 miles from the island.

(b)   Find the measure of the angle opposite the 200 side:

$$\cos\beta = \frac{a^2 + c^2 - b^2}{2ac} \Rightarrow \cos\beta = \frac{72^2 + 131.8^2 - 200^2}{2(72)(131.8)} = -\frac{17,444.76}{18,979.2} \Rightarrow \beta \approx 156.8°$$

The sailboat should turn through an angle of $180° - 156.8° = 23.2°$ to correct its course.

(c)   The original trip would have taken: $t = \frac{200}{18} \approx 11.1$ hours.

The actual trip takes: $t = 4 + \frac{131.8}{18} \approx 11.3$ hours.

The trip takes about 0.2 hour, that is, 12 minutes longer.

57.   Find the lengths of the two unknown sides of the middle triangle:

$$x^2 = 100^2 + 125^2 - 2(100)(125)\cos 50°$$

$x \approx 97.75$ feet

$$y^2 = 70^2 + 50^2 - 2(70)(50)\cos 100°$$

$y \approx 92.82$ feet

Find the areas of the three triangles:

$$A_1 = \frac{1}{2}(100)(125)\sin 50° \approx 4787.78 \qquad A_2 = \frac{1}{2}(50)(70)\sin 100° \approx 1723.41$$

$$s = \frac{1}{2}(50 + 97.75 + 92.82) = 120.285$$

$$A_3 = \sqrt{(120.285)(70.285)(22.535)(27.465)} \approx 2287.47$$

The approximate area of the lake is $4787.78 + 1723.41 + 2287.47 = 8798.67$ sq. ft.

59.   Area of the segment = area of the sector–area of the triangle.

$$\text{Area of sector} = \frac{1}{2}r^2\theta = \frac{1}{2}\cdot 6^2\left(50\cdot\frac{\pi}{180}\right) \approx 15.708 \text{ in}^2$$

$$\text{Area of triangle } = \frac{1}{2}ab\sin\theta = \frac{1}{2}\cdot 6\cdot 6\sin 50° \approx 13.789 \text{ in}^2$$

$$\text{Area of segment} = 15.708 - 13.789 \approx 1.92 \text{ in}^2$$

61.   Extend the tangent line until it meets a line extended through the centers of the pulleys. Label these extensions x and y. The distance between the points of tangency is z. Two similar triangles are formed.
Therefore:

$$\frac{24 + y}{y} = \frac{6.5}{2.5}$$

where $24 + y$ is the hypotenuse of the larger triangle and y is the hypotenuse of the smaller triangle .

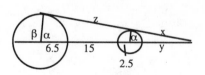

Solve for y:

$6.5y = 2.5(24 + y)$

$6.5y = 60 + 2.5y$

$4y = 60$

$y = 15$

Use the Pythagorean Theorem to find x:
$$x^2 + 2.5^2 = 15^2$$
$$x^2 = 225 - 6.25 = 218.75 \Rightarrow x \approx 14.79$$
Use the Pythagorean Theorem to find z:
$$(z + 14.79)^2 + 6.5^2 = (24 + 15)^2$$
$$(z + 14.79)^2 = 1521 - 42.25 = 1478.75$$
$$z + 14.79 \approx 38.45 \Rightarrow z \approx 23.66$$
Find α:
$$\cos\alpha = \frac{2.5}{15} \approx 0.1667 \Rightarrow \alpha \approx 1.4033 \text{ radians}$$
$$\beta \approx \pi - 1.4033 \approx 1.7383 \text{ radians}$$
The arc length on the top of the larger pulley is about: 6.5(1.7383) = 11.30 inches.
The arc length on the top of the smaller pulley is about: 2.5(1.4033) = 3.51 inches.
The distance between the points of tangency is about 23.66 inches.
The length of the belt is about: 2(11.30 + 3.51 + 23.66) = 76.94 inches.

63. $d = -3\cos\left(\dfrac{\pi}{2}t\right)$

65. $d = 6\sin(2t)$
    (a)   Simple harmonic
    (b)   6 feet
    (c) π seconds
    (d) $\dfrac{1}{\pi}$ oscillation/second

67. $d = -2\cos(\pi t)$
    (a)   Simple harmonic
    (b)   2 feet
    (c)   2 seconds
    (d) $\dfrac{1}{2}$ oscillation/second

69. $y = e^{-x/2\pi}\sin(2x), \qquad 0 \le x \le 2\pi$

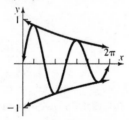

71. $y = x\cos x, \qquad 0 \le x \le 2\pi$

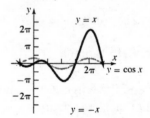

73. $y = 2\sin x + \cos(2x), \qquad 0 \le x \le 2\pi$

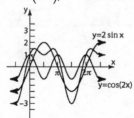

Applications of Trigonometric Functions

7.CR   Cumulative Review

1. $3x^2 + 1 = 4x$

$3x^2 - 4x + 1 = 0$

$(3x - 1)(x - 1) = 0$

$x = \dfrac{1}{3}$ or $x = 1$

The solution set is $\left\{ \dfrac{1}{3}, 1 \right\}$.

3. $f(x) = \sqrt{x^2 - 3x - 4}$

f will be defined provided $g(x) = x^2 - 3x - 4 \geq 0$.

$x^2 - 3x - 4 \geq 0$

$(x - 4)(x + 1) \geq 0$

$x = 4$, $x = -1$ are the zeros.

Interval	Test Number	$g(x)$	Positive/Negative
$-\infty < x < -1$	-2	6	Positive
$-1 < x < 4$	0	-4	Negative
$4 < x < \infty$	5	6	Positive

The domain of $f(x) = \sqrt{x^2 - 3x - 4}$ is $\left\{ x \mid -\infty < x \leq -1 \text{ or } 4 \leq x < \infty \right\}$.

5. $y = -2\cos(2x - \pi) = -2\cos\left(2\left(x - \dfrac{\pi}{2} \right) \right)$

Amplitude: $|A| = |-2| = 2$

Period: $T = \dfrac{2\pi}{2} = \pi$

Phase Shift: $\dfrac{\phi}{\omega} = \dfrac{\pi}{2}$

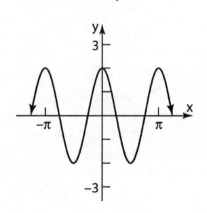

7. (a) $y = e^x,\ 0 \le x \le 4$

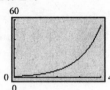

(b) $y = \sin x,\ 0 \le x \le 4$

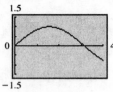

(c) $y = e^x \sin x,\ 0 \le x \le 4$

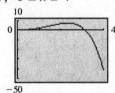

(d) $y = 2x + \sin x,\ 0 \le x \le 4$

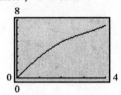

9. $a = 20,\ c = 15,\ \gamma = 40°$

$$\frac{\sin \gamma}{c} = \frac{\sin \alpha}{a} \Rightarrow \frac{\sin 40°}{15} = \frac{\sin \alpha}{20} \Rightarrow \sin \alpha = \frac{20 \sin 40°}{15}$$

$\alpha \approx 59.0°$ or $\alpha \approx 121.0°$

Case 1: $\alpha \approx 58.99° \Rightarrow \beta \approx 180° - \left(40° + 59.0°\right) = 81.0°$

$$\frac{\sin \gamma}{c} = \frac{\sin \beta}{b} \Rightarrow \frac{\sin 40°}{15} = \frac{\sin 81.0°}{b}$$

$$b \approx \frac{15 \sin 81.0°}{\sin 40°} \approx 23.05$$

Therefore, $\alpha \approx 59.0°,\ \beta \approx 81.0°,\ b \approx 23.05$.

Case 2: $\alpha \approx 121.0° \Rightarrow \beta \approx 180° - \left(40° + 121.0°\right) = 19.0°$

$$\frac{\sin \gamma}{c} = \frac{\sin \beta}{b} \Rightarrow \frac{\sin 40°}{15} = \frac{\sin 19.0°}{b}$$

$$b \approx \frac{15 \sin 19.0°}{\sin 40°} \approx 7.60$$

Therefore, $\alpha \approx 121.0°,\ \beta \approx 19.0°,\ b \approx 7.60$.

11. $R(x) = \dfrac{2x^2 - 7x - 4}{x^2 + 2x - 15} = \dfrac{(2x + 1)(x - 4)}{(x - 3)(x + 5)}$ $p(x) = 2x^2 - 7x - 4;\ q(x) = x^2 + 2x - 15;$

$$n = 2;\ m = 2$$

Step 1:   Domain: $\left\{x \mid x \ne -5,\ x \ne 3\right\}$

Step 2: R is in lowest terms.

Step 3:   (a)   The x-intercepts are the zeros of $p(x)$:

$$2x^2 - 7x - 4 = 0$$

$$(2x + 1)(x - 4) = 0$$

$$2x + 1 = 0 \Rightarrow x = -\frac{1}{2}$$

or $x - 4 = 0 \Rightarrow x = 4$

(b)   The y-intercept is $R(0) = \dfrac{2 \cdot 0^2 - 7 \cdot 0 - 4}{0^2 + 2 \cdot 0 - 15} = \dfrac{-4}{-15} = \dfrac{4}{15}$.

Step 4: $R(-x) = \dfrac{2(-x)^2 - 7(-x) - 4}{(-x)^2 + 2(-x) - 15} = \dfrac{2x^2 + 7x - 4}{x^2 - 2x - 15}$; this is neither $R(x)$ nor $-R(x)$, so there is no symmetry.

Step 5:   The vertical asymptotes are the zeros of $q(x)$:

$$x^2 + 2x - 15 = 0$$
$$(x + 5)(x - 3) = 0$$
$$x + 5 = 0 \Rightarrow x = -5$$
$$\text{or } x - 3 = 0 \Rightarrow x = 3$$

Step 6:   Since $n = m$, the line $y = 2$ is the horizontal asymptote.

$R(x)$ intersects $y = 2$ at $\left(\dfrac{26}{11}, 2\right)$, since:

$$\frac{2x^2 - 7x - 4}{x^2 + 2x - 15} = 2$$
$$2x^2 - 7x - 4 = 2\left(x^2 + 2x - 15\right)$$
$$2x^2 - 7x - 4 = 2x^2 + 4x - 30$$
$$-11x = -26$$
$$x = \frac{26}{11}$$

Step 7:   Graphing utility:

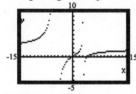

Step 8:   Graphing by hand:

Interval	$(-\infty, -5)$	$(-5, -0.5)$	$(-0.5, 3)$	$(3, 4)$	$(4, \infty)$
Number Chosen	-6	-1	0	3.5	5
Value of f	$f(-6) \approx 12.22$	$f(-1) = -0.3125$	$f(0) \approx 0.27$	$f(3.5) \approx -0.94$	$f(5) = 0.55$
Location of Graph	Above x-axis	Below x-axis	Above x-axis	Below x-axis	Above x-axis
Point on Graph	$(-6, 12.22)$	$(-1, -0.3125)$	$(0, 0.27)$	$(3.5, -0.94)$	$(5, 0.55)$

(marks above table: -5, -0.5, 3, 4)

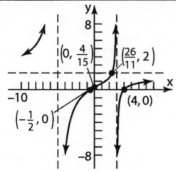

13. $\log_3(x+8) + \log_3 x = 2$

$\log_3\big((x+8)(x)\big) = 2$

$(x+8)(x) = 3^2$

$x^2 + 8x = 9$

$x^2 + 8x - 9 = 0$

$(x+9)(x-1) = 0$

$x = -9 \text{ or } x = 1$

Since $x = -9$ makes the original logarithms undefined, the only solution is $x = 1$.

Polar Coordinates; Vectors

8.1   Polar Coordinates

1.

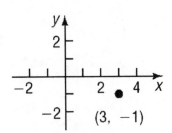

3. $\dfrac{b}{\sqrt{a^2 + b^2}}$

5.   pole, polar axis

7. $-\sqrt{3}, -1$

9.   True

11.   *A*   13.   *C*

15.   *B*   17.   *A*

19. $(3, 90°)$

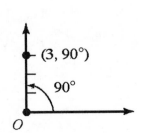

21. $(-2, 0)$

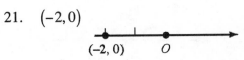

23. $\left(6, \dfrac{\pi}{6}\right)$

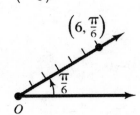

25. $(-2, 135°)$

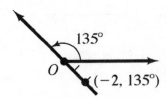

27. $\left(-1, -\dfrac{\pi}{3}\right)$

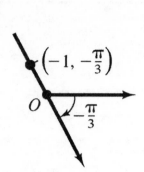

29. $(-2, -\pi)$

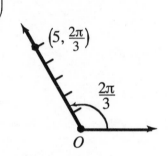

31. $\left(5, \dfrac{2\pi}{3}\right)$

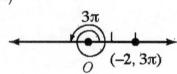

(a) $r > 0, \ -2\pi \le \theta < 0 \quad \left(5, -\dfrac{4\pi}{3}\right)$

(b) $r < 0, \ 0 \le \theta < 2\pi \quad \left(-5, \dfrac{5\pi}{3}\right)$

(c) $r > 0, \ 2\pi \le \theta < 4\pi \quad \left(5, \dfrac{8\pi}{3}\right)$

33. $(-2, 3\pi)$

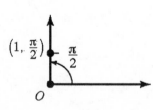

(a) $r > 0, \ -2\pi \le \theta < 0 \quad (2, -2\pi)$

(b) $r < 0, \ 0 \le \theta < 2\pi \quad (-2, \pi)$

(c) $r > 0, \ 2\pi \le \theta < 4\pi \quad (2, 2\pi)$

35. $\left(1, \dfrac{\pi}{2}\right)$

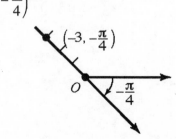

(a) $r > 0, \ -2\pi \le \theta < 0 \quad \left(1, -\dfrac{3\pi}{2}\right)$

(b) $r < 0, \ 0 \le \theta < 2\pi \quad \left(-1, \dfrac{3\pi}{2}\right)$

(c) $r > 0, \ 2\pi \le \theta < 4\pi \quad \left(1, \dfrac{5\pi}{2}\right)$

37. $\left(-3, -\dfrac{\pi}{4}\right)$

(a) $r > 0, \ -2\pi \le \theta < 0 \quad \left(3, -\dfrac{5\pi}{4}\right)$

(b) $r < 0, \ 0 \le \theta < 2\pi \quad \left(-3, \dfrac{7\pi}{4}\right)$

(c) $r > 0, \ 2\pi \le \theta < 4\pi \quad \left(3, \dfrac{11\pi}{4}\right)$

39. $x = r\cos\theta = 3\cos\dfrac{\pi}{2} = 3\cdot 0 = 0$

$y = r\sin\theta = 3\sin\dfrac{\pi}{2} = 3\cdot 1 = 3$

The rectangular coordinates of the point $\left(3, \dfrac{\pi}{2}\right)$ are $(0, 3)$.

41. $x = r\cos\theta = -2\cos 0 = -2\cdot 1 = -2$
$y = r\sin\theta = -2\sin 0 = -2\cdot 0 = 0$

The rectangular coordinates of the point $(-2, 0)$ are $(-2, 0)$.

43. $x = r\cos\theta = 6\cos 150° = 6\left(-\dfrac{\sqrt{3}}{2}\right) = -3\sqrt{3}$

$y = r\sin\theta = 6\sin 150° = 6\cdot\dfrac{1}{2} = 3$

The rectangular coordinates of the point $(6, 150°)$ are $\left(-3\sqrt{3}, 3\right)$.

45. $x = r\cos\theta = -2\cos\dfrac{3\pi}{4} = -2\left(-\dfrac{\sqrt{2}}{2}\right) = \sqrt{2}$

$y = r\sin\theta = -2\sin\dfrac{3\pi}{4} = -2\cdot\dfrac{\sqrt{2}}{2} = -\sqrt{2}$

The rectangular coordinates of the point $\left(-2, \dfrac{3\pi}{4}\right)$ are $\left(\sqrt{2}, -\sqrt{2}\right)$.

47. $x = r\cos\theta = -1\cos\left(-\dfrac{\pi}{3}\right) = -1\cdot\dfrac{1}{2} = -\dfrac{1}{2}$

$y = r\sin\theta = -1\sin\left(-\dfrac{\pi}{3}\right) = -1\left(-\dfrac{\sqrt{3}}{2}\right) = \dfrac{\sqrt{3}}{2}$

The rectangular coordinates of the point $\left(-1, -\dfrac{\pi}{3}\right)$ are $\left(-\dfrac{1}{2}, \dfrac{\sqrt{3}}{2}\right)$.

49. $x = r\cos\theta = -2\cos(-180°) = -2(-1) = 2$
$y = r\sin\theta = -2\sin(-180°) = -2\cdot 0 = 0$

The rectangular coordinates of the point $(-2, -180°)$ are $(2, 0)$.

51. $x = r\cos\theta = 7.5\cos 110° \approx 7.5(-0.3420) \approx -2.57$
$y = r\sin\theta = 7.5\sin 110° \approx 7.5(0.9397) \approx 7.05$

The rectangular coordinates of the point $(7.5, 110°)$ are $(-2.57, 7.05)$.

53. $x = r\cos\theta = 6.3\cos(3.8) \approx 6.3(-0.7910) \approx -4.98$
     $y = r\sin\theta = 6.3\sin(3.8) \approx 6.3(-0.6119) \approx -3.85$

     The rectangular coordinates of the point $(6.3, 3.8)$ are $(-4.98, -3.85)$.

55. $r = \sqrt{x^2 + y^2} = \sqrt{3^2 + 0^2} = \sqrt{9} = 3$ $\theta = \tan^{-1}\left(\dfrac{y}{x}\right) = \tan^{-1}\left(\dfrac{0}{3}\right) = \tan^{-1}0 = 0$

     Polar coordinates of the point $(3, 0)$ are $(3, 0)$.

57. $r = \sqrt{x^2 + y^2} = \sqrt{(-1)^2 + 0^2} = \sqrt{1} = 1$ $\theta = \tan^{-1}\left(\dfrac{y}{x}\right) = \tan^{-1}\left(\dfrac{0}{-1}\right) = \tan^{-1}0 = 0$

     The point lies on the negative x-axis thus $\theta = \pi$.
     Polar coordinates of the point $(-1, 0)$ are $(1, \pi)$.

59. The point $(1, -1)$ lies in quadrant IV.
     $r = \sqrt{x^2 + y^2} = \sqrt{1^2 + (-1)^2} = \sqrt{2}$ $\theta = \tan^{-1}\left(\dfrac{y}{x}\right) = \tan^{-1}\left(\dfrac{-1}{1}\right) = \tan^{-1}(-1) = -\dfrac{\pi}{4}$

     Polar coordinates of the point $(1, -1)$ are $\left(\sqrt{2}, -\dfrac{\pi}{4}\right)$.

61. The point $\left(\sqrt{3}, 1\right)$ lies in quadrant I.
     $r = \sqrt{x^2 + y^2} = \sqrt{\left(\sqrt{3}\right)^2 + 1^2} = \sqrt{4} = 2$ $\theta = \tan^{-1}\left(\dfrac{y}{x}\right) = \tan^{-1}\dfrac{1}{\sqrt{3}} = \dfrac{\pi}{6}$

     Polar coordinates of the point $\left(\sqrt{3}, 1\right)$ are $\left(2, \dfrac{\pi}{6}\right)$.

63. The point $(1.3, -2.1)$ lies in quadrant IV.
     $r = \sqrt{x^2 + y^2} = \sqrt{1.3^2 + (-2.1)^2} = \sqrt{6.1} \approx 2.47$

     $\theta = \tan^{-1}\left(\dfrac{y}{x}\right) = \tan^{-1}\left(\dfrac{-2.1}{1.3}\right) \approx \tan^{-1}(-1.6154) \approx -1.02$
     Polar coordinates of the point $(1.3, -2.1)$ are $(2.47, -1.02)$.

65. The point $(8.3, 4.2)$ lies in quadrant I.
     $r = \sqrt{x^2 + y^2} = \sqrt{8.3^2 + 4.2^2} = \sqrt{86.53} \approx 9.30$

     $\theta = \tan^{-1}\left(\dfrac{y}{x}\right) = \tan^{-1}\left(\dfrac{4.2}{8.3}\right) \approx \tan^{-1}(0.5060) \approx 0.47$
     Polar coordinates of the point $(8.3, 4.2)$ are $(9.30, 0.47)$.

67. $2x^2 + 2y^2 = 3$

$2(x^2 + y^2) = 3$

$2r^2 = 3$

$r^2 = \dfrac{3}{2}$

69. $x^2 = 4y$

$(r\cos\theta)^2 = 4r\sin\theta$

$r^2\cos^2\theta - 4r\sin\theta = 0$

71. $2xy = 1$

$2(r\cos\theta)(r\sin\theta) = 1$

$2r^2\sin\theta\cos\theta = 1$

$r^2\sin 2\theta = 1$

73. $x = 4$

$r\cos\theta = 4$

75. $r = \cos\theta$

$r^2 = r\cos\theta$

$x^2 + y^2 = x$

$x^2 - x + y^2 = 0$

77. $r^2 = \cos\theta$

$r^3 = r\cos\theta$

$\left(x^2 + y^2\right)^{3/2} = x$

$\left(x^2 + y^2\right)^{3/2} - x = 0$

79. $r = 2$

$\sqrt{x^2 + y^2} = 2$

$x^2 + y^2 = 4$

81. $r = \dfrac{4}{1 - \cos\theta}$

$r(1 - \cos\theta) = 4$

$r - r\cos\theta = 4$

$\sqrt{x^2 + y^2} - x = 4$

$\sqrt{x^2 + y^2} = x + 4$

$x^2 + y^2 = x^2 + 8x + 16$

$y^2 = 8(x + 2)$

83. Rewrite the polar coordinates in rectangular form:

$P_1 = (r_1, \theta_1) \Rightarrow P_1 = (r_1\cos\theta_1, r_1\sin\theta_1)$

$P_2 = (r_2, \theta_2) \Rightarrow P_2 = (r_2\cos\theta_2, r_2\sin\theta_2)$

$d = \sqrt{(r_2\cos\theta_2 - r_1\cos\theta_1)^2 + (r_2\sin\theta_2 - r_1\sin\theta_1)^2}$

$ = \sqrt{r_2^2\cos^2\theta_2 - 2r_1r_2\cos\theta_2\cos\theta_1 + r_1^2\cos^2\theta_1 + r_2^2\sin^2\theta_2 - 2r_1r_2\sin\theta_2\sin\theta_1 + r_1^2\sin^2\theta_1}$

$ = \sqrt{r_2^2\left(\cos^2\theta_2 + \sin^2\theta_2\right) + r_1^2\left(\cos^2\theta_1 + \sin^2\theta_1\right) - 2r_1r_2\left(\cos\theta_2\cos\theta_1 + \sin\theta_2\sin\theta_1\right)}$

$ = \sqrt{r_2^2 + r_1^2 - 2r_1r_2\cos(\theta_2 - \theta_1)}$

85. Answers will vary.

Chapter 8

Polar Coordinates; Vectors

8.2 Polar Equations and Graphs

1. $(-4,6)$

3. $\left(x-(-2)\right)^2 + (y-5)^2 = 3^2 \Rightarrow (x+2)^2 + (y-5)^2 = 9$

5. $-\dfrac{\sqrt{2}}{2}$

7. polar equation

9. $-r$

11. False

13. $r = 4$

The equation is of the form $r = a, \ a > 0$.
It is a circle, center at the pole and radius 4.
Transform to rectangular form:
$$r = 4$$
$$r^2 = 16$$
$$x^2 + y^2 = 16$$

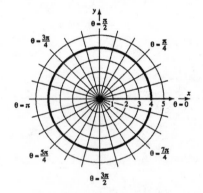

15. $\theta = \dfrac{\pi}{3}$

The equation is of the form $\theta = \alpha$. It is a
line, passing through the pole at an angle
of $\dfrac{\pi}{3}$.
Transform to rectangular form:
$$\theta = \frac{\pi}{3} \Rightarrow \tan\theta = \tan\frac{\pi}{3}$$
$$\frac{y}{x} = \sqrt{3}$$
$$y = \sqrt{3}x$$

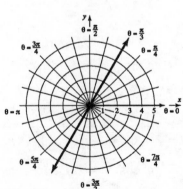

17. $r\sin\theta = 4$

The equation is of the form $r\sin\theta = b$. It is a horizontal line, 4 units above the pole. Transform to rectangular form:

$$r\sin\theta = 4$$

$$y = 4$$

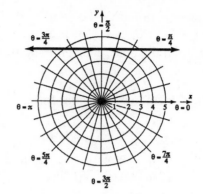

19. $r\cos\theta = -2$

The equation is of the form $r\cos\theta = a$. It is a vertical line, 2 units to the left of the pole. Transform to rectangular form:

$$r\cos\theta = -2$$

$$x = -2$$

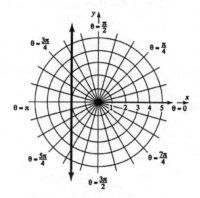

21. $r = 2\cos\theta$

The equation is of the form $r = \pm 2a\cos\theta$, $a > 0$. It is a circle, passing through the pole, and center on the polar axis.

Transform to rectangular form:

$$r = 2\cos\theta$$

$$r^2 = 2r\cos\theta$$

$$x^2 + y^2 = 2x$$

$$x^2 - 2x + y^2 = 0$$

$$(x-1)^2 + y^2 = 1$$

center $(1,0)$; radius 1

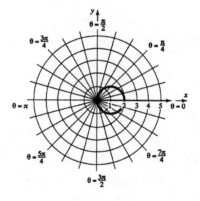

23. $r = -4\sin\theta$

The equation is of the form
$r = \pm 2a\sin\theta,\ a > 0$. It is a circle, passing
through the pole, and center on the line
$\theta = \dfrac{\pi}{2}$.

Transform to rectangular form:
$$r = -4\sin\theta$$
$$r^2 = -4r\sin\theta$$
$$x^2 + y^2 = -4y$$
$$x^2 + y^2 + 4y = 0$$
$$x^2 + (y+2)^2 = 4$$
center $(0,-2)$; radius 2

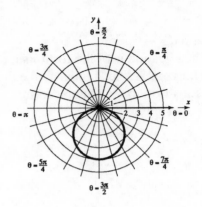

25. $r\sec\theta = 4$

The equation is a circle, passing through the
pole, center on the polar axis and radius 2.
Transform to rectangular form:
$$r\sec\theta = 4$$
$$r \cdot \frac{1}{\cos\theta} = 4$$
$$r = 4\cos\theta$$
$$r^2 = 4r\cos\theta$$
$$x^2 + y^2 = 4x$$
$$x^2 - 4x + y^2 = 0$$
$$(x-2)^2 + y^2 = 4$$
center $(2,0)$; radius 2

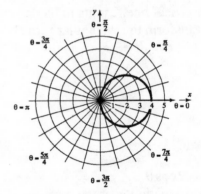

27. $r\csc\theta = -2$

The equation is a circle, passing through the
pole, center on the line $\theta = \dfrac{\pi}{2}$ and radius 1.
Transform to rectangular form:
$$r\csc\theta = -2$$
$$r \cdot \frac{1}{\sin\theta} = -2$$
$$r = -2\sin\theta$$
$$r^2 = -2r\sin\theta$$
$$x^2 + y^2 = -2y$$
$$x^2 + y^2 + 2y = 0$$
$$x^2 + (y+1)^2 = 1$$
center $(0,-1)$; radius 1

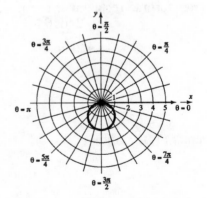

29.   E   31.   F   33.   H   35.   D

37. $r = 2 + 2\cos\theta$ The graph will be a cardioid. Check for symmetry:
Polar axis: Replace θ by $-\theta$. The result is $r = 2 + 2\cos(-\theta) = 2 + 2\cos\theta$.

        The graph is symmetric with respect to the polar axis.

The line $\theta = \dfrac{\pi}{2}$: Replace θ by $\pi - \theta$.

$$r = 2 + 2\cos(\pi - \theta) = 2 + 2(\cos(\pi)\cos\theta + \sin(\pi)\sin\theta)$$

$$= 2 + 2(-\cos\theta + 0) = 2 - 2\cos\theta$$

        The test fails.

The pole: Replace r by $-r$. $-r = 2 + 2\cos\theta$. The test fails.

Due to symmetry with respect to the polar axis, assign values to θ from 0 to π.

θ	0	$\dfrac{\pi}{6}$	$\dfrac{\pi}{3}$	$\dfrac{\pi}{2}$	$\dfrac{2\pi}{3}$	$\dfrac{5\pi}{6}$	π
$r = 2 + 2\cos\theta$	4	$2 + \sqrt{3} \approx 3.7$	3	2	1	$2 - \sqrt{3} \approx 0.3$	0

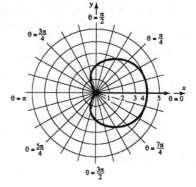

39. $r = 3 - 3\sin\theta$ The graph will be a cardioid. Check for symmetry:
Polar axis: Replace θ by $-\theta$. The result is $r = 3 - 3\sin(-\theta) = 3 + 3\sin\theta$. The test fails.

The line $\theta = \dfrac{\pi}{2}$: Replace θ by $\pi - \theta$.

$$r = 3 - 3\sin(\pi - \theta) = 3 - 3(\sin(\pi)\cos\theta - \cos(\pi)\sin\theta)$$

$$= 3 - 3(0 + \sin\theta) = 3 - 3\sin\theta$$

        The graph is symmetric with respect to the line $\theta = \dfrac{\pi}{2}$.

The pole: Replace r by $-r$. $-r = 3 - 3\sin\theta$. The test fails.

Due to symmetry with respect to the line $\theta = \dfrac{\pi}{2}$, assign values to θ from $-\dfrac{\pi}{2}$ to $\dfrac{\pi}{2}$.

θ	$-\dfrac{\pi}{2}$	$-\dfrac{\pi}{3}$	$-\dfrac{\pi}{6}$	0	$\dfrac{\pi}{6}$	$\dfrac{\pi}{3}$	$\dfrac{\pi}{2}$
$r = 3 - 3\sin\theta$	6	$3 + \dfrac{3\sqrt{3}}{2} \approx 5.6$	$\dfrac{9}{2}$	3	$\dfrac{3}{2}$	$3 - \dfrac{3\sqrt{3}}{2} \approx 0.4$	0

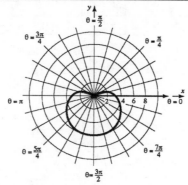

41. $r = 2 + \sin\theta$ The graph will be a limacon without an inner loop.
Check for symmetry:
Polar axis: Replace θ by $-\theta$. The result is $r = 2 + \sin(-\theta) = 2 - \sin\theta$. The test fails.

The line $\theta = \dfrac{\pi}{2}$: Replace θ by $\pi - \theta$.

$$r = 2 + \sin(\pi - \theta) = 2 + (\sin(\pi)\cos\theta - \cos(\pi)\sin\theta)$$

$$= 2 + (0 + \sin\theta) = 2 + \sin\theta$$

The graph is symmetric with respect to the line $\theta = \dfrac{\pi}{2}$.

The pole: Replace r by $-r$. $-r = 2 + \sin\theta$. The test fails.

Due to symmetry with respect to the line $\theta = \dfrac{\pi}{2}$, assign values to θ from $-\dfrac{\pi}{2}$ to $\dfrac{\pi}{2}$.

θ	$-\dfrac{\pi}{2}$	$-\dfrac{\pi}{3}$	$-\dfrac{\pi}{6}$	0	$\dfrac{\pi}{6}$	$\dfrac{\pi}{3}$	$\dfrac{\pi}{2}$
$r = 2 + \sin\theta$	1	$2 - \dfrac{\sqrt{3}}{2} \approx 1.1$	$\dfrac{3}{2}$	2	$\dfrac{5}{2}$	$2 + \dfrac{\sqrt{3}}{2} \approx 2.9$	3

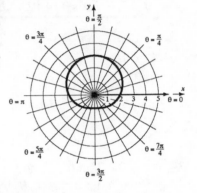

43. $r = 4 - 2\cos\theta$ The graph will be a limacon without an inner loop.
Check for symmetry:
Polar axis: Replace θ by $-\theta$. The result is $r = 4 - 2\cos(-\theta) = 4 - 2\cos\theta$.
The graph is symmetric with respect to the polar axis.

The line $\theta = \dfrac{\pi}{2}$: Replace θ by $\pi - \theta$.

$$r = 4 - 2\cos(\pi - \theta) = 4 - 2(\cos(\pi)\cos\theta + \sin(\pi)\sin\theta)$$

$$= 4 - 2(-\cos\theta + 0) = 4 + 2\cos\theta$$

The test fails.

The pole: Replace r by $-r$. $-r = 4 - 2\cos\theta$. The test fails.
Due to symmetry with respect to the polar axis, assign values to θ from 0 to π.

θ	0	$\dfrac{\pi}{6}$	$\dfrac{\pi}{3}$	$\dfrac{\pi}{2}$	$\dfrac{2\pi}{3}$	$\dfrac{5\pi}{6}$	π
$r = 4 - 2\cos\theta$	2	$4 - \sqrt{3} \approx 2.3$	3	4	5	$4 + \sqrt{3} \approx 5.7$	6

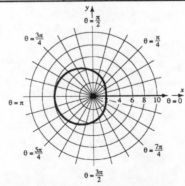

45. $r = 1 + 2\sin\theta$ The graph will be a limacon with an inner loop.
Check for symmetry:
Polar axis: Replace θ by $-\theta$. The result is $r = 1 + 2\sin(-\theta) = 1 - 2\sin\theta$. The test fails.

The line $\theta = \dfrac{\pi}{2}$: Replace θ by $\pi - \theta$.

$$r = 1 + 2\sin(\pi - \theta) = 1 + 2(\sin(\pi)\cos\theta - \cos(\pi)\sin\theta)$$

$$= 1 + 2(0 + \sin\theta) = 1 + 2\sin\theta$$

The graph is symmetric with respect to the line $\theta = \dfrac{\pi}{2}$.

The pole: Replace r by $-r$. $-r = 1 + 2\sin\theta$. The test fails.

Due to symmetry with respect to the line $\theta = \dfrac{\pi}{2}$, assign values to θ from $-\dfrac{\pi}{2}$ to $\dfrac{\pi}{2}$.

θ	$-\dfrac{\pi}{2}$	$-\dfrac{\pi}{3}$	$-\dfrac{\pi}{6}$	0	$\dfrac{\pi}{6}$	$\dfrac{\pi}{3}$	$\dfrac{\pi}{2}$
$r = 1 + 2\sin\theta$	-1	$1 - \sqrt{3} \approx -0.7$	0	1	2	$1 + \sqrt{3} \approx 2.7$	3

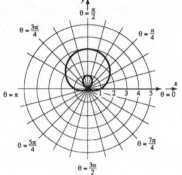

47. $r = 2 - 3\cos\theta$ The graph will be a limacon with an inner loop.
Check for symmetry:
Polar axis: Replace θ by $-\theta$. The result is $r = 2 - 3\cos(-\theta) = 2 - 3\cos\theta$.
        The graph is symmetric with respect to the polar axis.

The line $\theta = \dfrac{\pi}{2}$: Replace θ by $\pi - \theta$.

$$r = 2 - 3\cos(\pi - \theta) = 2 - 3(\cos(\pi)\cos\theta + \sin(\pi)\sin\theta)$$

$$= 2 - 3(-\cos\theta + 0) = 2 + 3\cos\theta$$

        The test fails.

The pole: Replace r by $-r$. $-r = 2 - 3\cos\theta$. The test fails.

Due to symmetry with respect to the polar axis, assign values to θ from 0 to π.

θ	0	$\dfrac{\pi}{6}$	$\dfrac{\pi}{3}$	$\dfrac{\pi}{2}$	$\dfrac{2\pi}{3}$	$\dfrac{5\pi}{6}$	π
$r = 2 - 3\cos\theta$	-1	$2 - \dfrac{3\sqrt{3}}{2} \approx -0.6$	$\dfrac{1}{2}$	2	$\dfrac{7}{2}$	$2 + \dfrac{3\sqrt{3}}{2} \approx 4.6$	5

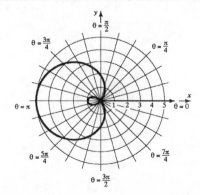

49. $r = 3\cos(2\theta)$ The graph will be a rose with four petals. Check for symmetry:

Polar axis: Replace θ by $-\theta$. $r = 3\cos(2(-\theta)) = 3\cos(-2\theta) = 3\cos(2\theta)$.

       The graph is symmetric with respect to the polar axis.

The line $\theta = \dfrac{\pi}{2}$: Replace θ by $\pi - \theta$.

$$r = 3\cos(2(\pi - \theta)) = 3\cos(2\pi - 2\theta)$$

$$= 3(\cos(2\pi)\cos(2\theta) + \sin(2\pi)\sin(2\theta)) = 3(\cos 2\theta + 0) = 3\cos(2\theta)$$

       The graph is symmetric with respect to the line $\theta = \dfrac{\pi}{2}$.

The pole: Since the graph is symmetric with respect to both the polar axis and the line $\theta = \dfrac{\pi}{2}$, it is also symmetric with respect to the pole.

Due to symmetry, assign values to θ from 0 to $\dfrac{\pi}{2}$.

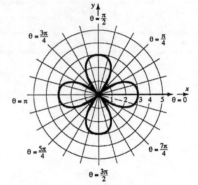

θ	0	$\dfrac{\pi}{6}$	$\dfrac{\pi}{4}$	$\dfrac{\pi}{3}$	$\dfrac{\pi}{2}$
$r = 3\cos(2\theta)$	3	$\dfrac{3}{2}$	0	$-\dfrac{3}{2}$	-3

51. $r = 4\sin(5\theta)$ The graph will be a rose with five petals. Check for symmetry:

Polar axis: Replace θ by $-\theta$. $r = 4\sin(5(-\theta)) = 4\sin(-5\theta) = -4\sin(5\theta)$. The test fails.

The line $\theta = \dfrac{\pi}{2}$: Replace θ by $\pi - \theta$.

$$r = 4\sin(5(\pi - \theta)) = 4\sin(5\pi - 5\theta)$$

$$= 4(\sin(5\pi)\cos(5\theta) - \cos(5\pi)\sin(5\theta)) = 4(0 + \sin(5\theta)) = 4\sin(5\theta)$$

       The graph is symmetric with respect to the line $\theta = \dfrac{\pi}{2}$.

The pole: Replace r by $-r$. $-r = 4\sin(5\theta)$. The test fails.

Due to symmetry with respect to the line $\theta = \dfrac{\pi}{2}$, assign values to θ from $-\dfrac{\pi}{2}$ to $\dfrac{\pi}{2}$.

θ	$-\dfrac{\pi}{2}$	$-\dfrac{\pi}{3}$	$-\dfrac{\pi}{4}$	$-\dfrac{\pi}{6}$	0	$\dfrac{\pi}{6}$	$\dfrac{\pi}{4}$	$\dfrac{\pi}{3}$	$\dfrac{\pi}{2}$
$r = 4\sin(5\theta)$	-4	$2\sqrt{3} \approx 3.5$	$2\sqrt{2} \approx 2.8$	-2	0	2	$-2\sqrt{2} \approx -2.8$	$-2\sqrt{3} \approx -3.5$	4

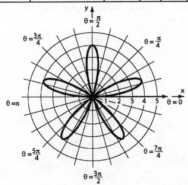

53. $r^2 = 9\cos(2\theta)$ The graph will be a lemniscate. Check for symmetry:
Polar axis: Replace θ by $-\theta$. $r^2 = 9\cos(2(-\theta)) = 9\cos(-2\theta) = 9\cos(2\theta)$.
        The graph is symmetric with respect to the polar axis.

The line $\theta = \dfrac{\pi}{2}$: Replace θ by $\pi - \theta$.

$$r^2 = 9\cos(2(\pi - \theta)) = 9\cos(2\pi - 2\theta)$$
$$= 9(\cos(2\pi)\cos 2\theta + \sin(2\pi)\sin 2\theta) = 9(\cos 2\theta + 0) = 9\cos(2\theta)$$

        The graph is symmetric with respect to the line $\theta = \dfrac{\pi}{2}$.

The pole: Since the graph is symmetric with respect to both the polar axis and the
        line $\theta = \dfrac{\pi}{2}$, it is also symmetric with respect to the pole.

Due to symmetry, assign values to θ from 0 to $\dfrac{\pi}{2}$.

θ	0	$\dfrac{\pi}{6}$	$\dfrac{\pi}{4}$	$\dfrac{\pi}{3}$	$\dfrac{\pi}{2}$
$r = \pm\sqrt{9\cos(2\theta)}$	± 3	$\pm\dfrac{3\sqrt{2}}{2} \approx \pm 2.1$	0	not defined	not defined

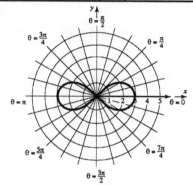

55. $r = 2^\theta$ The graph will be a spiral. Check for symmetry:
Polar axis: Replace θ by $-\theta$. $r = 2^{-\theta}$. The test fails.

The line $\theta = \dfrac{\pi}{2}$: Replace θ by $\pi - \theta$. $r = 2^{\pi-\theta}$. The test fails.

The pole: Replace r by $-r$. $-r = 2^\theta$. The test fails.

θ	$-\pi$	$-\dfrac{\pi}{2}$	$-\dfrac{\pi}{4}$	0	$\dfrac{\pi}{4}$	$\dfrac{\pi}{2}$	π	$\dfrac{3\pi}{2}$	2π
$r = 2^\theta$	0.1	0.3	0.6	1	1.7	3.0	8.8	26.2	77.9

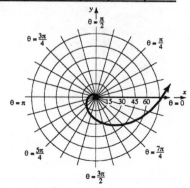

57.　$r = 1 - \cos\theta$　　The graph will be a cardioid.　Check for symmetry:
Polar axis: Replace θ by $-\theta$. The result is $r = 1 - \cos(-\theta) = 1 - \cos\theta$.
　　　　The graph is symmetric with respect to the polar axis.

The line $\theta = \dfrac{\pi}{2}$: Replace θ by $\pi - \theta$.

$$r = 1 - \cos(\pi - \theta) = 1 - (\cos(\pi)\cos\theta + \sin(\pi)\sin\theta)$$

$$= 1 - (-\cos\theta + 0) = 1 + \cos\theta.\quad \text{The test fails}$$

The pole:　Replace r by $-r$.　$-r = 1 - \cos\theta$.　The test fails.
Due to symmetry, assign values to θ from 0 to π.

θ	0	$\dfrac{\pi}{6}$	$\dfrac{\pi}{3}$	$\dfrac{\pi}{2}$	$\dfrac{2\pi}{3}$	$\dfrac{5\pi}{6}$	π
$r = 1 - \cos\theta$	0	$1 - \dfrac{\sqrt{3}}{2} \approx 0.1$	$\dfrac{1}{2}$	1	$\dfrac{3}{2}$	$1 + \dfrac{\sqrt{3}}{2} \approx 1.9$	2

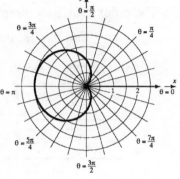

59.　$r = 1 - 3\cos\theta$　　The graph will be a limacon with an inner loop.
Check for symmetry:
Polar axis: Replace θ by $-\theta$. The result is $r = 1 - 3\cos(-\theta) = 1 - 3\cos\theta$.
　　　　The graph is symmetric with respect to the polar axis.

The line $\theta = \dfrac{\pi}{2}$: Replace θ by $\pi - \theta$.

$$r = 1 - 3\cos(\pi - \theta) = 1 - 3(\cos(\pi)\cos\theta + \sin(\pi)\sin\theta)$$

$$= 1 - 3(-\cos\theta + 0) = 1 + 3\cos\theta$$
　　　　The test fails.

The pole:　Replace r by $-r$.　$-r = 1 - 3\cos\theta$.　The test fails.
Due to symmetry, assign values to θ from 0 to π.

θ	0	$\dfrac{\pi}{6}$	$\dfrac{\pi}{3}$	$\dfrac{\pi}{2}$	$\dfrac{2\pi}{3}$	$\dfrac{5\pi}{6}$	π
$r = 1 - 3\cos\theta$	-2	$1 - \dfrac{3\sqrt{3}}{2} \approx -1.6$	$-\dfrac{1}{2}$	1	$\dfrac{5}{2}$	$1 + \dfrac{3\sqrt{3}}{2} \approx 3.6$	4

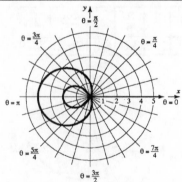

61. The graph is a cardioid whose equation is of the form $r = a + b\cos\theta$.
The graph contains the point $(6,0)$, so we have $6 = a + b\cos 0 \Rightarrow 6 = a + b(1) \Rightarrow 6 = a + b$.

The graph contains the point $\left(3, \dfrac{\pi}{2}\right)$, so we have $3 = a + b\cos\dfrac{\pi}{2} \Rightarrow 3 = a + b(0) \Rightarrow 3 = a$.

Substituting $a = 3$ into the first equation yields: $6 = a + b \Rightarrow 6 = 3 + b \Rightarrow 3 = b$.
Therefore, the graph has equation $r = 3 + 3\cos\theta$.

63. The graph is a limacon without inner loop whose equation is of the form $r = a + b\sin\theta$,
where $0 < b < a$.
The graph contains the point $(4,0)$, so we have $4 = a + b\sin 0 \Rightarrow 4 = a + b(0) \Rightarrow 4 = a$.

The graph contains the point $\left(5, \dfrac{\pi}{2}\right)$, so we have $5 = a + b\sin\dfrac{\pi}{2} \Rightarrow 5 = a + b(1) \Rightarrow 5 = a + b$.

Substituting $a = 4$ into the second equation yields: $5 = a + b \Rightarrow 5 = 4 + b \Rightarrow 1 = b$.
Therefore, the graph has equation $r = 4 + \sin\theta$.

65. $r = \dfrac{2}{1 - \cos\theta}$ Check for symmetry:

Polar axis: Replace θ by $-\theta$. The result is $r = \dfrac{2}{1 - \cos(-\theta)} = \dfrac{2}{1 - \cos\theta}$.

The graph is symmetric with respect to the polar axis.

The line $\theta = \dfrac{\pi}{2}$: Replace θ by $\pi - \theta$.

$$r = \dfrac{2}{1 - \cos(\pi - \theta)} = \dfrac{2}{1 - (\cos\pi\cos\theta + \sin\pi\sin\theta)}$$

$$= \dfrac{2}{1 - (-\cos\theta + 0)} = \dfrac{2}{1 + \cos\theta}$$

The test fails.

The pole: Replace r by $-r$. $-r = \dfrac{2}{1 - \cos\theta}$. The test fails.

Due to symmetry, assign values to θ from 0 to π.

θ	0	$\dfrac{\pi}{6}$	$\dfrac{\pi}{3}$	$\dfrac{\pi}{2}$	$\dfrac{2\pi}{3}$	$\dfrac{5\pi}{6}$	π
$r = \dfrac{2}{1 - \cos\theta}$	undefined	$\dfrac{2}{1 - \sqrt{3}/2} \approx 14.9$	4	2	$\dfrac{4}{3}$	$\dfrac{2}{1 + \sqrt{3}/2} \approx 1.1$	1

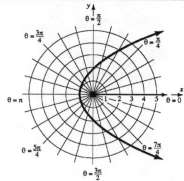

461

67. $r = \dfrac{1}{3 - 2\cos\theta}$ Check for symmetry:

Polar axis: Replace θ by $-\theta$. The result is $r = \dfrac{1}{3 - 2\cos(-\theta)} = \dfrac{1}{3 - 2\cos\theta}$.

The graph is symmetric with respect to the polar axis.

The line $\theta = \dfrac{\pi}{2}$: Replace θ by $\pi - \theta$.

$$r = \frac{1}{3 - 2\cos(\pi - \theta)} = \frac{1}{3 - 2(\cos\pi\cos\theta + \sin\pi\sin\theta)}$$

$$= \frac{1}{3 - 2(-\cos\theta + 0)} = \frac{1}{3 + 2\cos\theta}$$

The test fails.

The pole: Replace r by $-r$. $-r = \dfrac{1}{3 - 2\cos\theta}$. The test fails.

Due to symmetry, assign values to θ from 0 to π.

θ	0	$\dfrac{\pi}{6}$	$\dfrac{\pi}{3}$	$\dfrac{\pi}{2}$	$\dfrac{2\pi}{3}$	$\dfrac{5\pi}{6}$	π
$r = \dfrac{1}{3 - 2\cos\theta}$	1	$\dfrac{1}{3 - \sqrt{3}} \approx 0.8$	$\dfrac{1}{2}$	$\dfrac{1}{3}$	$\dfrac{1}{4}$	$\dfrac{1}{3 + \sqrt{3}} \approx 0.2$	$\dfrac{1}{5}$

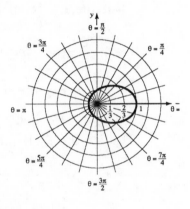

69. $r = \theta$, $\theta \geq 0$ Check for symmetry:

Polar axis: Replace θ by $-\theta$. $r = -\theta$. The test fails.

The line $\theta = \dfrac{\pi}{2}$: Replace θ by $\pi - \theta$. $r = \pi - \theta$. The test fails.

The pole: Replace r by $-r$. $-r = \theta$. The test fails.

θ	0	$\dfrac{\pi}{6}$	$\dfrac{\pi}{3}$	$\dfrac{\pi}{2}$	π	$\dfrac{3\pi}{2}$	2π
$r = \theta$	0	$\dfrac{\pi}{6} \approx 0.5$	$\dfrac{\pi}{3} \approx 1.0$	$\dfrac{\pi}{2} \approx 1.6$	$\pi \approx 3.1$	$\dfrac{3\pi}{2} \approx 4.7$	$2\pi \approx 6.3$

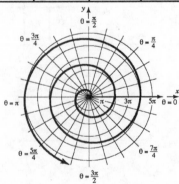

71. $r = \csc\theta - 2 = \dfrac{1}{\sin\theta} - 2, \quad 0 < \theta < \pi$ Check for symmetry:

Polar axis: Replace θ by $-\theta$. $r = \csc(-\theta) - 2 = -\csc\theta - 2$. The test fails.

The line $\theta = \dfrac{\pi}{2}$: Replace θ by $\pi - \theta$.

$$r = \csc(\pi - \theta) - 2 = \frac{1}{\sin(\pi - \theta)} - 2 = \frac{1}{\sin\pi\cos\theta - \cos\pi\sin\theta} - 2 = \frac{1}{\sin\theta} - 2 = \csc\theta - 2$$

The graph is symmetric with respect to the line $\theta = \dfrac{\pi}{2}$.

The pole: Replace r by $-r$. $-r = \csc\theta - 2$. The test fails.

Due to symmetry, assign values to θ from 0 to $\dfrac{\pi}{2}$.

θ	0	$\dfrac{\pi}{6}$	$\dfrac{\pi}{4}$	$\dfrac{\pi}{3}$	$\dfrac{\pi}{2}$
$r = \csc\theta - 2$	not defined	0	$\sqrt{2} - 2 \approx -0.6$	$\dfrac{2\sqrt{3}}{3} - 2 \approx -0.8$	-1

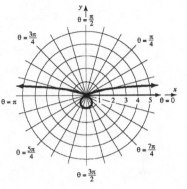

73. $r = \tan\theta, \quad -\dfrac{\pi}{2} < \theta < \dfrac{\pi}{2}$ Check for symmetry:

Polar axis: Replace θ by $-\theta$. $r = \tan(-\theta) = -\tan\theta$. The test fails.

The line $\theta = \dfrac{\pi}{2}$: Replace θ by $\pi - \theta$.

$$r = \tan(\pi - \theta) = \frac{\tan(\pi) - \tan\theta}{1 + \tan(\pi)\tan\theta} = \frac{-\tan\theta}{1} = -\tan\theta$$

The test fails.

The pole: Replace r by $-r$. $-r = \tan\theta$. The test fails.

θ	$-\dfrac{\pi}{3}$	$-\dfrac{\pi}{4}$	$-\dfrac{\pi}{6}$	0	$\dfrac{\pi}{6}$	$\dfrac{\pi}{4}$	$\dfrac{\pi}{3}$
$r = \tan\theta$	$-\sqrt{3} \approx -1.7$	-1	$-\dfrac{\sqrt{3}}{3} \approx -0.6$	0	$\dfrac{\sqrt{3}}{3} \approx 0.6$	1	$\sqrt{3} \approx 1.7$

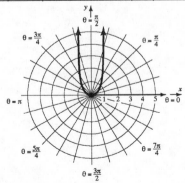

75. Convert the equation to rectangular form:
$$r \sin \theta = a \Rightarrow y = a$$
The graph of $r \sin \theta = a$ is a horizontal line a units above the pole if $a > 0$, and $|a|$ units below the pole if $a < 0$.

77. Convert the equation to rectangular form:
$$r = 2a \sin \theta, \ a > 0$$
$$r^2 = 2ar \sin \theta$$
$$x^2 + y^2 = 2ay \Rightarrow x^2 + y^2 - 2ay = 0 \Rightarrow x^2 + (y - a)^2 = a^2$$
Circle: radius a, center at rectangular coordinates $(0, a)$.

79. Convert the equation to rectangular form:
$$r = 2a \cos \theta, \ a > 0$$
$$r^2 = 2ar \cos \theta$$
$$x^2 + y^2 = 2ax \Rightarrow x^2 - 2ax + y^2 = 0 \Rightarrow (x - a)^2 + y^2 = a^2$$
Circle: radius a, center at rectangular coordinates $(a, 0)$.

81. (a) $r^2 = \cos \theta$: $\quad r^2 = \cos(\pi - \theta) \ \Rightarrow \ r^2 = -\cos \theta \quad$ Test fails.
$$(-r)^2 = \cos(-\theta) \ \Rightarrow \ r^2 = \cos \theta \quad \text{New test works.}$$
   (b) $r^2 = \sin \theta$: $\quad r^2 = \sin(\pi - \theta) \ \Rightarrow \ r^2 = \sin \theta \quad$ Test works.
$$(-r)^2 = \sin(-\theta) \ \Rightarrow \ r^2 = -\sin \theta \quad \text{New test fails.}$$

83. Answers will vary.

Polar Coordinates; Vectors

8.3 The Complex Plane; De Moivre's Theorem

1. $-4 + 3i$

3. $\cos\alpha\cos\beta - \sin\alpha\sin\beta$

5. magnitude, modulus, argument

7. three

9. False

11. $r = \sqrt{x^2 + y^2} = \sqrt{1^2 + 1^2} = \sqrt{2}$

$\tan\theta = \dfrac{y}{x} = 1 \implies \theta = 45°$

The polar form of $z = 1 + i$ is

$z = r(\cos\theta + i\sin\theta)$

$\quad = \sqrt{2}(\cos 45° + i\sin 45°)$

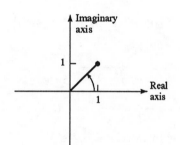

13. $r = \sqrt{x^2 + y^2} = \sqrt{\left(\sqrt{3}\right)^2 + (-1)^2} = \sqrt{4} = 2$

$\tan\theta = \dfrac{y}{x} = \dfrac{-1}{\sqrt{3}} = -\dfrac{\sqrt{3}}{3} \implies \theta = 330°$

The polar form of $z = \sqrt{3} - i$ is

$z = r(\cos\theta + i\sin\theta)$

$\quad = 2(\cos 330° + i\sin 330°)$

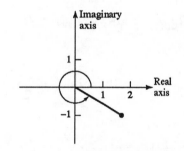

15. $r = \sqrt{x^2 + y^2} = \sqrt{0^2 + (-3)^2} = \sqrt{9} = 3$

$\tan\theta = \dfrac{y}{x} = \dfrac{-3}{0} \implies \theta = 270°$

The polar form of $z = -3i$ is

$z = r(\cos\theta + i\sin\theta)$

$\quad = 3(\cos 270° + i\sin 270°)$

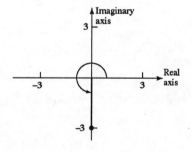

17. $r = \sqrt{x^2 + y^2} = \sqrt{4^2 + (-4)^2} = \sqrt{32} = 4\sqrt{2}$

$\tan\theta = \dfrac{y}{x} = \dfrac{-4}{4} = -1 \;\Rightarrow\; \theta = 315°$

The polar form of $z = 4 - 4i$ is

$z = r(\cos\theta + i\sin\theta)$

$\quad = 4\sqrt{2}\left(\cos 315° + i\sin 315°\right)$

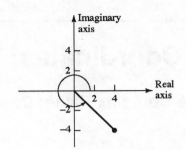

19. $r = \sqrt{x^2 + y^2} = \sqrt{3^2 + (-4)^2} = \sqrt{25} = 5$

$\tan\theta = \dfrac{y}{x} = \dfrac{-4}{3} \;\Rightarrow\; \theta \approx 306.87°$

The polar form of $z = 3 - 4i$ is

$z = r(\cos\theta + i\sin\theta)$

$\quad = 5\left(\cos(306.87°) + i\sin(306.87°)\right)$

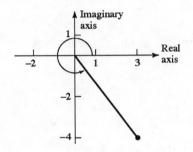

21. $r = \sqrt{x^2 + y^2} = \sqrt{(-2)^2 + 3^2} = \sqrt{13}$

$\tan\theta = \dfrac{y}{x} = \dfrac{3}{-2} = -\dfrac{3}{2} \;\Rightarrow\; \theta \approx 123.69°$

The polar form of $z = -2 + 3i$ is

$z = r(\cos\theta + i\sin\theta)$

$\quad = \sqrt{13}\left(\cos(123.69°) + i\sin(123.69°)\right)$

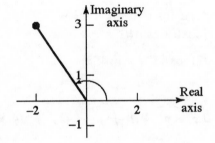

23. $2(\cos 120° + i\sin 120°) = 2\left(-\dfrac{1}{2} + \dfrac{\sqrt{3}}{2}i\right) = -1 + \sqrt{3}\,i$

25. $4\left(\cos\dfrac{7\pi}{4} + i\sin\dfrac{7\pi}{4}\right) = 4\left(\dfrac{\sqrt{2}}{2} - \dfrac{\sqrt{2}}{2}i\right) = 2\sqrt{2} - 2\sqrt{2}\,i$

27. $3\left(\cos\dfrac{3\pi}{2} + i\sin\dfrac{3\pi}{2}\right) = 3(0 - 1i) = -3i$

29. $0.2(\cos 100° + i\sin 100°) \approx 0.2(-0.1736 + 0.9848i) \approx -0.0347 + 0.1970\,i$

31. $2\left(\cos\dfrac{\pi}{18} + i\sin\dfrac{\pi}{18}\right) \approx 2(0.9848 + 0.1736i) = 1.9696 + 0.3472i$

33. $z \cdot w = 2(\cos 40° + i \sin 40°) \cdot 4(\cos 20° + i \sin 20°)$

$= 2 \cdot 4(\cos(40° + 20°) + i \sin(40° + 20°)) = 8(\cos 60° + i \sin 60°)$

$\dfrac{z}{w} = \dfrac{2(\cos 40° + i \sin 40°)}{4(\cos 20° + i \sin 20°)} = \dfrac{2}{4}(\cos(40° - 20°) + i \sin(40° - 20°))$

$= \dfrac{1}{2}(\cos 20° + i \sin 20°)$

35. $z \cdot w = 3(\cos 130° + i \sin 130°) \cdot 4(\cos 270° + i \sin 270°)$

$= 3 \cdot 4(\cos(130° + 270°) + i \sin(130° + 270°)) = 12(\cos 400° + i \sin 400°)$

$= 12(\cos(400° - 360°) + i \sin(400° - 360°)) = 12(\cos 40° + i \sin 40°)$

$\dfrac{z}{w} = \dfrac{3(\cos 130° + i \sin 130°)}{4(\cos 270° + i \sin 270°)} = \dfrac{3}{4}(\cos(130° - 270°) + i \sin(130° - 270°))$

$= \dfrac{3}{4}(\cos(-140°) + i \sin(-140°)) = \dfrac{3}{4}(\cos(360° - 140°) + i \sin(360° - 140°))$

$= \dfrac{3}{4}(\cos 220° + i \sin 220°)$

37. $z \cdot w = 2\left(\cos \dfrac{\pi}{8} + i \sin \dfrac{\pi}{8}\right) \cdot 2\left(\cos \dfrac{\pi}{10} + i \sin \dfrac{\pi}{10}\right) = 2 \cdot 2\left(\cos\left(\dfrac{\pi}{8} + \dfrac{\pi}{10}\right) + i \sin\left(\dfrac{\pi}{8} + \dfrac{\pi}{10}\right)\right)$

$= 4\left(\cos \dfrac{9\pi}{40} + i \sin \dfrac{9\pi}{40}\right)$

$\dfrac{z}{w} = \dfrac{2\left(\cos \dfrac{\pi}{8} + i \sin \dfrac{\pi}{8}\right)}{2\left(\cos \dfrac{\pi}{10} + i \sin \dfrac{\pi}{10}\right)} = \dfrac{2}{2}\left(\cos\left(\dfrac{\pi}{8} - \dfrac{\pi}{10}\right) + i \sin\left(\dfrac{\pi}{8} - \dfrac{\pi}{10}\right)\right) = \cos \dfrac{\pi}{40} + i \sin \dfrac{\pi}{40}$

39. $z = 2 + 2i$ $r = \sqrt{2^2 + 2^2} = \sqrt{8} = 2\sqrt{2}$ $\tan\theta = \dfrac{2}{2} = 1 \Rightarrow \theta = 45°$

$z = 2\sqrt{2}(\cos 45° + i \sin 45°)$

$w = \sqrt{3} - i$ $r = \sqrt{(\sqrt{3})^2 + (-1)^2} = \sqrt{4} = 2$ $\tan\theta = \dfrac{-1}{\sqrt{3}} = -\dfrac{\sqrt{3}}{3} \Rightarrow \theta = 330°$

$w = 2(\cos 330° + i \sin 330°)$

$z \cdot w = 2\sqrt{2}(\cos 45° + i \sin 45°) \cdot 2(\cos 330° + i \sin 330°)$

$= 2\sqrt{2} \cdot 2(\cos(45° + 330°) + i \sin(45° + 330°)) = 4\sqrt{2}(\cos 375° + i \sin 375°)$

$= 4\sqrt{2}(\cos(375° - 360°) + i \sin(375° - 360°)) = 4\sqrt{2}(\cos 15° + i \sin 15°)$

$\dfrac{z}{w} = \dfrac{2\sqrt{2}(\cos 45° + i \sin 45°)}{2(\cos 330° + i \sin 330°)} = \dfrac{2\sqrt{2}}{2}(\cos(45° - 330°) + i \sin(45° - 330°))$

$= \sqrt{2}(\cos(-285°) + i \sin(-285°)) = \sqrt{2}(\cos(360° - 285°) + i \sin(360° - 285°))$

$= \sqrt{2}(\cos 75° + i \sin 75°)$

41. $\left[4(\cos 40° + i\sin 40°)\right]^3 = 4^3(\cos(3 \cdot 40°) + i\sin(3 \cdot 40°)) = 64(\cos 120° + i\sin 120°)$

$$= 64\left(-\frac{1}{2} + \frac{\sqrt{3}}{2}i\right) = -32 + 32\sqrt{3}\,i$$

43. $\left[2\left(\cos\frac{\pi}{10} + i\sin\frac{\pi}{10}\right)\right]^5 = 2^5\left(\cos\left(5 \cdot \frac{\pi}{10}\right) + i\sin\left(5 \cdot \frac{\pi}{10}\right)\right) = 32\left(\cos\frac{\pi}{2} + i\sin\frac{\pi}{2}\right)$

$$= 32(0 + 1\,i) = 0 + 32\,i$$

45. $\left[\sqrt{3}(\cos 10° + i\sin 10°)\right]^6 = \left(\sqrt{3}\right)^6(\cos(6 \cdot 10°) + i\sin(6 \cdot 10°)) = 27(\cos 60° + i\sin 60°)$

$$= 27\left(\frac{1}{2} + \frac{\sqrt{3}}{2}i\right) = \frac{27}{2} + \frac{27\sqrt{3}}{2}i$$

47. $\left[\sqrt{5}\left(\cos\frac{3\pi}{16} + i\sin\frac{3\pi}{16}\right)\right]^4 = \left(\sqrt{5}\right)^4\left(\cos\left(4 \cdot \frac{3\pi}{16}\right) + i\sin\left(4 \cdot \frac{3\pi}{16}\right)\right)$

$$= 25\left(\cos\frac{3\pi}{4} + i\sin\frac{3\pi}{4}\right) = 25\left(-\frac{\sqrt{2}}{2} + \frac{\sqrt{2}}{2}i\right) = -\frac{25\sqrt{2}}{2} + \frac{25\sqrt{2}}{2}i$$

49. $1 - i \qquad r = \sqrt{1^2 + (-1)^2} = \sqrt{2} \qquad \tan\theta = \frac{-1}{1} = -1 \Rightarrow \theta = \frac{7\pi}{4}$

$$1 - i = \sqrt{2}\left(\cos\frac{7\pi}{4} + i\sin\frac{7\pi}{4}\right)$$

$$(1-i)^5 = \left[\sqrt{2}\left(\cos\frac{7\pi}{4} + i\sin\frac{7\pi}{4}\right)\right]^5 = \left(\sqrt{2}\right)^5\left(\cos\left(5 \cdot \frac{7\pi}{4}\right) + i\sin\left(5 \cdot \frac{7\pi}{4}\right)\right)$$

$$= 4\sqrt{2}\left(\cos\frac{35\pi}{4} + i\sin\frac{35\pi}{4}\right) = 4\sqrt{2}\left(-\frac{\sqrt{2}}{2} + \frac{\sqrt{2}}{2}i\right) = -4 + 4\,i$$

51. $\sqrt{2} - i \qquad r = \sqrt{\left(\sqrt{2}\right)^2 + (-1)^2} = \sqrt{3} \qquad \tan\theta = \frac{-1}{\sqrt{2}} = -\frac{\sqrt{2}}{2} \Rightarrow \theta \approx 324.7°$

$$\sqrt{2} - i \approx \sqrt{3}(\cos(324.7°) + i\sin(324.7°))$$

$$\left(\sqrt{2} - i\right)^6 \approx \left[\sqrt{3}(\cos(324.7°) + i\sin(324.7°))\right]^6 = \left(\sqrt{3}\right)^6(\cos(6 \cdot 324.7°) + i\sin(6 \cdot 324.7°))$$

$$= 27(\cos(1948.2°) + i\sin(1948.2°)) \approx 27(-0.8499 + 0.5270\,i)$$

$$= -22.9473 + 14.229\,i$$

53. $1+i$ $r = \sqrt{1^2 + 1^2} = \sqrt{2}$ $\tan\theta = \dfrac{1}{1} = 1 \Rightarrow \theta = 45°$

$1 + i = \sqrt{2}(\cos 45° + i\sin 45°)$

The three complex cube roots of $1 + i = \sqrt{2}(\cos 45° + i\sin 45°)$ are:

$z_k = \sqrt[3]{\sqrt{2}}\left[\cos\left(\dfrac{45°}{3} + \dfrac{360°k}{3}\right) + i\sin\left(\dfrac{45°}{3} + \dfrac{360°k}{3}\right)\right]$

$ = \sqrt[6]{2}\left[\cos(15° + 120°k) + i\sin(15° + 120°k)\right]$

$z_0 = \sqrt[6]{2}\left[\cos(15° + 120°\cdot 0) + i\sin(15° + 120°\cdot 0)\right] = \sqrt[6]{2}(\cos 15° + i\sin 15°)$

$z_1 = \sqrt[6]{2}\left[\cos(15° + 120°\cdot 1) + i\sin(15° + 120°\cdot 1)\right] = \sqrt[6]{2}(\cos 135° + i\sin 135°)$

$z_2 = \sqrt[6]{2}\left[\cos(15° + 120°\cdot 2) + i\sin(15° + 120°\cdot 2)\right] = \sqrt[6]{2}(\cos 255° + i\sin 255°)$

55. $4 - 4\sqrt{3}\,i$ $r = \sqrt{4^2 + \left(-4\sqrt{3}\right)^2} = \sqrt{64} = 8$ $\tan\theta = \dfrac{-4\sqrt{3}}{4} = -\sqrt{3} \Rightarrow \theta = 300°$

$4 - 4\sqrt{3}\,i = 8(\cos 300° + i\sin 300°)$

The four complex fourth roots of $4 - 4\sqrt{3}\,i = 8(\cos 300° + i\sin 300°)$ are:

$z_k = \sqrt[4]{8}\left[\cos\left(\dfrac{300°}{4} + \dfrac{360°k}{4}\right) + i\sin\left(\dfrac{300°}{4} + \dfrac{360°k}{4}\right)\right]$

$ = \sqrt[4]{8}\left[\cos(75° + 90°k) + i\sin(75° + 90°k)\right]$

$z_0 = \sqrt[4]{8}\left[\cos(75° + 90°\cdot 0) + i\sin(75° + 90°\cdot 0)\right] = \sqrt[4]{8}(\cos 75° + i\sin 75°)$

$z_1 = \sqrt[4]{8}\left[\cos(75° + 90°\cdot 1) + i\sin(75° + 90°\cdot 1)\right] = \sqrt[4]{8}(\cos 165° + i\sin 165°)$

$z_2 = \sqrt[4]{8}\left[\cos(75° + 90°\cdot 2) + i\sin(75° + 90°\cdot 2)\right] = \sqrt[4]{8}(\cos 255° + i\sin 255°)$

$z_3 = \sqrt[4]{8}\left[\cos(75° + 90°\cdot 3) + i\sin(75° + 90°\cdot 3)\right] = \sqrt[4]{8}(\cos 345° + i\sin 345°)$

57. $-16i$ $r = \sqrt{0^2 + (-16)^2} = \sqrt{256} = 16$ $\tan\theta = \dfrac{-16}{0} \Rightarrow \theta = 270°$

$-16i = 16(\cos 270° + i\sin 270°)$

The four complex fourth roots of $-16i = 16(\cos 270° + i\sin 270°)$ are:

$z_k = \sqrt[4]{16}\left[\cos\left(\dfrac{270°}{4} + \dfrac{360°k}{4}\right) + i\sin\left(\dfrac{270°}{4} + \dfrac{360°k}{4}\right)\right]$

$ = 2\left[\cos(67.5° + 90°k) + i\sin(67.5° + 90°k)\right]$

$z_0 = 2\left[\cos(67.5° + 90°\cdot 0) + i\sin(67.5° + 90°\cdot 0)\right] = 2\left(\cos(67.5°) + i\sin(67.5°)\right)$

$z_1 = 2\left[\cos(67.5° + 90°\cdot 1) + i\sin(67.5° + 90°\cdot 1)\right] = 2\left(\cos(157.5°) + i\sin(157.5°)\right)$

$z_2 = 2\left[\cos(67.5° + 90°\cdot 2) + i\sin(67.5° + 90°\cdot 2)\right] = 2\left(\cos(247.5°) + i\sin(247.5°)\right)$

$z_3 = 2\left[\cos(67.5° + 90°\cdot 3) + i\sin(67.5° + 90°\cdot 3)\right] = 2\left(\cos(337.5°) + i\sin(337.5°)\right)$

59. i $r = \sqrt{0^2 + 1^2} = \sqrt{1} = 1$ $\tan\theta = \dfrac{1}{0} \Rightarrow \theta = 90°$

$i = 1(\cos 90° + i\sin 90°)$

The five complex fifth roots of $i = 1(\cos 90° + i\sin 90°)$ are:

$$z_k = \sqrt[5]{1}\left[\cos\left(\frac{90°}{5} + \frac{360°k}{5}\right) + i\sin\left(\frac{90°}{5} + \frac{360°k}{5}\right)\right]$$

$$= 1\left[\cos(18° + 72°k) + i\sin(18° + 72°k)\right]$$

$z_0 = 1\left[\cos(18° + 72°\cdot 0) + i\sin(18° + 72°\cdot 0)\right] = \cos 18° + i\sin 18°$

$z_1 = 1\left[\cos(18° + 72°\cdot 1) + i\sin(18° + 72°\cdot 1)\right] = \cos 90° + i\sin 90°$

$z_2 = 1\left[\cos(18° + 72°\cdot 2) + i\sin(18° + 72°\cdot 2)\right] = \cos 162° + i\sin 162°$

$z_3 = 1\left[\cos(18° + 72°\cdot 3) + i\sin(18° + 72°\cdot 3)\right] = \cos 234° + i\sin 234°$

$z_4 = 1\left[\cos(18° + 72°\cdot 4) + i\sin(18° + 72°\cdot 4)\right] = \cos 306° + i\sin 306°$

61. $1 = 1 + 0i$ $r = \sqrt{1^2 + 0^2} = \sqrt{1} = 1$ $\tan\theta = \dfrac{0}{1} = 0 \Rightarrow \theta = 0°$

$1 + 0i = 1(\cos 0° + i\sin 0°)$

The four complex fourth roots of unity are:

$$z_k = \sqrt[4]{1}\left[\cos\left(\frac{0°}{4} + \frac{360°k}{4}\right) + i\sin\left(\frac{0°}{4} + \frac{360°k}{4}\right)\right]$$

$$= 1\left[\cos(90°k) + i\sin(90°k)\right]$$

$z_0 = \cos(90°\cdot 0) + i\sin(90°\cdot 0) = \cos 0° + i\sin 0° = 1 + 0i = 1$

$z_1 = \cos(90°\cdot 1) + i\sin(90°\cdot 1) = \cos 90° + i\sin 90° = 0 + 1i = i$

$z_2 = \cos(90°\cdot 2) + i\sin(90°\cdot 2) = \cos 180° + i\sin 180° = -1 + 0i = -1$

$z_3 = \cos(90°\cdot 3) + i\sin(90°\cdot 3) = \cos 270° + i\sin 270° = 0 - 1i = -i$

The complex fourth roots of unity are: $1,\ i,\ -1,\ -i$.

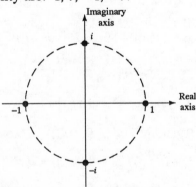

63. Let $w = r(\cos\theta + i\sin\theta)$ be a complex number. If $w \neq 0$, there are n distinct nth roots of w, given by the formula:

$$z_k = \sqrt[n]{r}\left(\cos\left(\frac{\theta}{n} + \frac{2k\pi}{n}\right) + i\sin\left(\frac{\theta}{n} + \frac{2k\pi}{n}\right)\right), \text{ where } k = 0, 1, 2, ..., n-1$$

$|z_k| = \sqrt[n]{r}$ for all k

65. Examining the formula for the distinct complex nth roots of the complex number $w = r(\cos\theta + i\sin\theta)$,

$$z_k = \sqrt[n]{r}\left(\cos\left(\frac{\theta}{n} + \frac{2k\pi}{n}\right) + i\sin\left(\frac{\theta}{n} + \frac{2k\pi}{n}\right)\right), \text{where } k = 0, 1, 2, ..., n-1$$

we see that the z_k are spaced apart by an angle of $\dfrac{2\pi}{n}$.

Polar Coordinates; Vectors

8.4   Vectors

1.   unit

3.   horizontal, vertical

5.   True

7.   **v + w**

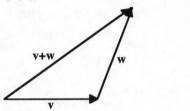

9.   **3v**

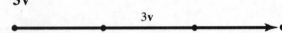

11.   **v − w**

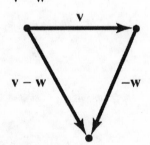

13. **3v + u − 2w**

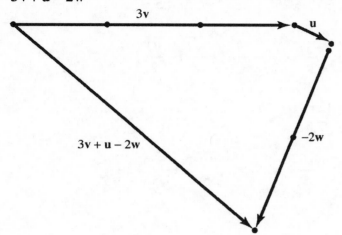

15. True

17. False   **C = −F + E − D**

19. False   **D − E = H + G**

21. True

23. If $\|\mathbf{v}\| = 4$, then $\|3\mathbf{v}\| = |3|\|\mathbf{v}\| = 3(4) = 12$.

25. $P = (0, 0)$, $Q = (3, 4)$ $\mathbf{v} = (3 - 0)\mathbf{i} + (4 - 0)\mathbf{j} = 3\mathbf{i} + 4\mathbf{j}$

27. $P = (3, 2)$, $Q = (5, 6)$ $\mathbf{v} = (5 - 3)\mathbf{i} + (6 - 2)\mathbf{j} = 2\mathbf{i} + 4\mathbf{j}$

29. $P = (-2, -1)$, $Q = (6, -2)$ $\mathbf{v} = (6 - (-2))\mathbf{i} + (-2 - (-1))\mathbf{j} = 8\mathbf{i} - \mathbf{j}$

31. $P = (1, 0)$, $Q = (0, 1)$ $\mathbf{v} = (0 - 1)\mathbf{i} + (1 - 0)\mathbf{j} = -\mathbf{i} + \mathbf{j}$

33. For $\mathbf{v} = 3\mathbf{i} - 4\mathbf{j}$, $\|\mathbf{v}\| = \sqrt{3^2 + (-4)^2} = \sqrt{25} = 5$

35. For $\mathbf{v} = \mathbf{i} - \mathbf{j}$, $\|\mathbf{v}\| = \sqrt{1^2 + (-1)^2} = \sqrt{2}$

37. For $\mathbf{v} = -2\mathbf{i} + 3\mathbf{j}$, $\|\mathbf{v}\| = \sqrt{(-2)^2 + 3^2} = \sqrt{13}$

39. $2\mathbf{v} + 3\mathbf{w} = 2(3\mathbf{i} - 5\mathbf{j}) + 3(-2\mathbf{i} + 3\mathbf{j}) = 6\mathbf{i} - 10\mathbf{j} - 6\mathbf{i} + 9\mathbf{j} = -\mathbf{j}$

41. $\|\mathbf{v} - \mathbf{w}\| = \|(3\mathbf{i} - 5\mathbf{j}) - (-2\mathbf{i} + 3\mathbf{j})\| = \|5\mathbf{i} - 8\mathbf{j}\| = \sqrt{5^2 + (-8)^2} = \sqrt{89}$

43. $\|\mathbf{v}\| - \|\mathbf{w}\| = \|3\mathbf{i} - 5\mathbf{j}\| - \|-2\mathbf{i} + 3\mathbf{j}\| = \sqrt{3^2 + (-5)^2} - \sqrt{(-2)^2 + 3^2} = \sqrt{34} - \sqrt{13}$

45. $\mathbf{u} = \dfrac{\mathbf{v}}{\|\mathbf{v}\|} = \dfrac{5\mathbf{i}}{\|5\mathbf{i}\|} = \dfrac{5\mathbf{i}}{\sqrt{25 + 0}} = \dfrac{5\mathbf{i}}{5} = \mathbf{i}$

47. $\mathbf{u} = \dfrac{\mathbf{v}}{\|\mathbf{v}\|} = \dfrac{3\mathbf{i} - 4\mathbf{j}}{\|3\mathbf{i} - 4\mathbf{j}\|} = \dfrac{3\mathbf{i} - 4\mathbf{j}}{\sqrt{3^2 + (-4)^2}} = \dfrac{3\mathbf{i} - 4\mathbf{j}}{\sqrt{25}} = \dfrac{3\mathbf{i} - 4\mathbf{j}}{5} = \dfrac{3}{5}\mathbf{i} - \dfrac{4}{5}\mathbf{j}$

49. $\mathbf{u} = \dfrac{\mathbf{v}}{\|\mathbf{v}\|} = \dfrac{\mathbf{i} - \mathbf{j}}{\|\mathbf{i} - \mathbf{j}\|} = \dfrac{\mathbf{i} - \mathbf{j}}{\sqrt{1^2 + (-1)^2}} = \dfrac{\mathbf{i} - \mathbf{j}}{\sqrt{2}} = \dfrac{1}{\sqrt{2}}\mathbf{i} - \dfrac{1}{\sqrt{2}}\mathbf{j} = \dfrac{\sqrt{2}}{2}\mathbf{i} - \dfrac{\sqrt{2}}{2}\mathbf{j}$

51. Let $\mathbf{v} = a\mathbf{i} + b\mathbf{j}$. We want $\|\mathbf{v}\| = 4$ and $a = 2b$.

$\|\mathbf{v}\| = \sqrt{a^2 + b^2} = \sqrt{(2b)^2 + b^2} = \sqrt{5b^2}$

$\sqrt{5b^2} = 4 \;\Rightarrow\; 5b^2 = 16 \;\Rightarrow\; b^2 = \dfrac{16}{5} \;\Rightarrow\; b = \pm\sqrt{\dfrac{16}{5}} = \pm\dfrac{4}{\sqrt{5}} = \pm\dfrac{4\sqrt{5}}{5}$

$a = 2b = \pm\dfrac{8\sqrt{5}}{5}$

$\mathbf{v} = \dfrac{8\sqrt{5}}{5}\mathbf{i} + \dfrac{4\sqrt{5}}{5}\mathbf{j}$ or $\mathbf{v} = -\dfrac{8\sqrt{5}}{5}\mathbf{i} - \dfrac{4\sqrt{5}}{5}\mathbf{j}$

53. $\mathbf{v} = 2\mathbf{i} - \mathbf{j}, \quad \mathbf{w} = x\mathbf{i} + 3\mathbf{j} \quad \|\mathbf{v} + \mathbf{w}\| = 5$

$\|\mathbf{v} + \mathbf{w}\| = \|2\mathbf{i} - \mathbf{j} + x\mathbf{i} + 3\mathbf{j}\| = \|(2 + x)\mathbf{i} + 2\mathbf{j}\| = \sqrt{(2 + x)^2 + 2^2}$

$\qquad\qquad = \sqrt{x^2 + 4x + 4 + 4} = \sqrt{x^2 + 4x + 8}$

Solve for x:

$\sqrt{x^2 + 4x + 8} = 5 \Rightarrow x^2 + 4x + 8 = 25 \Rightarrow x^2 + 4x - 17 = 0$

$\qquad\qquad x = \dfrac{-4 \pm \sqrt{16 - 4(1)(-17)}}{2(1)} = \dfrac{-4 \pm \sqrt{84}}{2} = \dfrac{-4 \pm 2\sqrt{21}}{2} = -2 \pm \sqrt{21}$

$\qquad\qquad x = -2 + \sqrt{21} \approx 2.58 \;\text{ or }\; x = -2 - \sqrt{21} \approx -6.58$

55. $\|\mathbf{v}\| = 5, \quad \alpha = 60°$

$\mathbf{v} = \|\mathbf{v}\|(\cos\alpha\,\mathbf{i} + \sin\alpha\,\mathbf{j}) = 5\left(\cos(60°)\mathbf{i} + \sin(60°)\mathbf{j}\right) = 5\left(\dfrac{1}{2}\mathbf{i} + \dfrac{\sqrt{3}}{2}\mathbf{j}\right) = \dfrac{5}{2}\mathbf{i} + \dfrac{5\sqrt{3}}{2}\mathbf{j}$

57. $\|\mathbf{v}\| = 14, \quad \alpha = 120°$

$\mathbf{v} = \|\mathbf{v}\|(\cos\alpha\,\mathbf{i} + \sin\alpha\,\mathbf{j}) = 14\left(\cos(120°)\mathbf{i} + \sin(120°)\mathbf{j}\right) = 14\left(-\dfrac{1}{2}\mathbf{i} + \dfrac{\sqrt{3}}{2}\mathbf{j}\right) = -7\mathbf{i} + 7\sqrt{3}\mathbf{j}$

59. $\|\mathbf{v}\| = 25, \quad \alpha = 330°$

$\mathbf{v} = \|\mathbf{v}\|(\cos\alpha\,\mathbf{i} + \sin\alpha\,\mathbf{j}) = 25\left(\cos(330°)\mathbf{i} + \sin(330°)\mathbf{j}\right) = 25\left(\dfrac{\sqrt{3}}{2}\mathbf{i} - \dfrac{1}{2}\mathbf{j}\right) = \dfrac{25\sqrt{3}}{2}\mathbf{i} - \dfrac{25}{2}\mathbf{j}$

61. $\mathbf{F} = 40\left(\cos(30°)\mathbf{i} + \sin(30°)\mathbf{j}\right) = 40\left(\dfrac{\sqrt{3}}{2}\mathbf{i} + \dfrac{1}{2}\mathbf{j}\right) = 20\sqrt{3}\mathbf{i} + 20\mathbf{j} = 20\left(\sqrt{3}\mathbf{i} + \mathbf{j}\right)$

63. $\mathbf{F_1} = 40\big(\cos(30°)\mathbf{i} + \sin(30°)\mathbf{j}\big) = 40\left(\dfrac{\sqrt{3}}{2}\mathbf{i} + \dfrac{1}{2}\mathbf{j}\right) = 20\sqrt{3}\,\mathbf{i} + 20\mathbf{j}$

$\mathbf{F_2} = 60\big(\cos(-45°)\mathbf{i} + \sin(-45°)\mathbf{j}\big) = 60\left(\dfrac{\sqrt{2}}{2}\mathbf{i} - \dfrac{\sqrt{2}}{2}\mathbf{j}\right) = 30\sqrt{2}\,\mathbf{i} - 30\sqrt{2}\,\mathbf{j}$

$\mathbf{F_1} + \mathbf{F_2} = 20\sqrt{3}\,\mathbf{i} + 20\mathbf{j} + 30\sqrt{2}\,\mathbf{i} - 30\sqrt{2}\,\mathbf{j} = \big(20\sqrt{3} + 30\sqrt{2}\big)\mathbf{i} + \big(20 - 30\sqrt{2}\big)\mathbf{j}$

magnitude of $\mathbf{F_1} + \mathbf{F_2} = \sqrt{\big(20\sqrt{3} + 30\sqrt{2}\big)^2 + \big(20 - 30\sqrt{2}\big)^2} \approx 80.26$ newtons

direction of $\mathbf{F_1} + \mathbf{F_2} = \tan^{-1}\left(\dfrac{20 - 30\sqrt{2}}{20\sqrt{3} + 30\sqrt{2}}\right) \approx -16.22°$

65. Let $\mathbf{F_1}$ be the tension on the left cable and $\mathbf{F_2}$ be the tension on the right cable.
Let $\mathbf{F_3}$ represent the force of the weight of the box.

$\mathbf{F_1} = \|\mathbf{F_1}\|\big(\cos(155°)\mathbf{i} + \sin(155°)\mathbf{j}\big) \approx \|\mathbf{F_1}\|(-0.9063\mathbf{i} + 0.4226\mathbf{j})$

$\mathbf{F_2} = \|\mathbf{F_2}\|\big(\cos(40°)\mathbf{i} + \sin(40°)\mathbf{j}\big) \approx \|\mathbf{F_2}\|(0.7660\mathbf{i} + 0.6428\mathbf{j})$

$\mathbf{F_3} = -1000\mathbf{j}$

For equilibrium, the sum of the force vectors must be zero.

$\mathbf{F_1} + \mathbf{F_2} + \mathbf{F_3} = -0.9063\|\mathbf{F_1}\|\mathbf{i} + 0.4226\|\mathbf{F_1}\|\mathbf{j} + 0.7660\|\mathbf{F_2}\|\mathbf{i} + 0.6428\|\mathbf{F_2}\|\mathbf{j} - 1000\mathbf{j}$

$\quad = \big(-0.9063\|\mathbf{F_1}\| + 0.7660\|\mathbf{F_2}\|\big)\mathbf{i} + \big(0.4226\|\mathbf{F_1}\| + 0.6428\|\mathbf{F_2}\| - 1000\big)\mathbf{j} = 0$

Set the $\mathbf{i}$ and $\mathbf{j}$ components equal to zero and solve:

$\begin{cases} -0.9063\|\mathbf{F_1}\| + 0.7660\|\mathbf{F_2}\| = 0 \ \Rightarrow\ \|\mathbf{F_2}\| = \dfrac{0.9063}{0.7660}\|\mathbf{F_1}\| = 1.1832\|\mathbf{F_1}\| \\[2mm] 0.4226\|\mathbf{F_1}\| + 0.6428\|\mathbf{F_2}\| - 1000 = 0 \end{cases}$

$0.4226\|\mathbf{F_1}\| + 0.6428\big(1.1832\|\mathbf{F_1}\|\big) - 1000 = 0 \Rightarrow 1.1832\|\mathbf{F_1}\| = 1000$

$\|\mathbf{F_1}\| = 845.17$ pounds

$\|\mathbf{F_2}\| = 1.1832(845.17) = 1000$ pounds

The tension in the left cable is about 845.17 pounds and the tension in the right cable is about 1000 pounds.

67. Let $\mathbf{F_1}$ be the tension on the left end of the rope and $\mathbf{F_2}$ be the tension on the right end of the rope. Let $\mathbf{F_3}$ represent the force of the weight of the tightrope walker.

$\mathbf{F_1} = \|\mathbf{F_1}\|\big(\cos(175.8°)\mathbf{i} + \sin(175.8°)\mathbf{j}\big) \approx \|\mathbf{F_1}\|(-0.9973\mathbf{i} + 0.0732\mathbf{j})$

$\mathbf{F_2} = \|\mathbf{F_2}\|\big(\cos(3.7°)\mathbf{i} + \sin(3.7°)\mathbf{j}\big) \approx \|\mathbf{F_2}\|(0.9979\mathbf{i} + 0.0645\mathbf{j})$

$\mathbf{F_3} = -150\mathbf{j}$

For equilibrium, the sum of the force vectors must be zero.

$\mathbf{F_1} + \mathbf{F_2} + \mathbf{F_3} = -0.9973\|\mathbf{F_1}\|\mathbf{i} + 0.0732\|\mathbf{F_1}\|\mathbf{j} + 0.9979\|\mathbf{F_2}\|\mathbf{i} + 0.0645\|\mathbf{F_2}\|\mathbf{j} - 150\mathbf{j}$

$\quad = \big(-0.9973\|\mathbf{F_1}\| + 0.9979\|\mathbf{F_2}\|\big)\mathbf{i} + \big(0.0732\|\mathbf{F_1}\| + 0.0645\|\mathbf{F_2}\| - 150\big)\mathbf{j} = 0$

Set the **i** and **j** components equal to zero and solve:

$$\begin{cases} -0.9973\|\mathbf{F_1}\| + 0.9979\|\mathbf{F_2}\| = 0 \;\Rightarrow\; \|\mathbf{F_2}\| = \dfrac{0.9973}{0.9979}\|\mathbf{F_1}\| = 0.9994\|\mathbf{F_1}\| \\[2mm] 0.0732\|\mathbf{F_1}\| + 0.0645\|\mathbf{F_2}\| - 150 = 0 \end{cases}$$

$$0.0732\|\mathbf{F_1}\| + 0.0645\big(0.9994\|\mathbf{F_1}\|\big) - 150 = 0 \Rightarrow 0.1377\|\mathbf{F_1}\| = 150$$

$$\|\mathbf{F_1}\| = 1089.3 \text{ pounds}; \quad \|\mathbf{F_2}\| = 0.9994(1089.3) = 1088.6 \text{ pounds}$$

The tension in the left end of the rope is about 1089.3 pounds and the tension in the right end of the rope is about 1088.6 pounds.

69. The given forces are: $\mathbf{F_1} = -3\mathbf{i}$; $\mathbf{F_2} = -\mathbf{i} + 4\mathbf{j}$; $\mathbf{F_3} = 4\mathbf{i} - 2\mathbf{j}$; $\mathbf{F_4} = -4\mathbf{j}$
A vector $\mathbf{v} = a\mathbf{i} + b\mathbf{j}$ needs to be added for equilibrium. Find vector $\mathbf{v} = a\mathbf{i} + b\mathbf{j}$:

$$\mathbf{F_1} + \mathbf{F_2} + \mathbf{F_3} + \mathbf{F_4} + \mathbf{v} = \mathbf{0}$$

$$-3\mathbf{i} + (-\mathbf{i} + 4\mathbf{j}) + (4\mathbf{i} - 2\mathbf{j}) + (-4\mathbf{j}) + (a\mathbf{i} + b\mathbf{j}) = \mathbf{0}$$

$$0\mathbf{i} - 2\mathbf{j} + (a\mathbf{i} + b\mathbf{j}) = \mathbf{0}$$

$$a\mathbf{i} + (-2 + b)\mathbf{j} = \mathbf{0}$$

$$a = 0$$

$$-2 + b = 0 \;\Rightarrow\; b = 2$$

Therefore, $\mathbf{v} = 2\mathbf{j}$.

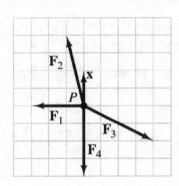

71. Answers will vary.

Polar Coordinates; Vectors

8.5 The Dot Product

1. $c^2 = a^2 + b^2 - 2ab\cos\gamma$

3. parallel

5. True

7. $\mathbf{v} = \mathbf{i} - \mathbf{j}, \quad \mathbf{w} = \mathbf{i} + \mathbf{j}$

   (a) $\mathbf{v} \cdot \mathbf{w} = 1(1) + (-1)(1) = 1 - 1 = 0$

   (b) $\cos\theta = \dfrac{\mathbf{v} \cdot \mathbf{w}}{\|\mathbf{v}\|\|\mathbf{w}\|} = \dfrac{0}{\sqrt{1^2 + (-1)^2}\sqrt{1^2 + 1^2}} = \dfrac{0}{\sqrt{2}\sqrt{2}} = \dfrac{0}{2} = 0 \Rightarrow \theta = 90°$

   (c) The vectors are orthogonal.

9. $\mathbf{v} = 2\mathbf{i} + \mathbf{j}, \quad \mathbf{w} = \mathbf{i} + 2\mathbf{j}$

   (a) $\mathbf{v} \cdot \mathbf{w} = 2(1) + 1(2) = 2 + 2 = 4$

   (b) $\cos\theta = \dfrac{\mathbf{v} \cdot \mathbf{w}}{\|\mathbf{v}\|\|\mathbf{w}\|} = \dfrac{4}{\sqrt{2^2 + 1^2}\sqrt{1^2 + 2^2}} = \dfrac{4}{\sqrt{5}\sqrt{5}} = \dfrac{4}{5} = 0.80 \Rightarrow \theta \approx 36.87°$

   (c) The vectors are neither parallel nor orthogonal.

11. $\mathbf{v} = \sqrt{3}\,\mathbf{i} - \mathbf{j}, \quad \mathbf{w} = \mathbf{i} + \mathbf{j}$

    (a) $\mathbf{v} \cdot \mathbf{w} = \sqrt{3}(1) + (-1)(1) = \sqrt{3} - 1$

    (b) $\cos\theta = \dfrac{\mathbf{v} \cdot \mathbf{w}}{\|\mathbf{v}\|\|\mathbf{w}\|} = \dfrac{\sqrt{3} - 1}{\sqrt{\left(\sqrt{3}\right)^2 + (-1)^2}\sqrt{1^2 + 1^2}} = \dfrac{\sqrt{3} - 1}{\sqrt{4}\sqrt{2}} = \dfrac{\sqrt{3} - 1}{2\sqrt{2}} = \dfrac{\sqrt{6} - \sqrt{2}}{4} \Rightarrow \theta = 75°$

    (c) The vectors are neither parallel nor orthogonal.

13. $\mathbf{v} = 3\mathbf{i} + 4\mathbf{j}, \quad \mathbf{w} = 4\mathbf{i} + 3\mathbf{j}$

    (a) $\mathbf{v} \cdot \mathbf{w} = 3(4) + 4(3) = 12 + 12 = 24$

    (b) $\cos\theta = \dfrac{\mathbf{v} \cdot \mathbf{w}}{\|\mathbf{v}\|\|\mathbf{w}\|} = \dfrac{24}{\sqrt{3^2 + 4^2}\sqrt{4^2 + 3^2}} = \dfrac{24}{\sqrt{25}\sqrt{25}} = \dfrac{24}{25} = 0.96 \Rightarrow \theta \approx 16.26°$

    (c) The vectors are neither parallel nor orthogonal.

15. $\mathbf{v} = 4\mathbf{i}, \quad \mathbf{w} = \mathbf{j}$

  (a) $\mathbf{v} \cdot \mathbf{w} = 4(0) + 0(1) = 0 + 0 = 0$

  (b) $\cos\theta = \dfrac{\mathbf{v} \cdot \mathbf{w}}{\|\mathbf{v}\|\|\mathbf{w}\|} = \dfrac{0}{\sqrt{4^2 + 0^2}\sqrt{0^2 + 1^2}} = \dfrac{0}{4 \cdot 1} = \dfrac{0}{4} = 0 \Rightarrow \theta = 90°$

  (c) The vectors are orthogonal.

17. $\mathbf{v} = \mathbf{i} - a\mathbf{j}, \quad \mathbf{w} = 2\mathbf{i} + 3\mathbf{j}$

  Two vectors are orthogonal if the dot product is zero. Solve for a:

  $\mathbf{v} \cdot \mathbf{w} = 1(2) + (-a)(3) = 2 - 3a$

  $2 - 3a = 0 \Rightarrow 3a = 2 \Rightarrow a = \dfrac{2}{3}$

19. $\mathbf{v} = 2\mathbf{i} - 3\mathbf{j}, \quad \mathbf{w} = \mathbf{i} - \mathbf{j}$

  $\mathbf{v}_1 = \dfrac{\mathbf{v} \cdot \mathbf{w}}{(\|\mathbf{w}\|)^2}\mathbf{w} = \dfrac{2(1) + (-3)(-1)}{\left(\sqrt{1^2 + (-1)^2}\right)^2}(\mathbf{i} - \mathbf{j}) = \dfrac{5}{2}(\mathbf{i} - \mathbf{j}) = \dfrac{5}{2}\mathbf{i} - \dfrac{5}{2}\mathbf{j}$

  $\mathbf{v}_2 = \mathbf{v} - \mathbf{v}_1 = (2\mathbf{i} - 3\mathbf{j}) - \left(\dfrac{5}{2}\mathbf{i} - \dfrac{5}{2}\mathbf{j}\right) = -\dfrac{1}{2}\mathbf{i} - \dfrac{1}{2}\mathbf{j}$

21. $\mathbf{v} = \mathbf{i} - \mathbf{j}, \quad \mathbf{w} = \mathbf{i} + 2\mathbf{j}$

  $\mathbf{v}_1 = \dfrac{\mathbf{v} \cdot \mathbf{w}}{(\|\mathbf{w}\|)^2}\mathbf{w} = \dfrac{1(1) + (-1)(2)}{\left(\sqrt{1^2 + 2^2}\right)^2}(\mathbf{i} + 2\mathbf{j}) = -\dfrac{1}{5}(\mathbf{i} + 2\mathbf{j}) = -\dfrac{1}{5}\mathbf{i} - \dfrac{2}{5}\mathbf{j}$

  $\mathbf{v}_2 = \mathbf{v} - \mathbf{v}_1 = (\mathbf{i} - \mathbf{j}) - \left(-\dfrac{1}{5}\mathbf{i} - \dfrac{2}{5}\mathbf{j}\right) = \dfrac{6}{5}\mathbf{i} - \dfrac{3}{5}\mathbf{j}$

23. $\mathbf{v} = 3\mathbf{i} + \mathbf{j}, \quad \mathbf{w} = -2\mathbf{i} - \mathbf{j}$

  $\mathbf{v}_1 = \dfrac{\mathbf{v} \cdot \mathbf{w}}{(\|\mathbf{w}\|)^2}\mathbf{w} = \dfrac{3(-2) + 1(-1)}{\left(\sqrt{(-2)^2 + (-1)^2}\right)^2}(-2\mathbf{i} - \mathbf{j}) = -\dfrac{7}{5}(-2\mathbf{i} - \mathbf{j}) = \dfrac{14}{5}\mathbf{i} + \dfrac{7}{5}\mathbf{j}$

  $\mathbf{v}_2 = \mathbf{v} - \mathbf{v}_1 = (3\mathbf{i} + \mathbf{j}) - \left(\dfrac{14}{5}\mathbf{i} + \dfrac{7}{5}\mathbf{j}\right) = \dfrac{1}{5}\mathbf{i} - \dfrac{2}{5}\mathbf{j}$

25. Let $\mathbf{v}_a$ = the velocity of the plane in still air.

  $\mathbf{v}_w$ = the velocity of the wind.

  $\mathbf{v}_g$ = the velocity of the plane relative to the ground.

  $\mathbf{v}_g = \mathbf{v}_a + \mathbf{v}_w$

  $\mathbf{v}_a = 550\left(\cos(225°)\mathbf{i} + \sin(225°)\mathbf{j}\right) = 550\left(-\dfrac{\sqrt{2}}{2}\mathbf{i} - \dfrac{\sqrt{2}}{2}\mathbf{j}\right) = -275\sqrt{2}\,\mathbf{i} - 275\sqrt{2}\,\mathbf{j}$

  $\mathbf{v}_w = 80\mathbf{i}$

  $\mathbf{v}_g = \mathbf{v}_a + \mathbf{v}_w = -275\sqrt{2}\,\mathbf{i} - 275\sqrt{2}\,\mathbf{j} + 80\mathbf{i} = \left(80 - 275\sqrt{2}\right)\mathbf{i} - 275\sqrt{2}\,\mathbf{j}$

The speed of the plane relative to the ground is:

$$\| \mathbf{v_g} \| = \sqrt{\left(80 - 275\sqrt{2}\right)^2 + \left(-275\sqrt{2}\right)^2} = \sqrt{6400 - 44{,}000\sqrt{2} + 151{,}250 + 151{,}250}$$

$$\approx \sqrt{246{,}674.6} \approx 496.66 \text{ miles per hour}$$

To find the direction, find the angle between $\mathbf{v_g}$ and a convenient vector such as due south, $-\mathbf{j}$.

$$\cos\theta = \frac{\mathbf{v_g} \cdot (-\mathbf{j})}{\|\mathbf{v_g}\|\|-\mathbf{j}\|} = \frac{\left(80 - 275\sqrt{2}\right)(0) + \left(-275\sqrt{2}\right)(-1)}{496.66\sqrt{0^2 + (-1)^2}} = \frac{275\sqrt{2}}{496.66} \approx 0.7830$$

$$\theta \approx 38.46°$$

The plane is traveling with a ground speed of about 496.66 miles per hour in an approximate direction of 38.46 degrees west of south.

27. Let the positive x-axis point downstream, so that the velocity of the current is $\mathbf{v_c} = 3\mathbf{i}$.
Let $\mathbf{v_w}$ = the velocity of the boat in the water.
Let $\mathbf{v_g}$ = the velocity of the boat relative to the land.
Then $\mathbf{v_g} = \mathbf{v_w} + \mathbf{v_c}$ and $\mathbf{v_g} = k_j$ since the boat is going directly across the river.
The speed of the boat is $|\mathbf{v_w}| = 20$; we need to find the direction.

Let $\mathbf{v_w} = a\mathbf{i} + b\mathbf{j}$ so $\|\mathbf{v_w}\| = \sqrt{a^2 + b^2} = 20 \Rightarrow a^2 + b^2 = 400$.
Since $\mathbf{v_g} = \mathbf{v_w} + \mathbf{v_c}$, $k\mathbf{j} = a\mathbf{i} + b\mathbf{j} + 3\mathbf{i} \Rightarrow k\mathbf{j} = (a+3)\mathbf{i} + b\mathbf{j}$

$a + 3 = 0$ and $k = b \Rightarrow a = -3$

$a^2 + b^2 = 400 \Rightarrow 9 + b^2 = 400 \Rightarrow b^2 = 391 \Rightarrow k = b \approx 19.77$

$\mathbf{v_w} = -3\mathbf{i} + 19.77\mathbf{j}$ and $\mathbf{v_g} = 19.77\mathbf{j}$

Find the angle between $\mathbf{v_w}$ and $\mathbf{j}$:

$$\cos\theta = \frac{\mathbf{v_w} \cdot \mathbf{j}}{\|\mathbf{v_w}\|\|\mathbf{j}\|} = \frac{-3(0) + 19.77(1)}{20\sqrt{0^2 + 1^2}} = \frac{19.77}{20} = 0.9885$$

$$\theta \approx 8.70°$$

The heading of the boat needs to be about 8.70 degrees upstream.
The velocity of the boat directly across the river is about 19.77 kilometers per hour. The time to cross the river is: $t = \dfrac{0.5}{19.77} \approx 0.025$ hours, that is, $t = \dfrac{0.5}{19.77} \approx 1.5$ minutes.

29. Split the force into the components going down the hill and perpendicular to the hill.

$\mathbf{F_d} = \mathbf{F}\sin 8° = 5300\sin 8°$

  ≈ 737.62 pounds

$\mathbf{F_p} = \mathbf{F}\cos 8° = 5300\cos 8°$

  ≈ 5248.42 pounds

The force required to keep the car from rolling down the hill is about 737.62 pounds.
The force perpendicular to the hill is approximately 5248.42 pounds.

31.   Let $\mathbf{v}_a$ = the velocity of the plane in still air.

     $\mathbf{v}_w$ = the velocity of the wind.

     $\mathbf{v}_g$ = the velocity of the plane relative to the ground.

     $\mathbf{v}_g = \mathbf{v}_a + \mathbf{v}_w$

$$\mathbf{v}_a = 500\left(\cos(45°)\mathbf{i} + \sin(45°)\mathbf{j}\right) = 500\left(\frac{\sqrt{2}}{2}\mathbf{i} + \frac{\sqrt{2}}{2}\mathbf{j}\right) = 250\sqrt{2}\,\mathbf{i} + 250\sqrt{2}\,\mathbf{j}$$

$$\mathbf{v}_w = 60\left(\cos(120°)\mathbf{i} + \sin(120°)\mathbf{j}\right) = 60\left(-\frac{1}{2}\mathbf{i} + \frac{\sqrt{3}}{2}\mathbf{j}\right) = -30\mathbf{i} + 30\sqrt{3}\,\mathbf{j}$$

$$\mathbf{v}_g = \mathbf{v}_a + \mathbf{v}_w = 250\sqrt{2}\,\mathbf{i} + 250\sqrt{2}\,\mathbf{j} - 30\mathbf{i} + 30\sqrt{3}\,\mathbf{j}$$

$$= \left(-30 + 250\sqrt{2}\right)\mathbf{i} + \left(250\sqrt{2} + 30\sqrt{3}\right)\mathbf{j}$$

The speed of the plane relative to the ground is:

$$\left\| \mathbf{v}_g \right\| = \sqrt{\left(-30 + 250\sqrt{2}\right)^2 + \left(250\sqrt{2} + 30\sqrt{3}\right)^2}$$

$$\approx \sqrt{269{,}129.1} \approx 518.8 \text{ kilometers per hour}$$

To find the direction, find the angle between $\mathbf{v}_g$ and a convenient vector such as due north, $\mathbf{j}$.

$$\cos\theta = \frac{\mathbf{v}_g \cdot \mathbf{j}}{\left\|\mathbf{v}_g\right\|\left\|\mathbf{j}\right\|} = \frac{\left(-30 + 250\sqrt{2}\right)(0) + \left(250\sqrt{2} + 30\sqrt{3}\right)(1)}{518.8\sqrt{0^2 + 1^2}} = \frac{250\sqrt{2} + 30\sqrt{3}}{518.8}$$

$$\approx \frac{405.5}{518.8} \approx 0.7816 \Rightarrow \theta \approx 38.59°$$

The plane is traveling with a ground speed of about 518.78 kilometers per hour in a direction of 38.59 degrees east of north.

33.   Let the positive x-axis point downstream, so that the velocity of the current is $\mathbf{v}_c = 3\mathbf{i}$.

Let $\mathbf{v}_w$ = the velocity of the boat in the water.

Let $\mathbf{v}_g$ = the velocity of the boat relative to the land.

Then $\mathbf{v}_g = \mathbf{v}_w + \mathbf{v}_c$

The speed of the boat is $\left\|\mathbf{v}_w\right\| = 20$; its heading is directly across the river, so $\mathbf{v}_w = 20\mathbf{j}$.

     Then $\mathbf{v}_g = \mathbf{v}_w + \mathbf{v}_c = 20\mathbf{j} + 3\mathbf{i} = 3\mathbf{i} + 20\mathbf{j}$

$$\left\|\mathbf{v}_g\right\| = \sqrt{3^2 + 20^2} = \sqrt{409} \approx 20.22 \text{ miles per hour}$$

Find the angle between $\mathbf{v}_g$ and $\mathbf{j}$:

$$\cos\theta = \frac{\mathbf{v}_g \cdot \mathbf{j}}{\left\|\mathbf{v}_g\right\|\left\|\mathbf{j}\right\|} = \frac{3(0) + 20(1)}{20.2\sqrt{0^2 + 1^2}} = \frac{20}{20.22} \approx 0.9891 \Rightarrow \theta \approx 8.47°$$

The approximate direction of the boat will be 8.47 degrees downstream.

35. $\mathbf{F} = 3\big(\cos(60°)\mathbf{i} + \sin(60°)\mathbf{j}\big) = 3\left(\dfrac{1}{2}\mathbf{i} + \dfrac{\sqrt{3}}{2}\mathbf{j}\right) = \dfrac{3}{2}\mathbf{i} + \dfrac{3\sqrt{3}}{2}\mathbf{j}$

$W = \mathbf{F} \cdot \overrightarrow{AB} = \left(\dfrac{3}{2}\mathbf{i} + \dfrac{3\sqrt{3}}{2}\mathbf{j}\right) \cdot 2\mathbf{i} = \dfrac{3}{2}(2) + \dfrac{3\sqrt{3}}{2}(0) = 3$ foot-pounds

37. $\mathbf{F} = 20\big(\cos(30°)\mathbf{i} + \sin(30°)\mathbf{j}\big) = 20\left(\dfrac{\sqrt{3}}{2}\mathbf{i} + \dfrac{1}{2}\mathbf{j}\right) = 10\sqrt{3}\mathbf{i} + 10\mathbf{j}$

$W = \mathbf{F} \cdot \overrightarrow{AB} = \big(10\sqrt{3}\mathbf{i} + 10\mathbf{j}\big) \cdot 100\mathbf{i} = 10\sqrt{3}(100) + 10(0) \approx 1732$ foot-pounds

39. Let $\mathbf{u} = a_1\mathbf{i} + b_1\mathbf{j}, \quad \mathbf{v} = a_2\mathbf{i} + b_2\mathbf{j}, \quad \mathbf{w} = a_3\mathbf{i} + b_3\mathbf{j}$

$\mathbf{u} \cdot (\mathbf{v} + \mathbf{w}) = \big(a_1\mathbf{i} + b_1\mathbf{j}\big) \cdot \big(a_2\mathbf{i} + b_2\mathbf{j} + a_3\mathbf{i} + b_3\mathbf{j}\big) = \big(a_1\mathbf{i} + b_1\mathbf{j}\big) \cdot \big(a_2\mathbf{i} + a_3\mathbf{i} + b_2\mathbf{j} + b_3\mathbf{j}\big)$

$\qquad = \big(a_1\mathbf{i} + b_1\mathbf{j}\big) \cdot \big((a_2 + a_3)\mathbf{i} + (b_2 + b_3)\mathbf{j}\big) = a_1(a_2 + a_3) + b_1(b_2 + b_3)$

$\qquad = a_1 a_2 + a_1 a_3 + b_1 b_2 + b_1 b_3 = a_1 a_2 + b_1 b_2 + a_1 a_3 + b_1 b_3$

$\qquad = \big(a_1\mathbf{i} + b_1\mathbf{j}\big) \cdot \big(a_2\mathbf{i} + b_2\mathbf{j}\big) + \big(a_1\mathbf{i} + b_1\mathbf{j}\big) \cdot \big(a_3\mathbf{i} + b_3\mathbf{j}\big) = \mathbf{u} \cdot \mathbf{v} + \mathbf{u} \cdot \mathbf{w}$

41. Let $\mathbf{v} = a\mathbf{i} + b\mathbf{j}$.

Since $\mathbf{v}$ is a unit vector, $\|\mathbf{v}\| = \sqrt{a^2 + b^2} = 1$ or $a^2 + b^2 = 1$

If α is the angle between $\mathbf{v}$ and $\mathbf{i}$, then $\cos\alpha = \dfrac{\mathbf{v} \cdot \mathbf{i}}{\|\mathbf{v}\|\|\mathbf{i}\|} \Rightarrow \cos\alpha = \dfrac{(a\mathbf{i} + b\mathbf{j}) \cdot \mathbf{i}}{1 \cdot 1} = a$.

$a^2 + b^2 = 1 \Rightarrow \cos^2\alpha + b^2 = 1 \Rightarrow b^2 = 1 - \cos^2\alpha \Rightarrow b^2 = \sin^2\alpha \Rightarrow b = \sin\alpha$

Thus, $\mathbf{v} = \cos\alpha\, \mathbf{i} + \sin\alpha\, \mathbf{j}$.

43. Let $\mathbf{v} = a\mathbf{i} + b\mathbf{j}$.

$\mathbf{v}_1 =$ The projection of $\mathbf{v}$ onto $\mathbf{i}$ $= \dfrac{\mathbf{v} \cdot \mathbf{i}}{\big(\|\mathbf{i}\|\big)^2}\mathbf{i} = \dfrac{(a\mathbf{i} + b\mathbf{j}) \cdot \mathbf{i}}{\left(\sqrt{1^2 + 0^2}\right)^2}\mathbf{i} = \dfrac{a(1) + b(0)}{1^2}\mathbf{i} = a\mathbf{i}$.

$\mathbf{v}_2 = \mathbf{v} - \mathbf{v}_1 = a\mathbf{i} + b\mathbf{j} - a\mathbf{i} = b\mathbf{j}$

Since $\mathbf{v} \cdot \mathbf{i} = (a\mathbf{i} + b\mathbf{j}) \cdot \mathbf{i} = (a)(1) + (b)(0) = a$ and $\mathbf{v} \cdot \mathbf{j} = (a\mathbf{i} + b\mathbf{j}) \cdot \mathbf{j} = (a)(0) + (b)(1) = b$,

$\mathbf{v} = \mathbf{v}_1 + \mathbf{v}_2 = (\mathbf{v} \cdot \mathbf{i})\mathbf{i} + (\mathbf{v} \cdot \mathbf{j})\mathbf{j}$.

45. $(\mathbf{v} - \alpha\mathbf{w}) \cdot \mathbf{w} = \mathbf{v} \cdot \mathbf{w} - \alpha\mathbf{w} \cdot \mathbf{w} = \mathbf{v} \cdot \mathbf{w} - \alpha\big(\|\mathbf{w}\|\big)^2 = \mathbf{v} \cdot \mathbf{w} - \dfrac{\mathbf{v} \cdot \mathbf{w}}{\big(\|\mathbf{w}\|\big)^2}\big(\|\mathbf{w}\|\big)^2 = 0$

Therefore the vectors are orthogonal.

47. If $\mathbf{F}$ is orthogonal to $\overrightarrow{AB}$, then $\mathbf{F} \cdot \overrightarrow{AB} = 0$. So $W = \mathbf{F} \cdot \overrightarrow{AB} = 0$.

49. Answers will vary.

Polar Coordinates; Vectors

8.6   Vectors in Space

1. $\sqrt{(x_2 - x_1)^2 + (y_2 - y_1)^2}$

3.   components

5.   False

7. $y = 0$ is the set of all points in the xz-plane, that is, all points of the form $(x, 0, z)$.

9. $z = 2$ is the set of all points of the form $(x, y, 2)$, that is, the plane two units above the xy-plane.

11. $x = -4$ is the set of all points of the form $(-4, y, z)$, that is, the plane four units to the right of the yz-plane.

13. $x = 1$ and $y = 2$ is the set of all points of the form $(1, 2, z)$, that is, a line parallel to the z-axis.

15. $d = \sqrt{(4-0)^2 + (1-0)^2 + (2-0)^2} = \sqrt{16+1+4} = \sqrt{21}$

17. $d = \sqrt{(0-(-1))^2 + (-2-2)^2 + (1-(-3))^2} = \sqrt{1+16+16} = \sqrt{33}$

19. $d = \sqrt{(3-4)^2 + (2-(-2))^2 + (1-(-2))^2} = \sqrt{1+16+9} = \sqrt{26}$

21.   The bottom of the box is formed by the vertices $(0, 0, 0)$, $(2, 0, 0)$, $(0, 1, 0)$, and $(2, 1, 0)$. The top of the box is formed by the vertices $(0, 0, 3)$, $(2, 0, 3)$, $(0, 1, 3)$, and $(2, 1, 3)$.

23.   The bottom of the box is formed by the vertices $(1, 2, 3)$, $(3, 2, 3)$, $(3, 4, 3)$, and $(1, 4, 3)$. The top of the box is formed by the vertices $(3, 4, 5)$, $(1, 2, 5)$, $(3, 2, 5)$, and $(1, 4, 5)$.

25.   The bottom of the box is formed by the vertices $(-1, 0, 2)$, $(4, 0, 2)$, $(-1, 2, 2)$, and $(4, 2, 2)$. The top of the box is formed by the vertices $(4, 2, 5)$, $(-1, 0, 5)$, $(4, 0, 5)$, and $(-1, 2, 5)$.

27. $\mathbf{v} = (3-0)\mathbf{i} + (4-0)\mathbf{j} + (-1-0)\mathbf{k} = 3\mathbf{i} + 4\mathbf{j} - \mathbf{k}$

29. $\mathbf{v} = (5-3)\mathbf{i} + (6-2)\mathbf{j} + (0-(-1))\mathbf{k} = 2\mathbf{i} + 4\mathbf{j} + \mathbf{k}$

31. $\mathbf{v} = (6-(-2))\mathbf{i} + (-2-(-1))\mathbf{j} + (4-4)\mathbf{k} = 8\mathbf{i} - \mathbf{j}$

33. $\|\mathbf{v}\| = \sqrt{3^2 + (-6)^2 + (-2)^2} = \sqrt{9 + 36 + 4} = \sqrt{49} = 7$

35. $\|\mathbf{v}\| = \sqrt{1^2 + (-1)^2 + 1^2} = \sqrt{1 + 1 + 1} = \sqrt{3}$

37. $\|\mathbf{v}\| = \sqrt{(-2)^2 + 3^2 + (-3)^2} = \sqrt{4 + 9 + 9} = \sqrt{22}$

39. $2\mathbf{v} + 3\mathbf{w} = 2(3\mathbf{i} - 5\mathbf{j} + 2\mathbf{k}) + 3(-2\mathbf{i} + 3\mathbf{j} - 2\mathbf{k}) = 6\mathbf{i} - 10\mathbf{j} + 4\mathbf{k} - 6\mathbf{i} + 9\mathbf{j} - 6\mathbf{k}$
$= 0\mathbf{i} - 1\mathbf{j} - 2\mathbf{k} = -\mathbf{j} - 2\mathbf{k}$

41. $\|\mathbf{v} - \mathbf{w}\| = \left\|(3\mathbf{i} - 5\mathbf{j} + 2\mathbf{k}) - (-2\mathbf{i} + 3\mathbf{j} - 2\mathbf{k})\right\| = \|3\mathbf{i} - 5\mathbf{j} + 2\mathbf{k} + 2\mathbf{i} - 3\mathbf{j} + 2\mathbf{k}\|$
$= \|5\mathbf{i} - 8\mathbf{j} + 4\mathbf{k}\| = \sqrt{5^2 + (-8)^2 + 4^2} = \sqrt{25 + 64 + 16} = \sqrt{105}$

43. $\|\mathbf{v}\| - \|\mathbf{w}\| = \|3\mathbf{i} - 5\mathbf{j} + 2\mathbf{k}\| - \|-2\mathbf{i} + 3\mathbf{j} - 2\mathbf{k}\|$
$= \sqrt{3^2 + (-5)^2 + 2^2} - \sqrt{(-2)^2 + 3^2 + (-2)^2} = \sqrt{38} - \sqrt{17}$

45. $\mathbf{u} = \dfrac{\mathbf{v}}{\|\mathbf{v}\|} = \dfrac{5\mathbf{i}}{\sqrt{5^2 + 0^2 + 0^2}} = \dfrac{5\mathbf{i}}{5} = \mathbf{i}$

47. $\mathbf{u} = \dfrac{\mathbf{v}}{\|\mathbf{v}\|} = \dfrac{3\mathbf{i} - 6\mathbf{j} - 2\mathbf{k}}{\sqrt{3^2 + (-6)^2 + (-2)^2}} = \dfrac{3\mathbf{i} - 6\mathbf{j} - 2\mathbf{k}}{7} = \dfrac{3}{7}\mathbf{i} - \dfrac{6}{7}\mathbf{j} - \dfrac{2}{7}\mathbf{k}$

49. $\mathbf{u} = \dfrac{\mathbf{v}}{\|\mathbf{v}\|} = \dfrac{\mathbf{i} + \mathbf{j} + \mathbf{k}}{\sqrt{1^2 + 1^2 + 1^2}} = \dfrac{\mathbf{i} + \mathbf{j} + \mathbf{k}}{\sqrt{3}} = \dfrac{1}{\sqrt{3}}\mathbf{i} + \dfrac{1}{\sqrt{3}}\mathbf{j} + \dfrac{1}{\sqrt{3}}\mathbf{k} = \dfrac{\sqrt{3}}{3}\mathbf{i} + \dfrac{\sqrt{3}}{3}\mathbf{j} + \dfrac{\sqrt{3}}{3}\mathbf{k}$

51. $\mathbf{v} \cdot \mathbf{w} = (\mathbf{i} - \mathbf{j}) \cdot (\mathbf{i} + \mathbf{j} + \mathbf{k}) = (1)(1) + (-1)(1) + (0)(1) = 1 - 1 + 0 = 0$
$\cos\theta = \dfrac{\mathbf{v} \cdot \mathbf{w}}{\|\mathbf{v}\|\|\mathbf{w}\|} = \dfrac{0}{\sqrt{1^2 + (-1)^2 + 0^2}\sqrt{1^2 + 1^2 + 1^2}} = \dfrac{0}{\sqrt{2}\sqrt{3}} = \dfrac{0}{\sqrt{6}} = 0$

$\theta = \dfrac{\pi}{2}$ radians $= 90°$

53. $\mathbf{v} \cdot \mathbf{w} = (2\mathbf{i} + \mathbf{j} - 3\mathbf{k}) \cdot (\mathbf{i} + 2\mathbf{j} + 2\mathbf{k}) = (2)(1) + (1)(2) + (-3)(2) = 2 + 2 - 6 = -2$
$\cos\theta = \dfrac{\mathbf{v} \cdot \mathbf{w}}{\|\mathbf{v}\|\|\mathbf{w}\|} = \dfrac{-2}{\sqrt{2^2 + 1^2 + (-3)^2}\sqrt{1^2 + 2^2 + 2^2}} = \dfrac{-2}{\sqrt{14}\sqrt{9}} = \dfrac{-2}{3\sqrt{14}}$
$\theta \approx 1.75$ radians $\approx 100.3°$

55. $\mathbf{v} \cdot \mathbf{w} = (3\mathbf{i} - \mathbf{j} + 2\mathbf{k}) \cdot (\mathbf{i} + \mathbf{j} - \mathbf{k}) = (3)(1) + (-1)(1) + (2)(-1) = 3 - 1 - 2 = 0$
$\cos\theta = \dfrac{\mathbf{v} \cdot \mathbf{w}}{\|\mathbf{v}\|\|\mathbf{w}\|} = \dfrac{0}{\sqrt{3^2 + (-1)^2 + 2^2}\sqrt{1^2 + 1^2 + (-1)^2}} = \dfrac{0}{\sqrt{14}\sqrt{3}} = 0$

$\theta = \dfrac{\pi}{2}$ radians $= 90°$

57. $\mathbf{v} \cdot \mathbf{w} = (3\mathbf{i} + 4\mathbf{j} + \mathbf{k}) \cdot (6\mathbf{i} + 8\mathbf{j} + 2\mathbf{k}) = (3)(6) + (4)(8) + (1)(2) = 18 + 32 + 2 = 52$

$\cos\theta = \dfrac{\mathbf{v} \cdot \mathbf{w}}{\|\mathbf{v}\|\,\|\mathbf{w}\|} = \dfrac{52}{\sqrt{3^2 + 4^2 + 1^2}\,\sqrt{6^2 + 8^2 + 2^2}} = \dfrac{52}{\sqrt{26}\,\sqrt{104}} = \dfrac{52}{52} = 1$

$\theta = 0$ radians $= 0°$

59. $\cos\alpha = \dfrac{a}{\|\mathbf{v}\|} = \dfrac{3}{\sqrt{3^2 + (-6)^2 + (-2)^2}} = \dfrac{3}{\sqrt{49}} = \dfrac{3}{7} \;\Rightarrow\; \alpha \approx 64.6°$

$\cos\beta = \dfrac{b}{\|\mathbf{v}\|} = \dfrac{-6}{\sqrt{3^2 + (-6)^2 + (-2)^2}} = \dfrac{-6}{\sqrt{49}} = -\dfrac{6}{7} \;\Rightarrow\; \beta \approx 149.0°$

$\cos\gamma = \dfrac{c}{\|\mathbf{v}\|} = \dfrac{-2}{\sqrt{3^2 + (-6)^2 + (-2)^2}} = \dfrac{-2}{\sqrt{49}} = -\dfrac{2}{7} \;\Rightarrow\; \gamma \approx 106.6°$

$\mathbf{v} = 7\big(\cos(64.6°)\mathbf{i} + \cos(149.0°)\mathbf{j} + \cos(106.6°)\mathbf{k}\big)$

61. $\cos\alpha = \dfrac{a}{\|\mathbf{v}\|} = \dfrac{1}{\sqrt{1^2 + 1^2 + 1^2}} = \dfrac{1}{\sqrt{3}} = \dfrac{\sqrt{3}}{3} \;\Rightarrow\; \alpha \approx 54.7°$

$\cos\beta = \dfrac{b}{\|\mathbf{v}\|} = \dfrac{1}{\sqrt{1^2 + 1^2 + 1^2}} = \dfrac{1}{\sqrt{3}} = \dfrac{\sqrt{3}}{3} \;\Rightarrow\; \beta \approx 54.7°$

$\cos\gamma = \dfrac{c}{\|\mathbf{v}\|} = \dfrac{1}{\sqrt{1^2 + 1^2 + 1^2}} = \dfrac{1}{\sqrt{3}} = \dfrac{\sqrt{3}}{3} \;\Rightarrow\; \gamma \approx 54.7°$

$\mathbf{v} = \sqrt{3}\big(\cos(54.7°)\mathbf{i} + \cos(54.7°)\mathbf{j} + \cos(54.7°)\mathbf{k}\big)$

63. $\cos\alpha = \dfrac{a}{\|\mathbf{v}\|} = \dfrac{1}{\sqrt{1^2 + 1^2 + 0^2}} = \dfrac{1}{\sqrt{2}} = \dfrac{\sqrt{2}}{2} \;\Rightarrow\; \alpha = 45°$

$\cos\beta = \dfrac{b}{\|\mathbf{v}\|} = \dfrac{1}{\sqrt{1^2 + 1^2 + 0^2}} = \dfrac{1}{\sqrt{2}} = \dfrac{\sqrt{2}}{2} \;\Rightarrow\; \beta = 45°$

$\cos\gamma = \dfrac{c}{\|\mathbf{v}\|} = \dfrac{0}{\sqrt{1^2 + 1^2 + 0^2}} = \dfrac{0}{\sqrt{2}} = 0 \;\Rightarrow\; \gamma = 90°$

$\mathbf{v} = \sqrt{2}\big(\cos(45°)\mathbf{i} + \cos(45°)\mathbf{j} + \cos(90°)\mathbf{k}\big)$

65. $\cos\alpha = \dfrac{a}{\|\mathbf{v}\|} = \dfrac{3}{\sqrt{3^2 + (-5)^2 + 2^2}} = \dfrac{3}{\sqrt{38}} \;\Rightarrow\; \alpha \approx 60.9°$

$\cos\beta = \dfrac{b}{\|\mathbf{v}\|} = \dfrac{-5}{\sqrt{3^2 + (-5)^2 + 2^2}} = -\dfrac{5}{\sqrt{38}} \;\Rightarrow\; \beta \approx 144.2°$

$\cos\gamma = \dfrac{c}{\|\mathbf{v}\|} = \dfrac{2}{\sqrt{3^2 + (-5)^2 + 2^2}} = \dfrac{2}{\sqrt{38}} \;\Rightarrow\; \gamma \approx 71.1°$

$\mathbf{v} = \sqrt{38}\big(\cos(60.9°)\mathbf{i} + \cos(144.2°)\mathbf{j} + \cos(71.1°)\mathbf{k}\big)$

67. $d(P_0, P) = \sqrt{(x - x_0)^2 + (y - y_0)^2 + (z - z_0)^2} = r$
$\Rightarrow (x - x_0)^2 + (y - y_0)^2 + (z - z_0)^2 = r^2$

69. $(x - 1)^2 + (y - 2)^2 + (z - 2)^2 = 4$

71. $x^2 + y^2 + z^2 + 2x - 2y = 2$
$x^2 + 2x + y^2 - 2y + z^2 = 2$
$x^2 + 2x + 1 + y^2 - 2y + 1 + z^2 = 2 + 1 + 1$
$(x + 1)^2 + (y - 1)^2 + (z - 0)^2 = 4$
$(x + 1)^2 + (y - 1)^2 + (z - 0)^2 = 2^2$
Center: $(-1, 1, 0)$;   Radius: 2

73. $x^2 + y^2 + z^2 - 4x + 4y + 2z = 0$
$x^2 - 4x + y^2 + 4y + z^2 + 2z = 0$
$x^2 - 4x + 4 + y^2 + 4y + 4 + z^2 + 2z + 1 = 4 + 4 + 1$
$(x - 2)^2 + (y + 2)^2 + (z + 1)^2 = 9$
$(x - 2)^2 + (y + 2)^2 + (z + 1)^2 = 3^2$
Center: $(2, -2, -1)$;   Radius: 3

75. $2x^2 + 2y^2 + 2z^2 - 8x + 4z = -1$
$x^2 - 4x + y^2 + z^2 + 2z = -\dfrac{1}{2}$
$x^2 - 4x + 4 + y^2 + z^2 + 2z + 1 = -\dfrac{1}{2} + 4 + 1$
$(x - 2)^2 + (y - 0)^2 + (z + 1)^2 = \dfrac{9}{2}$
$(x - 2)^2 + (y - 0)^2 + (z + 1)^2 = \left(\dfrac{3}{\sqrt{2}}\right)^2$
Center: $(2, 0, -1)$;   Radius: $\dfrac{3\sqrt{2}}{2}$

77. Write the force as a vector:
$\cos\alpha = \dfrac{2}{\sqrt{2^2 + 1^2 + 2^2}} = \dfrac{2}{\sqrt{9}} = \dfrac{2}{3};\quad \cos\beta = \dfrac{1}{3};\quad \cos\gamma = \dfrac{2}{3}$
$\mathbf{F} = 3\left(\dfrac{2}{3}\mathbf{i} + \dfrac{1}{3}\mathbf{j} + \dfrac{2}{3}\mathbf{k}\right)$
$W = 3\left(\dfrac{2}{3}\mathbf{i} + \dfrac{1}{3}\mathbf{j} + \dfrac{2}{3}\mathbf{k}\right) \cdot 2\mathbf{j} = (3)\left(\dfrac{1}{3}\right)(2) = 2$ joules

79. $W = \mathbf{F} \cdot \mathbf{u} = (2\mathbf{i} - \mathbf{j} - \mathbf{k}) \cdot (3\mathbf{i} + 2\mathbf{j} - 5\mathbf{k}) = (2)(3) + (-1)(2) + (-1)(-5) = 9$ joules

Polar Coordinates; Vectors

8.7   The Cross Product

1. True

3. True

5. False

7. $\begin{vmatrix} 3 & 4 \\ 1 & 2 \end{vmatrix} = 3 \cdot 2 - 1 \cdot 4 = 6 - 4 = 2$

9. $\begin{vmatrix} 6 & 5 \\ -2 & -1 \end{vmatrix} = 6(-1) - (-2)(5) = -6 + 10 = 4$

11. $\begin{vmatrix} A & B & C \\ 2 & 1 & 4 \\ 1 & 3 & 1 \end{vmatrix} = \begin{vmatrix} 1 & 4 \\ 3 & 1 \end{vmatrix} A - \begin{vmatrix} 2 & 4 \\ 1 & 1 \end{vmatrix} B + \begin{vmatrix} 2 & 1 \\ 1 & 3 \end{vmatrix} C = (1 - 12)A - (2 - 4)B + (6 - 1)C$

$\qquad = -11A + 2B + 5C$

13. $\begin{vmatrix} A & B & C \\ -1 & 3 & 5 \\ 5 & 0 & -2 \end{vmatrix} = \begin{vmatrix} 3 & 5 \\ 0 & -2 \end{vmatrix} A - \begin{vmatrix} -1 & 5 \\ 5 & -2 \end{vmatrix} B + \begin{vmatrix} -1 & 3 \\ 5 & 0 \end{vmatrix} C = (-6 - 0)A - (2 - 25)B + (0 - 15)C$

$\qquad = -6A + 23B - 15C$

15. (a) $\mathbf{v} \times \mathbf{w} = \begin{vmatrix} \mathbf{i} & \mathbf{j} & \mathbf{k} \\ 2 & -3 & 1 \\ 3 & -2 & -1 \end{vmatrix} = \begin{vmatrix} -3 & 1 \\ -2 & -1 \end{vmatrix} \mathbf{i} - \begin{vmatrix} 2 & 1 \\ 3 & -1 \end{vmatrix} \mathbf{j} + \begin{vmatrix} 2 & -3 \\ 3 & -2 \end{vmatrix} \mathbf{k} = 5\mathbf{i} + 5\mathbf{j} + 5\mathbf{k}$

(b) $\mathbf{w} \times \mathbf{v} = \begin{vmatrix} \mathbf{i} & \mathbf{j} & \mathbf{k} \\ 3 & -2 & -1 \\ 2 & -3 & 1 \end{vmatrix} = \begin{vmatrix} -2 & -1 \\ -3 & 1 \end{vmatrix} \mathbf{i} - \begin{vmatrix} 3 & -1 \\ 2 & 1 \end{vmatrix} \mathbf{j} + \begin{vmatrix} 3 & -2 \\ 2 & -3 \end{vmatrix} \mathbf{k} = -5\mathbf{i} - 5\mathbf{j} - 5\mathbf{k}$

(c) $\mathbf{w} \times \mathbf{w} = \begin{vmatrix} \mathbf{i} & \mathbf{j} & \mathbf{k} \\ 3 & -2 & -1 \\ 3 & -2 & -1 \end{vmatrix} = \begin{vmatrix} -2 & -1 \\ -2 & -1 \end{vmatrix} \mathbf{i} - \begin{vmatrix} 3 & -1 \\ 3 & -1 \end{vmatrix} \mathbf{j} + \begin{vmatrix} 3 & -2 \\ 3 & -2 \end{vmatrix} \mathbf{k} = 0\mathbf{i} + 0\mathbf{j} + 0\mathbf{k} = \mathbf{0}$

(d) $\mathbf{v} \times \mathbf{v} = \begin{vmatrix} \mathbf{i} & \mathbf{j} & \mathbf{k} \\ 2 & -3 & 1 \\ 2 & -3 & 1 \end{vmatrix} = \begin{vmatrix} -3 & 1 \\ -3 & 1 \end{vmatrix} \mathbf{i} - \begin{vmatrix} 2 & 1 \\ 2 & 1 \end{vmatrix} \mathbf{j} + \begin{vmatrix} 2 & -3 \\ 2 & -3 \end{vmatrix} \mathbf{k} = 0\mathbf{i} + 0\mathbf{j} + 0\mathbf{k} = \mathbf{0}$

17. (a) $\mathbf{v} \times \mathbf{w} = \begin{vmatrix} \mathbf{i} & \mathbf{j} & \mathbf{k} \\ 1 & 1 & 0 \\ 2 & 1 & 1 \end{vmatrix} = \begin{vmatrix} 1 & 0 \\ 1 & 1 \end{vmatrix} \mathbf{i} - \begin{vmatrix} 1 & 0 \\ 2 & 1 \end{vmatrix} \mathbf{j} + \begin{vmatrix} 1 & 1 \\ 2 & 1 \end{vmatrix} \mathbf{k} = \mathbf{i} - \mathbf{j} - \mathbf{k}$

(b) $\mathbf{w} \times \mathbf{v} = \begin{vmatrix} \mathbf{i} & \mathbf{j} & \mathbf{k} \\ 2 & 1 & 1 \\ 1 & 1 & 0 \end{vmatrix} = \begin{vmatrix} 1 & 1 \\ 1 & 0 \end{vmatrix} \mathbf{i} - \begin{vmatrix} 2 & 1 \\ 1 & 0 \end{vmatrix} \mathbf{j} + \begin{vmatrix} 2 & 1 \\ 1 & 1 \end{vmatrix} \mathbf{k} = -\mathbf{i} + \mathbf{j} + \mathbf{k}$

(c) $\mathbf{w} \times \mathbf{w} = \begin{vmatrix} \mathbf{i} & \mathbf{j} & \mathbf{k} \\ 2 & 1 & 1 \\ 2 & 1 & 1 \end{vmatrix} = \begin{vmatrix} 1 & 1 \\ 1 & 1 \end{vmatrix}\mathbf{i} - \begin{vmatrix} 2 & 1 \\ 2 & 1 \end{vmatrix}\mathbf{j} + \begin{vmatrix} 2 & 1 \\ 2 & 1 \end{vmatrix}\mathbf{k} = 0\mathbf{i} + 0\mathbf{j} + 0\mathbf{k} = \mathbf{0}$

(d) $\mathbf{v} \times \mathbf{v} = \begin{vmatrix} \mathbf{i} & \mathbf{j} & \mathbf{k} \\ 1 & 1 & 0 \\ 1 & 1 & 0 \end{vmatrix} = \begin{vmatrix} 1 & 0 \\ 1 & 0 \end{vmatrix}\mathbf{i} - \begin{vmatrix} 1 & 0 \\ 1 & 0 \end{vmatrix}\mathbf{j} + \begin{vmatrix} 1 & 1 \\ 1 & 1 \end{vmatrix}\mathbf{k} = 0\mathbf{i} + 0\mathbf{j} + 0\mathbf{k} = \mathbf{0}$

19. (a) $\mathbf{v} \times \mathbf{w} = \begin{vmatrix} \mathbf{i} & \mathbf{j} & \mathbf{k} \\ 2 & -1 & 2 \\ 0 & 1 & -1 \end{vmatrix} = \begin{vmatrix} -1 & 2 \\ 1 & -1 \end{vmatrix}\mathbf{i} - \begin{vmatrix} 2 & 2 \\ 0 & -1 \end{vmatrix}\mathbf{j} + \begin{vmatrix} 2 & -1 \\ 0 & 1 \end{vmatrix}\mathbf{k} = -\mathbf{i} + 2\mathbf{j} + 2\mathbf{k}$

(b) $\mathbf{w} \times \mathbf{v} = \begin{vmatrix} \mathbf{i} & \mathbf{j} & \mathbf{k} \\ 0 & 1 & -1 \\ 2 & -1 & 2 \end{vmatrix} = \begin{vmatrix} 1 & -1 \\ -1 & 2 \end{vmatrix}\mathbf{i} - \begin{vmatrix} 0 & -1 \\ 2 & 2 \end{vmatrix}\mathbf{j} + \begin{vmatrix} 0 & 1 \\ 2 & -1 \end{vmatrix}\mathbf{k} = \mathbf{i} - 2\mathbf{j} - 2\mathbf{k}$

(c) $\mathbf{w} \times \mathbf{w} = \begin{vmatrix} \mathbf{i} & \mathbf{j} & \mathbf{k} \\ 0 & 1 & -1 \\ 0 & 1 & -1 \end{vmatrix} = \begin{vmatrix} 1 & -1 \\ 1 & -1 \end{vmatrix}\mathbf{i} - \begin{vmatrix} 0 & -1 \\ 0 & -1 \end{vmatrix}\mathbf{j} + \begin{vmatrix} 0 & 1 \\ 0 & 1 \end{vmatrix}\mathbf{k} = 0\mathbf{i} + 0\mathbf{j} + 0\mathbf{k} = \mathbf{0}$

(d) $\mathbf{v} \times \mathbf{v} = \begin{vmatrix} \mathbf{i} & \mathbf{j} & \mathbf{k} \\ 2 & -1 & 2 \\ 2 & -1 & 2 \end{vmatrix} = \begin{vmatrix} -1 & 2 \\ -1 & 2 \end{vmatrix}\mathbf{i} - \begin{vmatrix} 2 & 2 \\ 2 & 2 \end{vmatrix}\mathbf{j} + \begin{vmatrix} 2 & -1 \\ 2 & -1 \end{vmatrix}\mathbf{k} = 0\mathbf{i} + 0\mathbf{j} + 0\mathbf{k} = \mathbf{0}$

21. (a) $\mathbf{v} \times \mathbf{w} = \begin{vmatrix} \mathbf{i} & \mathbf{j} & \mathbf{k} \\ 1 & -1 & -1 \\ 4 & 0 & -3 \end{vmatrix} = \begin{vmatrix} -1 & -1 \\ 0 & -3 \end{vmatrix}\mathbf{i} - \begin{vmatrix} 1 & -1 \\ 4 & -3 \end{vmatrix}\mathbf{j} + \begin{vmatrix} 1 & -1 \\ 4 & 0 \end{vmatrix}\mathbf{k} = 3\mathbf{i} - \mathbf{j} + 4\mathbf{k}$

(b) $\mathbf{w} \times \mathbf{v} = \begin{vmatrix} \mathbf{i} & \mathbf{j} & \mathbf{k} \\ 4 & 0 & -3 \\ 1 & -1 & -1 \end{vmatrix} = \begin{vmatrix} 0 & -3 \\ -1 & -1 \end{vmatrix}\mathbf{i} - \begin{vmatrix} 4 & -3 \\ 1 & -1 \end{vmatrix}\mathbf{j} + \begin{vmatrix} 4 & 0 \\ 1 & -1 \end{vmatrix}\mathbf{k} = -3\mathbf{i} + \mathbf{j} - 4\mathbf{k}$

(c) $\mathbf{w} \times \mathbf{w} = \begin{vmatrix} \mathbf{i} & \mathbf{j} & \mathbf{k} \\ 4 & 0 & -3 \\ 4 & 0 & -3 \end{vmatrix} = \begin{vmatrix} 0 & -3 \\ 0 & -3 \end{vmatrix}\mathbf{i} - \begin{vmatrix} 4 & -3 \\ 4 & -3 \end{vmatrix}\mathbf{j} + \begin{vmatrix} 4 & 0 \\ 4 & 0 \end{vmatrix}\mathbf{k} = 0\mathbf{i} + 0\mathbf{j} + 0\mathbf{k} = \mathbf{0}$

(d) $\mathbf{v} \times \mathbf{v} = \begin{vmatrix} \mathbf{i} & \mathbf{j} & \mathbf{k} \\ 1 & -1 & -1 \\ 1 & -1 & -1 \end{vmatrix} = \begin{vmatrix} -1 & -1 \\ -1 & -1 \end{vmatrix}\mathbf{i} - \begin{vmatrix} 1 & -1 \\ 1 & -1 \end{vmatrix}\mathbf{j} + \begin{vmatrix} 1 & -1 \\ 1 & -1 \end{vmatrix}\mathbf{k} = 0\mathbf{i} + 0\mathbf{j} + 0\mathbf{k} = \mathbf{0}$

23. $\mathbf{u} \times \mathbf{v} = \begin{vmatrix} \mathbf{i} & \mathbf{j} & \mathbf{k} \\ 2 & -3 & 1 \\ -3 & 3 & 2 \end{vmatrix} = \begin{vmatrix} -3 & 1 \\ 3 & 2 \end{vmatrix}\mathbf{i} - \begin{vmatrix} 2 & 1 \\ -3 & 2 \end{vmatrix}\mathbf{j} + \begin{vmatrix} 2 & -3 \\ -3 & 3 \end{vmatrix}\mathbf{k} = -9\mathbf{i} - 7\mathbf{j} - 3\mathbf{k}$

25. $\mathbf{v} \times \mathbf{u} = \begin{vmatrix} \mathbf{i} & \mathbf{j} & \mathbf{k} \\ -3 & 3 & 2 \\ 2 & -3 & 1 \end{vmatrix} = \begin{vmatrix} 3 & 2 \\ -3 & 1 \end{vmatrix}\mathbf{i} - \begin{vmatrix} -3 & 2 \\ 2 & 1 \end{vmatrix}\mathbf{j} + \begin{vmatrix} -3 & 3 \\ 2 & -3 \end{vmatrix}\mathbf{k} = 9\mathbf{i} + 7\mathbf{j} + 3\mathbf{k}$

27. $\mathbf{v} \times \mathbf{v} = \begin{vmatrix} \mathbf{i} & \mathbf{j} & \mathbf{k} \\ -3 & 3 & 2 \\ -3 & 3 & 2 \end{vmatrix} = \begin{vmatrix} 3 & 2 \\ 3 & 2 \end{vmatrix}\mathbf{i} - \begin{vmatrix} -3 & 2 \\ -3 & 2 \end{vmatrix}\mathbf{j} + \begin{vmatrix} -3 & 3 \\ -3 & 3 \end{vmatrix}\mathbf{k} = 0\mathbf{i} + 0\mathbf{j} + 0\mathbf{k} = \mathbf{0}$

29. $(3\mathbf{u}) \times \mathbf{v} = \begin{vmatrix} \mathbf{i} & \mathbf{j} & \mathbf{k} \\ 6 & -9 & 3 \\ -3 & 3 & 2 \end{vmatrix} = \begin{vmatrix} -9 & 3 \\ 3 & 2 \end{vmatrix}\mathbf{i} - \begin{vmatrix} 6 & 3 \\ -3 & 2 \end{vmatrix}\mathbf{j} + \begin{vmatrix} 6 & -9 \\ -3 & 3 \end{vmatrix}\mathbf{k} = -27\mathbf{i} - 21\mathbf{j} - 9\mathbf{k}$

31. $\mathbf{u} \times (2\mathbf{v}) = \begin{vmatrix} \mathbf{i} & \mathbf{j} & \mathbf{k} \\ 2 & -3 & 1 \\ -6 & 6 & 4 \end{vmatrix} = \begin{vmatrix} -3 & 1 \\ 6 & 4 \end{vmatrix}\mathbf{i} - \begin{vmatrix} 2 & 1 \\ -6 & 4 \end{vmatrix}\mathbf{j} + \begin{vmatrix} 2 & -3 \\ -6 & 6 \end{vmatrix}\mathbf{k} = -18\mathbf{i} - 14\mathbf{j} - 6\mathbf{k}$

33. $\mathbf{u} \cdot (\mathbf{u} \times \mathbf{v}) = \mathbf{u} \cdot \begin{vmatrix} \mathbf{i} & \mathbf{j} & \mathbf{k} \\ 2 & -3 & 1 \\ -3 & 3 & 2 \end{vmatrix} = \mathbf{u} \cdot \left(\begin{vmatrix} -3 & 1 \\ 3 & 2 \end{vmatrix}\mathbf{i} - \begin{vmatrix} 2 & 1 \\ -3 & 2 \end{vmatrix}\mathbf{j} + \begin{vmatrix} 2 & -3 \\ -3 & 3 \end{vmatrix}\mathbf{k} \right)$

   $= (2\mathbf{i} - 3\mathbf{j} + \mathbf{k}) \cdot (-9\mathbf{i} - 7\mathbf{j} - 3\mathbf{k}) = (2)(-9) + (-3)(-7) + (1)(-3) = -18 + 21 - 3 = 0$

35. $\mathbf{u} \cdot (\mathbf{v} \times \mathbf{w}) = \mathbf{u} \cdot \begin{vmatrix} \mathbf{i} & \mathbf{j} & \mathbf{k} \\ -3 & 3 & 2 \\ 1 & 1 & 3 \end{vmatrix} = \mathbf{u} \cdot \left(\begin{vmatrix} 3 & 2 \\ 1 & 3 \end{vmatrix}\mathbf{i} - \begin{vmatrix} -3 & 2 \\ 1 & 3 \end{vmatrix}\mathbf{j} + \begin{vmatrix} -3 & 3 \\ 1 & 1 \end{vmatrix}\mathbf{k} \right)$

   $= (2\mathbf{i} - 3\mathbf{j} + \mathbf{k}) \cdot (7\mathbf{i} + 11\mathbf{j} - 6\mathbf{k}) = (2)(7) + (-3)(11) + (1)(-6) = 14 - 33 - 6 = -25$

37. $\mathbf{v} \cdot (\mathbf{u} \times \mathbf{w}) = \mathbf{v} \cdot \begin{vmatrix} \mathbf{i} & \mathbf{j} & \mathbf{k} \\ 2 & -3 & 1 \\ 1 & 1 & 3 \end{vmatrix} = \mathbf{v} \cdot \left(\begin{vmatrix} -3 & 1 \\ 1 & 3 \end{vmatrix}\mathbf{i} - \begin{vmatrix} 2 & 1 \\ 1 & 3 \end{vmatrix}\mathbf{j} + \begin{vmatrix} 2 & -3 \\ 1 & 1 \end{vmatrix}\mathbf{k} \right)$

   $= (-3\mathbf{i} + 3\mathbf{j} + 2\mathbf{k}) \cdot (-10\mathbf{i} - 5\mathbf{j} + 5\mathbf{k}) = (-3)(-10) + (3)(-5) + (2)(5) = 30 - 15 + 10 = 25$

39. $\mathbf{u} \times (\mathbf{v} \times \mathbf{v}) = \mathbf{u} \times \begin{vmatrix} \mathbf{i} & \mathbf{j} & \mathbf{k} \\ -3 & 3 & 2 \\ -3 & 3 & 2 \end{vmatrix} = \mathbf{u} \times \left(\begin{vmatrix} 3 & 2 \\ 3 & 2 \end{vmatrix}\mathbf{i} - \begin{vmatrix} -3 & 2 \\ -3 & 2 \end{vmatrix}\mathbf{j} + \begin{vmatrix} -3 & 3 \\ -3 & 3 \end{vmatrix}\mathbf{k} \right)$

   $= (2\mathbf{i} - 3\mathbf{j} + \mathbf{k}) \times (0\mathbf{i} + 0\mathbf{j} + 0\mathbf{k}) = \begin{vmatrix} \mathbf{i} & \mathbf{j} & \mathbf{k} \\ 2 & -3 & 1 \\ 0 & 0 & 0 \end{vmatrix}$

   $= \begin{vmatrix} -3 & 1 \\ 0 & 0 \end{vmatrix}\mathbf{i} - \begin{vmatrix} 2 & 1 \\ 0 & 0 \end{vmatrix}\mathbf{j} + \begin{vmatrix} 2 & -3 \\ 0 & 0 \end{vmatrix}\mathbf{k} = 0\mathbf{i} + 0\mathbf{j} + 0\mathbf{k} = \mathbf{0}$

41. $\mathbf{u} \times \mathbf{v} = \begin{vmatrix} \mathbf{i} & \mathbf{j} & \mathbf{k} \\ 2 & -3 & 1 \\ -3 & 3 & 2 \end{vmatrix} = \begin{vmatrix} -3 & 1 \\ 3 & 2 \end{vmatrix}\mathbf{i} - \begin{vmatrix} 2 & 1 \\ -3 & 2 \end{vmatrix}\mathbf{j} + \begin{vmatrix} 2 & -3 \\ -3 & 3 \end{vmatrix}\mathbf{k} = -9\mathbf{i} - 7\mathbf{j} - 3\mathbf{k}$ is orthogonal to

   both $\mathbf{u}$ and $\mathbf{v}$, so choose any vector of the form $c(-9\mathbf{i} - 7\mathbf{j} - 3\mathbf{k})$, where c is a nonzero scalar.

43. A vector that is orthogonal to both $\mathbf{u}$ and $\mathbf{i} + \mathbf{j}$ is $\mathbf{u} \times (\mathbf{i} + \mathbf{j})$.

   $\mathbf{u} \times (\mathbf{i} + \mathbf{j}) = \begin{vmatrix} \mathbf{i} & \mathbf{j} & \mathbf{k} \\ 2 & -3 & 1 \\ 1 & 1 & 0 \end{vmatrix} = \begin{vmatrix} -3 & 1 \\ 1 & 0 \end{vmatrix}\mathbf{i} - \begin{vmatrix} 2 & 1 \\ 1 & 0 \end{vmatrix}\mathbf{j} + \begin{vmatrix} 2 & -3 \\ 1 & 1 \end{vmatrix}\mathbf{k} = -\mathbf{i} + \mathbf{j} + 5\mathbf{k}$, so choose any vector

   of the form $c(-1\mathbf{i} + 1\mathbf{j} + 5\mathbf{k})$, where c is a nonzero scalar.

45. $\mathbf{u} = \overrightarrow{P_1P_2} = \mathbf{i} + 2\mathbf{j} + 3\mathbf{k}$ $\mathbf{v} = \overrightarrow{P_1P_3} = -2\mathbf{i} + 3\mathbf{j} + 0\mathbf{k}$

$$\mathbf{u} \times \mathbf{v} = \begin{vmatrix} \mathbf{i} & \mathbf{j} & \mathbf{k} \\ 1 & 2 & 3 \\ -2 & 3 & 0 \end{vmatrix} = \begin{vmatrix} 2 & 3 \\ 3 & 0 \end{vmatrix}\mathbf{i} - \begin{vmatrix} 1 & 3 \\ -2 & 0 \end{vmatrix}\mathbf{j} + \begin{vmatrix} 1 & 2 \\ -2 & 3 \end{vmatrix}\mathbf{k} = -9\mathbf{i} - 6\mathbf{j} + 7\mathbf{k}$$

$$\text{Area} = \|\mathbf{u} \times \mathbf{v}\| = \sqrt{(-9)^2 + (-6)^2 + 7^2} = \sqrt{166}$$

47. $\mathbf{u} = \overrightarrow{P_1P_2} = -3\mathbf{i} + \mathbf{j} + 4\mathbf{k}$ $\mathbf{v} = \overrightarrow{P_1P_3} = -\mathbf{i} - 4\mathbf{j} + 3\mathbf{k}$

$$\mathbf{u} \times \mathbf{v} = \begin{vmatrix} \mathbf{i} & \mathbf{j} & \mathbf{k} \\ -3 & 1 & 4 \\ -1 & -4 & 3 \end{vmatrix} = \begin{vmatrix} 1 & 4 \\ -4 & 3 \end{vmatrix}\mathbf{i} - \begin{vmatrix} -3 & 4 \\ -1 & 3 \end{vmatrix}\mathbf{j} + \begin{vmatrix} -3 & 1 \\ -1 & -4 \end{vmatrix}\mathbf{k} = 19\mathbf{i} + 5\mathbf{j} + 13\mathbf{k}$$

$$\text{Area} = \|\mathbf{u} \times \mathbf{v}\| = \sqrt{19^2 + 5^2 + 13^2} = \sqrt{555}$$

49. $\mathbf{u} = \overrightarrow{P_1P_2} = 0\mathbf{i} + \mathbf{j} + 1\mathbf{k}$ $\mathbf{v} = \overrightarrow{P_1P_3} = -3\mathbf{i} + 2\mathbf{j} - 2\mathbf{k}$

$$\mathbf{u} \times \mathbf{v} = \begin{vmatrix} \mathbf{i} & \mathbf{j} & \mathbf{k} \\ 0 & 1 & 1 \\ -3 & 2 & -2 \end{vmatrix} = \begin{vmatrix} 1 & 1 \\ 2 & -2 \end{vmatrix}\mathbf{i} - \begin{vmatrix} 0 & 1 \\ -3 & -2 \end{vmatrix}\mathbf{j} + \begin{vmatrix} 0 & 1 \\ -3 & 2 \end{vmatrix}\mathbf{k} = -4\mathbf{i} - 3\mathbf{j} + 3\mathbf{k}$$

$$\text{Area} = \|\mathbf{u} \times \mathbf{v}\| = \sqrt{(-4)^2 + (-3)^2 + 3^2} = \sqrt{34}$$

51. $\mathbf{u} = \overrightarrow{P_1P_2} = 3\mathbf{i} + 0\mathbf{j} - 2\mathbf{k}$ $\mathbf{v} = \overrightarrow{P_1P_3} = 5\mathbf{i} - 7\mathbf{j} + 3\mathbf{k}$

$$\mathbf{u} \times \mathbf{v} = \begin{vmatrix} \mathbf{i} & \mathbf{j} & \mathbf{k} \\ 3 & 0 & -2 \\ 5 & -7 & 3 \end{vmatrix} = \begin{vmatrix} 0 & -2 \\ -7 & 3 \end{vmatrix}\mathbf{i} - \begin{vmatrix} 3 & -2 \\ 5 & 3 \end{vmatrix}\mathbf{j} + \begin{vmatrix} 3 & 0 \\ 5 & -7 \end{vmatrix}\mathbf{k} = -14\mathbf{i} - 19\mathbf{j} - 21\mathbf{k}$$

$$\text{Area} = \|\mathbf{u} \times \mathbf{v}\| = \sqrt{(-14)^2 + (-19)^2 + (-21)^2} = \sqrt{998}$$

53. $\mathbf{v} \times \mathbf{w} = \begin{vmatrix} \mathbf{i} & \mathbf{j} & \mathbf{k} \\ 1 & 3 & -2 \\ -2 & 1 & 3 \end{vmatrix} = \begin{vmatrix} 3 & -2 \\ 1 & 3 \end{vmatrix}\mathbf{i} - \begin{vmatrix} 1 & -2 \\ -2 & 3 \end{vmatrix}\mathbf{j} + \begin{vmatrix} 1 & 3 \\ -2 & 1 \end{vmatrix}\mathbf{k} = 11\mathbf{i} + \mathbf{j} + 7\mathbf{k}$

$$\|\mathbf{v} \times \mathbf{w}\| = \sqrt{11^2 + 1^2 + 7^2} = \sqrt{171} = 3\sqrt{19}$$

$$\mathbf{u} = \pm\frac{\mathbf{v} \times \mathbf{w}}{\|\mathbf{v} \times \mathbf{w}\|} = \pm\frac{11\mathbf{i} + \mathbf{j} + 7\mathbf{k}}{3\sqrt{19}} = \pm\left(\frac{11}{3\sqrt{19}}\mathbf{i} + \frac{1}{3\sqrt{19}}\mathbf{j} + \frac{7}{3\sqrt{19}}\mathbf{k}\right)$$

$$= \pm\left(\frac{11\sqrt{19}}{57}\mathbf{i} + \frac{\sqrt{19}}{57}\mathbf{j} + \frac{7\sqrt{19}}{57}\mathbf{k}\right)$$

55. Prove: $\mathbf{u} \times \mathbf{v} = -(\mathbf{v} \times \mathbf{u})$

Let $\mathbf{u} = a_1\mathbf{i} + b_1\mathbf{j} + c_1\mathbf{k}$ and $\mathbf{v} = a_2\mathbf{i} + b_2\mathbf{j} + c_2\mathbf{k}$

$$\mathbf{u} \times \mathbf{v} = \begin{vmatrix} \mathbf{i} & \mathbf{j} & \mathbf{k} \\ a_1 & b_1 & c_1 \\ a_2 & b_2 & c_2 \end{vmatrix} = \begin{vmatrix} b_1 & c_1 \\ b_2 & c_2 \end{vmatrix}\mathbf{i} - \begin{vmatrix} a_1 & c_1 \\ a_2 & c_2 \end{vmatrix}\mathbf{j} + \begin{vmatrix} a_1 & b_1 \\ a_2 & b_2 \end{vmatrix}\mathbf{k}$$

$$= (b_1c_2 - b_2c_1)\mathbf{i} - (a_1c_2 - a_2c_1)\mathbf{j} + (a_1b_2 - a_2b_1)\mathbf{k}$$

$$= -(b_2c_1 - b_1c_2)\mathbf{i} + (a_2c_1 - a_1c_2)\mathbf{j} - (a_2b_1 - a_1b_2)\mathbf{k}$$

$$= -\big((b_2c_1 - b_1c_2)\mathbf{i} - (a_2c_1 - a_1c_2)\mathbf{j} + (a_2b_1 - a_1b_2)\mathbf{k}\big)$$

$$= -\left(\begin{vmatrix} b_2 & c_2 \\ b_1 & c_1 \end{vmatrix}\mathbf{i} - \begin{vmatrix} a_2 & c_2 \\ a_1 & c_1 \end{vmatrix}\mathbf{j} + \begin{vmatrix} a_2 & b_2 \\ a_1 & b_1 \end{vmatrix}\mathbf{k}\right) = -\begin{vmatrix} \mathbf{i} & \mathbf{j} & \mathbf{k} \\ a_2 & b_2 & c_2 \\ a_1 & b_1 & c_1 \end{vmatrix} = -(\mathbf{v} \times \mathbf{u})$$

57. Prove: $\|\mathbf{u} \times \mathbf{v}\|^2 = \|\mathbf{u}\|^2\|\mathbf{v}\|^2 - (\mathbf{u} \cdot \mathbf{v})^2$

Let $\mathbf{u} = a_1\mathbf{i} + b_1\mathbf{j} + c_1\mathbf{k}$ and $\mathbf{v} = a_2\mathbf{i} + b_2\mathbf{j} + c_2\mathbf{k}$

$$\mathbf{u} \times \mathbf{v} = \begin{vmatrix} \mathbf{i} & \mathbf{j} & \mathbf{k} \\ a_1 & b_1 & c_1 \\ a_2 & b_2 & c_2 \end{vmatrix} = \begin{vmatrix} b_1 & c_1 \\ b_2 & c_2 \end{vmatrix}\mathbf{i} - \begin{vmatrix} a_1 & c_1 \\ a_2 & c_2 \end{vmatrix}\mathbf{j} + \begin{vmatrix} a_1 & b_1 \\ a_2 & b_2 \end{vmatrix}\mathbf{k}$$

$$= (b_1c_2 - b_2c_1)\mathbf{i} - (a_1c_2 - a_2c_1)\mathbf{j} + (a_1b_2 - a_2b_1)\mathbf{k}$$

$$\|\mathbf{u} \times \mathbf{v}\|^2 = (b_1c_2 - b_2c_1)^2 + (a_1c_2 - a_2c_1)^2 + (a_1b_2 - a_2b_1)^2$$

$$= b_1^2c_2^2 - 2b_1b_2c_1c_2 + b_2^2c_1^2 + a_1^2c_2^2 - 2a_1a_2c_1c_2 + a_2^2c_1^2$$
$$+ a_1^2b_2^2 - 2a_1a_2b_1b_2 + a_2^2b_1^2$$

$$= a_1^2b_2^2 + a_1^2c_2^2 + a_2^2b_1^2 + a_2^2c_1^2 + b_1^2c_2^2 + b_2^2c_1^2 - 2a_1a_2b_1b_2$$
$$- 2a_1a_2c_1c_2 - 2b_1b_2c_1c_2$$

$$\|\mathbf{u}\|^2 = a_1^2 + b_1^2 + c_1^2; \qquad \|\mathbf{v}\|^2 = a_2^2 + b_2^2 + c_2^2; \qquad (\mathbf{u} \cdot \mathbf{v})^2 = (a_1a_2 + b_1b_2 + c_1c_2)^2$$

$$\|\mathbf{u}\|^2\|\mathbf{v}\|^2 - (\mathbf{u} \cdot \mathbf{v})^2$$

$$= (a_1^2 + b_1^2 + c_1^2)(a_2^2 + b_2^2 + c_2^2) - (a_1a_2 + b_1b_2 + c_1c_2)^2$$

$$= a_1^2a_2^2 + a_1^2b_2^2 + a_1^2c_2^2 + b_1^2a_2^2 + b_1^2b_2^2 + b_1^2c_2^2 + c_1^2a_2^2 + c_1^2b_2^2 + c_1^2c_2^2$$
$$- a_1^2a_2^2 + a_1a_2b_1b_2 + a_1a_2c_1c_2 + a_1a_2b_1b_2 + b_1^2b_2^2$$
$$+ b_1b_2c_1c_2 + a_1a_2c_1c_2 + b_1b_2c_1c_2 + c_1^2c_2^2$$

$$= a_1^2b_2^2 + a_1^2c_2^2 + a_2^2b_1^2 + a_2^2c_1^2 + b_1^2c_2^2 + b_2^2c_1^2 - 2a_1a_2b_1b_2$$
$$- 2a_1a_2c_1c_2 - 2b_1b_2c_1c_2$$

$$= \|\mathbf{u} \times \mathbf{v}\|^2$$

59. If $\mathbf{u}$ and $\mathbf{v}$ are orthogonal, then $\mathbf{u} \cdot \mathbf{v} = 0$. From problem 57, then:

$$\|\mathbf{u} \times \mathbf{v}\|^2 = \|\mathbf{u}\|^2 \|\mathbf{v}\|^2 - (\mathbf{u} \bullet \mathbf{v})^2 = \|\mathbf{u}\|^2 \|\mathbf{v}\|^2 - (0)^2 = \|\mathbf{u}\|^2 \|\mathbf{v}\|^2$$

$$\|\mathbf{u} \times \mathbf{v}\| = \|\mathbf{u}\| \|\mathbf{v}\|$$

61. $\mathbf{u} \cdot \mathbf{v} = 0 \Rightarrow \mathbf{u}$ and $\mathbf{v}$ are orthogonal.

$\mathbf{u} \times \mathbf{v} = \mathbf{0} \Rightarrow \mathbf{u}$ and $\mathbf{v}$ are parallel.

Therefore, if $\mathbf{u} \cdot \mathbf{v} = 0$ and $\mathbf{u} \times \mathbf{v} = \mathbf{0}$, then either $\mathbf{u} = \mathbf{0}$ or $\mathbf{v} = \mathbf{0}$.

Polar Coordinates; Vectors

8.R   Chapter Review

1. $\left(3, \dfrac{\pi}{6}\right)$

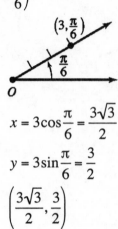

$$x = 3\cos\dfrac{\pi}{6} = \dfrac{3\sqrt{3}}{2}$$

$$y = 3\sin\dfrac{\pi}{6} = \dfrac{3}{2}$$

$$\left(\dfrac{3\sqrt{3}}{2}, \dfrac{3}{2}\right)$$

3. $\left(-2, \dfrac{4\pi}{3}\right)$

$$x = -2\cos\dfrac{4\pi}{3} = 1$$

$$y = -2\sin\dfrac{4\pi}{3} = \sqrt{3}$$

$$\left(1, \sqrt{3}\right)$$

5. $\left(-3, -\dfrac{\pi}{2}\right)$

$$x = -3\cos\left(-\dfrac{\pi}{2}\right) = 0$$

$$y = -3\sin\left(-\dfrac{\pi}{2}\right) = 3$$

$$(0, 3)$$

7. The point (–3, 3) lies in quadrant II.

$$r = \sqrt{x^2 + y^2} = \sqrt{(-3)^2 + 3^2} = 3\sqrt{2}$$

$$\theta = \tan^{-1}\left(\dfrac{y}{x}\right) = \tan^{-1}\left(\dfrac{3}{-3}\right) = \tan^{-1}(-1) = -\dfrac{\pi}{4}$$

Polar coordinates of the point

$(-3, 3)$ are $\left(-3\sqrt{2}, -\dfrac{\pi}{4}\right)$ or $\left(3\sqrt{2}, \dfrac{3\pi}{4}\right)$.

9. The point (0, –2) lies on the negative y-axis.

$$r = \sqrt{x^2 + y^2} = \sqrt{0^2 + (-2)^2} = 2 \qquad \theta = \tan^{-1}\left(\dfrac{y}{x}\right) = \tan^{-1}\left(\dfrac{-2}{0}\right), \ \dfrac{-2}{0} \text{ is not defined} \Rightarrow \theta = -\dfrac{\pi}{2}$$

Polar coordinates of the point $(0, -2)$ are $\left(2, -\dfrac{\pi}{2}\right)$ or $\left(-2, \dfrac{\pi}{2}\right)$.

11. The point $(3, 4)$ lies in quadrant I.

$r = \sqrt{x^2 + y^2} = \sqrt{3^2 + 4^2} = 5 \qquad \theta = \tan^{-1}\left(\frac{y}{x}\right) = \tan^{-1}\frac{4}{3} \approx 0.93$

Polar coordinates of the point $(3, 4)$ are $(5, 0.93)$ or $(-5, 4.07)$.

13.
$$r = 2\sin\theta$$
$$r^2 = 2r\sin\theta$$
$$x^2 + y^2 = 2y$$
$$x^2 + y^2 - 2y = 0$$
$$x^2 + y^2 - 2y + 1 = 1$$
$$x^2 + (y - 1)^2 = 1^2$$

circle with center $(0, 1)$, radius 1.

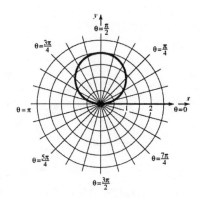

15.
$$r = 5$$
$$r^2 = 25$$
$$x^2 + y^2 = 5^2$$

circle with center $(0, 0)$, radius 5.

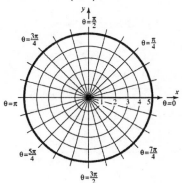

17. $r\cos\theta + 3r\sin\theta = 6$
$$x + 3y = 6$$
$$3y = -x + 6$$
$$y = -\frac{1}{3}x + 2$$

line through $(0, 2)$ and $(6, 2)$, slope $-\frac{1}{3}$

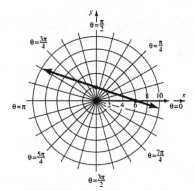

19. $r = 4\cos\theta$ The graph will be a circle, radius 2, center $(2, 0)$. Check for symmetry:
Polar axis: Replace θ by $-\theta$. The result is $r = 4\cos(-\theta) = 4\cos\theta$.

The graph is symmetric with respect to the polar axis.

The line $\theta = \frac{\pi}{2}$: Replace θ by $\pi - \theta$.
$$r = 4\cos(\pi - \theta) = 4(\cos(\pi)\cos\theta + \sin(\pi)\sin\theta)$$
$$= 4(-\cos\theta + 0) = -4\cos\theta$$

The test fails.

The pole: Replace r by $-r$. $-r = 4\cos\theta$. The test fails.

Due to symmetry with respect to the polar axis, assign values to θ from 0 to π.

θ	0	$\dfrac{\pi}{6}$	$\dfrac{\pi}{3}$	$\dfrac{\pi}{2}$	$\dfrac{2\pi}{3}$	$\dfrac{5\pi}{6}$	π
$r = 4\cos\theta$	4	$2\sqrt{3} \approx 3.5$	2	0	-2	$-2\sqrt{3} \approx -3.5$	-4

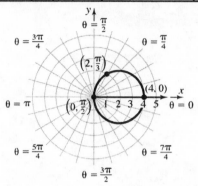

21. $r = 3 - 3\sin\theta$ The graph will be a cardioid. Check for symmetry:
Polar axis: Replace θ by $-\theta$. The result is $r = 3 - 3\sin(-\theta) = 3 + 3\sin\theta$. The test fails.

The line $\theta = \dfrac{\pi}{2}$: Replace θ by $\pi - \theta$.

$$r = 3 - 3\sin(\pi - \theta) = 3 - 3(\sin(\pi)\cos\theta - \cos(\pi)\sin\theta)$$
$$= 3 - 3(0 + \sin\theta) = 3 - 3\sin\theta$$

The graph is symmetric with respect to the line $\theta = \dfrac{\pi}{2}$.

The pole: Replace r by $-r$. $-r = 3 - 3\sin\theta$. The test fails.

Due to symmetry with respect to the line $\theta = \dfrac{\pi}{2}$, assign values to θ from $-\dfrac{\pi}{2}$ to $\dfrac{\pi}{2}$.

θ	$-\dfrac{\pi}{2}$	$-\dfrac{\pi}{3}$	$-\dfrac{\pi}{6}$	0	$\dfrac{\pi}{6}$	$\dfrac{\pi}{3}$	$\dfrac{\pi}{2}$
$r = 3 - 3\sin\theta$	6	$3 + \dfrac{3\sqrt{3}}{2} \approx 5.6$	$\dfrac{9}{2}$	3	$\dfrac{3}{2}$	$3 - \dfrac{3\sqrt{3}}{2} \approx 0.4$	0

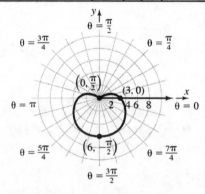

23. $r = 4 - \cos\theta$ The graph will be a limacon without inner loop.
Check for symmetry:
Polar axis: Replace θ by $-\theta$. The result is $r = 4 - \cos(-\theta) = 4 - \cos\theta$.
The graph is symmetric with respect to the polar axis.

The line $\theta = \frac{\pi}{2}$: Replace θ by $\pi - \theta$.

$$r = 4 - \cos(\pi - \theta) = 4 - (\cos(\pi)\cos\theta + \sin(\pi)\sin\theta)$$

$$= 4 - (-\cos\theta + 0) = 4 + \cos\theta$$

The test fails.

The pole: Replace r by $-r$. $-r = 4 - \cos\theta$. The test fails.

Due to symmetry with respect to the polar axis, assign values to θ from 0 to π.

θ	0	$\frac{\pi}{6}$	$\frac{\pi}{3}$	$\frac{\pi}{2}$	$\frac{2\pi}{3}$	$\frac{5\pi}{6}$	π
$r = 4 - \cos\theta$	3	$4 - \frac{\sqrt{3}}{2} \approx 3.1$	$\frac{7}{2}$	4	$\frac{9}{2}$	$4 + \frac{\sqrt{3}}{2} \approx 4.9$	5

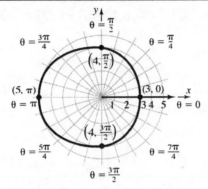

25. $r = \sqrt{x^2 + y^2} = \sqrt{(-1)^2 + (-1)^2} = \sqrt{2}$

$\tan\theta = \frac{y}{x} = \frac{-1}{-1} = 1 \implies \theta = 225°$

The polar form of $z = -1 - i$ is $z = r(\cos\theta + i\sin\theta) = \sqrt{2}(\cos 225° + i\sin 225°)$.

27. $r = \sqrt{x^2 + y^2} = \sqrt{4^2 + (-3)^2} = \sqrt{25} = 5$

$\tan\theta = \frac{y}{x} = -\frac{3}{4} \implies \theta \approx 323.1°$

The polar form of $z = 4 - 3i$ is $z = r(\cos\theta + i\sin\theta) = 5(\cos(323.1°) + i\sin(323.1°))$.

29. $2(\cos 150° + i\sin 150°) = 2\left(-\frac{\sqrt{3}}{2} + \frac{1}{2}i\right) = -\sqrt{3} + i$

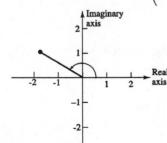

31. $3\left(\cos\dfrac{2\pi}{3} + i\sin\dfrac{2\pi}{3}\right) = 3\left(-\dfrac{1}{2} + \dfrac{\sqrt{3}}{2}i\right) = -\dfrac{3}{2} + \dfrac{3\sqrt{3}}{2}i$

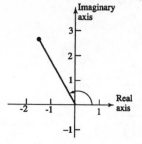

33. $0.1(\cos 350° + i\sin 350°) \approx 0.1(0.9848 - 0.1736i) = 0.0985 - 0.0174i$

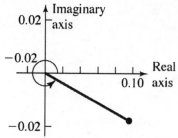

35. $z \cdot w = (\cos 80° + i\sin 80°) \cdot (\cos 50° + i\sin 50°)$

$= 1 \cdot 1(\cos(80° + 50°) + i\sin(80° + 50°)) = \cos 130° + i\sin 130°$

$\dfrac{z}{w} = \dfrac{\cos 80° + i\sin 80°}{\cos 50° + i\sin 50°} = \dfrac{1}{1}(\cos(80° - 50°) + i\sin(80° - 50°))$

$= \cos 30° + i\sin 30°$

37. $z \cdot w = 3\left(\cos\dfrac{9\pi}{5} + i\sin\dfrac{9\pi}{5}\right) \cdot 2\left(\cos\dfrac{\pi}{5} + i\sin\dfrac{\pi}{5}\right) = 3 \cdot 2\left(\cos\left(\dfrac{9\pi}{5} + \dfrac{\pi}{5}\right) + i\sin\left(\dfrac{9\pi}{5} + \dfrac{\pi}{5}\right)\right)$

$= 6 \cdot (\cos 2\pi + i\sin 2\pi) = 6(\cos 0 + i\sin 0) = 6$

$\dfrac{z}{w} = \dfrac{3\cdot\left(\cos\dfrac{9\pi}{5} + i\sin\dfrac{9\pi}{5}\right)}{2\cdot\left(\cos\dfrac{\pi}{5} + i\sin\dfrac{\pi}{5}\right)} = \dfrac{3}{2}\left(\cos\left(\dfrac{9\pi}{5} - \dfrac{\pi}{5}\right) + i\sin\left(\dfrac{9\pi}{5} - \dfrac{\pi}{5}\right)\right) = \dfrac{3}{2}\left(\cos\dfrac{8\pi}{5} + i\sin\dfrac{8\pi}{5}\right)$

39. $z \cdot w = 5(\cos 10° + i\sin 10°) \cdot (\cos 355° + i\sin 355°)$

$= 5 \cdot 1(\cos(10° + 355°) + i\sin(10° + 355°)) = 5 \cdot (\cos 365° + i\sin 365°)$

$= 5 \cdot (\cos 5° + i\sin 5°)$

$\dfrac{z}{w} = \dfrac{5(\cos 10° + i\sin 10°)}{\cos 355° + i\sin 355°} = \dfrac{5}{1}(\cos(10° - 355°) + i\sin(10° - 355°))$

$= 5 \cdot (\cos(-345°) + i\sin(-345°)) = 5 \cdot (\cos 15° + i\sin 15°)$

41. $\left[3(\cos 20° + i \sin 20°)\right]^3 = 3^3(\cos(3 \cdot 20°) + i \sin(3 \cdot 20°)) = 27(\cos 60° + i \sin 60°)$

$$= 27\left(\frac{1}{2} + \frac{\sqrt{3}}{2} i\right) = \frac{27}{2} + \frac{27\sqrt{3}}{2} i$$

43. $\left[\sqrt{2}\left(\cos\frac{5\pi}{8} + i \sin\frac{5\pi}{8}\right)\right]^4 = \left(\sqrt{2}\right)^4\left(\cos\left(4 \cdot \frac{5\pi}{8}\right) + i \sin\left(4 \cdot \frac{5\pi}{8}\right)\right)$

$$= 4 \cdot \left(\cos\frac{5\pi}{2} + i \sin\frac{5\pi}{2}\right) = 4(0 + 1\,i) = 4i$$

45. $1 - \sqrt{3}\,i$ $r = \sqrt{1^2 + \left(-\sqrt{3}\right)^2} = 2$ $\tan\theta = \frac{-\sqrt{3}}{1} = -\sqrt{3} \Rightarrow \theta = 300°$

$1 - \sqrt{3}\,i = 2(\cos 300° + i \sin 300°)$

$\left(1 - \sqrt{3}\,i\right)^6 = \left[2(\cos 300° + i \sin 300°)\right]^6 = 2^6 \cdot (\cos(6 \cdot 300°) + i \sin(6 \cdot 300°))$

$$= 64 \cdot (\cos 1800° + i \sin 1800°) = 64 \cdot (\cos 0° + i \sin 0°) = 64 + 0i = 64$$

47. $3 + 4i$ $r = \sqrt{3^2 + 4^2} = 5$ $\tan\theta = \frac{4}{3} \Rightarrow \theta \approx 53.1°$

$3 + 4i = 5 \cdot (\cos(53.1°) + i \sin(53.1°))$

$(3 + 4i)^4 = \left[5^4 \cdot (\cos(4 \cdot 53.1°) + i \sin(4 \cdot 53.1°))\right]^4 = 625 \cdot (\cos(212.4°) + i \sin(212.4°))$

$$\approx 625 \cdot (-0.8443 + i(-0.5358)) = -527.7 - 334.9i$$

49. $27 + 0i$ $r = \sqrt{27^2 + 0^2} = 27$ $\tan\theta = \frac{0}{27} = 0 \Rightarrow \theta = 0°$

$27 + 0i = 27(\cos 0° + i \sin 0°)$

The three complex cube roots of $27 + 0i = 27(\cos 0° + i \sin 0°)$ are:

$$z_k = \sqrt[3]{27} \cdot \left[\cos\left(\frac{0°}{3} + \frac{360° \cdot k}{3}\right) + i \sin\left(\frac{0°}{3} + \frac{360° \cdot k}{3}\right)\right]$$

$$= 3\left[\cos(120° k) + i \sin(120° \cdot k)\right]$$

$$z_0 = 3\left[\cos(120° \cdot 0) + i \sin(120° \cdot 0)\right] = 3 \cdot (\cos 0° + i \sin 0°) = 3$$

$$z_1 = 3\left[\cos(120° \cdot 1) + i \sin(120° \cdot 1)\right] = 3 \cdot (\cos 120° + i \sin 120°) = -\frac{3}{2} + \frac{3\sqrt{3}}{2} i$$

$$z_2 = 3\left[\cos(120° \cdot 2) + i \sin(120° \cdot 2)\right] = 3 \cdot (\cos 240° + i \sin 240°) = -\frac{3}{2} - \frac{3\sqrt{3}}{2} i$$

51.

53.

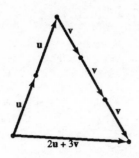

55. $P = (1, -2)$, $Q = (3, -6)$ $\mathbf{v} = (3 - 1)\mathbf{i} + (-6 - (-2))\mathbf{j} = 2\mathbf{i} - 4\mathbf{j}$

$\|\mathbf{v}\| = \sqrt{2^2 + (-4)^2} = \sqrt{20} = 2\sqrt{5}$

57. $P = (0, -2)$, $Q = (-1, 1)$ $\mathbf{v} = (-1 - 0)\mathbf{i} + (1 - (-2))\mathbf{j} = -1\mathbf{i} + 3\mathbf{j}$

$\|\mathbf{v}\| = \sqrt{(-1)^2 + 3^2} = \sqrt{10}$

59. $\mathbf{v} + \mathbf{w} = (-2\mathbf{i} + \mathbf{j}) + (4\mathbf{i} - 3\mathbf{j}) = 2\mathbf{i} - 2\mathbf{j}$

61. $4\mathbf{v} - 3\mathbf{w} = 4(-2\mathbf{i} + \mathbf{j}) - 3(4\mathbf{i} - 3\mathbf{j}) = -8\mathbf{i} + 4\mathbf{j} - 12\mathbf{i} + 9\mathbf{j} = -20\mathbf{i} + 13\mathbf{j}$

63. $\|\mathbf{v}\| = \|-2\mathbf{i} + \mathbf{j}\| = \sqrt{(-2)^2 + 1^2} = \sqrt{5}$

65. $\|\mathbf{v}\| + \|\mathbf{w}\| = \|-2\mathbf{i} + \mathbf{j}\| + \|4\mathbf{i} - 3\mathbf{j}\| = \sqrt{(-2)^2 + 1^2} + \sqrt{4^2 + (-3)^2} = \sqrt{5} + 5 \approx 7.24$

67. $\mathbf{u} = \dfrac{\mathbf{v}}{\|\mathbf{v}\|} = \dfrac{-2\mathbf{i} + \mathbf{j}}{\|-2\mathbf{i} + \mathbf{j}\|} = \dfrac{-2\mathbf{i} + \mathbf{j}}{\sqrt{(-2)^2 + 1^2}} = \dfrac{-2\mathbf{i} + \mathbf{j}}{\sqrt{5}} = -\dfrac{2\sqrt{5}}{5}\mathbf{i} + \dfrac{\sqrt{5}}{5}\mathbf{j}$

69. Let $\mathbf{v} = x\mathbf{i} + y\mathbf{j}$, $\|\mathbf{v}\| = 3$, with the angle between $\mathbf{v}$ and $\mathbf{i}$ equal to $60°$.

$\|\mathbf{v}\| = 3 \Rightarrow \sqrt{x^2 + y^2} = 3 \Rightarrow x^2 + y^2 = 9$

The angle between $\mathbf{v}$ and $\mathbf{i}$ equals $60°$. Thus, $\cos 60° = \dfrac{\mathbf{v} \cdot \mathbf{i}}{\|\mathbf{v}\| \|\mathbf{i}\|} = \dfrac{x}{3(1)} = \dfrac{x}{3}$.

We also conclude that $\mathbf{v}$ lies in Quadrant I.

$\cos 60° = \dfrac{x}{3}$

$\dfrac{1}{2} = \dfrac{x}{3} \Rightarrow x = \dfrac{3}{2}$

$x^2 + y^2 = 9 \Rightarrow \left(\dfrac{3}{2}\right)^2 + y^2 = 9$

$y^2 = 9 - \left(\dfrac{3}{2}\right)^2 \Rightarrow y = \pm\sqrt{9 - \left(\dfrac{3}{2}\right)^2} = \pm\sqrt{9 - \dfrac{9}{4}} = \pm\sqrt{\dfrac{36 - 9}{4}} = \pm\sqrt{\dfrac{27}{4}} = \pm\dfrac{3\sqrt{3}}{2}$

Since $\mathbf{v}$ lies in Quadrant I, $y = \dfrac{3\sqrt{3}}{2}$. So $\mathbf{v} = x\mathbf{i} + y\mathbf{j} = \dfrac{3}{2}\mathbf{i} + \dfrac{3\sqrt{3}}{2}\mathbf{j}$.

71. $d(P_1, P_2) = \sqrt{(4-1)^2 + (-2-3)^2 + (1-(-2))^2} = \sqrt{9+25+9} = \sqrt{43} \approx 6.56$

73. $\mathbf{v} = (4-1)\mathbf{i} + (-2-3)\mathbf{j} + (1-(-2))\mathbf{k} = 3\mathbf{i} - 5\mathbf{j} + 3\mathbf{k}$

75. $4\mathbf{v} - 3\mathbf{w} = 4(3\mathbf{i} + \mathbf{j} - 2\mathbf{k}) - 3(-3\mathbf{i} + 2\mathbf{j} - \mathbf{k}) = 12\mathbf{i} + 4\mathbf{j} - 8\mathbf{k} + 9\mathbf{i} - 6\mathbf{j} + 3\mathbf{k}$
$$= 21\mathbf{i} - 2\mathbf{j} - 5\mathbf{k}$$

77. $\|\mathbf{v} - \mathbf{w}\| = \|(3\mathbf{i} + \mathbf{j} - 2\mathbf{k}) - (-3\mathbf{i} + 2\mathbf{j} - \mathbf{k})\| = \|3\mathbf{i} + \mathbf{j} - 2\mathbf{k} + 3\mathbf{i} - 2\mathbf{j} + \mathbf{k}\|$
$$= \|6\mathbf{i} - \mathbf{j} - \mathbf{k}\| = \sqrt{6^2 + (-1)^2 + (-1)^2} = \sqrt{36 + 1 + 1} = \sqrt{38}$$

79. $\|\mathbf{v}\| - \|\mathbf{w}\| = \|3\mathbf{i} + \mathbf{j} - 2\mathbf{k}\| - \|-3\mathbf{i} + 2\mathbf{j} - \mathbf{k}\|$
$$= \sqrt{3^2 + 1^2 + (-2)^2} - \sqrt{(-3)^2 + 2^2 + (-1)^2} = \sqrt{14} - \sqrt{14} = 0$$

81. $\mathbf{v} \times \mathbf{w} = \begin{vmatrix} \mathbf{i} & \mathbf{j} & \mathbf{k} \\ 3 & 1 & -2 \\ -3 & 2 & -1 \end{vmatrix} = \begin{vmatrix} 1 & -2 \\ 2 & -1 \end{vmatrix}\mathbf{i} - \begin{vmatrix} 3 & -2 \\ -3 & -1 \end{vmatrix}\mathbf{j} + \begin{vmatrix} 3 & 1 \\ -3 & 2 \end{vmatrix}\mathbf{k} = 3\mathbf{i} + 9\mathbf{j} + 9\mathbf{k}$

83. Same direction:
$$\frac{\mathbf{v}}{\|\mathbf{v}\|} = \frac{3\mathbf{i} + \mathbf{j} - 2\mathbf{k}}{\sqrt{3^2 + 1^2 + (-2)^2}} = \frac{3\mathbf{i} + \mathbf{j} - 2\mathbf{k}}{\sqrt{14}} = \frac{3\sqrt{14}}{14}\mathbf{i} + \frac{\sqrt{14}}{14}\mathbf{j} - \frac{\sqrt{14}}{7}\mathbf{k}$$

Opposite direction:
$$\frac{-\mathbf{v}}{\|\mathbf{v}\|} = -\frac{3\sqrt{14}}{14}\mathbf{i} - \frac{\sqrt{14}}{14}\mathbf{j} + \frac{\sqrt{14}}{7}\mathbf{k}$$

85. $\mathbf{v} = -2\mathbf{i} + \mathbf{j}, \quad \mathbf{w} = 4\mathbf{i} - 3\mathbf{j}$
$\mathbf{v} \cdot \mathbf{w} = (-2)(4) + (1)(-3) = -8 - 3 = -11$
$$\cos\theta = \frac{\mathbf{v} \cdot \mathbf{w}}{\|\mathbf{v}\| \|\mathbf{w}\|} = \frac{-11}{\sqrt{(-2)^2 + 1^2}\sqrt{4^2 + (-3)^2}} = \frac{-11}{\sqrt{5}(5)} \approx -0.9839 \Rightarrow \theta \approx 169.7°$$

87. $\mathbf{v} = \mathbf{i} - 3\mathbf{j}, \quad \mathbf{w} = -\mathbf{i} + \mathbf{j}$
$\mathbf{v} \bullet \mathbf{w} = 1(-1) + (-3)(1) = -1 - 3 = -4$
$$\cos\theta = \frac{\mathbf{v} \cdot \mathbf{w}}{\|\mathbf{v}\| \|\mathbf{w}\|} = \frac{-4}{\sqrt{1^2 + (-3)^2}\sqrt{(-1)^2 + 1^2}} = \frac{-4}{\sqrt{10}\sqrt{2}} = \frac{-2}{\sqrt{5}} \approx -0.8944 \Rightarrow \theta \approx 153.4°$$

89. $\mathbf{v} \cdot \mathbf{w} = (\mathbf{i} + \mathbf{j} + \mathbf{k}) \cdot (\mathbf{i} - \mathbf{j} + \mathbf{k}) = (1)(1) + (1)(-1) + (1)(1) = 1 - 1 + 1 = 1$
$$\cos\theta = \frac{\mathbf{v} \cdot \mathbf{w}}{\|\mathbf{v}\| \|\mathbf{w}\|} = \frac{1}{\sqrt{1^2 + 1^2 + 1^2}\sqrt{1^2 + (-1)^2 + 1^2}} = \frac{1}{\sqrt{3}\sqrt{3}} = \frac{1}{3} \Rightarrow \theta \approx 70.5°$$

91. $\mathbf{v} \cdot \mathbf{w} = (4\mathbf{i} - \mathbf{j} + 2\mathbf{k}) \cdot (\mathbf{i} - 2\mathbf{j} - 3\mathbf{k}) = (4)(1) + (-1)(-2) + (2)(-3) = 4 + 2 - 6 = 0$

$\cos\theta = \dfrac{\mathbf{v} \cdot \mathbf{w}}{\|\mathbf{v}\|\|\mathbf{w}\|} = \dfrac{0}{\sqrt{4^2 + (-1)^2 + 2^2}\sqrt{1^2 + (-2)^2 + (-3)^2}} = 0 \Rightarrow \theta = 90°$

93. $\mathbf{v} = 2\mathbf{i} + 3\mathbf{j}, \quad \mathbf{w} = -4\mathbf{i} - 6\mathbf{j}$

$\mathbf{v} \cdot \mathbf{w} = (2)(-4) + (3)(-6) = -8 - 18 = -26$

$\cos\theta = \dfrac{\mathbf{v} \cdot \mathbf{w}}{\|\mathbf{v}\|\|\mathbf{w}\|} = \dfrac{-26}{\sqrt{2^2 + 3^2}\sqrt{(-4)^2 + (-6)^2}} = \dfrac{-5}{\sqrt{13}\sqrt{52}} = \dfrac{-26}{\sqrt{676}} = -1$

$\theta = \cos^{-1}(-1) = 180°$ Therefore, the vectors are parallel.

95. $\mathbf{v} = 3\mathbf{i} - 4\mathbf{j}, \quad \mathbf{w} = -3\mathbf{i} + 4\mathbf{j}$

$\mathbf{v} \cdot \mathbf{w} = (3)(-3) + (-4)(4) = -9 - 16 = -25$

$\cos\theta = \dfrac{\mathbf{v} \cdot \mathbf{w}}{\|\mathbf{v}\|\|\mathbf{w}\|} = \dfrac{-25}{\sqrt{3^2 + (-4)^2}\sqrt{(-3)^2 + 4^2}} = \dfrac{-25}{\sqrt{25}\sqrt{25}} = \dfrac{-25}{25} = -1$

$\theta = \cos^{-1}(-1) = 180°$ Therefore, the vectors are parallel.

97. $\mathbf{v} = 3\mathbf{i} - 2\mathbf{j}, \quad \mathbf{w} = 4\mathbf{i} + 6\mathbf{j}$

$\mathbf{v} \cdot \mathbf{w} = (3)(4) + (-2)(6) = 12 - 12 = 0$

Therefore, the vectors are orthogonal.

99. $\mathbf{v} = 2\mathbf{i} + \mathbf{j}, \quad \mathbf{w} = -4\mathbf{i} + 3\mathbf{j}$

The decomposition of $\mathbf{v}$ into 2 vectors $\mathbf{v}_1$ and $\mathbf{v}_2$ so that $\mathbf{v}_1$ is parallel to $\mathbf{w}$ and $\mathbf{v}_2$ is perpendicular to $\mathbf{w}$ is given by:

$\mathbf{v}_1 = \dfrac{\mathbf{v} \cdot \mathbf{w}}{\|\mathbf{w}\|^2}\mathbf{w}$ and $\mathbf{v}_2 = \mathbf{v} - \mathbf{v}_1$

$\mathbf{v}_1 = \dfrac{\mathbf{v} \cdot \mathbf{w}}{\|\mathbf{w}\|^2} = \dfrac{(2\mathbf{i} + \mathbf{j}) \cdot (-4\mathbf{i} + 3\mathbf{j})}{\left(\sqrt{(-4)^2 + 3^2}\right)^2}(-4\mathbf{i} + 3\mathbf{j}) = \dfrac{(2)(-4) + (1)(3)}{25}(-4\mathbf{i} + 3\mathbf{j})$

$= \left(-\dfrac{1}{5}\right)(-4\mathbf{i} + 3\mathbf{j}) = \dfrac{4}{5}\mathbf{i} - \dfrac{3}{5}\mathbf{j}$

$\mathbf{v}_2 = \mathbf{v} - \mathbf{v}_1 = 2\mathbf{i} + \mathbf{j} - \left(\dfrac{4}{5}\mathbf{i} - \dfrac{3}{5}\mathbf{j}\right) = \dfrac{6}{5}\mathbf{i} + \dfrac{8}{5}\mathbf{j}$

101. $\mathbf{v} = 2\mathbf{i} + 3\mathbf{j}, \quad \mathbf{w} = 3\mathbf{i} + \mathbf{j}$

The projection of $\mathbf{v}$ onto $\mathbf{w}$ is given by:

$\dfrac{\mathbf{v} \cdot \mathbf{w}}{\|\mathbf{w}\|^2}\mathbf{w} = \dfrac{(2\mathbf{i} + 3\mathbf{j}) \cdot (3\mathbf{i} + \mathbf{j})}{\left(\sqrt{3^2 + 1^2}\right)^2}(3\mathbf{i} + \mathbf{j}) = \dfrac{(2)(3) + (3)(1)}{10}(3\mathbf{i} + \mathbf{j}) = \left(\dfrac{9}{10}\right)(3\mathbf{i} + \mathbf{j}) = \dfrac{27}{10}\mathbf{i} + \dfrac{9}{10}\mathbf{j}$

103. $\cos\alpha = \dfrac{a}{\|\mathbf{v}\|} = \dfrac{3}{\sqrt{3^2 + (-4)^2 + 2^2}} = \dfrac{3}{\sqrt{29}} \Rightarrow \alpha \approx 56.1°$

$\cos\beta = \dfrac{b}{\|\mathbf{v}\|} = \dfrac{-4}{\sqrt{3^2 + (-4)^2 + 2^2}} = -\dfrac{4}{\sqrt{29}} \Rightarrow \beta \approx 138.0°$

$\cos\gamma = \dfrac{c}{\|\mathbf{v}\|} = \dfrac{2}{\sqrt{3^2 + (-4)^2 + 2^2}} = \dfrac{2}{\sqrt{29}} \Rightarrow \gamma \approx 68.2°$

105. $\mathbf{u} = \overrightarrow{P_1P_2} = \mathbf{i} + 2\mathbf{j} + 3\mathbf{k} \qquad \mathbf{v} = \overrightarrow{P_1P_3} = 5\mathbf{i} + 4\mathbf{j} + \mathbf{k}$

$\mathbf{u} \times \mathbf{v} = \begin{vmatrix} \mathbf{i} & \mathbf{j} & \mathbf{k} \\ 1 & 2 & 3 \\ 5 & 4 & 1 \end{vmatrix} = \begin{vmatrix} 2 & 3 \\ 4 & 1 \end{vmatrix}\mathbf{i} - \begin{vmatrix} 1 & 3 \\ 5 & 1 \end{vmatrix}\mathbf{j} + \begin{vmatrix} 1 & 2 \\ 5 & 4 \end{vmatrix}\mathbf{k} = -10\mathbf{i} + 14\mathbf{j} - 6\mathbf{k}$

Area $= \|\mathbf{u} \times \mathbf{v}\| = \sqrt{(-10)^2 + 14^2 + (-6)^2} = \sqrt{332} = 2\sqrt{83} \approx 18.2$ sq. units

107. $\mathbf{v} \times \mathbf{u} = -(\mathbf{u} \times \mathbf{v}) = -(2\mathbf{i} - 3\mathbf{j} + \mathbf{k}) = -2\mathbf{i} + 3\mathbf{j} - \mathbf{k}$

109. Let the positive x-axis point downstream, so that the velocity of the current is $\mathbf{v}_c = 2\mathbf{i}$.

Let $\mathbf{v}_w =$ the velocity of the swimmer in the water.

Let $\mathbf{v}_g =$ the velocity of the swimmer relative to the land.

Then $\mathbf{v}_g = \mathbf{v}_w + \mathbf{v}_c$

The speed of the swimmer is $\|\mathbf{v}_w\| = 5$; the heading is directly across the river, so $\mathbf{v}_w = 5\mathbf{j}$.

Then $\mathbf{v}_g = \mathbf{v}_w + \mathbf{v}_c = 5\mathbf{j} + 2\mathbf{i} = 2\mathbf{i} + 5\mathbf{j}$

$\Rightarrow \|\mathbf{v}_g\| = \sqrt{2^2 + 5^2} = \sqrt{29} \approx 5.39$ miles per hour

Since the river is 1 mile wide, it takes the swimmer about 0.2 hour to cross the river. The swimmer will end up $(0.2)(2) = 0.4$ mile downstream.

111. Let $\mathbf{F}_1$ be the tension on the left cable and $\mathbf{F}_2$ be the tension on the right cable.

Let $\mathbf{F}_3$ represent the force of the weight of the box.

$\mathbf{F}_1 = \|\mathbf{F}_1\|\left(\cos(140°)\mathbf{i} + \sin(140°)\mathbf{j}\right) \approx \|\mathbf{F}_1\|(-0.7660\mathbf{i} + 0.6428\mathbf{j})$

$\mathbf{F}_2 = \|\mathbf{F}_2\|\left(\cos(30°)\mathbf{i} + \sin(30°)\mathbf{j}\right) \approx \|\mathbf{F}_2\|(0.8660\mathbf{i} + 0.5000\mathbf{j})$

$\mathbf{F}_3 = -2000\mathbf{j}$

For equilibrium, the sum of the force vectors must be zero.

$\mathbf{F}_1 + \mathbf{F}_2 + \mathbf{F}_3 = -0.7660\|\mathbf{F}_1\|\mathbf{i} + 0.6428\|\mathbf{F}_1\|\mathbf{j} + 0.8660\|\mathbf{F}_2\|\mathbf{i} + 0.5000\|\mathbf{F}_2\|\mathbf{j} - 2000\mathbf{j}$

$= \left(-0.7660\|\mathbf{F}_1\| + 0.8660\|\mathbf{F}_2\|\right)\mathbf{i} + \left(0.6428\|\mathbf{F}_1\| + 0.5000\|\mathbf{F}_2\| - 2000\right)\mathbf{j}$

$= 0$

Set the $\mathbf{i}$ and $\mathbf{j}$ components equal to zero and solve:

$\begin{cases} -0.7660\|\mathbf{F}_1\| + 0.8660\|\mathbf{F}_2\| = 0 \Rightarrow \|\mathbf{F}_2\| = \dfrac{0.7660}{0.8660}\|\mathbf{F}_1\| = 0.8845\|\mathbf{F}_1\| \\ 0.6428\|\mathbf{F}_1\| + 0.5000\|\mathbf{F}_2\| - 2000 = 0 \end{cases}$

$$0.6428\|\mathbf{F}_1\| + 0.5000\big(0.8845\|\mathbf{F}_1\|\big) - 2000 = 0$$

$$1.0851\|\mathbf{F}_1\| = 2000 \Rightarrow \|\mathbf{F}_1\| \approx 1843 \text{ pounds}; \quad \|\mathbf{F}_2\| = 0.8845(1843) = 1630 \text{ pounds}$$

The tension in the left cable is about 1843 pounds and the tension in the right cable is about 1630 pounds.

113. $\mathbf{F} = 5\big(\cos(60°)\mathbf{i} + \sin(60°)\mathbf{j}\big) = 5\left(\dfrac{1}{2}\mathbf{i} + \dfrac{\sqrt{3}}{2}\mathbf{j}\right) = \dfrac{5}{2}\mathbf{i} + \dfrac{5\sqrt{3}}{2}\mathbf{j}$

$\overrightarrow{AB} = 20\mathbf{i}$

$W = \mathbf{F} \cdot \overrightarrow{AB} = \left(\dfrac{5}{2}\mathbf{i} + \dfrac{5\sqrt{3}}{2}\mathbf{j}\right) \cdot 20\mathbf{i} = \left(\dfrac{5}{2}\right)(20) + \left(\dfrac{5\sqrt{3}}{2}\right)(0) = 50$ foot-pounds

Polar Coordinates; Vectors

8.CR Cumulative Review

1.
$$e^{x^2-9} = 1$$
$$\ln\left(e^{x^2-9}\right) = \ln(1)$$
$$x^2 - 9 = 0$$
$$(x+3)(x-3) = 0$$
$$x = -3 \text{ or } x = 3$$
The solution set is $\{-3,3\}$.

3. The circle with center point (0,1) and radius has equation:
$$(x-h)^2 + (y-k)^2 = r^2$$
$$(x-0)^2 + (y-1)^2 = 3^2$$
$$x^2 + (y-1)^2 = 9$$
$$x^2 + y^2 - 2y + 1 = 9$$
$$x^2 + y^2 - 2y - 8 = 0$$

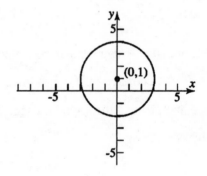

5. $x^2 + y^3 = 2x^4$
Test for symmetry:

  x-axis: Replace y by $-y$: $x^2 + (-y)^3 = 2x^4 \Rightarrow x^2 - y^3 = 2x^4$
  which is not equivalent to $x^2 + y^3 = 2x^4$.

  y-axis: Replace x by $-x$: $(-x)^2 + y^3 = 2(-x)^4 \Rightarrow x^2 + y^3 = 2x^4$,
  which is equivalent to $x^2 + y^3 = 2x^4$.

  Origin: Replace x by $-x$ and y by $-y$: $(-x)^2 + (-y)^3 = 2(-x)^4 \Rightarrow x^2 - y^3 = 2x^4$,
  which is not equivalent to $x^2 + y^3 = 2x^4$.

Therefore, the graph is symmetric with respect to the y-axis.

7. $y = |\sin x| = \begin{cases} \sin x, & \text{when } \sin x \geq 0 \\ -\sin x, & \text{when } \sin x < 0 \end{cases}$

$= \begin{cases} \sin x, & \text{when } 0 \leq x \leq \pi \\ -\sin x, & \text{when } \pi < x < 2\pi \end{cases}$

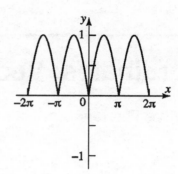

9. $\sin^{-1}\left(-\dfrac{1}{2}\right)$

We are finding the angle θ, $-\dfrac{\pi}{2} \leq \theta \leq \dfrac{\pi}{2}$, whose sine equals $-\dfrac{1}{2}$.

$\sin \theta = -\dfrac{1}{2} \qquad -\dfrac{\pi}{2} \leq \theta \leq \dfrac{\pi}{2}$

$\theta = -\dfrac{\pi}{6} \Rightarrow \sin^{-1}\left(-\dfrac{1}{2}\right) = -\dfrac{\pi}{6}$

11.   Graphing $r = 2$ and $\theta = \dfrac{\pi}{3}$ using polar coordinates:

$r = 2$ yields a circle, centered at $(0,0)$, with radius $= 2$.

$\theta = \dfrac{\pi}{3}$ yields a line passing through the point $(0,0)$, forming an angle of $\theta = \dfrac{\pi}{3}$

with the positive x-axis.

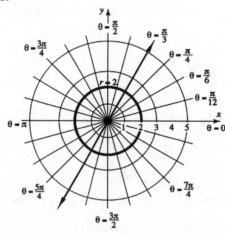

Analytic Geometry

9.2   The Parabola

1. $\sqrt{\left(x_2 - x_1\right)^2 + \left(y_2 - y_1\right)^2}$

3. $(x + 4)^2 = 9$

 $x + 4 = \pm 3$

 $x + 4 = 3 \Rightarrow x = -1$

 or $x + 4 = -3 \Rightarrow x = -7$
 The solution set is $\{-7, -1\}$

5. 3, up

7. paraboloid of revolution

9. True

11. *B*   13. *E*   15. *H*   17. *C*

19. The focus is (4, 0) and the vertex is (0, 0). Both lie
 on the horizontal line $y = 0$. $a = 4$ and since (4, 0)
 is to the right of (0, 0), the parabola opens to the
 right. The equation of the parabola is:

 $y^2 = 4ax$

 $y^2 = 4 \cdot 4 \cdot x$

 $y^2 = 16x$

 Letting $x = 4$, we find $y^2 = 64$ or $y = \pm 8$.
 The points (4, 8) and (4, –8) define the latus rectum.

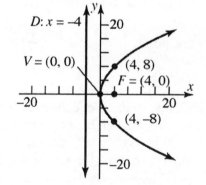

21. The focus is (0, –3) and the vertex is (0, 0). Both lie
 on the vertical line $x = 0$. $a = 3$ and since (0, –3) is
 below (0, 0), the parabola opens down. The
 equation of the parabola is:

 $x^2 = -4ay$

 $x^2 = -4 \cdot 3 \cdot y$

 $x^2 = -12y$

 Letting $y = -3$, we find $x^2 = 36$ or $x = \pm 6$.
 The points (6, 3) and (6, –3) define the latus rectum.

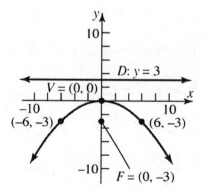

23.   The focus is $(-2, 0)$ and the directrix is $x = 2$. The vertex is $(0, 0)$. $a = 2$ and since $(-2, 0)$ is to the left of $(0, 0)$, the parabola opens to the left. The equation of the parabola is:

$$y^2 = -4ax$$
$$y^2 = -4 \cdot 2 \cdot x$$
$$y^2 = -8x$$

Letting $x = -2$, we find $y^2 = 16$ or $y = \pm 4$. The points $(-2, 4)$ and $(-2, -4)$ define the latus rectum.

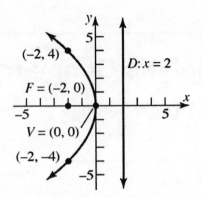

25.   The directrix is $y = -\dfrac{1}{2}$ and the vertex is $(0, 0)$. The focus is $\left(0, \dfrac{1}{2}\right)$. $a = \dfrac{1}{2}$ and since $\left(0, \dfrac{1}{2}\right)$ is above $(0, 0)$, the parabola opens up. The equation of the parabola is:

$$x^2 = 4ay$$
$$x^2 = 4 \cdot \frac{1}{2} \cdot y \Rightarrow x^2 = 2y$$

Letting $y = \dfrac{1}{2}$, we find $x^2 = 1$ or $x = \pm 1$.

The points $\left(1, \dfrac{1}{2}\right)$ and $\left(-1, \dfrac{1}{2}\right)$ define the latus rectum.

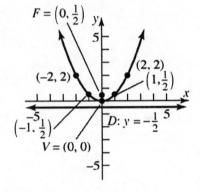

27.   Vertex: $(0,0)$. Since the axis of symmetry is vertical, the parabola opens up or down. Since $(2, 3)$ is above $(0, 0)$, the parabola opens up. The equation has the form $x^2 = 4ay$. Substitute the coordinates of $(2, 3)$ into the equation to find a:

$$2^2 = 4a \cdot 3 \Rightarrow 4 = 12a \Rightarrow a = \frac{1}{3}$$

The equation of the parabola is: $x^2 = \dfrac{4}{3}y$. The focus is $\left(0, \dfrac{1}{3}\right)$. Letting $y = \dfrac{1}{3}$,

we find $x^2 = \dfrac{4}{9}$ or $x = \pm\dfrac{2}{3}$. The points $\left(\dfrac{2}{3}, \dfrac{1}{3}\right)$ and

$\left(-\dfrac{2}{3}, \dfrac{1}{3}\right)$ define the latus rectum.

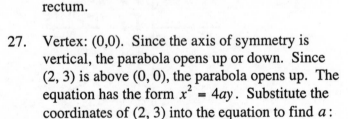

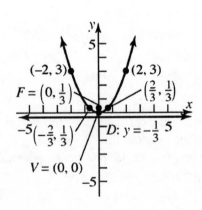

29.   The vertex is (2, –3) and the focus is (2, –5). Both
       lie on the vertical line $x = 2$. $a = |-5 - (-3)| = 2$
       and since (2, –5) is below (2, –3), the parabola
       opens down. The equation of the parabola is:

$$(x - h)^2 = -4a(y - k)$$

$$(x - 2)^2 = -4(2)(y - (-3))$$

$$(x - 2)^2 = -8(y + 3)$$

Letting $y = -5$, we find

$$(x - 2)^2 = 16$$

$$x - 2 = \pm 4 \Rightarrow x = -2 \text{ or } x = 6$$

The points (–2, –5) and (6, –5) define the latus
rectum.

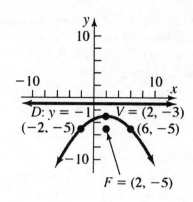

31.   The vertex is (–1, –2) and the focus is (0, –2).
       Both lie on the horizontal line $y = -2$.
       $a = |-1 - 0| = 1$ and since (0, –2) is to the right of
       (–1, –2), the parabola opens to the right. The
       equation of the parabola is:

$$(y - k)^2 = 4a(x - h)$$

$$(y - (-2))^2 = 4(1)(x - (-1))$$

$$(y + 2)^2 = 4(x + 1)$$

Letting $x = 0$, we find

$$(y + 2)^2 = 4$$

$$y + 2 = \pm 2 \Rightarrow y = -4 \text{ or } y = 0$$

The points (0, –4) and (0, 0) define the latus
rectum.

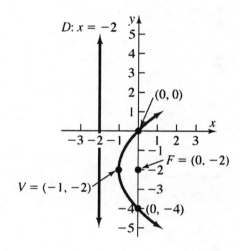

33.   The directrix is $y = 2$ and the focus is (–3, 4).
       This is a vertical case, so the vertex is (–3, 3).
       $a = 1$ and since (–3, 4) is above $y = 2$, the
       parabola opens up. The equation of the parabola
       is: $(x - h)^2 = 4a(y - k)$

$$(x - (-3))^2 = 4 \cdot 1 \cdot (y - 3)$$

$$(x + 3)^2 = 4(y - 3)$$

Letting $y = 4$, we find $(x + 3)^2 = 4$ or $x + 3 = \pm 2$.
So, $x = -1$ or $x = -5$. The points (–1, 4) and
(–5, 4) define the latus rectum.

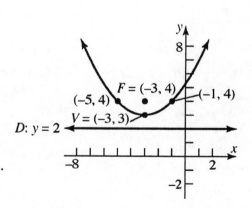

35. The directrix is $x = 1$ and the focus is $(-3, -2)$. This is a horizontal case, so the vertex is $(-1, -2)$. $a = 2$ and since $(-3, -2)$ is to the left of $x = 1$, the parabola opens to the left. The equation of the parabola is: $(y - k)^2 = -4a(x - h)$

$$(y - (-2))^2 = -4 \cdot 2 \cdot (x - (-1))$$

$$(y + 2)^2 = -8(x + 1)$$

Letting $x = -3$, we find $(y + 2)^2 = 16$ or $y + 2 = \pm 4$. So, $y = 2$ or $y = -6$. The points $(-3, 2)$ and $(-3, -6)$ define the latus rectum.

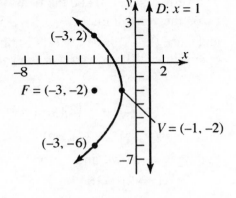

37. The equation $x^2 = 4y$ is in the form $x^2 = 4ay$ where $4a = 4$ or $a = 1$. Thus, we have:
Vertex: $(0, 0)$
Focus: $(0, 1)$
Directrix: $y = -1$

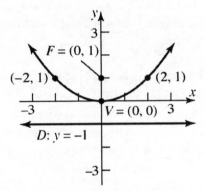

39. The equation $y^2 = -16x$ is in the form $y^2 = -4ax$ where $-4a = -16$ or $a = 4$. Thus, we have:
Vertex: $(0, 0)$
Focus: $(-4, 0)$
Directrix: $x = 4$

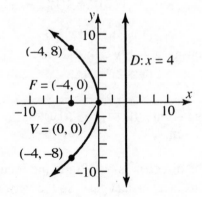

41. The equation $(y - 2)^2 = 8(x + 1)$ is in the form $(y - k)^2 = 4a(x - h)$ where $4a = 8$ or $a = 2$, $h = -1$, and $k = 2$. Thus, we have:
Vertex: $(-1, 2)$
Focus: $(1, 2)$
Directrix: $x = -3$

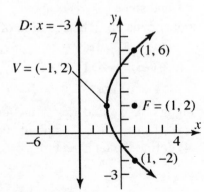

43. The equation $(x - 3)^2 = -(y + 1)$ is in the form

$(x - h)^2 = -4a(y - k)$ where $-4a = -1$ or $a = \dfrac{1}{4}$,

$h = 3$, and $k = -1$. Thus, we have:

Vertex: $(3, -1)$

Focus: $\left(3, -\dfrac{5}{4}\right)$

Directrix: $y = -\dfrac{3}{4}$

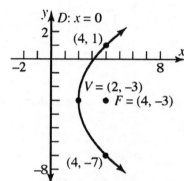

45. The equation $(y + 3)^2 = 8(x - 2)$ is in the form

$(y - k)^2 = 4a(x - h)$ where $4a = 8$ or $a = 2$,

$h = 2$, and $k = -3$. Thus, we have:

Vertex: $(2, -3)$

Focus: $(4, -3)$

Directrix: $x = 0$

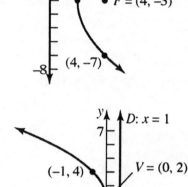

47. Complete the square to put in standard form:

$y^2 - 4y + 4x + 4 = 0$

$\qquad y^2 - 4y + 4 = -4x$

$\qquad\qquad (y - 2)^2 = -4x$

The equation is in the form $(y - k)^2 = -4a(x - h)$

where $-4a = -4$ or $a = 1$, $h = 0$, and $k = 2$.

Thus, we have:

Vertex: $(0, 2)$

Focus: $(-1, 2)$

Directrix: $x = 1$

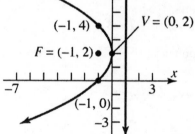

49. Complete the square to put in standard form:

$\qquad x^2 + 8x = 4y - 8$

$x^2 + 8x + 16 = 4y - 8 + 16$

$\qquad (x + 4)^2 = 4(y + 2)$

The equation is in the form $(x - h)^2 = 4a(y - k)$

where $4a = 4$ or $a = 1$, $h = -4$, and $k = -2$.

Thus, we have:

Vertex: $(-4, -2)$

Focus: $(-4, -1)$

Directrix: $y = -3$

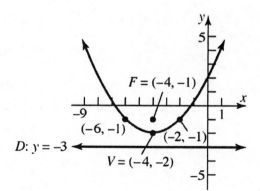

51. Complete the square to put in standard form:
$y^2 + 2y - x = 0$

$y^2 + 2y + 1 = x + 1$

$(y + 1)^2 = x + 1$

The equation is in the form

$(y - k)^2 = 4a(x - h)$ where $4a = 1$ or $a = \dfrac{1}{4}$,

$h = -1$, and $k = -1$.
Thus, we have:
Vertex: $(-1, -1)$

Focus: $\left(-\dfrac{3}{4}, -1\right)$

Directrix: $x = -\dfrac{5}{4}$

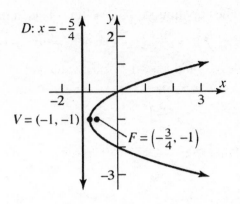

53. Complete the square to put in standard form:
$x^2 - 4x = y + 4$

$x^2 - 4x + 4 = y + 4 + 4$

$(x - 2)^2 = y + 8$

The equation is in the form

$(x - h)^2 = 4a(y - k)$ where $4a = 1$ or $a = \dfrac{1}{4}$,

$h = 2$, and $k = -8$.
Thus, we have:
Vertex: $(2, -8)$

Focus: $\left(2, -\dfrac{31}{4}\right)$

Directrix: $y = -\dfrac{33}{4}$

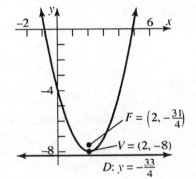

55. $(y - 1)^2 = c(x - 0)$
$(y - 1)^2 = cx$

$(2 - 1)^2 = c(1) \Rightarrow 1 = c$

$(y - 1)^2 = x$

57. $(y - 1)^2 = c(x - 2)$
$(0 - 1)^2 = c(1 - 2)$

$1 = -c \Rightarrow c = -1$

$(y - 1)^2 = -(x - 2)$

59. $(x - 0)^2 = c(y - 1)$
$x^2 = c(y - 1)$

$2^2 = c(2 - 1) \Rightarrow 4 = c$

$x^2 = 4(y - 1)$

61. $(y - 0)^2 = c(x - (-2))$
$y^2 = c(x + 2)$

$1^2 = c(0 + 2) \Rightarrow 1 = 2c \Rightarrow c = \dfrac{1}{2}$

$y^2 = \dfrac{1}{2}(x + 2)$

63. Set up the problem so that the vertex of the parabola is at $(0, 0)$ and it opens up. Then the equation of the parabola has the form: $x^2 = 4ay$. Since the parabola is 10 feet across and 4 feet deep, the points $(5, 4)$ and $(-5, 4)$ are on the parabola.

Substitute and solve for a: $5^2 = 4a(4) \Rightarrow 25 = 16a \Rightarrow a = \dfrac{25}{16}$

a is the distance from the vertex to the focus. Thus, the receiver (located at the focus) is $\dfrac{25}{16} = 1.5625$ feet, or 18.75 inches from the base of the dish, along the axis of the parabola.

65. Set up the problem so that the vertex of the parabola is at $(0, 0)$ and it opens up. Then the equation of the parabola has the form: $x^2 = 4ay$. Since the parabola is 4 inches across and 1 inch deep, the points $(2, 1)$ and $(-2, 1)$ are on the parabola.

Substitute and solve for a: $2^2 = 4a(1) \Rightarrow 4 = 4a \Rightarrow a = 1$, a is the distance from the vertex to the focus. Thus, the bulb (located at the focus) should be 1 inch from the vertex.

67. Set up the problem so that the vertex of the parabola is at $(0, 0)$ and it opens up. Then the equation of the parabola has the form: $x^2 = cy$.
The point $(300, 80)$ is a point on the parabola.
Solve for c and find the equation:
$300^2 = c(80) \Rightarrow c = 1125$

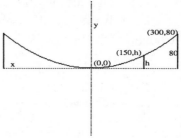

$$x^2 = 1125y$$

Since the height of the cable 150 feet from the center is to be found, the point $(150, h)$ is a point on the parabola. Solve for h:
$150^2 = 1125h \Rightarrow 22,500 = 1125h \Rightarrow h = 20$
The height of the cable 150 feet from the center is 20 feet.

69. Set up the problem so that the vertex of the parabola is at $(0, 0)$ and it opens up. Then the equation of the parabola has the form: $x^2 = 4ay$. a is the distance from the vertex to the focus (where the source is located), so $a = 2$. Since the opening is 5 feet across, there is a point $(2.5, y)$ on the parabola.
Solve for y: $x^2 = 8y \Rightarrow 2.5^2 = 8y \Rightarrow 6.25 = 8y \Rightarrow y = 0.78125$ feet
The depth of the searchlight should be 0.78125 foot.

71. Set up the problem so that the vertex of the parabola is at $(0, 0)$ and it opens up. Then the equation of the parabola has the form: $x^2 = 4ay$. Since the parabola is 20 feet across and 6 feet deep, the points $(10, 6)$ and $(-10, 6)$ are on the parabola.
Substitute and solve for a: $10^2 = 4a(6) \Rightarrow 100 = 24a \Rightarrow a \approx 4.17$ feet
The heat will be concentrated about 4.17 feet from the base, along the axis of symmetry.

73. Set up the problem so that the vertex of the parabola is at $(0, 0)$ and it opens down. Then the equation of the parabola has the form: $x^2 = cy$.
The point $(60, -25)$ is a point on the parabola.
Solve for c and find the equation:
$$60^2 = c(-25) \Rightarrow c = -144$$
$$x^2 = -144y$$

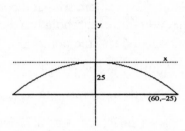

To find the height of the bridge 10 feet from the center the point $(10, y)$ is a point on the parabola. Solve for y:
$$10^2 = -144y \Rightarrow 100 = -144y \Rightarrow y \approx -0.69$$
The height of the bridge 10 feet from the center is about $25 - 0.69 = 24.31$ feet.
To find the height of the bridge 30 feet from the center the point $(30, y)$ is a point on the parabola. Solve for y:
$$30^2 = -144y \Rightarrow 900 = -144y \Rightarrow y \approx -6.25$$
The height of the bridge 30 feet from the center is $25 - 6.25 = 18.75$ feet.
To find the height of the bridge, 50 feet from the center, the point $(50, y)$ is a point on the parabola. Solve for y: $50^2 = -144y \Rightarrow 2500 = -144y \Rightarrow y = -17.36$
The height of the bridge 50 feet from the center is about $25 - 17.36 = 7.64$ feet.

75. $Ax^2 + Ey = 0 \quad A \neq 0, \ E \neq 0$
$$Ax^2 = -Ey \Rightarrow x^2 = -\frac{E}{A}y$$
This is the equation of a parabola with vertex at $(0, 0)$ and axis of symmetry being the y-axis. The focus is $\left(0, -\frac{E}{4A}\right)$. The directrix is $y = \frac{E}{4A}$.
The parabola opens up if $-\frac{E}{A} > 0$ and down if $-\frac{E}{A} < 0$.

77. $Ax^2 + Dx + Ey + F = 0 \quad A \neq 0$
(a) If $E \neq 0$, then:
$$Ax^2 + Dx = -Ey - F$$
$$A\left(x^2 + \frac{D}{A}x + \frac{D^2}{4A^2}\right) = -Ey - F + \frac{D^2}{4A}$$
$$\left(x + \frac{D}{2A}\right)^2 = \frac{1}{A}\left(-Ey - F + \frac{D^2}{4A}\right)$$
$$\left(x + \frac{D}{2A}\right)^2 = \frac{-E}{A}\left(y + \frac{F}{E} - \frac{D^2}{4AE}\right)$$
$$\left(x + \frac{D}{2A}\right)^2 = \frac{-E}{A}\left(y - \frac{D^2 - 4AF}{4AE}\right)$$

This is the equation of a parabola whose vertex is $\left(-\frac{D}{2A}, \frac{D^2 - 4AF}{4AE}\right)$ and whose axis of symmetry is parallel to the y-axis.

(b)   If $E = 0$, then $Ax^2 + Dx + F = 0 \Rightarrow x = \dfrac{-D \pm \sqrt{D^2 - 4AF}}{2A}$

If $D^2 - 4AF = 0$, then $x = -\dfrac{D}{2A}$ is a single vertical line.

(c)   If $E = 0$, then $Ax^2 + Dx + F = 0 \Rightarrow x = \dfrac{-D \pm \sqrt{D^2 - 4AF}}{2A}$

If $D^2 - 4AF > 0$,

then $x = \dfrac{-D + \sqrt{D^2 - 4AF}}{2A}$ and $x = \dfrac{-D - \sqrt{D^2 - 4AF}}{2A}$ are two vertical lines.

(d)   If $E = 0$, then $Ax^2 + Dx + F = 0 \Rightarrow x = \dfrac{-D \pm \sqrt{D^2 - 4AF}}{2A}$

If $D^2 - 4AF < 0$, there is no real solution. The graph contains no points.

Analytic Geometry

9.3　The Ellipse

1.　$\sqrt{(4-2)^2 + (-2-(-5))^2} = \sqrt{2^2 + 3^2} = \sqrt{4+9} = \sqrt{13}$

3.　$y^2 = 16 - 4x^2$
　　x-intercept(s):
　　　$0^2 = 16 - 4x^2$
　　$4x^2 = 16$
　　　$x^2 = 4 \Rightarrow x = -2, x = 2$
　　　　　$(-2,0)$ and $(2,0)$
　　y-intercept(s):
　　$y^2 = 16 - 4(0)^2$
　　$y^2 = 16 \Rightarrow y = -4, y = 4$
　　　　$(0,-4)$ and $(0,4)$

5.　left, 1; down, 4

7.　ellipse

9.　$(0,-5), (0,5)$

11.　True

13.　C

15.　B

17.　$\dfrac{x^2}{25} + \dfrac{y^2}{4} = 1$
　　The center of the ellipse is at the origin.
　　$a = 5$, $b = 2$. The vertices are $(5, 0)$ and
　　$(-5, 0)$. Find the value of c:
　　$c^2 = a^2 - b^2 = 25 - 4 = 21 \Rightarrow c = \sqrt{21}$
　　The foci are $\left(\sqrt{21},0\right)$ and $\left(-\sqrt{21},0\right)$.

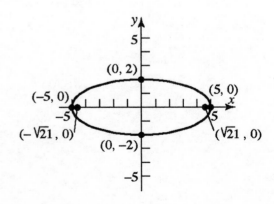

19. $\dfrac{x^2}{9} + \dfrac{y^2}{25} = 1$

The center of the ellipse is at the origin.
$a = 5,\ b = 3$. The vertices are $(0, 5)$ and $(0, -5)$.
Find the value of c:
$c^2 = a^2 - b^2 = 25 - 9 = 16$
$c = 4$
The foci are $(0, 4)$ and $(0, -4)$.

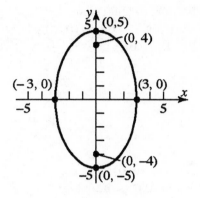

21. $4x^2 + y^2 = 16$

Divide by 16 to put in standard form:
$\dfrac{4x^2}{16} + \dfrac{y^2}{16} = \dfrac{16}{16} \Rightarrow \dfrac{x^2}{4} + \dfrac{y^2}{16} = 1$
The center of the ellipse is at the origin.
$a = 4,\ b = 2$.
The vertices are $(0, 4)$ and $(0, -4)$.
Find the value of c:
$c^2 = a^2 - b^2 = 16 - 4 = 12$
$c = \sqrt{12} = 2\sqrt{3}$
The foci are $\left(0,\ 2\sqrt{3}\right)$ and $\left(0,\ -2\sqrt{3}\right)$.

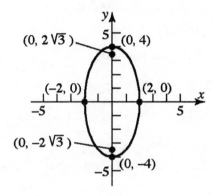

23. $4y^2 + x^2 = 8$

Divide by 8 to put in standard form:
$\dfrac{4y^2}{8} + \dfrac{x^2}{8} = \dfrac{8}{8} \Rightarrow \dfrac{x^2}{8} + \dfrac{y^2}{2} = 1$
The center of the ellipse is at the origin.
$a = \sqrt{8} = 2\sqrt{2},\ b = \sqrt{2}$.
The vertices are $\left(2\sqrt{2}, 0\right)$ and $\left(-2\sqrt{2}, 0\right)$.
Find the value of c:
$c^2 = a^2 - b^2 = 8 - 2 = 6$
$c = \sqrt{6}$
The foci are $\left(\sqrt{6}, 0\right)$ and $\left(-\sqrt{6}, 0\right)$.

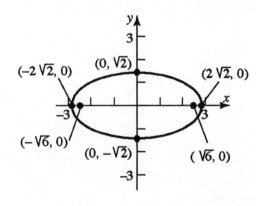

25. $x^2 + y^2 = 16$
This is the equation of a circle whose center
is at $(0, 0)$ and radius 4.
Vertices: $(-4,0)$, $(4,0)$, $(0,-4)$, $(0,4)$
Focus: $(0,0)$

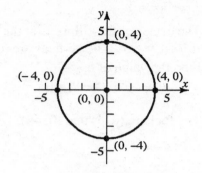

27. Center: $(0, 0)$; Focus: $(3, 0)$; Vertex: $(5, 0)$;
Major axis is the x-axis; $a = 5$; $c = 3$. Find b:
$b^2 = a^2 - c^2 = 25 - 9 = 16$
$b = 4$

Write the equation: $\dfrac{x^2}{25} + \dfrac{y^2}{16} = 1$

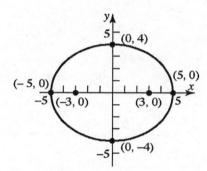

29. Center: $(0, 0)$; Focus: $(0, -4)$; Vertex: $(0, 5)$;
Major axis is the y-axis; $a = 5$; $c = 4$. Find b:
$b^2 = a^2 - c^2 = 25 - 16 = 9$
$b = 3$

Write the equation: $\dfrac{x^2}{9} + \dfrac{y^2}{25} = 1$

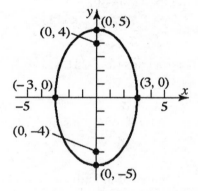

31. Foci: $(\pm2, 0)$; Length of major axis is 6.
Center: $(0, 0)$; Major axis is the x-axis;
$a = 3$; $c = 2$. Find b:
$b^2 = a^2 - c^2 = 9 - 4 = 5$
$b = \sqrt{5}$

Write the equation: $\dfrac{x^2}{9} + \dfrac{y^2}{5} = 1$

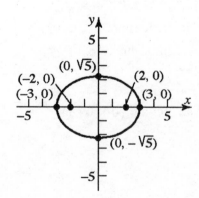

33. Focus: $(-4, 0)$; Vertices: $(\pm 5, 0)$.
Center: $(0, 0)$; Major axis is the x-axis;
$a = 5$; $c = 4$. Find b:
$$b^2 = a^2 - c^2 = 25 - 16 = 9$$
$$b = 3$$
Write the equation: $\dfrac{x^2}{25} + \dfrac{y^2}{9} = 1$

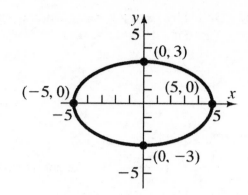

35. Foci: $(0, \pm 3)$; x-intercepts are ± 2. Center: $(0, 0)$;
Major axis is the y-axis; $c = 3$; $b = 2$. Find a:
$$a^2 = b^2 + c^2 = 4 + 9 = 13$$
$$a = \sqrt{13}$$
Write the equation: $\dfrac{x^2}{4} + \dfrac{y^2}{13} = 1$

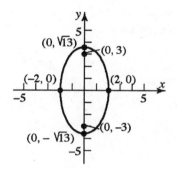

37. Center: $(0, 0)$; Vertex: $(0, 4)$; $b = 1$; Major axis is
the y-axis; $a = 4$; $b = 1$. Find c:
$$c^2 = a^2 - b^2 = 16 - 1 = 15$$
$$c = \sqrt{15}$$
Write the equation: $\dfrac{x^2}{1} + \dfrac{y^2}{16} = 1$

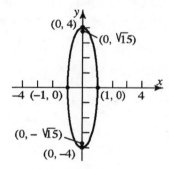

39. $\dfrac{(x+1)^2}{4} + \dfrac{(y-1)^2}{1} = 1$ 41. $\dfrac{(x-1)^2}{1} + \dfrac{y^2}{4} = 1$

43. The equation $\dfrac{(x-3)^2}{4} + \dfrac{(y+1)^2}{9} = 1$ is in the form

$\dfrac{(x-h)^2}{b^2} + \dfrac{(y-k)^2}{a^2} = 1$

(major axis parallel to the y-axis) where
$a = 3$, $b = 2$, $h = 3$, and $k = -1$.
Solving for c: $c^2 = a^2 - b^2 = 9 - 4 = 5 \Rightarrow c = \sqrt{5}$
Thus, we have:
Center: $(3, -1)$
Foci: $\left(3, -1 + \sqrt{5}\right), \left(3, -1 - \sqrt{5}\right)$
Vertices: $(3, 2), (3, -4)$

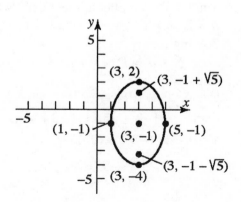

45.   Divide by 16 to put the equation in standard form:
$$(x+5)^2 + 4(y-4)^2 = 16$$

$$\frac{(x+5)^2}{16} + \frac{4(y-4)^2}{16} = \frac{16}{16}$$

$$\frac{(x+5)^2}{16} + \frac{(y-4)^2}{4} = 1$$

The equation is in the form
$\frac{(x-h)^2}{a^2} + \frac{(y-k)^2}{b^2} = 1$ (major axis parallel to
the x-axis) where $a = 4$, $b = 2$,
$h = -5$, and $k = 4$.
Solving for c:
$$c^2 = a^2 - b^2 = 16 - 4 = 12 \Rightarrow c = \sqrt{12} = 2\sqrt{3}$$
Thus, we have:
Center: $(-5, 4)$
Foci: $\left(-5 - 2\sqrt{3}, 4\right), \left(-5 + 2\sqrt{3}, 4\right)$
Vertices: $(-9, 4)$, $(-1, 4)$

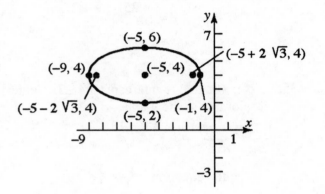

47.   Complete the squares to put the equation in standard form:
$$x^2 + 4x + 4y^2 - 8y + 4 = 0$$

$$(x^2 + 4x + 4) + 4(y^2 - 2y + 1) = -4 + 4 + 4$$

$$(x+2)^2 + 4(y-1)^2 = 4$$

$$\frac{(x+2)^2}{4} + \frac{4(y-1)^2}{4} = \frac{4}{4}$$

$$\frac{(x+2)^2}{4} + \frac{(y-1)^2}{1} = 1$$

The equation is in the form
$$\frac{(x-h)^2}{a^2} + \frac{(y-k)^2}{b^2} = 1$$
(major axis parallel to the x-axis) where
$a = 2$, $b = 1$, $h = -2$, and $k = 1$.
Solving for c:
$$c^2 = a^2 - b^2 = 4 - 1 = 3 \Rightarrow c = \sqrt{3}$$
Thus, we have:
Center: $(-2, 1)$
Foci: $\left(-2 - \sqrt{3}, 1\right), \left(-2 + \sqrt{3}, 1\right)$
Vertices: $(-4, 1)$, $(0, 1)$

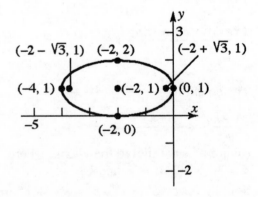

49.   Complete the squares to put the equation in standard form:

$$2x^2 + 3y^2 - 8x + 6y + 5 = 0$$

$$2(x^2 - 4x) + 3(y^2 + 2y) = -5$$

$$2(x^2 - 4x + 4) + 3(y^2 + 2y + 1) = -5 + 8 + 3$$

$$2(x - 2)^2 + 3(y + 1)^2 = 6$$

$$\frac{2(x - 2)^2}{6} + \frac{3(y + 1)^2}{6} = \frac{6}{6}$$

$$\frac{(x - 2)^2}{3} + \frac{(y + 1)^2}{2} = 1$$

The equation is in the form $\dfrac{(x - h)^2}{a^2} + \dfrac{(y - k)^2}{b^2} = 1$

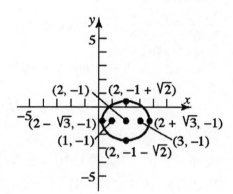

(major axis parallel to the x-axis) where
$a = \sqrt{3}$, $b = \sqrt{2}$, $h = 2$, and $k = -1$.
Solving for c: $c^2 = a^2 - b^2 = 3 - 2 = 1 \implies c = 1$
Thus, we have:
Center: $(2, -1)$
Foci: $(1, -1)$, $(3, -1)$
Vertices: $\left(2 - \sqrt{3}, -1\right)$, $\left(2 + \sqrt{3}, -1\right)$

51.   Complete the squares to put the equation in standard form:

$$9x^2 + 4y^2 - 18x + 16y - 11 = 0$$

$$9(x^2 - 2x) + 4(y^2 + 4y) = 11$$

$$9(x^2 - 2x + 1) + 4(y^2 + 4y + 4) = 11 + 9 + 16$$

$$9(x - 1)^2 + 4(y + 2)^2 = 36$$

$$\frac{9(x - 1)^2}{36} + \frac{4(y + 2)^2}{36} = \frac{36}{36}$$

$$\frac{(x - 1)^2}{4} + \frac{(y + 2)^2}{9} = 1$$

The equation is in the form $\dfrac{(x - h)^2}{b^2} + \dfrac{(y - k)^2}{a^2} = 1$

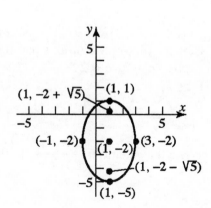

(major axis parallel to the y-axis) where
$a = 3$, $b = 2$, $h = 1$, and $k = -2$.
Solving for c: $c^2 = a^2 - b^2 = 9 - 4 = 5 \Rightarrow c = \sqrt{5}$
Thus, we have:
Center: $(1, -2)$
Foci: $\left(1, -2 + \sqrt{5}\right)$, $\left(1, -2 - \sqrt{5}\right)$
Vertices: $(1, 1)$, $(1, -5)$

53. Complete the square to put the equation in standard form:

$$4x^2 + y^2 + 4y = 0$$

$$4x^2 + y^2 + 4y + 4 = 4$$

$$4x^2 + (y+2)^2 = 4$$

$$\frac{4x^2}{4} + \frac{(y+2)^2}{4} = \frac{4}{4}$$

$$\frac{x^2}{1} + \frac{(y+2)^2}{4} = 1$$

The equation is in the form

$$\frac{(x-h)^2}{b^2} + \frac{(y-k)^2}{a^2} = 1 \text{ (major axis parallel}$$

to the y-axis) where

$a = 2$, $b = 1$, $h = 0$, and $k = -2$.

Solving for c:

$$c^2 = a^2 - b^2 = 4 - 1 = 3 \implies c = \sqrt{3}$$

Thus, we have:

Center: $(0, -2)$

Foci: $\left(0, -2 + \sqrt{3}\right), \left(0, -2 - \sqrt{3}\right)$

Vertices: $(0, 0), (0, -4)$

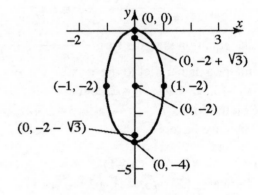

55. Center: $(2, -2)$; Vertex: $(7, -2)$;
Focus: $(4, -2)$; Major axis parallel to the
x-axis; $a = 5$; $c = 2$. Find b:

$$b^2 = a^2 - c^2 = 25 - 4 = 21$$

$$b = \sqrt{21}$$

Write the equation:

$$\frac{(x-2)^2}{25} + \frac{(y+2)^2}{21} = 1$$

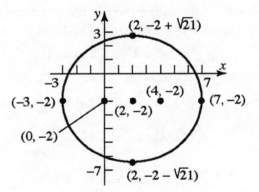

57. Vertices: $(4, 3), (4, 9)$; Focus: $(4, 8)$;
Center: $(4, 6)$; Major axis parallel to the
y-axis; $a = 3$; $c = 2$.

Find b:

$$b^2 = a^2 - c^2 = 9 - 4 = 5$$

$$b = \sqrt{5}$$

Write the equation: $\dfrac{(x-4)^2}{5} + \dfrac{(y-6)^2}{9} = 1$

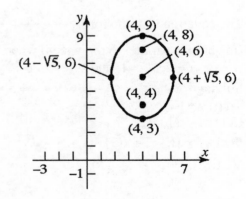

59. Foci: (5, 1), (–1, 1); length of the major axis = 8;
     Center: (2, 1); Major axis parallel to the x-axis;
     $a = 4$; $c = 3$.   Find b:
     $$b^2 = a^2 - c^2 = 16 - 9 = 7$$
     $$b = \sqrt{7}$$
     Write the equation: $\dfrac{(x-2)^2}{16} + \dfrac{(y-1)^2}{7} = 1$

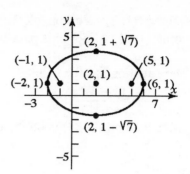

61. Center: (1, 2); Focus: (4, 2); contains the point (1, 3);
     Major axis parallel to the x-axis; $c = 3$.

     The equation has the form: $\dfrac{(x-1)^2}{a^2} + \dfrac{(y-2)^2}{b^2} = 1$.

     Since the point (1, 3) is on the curve:
     $$\frac{0}{a^2} + \frac{1}{b^2} = 1$$
     $$\frac{1}{b^2} = 1$$
     $$b^2 = 1 \Rightarrow b = 1$$
     Find a:
     $$a^2 = b^2 + c^2 = 1 + 9 = 10 \Rightarrow a = \sqrt{10}$$
     Write the equation: $\dfrac{(x-1)^2}{10} + \dfrac{(y-2)^2}{1} = 1$

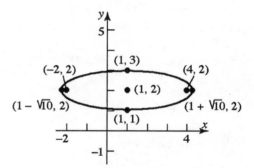

63.   Center: $(1, 2)$; Vertex: $(4, 2)$; contains the
point $(1, 3)$; Major axis parallel to the
x-axis; $a = 3$.
The equation has the form:
$$\frac{(x-1)^2}{a^2} + \frac{(y-2)^2}{b^2} = 1$$
Since the point $(1, 3)$ is on the curve:
$$\frac{0}{a^2} + \frac{1}{b^2} = 1$$
$$\frac{1}{b^2} = 1$$
$$b^2 = 1 \Rightarrow b = 1$$
Find c:
$$c^2 = a^2 - b^2 = 9 - 1 = 8 \;\Rightarrow\; c = \sqrt{8} = 2\sqrt{2}$$
Write the equation: $\dfrac{(x-1)^2}{9} + \dfrac{(y-2)^2}{1} = 1$

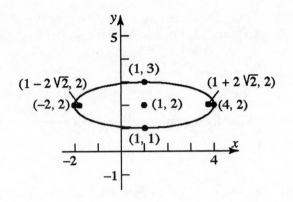

65.   Rewrite the equation:
$$y = \sqrt{16 - 4x^2}$$
$$y^2 = 16 - 4x^2, \quad y \geq 0$$
$$4x^2 + y^2 = 16, \qquad y \geq 0$$
$$\frac{x^2}{4} + \frac{y^2}{16} = 1, \qquad y \geq 0$$

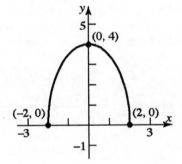

67.   Rewrite the equation:
$$y = -\sqrt{64 - 16x^2}$$
$$y^2 = 64 - 16x^2, \quad y \leq 0$$
$$16x^2 + y^2 = 64, \qquad y \leq 0$$
$$\frac{x^2}{4} + \frac{y^2}{64} = 1, \qquad y \leq 0$$

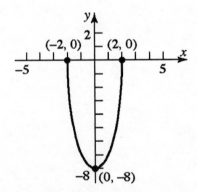

69.   The center of the ellipse is $(0, 0)$. The length of the major axis is 20, so $a = 10$. The length
of half the minor axis is 6, so $b = 6$. The ellipse is situated with its major axis on the

x-axis. The equation is: $\dfrac{x^2}{100} + \dfrac{y^2}{36} = 1$.

71. Assume that the half ellipse formed by the gallery is centered at (0, 0). Since the hall is 100 feet long, $2a = 100$ or $a = 50$. The distance from the center to the foci is 25 feet, so $c = 25$. Find the height of the gallery, which is b:
$b^2 = a^2 - c^2 = 2500 - 625 = 1875 \Rightarrow b = \sqrt{1875} \approx 43.3$
The ceiling is about 43.3 feet high in the center.

73. Place the semielliptical arch so that the x-axis coincides with the major axis and the y-axis passes through the center of the arch. Since the bridge has a span of 120 feet, the length of the major axis is 120, or $2a = 120$ or $a = 60$. The maximum height of the bridge is 25 feet, so $b = 25$. The equation is: $\dfrac{x^2}{3600} + \dfrac{y^2}{625} = 1$.

The height 10 feet from the center:
$\dfrac{10^2}{3600} + \dfrac{y^2}{625} = 1 \Rightarrow \dfrac{y^2}{625} = 1 - \dfrac{100}{3600} \Rightarrow y^2 = 625 \cdot \dfrac{3500}{3600} \Rightarrow y \approx 24.65$ feet

The height 30 feet from the center:
$\dfrac{30^2}{3600} + \dfrac{y^2}{625} = 1 \Rightarrow \dfrac{y^2}{625} = 1 - \dfrac{900}{3600} \Rightarrow y^2 = 625 \cdot \dfrac{2700}{3600} \Rightarrow y \approx 21.65$ feet

The height 50 feet from the center:
$\dfrac{50^2}{3600} + \dfrac{y^2}{625} = 1 \Rightarrow \dfrac{y^2}{625} = 1 - \dfrac{2500}{3600} \Rightarrow y^2 = 625 \cdot \dfrac{1100}{3600} \Rightarrow y \approx 13.82$ feet

75. Place the semielliptical arch so that the x-axis coincides with the major axis and the y-axis passes through the center of the arch. Since the ellipse is 40 feet wide, the length of the major axis is 40, or $2a = 40$ or $a = 20$. The height is 15 feet at the center, so $b = 15$. The equation is: $\dfrac{x^2}{400} + \dfrac{y^2}{225} = 1$.

The height 10 feet either side of the center:
$\dfrac{10^2}{400} + \dfrac{y^2}{225} = 1 \Rightarrow \dfrac{y^2}{225} = 1 - \dfrac{100}{400} \Rightarrow y^2 = 225 \cdot \dfrac{3}{4} \Rightarrow y \approx 12.99$ feet

The height 20 feet either side of the center:
$\dfrac{20^2}{400} + \dfrac{y^2}{225} = 1 \Rightarrow \dfrac{y^2}{225} = 1 - \dfrac{400}{400} \Rightarrow y^2 = 225 \cdot 0 \Rightarrow y \approx 0$ feet

77. Since the mean distance is 93 million miles, $a = 93$ million. The length of the major axis is 186 million. The perihelion is 186 million – 94.5 million = 91.5 million miles. The distance from the center of the ellipse to the sun (focus) is
93 million – 91.5 million = 1.5 million miles; therefore, $c = 1.5$ million. Find b:
$b^2 = a^2 - c^2 = (93 \times 10^6)^2 - (1.5 \times 10^6)^2 = 8.64675 \times 10^{15} \Rightarrow b \approx 92.99 \times 10^6$

The equation of the orbit is: $\dfrac{x^2}{(93 \times 10^6)^2} + \dfrac{y^2}{(92.99 \times 10^6)^2} = 1$.

79.   The mean distance is 507 million – 23.2 million = 483.8 million miles.
The perihelion is 483.8 million – 23.2 million = 460.6 million miles.
Since $a = 483.8 \times 10^6$ and $c = 23.2 \times 10^6$, we can find b:

$$b^2 = a^2 - c^2 = \left(483.8 \times 10^6\right)^2 - \left(23.2 \times 10^6\right)^2 = 2.335242 \times 10^{17} \Rightarrow b \approx 483.2 \times 10^6$$

The equation of the orbit of Jupiter is: $\dfrac{x^2}{\left(483.8 \times 10^6\right)^2} + \dfrac{y^2}{\left(483.2 \times 10^6\right)^2} = 1.$

81.   If the x-axis is placed along the 100 foot length and the y-axis is placed along the 50-foot width, the equation for the ellipse is: $\dfrac{x^2}{50^2} + \dfrac{y^2}{25^2} = 1.$

Find y when $x = 40$:

$$\frac{40^2}{50^2} + \frac{y^2}{25^2} = 1 \Rightarrow \frac{y^2}{625} = 1 - \frac{1600}{2500} \Rightarrow y^2 = 625 \cdot \frac{9}{25} \Rightarrow y = 15 \text{ feet}$$

The width 10 feet from a vertex is about 30 feet.

83.   (a)   Put the equation in standard ellipse form:

$$Ax^2 + Cy^2 + F = 0 \qquad A \neq 0, C \neq 0, F \neq 0$$

$$Ax^2 + Cy^2 = -F$$

$$\frac{Ax^2}{-F} + \frac{Cy^2}{-F} = 1$$

$$\frac{x^2}{(-F/A)} + \frac{y^2}{(-F/C)} = 1$$

Since the sign of F is opposite that of A and C, $-F/A$ and $-F/C$ are positive.
This is the equation of an ellipse with center at $(0, 0)$.

(b)   If $A = C$, the equation becomes:

$$Ax^2 + Ay^2 = -F \Rightarrow x^2 + y^2 = -\frac{F}{A}$$

This is the equation of a circle with center at $(0, 0)$ and radius of $\sqrt{-\dfrac{F}{A}}$.

85.   Answers will vary.

Analytic Geometry
9.4 The Hyperbola

1. $\sqrt{(-2-3)^2 + (1-(-4))^2} = \sqrt{(-5)^2 + 5^2} = \sqrt{25 + 25} = \sqrt{50} = 5\sqrt{2}$

3. $y^2 = 9 + 4x^2$

   x-intercept(s):

   $\qquad 0^2 = 9 + 4x^2$

   $\qquad -4x^2 = 9$

   $\qquad x^2 = -\dfrac{9}{4} \Rightarrow$ no real solution

   No x-intercepts.

   y-intercept(s):

   $\qquad y^2 = 9 + 4(0)^2$

   $\qquad y^2 = 9 \Rightarrow y = -3, y = 3$

   $\qquad (0, -3)$ and $(0, 3)$

5. right, 5; down, 4

7. hyperbola

9. $y = \dfrac{3}{2}x, \ y = -\dfrac{3}{2}x$

11. True

13. B

15. A

17. Center: $(0, 0)$; Focus: $(3, 0)$; Vertex: $(1, 0)$;
    Transverse axis is the x-axis; $a = 1$; $c = 3$.
    Find the value of b:
    $b^2 = c^2 - a^2 = 9 - 1 = 8$
    $b = \sqrt{8} = 2\sqrt{2}$
    Write the equation: $\dfrac{x^2}{1} - \dfrac{y^2}{8} = 1$.

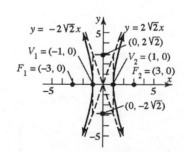

19. Center: $(0, 0)$; Focus: $(0, -6)$; Vertex: $(0, 4)$
    Transverse axis is the y-axis; $a = 4$; $c = 6$.
    Find the value of b:
    $b^2 = c^2 - a^2 = 36 - 16 = 20$
    $b = \sqrt{20} = 2\sqrt{5}$
    Write the equation: $\dfrac{y^2}{16} - \dfrac{x^2}{20} = 1$.

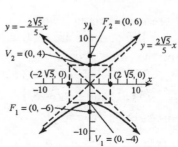

21. Foci: $(-5, 0)$, $(5, 0)$; Vertex: $(3, 0)$
Center: $(0, 0)$; Transverse axis is the
x-axis; $a = 3$; $c = 5$.
Find the value of b:
$b^2 = c^2 - a^2 = 25 - 9 = 16 \Rightarrow b = 4$
Write the equation: $\dfrac{x^2}{9} - \dfrac{y^2}{16} = 1$.

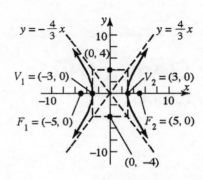

23. Vertices: $(0, -6)$, $(0, 6)$; Asymptote: $y = 2x$;
Center: $(0, 0)$; Transverse axis is the y-axis;
$a = 6$. Find the value of b using the slope of
the asymptote: $\dfrac{a}{b} = \dfrac{6}{b} = 2 \Rightarrow 2b = 6 \Rightarrow b = 3$
Find the value of c:
$c^2 = a^2 + b^2 = 36 + 9 = 45 \Rightarrow c = 3\sqrt{5}$
Write the equation: $\dfrac{y^2}{36} - \dfrac{x^2}{9} = 1$.

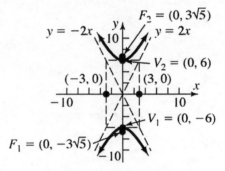

25. Foci: $(-4, 0)$, $(4, 0)$; Asymptote: $y = -x$;
Center: $(0, 0)$; Transverse axis is the
x-axis; $c = 4$. Using the slope of the
asymptote: $-\dfrac{b}{a} = -1 \Rightarrow -b = -a \Rightarrow b = a$.
Find the value of b:
$$b^2 = c^2 - a^2 \Rightarrow a^2 + b^2 = c^2 \quad (c = 4)$$
$$b^2 + b^2 = 16 \Rightarrow 2b^2 = 16 \Rightarrow b^2 = 8$$
$$b = \sqrt{8} = 2\sqrt{2}$$
$$a = \sqrt{8} = 2\sqrt{2} \quad (a = b)$$
Write the equation: $\dfrac{x^2}{8} - \dfrac{y^2}{8} = 1$.

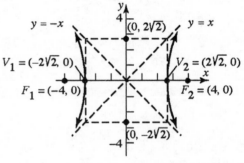

27. $\dfrac{x^2}{25} - \dfrac{y^2}{9} = 1$
The center of the hyperbola is at $(0, 0)$.
$a = 5$, $b = 3$. The vertices are $(5, 0)$ and
$(-5, 0)$. Find the value of c:
$c^2 = a^2 + b^2 = 25 + 9 = 34 \Rightarrow c = \sqrt{34}$
The foci are $\left(\sqrt{34}, 0\right)$ and $\left(-\sqrt{34}, 0\right)$.
The transverse axis is the x-axis.
The asymptotes are $y = \dfrac{3}{5}x$; $y = -\dfrac{3}{5}x$.

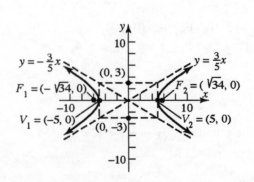

29. $4x^2 - y^2 = 16$

Divide both sides by 16 to put in standard form: $\dfrac{4x^2}{16} - \dfrac{y^2}{16} = \dfrac{16}{16} \Rightarrow \dfrac{x^2}{4} - \dfrac{y^2}{16} = 1$.

The center of the hyperbola is at $(0, 0)$.
$a = 2, \ b = 4$.
The vertices are $(2, 0)$ and $(-2, 0)$.

Find the value of c:
$c^2 = a^2 + b^2 = 4 + 16 = 20$

$c = \sqrt{20} = 2\sqrt{5}$

The foci are $\left(2\sqrt{5}, 0\right)$ and $\left(-2\sqrt{5}, 0\right)$.
The transverse axis is the x-axis.
The asymptotes are $y = 2x; \ y = -2x$.

31. $y^2 - 9x^2 = 9$

Divide both sides by 9 to put in standard form: $\dfrac{y^2}{9} - \dfrac{9x^2}{9} = \dfrac{9}{9} \Rightarrow \dfrac{y^2}{9} - \dfrac{x^2}{1} = 1$.

The center of the hyperbola is at $(0, 0)$.
$a = 3, \ b = 1$.
The vertices are $(0, 3)$ and $(0, -3)$.
Find the value of c:
$c^2 = a^2 + b^2 = 9 + 1 = 10$

$c = \sqrt{10}$

The foci are $\left(0, \sqrt{10}\right)$ and $\left(0, -\sqrt{10}\right)$.

The transverse axis is the y-axis.
The asymptotes are $y = 3x; \ y = -3x$.

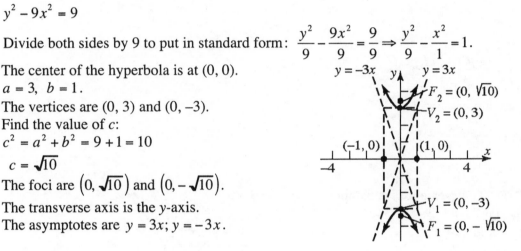

33. $y^2 - x^2 = 25$

Divide both sides by 25 to put in standard form: $\dfrac{y^2}{25} - \dfrac{x^2}{25} = 1$.

The center of the hyperbola is at $(0, 0)$.
$a = 5, \ b = 5$. The vertices are $(0, 5)$ and $(0, -5)$.
Find the value of c:
$c^2 = a^2 + b^2 = 25 + 25 = 50$

$c = \sqrt{50} = 5\sqrt{2}$

The foci are $\left(0, 5\sqrt{2}\right)$ and $\left(0, -5\sqrt{2}\right)$.

The transverse axis is the y-axis.
The asymptotes are $y = x; \ y = -x$.

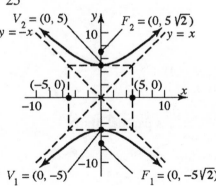

35. The center of the hyperbola is at $(0, 0)$.

$a = 1,\ b = 1$. The vertices are $(1, 0)$ and $(-1, 0)$.

Find the value of c:

$c^2 = a^2 + b^2 = 1 + 1 = 2$

$c = \sqrt{2}$

The foci are $\left(\sqrt{2}, 0\right)$ and $\left(-\sqrt{2}, 0\right)$.

The transverse axis is the x-axis.

The asymptotes are $y = x;\ y = -x$.

The equation is: $x^2 - y^2 = 1$.

37. The center of the hyperbola is at $(0, 0)$.

$a = 6,\ b = 3$. The vertices are $(0, -6)$ and $(0, 6)$.

Find the value of c:

$c^2 = a^2 + b^2 = 36 + 9 = 45$

$c = \sqrt{45} = 3\sqrt{5}$

The foci are $\left(0, -3\sqrt{5}\right)$ and $\left(0, 3\sqrt{5}\right)$.

The transverse axis is the y-axis.

The asymptotes are $y = 2x;\ y = -2x$.

The equation is: $\dfrac{y^2}{36} - \dfrac{x^2}{9} = 1$.

39. Center: $(4, -1)$; Focus: $(7, -1)$;
Vertex: $(6, -1)$; Transverse axis is parallel
to the x-axis; $a = 2;\ c = 3$.

Find the value of b:

$b^2 = c^2 - a^2 = 9 - 4 = 5 \Rightarrow b = \sqrt{5}$

Write the equation: $\dfrac{(x-4)^2}{4} - \dfrac{(y+1)^2}{5} = 1$.

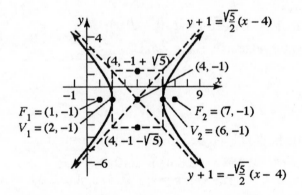

41. Center: $(-3, -4)$;
Focus: $(-3, -8)$;
Vertex: $(-3, -2)$;
Transverse axis is parallel to the
y-axis; $a = 2;\ c = 4$.

Find the value of b:

$b^2 = c^2 - a^2 = 16 - 4 = 12$

$b = \sqrt{12} = 2\sqrt{3}$

Write the equation:

$\dfrac{(y+4)^2}{4} - \dfrac{(x+3)^2}{12} = 1$.

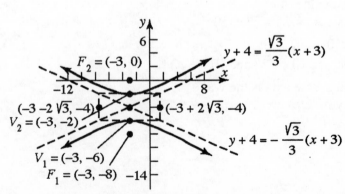

43. Foci: $(3, 7)$, $(7, 7)$; Vertex: $(6, 7)$;
Center: $(5, 7)$; Transverse axis is parallel to the x-axis; $a = 1$; $c = 2$.
Find the value of b:
$$b^2 = c^2 - a^2 = 4 - 1 = 3$$
$$b = \sqrt{3}$$
Write the equation: $\dfrac{(x-5)^2}{1} - \dfrac{(y-7)^2}{3} = 1$.

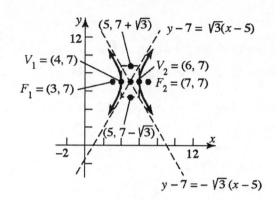

45. Vertices: $(-1, -1)$, $(3, -1)$;
Center: $(1, -1)$; Transverse axis is parallel to the x-axis; $a = 2$.

Asymptote: $y + 1 = \dfrac{3}{2}(x - 1)$

Using the slope of the asymptote, find the value of b:
$$\frac{b}{a} = \frac{b}{2} = \frac{3}{2} \Rightarrow b = 3$$
Find the value of c:
$$c^2 = a^2 + b^2 = 4 + 9 = 13$$
$$c = \sqrt{13}$$
Write the equation: $\dfrac{(x-1)^2}{4} - \dfrac{(y+1)^2}{9} = 1$.

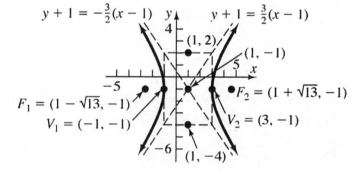

47. $\dfrac{(x-2)^2}{4} - \dfrac{(y+3)^2}{9} = 1$
The center of the hyperbola is at $(2, -3)$.
$a = 2$, $b = 3$.
The vertices are $(0, -3)$ and $(4, -3)$.
Find the value of c:
$$c^2 = a^2 + b^2 = 4 + 9 = 13 \Rightarrow c = \sqrt{13}$$
Foci: $\left(2 - \sqrt{13}, -3\right)$ and $\left(2 + \sqrt{13}, -3\right)$.
Transverse axis: $y = -3$,
parallel to the x-axis.

Asymptotes: $y + 3 = \dfrac{3}{2}(x - 2)$;

$$y + 3 = -\frac{3}{2}(x - 2).$$

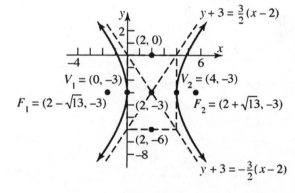

49. $(y-2)^2 - 4(x+2)^2 = 4$

Divide both sides by 4 to put in standard form: $\dfrac{(y-2)^2}{4} - \dfrac{(x+2)^2}{1} = 1.$

The center of the hyperbola is at $(-2, 2)$.
$a = 2, \ b = 1.$
The vertices are $(-2, 4)$ and $(-2, 0)$.
Find the value of c:
$c^2 = a^2 + b^2 = 4 + 1 = 5 \Rightarrow c = \sqrt{5}$
Foci: $\left(-2, 2 - \sqrt{5}\right)$ and $\left(-2, 2 + \sqrt{5}\right)$.
Transverse axis: $x = -2$, parallel to the y-axis.
Asymptotes:
$y - 2 = 2(x + 2); \ y - 2 = -2(x + 2).$

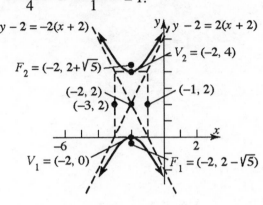

51. $(x+1)^2 - (y+2)^2 = 4$

Divide both sides by 4 to put in standard form: $\dfrac{(x+1)^2}{4} - \dfrac{(y+2)^2}{4} = 1.$

The center of the hyperbola is $(-1, -2)$.
$a = 2, \ b = 2.$
The vertices are $(-3, -2)$ and $(1, -2)$.
Find the value of c:
$c^2 = a^2 + b^2 = 4 + 4 = 8$

$c = \sqrt{8} = 2\sqrt{2}$
Foci: $\left(-1 - 2\sqrt{2}, -2\right)$ and $\left(-1 + 2\sqrt{2}, -2\right)$.
Transverse axis: $y = -2$, parallel to the x-axis.
Asymptotes: $y + 2 = x + 1;$
$\qquad\qquad y + 2 = -(x + 1).$

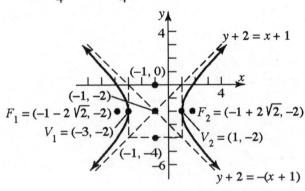

53. Complete the squares to put in standard form:
$$x^2 - y^2 - 2x - 2y - 1 = 0$$
$$(x^2 - 2x + 1) - (y^2 + 2y + 1) = 1 + 1 - 1$$
$$(x - 1)^2 - (y + 1)^2 = 1$$
The center of the hyperbola is $(1, -1)$.
$a = 1, \ b = 1.$
The vertices are $(0, -1)$ and $(2, -1)$.
Find the value of c:
$c^2 = a^2 + b^2 = 1 + 1 = 2 \Rightarrow c = \sqrt{2}$
Foci: $\left(1 - \sqrt{2}, -1\right)$ and $\left(1 + \sqrt{2}, -1\right)$.
Transverse axis: $y = -1$, parallel to the x-axis.
Asymptotes: $y + 1 = x - 1; \ y + 1 = -(x - 1).$

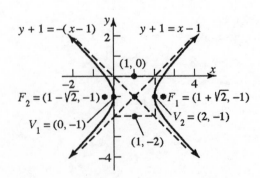

55. Complete the squares to put in standard form:
$$y^2 - 4x^2 - 4y - 8x - 4 = 0$$
$$(y^2 - 4y + 4) - 4(x^2 + 2x + 1) = 4 + 4 - 4$$
$$(y - 2)^2 - 4(x + 1)^2 = 4$$
$$\frac{(y - 2)^2}{4} - \frac{(x + 1)^2}{1} = 1$$

The center of the hyperbola is $(-1, 2)$.
$a = 2, \ b = 1$.
The vertices are $(-1, 4)$ and $(-1, 0)$.
Find the value of c:
$$c^2 = a^2 + b^2 = 4 + 1 = 5 \Rightarrow c = \sqrt{5}$$
Foci: $\left(-1, 2 - \sqrt{5}\right)$ and $\left(-1, 2 + \sqrt{5}\right)$.

Transverse axis: $x = -1$, parallel to the y-axis.
Asymptotes: $y - 2 = 2(x + 1); \ y - 2 = -2(x + 1)$.

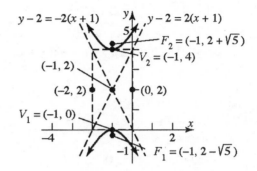

57. Complete the squares to put in standard form:
$$4x^2 - y^2 - 24x - 4y + 16 = 0$$
$$4(x^2 - 6x + 9) - (y^2 + 4y + 4) = -16 + 36 - 4$$
$$4(x - 3)^2 - (y + 2)^2 = 16$$
$$\frac{(x - 3)^2}{4} - \frac{(y + 2)^2}{16} = 1$$

The center of the hyperbola
is $(3, -2)$. $a = 2, \ b = 4$.
The vertices are $(1, -2)$
and $(5, -2)$. Find the value of c:
$$c^2 = a^2 + b^2 = 4 + 16 = 20$$
$$c = \sqrt{20} = 2\sqrt{5}$$
Foci:
$\left(3 - 2\sqrt{5}, -2\right)$ and $\left(3 + 2\sqrt{5}, -2\right)$.

Transverse axis: $y = -2$,
parallel to the x-axis.
Asymptotes: $y + 2 = 2(x - 3);$
$$y + 2 = -2(x - 3)$$

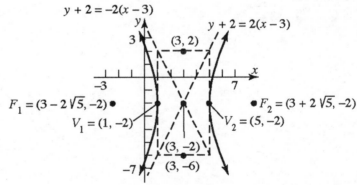

59. Complete the squares to put in standard form:
$$y^2 - 4x^2 - 16x - 2y - 19 = 0$$
$$(y^2 - 2y + 1) - 4(x^2 + 4x + 4) = 19 + 1 - 16$$
$$(y - 1)^2 - 4(x + 2)^2 = 4$$
$$\frac{(y - 1)^2}{4} - \frac{(x + 2)^2}{1} = 1$$

The center of the hyperbola is $(-2, 1)$.
  $a = 2, \quad b = 1$.
The vertices are $(-2, 3)$ and $(-2, -1)$. Find the value of c:
$$c^2 = a^2 + b^2 = 4 + 1 = 5$$
$$c = \sqrt{5}$$
Foci: $\left(-2, 1 - \sqrt{5}\right)$ and $\left(-2, 1 + \sqrt{5}\right)$.

Transverse axis: $x = -2$, parallel to the y-axis.
Asymptotes: $y - 1 = 2(x + 2); y - 1 = -2(x + 2)$.

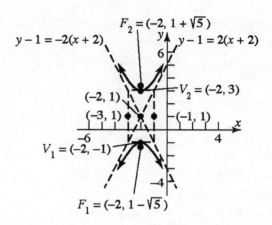

61. Rewrite the equation:
$$y = \sqrt{16 + 4x^2}$$
$$y^2 = 16 + 4x^2, \quad y \geq 0$$
$$y^2 - 4x^2 = 16, \quad y \geq 0$$
$$\frac{y^2}{16} - \frac{x^2}{4} = 1, \quad y \geq 0$$

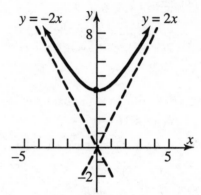

63. Rewrite the equation:
$$y = -\sqrt{-25 + x^2}$$
$$y^2 = -25 + x^2, \quad y \leq 0$$
$$x^2 - y^2 = 25, \quad y \leq 0$$
$$\frac{x^2}{25} - \frac{y^2}{25} = 1, \quad y \leq 0$$

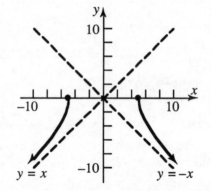

65. (a) Set up a coordinate system so that the two stations lie on the x-axis and the origin is midway between them. The ship lies on a hyperbola whose foci are the locations of the two stations. Since the time difference is 0.00038 second and the speed of the signal is 186,000 miles per second, the difference in the distances of the ships from each station is: $(186,000)(0.00038) = 70.68$ miles

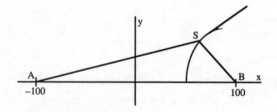

The difference of the distances from the ship to each station, 70.68, equals $2a$, so $a = 35.34$ and the vertex of the corresponding hyperbola is at $(35.34, 0)$. Since the focus is at $(100, 0)$, following this hyperbola, the ship would reach shore 64.66 miles from the master station.

(b) The ship should follow a hyperbola with a vertex at $(80, 0)$. For this hyperbola, $a = 80$, so the constant difference of the distances from the ship to each station is 160. The time difference the ship should look for is: time $= \dfrac{160}{186,000} \approx 0.00086$ second.

(c) Find the equation of the hyperbola with vertex at $(80, 0)$ and a focus at $(100, 0)$. The form of the equation of the hyperbola is $\dfrac{x^2}{a^2} - \dfrac{y^2}{b^2} = 1$ where $a = 80$.

Since $c = 100$ and $b^2 = c^2 - a^2 \Rightarrow b^2 = 100^2 - 80^2 = 3600$.

The equation of the hyperbola is: $\dfrac{x^2}{6400} - \dfrac{y^2}{3600} = 1$.

Since the ship is 50 miles off shore, we have $y = 50$. Solve the equation for x:

$$\frac{x^2}{6400} - \frac{50^2}{3600} = 1 \Rightarrow \frac{x^2}{6400} = 1 + \frac{2500}{3600} = \frac{61}{36}$$

$$x^2 = 6400 \cdot \frac{61}{36} \Rightarrow x \approx 104 \text{ miles}$$

The ship's location is approximately $(104, 50)$.

67. (a) Set up a rectangular coordinate system so that the two devices lie on the x-axis and the origin is midway between them. The devices serve as foci to the hyperbola so $c = \dfrac{2000}{2} = 1000$. Since the explosion occurs 200 feet from point B, the vertex of the hyperbola is $(800, 0)$; therefore, $a = 800$. Finding b:

$b^2 = c^2 - a^2 \Rightarrow b^2 = 1000^2 - 800^2 = 360,000 \Rightarrow b = 600$

The equation of the hyperbola is: $\dfrac{x^2}{800^2} - \dfrac{y^2}{600^2} = 1$

If $x = 1000$, find y:

$$\frac{1000^2}{800^2} - \frac{y^2}{600^2} = 1 \Rightarrow \frac{y^2}{600^2} = \frac{1000^2}{800^2} - 1 = \frac{600^2}{800^2}$$

$$y^2 = 600^2 \cdot \frac{600^2}{800^2} \Rightarrow y = 450 \text{ feet}$$

The second detonation should take place 450 feet north of point B.

(b) Answers will vary.

69. If the eccentricity is close to 1, then $c \approx a$ and $b \approx 0$. When b is close to 0, the hyperbola is very narrow, because the slopes of the asymptotes are close to 0.
If the eccentricity is very large, then c is much larger than a and b is very large. The result is a hyperbola that is very wide.

71. $\dfrac{x^2}{4} - y^2 = 1$ $(a = 2,\ b = 1)$

This is a hyperbola with horizontal transverse axis, centered at $(0, 0)$ and

has asymptotes: $y = \pm\dfrac{1}{2}x$

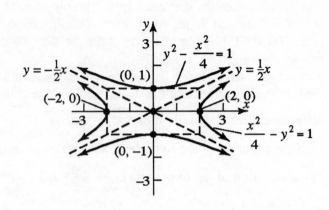

$y^2 - \dfrac{x^2}{4} = 1$ $(a = 1,\ b = 2)$

This is a hyperbola with vertical transverse axis, centered at $(0, 0)$ and

has asymptotes: $y = \pm\dfrac{1}{2}x$.

Since the two hyperbolas have the same asymptotes, they are conjugate.

73. Put the equation in standard hyperbola form:

$Ax^2 + Cy^2 + F = 0$ $A \neq 0,\ C \neq 0,\ F \neq 0$

$$Ax^2 + Cy^2 = -F$$

$$\frac{Ax^2}{-F} + \frac{Cy^2}{-F} = 1$$

$$\frac{x^2}{\left(-\dfrac{F}{A}\right)} + \frac{y^2}{\left(-\dfrac{F}{C}\right)} = 1$$

Since $-F / A$ and $-F / C$ have opposite signs, this is a hyperbola with center $(0, 0)$.

The transverse axis is the x-axis if $-F / A > 0$.

The transverse axis is the y-axis if $-F / A < 0$.

Chapter 9

Analytic Geometry

9.5　Rotation of Axes; General Form of a Conic

1.　$\sin\alpha\cos\beta + \cos\alpha\sin\beta$

3.　$\sqrt{\dfrac{1-\cos\theta}{2}}$

5.　$\cot(2\theta) = \dfrac{A-C}{B}$

7.　$x^2 + 2xy + 3y^2 - 2x + 4y + 10 = 0$
$A = 1$, $B = 2$, and $C = 3$; $B^2 - 4AC = 2^2 - 4(1)(3) = -8$. Since $B^2 - 4AC < 0$, the equation defines an ellipse.

9.　True

11.　$x^2 + 4x + y + 3 = 0$
$A = 1$ and $C = 0$; $AC = (1)(0) = 0$. Since $AC = 0$, the equation defines a parabola.

13.　$6x^2 + 3y^2 - 12x + 6y = 0$
$A = 6$ and $C = 3$; $AC = (6)(3) = 18$. Since $AC > 0$ and $A \neq C$, the equation defines an ellipse.

15.　$3x^2 - 2y^2 + 6x + 4 = 0$
$A = 3$ and $C = -2$; $AC = (3)(-2) = -6$. Since $AC < 0$, the equation defines a hyperbola.

17.　$2y^2 - x^2 - y + x = 0$
$A = -1$ and $C = 2$; $AC = (-1)(2) = -2$. Since $AC < 0$, the equation defines a hyperbola.

19.　$x^2 + y^2 - 8x + 4y = 0$
$A = 1$ and $C = 1$; $AC = (1)(1) = 1$. Since $AC > 0$ and $A = C$, the equation defines a circle.

21.　$x^2 + 4xy + y^2 - 3 = 0$
$\quad A = 1$, $B = 4$, and $C = 1$;

$$\cot(2\theta) = \frac{A-C}{B} = \frac{1-1}{4} = \frac{0}{4} = 0 \Rightarrow 2\theta = \frac{\pi}{2} \Rightarrow \theta = \frac{\pi}{4}$$

$$x = x'\cos\frac{\pi}{4} - y'\sin\frac{\pi}{4} = \frac{\sqrt{2}}{2}x' - \frac{\sqrt{2}}{2}y' = \frac{\sqrt{2}}{2}(x' - y')$$

$$y = x'\sin\frac{\pi}{4} + y'\cos\frac{\pi}{4} = \frac{\sqrt{2}}{2}x' + \frac{\sqrt{2}}{2}y' = \frac{\sqrt{2}}{2}(x' + y')$$

23. $5x^2 + 6xy + 5y^2 - 8 = 0$

$A = 5, B = 6,$ and $C = 5;$

$$\cot(2\theta) = \frac{A-C}{B} = \frac{5-5}{6} = \frac{0}{6} = 0 \Rightarrow 2\theta = \frac{\pi}{2} \Rightarrow \theta = \frac{\pi}{4}$$

$$x = x'\cos\frac{\pi}{4} - y'\sin\frac{\pi}{4} = \frac{\sqrt{2}}{2}x' - \frac{\sqrt{2}}{2}y' = \frac{\sqrt{2}}{2}(x' - y')$$

$$y = x'\sin\frac{\pi}{4} + y'\cos\frac{\pi}{4} = \frac{\sqrt{2}}{2}x' + \frac{\sqrt{2}}{2}y' = \frac{\sqrt{2}}{2}(x' + y')$$

25. $13x^2 - 6\sqrt{3}xy + 7y^2 - 16 = 0$

$A = 13, B = -6\sqrt{3},$ and $C = 7;$

$$\cot(2\theta) = \frac{A-C}{B} = \frac{13-7}{-6\sqrt{3}} = \frac{6}{-6\sqrt{3}} = -\frac{\sqrt{3}}{3} \Rightarrow 2\theta = \frac{2\pi}{3} \Rightarrow \theta = \frac{\pi}{3}$$

$$x = x'\cos\frac{\pi}{3} - y'\sin\frac{\pi}{3} = \frac{1}{2}x' - \frac{\sqrt{3}}{2}y' = \frac{1}{2}\left(x' - \sqrt{3}y'\right)$$

$$y = x'\sin\frac{\pi}{3} + y'\cos\frac{\pi}{3} = \frac{\sqrt{3}}{2}x' + \frac{1}{2}y' = \frac{1}{2}\left(\sqrt{3}x' + y'\right)$$

27. $4x^2 - 4xy + y^2 - 8\sqrt{5}x - 16\sqrt{5}y = 0$

$A = 4, B = -4,$ and $C = 1;$ $\cot(2\theta) = \dfrac{A-C}{B} = \dfrac{4-1}{-4} = -\dfrac{3}{4};$ $\cos 2\theta = -\dfrac{3}{5}$

$$\sin\theta = \sqrt{\frac{1-\left(-\frac{3}{5}\right)}{2}} = \sqrt{\frac{4}{5}} = \frac{2}{\sqrt{5}} = \frac{2\sqrt{5}}{5};\quad \cos\theta = \sqrt{\frac{1+\left(-\frac{3}{5}\right)}{2}} = \sqrt{\frac{1}{5}} = \frac{1}{\sqrt{5}} = \frac{\sqrt{5}}{5}$$

$$x = x'\cos\theta - y'\sin\theta = \frac{\sqrt{5}}{5}x' - \frac{2\sqrt{5}}{5}y' = \frac{\sqrt{5}}{5}(x' - 2y')$$

$$y = x'\sin\theta + y'\cos\theta = \frac{2\sqrt{5}}{5}x' + \frac{\sqrt{5}}{5}y' = \frac{\sqrt{5}}{5}(2x' + y')$$

29. $25x^2 - 36xy + 40y^2 - 12\sqrt{13}x - 8\sqrt{13}y = 0$

$A = 25, B = -36,$ and $C = 40;$ $\cot(2\theta) = \dfrac{A-C}{B} = \dfrac{25-40}{-36} = \dfrac{5}{12};$ $\cos 2\theta = \dfrac{5}{13}$

$$\sin\theta = \sqrt{\frac{1-\frac{5}{13}}{2}} = \sqrt{\frac{4}{13}} = \frac{2}{\sqrt{13}} = \frac{2\sqrt{13}}{13};\quad \cos\theta = \sqrt{\frac{1+\frac{5}{13}}{2}} = \sqrt{\frac{9}{13}} = \frac{3}{\sqrt{13}} = \frac{3\sqrt{13}}{13}$$

$$x = x'\cos\theta - y'\sin\theta = \frac{3\sqrt{13}}{13}x' - \frac{2\sqrt{13}}{13}y' = \frac{\sqrt{13}}{13}(3x' - 2y')$$

$$y = x'\sin\theta + y'\cos\theta = \frac{2\sqrt{13}}{13}x' + \frac{3\sqrt{13}}{13}y' = \frac{\sqrt{13}}{13}(2x' + 3y')$$

31. $x^2 + 4xy + y^2 - 3 = 0$; $\theta = 45°$ (see Problem 21)

$$\left(\frac{\sqrt{2}}{2}(x' - y')\right)^2 + 4\left(\frac{\sqrt{2}}{2}(x' - y')\right)\left(\frac{\sqrt{2}}{2}(x' + y')\right) + \left(\frac{\sqrt{2}}{2}(x' + y')\right)^2 - 3 = 0$$

$$\frac{1}{2}\left(x'^2 - 2x'y' + y'^2\right) + 2\left(x'^2 - y'^2\right) + \frac{1}{2}\left(x'^2 + 2x'y' + y'^2\right) - 3 = 0$$

$$\frac{1}{2}x'^2 - x'y' + \frac{1}{2}y'^2 + 2x'^2 - 2y'^2 + \frac{1}{2}x'^2 + x'y' + \frac{1}{2}y'^2 = 3$$

$$3x'^2 - y'^2 = 3$$

$$\frac{x'^2}{1} - \frac{y'^2}{3} = 1$$

Hyperbola; center at the origin, transverse axis is the x'-axis, vertices $(\pm1, 0)$.

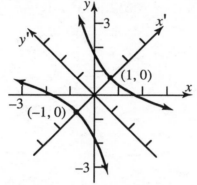

33. $5x^2 + 6xy + 5y^2 - 8 = 0$; $\theta = 45°$ (see Problem 23)

$$5\left(\frac{\sqrt{2}}{2}(x' - y')\right)^2 + 6\left(\frac{\sqrt{2}}{2}(x' - y')\right)\left(\frac{\sqrt{2}}{2}(x' + y')\right) + 5\left(\frac{\sqrt{2}}{2}(x' + y')\right)^2 - 8 = 0$$

$$\frac{5}{2}\left(x'^2 - 2x'y' + y'^2\right) + 3\left(x'^2 - y'^2\right) + \frac{5}{2}\left(x'^2 + 2x'y' + y'^2\right) - 8 = 0$$

$$\frac{5}{2}x'^2 - 5x'y' + \frac{5}{2}y'^2 + 3x'^2 - 3y'^2 + \frac{5}{2}x'^2 + 5x'y' + \frac{5}{2}y'^2 = 8$$

$$8x'^2 + 2y'^2 = 8$$

$$\frac{x'^2}{1} + \frac{y'^2}{4} = 1$$

Ellipse; center at the origin, major axis is the y'-axis, vertices $(0, \pm2)$.

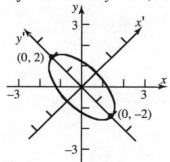

35. $13x^2 - 6\sqrt{3}\,xy + 7y^2 - 16 = 0$; $\theta = 60°$ (see Problem 25)

$$13\left(\frac{1}{2}\left(x' - \sqrt{3}y'\right)\right)^2 - 6\sqrt{3}\left(\frac{1}{2}\left(x' - \sqrt{3}y'\right)\right)\left(\frac{1}{2}\left(\sqrt{3}x' + y'\right)\right) + 7\left(\frac{1}{2}\left(\sqrt{3}x' + y'\right)\right)^2 - 16 = 0$$

$$\frac{13}{4}\left(x'^2 - 2\sqrt{3}x'y' + 3y'^2\right) - \frac{3\sqrt{3}}{2}\left(\sqrt{3}x'^2 - 2x'y' - \sqrt{3}y'^2\right) + \frac{7}{4}\left(3x'^2 + 2\sqrt{3}x'y' + y'^2\right) = 16$$

$$\frac{13}{4}x'^2 - \frac{13\sqrt{3}}{2}x'y' + \frac{39}{4}y'^2 - \frac{9}{2}x'^2 + 3\sqrt{3}x'y' + \frac{9}{2}y'^2 + \frac{21}{4}x'^2 + \frac{7\sqrt{3}}{2}x'y' + \frac{7}{4}y'^2 = 16$$

$$4x'^2 + 16y'^2 = 16$$

$$\frac{x'^2}{4} + \frac{y'^2}{1} = 1$$

Ellipse; center at the origin, major axis is the x'-axis, vertices $(\pm 2, 0)$.

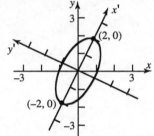

37. $4x^2 - 4xy + y^2 - 8\sqrt{5}\,x - 16\sqrt{5}\,y = 0$; $\theta \approx 63.4°$ (see Problem 27)

$$4\left(\frac{\sqrt{5}}{5}(x' - 2y')\right)^2 - 4\left(\frac{\sqrt{5}}{5}(x' - 2y')\right)\left(\frac{\sqrt{5}}{5}(2x' + y')\right) + \left(\frac{\sqrt{5}}{5}(2x' + y')\right)^2$$

$$-8\sqrt{5}\left(\frac{\sqrt{5}}{5}(x' - 2y')\right) - 16\sqrt{5}\left(\frac{\sqrt{5}}{5}(2x' + y')\right) = 0$$

$$\frac{4}{5}\left(x'^2 - 4x'y' + 4y'^2\right) - \frac{4}{5}\left(2x'^2 - 3x'y' - 2y'^2\right) + \frac{1}{5}\left(4x'^2 + 4x'y' + y'^2\right)$$

$$-8x' + 16y' - 32x' - 16y' = 0$$

$$\frac{4}{5}x'^2 - \frac{16}{5}x'y' + \frac{16}{5}y'^2 - \frac{8}{5}x'^2 + \frac{12}{5}x'y' + \frac{8}{5}y'^2 + \frac{4}{5}x'^2 + \frac{4}{5}x'y' + \frac{1}{5}y'^2 - 40x' = 0$$

$$5y'^2 - 40x' = 0$$

$$y'^2 = 8x'$$

Parabola; vertex at the origin, focus at $(2, 0)$.

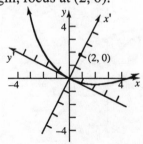

39. $25x^2 - 36xy + 40y^2 - 12\sqrt{13}x - 8\sqrt{13}y = 0$; $\theta \approx 33.7°$ (see Problem 29)

$$25\left(\frac{\sqrt{13}}{13}(3x' - 2y')\right)^2 - 36\left(\frac{\sqrt{13}}{13}(3x' - 2y')\right)\left(\frac{\sqrt{13}}{13}(2x' + 3y')\right) + 40\left(\frac{\sqrt{13}}{13}(2x' + 3y')\right)^2$$

$$-12\sqrt{13}\left(\frac{\sqrt{13}}{13}(3x' - 2y')\right) - 8\sqrt{13}\left(\frac{\sqrt{13}}{13}(2x' + 3y')\right) = 0$$

$$\frac{25}{13}(9x'^2 - 12x'y' + 4y'^2) - \frac{36}{13}(6x'^2 + 5x'y' - 6y'^2) + \frac{40}{13}(4x'^2 + 12x'y' + 9y'^2)$$

$$- 36x' + 24y' - 16x' - 24y' = 0$$

$$\frac{225}{13}x'^2 - \frac{300}{13}x'y' + \frac{100}{13}y'^2 - \frac{216}{13}x'^2 - \frac{180}{13}x'y' + \frac{216}{13}y'^2$$

$$+ \frac{160}{13}x'^2 + \frac{480}{13}x'y' + \frac{360}{13}y'^2 - 52x' = 0$$

$$13x'^2 + 52y'^2 - 52x' = 0$$

$$x'^2 - 4x' + 4y'^2 = 0$$

$$(x' - 2)^2 + 4y'^2 = 4$$

$$\frac{(x' - 2)^2}{4} + \frac{y'^2}{1} = 1$$

Ellipse; center at $(2, 0)$, major axis is the x'-axis, vertices $(4, 0)$ and $(0, 0)$.

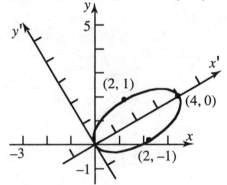

41. $16x^2 + 24xy + 9y^2 - 130x + 90y = 0$

$$A = 16, B = 24, \text{ and } C = 9; \ \cot(2\theta) = \frac{A - C}{B} = \frac{16 - 9}{24} = \frac{7}{24} \Rightarrow \cos(2\theta) = \frac{7}{25}$$

$$\sin\theta = \sqrt{\frac{1 - \frac{7}{25}}{2}} = \sqrt{\frac{9}{25}} = \frac{3}{5}; \ \cos\theta = \sqrt{\frac{1 + \frac{7}{25}}{2}} = \sqrt{\frac{16}{25}} = \frac{4}{5} \Rightarrow \theta \approx 36.9°$$

$$x = x'\cos\theta - y'\sin\theta = \frac{4}{5}x' - \frac{3}{5}y' = \frac{1}{5}(4x' - 3y')$$

$$y = x'\sin\theta + y'\cos\theta = \frac{3}{5}x' + \frac{4}{5}y' = \frac{1}{5}(3x' + 4y')$$

$$16\left(\frac{1}{5}(4x'-3y')\right)^2 + 24\left(\frac{1}{5}(4x'-3y')\right)\left(\frac{1}{5}(3x'+4y')\right) + 9\left(\frac{1}{5}(3x'+4y')\right)^2$$

$$-130\left(\frac{1}{5}(4x'-3y')\right) + 90\left(\frac{1}{5}(3x'+4y')\right) = 0$$

$$\frac{16}{25}\left(16x'^2 - 24x'y' + 9y'^2\right) + \frac{24}{25}\left(12x'^2 + 7x'y' - 12y'^2\right)$$

$$+ \frac{9}{25}\left(9x'^2 + 24x'y' + 16y'^2\right) - 104x' + 78y' + 54x' + 72y' = 0$$

$$\frac{256}{25}x'^2 - \frac{384}{25}x'y' + \frac{144}{25}y'^2 + \frac{288}{25}x'^2 + \frac{168}{25}x'y' - \frac{288}{25}y'^2$$

$$+ \frac{81}{25}x'^2 + \frac{216}{25}x'y' + \frac{144}{25}y'^2 - 50x' + 150y' = 0$$

$$25x'^2 - 50x' + 150y' = 0$$

$$x'^2 - 2x' = -6y'$$

$$(x'-1)^2 = -6y' + 1$$

$$(x'-1)^2 = -6\left(y' - \frac{1}{6}\right)$$

Parabola; vertex $\left(1, \frac{1}{6}\right)$, focus $\left(1, -\frac{4}{3}\right)$.

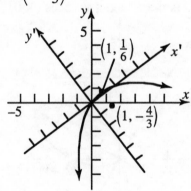

43. $A = 1,\ B = 3,\ C = -2 \quad B^2 - 4AC = 3^2 - 4(1)(-2) = 17 > 0;$ hyperbola

45. $A = 1,\ B = -7,\ C = 3 \quad B^2 - 4AC = (-7)^2 - 4(1)(3) = 37 > 0;$ hyperbola

47. $A = 9,\ B = 12,\ C = 4 \quad B^2 - 4AC = 12^2 - 4(9)(4) = 0;$ parabola

49. $A = 10,\ B = -12,\ C = 4 \quad B^2 - 4AC = (-12)^2 - 4(10)(4) = -16 < 0;$ ellipse

51. $A = 3,\ B = -2,\ C = 1 \quad B^2 - 4AC = (-2)^2 - 4(3)(1) = -8 < 0;$ ellipse

53. See equation 6 on page 808.

$A' = A\cos^2\theta + B\sin\theta\cos\theta + C\sin^2\theta$

$B' = B(\cos^2\theta - \sin^2\theta) + 2(C - A)(\sin\theta\cos\theta)$

$C' = A\sin^2\theta - B\sin\theta\cos\theta + C\cos^2\theta$

$D' = D\cos\theta + E\sin\theta$

$E' = -D\sin\theta + E\cos\theta$

$F' = F$

55. $B'^2 - 4A'C'$

$= \left[B(\cos^2\theta - \sin^2\theta) + 2(C - A)\sin\theta\cos\theta\right]^2$

$\quad - 4\left(A\cos^2\theta + B\sin\theta\cos\theta + C\sin^2\theta\right)\left(A\sin^2\theta - B\sin\theta\cos\theta + C\cos^2\theta\right)$

$= B^2\left(\cos^4\theta - 2\cos^2\theta\sin^2\theta + \sin^4\theta\right) + 4B(C - A)\sin\theta\cos\theta(\cos^2\theta - \sin^2\theta)$

$\quad + 4(C - A)^2\sin^2\theta\cos^2\theta - 4\left[A^2\sin^2\theta\cos^2\theta - AB\sin\theta\cos^3\theta + AC\cos^4\theta\right.$

$\quad + AB\sin^3\theta\cos\theta - B^2\sin^2\theta\cos^2\theta + BC\sin\theta\cos^3\theta + AC\sin^4\theta$

$\quad \left. - BC\sin^3\theta\cos\theta + C^2\sin^2\theta\cos^2\theta\right]$

$= B^2\left(\cos^4\theta - 2\cos^2\theta\sin^2\theta + \sin^4\theta + 4\sin^2\theta\cos^2\theta\right)$

$\quad + BC\left(4\sin\theta\cos\theta(\cos^2\theta - \sin^2\theta) - 4\sin\theta\cos^3\theta + 4\sin^3\theta\cos\theta\right)$

$\quad - AB\left(4\sin\theta\cos\theta(\cos^2\theta - \sin^2\theta) - 4\sin\theta\cos^3\theta + 4\sin^3\theta\cos\theta\right)$

$\quad + 4C^2\left(\sin^2\theta\cos^2\theta - \sin^2\theta\cos^2\theta\right) - 4AC\left(2\sin^2\theta\cos^2\theta + \cos^4\theta + \sin^4\theta\right)$

$\quad + 4A^2\left(\sin^2\theta\cos^2\theta - \sin^2\theta\cos^2\theta\right)$

$= B^2\left(\cos^4\theta + 2\sin^2\theta\cos^2\theta + \sin^4\theta\right) - 4AC\left(\cos^4\theta + 2\sin^2\theta\cos^2\theta + \sin^4\theta\right)$

$= B^2\left(\cos^2\theta + \sin^2\theta\right)^2 - 4AC\left(\cos^2\theta + \sin^2\theta\right)^2 = B^2 - 4AC$

57. $d^2 = (y_2 - y_1)^2 + (x_2 - x_1)^2$

$= \left(x_2'\sin\theta + y_2'\cos\theta - x_1'\sin\theta - y_1'\cos\theta\right)^2$

$\qquad\qquad + \left(x_2'\cos\theta - y_2'\sin\theta - x_1'\cos\theta + y_1'\sin\theta\right)^2$

$= \left((x_2' - x_1')\sin\theta + (y_2' - y_1')\cos\theta\right)^2 + \left((x_2' - x_1')\cos\theta - (y_2' - y_1')\sin\theta\right)^2$

$= (x_2' - x_1')^2\sin^2\theta + 2(x_2' - x_1')(y_2' - y_1')\sin\theta\cos\theta + (y_2' - y_1')^2\cos^2\theta$

$\qquad + (x_2' - x_1')^2\cos^2\theta - 2(x_2' - x_1')(y_2' - y_1')\sin\theta\cos\theta + (y_2' - y_1')^2\sin^2\theta$

$= (x_2' - x_1')^2\sin^2\theta + (x_2' - x_1')^2\cos^2\theta + (y_2' - y_1')^2\cos^2\theta + (y_2' - y_1')^2\sin^2\theta$

$= (x_2' - x_1')^2 + (y_2' - y_1')^2$

59. Answers will vary.

Chapter 9

Analytic Geometry

9.6   Polar Equations of Conics

1. $r\cos\theta,\ r\sin\theta$

3. $\dfrac{1}{2}$, ellipse, parallel, 4, below

5.   True

7. $e = 1;\quad p = 1;$ parabola; directrix is perpendicular to the polar axis and 1 unit to the right of the pole.

9. $r = \dfrac{4}{2 - 3\sin\theta} = \dfrac{4}{2\left(1 - \dfrac{3}{2}\sin\theta\right)} = \dfrac{2}{1 - \dfrac{3}{2}\sin\theta};\quad ep = 2,\ e = \dfrac{3}{2};\ p = \dfrac{4}{3}$

Hyperbola; directrix is parallel to the polar axis and $\dfrac{4}{3}$ units below the pole.

11. $r = \dfrac{3}{4 - 2\cos\theta} = \dfrac{3}{4\left(1 - \dfrac{1}{2}\cos\theta\right)} = \dfrac{\dfrac{3}{4}}{1 - \dfrac{1}{2}\cos\theta};\quad ep = \dfrac{3}{4},\ e = \dfrac{1}{2};\ p = \dfrac{3}{2}$

Ellipse; directrix is perpendicular to the polar axis and $\dfrac{3}{2}$ units to the left of the pole.

13. $r = \dfrac{1}{1 + \cos\theta}$
$ep = 1,\ e = 1,\ p = 1$
Parabola; directrix is perpendicular to the polar axis 1 unit to the right of the pole;
vertex is $\left(\dfrac{1}{2}, 0\right)$.

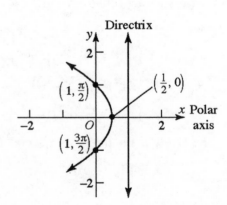

15. $r = \dfrac{8}{4 + 3\sin\theta}$

$r = \dfrac{8}{4\left(1 + \dfrac{3}{4}\sin\theta\right)} = \dfrac{2}{1 + \dfrac{3}{4}\sin\theta}$

$ep = 2,\ e = \dfrac{3}{4},\ p = \dfrac{8}{3}$

Ellipse; directrix is parallel to the polar

axis $\dfrac{8}{3}$ units above the pole; vertices are $\left(\dfrac{8}{7}, \dfrac{\pi}{2}\right)$ and $\left(8, \dfrac{3\pi}{2}\right)$.

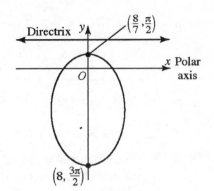

17. $r = \dfrac{9}{3 - 6\cos\theta}$

$r = \dfrac{9}{3(1 - 2\cos\theta)} = \dfrac{3}{1 - 2\cos\theta}$

$ep = 3,\ e = 2,\ p = \dfrac{3}{2}$

Hyperbola; directrix is perpendicular to the

polar axis $\dfrac{3}{2}$ units to the left of the pole;

vertices are $(-3, 0)$ and $(1, \pi)$.

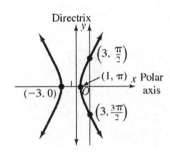

19. $r = \dfrac{8}{2 - \sin\theta}$

$r = \dfrac{8}{2\left(1 - \dfrac{1}{2}\sin\theta\right)} = \dfrac{4}{1 - \dfrac{1}{2}\sin\theta}$

$ep = 4,\ e = \dfrac{1}{2},\ p = 8$

Ellipse; directrix is parallel to the polar
axis 8 units below the pole; vertices are

$\left(8, \dfrac{\pi}{2}\right)$ and $\left(\dfrac{8}{3}, \dfrac{3\pi}{2}\right)$.

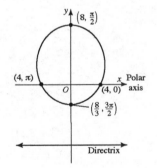

21. $r(3 - 2\sin\theta) = 6 \Rightarrow r = \dfrac{6}{3 - 2\sin\theta}$

$r = \dfrac{6}{3\left(1 - \dfrac{2}{3}\sin\theta\right)} = \dfrac{2}{1 - \dfrac{2}{3}\sin\theta}$

$ep = 2,\ e = \dfrac{2}{3},\ p = 3$

Ellipse; directrix is parallel to the polar
axis 3 units below the pole; vertices are

$\left(6, \dfrac{\pi}{2}\right)$ and $\left(\dfrac{6}{5}, \dfrac{3\pi}{2}\right)$.

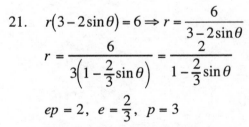

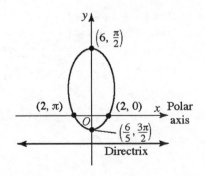

23. $r = \dfrac{6\sec\theta}{2\sec\theta - 1} = \dfrac{6\left(\dfrac{1}{\cos\theta}\right)}{2\left(\dfrac{1}{\cos\theta}\right) - 1} = \dfrac{\dfrac{6}{\cos\theta}}{\dfrac{2 - \cos\theta}{\cos\theta}} = \left(\dfrac{6}{\cos\theta}\right)\left(\dfrac{\cos\theta}{2 - \cos\theta}\right) = \dfrac{6}{2 - \cos\theta}$

$r = \dfrac{6}{2\left(1 - \dfrac{1}{2}\cos\theta\right)} = \dfrac{3}{1 - \dfrac{1}{2}\cos\theta}$

$ep = 3, \quad e = \dfrac{1}{2}, \quad p = 6$

Ellipse; directrix is perpendicular to the polar axis 6 units to the left of the pole; vertices are $(6, 0)$ and $(2, \pi)$.

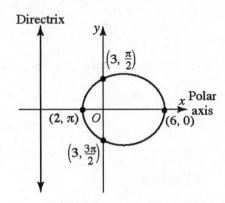

25. $r = \dfrac{1}{1 + \cos\theta}$

$r + r\cos\theta = 1$

$r = 1 - r\cos\theta$

$r^2 = (1 - r\cos\theta)^2$

$x^2 + y^2 = (1 - x)^2$

$x^2 + y^2 = 1 - 2x + x^2$

$y^2 + 2x - 1 = 0$

27. $r = \dfrac{8}{4 + 3\sin\theta}$

$4r + 3r\sin\theta = 8$

$4r = 8 - 3r\sin\theta$

$16r^2 = (8 - 3r\sin\theta)^2$

$16(x^2 + y^2) = (8 - 3y)^2$

$16x^2 + 16y^2 = 64 - 48y + 9y^2$

$16x^2 + 7y^2 + 48y - 64 = 0$

29. $r = \dfrac{9}{3 - 6\cos\theta}$

$3r - 6r\cos\theta = 9$

$3r = 9 + 6r\cos\theta$

$r = 3 + 2r\cos\theta$

$r^2 = (3 + 2r\cos\theta)^2$

$x^2 + y^2 = (3 + 2x)^2$

$x^2 + y^2 = 9 + 12x + 4x^2$

$3x^2 - y^2 + 12x + 9 = 0$

31. $r = \dfrac{8}{2 - \sin\theta}$

$2r - r\sin\theta = 8$

$2r = 8 + r\sin\theta$

$4r^2 = (8 + r\sin\theta)^2$

$4(x^2 + y^2) = (8 + y)^2$

$4x^2 + 4y^2 = 64 + 16y + y^2$

$4x^2 + 3y^2 - 16y - 64 = 0$

33. $r(3 - 2\sin\theta) = 6$

$3r - 2r\sin\theta = 6$

$3r = 6 + 2r\sin\theta$

$9r^2 = (6 + 2r\sin\theta)^2$

$9(x^2 + y^2) = (6 + 2y)^2$

$9x^2 + 9y^2 = 36 + 24y + 4y^2$

$9x^2 + 5y^2 - 24y - 36 = 0$

35. $r = \dfrac{6\sec\theta}{2\sec\theta - 1}$

$r = \dfrac{6}{2 - \cos\theta}$

$2r - r\cos\theta = 6$

$2r = 6 + r\cos\theta$

$4r^2 = (6 + r\cos\theta)^2$

$4(x^2 + y^2) = (6 + x)^2$

$4x^2 + 4y^2 = 36 + 12x + x^2$

$3x^2 + 4y^2 - 12x - 36 = 0$

37. $r = \dfrac{ep}{1 + e\sin\theta}$

$e = 1; \quad p = 1$

$r = \dfrac{1}{1 + \sin\theta}$

39. $r = \dfrac{ep}{1 - e\cos\theta}$

$e = \dfrac{4}{5}; \quad p = 3$

$r = \dfrac{\dfrac{12}{5}}{1 - \dfrac{4}{5}\cos\theta} = \dfrac{12}{5 - 4\cos\theta}$

41. $r = \dfrac{ep}{1 - e\sin\theta}$

$e = 6; \quad p = 2$

$r = \dfrac{12}{1 - 6\sin\theta}$

43.

$$d(F, P) = e \cdot d(D, P) \qquad d(D, P) = p - r\cos\theta$$

$$r = e(p - r\cos\theta)$$

$$r = ep - er\cos\theta$$

$$r + er\cos\theta = ep$$

$$r(1 + e\cos\theta) = ep$$

$$r = \frac{ep}{1 + e\cos\theta}$$

45.

$$d(F, P) = e \cdot d(D, P) \qquad d(D, P) = p + r\sin\theta$$

$$r = e(p + r\sin\theta)$$

$$r = ep + er\sin\theta$$

$$r - er\sin\theta = ep$$

$$r(1 - e\sin\theta) = ep$$

$$r = \frac{ep}{1 - e\sin\theta}$$

Chapter 9

Analytic Geometry

9.7 Plane Curves and Parametric Equations

1. $|3| = 3, \dfrac{2\pi}{4} = \dfrac{\pi}{2}$

3. ellipse

5. False

7. $x = 3t + 2, \ y = t + 1, \ 0 \le t \le 4$

$x = 3(y - 1) + 2$
$x = 3y - 3 + 2$
$x = 3y - 1$
$x - 3y + 1 = 0$

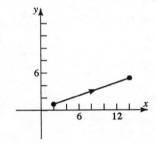

9. $x = t + 2, \ y = \sqrt{t}, \ t \ge 0$

$y = \sqrt{x - 2}$

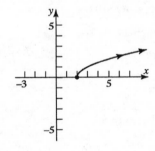

11. $x = t^2 + 4, \ y = t^2 - 4, \ -\infty < t < \infty$

$y = (x - 4) - 4$
$y = x - 8$
For $-\infty < t < 0$ the movement is to the left. For $0 < t < \infty$ the movement is to the right.

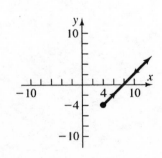

13. $x = 3t^2$, $y = t + 1$, $-\infty < t < \infty$

$x = 3(y - 1)^2$

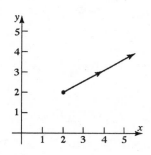

15. $x = 2e^t$, $y = 1 + e^t$, $t \geq 0$

$y = 1 + \dfrac{x}{2}$
$2y = 2 + x$

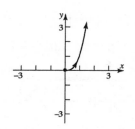

17. $x = \sqrt{t}$, $y = t^{3/2}$, $t \geq 0$

$y = \left(x^2\right)^{3/2}$
$y = x^3$

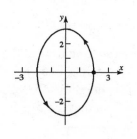

19. $x = 2\cos t$, $y = 3\sin t$, $0 \leq t \leq 2\pi$

$\dfrac{x}{2} = \cos t \qquad \dfrac{y}{3} = \sin t$

$\left(\dfrac{x}{2}\right)^2 + \left(\dfrac{y}{3}\right)^2 = \cos^2 t + \sin^2 t = 1$

$\dfrac{x^2}{4} + \dfrac{y^2}{9} = 1$

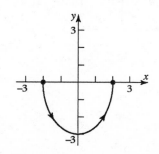

21. $x = 2\cos t$, $y = 3\sin t$, $-\pi \leq t \leq 0$

$\dfrac{x}{2} = \cos t \qquad \dfrac{y}{3} = \sin t$

$\left(\dfrac{x}{2}\right)^2 + \left(\dfrac{y}{3}\right)^2 = \cos^2 t + \sin^2 t = 1$

$\dfrac{x^2}{4} + \dfrac{y^2}{9} = 1 \quad y \leq 0$

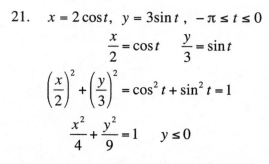

23. $x = \sec t,\ \ y = \tan t,\ \ 0 \le t \le \dfrac{\pi}{4}$

$\sec^2 t = 1 + \tan^2 t$

$x^2 = 1 + y^2$

$x^2 - y^2 = 1 \quad 1 \le x \le \sqrt{2},\ \ 0 \le y \le 1$

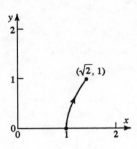

25. $x = \sin^2 t,\ \ y = \cos^2 t,\ \ 0 \le t \le 2\pi$

$\sin^2 t + \cos^2 t = 1$

$x + y = 1$

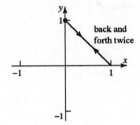

27. (a)   Use equations (2):
$x = (50\cos 90°)t = 0$

$y = -\dfrac{1}{2}(32)t^2 + (50\sin 90°)t + 6 = -16t^2 + 50t + 6$

(b)   The ball is in the air until $y = 0$. Solve:
$-16t^2 + 50t + 6 = 0$

$t = \dfrac{-50 \pm \sqrt{50^2 - 4(-16)(6)}}{2(-16)} = \dfrac{-50 \pm \sqrt{2884}}{-32} \approx -0.12 \text{ or } 3.24$

The ball is in the air for about 3.24 seconds. (The negative solution is extraneous.)

(c)   The maximum height occurs at the vertex of the quadratic function.

$t = -\dfrac{b}{2a} = -\dfrac{50}{2(-16)} = 1.5625 \text{ seconds}$

Evaluate the function to find the maximum height:
$-16(1.5625)^2 + 50(1.5625) + 6 = 45.0625$

The maximum height is 45.0625 feet.

(d)   We use $x = 3$ so that the line is not on top of the y-axis.

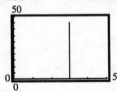

29. Let $y_1 = 1$ be the train's path and $y_2 = 3$ be Bill's path.
   (a) Train: Using the hint,
   $$x_1 = \frac{1}{2}(2)t^2 = t^2$$
   $$y_1 = 1$$
   Bill:
   $$x_2 = 5(t-5)$$
   $$y_2 = 3$$
   (b) Bill will catch the train if $x_1 = x_2$.
   $$t^2 = 5(t-5)$$
   $$t^2 = 5t - 25 \Rightarrow t^2 - 5t + 25 = 0$$
   Since $b^2 - 4ac = (-5)^2 - 4(1)(25) = 25 - 100 = -75 < 0$, the equation has no real solution. Thus, Bill will not catch the train.
   (c)

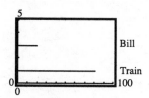

31. (a) Use equations (2):
   $$x = (145\cos 20°)t$$

   $$y = -\frac{1}{2}(32)t^2 + (145\sin 20°)t + 5$$
   (b) The ball is in the air until $y = 0$. Solve:
   $$-16t^2 + (145\sin 20°)t + 5 = 0$$

   $$t = \frac{-145\sin 20° \pm \sqrt{(145\sin 20°)^2 - 4(-16)(5)}}{2(-16)} \approx -0.10 \text{ or } 3.20$$
   The ball is in the air for about 3.20 seconds. (The negative solution is extraneous.)
   (c) The maximum height occurs at the vertex of the quadratic function.
   $$t = -\frac{b}{2a} = -\frac{145\sin 20°}{2(-16)} \approx 1.55 \text{ seconds}$$
   Evaluate the function to find the maximum height:
   $$-16(1.55)^2 + (145\sin 20°)(1.55) + 5 \approx 43.43$$
   The maximum height is about 43.43 feet.
   (d) Find the horizontal displacement:
   $$x = (145\cos 20°)(3.20) \approx 436 \text{ feet}$$
   (e)

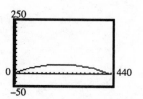

33. (a) Use equations (2):

$$x = (40\cos 45°)t = 20\sqrt{2}t$$

$$y = -\frac{1}{2}(9.8)t^2 + (40\sin 45°)t + 300 = -4.9t^2 + 20\sqrt{2}t + 300$$

(b) The ball is in the air until $y = 0$. Solve:

$$-4.9t^2 + 20\sqrt{2}t + 300 = 0$$

$$t = \frac{-20\sqrt{2} \pm \sqrt{\left(20\sqrt{2}\right)^2 - 4(-4.9)(300)}}{2(-4.9)} \approx -5.45 \text{ or } 11.23$$

The ball is in the air for about 11.23 seconds. (The negative solution is extraneous.)

(c) The maximum height occurs at the vertex of the quadratic function.

$$t = -\frac{b}{2a} = -\frac{20\sqrt{2}}{2(-4.9)} \approx 2.89 \text{ seconds}$$

Evaluate the function to find the maximum height:

$$-4.9(2.89)^2 + 20\sqrt{2}(2.89) + 300 = 340.8 \text{ meters}$$

(d) Find the horizontal displacement:

$$x = \left(20\sqrt{2}t\right)(11.23) \approx 317.6 \text{ meters}$$

(e)

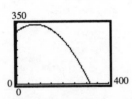

35. (a) At $t = 0$, the Paseo is 5 miles from the intersection (at $(0, 0)$) traveling east (along the x-axis) at 40 mph. Thus, $x = 40t - 5$ describes the position of the Paseo as a function of time. The Bonneville, at $t = 0$, is 4 miles from the intersection traveling north (along the y-axis) at 30 mph. Thus, $y = 30t - 4$ describes the position of the Bonneville as a function of time.

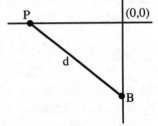

(b) Let d represent the distance between the cars. Use the Pythagorean Theorem to find the distance: $d = \sqrt{(40t - 5)^2 + (30t - 4)^2}$.

(c) Note this is a function graph not a parametric graph.

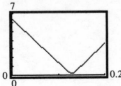

(d) The minimum distance between the cars is 0.2 mile and occurs at 0.128 second.
(e) The axes were turned off to get this graph.

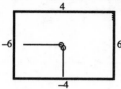

37. $x = t, \ y = 4t - 1$ $x = t + 1, \ y = 4t + 3$

39. $x = t, \ y = t^2 + 1$ $x = t - 1, \ y = t^2 - 2t + 2$

41. $x = t, \ y = t^3$ $x = \sqrt[3]{t}, \ y = t$

43. $x = t^{3/2}, \ y = t$ $x = t, \ y = t^{2/3}$

45. $x = t + 2, \ y = t; \ \ 0 \le t \le 5$

47. $x = 3\cos t, \ y = 2\sin t; \ \ 0 \le t \le 2\pi$

49. Since the motion begins at (2, 0), we want $x = 2$ and $y = 0$ when $t = 0$. For the motion to be clockwise, both x and y must be decreasing.
$$x = 2\cos(\omega t), \ y = -3\sin(\omega t)$$
$$\frac{2\pi}{\omega} = 2 \Rightarrow \omega = \pi$$
$$x = 2\cos(\pi t), \ y = -3\sin(\pi t), \ \ 0 \le t \le 2$$

51. Since the motion begins at (0, 3), we want $x = 0$ and $y = 3$ when $t = 0$. For the motion to be clockwise, x must be increasing and y must be decreasing.
$$x = -2\sin(\omega t), \ y = 3\cos(\omega t)$$
$$\frac{2\pi}{\omega} = 1 \Rightarrow \omega = 2\pi$$
$$x = -2\sin(2\pi t), \ y = 3\cos(2\pi t), \ \ 0 \le t \le 1$$

53. C_1

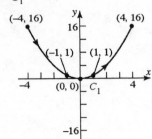

C_2

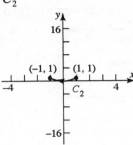

C_3

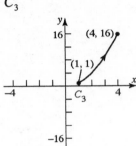

C_4

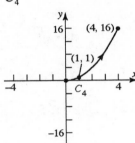

55. $x = (x_2 - x_1)t + x_1, \quad y = (y_2 - y_1)t + y_1, \quad -\infty < t < \infty$

$$\frac{x - x_1}{x_2 - x_1} = t$$

$$y = (y_2 - y_1)\left(\frac{x - x_1}{x_2 - x_1}\right) + y_1$$

$$y - y_1 = \left(\frac{y_2 - y_1}{x_2 - x_1}\right)(x - x_1)$$

This is the two-point form for the equation of a line.
Its orientation is from (x_1, y_1) to (x_2, y_2).

57. $x = t\sin t, \quad y = t\cos t$

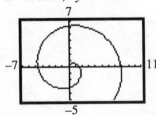

59. $x = 4\sin t - 2\sin(2t)$
$y = 4\cos t - 2\cos(2t)$

61. (a) $x(t) = \cos^3 t,\ \ y(t) = \sin^3 t,\ \ 0 \le t \le 2\pi$

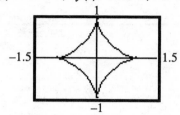

(b) $\cos^2 t + \sin^2 t = \left(x^{1/3}\right)^2 + \left(y^{1/3}\right)^2$

$x^{2/3} + y^{2/3} = 1$

63. Answers will vary.

Chapter 9

Analytic Geometry

9.R Chapter Review

1. $y^2 = -16x$
   This is a parabola.
   $a = 4$
   Vertex: $(0, 0)$
   Focus: $(-4, 0)$
   Directrix: $x = 4$

3. $\dfrac{x^2}{25} - y^2 = 1$
   This is a hyperbola.
   $a = 5, \quad b = 1$.

   Find the value of c:
   $c^2 = a^2 + b^2 = 25 + 1 = 26$
   $c = \sqrt{26}$
   Center: $(0, 0)$
   Vertices: $(-5, 0), (5, 0)$
   Foci: $\left(-\sqrt{26}, 0\right), \left(\sqrt{26}, 0\right)$
   Asymptotes: $y = \dfrac{1}{5}x; \quad y = -\dfrac{1}{5}x$

5. $\dfrac{y^2}{25} + \dfrac{x^2}{16} = 1$
   This is an ellipse.
   $a = 5, \quad b = 4$.

   Find the value of c:
   $c^2 = a^2 - b^2 = 25 - 16 = 9$
   $c = 3$
   Center: $(0, 0)$
   Vertices: $(0, -5), (0, 5)$
   Foci: $(0, -3), (0, 3)$

7. $x^2 + 4y = 4$
   This is a parabola.
   Write in standard form:
   $x^2 = -4y + 4$
   $x^2 = -4(y - 1)$

   $a = 1$
   Vertex: $(0, 1)$
   Focus: $(0, 0)$
   Directrix: $y = 2$

9. $4x^2 - y^2 = 8$
This is a hyperbola.
Write in standard form:
$$\frac{x^2}{2} - \frac{y^2}{8} = 1$$
$a = \sqrt{2}, \ b = \sqrt{8} = 2\sqrt{2}$

Find the value of c:
$c^2 = a^2 + b^2 = 2 + 8 = 10$
$c = \sqrt{10}$
Center: $(0, 0)$
Vertices: $\left(-\sqrt{2}, 0\right), \left(\sqrt{2}, 0\right)$
Foci: $\left(-\sqrt{10}, 0\right), \left(\sqrt{10}, 0\right)$
Asymptotes: $y = 2x; \ \ y = -2x$

11. $x^2 - 4x = 2y$
This is a parabola.
Write in standard form:
$x^2 - 4x + 4 = 2y + 4$
$\ \ \ \ (x - 2)^2 = 2(y + 2)$

$a = \dfrac{1}{2}$
Vertex: $(2, -2)$
Focus: $\left(2, -\dfrac{3}{2}\right)$
Directrix: $y = -\dfrac{5}{2}$

13. $y^2 - 4y - 4x^2 + 8x = 4$
This is a hyperbola.
Write in standard form:
$(y^2 - 4y + 4) - 4(x^2 - 2x + 1) = 4 + 4 - 4$
$\ \ \ \ \ \ \ \ (y - 2)^2 - 4(x - 1)^2 = 4$
$\ \ \ \ \ \ \ \ \dfrac{(y - 2)^2}{4} - \dfrac{(x - 1)^2}{1} = 1$
$a = 2, \ b = 1.$

Find the value of c:
$c^2 = a^2 + b^2 = 4 + 1 = 5$
$c = \sqrt{5}$
Center: $(1, 2)$
Vertices: $(1, 0), (1, 4)$
Foci: $\left(1, 2 - \sqrt{5}\right), \left(1, 2 + \sqrt{5}\right)$
Asymptotes:
$y - 2 = 2(x - 1); \ \ y - 2 = -2(x - 1)$

15. $4x^2 + 9y^2 - 16x - 18y = 11$
This is an ellipse.
Write in standard form:
$4(x^2 - 4x + 4) + 9(y^2 - 2y + 1) = 11 + 16 + 9$
$\ \ \ \ \ \ \ \ \ \ 4(x - 2)^2 + 9(y - 1)^2 = 36$
$\ \ \ \ \ \ \ \ \ \ \dfrac{(x - 2)^2}{9} + \dfrac{(y - 1)^2}{4} = 1$
$a = 3, \ b = 2.$

Find the value of c:
$c^2 = a^2 - b^2 = 9 - 4 = 5$
$c = \sqrt{5}$
Center: $(2, 1)$
Vertices: $(-1, 1), (5, 1)$
Foci: $\left(2 - \sqrt{5}, 1\right), \left(2 + \sqrt{5}, 1\right)$

17. $4x^2 - 16x + 16y + 32 = 0$
This is a parabola.
Write in standard form:
$4(x^2 - 4x + 4) = -16y - 32 + 16$
$\ \ \ \ \ \ \ 4(x - 2)^2 = -16(y + 1)$
$\ \ \ \ \ \ \ \ (x - 2)^2 = -4(y + 1)$

$a = 1$
Vertex: $(2, -1)$
Focus: $(2, -2)$
Directrix: $y = 0$

19. $9x^2 + 4y^2 - 18x + 8y = 23$
This is an ellipse.
Write in standard form:
$9(x^2 - 2x + 1) + 4(y^2 + 2y + 1) = 23 + 9 + 4$
$$9(x-1)^2 + 4(y+1)^2 = 36$$
$$\frac{(x-1)^2}{4} + \frac{(y+1)^2}{9} = 1$$
$a = 3,\ b = 2$.

Find the value of c:
$c^2 = a^2 - b^2 = 9 - 4 = 5$

$c = \sqrt{5}$
Center: $(1, -1)$
Vertices: $(1, -4), (1, 2)$
Foci: $\left(1, -1 - \sqrt{5}\right), \left(1, -1 + \sqrt{5}\right)$

21. Parabola: The focus is $(-2, 0)$ and the directrix is
$x = 2$. The vertex is $(0, 0)$. $a = 2$ and since $(-2, 0)$
is to the left of $(0, 0)$, the parabola opens to the left.
Write the equation:
$y^2 = -4ax$
$y^2 = -4 \cdot 2 \cdot x$
$y^2 = -8x$

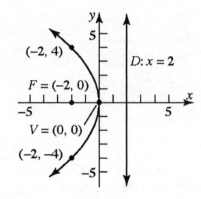

23. Hyperbola: Center: $(0, 0)$;
Focus: $(0, 4)$; Vertex: $(0, -2)$;
Transverse axis is the y-axis;
$a = 2;\ c = 4$.
Find the value of b:
$b^2 = c^2 - a^2 = 16 - 4 = 12$
$b = \sqrt{12} = 2\sqrt{3}$

Write the equation: $\dfrac{y^2}{4} - \dfrac{x^2}{12} = 1$

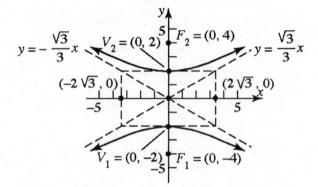

25. Ellipse: Foci: $(-3, 0), (3, 0)$; Vertex: $(4, 0)$;
Center: $(0, 0)$; Major axis is the x-axis;
$a = 4;\ c = 3$.
Find the value of b:
$b^2 = a^2 - c^2 = 16 - 9 = 7$
$b = \sqrt{7}$

Write the equation: $\dfrac{x^2}{16} + \dfrac{y^2}{7} = 1$

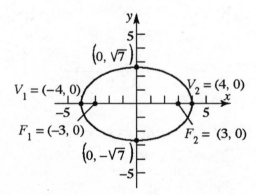

27. Parabola: The focus is $(2, -4)$ and the vertex is $(2, -3)$. Both lie on the vertical line $x = 2$. $a = 1$ and since $(2, -4)$ is below $(2, -3)$, the parabola opens down. Directrix is $y = -2$.
    Write the equation:
    $$(x - 2)^2 = -4 \cdot 1 \cdot (y - (-3))$$
    $$(x - 2)^2 = -4(y + 3)$$

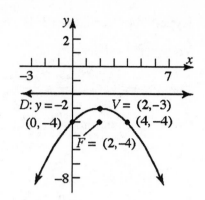

29. Hyperbola: Center: $(-2, -3)$;
    Focus: $(-4, -3)$; Vertex: $(-3, -3)$;
    Transverse axis is parallel to the x-axis;
    $a = 1; \; c = 2$.
    Find the value of b:
    $$b^2 = c^2 - a^2 = 4 - 1 = 3$$
    $$b = \sqrt{3}$$
    Write the equation:
    $$\frac{(x + 2)^2}{1} - \frac{(y + 3)^2}{3} = 1$$

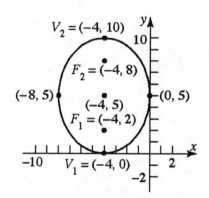

31. Ellipse: Foci: $(-4, 2), (-4, 8)$; Vertex: $(-4, 10)$;
    Center: $(-4, 5)$; Major axis is parallel to the y-axis;
    $a = 5; \; c = 3$.
    Find the value of b:
    $$b^2 = a^2 - c^2 = 25 - 9 = 16$$
    $$b = 4$$
    Write the equation: $\dfrac{(x + 4)^2}{16} + \dfrac{(y - 5)^2}{25} = 1$

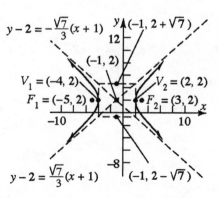

33. Hyperbola: Center: $(-1, 2)$;
    $a = 3; \; c = 4$; Transverse axis parallel to the x-axis;
    Find the value of b:
    $$b^2 = c^2 - a^2 = 16 - 9 = 7$$
    $$b = \sqrt{7}$$
    Write the equation:
    $$\frac{(x + 1)^2}{9} - \frac{(y - 2)^2}{7} = 1$$

35. Hyperbola: Vertices: $(0, 1), (6, 1)$; Asymptote: $3y + 2x = 9$; Center: $(3, 1)$;

Transverse axis is parallel to the x-axis; $a = 3$; The slope of the asymptote is $-\dfrac{2}{3}$;

Find the value of b: $\dfrac{-b}{a} = \dfrac{-b}{3} = \dfrac{-2}{3} \Rightarrow -3b = -6 \Rightarrow b = 2$

Write the equation: $\dfrac{(x-3)^2}{9} - \dfrac{(y-1)^2}{4} = 1$

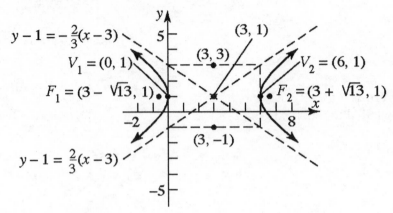

37. $y^2 + 4x + 3y - 8 = 0$
$A = 0$ and $C = 1$; $AC = (0)(1) = 0$. Since $AC = 0$, the equation defines a parabola.

39. $x^2 + 2y^2 + 4x - 8y + 2 = 0$
$A = 1$ and $C = 2$; $AC = (1)(2) = 2$. Since $AC > 0$ and $A \neq C$, the equation defines an ellipse.

41. $9x^2 - 12xy + 4y^2 + 8x + 12y = 0$
$A = 9$, $B = -12$, $C = 4$ $B^2 - 4AC = (-12)^2 - 4(9)(4) = 0$; The equation defines a parabola.

43. $4x^2 + 10xy + 4y^2 - 9 = 0$
$A = 4$, $B = 10$, $C = 4$ $B^2 - 4AC = 10^2 - 4(4)(4) = 36 > 0$; The equation defines a hyperbola.

45. $x^2 - 2xy + 3y^2 + 2x + 4y - 1 = 0$
$A = 1$, $B = -2$, $C = 3$ $B^2 - 4AC = (-2)^2 - 4(1)(3) = -8 < 0$; The equation defines an ellipse.

47. $2x^2 + 5xy + 2y^2 - \dfrac{9}{2} = 0$

$A = 2, B = 5,$ and $C = 2;$ $\cot(2\theta) = \dfrac{A-C}{B} = \dfrac{2-2}{5} = 0 \Rightarrow 2\theta = \dfrac{\pi}{2} \Rightarrow \theta = \dfrac{\pi}{4}$

$x = x'\cos\theta - y'\sin\theta = x'\cos\dfrac{\pi}{4} - y'\sin\dfrac{\pi}{4} = \dfrac{\sqrt{2}}{2}x' - \dfrac{\sqrt{2}}{2}y' = \dfrac{\sqrt{2}}{2}(x' - y')$

$y = x'\sin\theta + y'\cos\theta = x'\sin\dfrac{\pi}{4} + y'\cos\dfrac{\pi}{4} = \dfrac{\sqrt{2}}{2}x' + \dfrac{\sqrt{2}}{2}y' = \dfrac{\sqrt{2}}{2}(x' + y')$

$2\left(\dfrac{\sqrt{2}}{2}(x' - y')\right)^2 + 5\left(\dfrac{\sqrt{2}}{2}(x' - y')\right)\left(\dfrac{\sqrt{2}}{2}(x' + y')\right) + 2\left(\dfrac{\sqrt{2}}{2}(x' + y')\right)^2 - \dfrac{9}{2} = 0$

$\left(x'^2 - 2x'y' + y'^2\right) + \dfrac{5}{2}\left(x'^2 - y'^2\right) + \left(x'^2 + 2x'y' + y'^2\right) - \dfrac{9}{2} = 0$

$\dfrac{9}{2}x'^2 - \dfrac{1}{2}y'^2 = \dfrac{9}{2}$

$9x'^2 - y'^2 = 9$

$\dfrac{x'^2}{1} - \dfrac{y'^2}{9} = 1$

Hyperbola; center $(0, 0)$, transverse axis is the x'-axis, vertices $(\pm 1, 0)$.

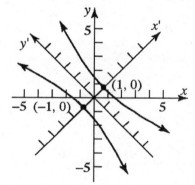

49. $6x^2 + 4xy + 9y^2 - 20 = 0$

$A = 6, B = 4,$ and $C = 9;$ $\cot(2\theta) = \dfrac{A-C}{B} = \dfrac{6-9}{4} = -\dfrac{3}{4} \Rightarrow \cos(2\theta) = -\dfrac{3}{5}$

$\sin\theta = \sqrt{\dfrac{1 - \left(-\dfrac{3}{5}\right)}{2}} = \sqrt{\dfrac{4}{5}} = \dfrac{2\sqrt{5}}{5};$ $\cos\theta = \sqrt{\dfrac{1 + \left(-\dfrac{3}{5}\right)}{2}} = \sqrt{\dfrac{1}{5}} = \dfrac{\sqrt{5}}{5} \Rightarrow \theta \approx 63.4°$

$x = x'\cos\theta - y'\sin\theta = \dfrac{\sqrt{5}}{5}x' - \dfrac{2\sqrt{5}}{5}y' = \dfrac{\sqrt{5}}{5}(x' - 2y')$

$y = x'\sin\theta + y'\cos\theta = \dfrac{2\sqrt{5}}{5}x' + \dfrac{\sqrt{5}}{5}y' = \dfrac{\sqrt{5}}{5}(2x' + y')$

$$6\left(\frac{\sqrt{5}}{5}(x'-2y')\right)^2 + 4\left(\frac{\sqrt{5}}{5}(x'-2y')\right)\left(\frac{\sqrt{5}}{5}(2x'+y')\right) + 9\left(\frac{\sqrt{5}}{5}(2x'+y')\right)^2 - 20 = 0$$

$$\frac{6}{5}\left(x'^2 - 4x'y' + 4y'^2\right) + \frac{4}{5}\left(2x'^2 - 3x'y' - 2y'^2\right) + \frac{9}{5}\left(4x'^2 + 4x'y' + y'^2\right) - 20 = 0$$

$$\frac{6}{5}x'^2 - \frac{24}{5}x'y' + \frac{24}{5}y'^2 + \frac{8}{5}x'^2 - \frac{12}{5}x'y' - \frac{8}{5}y'^2 + \frac{36}{5}x'^2 + \frac{36}{5}x'y' + \frac{9}{5}y'^2 = 20$$

$$10x'^2 + 5y'^2 = 20$$

$$\frac{x'^2}{2} + \frac{y'^2}{4} = 1$$

Ellipse; center at the origin, major axis is the y'-axis, vertices $(0, \pm 2)$.

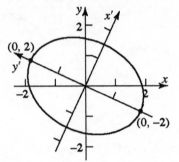

51. $4x^2 - 12xy + 9y^2 + 12x + 8y = 0$

$A = 4, B = -12,$ and $C = 9;$ $\cot(2\theta) = \frac{A-C}{B} = \frac{4-9}{-12} = \frac{5}{12} \Rightarrow \cos(2\theta) = \frac{5}{13}$

$\sin\theta = \sqrt{\dfrac{1 - \dfrac{5}{13}}{2}} = \sqrt{\dfrac{4}{13}} = \dfrac{2\sqrt{13}}{13};$ $\cos\theta = \sqrt{\dfrac{1 + \dfrac{5}{13}}{2}} = \sqrt{\dfrac{9}{13}} = \dfrac{3\sqrt{13}}{13} \Rightarrow \theta \approx 33.7°$

$x = x'\cos\theta - y'\sin\theta = \dfrac{3\sqrt{13}}{13}x' - \dfrac{2\sqrt{13}}{13}y' = \dfrac{\sqrt{13}}{13}(3x' - 2y')$

$y = x'\sin\theta + y'\cos\theta = \dfrac{2\sqrt{13}}{13}x' + \dfrac{3\sqrt{13}}{13}y' = \dfrac{\sqrt{13}}{13}(2x' + 3y')$

$$4\left(\frac{\sqrt{13}}{13}(3x'-2y')\right)^2 - 12\left(\frac{\sqrt{13}}{13}(3x'-2y')\right)\left(\frac{\sqrt{13}}{13}(2x'+3y')\right)$$

$$+9\left(\frac{\sqrt{13}}{13}(2x'+3y')\right)^2 + 12\left(\frac{\sqrt{13}}{13}(3x'-2y')\right) + 8\left(\frac{\sqrt{13}}{13}(2x'+3y')\right) = 0$$

$$\frac{4}{13}\left(9x'^2 - 12x'y' + 4y'^2\right) - \frac{12}{13}\left(6x'^2 + 5x'y' - 6y'^2\right)$$

$$+\frac{9}{13}\left(4x'^2 + 12x'y' + 9y'^2\right) + \frac{36\sqrt{13}}{13}x' - \frac{24\sqrt{13}}{13}y' + \frac{16\sqrt{13}}{13}x' + \frac{24\sqrt{13}}{13}y' = 0$$

$$\frac{36}{13}x'^2 - \frac{48}{13}x'y' + \frac{16}{13}y'^2 - \frac{72}{13}x'^2 - \frac{60}{13}x'y' + \frac{72}{13}y'^2$$

$$+\frac{36}{13}x'^2 + \frac{108}{13}x'y' + \frac{81}{13}y'^2 + 4\sqrt{13}x' = 0$$

$$13y'^2 + 4\sqrt{13}x' = 0$$

$$y'^2 = -\frac{4\sqrt{13}}{13}x'$$

Parabola; vertex at the origin, focus $\left(-\frac{\sqrt{13}}{13}, 0\right)$.

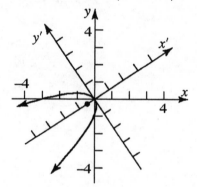

53. $r = \dfrac{4}{1 - \cos\theta}$

$ep = 4,\ \ e = 1,\ \ p = 4$

Parabola; directrix is perpendicular to the polar axis 4 units to the left of the pole; vertex is $(2, \pi)$.

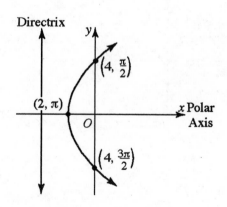

55. $r = \dfrac{6}{2 - \sin\theta} = \dfrac{3}{1 - \dfrac{1}{2}\sin\theta}$

$ep = 3, \ e = \dfrac{1}{2}, \ p = 6$

Ellipse; directrix is parallel to the polar axis, 6 units below the pole; vertices are $\left(6, \dfrac{\pi}{2}\right)$ and $\left(2, \dfrac{3\pi}{2}\right)$.

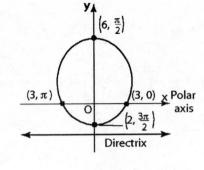

57. $r = \dfrac{8}{4 + 8\cos\theta} = \dfrac{2}{1 + 2\cos\theta}$

$ep = 2, \ e = 2, \ p = 1$

Hyperbola; directrix is perpendicular to the polar axis, 1 unit to the right of the pole; vertices are $\left(\dfrac{2}{3}, 0\right)$ and $(-2, \pi)$.

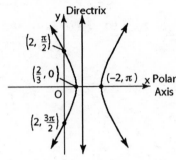

59. $r = \dfrac{4}{1 - \cos\theta}$

$r - r\cos\theta = 4$

$r = 4 + r\cos\theta$

$r^2 = (4 + r\cos\theta)^2$

$x^2 + y^2 = (4 + x)^2$

$x^2 + y^2 = 16 + 8x + x^2$

$y^2 - 8x - 16 = 0$

61. $r = \dfrac{8}{4 + 8\cos\theta}$

$4r + 8r\cos\theta = 8$

$4r = 8 - 8r\cos\theta$

$r = 2 - 2r\cos\theta$

$r^2 = (2 - 2r\cos\theta)^2$

$x^2 + y^2 = (2 - 2x)^2$

$x^2 + y^2 = 4 - 8x + 4x^2$

$3x^2 - y^2 - 8x + 4 = 0$

63. $x = 4t - 2, \ y = 1 - t, \ -\infty < t < \infty$

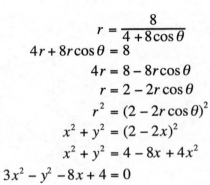

$x = 4(1 - y) - 2$

$x = 4 - 4y - 2$

$x + 4y = 2$

65. $x = 3\sin t$, $y = 4\cos t + 2$, $0 \le t \le 2\pi$

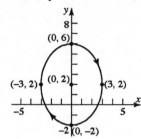

$$\frac{x}{3} = \sin t, \quad \frac{y-2}{4} = \cos t$$

$$\sin^2 t + \cos^2 t = 1$$

$$\left(\frac{x}{3}\right)^2 + \left(\frac{y-2}{4}\right)^2 = 1$$

$$\frac{x^2}{9} + \frac{(y-2)^2}{16} = 1$$

67. $x = \sec^2 t$, $y = \tan^2 t$, $0 \le t \le \dfrac{\pi}{4}$

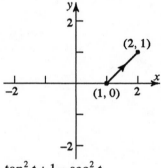

$$\tan^2 t + 1 = \sec^2 t$$

$$y + 1 = x$$

69. Answers will vary.
Two possible answers:
$y = -2x + 4$

$x = t$, $y = -2t + 4$, $-\infty < t < \infty$

or
$2x = -y + 4$

$$x = \frac{-y+4}{2} = \frac{y-4}{-2}$$

so

$$x = \frac{t-4}{-2}, \quad y = t, \quad -\infty < t < \infty$$

71. Since the motion begins at $(4, 0)$, we want $x = 4$ and $y = 0$ when $t = 0$. Furthermore, since the given equation is an ellipse, we begin by letting $\dfrac{x}{4} = \cos(\omega t)$ and $\dfrac{y}{3} = \sin(\omega t)$ for some constant ω. With this choice, when $t = 0$, we have $x = 4$ and $y = 0$. For the motion to be counterclockwise, the motion will have to begin with x decreasing and y increasing as t increases. This requires that $\omega > 0$. Since 1 revolution requires 4 seconds, the period is $\dfrac{2\pi}{\omega} = 4 \Rightarrow 4\omega = 2\pi \Rightarrow \omega = \dfrac{\pi}{2}$. Thus, the desired parametric equations are:

$$x = 4\cos\left(\frac{\pi}{2}t\right), \quad y = 3\sin\left(\frac{\pi}{2}t\right), \quad 0 \le t \le 4.$$

73. Write the equation of the ellipse in standard form:

$$4x^2 + 9y^2 = 36 \Rightarrow \frac{x^2}{9} + \frac{y^2}{4} = 1$$

The center of the ellipse is $(0, 0)$.

The major axis is the x-axis.

$a = 3;\ b = 2;$

$c^2 = a^2 - b^2 = 9 - 4 = 5$

$c = \sqrt{5}$

For the ellipse:

Vertices: $(-3, 0),\ (3, 0)$

Foci: $\left(-\sqrt{5}, 0\right), \left(\sqrt{5}, 0\right)$

For the hyperbola:

Foci: $(-3, 0),\ (3, 0)$

Vertices: $\left(-\sqrt{5}, 0\right), \left(\sqrt{5}, 0\right)$

Center: $(0, 0)$

$a = \sqrt{5};\ c = 3;$

$b^2 = c^2 - a^2 = 9 - 5 = 4$

The equation of the hyperbola is: $\frac{x^2}{5} - \frac{y^2}{4} = 1$.

75. Let (x, y) be any point in the collection of points.

The distance from (x, y) to $(3, 0) = \sqrt{(x - 3)^2 + y^2}$.

The distance from (x, y) to the line $x = \frac{16}{3}$ is $\left| x - \frac{16}{3} \right|$.

Relating the distances, we have:

$$\sqrt{(x - 3)^2 + y^2} = \frac{3}{4} \left| x - \frac{16}{3} \right|$$

$$(x - 3)^2 + y^2 = \frac{9}{16} \left(x - \frac{16}{3} \right)^2$$

$$x^2 - 6x + 9 + y^2 = \frac{9}{16} \left(x^2 - \frac{32}{3}x + \frac{256}{9} \right)$$

$$16x^2 - 96x + 144 + 16y^2 = 9x^2 - 96x + 256$$

$$7x^2 + 16y^2 = 112$$

$$\frac{7x^2}{112} + \frac{16y^2}{112} = 1$$

$$\frac{x^2}{16} + \frac{y^2}{7} = 1$$

The set of points is an ellipse.

77.   Locate the parabola so that the vertex is at $(0, 0)$ and opens up. It then has the equation:
$x^2 = 4ay$. Since the light source is located at the focus and is 1 foot from the base, $a = 1$.
The diameter is 2, so the point $(1, y)$ is located on the parabola.
Solve for y: $1^2 = 4(1)y \Rightarrow 1 = 4y \Rightarrow y = 0.25$ foot
The mirror should be 0.25 foot deep, that is, 3 inches deep.

79.   Place the semielliptical arch so that the x-axis coincides with the major axis and the y-axis
passes through the center of the arch. Since the bridge has a span of 60 feet, the length of
the major axis is 60, or $2a = 60$, or $a = 30$. The maximum height of the bridge is 20 feet,
so $b = 20$. The equation is: $\dfrac{x^2}{900} + \dfrac{y^2}{400} = 1$.
The height 5 feet from the center:
$$\frac{5^2}{900} + \frac{y^2}{400} = 1 \Rightarrow \frac{y^2}{400} = 1 - \frac{25}{900} \Rightarrow y^2 = 400 \cdot \frac{875}{900} \Rightarrow y \approx 19.72 \text{ feet}$$
The height 10 feet from the center:
$$\frac{10^2}{900} + \frac{y^2}{400} = 1 \Rightarrow \frac{y^2}{400} = 1 - \frac{100}{900} \Rightarrow y^2 = 400 \cdot \frac{800}{900} \Rightarrow y \approx 18.86 \text{ feet}$$
The height 20 feet from the center:
$$\frac{20^2}{900} + \frac{y^2}{400} = 1 \Rightarrow \frac{y^2}{400} = 1 - \frac{400}{900} \Rightarrow y^2 = 400 \cdot \frac{500}{900} \Rightarrow y \approx 14.91 \text{ feet}$$

81.   (a)   Set up a coordinate system so that the two stations
lie on the x-axis and the origin is midway between
them. The ship lies on a hyperbola whose foci are
the locations of the two stations. Since the time
difference is 0.00032 second and the speed of the
signal is 186,000 miles per second, the difference
in the distances of the ships from each station is:
distance $= (186{,}000)(0.00032) = 59.52$ miles

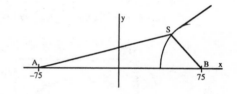

The difference of the distances from the ship to each station, 59.52, equals $2a$, so
$a = 29.76$ and the vertex of the corresponding hyperbola is at $(29.76, 0)$. Since the
focus is at $(75, 0)$, following this hyperbola, the ship would reach shore 45.24 miles
from the master station.

(b)   The ship should follow a hyperbola with a vertex at $(60, 0)$. For this hyperbola, $a = 60$, so
the constant difference of the distances from the ship to each station is 120. The time
difference the ship should look for is: time $= \dfrac{120}{186{,}000} = 0.000645$ second.

(c)   Find the equation of the hyperbola with vertex at $(60, 0)$ and a focus at $(75, 0)$. The
form of the equation of the hyperbola is: $\dfrac{x^2}{a^2} - \dfrac{y^2}{b^2} = 1$ where $a = 60$.
Since $c = 75$ and $b^2 = c^2 - a^2 \Rightarrow b^2 = 75^2 - 60^2 = 2025$.
The equation of the hyperbola is: $\dfrac{x^2}{3600} - \dfrac{y^2}{2025} = 1$.
Since the ship is 20 miles off shore, we have $y = 20$.

Solve the equation for x:

$$\frac{x^2}{3600} - \frac{20^2}{2025} = 1 \Rightarrow \frac{x^2}{3600} = 1 + \frac{400}{2025} = \frac{97}{81} \Rightarrow x^2 = 3600 \cdot \frac{97}{81}$$

$$x \approx 66 \text{ miles}$$

The ship's location is approximately (66, 20).

83. (a) Use equations (2):

$$x = (100\cos 35°)t; \qquad y = -\frac{1}{2}(32)t^2 + (100\sin 35°)t + 6$$

(b) The ball is in the air until $y = 0$. Solve:

$$-16t^2 + (100\sin 35°)t + 6 = 0$$

$$t = \frac{-100\sin 35° \pm \sqrt{(100\sin 35°)^2 - 4(-16)(6)}}{2(-16)} \approx -0.10 \text{ or } 3.69$$

The ball is in the air for about 3.69 seconds. (The negative solution is extraneous.)

(c) The maximum height occurs at the vertex of the quadratic function.

$$t = -\frac{b}{2a} = -\frac{100\sin 35°}{2(-16)} \approx 1.79 \text{ seconds}$$

Evaluate the function to find the maximum height:

$$-16(1.79)^2 + 100(\sin 35°)(1.79) + 6 \approx 57.4 \text{ feet}$$

(d) Find the horizontal displacement:

$$x = 100(\cos 35°)(3.69) \approx 302 \text{ feet}$$

(e)

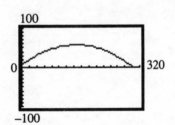

Chapter 9

Analytic Geometry

9.CR Cumulative Review

1. $\sin(2\theta) = 0.5$

$$2\theta = \frac{\pi}{6} + 2k\pi \quad \Rightarrow \quad \theta = \frac{\pi}{12} + k\pi$$

or $2\theta = \frac{5\pi}{6} + 2k\pi \quad \Rightarrow \quad \theta = \frac{5\pi}{12} + k\pi$,

where k is any integer.

3. Using rectangular coordinates, the circle with center point (0,4) and radius 4 has the equation:

$$(x-h)^2 + (y-k)^2 = r^2$$

$$(x-0)^2 + (y-4)^2 = 4^2$$

$$x^2 + y^2 - 8y + 16 = 16$$

$$x^2 + y^2 - 8y = 0$$

Converting to polar coordinates:

$r^2 - 8r\sin\theta = 0$

$$r^2 = 8r\sin\theta$$

$$r = 8\sin\theta$$

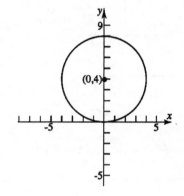

5.

$$\frac{f(x+h) - f(x)}{h} = \frac{-3(x+h)^2 + 5(x+h) - 2 - \left(-3x^2 + 5x - 2\right)}{h}$$

$$= \frac{-3\left(x^2 + 2xh + h^2\right) + 5x + 5h - 2 + 3x^2 - 5x + 2}{h}$$

$$= \frac{-3x^2 - 6xh - 3h^2 + 5x + 5h - 2 + 3x^2 - 5x + 2}{h}$$

$$= \frac{-6xh - 3h^2 + 5h}{h} = \frac{h(-6x - 3h + 5)}{h} = -6x - 3h + 5$$

7. $9x^4 + 33x^3 - 71x^2 - 57x - 10 = 0$

There are at most 4 real zeros.

Possible rational zeros:

$p = \pm 1, \pm 2, \pm 5, \pm 10; \quad q = \pm 1, \pm 3, \pm 9;$

$\dfrac{p}{q} = \pm 1, \pm \dfrac{1}{3}, \pm \dfrac{1}{9}, \pm 2, \pm \dfrac{2}{3}, \pm \dfrac{2}{9}, \pm 5, \pm \dfrac{5}{3}, \pm \dfrac{5}{9}, \pm 10, \pm \dfrac{10}{3}, \pm \dfrac{10}{9}$

Graphing $y_1 = 9x^4 + 33x^3 - 71x^2 - 57x - 10$ indicates that there appear to be zeros at $x = -5$ and at $x = 2$.

Using synthetic division with $x = -5$:

$$
\begin{array}{r|rrrrr}
-5 & 9 & 33 & -71 & -57 & -10 \\
   &   & -45 & 60 & 55 & 10 \\
\hline
   & 9 & -12 & -11 & -2 & 0
\end{array}
$$

Since the remainder is 0, -5 is a zero for f. So $x - (-5) = x + 5$ is a factor.

The other factor is the quotient: $9x^3 - 12x^2 - 11x - 2$.

Thus, $f(x) = (x + 5)\left(9x^3 - 12x^2 - 11x - 2\right)$.

Using synthetic division on the quotient and $x = 2$:

$$
\begin{array}{r|rrrr}
2 & 9 & -12 & -11 & -2 \\
  &   & 18 & 12 & 2 \\
\hline
  & 9 & 6 & 1 & 0
\end{array}
$$

Since the remainder is 0, 2 is a zero for f. So $x - 2$ is a factor

The other factor is the quotient: $9x^2 + 6x + 1$.

Thus, $f(x) = (x + 5)(x - 2)\left(9x^2 + 6x + 1\right) = (x + 5)(x - 2)(3x + 1)(3x + 1)$.

Therefore, $x = -\dfrac{1}{3}$ is also a zero for f (with multiplicity 2).

Solution set: $\left\{-5, -\dfrac{1}{3}, 2\right\}$.

9. $\cot(2\theta) = 1$, where $0° < \theta < 90°$

$2\theta = \dfrac{\pi}{4} + k\pi \Rightarrow \theta = \dfrac{\pi}{8} + \dfrac{k\pi}{2}$, where k is any integer.

On the interval $0° < \theta < 90°$, the solution is $\theta = \dfrac{\pi}{8} = 22.5°$.

11. $f(x) = \log_4(x - 2)$

(a) $f(x) = \log_4(x - 2) = 2$

      $x - 2 = 4^2$

      $x - 2 = 16$

        $x = 18$

      The solution set is $\{18\}$.

(b) $f(x) = \log_4(x - 2) \le 2$

      $x - 2 \le 4^2$ and $x - 2 > 0$

      $x - 2 \le 16$ and $x > 2$

      $x \le 18$ and $x > 2$

         $2 < x \le 18$

          $(2, 18]$

Chapter 10

Systems of Equations and Inequalities

10.1 Systems of Linear Equations: Substitution and Elimination

1. $3x + 4 = 8 - x$

    $4x = 4$

    $x = 1$

    The solution set is $\{1\}$.

3. inconsistent

5. False

7. Substituting the values of the variables:
    $$\begin{cases} 2x - y = 5 & \Rightarrow & 2(2) - (-1) = 4 + 1 = 5 \\ 5x + 2y = 8 & \Rightarrow & 5(2) + 2(-1) = 10 - 2 = 8 \end{cases}$$
    Each equation is satisfied, so $x = 2$, $y = -1$ is a solution of the system of equations.

9. Substituting the values of the variables:
    $$\begin{cases} 3x - 4y = 4 & \Rightarrow & 3(2) - 4\left(\dfrac{1}{2}\right) = 6 - 2 = 4 \\ \dfrac{1}{2}x - 3y = -\dfrac{1}{2} & \Rightarrow & \dfrac{1}{2}(2) - 3\left(\dfrac{1}{2}\right) = 1 - \dfrac{3}{2} = -\dfrac{1}{2} \end{cases}$$
    Each equation is satisfied, so $x = 2$, $y = \dfrac{1}{2}$ is a solution of the system of equations.

11. Substituting the values of the variables:
    $$\begin{cases} x - y = 3 & \Rightarrow & 4 - 1 = 3 \\ \dfrac{1}{2}x + y = 3 & \Rightarrow & \dfrac{1}{2}(4) + 1 = 2 + 1 = 3 \end{cases}$$
    Each equation is satisfied, so $x = 4$, $y = 1$ is a solution of the system of equations.

13. Substituting the values of the variables:
    $$\begin{cases} 3x + 3y + 2z = 4 & \Rightarrow & 3(1) + 3(-1) + 2(2) = 3 - 3 + 4 = 4 \\ x - y - z = 0 & \Rightarrow & 1 - (-1) - 2 = 1 + 1 - 2 = 0 \\ 2y - 3z = -8 & \Rightarrow & 2(-1) - 3(2) = -2 - 6 = -8 \end{cases}$$
    Each equation is satisfied, so $x = 1$, $y = -1$, $z = 2$ is a solution of the system of equations.

15.   Substituting the values of the variables:
$$\begin{cases} 3x + 3y + 2z = 4 & \Rightarrow & 3(2) + 3(-2) + 2(2) = 6 - 6 + 4 = 4 \\ x - 3y + z = 10 & \Rightarrow & 2 - 3(-2) + 2 = 2 + 6 + 2 = 10 \\ 5x - 2y - 3z = 8 & \Rightarrow & 5(2) - 2(-2) - 3(2) = 10 + 4 - 6 = 8 \end{cases}$$
Each equation is satisfied, so $x = 2, y = -2, z = 2$ is a solution of the system of equations.

17.   Solve the first equation for y, substitute into the second equation and solve:
$$\begin{cases} x + y = 8 & \Rightarrow & y = 8 - x \\ x - y = 4 \end{cases}$$
$$x - (8 - x) = 4 \Rightarrow x - 8 + x = 4 \Rightarrow 2x = 12 \Rightarrow x = 6$$
Since $x = 6, \quad y = 8 - 6 = 2$
The solution of the system is $x = 6, \ y = 2$.

19.   Multiply each side of the first equation by 3 and add the equations to eliminate y:
$$\begin{cases} 5x - y = 13 & \overset{3}{\longrightarrow} & 15x - 3y = 39 \\ 2x + 3y = 12 & \longrightarrow & \underline{2x + 3y = 12} \end{cases}$$
$$17x \qquad = 51 \Rightarrow x = 3$$
Substitute and solve for y:
$$5(3) - y = 13 \Rightarrow 15 - y = 13 \Rightarrow -y = -2 \Rightarrow y = 2$$
The solution of the system is $x = 3, \ y = 2$.

21.   Solve the first equation for x and substitute into the second equation:
$$\begin{cases} 3x \qquad = 24 & \Rightarrow & x = 8 \\ x + 2y = 0 \end{cases}$$
$$8 + 2y = 0 \Rightarrow 2y = -8 \Rightarrow y = -4$$
The solution of the system is $x = 8, \ y = -4$.

23.   Multiply each side of the first equation by 2 and each side of the second equation by 3, then add to eliminate y:
$$\begin{cases} 3x - 6y = 2 & \overset{2}{\longrightarrow} & 6x - 12y = 4 \\ 5x + 4y = 1 & \overset{3}{\longrightarrow} & \underline{15x + 12y = 3} \end{cases}$$
$$21x \qquad = 7 \Rightarrow x = \frac{1}{3}$$
Substitute and solve for y: $3(1/3) - 6y = 2 \Rightarrow 1 - 6y = 2 \Rightarrow -6y = 1 \Rightarrow y = -\frac{1}{6}$

The solution of the system is $x = \frac{1}{3}, \ y = -\frac{1}{6}$.

25. Solve the first equation for y, substitute into the second equation and solve:

$$\begin{cases} 2x + y = 1 & \Rightarrow \quad y = 1 - 2x \\ 4x + 2y = 3 \end{cases}$$

$$4x + 2(1 - 2x) = 3 \Rightarrow 4x + 2 - 4x = 3 \Rightarrow 0x = 1$$

This has no solution, so the system is inconsistent.

27. Solve the first equation for y, substitute into the second equation and solve:

$$\begin{cases} 2x - y = 0 & \Rightarrow \quad y = 2x \\ 3x + 2y = 7 \end{cases}$$

$$3x + 2(2x) = 7 \Rightarrow 3x + 4x = 7 \Rightarrow 7x = 7 \Rightarrow x = 1$$

Since $x = 1$, $y = 2(1) = 2$

The solution of the system is $x = 1$, $y = 2$.

29. Solve the first equation for x, substitute into the second equation and solve:

$$\begin{cases} x + 2y = 4 & \Rightarrow \quad x = 4 - 2y \\ 2x + 4y = 8 \end{cases}$$

$$2(4 - 2y) + 4y = 8 \Rightarrow 8 - 4y + 4y = 8 \Rightarrow 0y = 0$$

These equations are dependent. Any real number is a solution for y.

The solution of the system is $x = 4 - 2y$, where y is any real number.

31. Multiply each side of the first equation by -5, and add the equations to eliminate x:

$$\begin{cases} 2x - 3y = -1 & \xrightarrow{\ -5\ } & -10x + 15y = 5 \\ 10x + y = 11 & \longrightarrow & \underline{10x + \quad y = 11} \end{cases}$$

$$16y = 16 \Rightarrow y = 1$$

Substitute and solve for x:

$$2x - 3(1) = -1 \Rightarrow 2x - 3 = -1 \Rightarrow 2x = 2 \Rightarrow x = 1$$

The solution of the system is $x = 1$, $y = 1$.

33. Solve the second equation for x, substitute into the first equation and solve:

$$\begin{cases} 2x + 3y = 6 \\ x - y = \dfrac{1}{2} & \Rightarrow \quad x = y + \dfrac{1}{2} \end{cases}$$

$$2\left(y + \frac{1}{2}\right) + 3y = 6 \Rightarrow 2y + 1 + 3y = 6 \Rightarrow 5y = 5 \Rightarrow y = 1$$

Since $y = 1$, $x = 1 + \dfrac{1}{2} = \dfrac{3}{2}$.

The solution of the system is $x = \dfrac{3}{2}$, $y = 1$.

35. Multiply each side of the first equation by –6 and each side of the second equation by 12, then add to eliminate x:

$$\begin{cases} \dfrac{1}{2}x + \dfrac{1}{3}y = 3 \\[2mm] \dfrac{1}{4}x - \dfrac{2}{3}y = -1 \end{cases} \quad \begin{aligned} \xrightarrow{-6} \quad & -3x - 2y = -18 \\ \xrightarrow{12} \quad & \underline{3x - 8y = -12} \end{aligned}$$

$$-10y = -30 \Rightarrow y = 3$$

Substitute and solve for x:

$$\frac{1}{2}x + \frac{1}{3}(3) = 3 \Rightarrow \frac{1}{2}x + 1 = 3 \Rightarrow \frac{1}{2}x = 2 \Rightarrow x = 4$$

The solution of the system is $x = 4,\ y = 3$.

37. Add the equations to eliminate y:

$$\begin{cases} 3x - 5y = 3 \\ 15x + 5y = 21 \end{cases}$$

$$\overline{18x \qquad = 24} \Rightarrow x = \frac{4}{3}$$

Substitute and solve for y: $\quad 3\left(\dfrac{4}{3}\right) - 5y = 3 \Rightarrow 4 - 5y = 3 \Rightarrow -5y = -1 \Rightarrow y = \dfrac{1}{5}$

The solution of the system is $x = \dfrac{4}{3},\ y = \dfrac{1}{5}$.

39. Rewrite letting $u = \dfrac{1}{x},\ v = \dfrac{1}{y}$:

$$\begin{cases} \dfrac{1}{x} + \dfrac{1}{y} = 8 \\[2mm] \dfrac{3}{x} - \dfrac{5}{y} = 0 \end{cases} \begin{aligned} \Rightarrow \quad & u + v = 8 \\ \Rightarrow \quad & 3u - 5v = 0 \end{aligned}$$

Solve the first equation for u, substitute into the second equation and solve:

$$\begin{cases} u + v = 8 \quad \Rightarrow \quad u = 8 - v \\ 3u - 5v = 0 \end{cases}$$

$$3(8 - v) - 5v = 0 \Rightarrow 24 - 3v - 5v = 0 \Rightarrow -8v = -24 \Rightarrow v = 3$$

Since $v = 3,\ u = 8 - 3 = 5$. Thus, $x = \dfrac{1}{u} = \dfrac{1}{5},\ y = \dfrac{1}{v} = \dfrac{1}{3}$.

The solution of the system is $x = \dfrac{1}{5},\ y = \dfrac{1}{3}$.

41. Multiply each side of the first equation by -2 and add to the second equation to eliminate x:

$$\begin{cases} x- \ y= 6 \\ 2x-3z=16 \\ 2y+ \ z=4 \end{cases} \xrightarrow{-2} \quad \begin{array}{l} -2x+2y \quad\quad =-12 \\ \underline{2x \quad\quad -3z= \ \ 16} \\ 2y-3z= \ \ 4 \end{array}$$

Multiply each side of the result by -1 and add to the original third equation to eliminate y:

$$\begin{array}{l} 2y-3z=4 \xrightarrow{-1} \quad -2y+3z=-4 \\ 2y+ \ z=4 \longrightarrow \quad \underline{2y+ \ z= \ \ 4} \\ 4z= \ \ 0 \\ z=0 \end{array}$$

Substituting and solving for the other variables:

$$\begin{array}{ll} 2y+0=4 & 2x-3(0)=16 \\ 2y=4 & 2x=16 \\ y=2 & x=8 \end{array}$$

The solution is $x=8, \ y=2, \ z=0$.

43. Multiply each side of the first equation by -2 and add to the second equation to eliminate x; and multiply each side of the first equation by 3 and add to the third equation to eliminate x:

$$\begin{cases} x-2y+3z= \ \ 7 \\ 2x+ \ y+ \ z= \ \ 4 \\ -3x+2y-2z=-10 \end{cases} \xrightarrow{-2} \quad \begin{array}{l} -2x+4y-6z=-14 \\ \underline{2x+ \ y+ \ z= \ \ 4} \\ 5y-5z=-10 \xrightarrow{1/5} \ y-z=-2 \end{array}$$

$$\xrightarrow{3} \quad \begin{array}{l} 3x-6y+9z= \ \ 21 \\ \underline{-3x+2y-2z=-10} \\ -4y+7z= \ \ 11 \end{array}$$

Multiply each side of the first result by 4 and add to the second result to eliminate y:

$$\begin{array}{l} y- \ z=-2 \xrightarrow{4} \quad 4y-4z=-8 \\ -4y+7z=11 \longrightarrow \quad \underline{-4y+7z= \ 11} \\ 3z= \ \ 3 \Rightarrow z=1 \end{array}$$

Substituting and solving for the other variables:

$$\begin{array}{ll} y-1=-2 & x-2(-1)+3(1)=7 \\ y=-1 & x+2+3=7 \Rightarrow x=2 \end{array}$$

The solution is $x=2, \ y=-1, \ z=1$.

45. Add the first and second equations to eliminate z:

$$\begin{cases} x - y - z = 1 \\ 2x + 3y + z = 2 \\ 3x + 2y \quad = 0 \end{cases} \Rightarrow \begin{array}{l} x - y - z = 1 \\ \underline{2x + 3y + z = 2} \\ 3x + 2y \quad = 3 \end{array}$$

Multiply each side of the result by -1 and add to the original third equation to eliminate y:

$$\begin{array}{lcl} 3x + 2y = 3 & \xrightarrow{-1} & -3x - 2y = -3 \\ 3x + 2y = 0 & \longrightarrow & \underline{3x + 2y = 0} \\ & & 0 = -3 \end{array}$$

This result has no solution, so the system is inconsistent.

47. Add the first and second equations to eliminate x; multiply the first equation by -3 and add to the third equation to eliminate x:

$$\begin{cases} x - y - z = 1 \\ -x + 2y - 3z = -4 \\ 3x - 2y - 7z = 0 \end{cases} \Rightarrow \begin{array}{l} x - y - z = 1 \\ \underline{-x + 2y - 3z = -4} \\ y - 4z = -3 \\ \xrightarrow{-3} \quad -3x + 3y + 3z = -3 \\ \longrightarrow \quad \underline{3x - 2y - 7z = 0} \\ y - 4z = -3 \end{array}$$

Multiply each side of the first result by -1 and add to the second result to eliminate y:

$$\begin{array}{lcl} y - 4z = -3 & \xrightarrow{-1} & -y + 4z = 3 \\ y - 4z = -3 & \longrightarrow & \underline{y - 4z = -3} \\ & & 0 = 0 \end{array}$$

The system is dependent. If z is any real number, then $y = 4z - 3$.
Solving for x in terms of z in the first equation:

$$\begin{aligned} x - (4z - 3) - z &= 1 \\ x - 4z + 3 - z &= 1 \\ x - 5z + 3 &= 1 \\ x &= 5z - 2 \end{aligned}$$

The solution is $x = 5z - 2$, $y = 4z - 3$, z is any real number.

49. Multiply the first equation by -2 and add to the second equation to eliminate x; add the first and third equations to eliminate x:

$$\begin{cases} 2x - 2y + 3z = 6 \\ 4x - 3y + 2z = 0 \\ -2x + 3y - 7z = 1 \end{cases} \begin{array}{l} \xrightarrow{-2} \quad -4x + 4y - 6z = -12 \\ \longrightarrow \quad \underline{4x - 3y + 2z = 0} \\ y - 4z = -12 \\ \Rightarrow \quad 2x - 2y + 3z = 6 \\ \Rightarrow \quad \underline{-2x + 3y - 7z = 1} \\ y - 4z = 7 \end{array}$$

Multiply each side of the first result by –1 and add to the second result to eliminate y:

$$y - 4z = -12 \quad \xrightarrow{-1} \quad -y + 4z = 12$$
$$y - 4z = \quad 7 \quad \longrightarrow \quad \underline{y - 4z = 7}$$
$$0 = 19$$

This result has no solution, so the system is inconsistent.

51. Add the first and second equations to eliminate z; multiply the second equation by 2 and add to the third equation to eliminate z:

$$\begin{cases} x + y - z = 6 \\ 3x - 2y + z = -5 \\ x + 3y - 2z = 14 \end{cases}$$

$$\begin{aligned} \Rightarrow \quad & x + y - z = 6 \\ \Rightarrow \quad & \underline{3x - 2y + z = -5} \\ & 4x - y \quad\quad = 1 \\ \xrightarrow{2} \quad & 6x - 4y + 2z = -10 \\ \longrightarrow \quad & \underline{x + 3y - 2z = \quad 14} \\ & 7x - y \quad\quad = 4 \end{aligned}$$

Multiply each side of the first result by –1 and add to the second result to eliminate y:

$$4x - y = 1 \quad \xrightarrow{-1} \quad -4x + y = -1$$
$$7x - y = 4 \quad \longrightarrow \quad \underline{7x - y = \quad 4}$$
$$3x \quad\quad = 3$$
$$x \quad\quad = 1$$

Substituting and solving for the other variables:

$$4(1) - y = 1 \quad\quad\quad 3(1) - 2(3) + z = -5$$
$$-y = -3 \quad\quad\quad\quad 3 - 6 + z = -5$$
$$y = 3 \quad\quad\quad\quad\quad z = -2$$

The solution is $x = 1$, $y = 3$, $z = -2$.

53. Add the first and second equations to eliminate z; multiply the second equation by 3 and add to the third equation to eliminate z:

$$\begin{cases} x + 2y - z = -3 \\ 2x - 4y + z = -7 \\ -2x + 2y - 3z = 4 \end{cases}$$

$$\begin{aligned} \Rightarrow \quad & x + 2y - z = -3 \\ \Rightarrow \quad & \underline{2x - 4y + z = -7} \\ & 3x - 2y \quad\quad = -10 \\ \xrightarrow{3} \quad & 6x - 12y + 3z = -21 \\ \longrightarrow \quad & \underline{-2x + 2y - 3z = \quad 4} \\ & 4x - 10y \quad\quad = -17 \end{aligned}$$

Multiply each side of the first result by –5 and add to the second result to eliminate y:

$$3x - 2y = -10 \quad \xrightarrow{-5} \quad -15x + 10y = 50$$
$$4x - 10y = -17 \quad \longrightarrow \quad \underline{4x - 10y = -17}$$
$$-11x \quad\quad = 33 \Rightarrow x = -3$$

Substituting and solving for the other variables:

$$3(-3) - 2y = -10$$
$$-9 - 2y = -10$$
$$-2y = -1$$
$$y = \frac{1}{2}$$

$$-3 + 2\left(\frac{1}{2}\right) - z = -3$$
$$-3 + 1 - z = -3$$
$$-z = -1$$
$$z = 1$$

The solution is $x = -3$, $y = \frac{1}{2}$, $z = 1$.

55. Let l be the length of the rectangle and w be the width of the rectangle. Then:
$l = 2w$ and $2l + 2w = 90$
Solve by substitution:

$$2(2w) + 2w = 90 \Rightarrow 4w + 2w = 90 \Rightarrow 6w = 90 \Rightarrow w = 15 \text{ feet}$$
$$l = 2(15) = 30 \text{ feet}$$

The dimensions of the floor are 15 feet by 30 feet.

57. Let x = the cost of one cheeseburger and y = the cost of one shake. Then:
$4x + 2y = 790$ and $2y = x + 15$
Solve by substitution:

$$4x + x + 15 = 790 \Rightarrow 5x = 775 \Rightarrow x = 155$$
$$2y = 155 + 15 \Rightarrow 2y = 170 \Rightarrow y = 85$$

A cheeseburger cost \$1.55 and a shake costs \$0.85.

59. Let x = the number of pounds of cashews.
Then $x + 30$ is the number of pounds in the mixture.
The value of the cashews is $5x$.
The value of the peanuts is $1.50(30) = 45$.
The value of the mixture is $3(x + 30)$.
Setting up a value equation:

$$5x + 45 = 3(x + 30) \Rightarrow 5x + 45 = 3x + 90$$
$$2x = 45 \Rightarrow x = 22.5$$

22.5 pounds of cashews should be used in the mixture.

61. Let x = the plane's air speed and y = the wind speed.

	Rate	Time	Distance
With Wind	$x + y$	3	600
Against	$x - y$	4	600

$$(x + y)(3) = 600 \quad \Rightarrow \quad x + y = 200$$
$$(x - y)(4) = 600 \quad \Rightarrow \quad x - y = 150$$

Solving by elimination:

$$2x = 350 \Rightarrow x = 175$$
$$y = 200 - x = 200 - 175 = 25$$

The airspeed of the plane is 175 mph, and the wind speed is 25 mph.

63. Let x = the number of $25-design.
     Let y = the number of $45-design.
     Then $x + y$ = the total number of sets of dishes.
     $25x + 45y$ = the cost of the dishes.
     Setting up the equations and solving by substitution:
     $$\begin{cases} x + y = 200 \implies y = 200 - x \\ 25x + 45y = 7400 \end{cases}$$
     $$25x + 45(200 - x) = 7400$$
     $$25x + 9000 - 45x = 7400$$
     $$-20x = -1600 \implies x = 80$$
     $$y = 200 - 80 = 120$$
     Thus, 80 sets of the $25 dishes and 120 sets of the $45 dishes should be ordered.

65. Let x = the cost per package of bacon.
     Let y = the cost of a carton of eggs.
     Set up a system of equations for the problem:
     $$\begin{cases} 3x + 2y = 7.45 \\ 2x + 3y = 6.45 \end{cases}$$
     Multiply each side of the first equation by 3 and each side of the second equation by –2 and solve by elimination:
     $$\begin{cases} 3x + 2y = 7.45 & \xrightarrow{\ 3\ } & 9x + 6y = 22.35 \\ 2x + 3y = 6.45 & \xrightarrow{\ -2\ } & -4x - 6y = -12.90 \end{cases}$$
     $$5x \quad = \quad 9.45 \implies x = 1.89$$
     Substitute and solve for y:
     $$3(1.89) + 2y = 7.45$$
     $$5.67 + 2y = 7.45 \implies 2y = 1.78 \implies y = 0.89$$
     A package of bacon costs $1.89 and a carton of eggs cost $0.89.
     The refund for 2 packages of bacon and 2 cartons of eggs will
     be 2($1.89) + 2($0.89) =$5.56.

67. Let x = the # of mg of compound 1.
     Let y = the # of mg of compound 2.
     Setting up the equations and solving by substitution:
     $$\begin{cases} 0.2x + 0.4y = 40 & \text{vitamin C} \\ 0.3x + 0.2y = 30 & \text{vitamin D} \end{cases}$$
     Multiplying each equation by 10 yields
     $$\begin{cases} 2x + 4y = 400 & \longrightarrow & 2x + 4y = 400 \\ 3x + 2y = 300 & \xrightarrow{\ 2\ } & 6x + 4y = 600 \end{cases}$$

Subtracting the bottom equation from the top equation yields

$$2x + 4y - (6x + 4y) = -200$$

$$2x - 6x = -200$$

$$-4x = -200 \Rightarrow x = 50$$

$$\Rightarrow 2(50) + 4y = 400 \Rightarrow 100 + 4y = 400$$

$$\Rightarrow 4y = 300 \Rightarrow y = \frac{300}{4} = 75$$

So 50 mg of compound 1 should be mixed with 75 mg of compound 2.

69. $y = ax^2 + bx + c$

At $(-1, 4)$ the equation becomes:

$$4 = a(-1)^2 + b(-1) + c$$

$$4 = a - b + c \Rightarrow a - b + c = 4$$

At $(2, 3)$ the equation becomes:

$$3 = a(2)^2 + b(2) + c$$

$$3 = 4a + 2b + c \Rightarrow 4a + 2b + c = 3$$

At $(0, 1)$ the equation becomes:

$$1 = a(0)^2 + b(0) + c$$

$$c = 1$$

The system of equations is:

$$\begin{cases} a - b + c = 4 \\ 4a + 2b + c = 3 \\ c = 1 \end{cases}$$

Substitute $c = 1$ into the first and second equations and simplify:

$$\begin{cases} a - b + 1 = 4 & \Rightarrow & a - b = 3 & \Rightarrow & a = b + 3 \\ 4a + 2b + 1 = 3 & \Rightarrow & 4a + 2b = 2 \end{cases}$$

Solve the first equation for a, substitute into the second equation and solve:

$$4(b + 3) + 2b = 2$$

$$4b + 12 + 2b = 2 \Rightarrow 6b = -10 \Rightarrow b = -\frac{5}{3}$$

$$a = -\frac{5}{3} + 3 = \frac{4}{3}$$

The solution is $a = \frac{4}{3}$, $b = -\frac{5}{3}$, $c = 1$.

The equation is $y = \frac{4}{3}x^2 - \frac{5}{3}x + 1$.

71. Substitute the expression for I_2 into the second and third equations and simplify:

$$\begin{cases} I_2 = I_1 + I_3 \\ 5 - 3I_1 - 5I_2 = 0 \\ 10 - 5I_2 - 7I_3 = 0 \end{cases}$$

$$5 - 3I_1 - 5(I_1 + I_3) = 0 \implies -8I_1 - 5I_3 = -5$$
$$10 - 5(I_1 + I_3) - 7I_3 = 0 \implies -5I_1 - 12I_3 = -10$$

Multiply both sides of the second equation by 5 and multiply both sides of the third equation by -8 to eliminate I_1:

$$-8I_1 - 5I_3 = -5 \xrightarrow{\ 5\ } -40I_1 - 25I_3 = -25$$
$$-5I_1 - 12I_3 = -10 \xrightarrow{\ -8\ } \underline{\quad 40I_1 + 96I_3 = \ \ 80\quad}$$
$$71I_3 = \ \ 55 \implies I_3 = \frac{55}{71}$$

Substituting and solving for the other variables:

$$-8I_1 - 5\left(\frac{55}{71}\right) = -5 \qquad\qquad I_2 = \frac{10}{71} + \frac{55}{71}$$
$$-8I_1 - \frac{275}{71} = -5 \qquad\qquad I_2 = \frac{65}{71}$$
$$-8I_1 = -\frac{80}{71}$$
$$I_1 = \frac{10}{71}$$

The solution is $I_1 = \dfrac{10}{71}$, $I_2 = \dfrac{65}{71}$, $I_3 = \dfrac{55}{71}$.

73. Let x = the number of orchestra seats.
Let y = the number of main seats.
Let z = the number of balcony seats.
Since the total number of seats is 500, $x + y + z = 500$.
Since the total revenue is \$17,100 if all seats are sold, $50x + 35y + 25z = 17,100$.
If only half of the orchestra seats are sold, the revenue is \$14,600.
So, $50\left(\dfrac{1}{2}x\right) + 35y + 25z = 14,600$.

Multiply each side of the first equation by -25 and add to the second equation to eliminate z; multiply each side of the third equation by -1 and add to the second equation to eliminate z:

$$\begin{cases} x + y + z = 500 \\ 50x + 35y + 25z = 17,100 \\ 25x + 35y + 25z = 14,600 \end{cases}$$

$$x + y + z = 500 \xrightarrow{\ -25\ } -25x - 25y - 25z = -12,500$$
$$50x + 35y + 25z = 17,100 \xrightarrow{\quad} \underline{\quad 50x + 35y + 25z = \ \ 17,100\quad}$$
$$25x + 10y \qquad\quad = \ \ 4600$$

$$\xrightarrow{\quad} 50x + 35y + 25z = \ \ 17,100$$
$$\xrightarrow{\ -1\ } \underline{\quad -25x - 35y - 25z = -14,600\quad}$$
$$25x \qquad\qquad\quad = \ \ 2500 \implies x = \ \ 100$$

Substituting and solving for the other variables:

$$25(100) + 10y = 4600 \qquad 100 + 210 + z = 500$$
$$2500 + 10y = 4600 \qquad 310 + z = 500$$
$$10y = 2100 \qquad z = 190$$
$$y = 210$$

There are 100 orchestra seats, 210 main seats, and 190 balcony seats.

75. Let x = the number of servings of chicken.
Let y = the number of servings of corn.
Let z = the number of servings of 2% milk.
Protein equation: $30x + 3y + 9z = 66$
Carbohydrate equation: $35x + 16y + 13z = 94.5$
Calcium equation: $200x + 10y + 300z = 910$
Multiply each side of the first equation by –16 and multiply each side of the second
equation by 3 and add them to eliminate y; multiply each side of the second equation
by –5 and multiply each side of the third equation by 8 and add to eliminate y:

$$\begin{cases} 30x + 3y + 9z = 66 \\ 35x + 16y + 13z = 94.5 \\ 200x + 10y + 300z = 910 \end{cases}$$

$$\xrightarrow{-16} \quad -480x - 48y - 144z = -1056$$
$$\xrightarrow{3} \quad \underline{105x + 48y + 39z = 283.5}$$
$$-375x \qquad -105z = -772.5$$

$$\xrightarrow{-5} \quad -175x - 80y - 65z = -472.5$$
$$\xrightarrow{8} \quad \underline{1600x + 80y + 2400z = 7280}$$
$$1425x \qquad + 2335z = 6807.5$$

Multiply each side of the first result by 19 and multiply each side of the second result by 5
to eliminate x:

$$-375x - 105z = -772.5 \quad \xrightarrow{19} \quad -7125x - 1995z = -14{,}677.5$$
$$1425x + 2335z = 6807.5 \quad \xrightarrow{5} \quad \underline{7125x + 11{,}675z = 34{,}037.5}$$
$$9680z = 19{,}360 \Rightarrow z = 2$$

Substituting and solving for the other variables:

$$-375x - 105(2) = -772.5 \qquad 30(1.5) + 3y + 9(2) = 66$$
$$-375x - 210 = -772.5 \qquad 45 + 3y + 18 = 66$$
$$-375x = -562.5 \qquad 3y = 3$$
$$x = 1.5 \qquad y = 1$$

The dietitian should serve 1.5 servings of chicken, 1 serving of corn, and 2 servings of 2%
milk.

77. Let x = the price of 1 hamburger.
Let y = the price of 1 order of fries.
Let z = the price of 1 drink.
We can construct the system

$$\begin{cases} 8x + 6y + 6z = 26.10 \\ 10x + 6y + 8z = 31.60 \end{cases}$$

A system involving only 2 equations that contain 3 or more unknowns cannot be solved
uniquely.

Multiply the first equation by $-\dfrac{1}{2}$ and the second equation by $\dfrac{1}{2}$, then add to eliminate y:

$$\begin{cases} 8x + 6y + 6z = 26.10 \xrightarrow{\;-1/2\;} -4x - 3y - 3z = -13.05 \\ 10x + 6y + 8z = 31.60 \xrightarrow{\;1/2\;} \underline{\quad 5x + 3y + 4z = 15.80} \end{cases}$$

$$x \qquad\quad + z = 2.75 \quad\Rightarrow x = 2.75 - z$$

Substitute and solve for y in terms of z:
$$5(2.75 - z) + 3y + 4z = 15.80$$

$$\Rightarrow 13.75 + 3y - z = 15.80$$

$$\Rightarrow 3y = z + 2.05 \Rightarrow y = \frac{1}{3}z + \frac{41}{60}$$

Solutions of the system are: $x = 2.75 - z$, $y = \dfrac{1}{3}z + \dfrac{41}{60}$.

Since we are given that $0.60 \le z \le 0.90$, we choose values of z that give two-decimal-place values of x and y with $1.75 \le x \le 2.25$ and $0.75 \le y \le 1.00$.

The possible values of x, y, and z are shown in the table.

x	y	z
2.13	0.89	0.62
2.10	0.90	0.65
2.07	.091	0.68
2.04	.092	0.71
2.01	.093	0.74
1.98	.094	0.77
1.95	0.95	0.80
1.92	0.96	0.83
1.89	0.97	0.86
1.86	0.98	0.89
1.83	0.99	0.92
1.80	1.00	0.95

79. Let $x =$ Beth's time working alone.
    Let $y =$ Bill's time working alone.
    Let $z =$ Edie's time working alone.
    We can use the following tables to organize our work:

	Beth	Bill	Edie
Hours to do job	x	y	z
Part of job done in 1 hour	$\dfrac{1}{x}$	$\dfrac{1}{y}$	$\dfrac{1}{z}$

In 10 hours they complete 1 entire job, so $10\left(\dfrac{1}{x} + \dfrac{1}{y} + \dfrac{1}{z}\right) = 1 \Rightarrow \dfrac{1}{x} + \dfrac{1}{y} + \dfrac{1}{z} = \dfrac{1}{10}$.

	Bill	Edie
Hours to do job	y	z
Part of job done in 1 hour	$\dfrac{1}{y}$	$\dfrac{1}{z}$

In 15 hours they complete 1 entire job, so $15\left(\dfrac{1}{y}+\dfrac{1}{z}\right)=1\Rightarrow\dfrac{1}{y}+\dfrac{1}{z}=\dfrac{1}{15}$.

	Beth	Bill	Edie
Hours to do job	x	y	z
Part of job done in 1 hour	$\dfrac{1}{x}$	$\dfrac{1}{y}$	$\dfrac{1}{z}$

With all 3 working for 4 hours and Beth and Bill working for an additional 8 hours, they complete 1 entire job, so $4\left(\dfrac{1}{x}+\dfrac{1}{y}+\dfrac{1}{z}\right)+8\left(\dfrac{1}{y}+\dfrac{1}{z}\right)=1\Rightarrow\dfrac{12}{x}+\dfrac{12}{y}+\dfrac{4}{z}=1.$

We have the system

$$\begin{cases}\dfrac{1}{x}+\dfrac{1}{y}+\dfrac{1}{z}=\dfrac{1}{10}\Rightarrow\\[2mm]\dfrac{1}{y}+\dfrac{1}{z}=\dfrac{1}{15}\Rightarrow\\[2mm]\dfrac{12}{x}+\dfrac{12}{y}+\dfrac{4}{z}=1\end{cases}$$

Subtract the second equation from the first equation

$$\dfrac{1}{x}+\dfrac{1}{y}+\dfrac{1}{z}=\dfrac{1}{10}$$
$$\dfrac{1}{y}+\dfrac{1}{z}=\dfrac{1}{15}$$
$$\dfrac{1}{x}\qquad-\dfrac{1}{z}=\dfrac{1}{10}-\dfrac{1}{15}\Rightarrow\dfrac{1}{x}=\dfrac{1}{30}\Rightarrow x=30$$

Substitute $x=30$ into the third equation: $\dfrac{12}{30}+\dfrac{12}{y}+\dfrac{4}{z}=1\Rightarrow\dfrac{12}{y}+\dfrac{4}{z}=\dfrac{3}{5}.$

Now consider the system consisting of the last result and the second original equation. Multiply the second original equation by -12 and add it to the last result to eliminate y:

$$\begin{cases}\dfrac{1}{y}+\dfrac{1}{z}=\dfrac{1}{15}&\xrightarrow{-12}&\dfrac{-12}{y}+\dfrac{-12}{z}=\dfrac{-12}{15}\\[2mm]\dfrac{12}{y}+\dfrac{4}{z}=\dfrac{3}{5}&\longrightarrow&\dfrac{12}{y}+\dfrac{4}{z}=\dfrac{3}{5}\end{cases}$$

$$-\dfrac{12}{z}+\dfrac{4}{z}=-\dfrac{12}{15}+\dfrac{3}{5}\Rightarrow-\dfrac{8}{z}=-\dfrac{3}{15}\Rightarrow z=40$$

Plugging $z=40$ to find y:

$$\dfrac{12}{y}+\dfrac{4}{z}=\dfrac{3}{5}\Rightarrow\dfrac{12}{y}+\dfrac{4}{40}=\dfrac{3}{5}$$
$$\dfrac{12}{y}+\dfrac{1}{10}=\dfrac{3}{5}\Rightarrow\dfrac{12}{y}=\dfrac{3}{5}-\dfrac{1}{10}\Rightarrow\dfrac{12}{y}=\dfrac{1}{2}\Rightarrow y=24$$

Working alone, it would take Beth 30 hours, Bill 24 hours, and Edie 40 hours to complete the job.

81. Answers will vary.

Systems of Equations and Inequalities

10.2 Systems of Linear Equations: Matrices

1. matrix   3. True

5. Writing the augmented matrix for the system of equations:
$$\begin{cases} x - 5y = 5 \\ 4x + 3y = 6 \end{cases} \rightarrow \begin{bmatrix} 1 & -5 & | & 5 \\ 4 & 3 & | & 6 \end{bmatrix}$$

7. Writing the augmented matrix for the system of equations:
$$\begin{cases} 2x + 3y - 6 = 0 \\ 4x - 6y + 2 = 0 \end{cases} \Rightarrow \begin{cases} 2x + 3y = 6 \\ 4x - 6y = -2 \end{cases} \rightarrow \begin{bmatrix} 2 & 3 & | & 6 \\ 4 & -6 & | & -2 \end{bmatrix}$$

9. Writing the augmented matrix for the system of equations:
$$\begin{cases} 0.01x - 0.03y = 0.06 \\ 0.13x + 0.10y = 0.20 \end{cases} \rightarrow \begin{bmatrix} 0.01 & -0.03 & | & 0.06 \\ 0.13 & 0.10 & | & 0.20 \end{bmatrix}$$

11. Writing the augmented matrix for the system of equations:
$$\begin{cases} x - y + z = 10 \\ 3x + 3y \quad = 5 \\ x + y + 2z = 2 \end{cases} \rightarrow \begin{bmatrix} 1 & -1 & 1 & | & 10 \\ 3 & 3 & 0 & | & 5 \\ 1 & 1 & 2 & | & 2 \end{bmatrix}$$

13. Writing the augmented matrix for the system of equations:
$$\begin{cases} x + y - z = 2 \\ 3x - 2y \quad = 2 \\ 5x + 3y - z = 1 \end{cases} \rightarrow \begin{bmatrix} 1 & 1 & -1 & | & 2 \\ 3 & -2 & 0 & | & 2 \\ 5 & 3 & -1 & | & 1 \end{bmatrix}$$

15. Writing the augmented matrix for the system of equations:
$$\begin{cases} x - y - z = 10 \\ 2x + y + 2z = -1 \\ -3x + 4y = 5 \\ 4x - 5y + z = 0 \end{cases} \rightarrow \begin{bmatrix} 1 & -1 & -1 & | & 10 \\ 2 & 1 & 2 & | & -1 \\ -3 & 4 & 0 & | & 5 \\ 4 & -5 & 1 & | & 0 \end{bmatrix}$$

17. $\begin{bmatrix} 1 & -3 & | & -2 \\ 2 & -5 & | & 5 \end{bmatrix} \rightarrow \begin{bmatrix} 1 & -3 & | & -2 \\ 0 & 1 & | & 9 \end{bmatrix}$

$$R_2 = -2r_1 + r_2$$

583

19.
$$\begin{bmatrix} 1 & -3 & 4 & | & 3 \\ 2 & -5 & 6 & | & 6 \\ -3 & 3 & 4 & | & 6 \end{bmatrix} \rightarrow \begin{bmatrix} 1 & -3 & 4 & | & 3 \\ 0 & 1 & -2 & | & 0 \\ -3 & 3 & 4 & | & 6 \end{bmatrix}$$

(a) $R_2 = -2r_1 + r_2$

$$\begin{bmatrix} 1 & -3 & 4 & | & 3 \\ 2 & -5 & 6 & | & 6 \\ -3 & 3 & 4 & | & 6 \end{bmatrix} \rightarrow \begin{bmatrix} 1 & -3 & 4 & | & 3 \\ 2 & -5 & 6 & | & 6 \\ 0 & -6 & 16 & | & 15 \end{bmatrix}$$

(b) $R_3 = 3r_1 + r_3$

21.
$$\begin{bmatrix} 1 & -3 & 2 & | & -6 \\ 2 & -5 & 3 & | & -4 \\ -3 & -6 & 4 & | & 6 \end{bmatrix} \rightarrow \begin{bmatrix} 1 & -3 & 2 & | & -6 \\ 0 & 1 & -1 & | & 8 \\ -3 & -6 & 4 & | & 6 \end{bmatrix}$$

(a) $R_2 = -2r_1 + r_2$

$$\begin{bmatrix} 1 & -3 & 2 & | & -6 \\ 2 & -5 & 3 & | & -4 \\ -3 & -6 & 4 & | & 6 \end{bmatrix} \rightarrow \begin{bmatrix} 1 & -3 & 2 & | & -6 \\ 2 & -5 & 3 & | & -4 \\ 0 & -15 & 10 & | & -12 \end{bmatrix}$$

(b) $R_3 = 3r_1 + r_3$

23.
$$\begin{bmatrix} 1 & -3 & 1 & | & -2 \\ 2 & -5 & 6 & | & -2 \\ -3 & 1 & 4 & | & 6 \end{bmatrix} \rightarrow \begin{bmatrix} 1 & -3 & 1 & | & -2 \\ 0 & 1 & 4 & | & 2 \\ -3 & 1 & 4 & | & 6 \end{bmatrix}$$

(a) $R_2 = -2r_1 + r_2$

$$\begin{bmatrix} 1 & -3 & 1 & | & -2 \\ 2 & -5 & 6 & | & -2 \\ -3 & 1 & 4 & | & 6 \end{bmatrix} \rightarrow \begin{bmatrix} 1 & -3 & 1 & | & -2 \\ 2 & -5 & 6 & | & -2 \\ 0 & -8 & 7 & | & 0 \end{bmatrix}$$

(b) $R_3 = 3r_1 + r_3$

25. $\begin{cases} x = 5 \\ y = -1 \end{cases}$ consistent; $x = 5, y = -1$

27. $\begin{cases} x = 1 \\ y = 2 \\ 0 = 3 \end{cases}$ inconsistent

29. $\begin{cases} x + 2z = -1 \\ y - 4z = -2 \\ \qquad 0 = 0 \end{cases}$ consistent; $x = -1 - 2z, y = -2 + 4z, z$ is any real number

31. $\begin{cases} x_1 = 1 \\ x_2 + x_4 = 2 \\ x_3 + 2x_4 = 3 \end{cases}$ consistent; $x_1 = 1$, $x_2 = 2 - x_4$, $x_3 = 3 - 2x_4$, x_4 is any real number

33. $\begin{cases} x_1 + 4x_4 = 2 \\ x_2 + x_3 + 3x_4 = 3 \\ 0 = 0 \end{cases}$ consistent; $x_1 = 2 - 4x_4$, $x_2 = 3 - x_3 - 3x_4$, x_3, x_4 are any real numbers

35. $\begin{cases} x_1 + x_4 = -2 \\ x_2 + 2x_4 = 2 \\ x_3 - x_4 = 0 \\ 0 = 0 \end{cases}$ consistent; $x_1 = -2 - x_4$, $x_2 = 2 - 2x_4$, $x_3 = x_4$, x_4 is any real number

37. $\begin{cases} x + y = 8 \\ x - y = 4 \end{cases}$ Write the augmented matrix: $\begin{bmatrix} 1 & 1 & | & 8 \\ 1 & -1 & | & 4 \end{bmatrix}$

$\rightarrow \begin{bmatrix} 1 & 1 & | & 8 \\ 0 & -2 & | & -4 \end{bmatrix} \rightarrow \begin{bmatrix} 1 & 1 & | & 8 \\ 0 & 1 & | & 2 \end{bmatrix} \rightarrow \begin{bmatrix} 1 & 0 & | & 6 \\ 0 & 1 & | & 2 \end{bmatrix}$

$R_2 = -r_1 + r_2 \qquad R_2 = -\frac{1}{2}r_2 \quad R_1 = -r_2 + r_1$

The solution is $x = 6$, $y = 2$.

39. $\begin{cases} 2x - 4y = -2 \\ 3x + 2y = 3 \end{cases}$ Write the augmented matrix: $\begin{bmatrix} 2 & -4 & | & -2 \\ 3 & 2 & | & 3 \end{bmatrix}$

$\rightarrow \begin{bmatrix} 1 & -2 & | & -1 \\ 3 & 2 & | & 3 \end{bmatrix} \rightarrow \begin{bmatrix} 1 & -2 & | & -1 \\ 0 & 8 & | & 6 \end{bmatrix} \rightarrow \begin{bmatrix} 1 & -2 & | & -1 \\ 0 & 1 & | & \frac{3}{4} \end{bmatrix} \rightarrow \begin{bmatrix} 1 & 0 & | & \frac{1}{2} \\ 0 & 1 & | & \frac{3}{4} \end{bmatrix}$

$R_1 = \frac{1}{2}r_1 \qquad R_2 = -3r_1 + r_2 \quad R_2 = \frac{1}{8}r_2 \qquad R_1 = 2r_2 + r_1$

The solution is $x = \dfrac{1}{2}$, $y = \dfrac{3}{4}$.

41. $\begin{cases} x + 2y = 4 \\ 2x + 4y = 8 \end{cases}$ Write the augmented matrix: $\begin{bmatrix} 1 & 2 & | & 4 \\ 2 & 4 & | & 8 \end{bmatrix}$

$\rightarrow \begin{bmatrix} 1 & 2 & | & 4 \\ 0 & 0 & | & 0 \end{bmatrix}$

$R_2 = -2r_1 + r_2$

This is a dependent system and the solution is $x = 4 - 2y$, y is any real number.

43. $\begin{cases} 2x + 3y = 6 \\ x - y = \dfrac{1}{2} \end{cases}$ Write the augmented matrix: $\begin{bmatrix} 2 & 3 & | & 6 \\ 1 & -1 & | & \frac{1}{2} \end{bmatrix}$

$$\rightarrow \begin{bmatrix} 1 & \frac{3}{2} & | & 3 \\ 1 & -1 & | & \frac{1}{2} \end{bmatrix} \rightarrow \begin{bmatrix} 1 & \frac{3}{2} & | & 3 \\ 0 & -\frac{5}{2} & | & -\frac{5}{2} \end{bmatrix} \rightarrow \begin{bmatrix} 1 & \frac{3}{2} & | & 3 \\ 0 & 1 & | & 1 \end{bmatrix} \rightarrow \begin{bmatrix} 1 & 0 & | & \frac{3}{2} \\ 0 & 1 & | & 1 \end{bmatrix}$$

$\qquad R_1 = \frac{1}{2}r_1 \qquad R_2 = -r_1 + r_2 \qquad R_2 = -\frac{2}{5}r_2 \qquad R_1 = -\frac{3}{2}r_2 + r_1$

The solution is $x = \dfrac{3}{2}, y = 1$.

45. $\begin{cases} 3x - 5y = 3 \\ 15x + 5y = 21 \end{cases}$ Write the augmented matrix: $\begin{bmatrix} 3 & -5 & | & 3 \\ 15 & 5 & | & 21 \end{bmatrix}$

$$\rightarrow \begin{bmatrix} 1 & -\frac{5}{3} & | & 1 \\ 15 & 5 & | & 21 \end{bmatrix} \rightarrow \begin{bmatrix} 1 & -\frac{5}{3} & | & 1 \\ 0 & 30 & | & 6 \end{bmatrix} \rightarrow \begin{bmatrix} 1 & -\frac{5}{3} & | & 1 \\ 0 & 1 & | & \frac{1}{5} \end{bmatrix} \rightarrow \begin{bmatrix} 1 & 0 & | & \frac{4}{3} \\ 0 & 1 & | & \frac{1}{5} \end{bmatrix}$$

$\qquad R_1 = \frac{1}{3}r_1 \qquad R_2 = -15r_1 + r_2 \quad R_2 = \frac{1}{30}r_2 \qquad R_1 = \frac{5}{3}r_2 + r_1$

The solution is $x = \dfrac{4}{3}, y = \dfrac{1}{5}$.

47. $\begin{cases} x - y = 6 \\ 2x - 3z = 16 \\ 2y + z = 4 \end{cases}$ Write the augmented matrix: $\begin{bmatrix} 1 & -1 & 0 & | & 6 \\ 2 & 0 & -3 & | & 16 \\ 0 & 2 & 1 & | & 4 \end{bmatrix}$

$$\rightarrow \begin{bmatrix} 1 & -1 & 0 & | & 6 \\ 0 & 2 & -3 & | & 4 \\ 0 & 2 & 1 & | & 4 \end{bmatrix} \rightarrow \begin{bmatrix} 1 & -1 & 0 & | & 6 \\ 0 & 1 & -\frac{3}{2} & | & 2 \\ 0 & 2 & 1 & | & 4 \end{bmatrix} \rightarrow \begin{bmatrix} 1 & 0 & -\frac{3}{2} & | & 8 \\ 0 & 1 & -\frac{3}{2} & | & 2 \\ 0 & 0 & 4 & | & 0 \end{bmatrix} \rightarrow \begin{bmatrix} 1 & 0 & -\frac{3}{2} & | & 8 \\ 0 & 1 & -\frac{3}{2} & | & 2 \\ 0 & 0 & 1 & | & 0 \end{bmatrix}$$

$\quad R_2 = -2r_1 + r_2 \qquad R_2 = \frac{1}{2}r_2 \qquad\quad R_1 = r_2 + r_1 \qquad\quad R_3 = \frac{1}{4}r_3$

$\qquad\qquad\qquad\qquad\qquad\qquad\qquad\qquad\qquad R_3 = -2r_2 + r_3$

$$\rightarrow \begin{bmatrix} 1 & 0 & 0 & | & 8 \\ 0 & 1 & 0 & | & 2 \\ 0 & 0 & 1 & | & 0 \end{bmatrix}$$

$R_1 = \frac{3}{2}r_3 + r_1$

$R_2 = \frac{3}{2}r_3 + r_2$

The solution is $x = 8, y = 2, z = 0$.

49. $\begin{cases} x-2y+3z= 7 \\ 2x+ y+ z= 4 \\ -3x+2y-2z=-10 \end{cases}$ Write the augmented matrix: $\begin{bmatrix} 1 & -2 & 3 & | & 7 \\ 2 & 1 & 1 & | & 4 \\ -3 & 2 & -2 & | & -10 \end{bmatrix}$

$\rightarrow \begin{bmatrix} 1 & -2 & 3 & | & 7 \\ 0 & 5 & -5 & | & -10 \\ 0 & -4 & 7 & | & 11 \end{bmatrix} \rightarrow \begin{bmatrix} 1 & -2 & 3 & | & 7 \\ 0 & 1 & -1 & | & -2 \\ 0 & -4 & 7 & | & 11 \end{bmatrix} \rightarrow \begin{bmatrix} 1 & 0 & 1 & | & 3 \\ 0 & 1 & -1 & | & -2 \\ 0 & 0 & 3 & | & 3 \end{bmatrix}$

$\begin{array}{ll} R_2 = -2r_1 + r_2 & R_2 = \frac{1}{5}r_2 & R_1 = 2r_2 + r_1 \\ R_3 = 3r_1 + r_3 & & R_3 = 4r_2 + r_3 \end{array}$

$\rightarrow \begin{bmatrix} 1 & 0 & 1 & | & 3 \\ 0 & 1 & -1 & | & -2 \\ 0 & 0 & 1 & | & 1 \end{bmatrix} \rightarrow \begin{bmatrix} 1 & 0 & 0 & | & 2 \\ 0 & 1 & 0 & | & -1 \\ 0 & 0 & 1 & | & 1 \end{bmatrix}$

$\begin{array}{ll} R_3 = \frac{1}{3}r_3 & R_1 = -r_3 + r_1 \\ & R_2 = r_3 + r_2 \end{array}$

The solution is $x = 2$, $y = -1$, $z = 1$.

51. $\begin{cases} 2x-2y-2z= 2 \\ 2x+3y+ z= 2 \\ 3x+2y = 0 \end{cases}$ Write the augmented matrix: $\begin{bmatrix} 2 & -2 & -2 & | & 2 \\ 2 & 3 & 1 & | & 2 \\ 3 & 2 & 0 & | & 0 \end{bmatrix}$

$\rightarrow \begin{bmatrix} 1 & -1 & -1 & | & 1 \\ 2 & 3 & 1 & | & 2 \\ 3 & 2 & 0 & | & 0 \end{bmatrix} \rightarrow \begin{bmatrix} 1 & -1 & -1 & | & 1 \\ 0 & 5 & 3 & | & 0 \\ 0 & 5 & 3 & | & -3 \end{bmatrix} \rightarrow \begin{bmatrix} 1 & -1 & -1 & | & 1 \\ 0 & 5 & 3 & | & 0 \\ 0 & 0 & 0 & | & -3 \end{bmatrix}$

$\begin{array}{lll} R_1 = \frac{1}{2}r_1 & R_2 = -2r_1 + r_2 & R_3 = -r_2 + r_3 \\ & R_3 = -3r_1 + r_3 & \end{array}$

There is no solution. The system is inconsistent.

53. $\begin{cases} -x+ y+ z= -1 \\ -x+2y-3z=-4 \\ 3x-2y-7z= 0 \end{cases}$ Write the augmented matrix: $\begin{bmatrix} -1 & 1 & 1 & | & -1 \\ -1 & 2 & -3 & | & -4 \\ 3 & -2 & -7 & | & 0 \end{bmatrix}$

$\rightarrow \begin{bmatrix} 1 & -1 & -1 & | & 1 \\ -1 & 2 & -3 & | & -4 \\ 3 & -2 & -7 & | & 0 \end{bmatrix} \rightarrow \begin{bmatrix} 1 & -1 & -1 & | & 1 \\ 0 & 1 & -4 & | & -3 \\ 0 & 1 & -4 & | & -3 \end{bmatrix} \rightarrow \begin{bmatrix} 1 & 0 & -5 & | & -2 \\ 0 & 1 & -4 & | & -3 \\ 0 & 0 & 0 & | & 0 \end{bmatrix} \Rightarrow \begin{array}{l} x-5z=-2 \\ y-4z=-3 \end{array}$

$\begin{array}{lll} R_1 = -r_1 & R_2 = r_1 + r_2 & R_1 = r_2 + r_1 \\ & R_3 = -3r_1 + r_3 & R_3 = -r_2 + r_3 \end{array}$

The solution is $x = 5z-2$, $y = 4z-3$, z is any real number .

55. $\begin{cases} 2x - 2y + 3z = 6 \\ 4x - 3y + 2z = 0 \\ -2x + 3y - 7z = 1 \end{cases}$ Write the augmented matrix: $\begin{bmatrix} 2 & -2 & 3 & | & 6 \\ 4 & -3 & 2 & | & 0 \\ -2 & 3 & -7 & | & 1 \end{bmatrix}$

$$\rightarrow \begin{bmatrix} 1 & -1 & \frac{3}{2} & | & 3 \\ 4 & -3 & 2 & | & 0 \\ -2 & 3 & -7 & | & 1 \end{bmatrix} \rightarrow \begin{bmatrix} 1 & -1 & \frac{3}{2} & | & 3 \\ 0 & 1 & -4 & | & -12 \\ 0 & 1 & -4 & | & 7 \end{bmatrix} \rightarrow \begin{bmatrix} 1 & 0 & -\frac{5}{2} & | & -9 \\ 0 & 1 & -4 & | & -12 \\ 0 & 0 & 0 & | & 19 \end{bmatrix}$$

$$R_1 = \tfrac{1}{2}r_1 \qquad\qquad\qquad R_2 = -4r_1 + r_2 \qquad\qquad R_1 = r_2 + r_1$$
$$R_3 = 2r_1 + r_3 \qquad\qquad R_3 = -r_2 + r_3$$

There is no solution. The system is inconsistent.

57. $\begin{cases} x + y - z = 6 \\ 3x - 2y + z = -5 \\ x + 3y - 2z = 14 \end{cases}$ Write the augmented matrix: $\begin{bmatrix} 1 & 1 & -1 & | & 6 \\ 3 & -2 & 1 & | & -5 \\ 1 & 3 & -2 & | & 14 \end{bmatrix}$

$$\rightarrow \begin{bmatrix} 1 & 1 & -1 & | & 6 \\ 0 & -5 & 4 & | & -23 \\ 0 & 2 & -1 & | & 8 \end{bmatrix} \rightarrow \begin{bmatrix} 1 & 1 & -1 & | & 6 \\ 0 & 1 & -\frac{4}{5} & | & \frac{23}{5} \\ 0 & 2 & -1 & | & 8 \end{bmatrix} \rightarrow \begin{bmatrix} 1 & 0 & -\frac{1}{5} & | & \frac{7}{5} \\ 0 & 1 & -\frac{4}{5} & | & \frac{23}{5} \\ 0 & 0 & \frac{3}{5} & | & -\frac{6}{5} \end{bmatrix}$$

$$R_2 = -3r_1 + r_2 \qquad\qquad R_2 = -\tfrac{1}{5}r_2 \qquad\qquad R_1 = -r_2 + r_1$$
$$R_3 = -r_1 + r_3 \qquad\qquad\qquad\qquad\qquad\qquad R_3 = -2r_2 + r_3$$

$$\rightarrow \begin{bmatrix} 1 & 0 & -\frac{1}{5} & | & \frac{7}{5} \\ 0 & 1 & -\frac{4}{5} & | & \frac{23}{5} \\ 0 & 0 & 1 & | & -2 \end{bmatrix} \rightarrow \begin{bmatrix} 1 & 0 & 0 & | & 1 \\ 0 & 1 & 0 & | & 3 \\ 0 & 0 & 1 & | & -2 \end{bmatrix}$$

$$R_3 = \tfrac{5}{3}r_3 \qquad\qquad R_1 = \tfrac{1}{5}r_3 + r_1$$
$$R_2 = \tfrac{4}{5}r_3 + r_2$$

The solution is $x = 1$, $y = 3$, $z = -2$.

59. $\begin{cases} x + 2y - z = -3 \\ 2x - 4y + z = -7 \\ -2x + 2y - 3z = 4 \end{cases}$ Write the augmented matrix: $\begin{bmatrix} 1 & 2 & -1 & | & -3 \\ 2 & -4 & 1 & | & -7 \\ -2 & 2 & -3 & | & 4 \end{bmatrix}$

$$\rightarrow \begin{bmatrix} 1 & 2 & -1 & | & -3 \\ 0 & -8 & 3 & | & -1 \\ 0 & 6 & -5 & | & -2 \end{bmatrix} \rightarrow \begin{bmatrix} 1 & 2 & -1 & | & -3 \\ 0 & 1 & -\frac{3}{8} & | & \frac{1}{8} \\ 0 & 6 & -5 & | & -2 \end{bmatrix} \rightarrow \begin{bmatrix} 1 & 0 & -\frac{1}{4} & | & -\frac{13}{4} \\ 0 & 1 & -\frac{3}{8} & | & \frac{1}{8} \\ 0 & 0 & -\frac{11}{4} & | & -\frac{11}{4} \end{bmatrix}$$

$$R_2 = -2r_1 + r_2 \qquad\qquad R_2 = -\tfrac{1}{8}r_2 \qquad\qquad R_1 = -2r_2 + r_1$$
$$R_3 = 2r_1 + r_3 \qquad\qquad\qquad\qquad\qquad\qquad R_3 = -6r_2 + r_3$$

$$\rightarrow \begin{bmatrix} 1 & 0 & -\frac{1}{4} & | & -\frac{13}{4} \\ 0 & 1 & -\frac{3}{8} & | & \frac{1}{8} \\ 0 & 0 & 1 & | & 1 \end{bmatrix} \rightarrow \begin{bmatrix} 1 & 0 & 0 & | & -3 \\ 0 & 1 & 0 & | & \frac{1}{2} \\ 0 & 0 & 1 & | & 1 \end{bmatrix}$$

$$R_3 = -\frac{4}{11}r_3 \qquad\qquad R_1 = \frac{1}{4}r_3 + r_1$$
$$R_2 = \frac{3}{8}r_3 + r_2$$

The solution is $x = -3,\ y = \dfrac{1}{2},\ z = 1.$

61. $\begin{cases} 3x + y - z = \dfrac{2}{3} \\ 2x - y + z = 1 \\ 4x + 2y = \dfrac{8}{3} \end{cases}$ Write the augmented matrix : $\begin{bmatrix} 3 & 1 & -1 & | & \frac{2}{3} \\ 2 & -1 & 1 & | & 1 \\ 4 & 2 & 0 & | & \frac{8}{3} \end{bmatrix}$

$$\rightarrow \begin{bmatrix} 1 & \frac{1}{3} & -\frac{1}{3} & | & \frac{2}{9} \\ 2 & -1 & 1 & | & 1 \\ 4 & 2 & 0 & | & \frac{8}{3} \end{bmatrix} \rightarrow \begin{bmatrix} 1 & \frac{1}{3} & -\frac{1}{3} & | & \frac{2}{9} \\ 0 & -\frac{5}{3} & \frac{5}{3} & | & \frac{5}{9} \\ 0 & \frac{2}{3} & \frac{4}{3} & | & \frac{16}{9} \end{bmatrix} \rightarrow \begin{bmatrix} 1 & \frac{1}{3} & -\frac{1}{3} & | & \frac{2}{9} \\ 0 & 1 & -1 & | & -\frac{1}{3} \\ 0 & \frac{2}{3} & \frac{4}{3} & | & \frac{16}{9} \end{bmatrix} \rightarrow \begin{bmatrix} 1 & 0 & 0 & | & \frac{1}{3} \\ 0 & 1 & -1 & | & -\frac{1}{3} \\ 0 & 0 & 2 & | & 2 \end{bmatrix}$$

$$R_1 = \frac{1}{3}r_1 \qquad R_2 = -2r_1 + r_2 \qquad R_2 = -\frac{3}{5}r_2 \qquad R_1 = -\frac{1}{3}r_2 + r_1$$
$$R_3 = -4r_1 + r_3 \qquad\qquad\qquad R_3 = -\frac{2}{3}r_2 + r_3$$

$$\rightarrow \begin{bmatrix} 1 & 0 & 0 & | & \frac{1}{3} \\ 0 & 1 & -1 & | & -\frac{1}{3} \\ 0 & 0 & 1 & | & 1 \end{bmatrix} \rightarrow \begin{bmatrix} 1 & 0 & 0 & | & \frac{1}{3} \\ 0 & 1 & 0 & | & \frac{2}{3} \\ 0 & 0 & 1 & | & 1 \end{bmatrix}$$

$$R_3 = \frac{1}{2}r_3 \qquad\qquad R_2 = r_3 + r_2$$

The solution is $x = \dfrac{1}{3},\ y = \dfrac{2}{3},\ z = 1.$

63. $\begin{cases} x + y + z + w = 4 \\ 2x - y + z = 0 \\ 3x + 2y + z - w = 6 \\ x - 2y - 2z + 2w = -1 \end{cases}$ Write the augmented matrix : $\begin{bmatrix} 1 & 1 & 1 & 1 & | & 4 \\ 2 & -1 & 1 & 0 & | & 0 \\ 3 & 2 & 1 & -1 & | & 6 \\ 1 & -2 & -2 & 2 & | & -1 \end{bmatrix}$

$$\rightarrow \begin{bmatrix} 1 & 1 & 1 & 1 & | & 4 \\ 0 & -3 & -1 & -2 & | & -8 \\ 0 & -1 & -2 & -4 & | & -6 \\ 0 & -3 & -3 & 1 & | & -5 \end{bmatrix} \rightarrow \begin{bmatrix} 1 & 1 & 1 & 1 & | & 4 \\ 0 & -1 & -2 & -4 & | & -6 \\ 0 & -3 & -1 & -2 & | & -8 \\ 0 & -3 & -3 & 1 & | & -5 \end{bmatrix} \rightarrow \begin{bmatrix} 1 & 1 & 1 & 1 & | & 4 \\ 0 & 1 & 2 & 4 & | & 6 \\ 0 & -3 & -1 & -2 & | & -8 \\ 0 & -3 & -3 & 1 & | & -5 \end{bmatrix}$$

$$R_2 = -2r_1 + r_2 \qquad\qquad \text{Interchange } r_2 \text{ and } r_3 \qquad R_2 = -r_2$$
$$R_3 = -3r_1 + r_3$$
$$R_4 = -r_1 + r_4$$

$$\rightarrow \begin{bmatrix} 1 & 0 & -1 & -3 & | & -2 \\ 0 & 1 & 2 & 4 & | & 6 \\ 0 & 0 & 5 & 10 & | & 10 \\ 0 & 0 & 3 & 13 & | & 13 \end{bmatrix} \rightarrow \begin{bmatrix} 1 & 0 & -1 & -3 & | & -2 \\ 0 & 1 & 2 & 4 & | & 6 \\ 0 & 0 & 1 & 2 & | & 2 \\ 0 & 0 & 3 & 13 & | & 13 \end{bmatrix} \rightarrow \begin{bmatrix} 1 & 0 & 0 & -1 & | & 0 \\ 0 & 1 & 0 & 0 & | & 2 \\ 0 & 0 & 1 & 2 & | & 2 \\ 0 & 0 & 0 & 7 & | & 7 \end{bmatrix}$$

$R_1 = -r_2 + r_1$ $R_3 = \frac{1}{5}r_3$ $R_1 = r_3 + r_1$

$R_3 = 3r_2 + r_3$ $R_2 = -2r_3 + r_2$

$R_4 = 3r_2 + r_4$ $R_4 = -3r_3 + r_4$

$$\rightarrow \begin{bmatrix} 1 & 0 & 0 & -1 & | & 0 \\ 0 & 1 & 0 & 0 & | & 2 \\ 0 & 0 & 1 & 2 & | & 2 \\ 0 & 0 & 0 & 1 & | & 1 \end{bmatrix} \rightarrow \begin{bmatrix} 1 & 0 & 0 & 0 & | & 1 \\ 0 & 1 & 0 & 0 & | & 2 \\ 0 & 0 & 1 & 0 & | & 0 \\ 0 & 0 & 0 & 1 & | & 1 \end{bmatrix}$$

$R_4 = \frac{1}{7}r_4$ $R_1 = r_4 + r_1$

                       $R_3 = -2r_4 + r_3$

The solution is $x = 1$, $y = 2$, $z = 0$, $w = 1$.

65. $\begin{cases} x + 2y + z = 1 \\ 2x - y + 2z = 2 \\ 3x + y + 3z = 3 \end{cases}$ Write the augmented matrix: $\begin{bmatrix} 1 & 2 & 1 & | & 1 \\ 2 & -1 & 2 & | & 2 \\ 3 & 1 & 3 & | & 3 \end{bmatrix}$

$$\rightarrow \begin{bmatrix} 1 & 2 & 1 & | & 1 \\ 0 & -5 & 0 & | & 0 \\ 0 & -5 & 0 & | & 0 \end{bmatrix} \rightarrow \begin{bmatrix} 1 & 2 & 1 & | & 1 \\ 0 & -5 & 0 & | & 0 \\ 0 & 0 & 0 & | & 0 \end{bmatrix} \Rightarrow \begin{array}{l} x + 2y + z = 1 \\ -5y \quad = 0 \end{array}$$

$R_2 = -2r_1 + r_2$ $R_3 = -r_2 + r_3$

$R_3 = -3r_1 + r_3$

Substitute and solve:

$$y = 0$$

$$x + 2(0) + z = 1$$

$$x + z = 1 \Rightarrow z = 1 - x$$

The solution is $y = 0$, $z = 1 - x$, x is any real number.

67. $\begin{cases} x - y + z = 5 \\ 3x + 2y - 2z = 0 \end{cases}$ Write the augmented matrix: $\begin{bmatrix} 1 & -1 & 1 & | & 5 \\ 3 & 2 & -2 & | & 0 \end{bmatrix}$

$$\rightarrow \begin{bmatrix} 1 & -1 & 1 & | & 5 \\ 0 & 5 & -5 & | & -15 \end{bmatrix} \rightarrow \begin{bmatrix} 1 & -1 & 1 & | & 5 \\ 0 & 1 & -1 & | & -3 \end{bmatrix} \rightarrow \begin{bmatrix} 1 & 0 & 0 & | & 2 \\ 0 & 1 & -1 & | & -3 \end{bmatrix}$$

$R_2 = -3r_1 + r_2$ $R_2 = \frac{1}{5}r_2$ $R_1 = r_2 + r_1$

The matrix in the third step represents the system $\begin{cases} x = 2 \\ y - z = -3 \end{cases}$

Therefore the solution is $x = 2; y = -3 + z;$ z is any real number

or

$x = 2; z = y + 3;$ y is any real number

69. $\begin{cases} 2x + 3y - z = 3 \\ x - y - z = 0 \\ -x + y + z = 0 \\ x + y + 3z = 5 \end{cases}$ Write the augmented matrix : $\begin{bmatrix} 2 & 3 & -1 & 3 \\ 1 & -1 & -1 & 0 \\ -1 & 1 & 1 & 0 \\ 1 & 1 & 3 & 5 \end{bmatrix}$

$\rightarrow \begin{bmatrix} 1 & -1 & -1 & 0 \\ 2 & 3 & -1 & 3 \\ -1 & 1 & 1 & 0 \\ 1 & 1 & 3 & 5 \end{bmatrix} \longrightarrow \begin{bmatrix} 1 & -1 & -1 & 0 \\ 0 & 5 & 1 & 3 \\ 0 & 0 & 0 & 0 \\ 0 & 2 & 4 & 5 \end{bmatrix} \rightarrow \begin{bmatrix} 1 & -1 & -1 & 0 \\ 0 & 5 & 1 & 3 \\ 0 & 2 & 4 & 5 \\ 0 & 0 & 0 & 0 \end{bmatrix}$

interchange r_1 and r_2 $R_2 = -2r_1 + r_2$ interchange r_3 and r_4

$R_3 = r_1 + r_3$

$R_4 = -r_1 + r_4$

$\rightarrow \begin{bmatrix} 1 & -1 & -1 & 0 \\ 0 & 1 & -7 & -7 \\ 0 & 2 & 4 & 5 \\ 0 & 0 & 0 & 0 \end{bmatrix} \longrightarrow \begin{bmatrix} 1 & 0 & -8 & -7 \\ 0 & 1 & -7 & -7 \\ 0 & 0 & 18 & 19 \\ 0 & 0 & 0 & 0 \end{bmatrix} \rightarrow \begin{bmatrix} 1 & 0 & -8 & -7 \\ 0 & 1 & -7 & -7 \\ 0 & 0 & 1 & \frac{19}{18} \\ 0 & 0 & 0 & 0 \end{bmatrix}$

$R_2 = -2r_3 + r_2$ $R_1 = r_2 + r_1$ $R_3 = \frac{1}{18}r_3$

$R_3 = -2r_2 + r_3$

The matrix in the last step represents the system $\begin{cases} x - 8z = -7 \\ y - 7z = -7 \\ z = \dfrac{19}{18} \end{cases}$

Therefore the solution is

$z = \dfrac{19}{18};$ $x = -7 + 8z = -7 + 8\left(\dfrac{19}{18}\right) = \dfrac{13}{9};$ $y = -7 + 7z = -7 + 7\left(\dfrac{19}{18}\right) = \dfrac{7}{18}$

71. $\begin{cases} 4x + y + z - w = 4 \\ x - y + 2z + 3w = 3 \end{cases}$ Write the augmented matrix : $\begin{bmatrix} 4 & 1 & 1 & -1 & 4 \\ 1 & -1 & 2 & 3 & 3 \end{bmatrix}$

$\rightarrow \begin{bmatrix} 1 & -1 & 2 & 3 & 3 \\ 4 & 1 & 1 & -1 & 4 \end{bmatrix} \longrightarrow \begin{bmatrix} 1 & -1 & 2 & 3 & 3 \\ 0 & 5 & -7 & -13 & -8 \end{bmatrix}$

interchange r_1 and r_2 $R_2 = -4r_1 + r_2$

The matrix in the last step represents the system $\begin{cases} x - y + 2z + 3w = 3 \\ 5y - 7z - 13w = -8 \end{cases}$

The second equation yields

$$5y - 7z - 13w = -8 \Rightarrow 5y = -8 + 7z + 13w \Rightarrow y = -\frac{8}{5} + \frac{7}{5}z + \frac{13}{5}w$$

The first equation yields

$$x - y + 2z + 3w = 3 \Rightarrow x = 3 + y - 2z - 3w$$

substituting for y

$$x = 3 + \left(-\frac{8}{5} + \frac{7}{5}z + \frac{13}{5}w\right) - 2z - 3w$$

$$x = -\frac{3}{5}z - \frac{2}{5}w + \frac{7}{5}$$

Therefore the solution is

$$x = -\frac{3}{5}z - \frac{2}{5}w + \frac{7}{5}; \quad y = -\frac{8}{5} + \frac{7}{5}z + \frac{13}{5}w; \quad z \text{ and } w \text{ are any real numbers.}$$

73. Each of the points must satisfy the equation $y = ax^2 + bx + c$.

$(1,2):$ $2 = a + b + c$

$(-2,-7):$ $-7 = 4a - 2b + c$

$(2,-3):$ $-3 = 4a + 2b + c$

Set up a matrix and solve:

$$\begin{bmatrix} 1 & 1 & 1 & | & 2 \\ 4 & -2 & 1 & | & -7 \\ 4 & 2 & 1 & | & -3 \end{bmatrix} \rightarrow \begin{bmatrix} 1 & 1 & 1 & | & 2 \\ 0 & -6 & -3 & | & -15 \\ 0 & -2 & -3 & | & -11 \end{bmatrix} \rightarrow \begin{bmatrix} 1 & 1 & 1 & | & 2 \\ 0 & 1 & \frac{1}{2} & | & \frac{5}{2} \\ 0 & -2 & -3 & | & -11 \end{bmatrix} \rightarrow \begin{bmatrix} 1 & 0 & \frac{1}{2} & | & -\frac{1}{2} \\ 0 & 1 & \frac{1}{2} & | & \frac{5}{2} \\ 0 & 0 & -2 & | & -6 \end{bmatrix}$$

$$\begin{array}{cccc} R_2 = -4r_1 + r_2 & R_2 = -\frac{1}{6}r_2 & R_1 = -r_2 + r_1 \\ R_3 = -4r_1 + r_3 & & R_3 = 2r_2 + r_3 \end{array}$$

$$\rightarrow \begin{bmatrix} 1 & 0 & \frac{1}{2} & | & -\frac{1}{2} \\ 0 & 1 & \frac{1}{2} & | & \frac{5}{2} \\ 0 & 0 & 1 & | & 3 \end{bmatrix} \rightarrow \begin{bmatrix} 1 & 0 & 0 & | & -2 \\ 0 & 1 & 0 & | & 1 \\ 0 & 0 & 1 & | & 3 \end{bmatrix}$$

$$\begin{array}{cc} R_3 = -\frac{1}{2}r_3 & R_1 = -\frac{1}{2}r_3 + r_1 \\ & R_2 = -\frac{1}{2}r_3 + r_2 \end{array}$$

The solution is $a = -2$, $b = 1$, $c = 3$; so the equation is $y = -2x^2 + x + 3$.

75. Each of the points must satisfy the equation $f(x) = ax^3 + bx^2 + cx + d$.

$f(-3) = -112:$ $-27a + 9b - 3c + d = -112$

$f(-1) = -2:$ $-a + b - c + d = -2$

$f(1) = 4:$ $a + b + c + d = 4$

$f(2) = 13:$ $8a + 4b + 2c + d = 13$

Set up a matrix and solve:

$$\begin{bmatrix} -27 & 9 & -3 & 1 & | & -112 \\ -1 & 1 & -1 & 1 & | & -2 \\ 1 & 1 & 1 & 1 & | & 4 \\ 8 & 4 & 2 & 1 & | & 13 \end{bmatrix} \rightarrow \begin{bmatrix} 1 & 1 & 1 & 1 & | & 4 \\ -1 & 1 & -1 & 1 & | & -2 \\ -27 & 9 & -3 & 1 & | & -112 \\ 8 & 4 & 2 & 1 & | & 13 \end{bmatrix} \rightarrow \begin{bmatrix} 1 & 1 & 1 & 1 & | & 4 \\ 0 & 2 & 0 & 2 & | & 2 \\ 0 & 36 & 24 & 28 & | & -4 \\ 0 & -4 & -6 & -7 & | & -19 \end{bmatrix}$$

Interchange r_3 and r_1 $R_2 = r_1 + r_2$

$R_3 = 27 r_1 + r_3$

$R_4 = -8 r_1 + r_4$

$$\rightarrow \begin{bmatrix} 1 & 1 & 1 & 1 & | & 4 \\ 0 & 1 & 0 & 1 & | & 1 \\ 0 & 36 & 24 & 28 & | & -4 \\ 0 & -4 & -6 & -7 & | & -19 \end{bmatrix} \rightarrow \begin{bmatrix} 1 & 0 & 1 & 0 & | & 3 \\ 0 & 1 & 0 & 1 & | & 1 \\ 0 & 0 & 24 & -8 & | & -40 \\ 0 & 0 & -6 & -3 & | & -15 \end{bmatrix} \rightarrow \begin{bmatrix} 1 & 0 & 1 & 0 & | & 3 \\ 0 & 1 & 0 & 1 & | & 1 \\ 0 & 0 & 1 & -\frac{5}{3} & | & -\frac{5}{3} \\ 0 & 0 & -6 & -3 & | & -15 \end{bmatrix}$$

$R_2 = \frac{1}{2} r_2$ $R_1 = -r_2 + r_1$ $R_3 = \frac{1}{24} r_3$

$R_3 = -36 r_2 + r_3$

$R_4 = 4 r_2 + r_4$

$$\rightarrow \begin{bmatrix} 1 & 0 & 0 & \frac{1}{3} & | & \frac{14}{3} \\ 0 & 1 & 0 & 1 & | & 1 \\ 0 & 0 & 1 & -\frac{1}{3} & | & -\frac{5}{3} \\ 0 & 0 & 0 & -5 & | & -25 \end{bmatrix} \rightarrow \begin{bmatrix} 1 & 0 & 0 & \frac{1}{3} & | & \frac{14}{3} \\ 0 & 1 & 0 & 1 & | & 1 \\ 0 & 0 & 1 & -\frac{1}{3} & | & -\frac{5}{3} \\ 0 & 0 & 0 & 1 & | & 5 \end{bmatrix} \rightarrow \begin{bmatrix} 1 & 0 & 0 & 0 & | & 3 \\ 0 & 1 & 0 & 0 & | & -4 \\ 0 & 0 & 1 & 0 & | & 0 \\ 0 & 0 & 0 & 1 & | & 5 \end{bmatrix}$$

$R_1 = -r_3 + r_1$ $R_4 = -\frac{1}{5} r_4$ $R_1 = -\frac{1}{3} r_4 + r_1$

$R_4 = 6 r_3 + r_4$ $R_2 = -r_4 + r_2$

$R_3 = \frac{1}{3} r_4 + r_3$

The solution is $a = 3$, $b = -4$, $c = 0$, $d = 5$; so the equation is $f(x) = 3x^3 - 4x^2 + 5$.

77. Let x = the number of servings of salmon steak.
Let y = the number of servings of baked eggs.
Let z = the number of servings of acorn squash.
Protein equation: $30x + 15y + 3z = 78$
Carbohydrate equation: $20x + 2y + 25z = 59$
Vitamin A equation: $2x + 20y + 32z = 75$
Set up a matrix and solve:

$$\begin{bmatrix} 30 & 15 & 3 & | & 78 \\ 20 & 2 & 25 & | & 59 \\ 2 & 20 & 32 & | & 75 \end{bmatrix} \rightarrow \begin{bmatrix} 2 & 20 & 32 & | & 75 \\ 20 & 2 & 25 & | & 59 \\ 30 & 15 & 3 & | & 78 \end{bmatrix} \rightarrow \begin{bmatrix} 1 & 10 & 16 & | & 37.5 \\ 20 & 2 & 25 & | & 59 \\ 30 & 15 & 3 & | & 78 \end{bmatrix}$$

Interchange r_3 and r_1 $R_1 = \frac{1}{2} r_1$

$$\rightarrow \begin{bmatrix} 1 & 10 & 16 & | & 37.5 \\ 0 & -198 & -295 & | & -691 \\ 0 & -285 & -477 & | & -1047 \end{bmatrix} \rightarrow \begin{bmatrix} 1 & 10 & 16 & | & 37.5 \\ 0 & -198 & -295 & | & -691 \\ 0 & 0 & -\frac{3457}{66} & | & -\frac{3457}{66} \end{bmatrix}$$

$$R_2 = -20\,r_1 + r_2 \qquad\qquad R_3 = -\frac{95}{66}r_2 + r_3$$

$$R_3 = -30\,r_1 + r_3$$

$$\rightarrow \begin{bmatrix} 1 & 10 & 16 & | & 37.5 \\ 0 & -198 & -295 & | & -691 \\ 0 & 0 & 1 & | & 1 \end{bmatrix}$$

$$R_3 = -\frac{66}{3457}r_3$$

Substitute $z = 1$ and solve:

$$-198y - 295(1) = -691 \qquad\qquad x + 10(2) + 16(1) = 37.5$$

$$-198y = -396 \qquad\qquad x + 36 = 37.5$$

$$y = 2 \qquad\qquad x = 1.5$$

The dietitian should serve 1.5 servings of salmon steak, 2 servings of baked eggs, and 1 serving of acorn squash.

79. Let x = the amount invested in Treasury bills.
Let y = the amount invested in Treasury bonds.
Let z = the amount invested in corporate bonds.

Total investment equation: $x + y + z = 10,000$
Annual income equation: $0.06x + 0.07y + 0.08z = 680$
Condition on investment equation: $z = 0.5x \Rightarrow x - 2z = 0$

Set up a matrix and solve:

$$\begin{bmatrix} 1 & 1 & 1 & | & 10,000 \\ 0.06 & 0.07 & 0.08 & | & 680 \\ 1 & 0 & -2 & | & 0 \end{bmatrix} \rightarrow \begin{bmatrix} 1 & 1 & 1 & | & 10,000 \\ 0 & 0.01 & 0.02 & | & 80 \\ 0 & -1 & -3 & | & -10,000 \end{bmatrix} \rightarrow \begin{bmatrix} 1 & 1 & 1 & | & 10,000 \\ 0 & 1 & 2 & | & 8000 \\ 0 & -1 & -3 & | & -10,000 \end{bmatrix}$$

$$R_2 = -0.06\,r_1 + r_2 \qquad\qquad R_2 = 100\,r_2$$
$$R_3 = -r_1 + r_3$$

$$\rightarrow \begin{bmatrix} 1 & 0 & -1 & | & 2000 \\ 0 & 1 & 2 & | & 8000 \\ 0 & 0 & -1 & | & -2000 \end{bmatrix} \rightarrow \begin{bmatrix} 1 & 0 & -1 & | & 2000 \\ 0 & 1 & 2 & | & 8000 \\ 0 & 0 & 1 & | & 2000 \end{bmatrix} \rightarrow \begin{bmatrix} 1 & 0 & 0 & | & 4000 \\ 0 & 1 & 0 & | & 4000 \\ 0 & 0 & 1 & | & 2000 \end{bmatrix}$$

$$R_1 = -r_2 + r_1 \qquad\quad R_3 = -r_3 \qquad\quad R_1 = r_3 + r_1$$
$$R_3 = r_2 + r_3 \qquad\qquad\qquad\qquad R_2 = -2r_3 + r_2$$

Carletta should invest \$4000 in Treasury bills, \$4000 in Treasury bonds, and \$2000 in corporate bonds.

81.   Let x = the number of Deltas produced.
Let y = the number of Betas produced.
Let z = the number of Sigmas produced.
Painting equation: $10x + 16y + 8z = 240$
Drying equation: $3x + 5y + 2z = 69$
Polishing equation: $2x + 3y + z = 41$
Set up a matrix and solve:

$$\begin{bmatrix} 10 & 16 & 8 & | & 240 \\ 3 & 5 & 2 & | & 69 \\ 2 & 3 & 1 & | & 41 \end{bmatrix} \rightarrow \begin{bmatrix} 1 & 1 & 2 & | & 33 \\ 3 & 5 & 2 & | & 69 \\ 2 & 3 & 1 & | & 41 \end{bmatrix} \rightarrow \begin{bmatrix} 1 & 1 & 2 & | & 33 \\ 0 & 2 & -4 & | & -30 \\ 0 & 1 & -3 & | & -25 \end{bmatrix} \rightarrow \begin{bmatrix} 1 & 1 & 2 & | & 33 \\ 0 & 1 & -2 & | & -15 \\ 0 & 1 & -3 & | & -25 \end{bmatrix}$$

$$R_1 = -3r_2 + r_1 \qquad R_2 = -3r_1 + r_2 \qquad R_2 = \tfrac{1}{2}r_2$$
$$R_3 = -2r_1 + r_3$$

$$\rightarrow \begin{bmatrix} 1 & 0 & 4 & | & 48 \\ 0 & 1 & -2 & | & -15 \\ 0 & 0 & -1 & | & -10 \end{bmatrix} \rightarrow \begin{bmatrix} 1 & 0 & 4 & | & 48 \\ 0 & 1 & -2 & | & -15 \\ 0 & 0 & 1 & | & 10 \end{bmatrix} \rightarrow \begin{bmatrix} 1 & 0 & 0 & | & 8 \\ 0 & 1 & 0 & | & 5 \\ 0 & 0 & 1 & | & 10 \end{bmatrix}$$

$$R_1 = -r_2 + r_1 \qquad\qquad R_3 = -r_3 \qquad\qquad R_1 = -4r_3 + r_1$$
$$R_3 = -r_2 + r_3 \qquad\qquad\qquad\qquad\qquad R_2 = 2r_3 + r_2$$

The company should produce 8 Deltas, 5 Betas, and 10 Sigmas.

83.   Rewrite the system to set up the matrix and solve:

$$\begin{cases} -4 + 8 - 2I_2 = 0 \\ 8 = 5I_4 + I_1 \\ 4 = 3I_3 + I_1 \\ I_3 + I_4 = I_1 \end{cases} \rightarrow \begin{cases} 2I_2 = 4 \\ I_1 + 5I_4 = 8 \\ I_1 + 3I_3 = 4 \\ I_1 - I_3 - I_4 = 0 \end{cases}$$

$$\begin{bmatrix} 0 & 2 & 0 & 0 & | & 4 \\ 1 & 0 & 0 & 5 & | & 8 \\ 1 & 0 & 3 & 0 & | & 4 \\ 1 & 0 & -1 & -1 & | & 0 \end{bmatrix} \rightarrow \begin{bmatrix} 1 & 0 & 0 & 5 & | & 8 \\ 0 & 2 & 0 & 0 & | & 4 \\ 1 & 0 & 3 & 0 & | & 4 \\ 1 & 0 & -1 & -1 & | & 0 \end{bmatrix} \rightarrow \begin{bmatrix} 1 & 0 & 0 & 5 & | & 8 \\ 0 & 1 & 0 & 0 & | & 2 \\ 0 & 0 & 3 & -5 & | & -4 \\ 0 & 0 & -1 & -6 & | & -8 \end{bmatrix}$$

Interchange r_2 and r_1 $R_2 = \tfrac{1}{2}r_2$
$$R_3 = -r_1 + r_3$$
$$R_4 = -r_1 + r_4$$

$$\rightarrow \begin{bmatrix} 1 & 0 & 0 & 5 & | & 8 \\ 0 & 1 & 0 & 0 & | & 2 \\ 0 & 0 & -1 & -6 & | & -8 \\ 0 & 0 & 3 & -5 & | & -4 \end{bmatrix} \rightarrow \begin{bmatrix} 1 & 0 & 0 & 5 & | & 8 \\ 0 & 1 & 0 & 0 & | & 2 \\ 0 & 0 & 1 & 6 & | & 8 \\ 0 & 0 & 0 & -23 & | & -28 \end{bmatrix} \rightarrow \begin{bmatrix} 1 & 0 & 0 & 5 & | & 8 \\ 0 & 1 & 0 & 0 & | & 2 \\ 0 & 0 & 1 & 6 & | & 8 \\ 0 & 0 & 0 & 1 & | & \tfrac{28}{23} \end{bmatrix}$$

Interchange r_3 and r_4 $R_3 = -r_3$ $\qquad\qquad\qquad\qquad R_4 = -\tfrac{1}{23}r_4$
$$R_4 = -3r_3 + r_4$$

$$\rightarrow \begin{bmatrix} 1 & 0 & 0 & 0 & \frac{44}{23} \\ 0 & 1 & 0 & 0 & 2 \\ 0 & 0 & 1 & 0 & \frac{16}{23} \\ 0 & 0 & 0 & 1 & \frac{28}{23} \end{bmatrix}$$

$$R_1 = -5r_4 + r_1$$
$$R_3 = -6r_4 + r_3$$

The solution is $I_1 = \dfrac{44}{23}$, $I_2 = 2$, $I_3 = \dfrac{16}{23}$, $I_4 = \dfrac{28}{23}$.

85.　Let $x =$ the amount invested in Treasury bills.
　　Let $y =$ the amount invested in corporate bonds.
　　Let $z =$ the amount invested in junk bonds.

(a)　Total investment equation:　　　　　　$x + y + z = 20,000$
　　　Annual income equation:　　　　　　　$0.07x + 0.09y + 0.11z = 2000$

　　　Set up a matrix and solve:

$$\begin{bmatrix} 1 & 1 & 1 & 20{,}000 \\ 0.07 & 0.09 & 0.11 & 2000 \end{bmatrix} \rightarrow \begin{bmatrix} 1 & 1 & 1 & 20{,}000 \\ 7 & 9 & 11 & 200{,}000 \end{bmatrix} \rightarrow \begin{bmatrix} 1 & 1 & 1 & 20{,}000 \\ 0 & 2 & 4 & 60{,}000 \end{bmatrix}$$

$$R_2 = 100r_2 \qquad\qquad R_2 = r_2 - 7r_1$$

$$\rightarrow \begin{bmatrix} 1 & 1 & 1 & 20{,}000 \\ 0 & 1 & 2 & 30{,}000 \end{bmatrix} \rightarrow \begin{bmatrix} 1 & 0 & -1 & -10{,}000 \\ 0 & 1 & 2 & 30{,}000 \end{bmatrix}$$

$$R_2 = \tfrac{1}{2}r_2 \qquad\qquad R_1 = r_1 - r_2$$

The matrix in the last step represents the system $\begin{cases} x - z = -10{,}000 \\ y + 2z = 30{,}000 \end{cases}$

Therefore the solution is $x = -10{,}000 + z$; $y = 30{,}000 - 2z$; z is any real number.
Possible investment strategies:

Amount invested at

7%	9%	11%
0	10,000	10,000
1000	8000	11,000
2000	6000	12,000
3000	4000	13,000
4000	2000	14,000
5000	0	15,000

(b)　Total investment equation:　　　　$x + y + z = 25,000$
　　　Annual income equation:　　　　　$0.07x + 0.09y + 0.11z = 2000$

Set up a matrix and solve:

$$\begin{bmatrix} 1 & 1 & 1 & | & 25{,}000 \\ 0.07 & 0.09 & 0.11 & | & 2000 \end{bmatrix} \rightarrow \begin{bmatrix} 1 & 1 & 1 & | & 25{,}000 \\ 7 & 9 & 11 & | & 200{,}000 \end{bmatrix} \rightarrow \begin{bmatrix} 1 & 1 & 1 & | & 25{,}000 \\ 0 & 2 & 4 & | & 25{,}000 \end{bmatrix}$$

$$R_2 = 100r_2 \qquad\qquad R_2 = r_2 - 7r_1$$

$$\rightarrow \begin{bmatrix} 1 & 1 & 1 & | & 25{,}000 \\ 0 & 1 & 2 & | & 12{,}500 \end{bmatrix} \rightarrow \begin{bmatrix} 1 & 0 & -1 & | & 12{,}500 \\ 0 & 1 & 2 & | & 12{,}500 \end{bmatrix}$$

$$R_2 = \tfrac{1}{2}r_2 \qquad\qquad R_1 = r_1 - r_2$$

The matrix in the last step represents the system $\quad \begin{cases} x - z = 12{,}500 \\ y + 2z = 12{,}500 \end{cases}$

Therefore the solution is $x = 12{,}500 + z$; $y = 12{,}500 - 2z$; z is any real number.
Possible investment strategies:

Amount invested at

7%	9%	11%
12,500	12,500	0
14,500	8500	2000
16,500	4500	4000
18,750	0	6250

(c)   Total investment equation: $\qquad x + y + z = 30{,}000$
Annual income equation: $\qquad 0.07x + 0.09y + 0.11z = 2000$
Set up a matrix and solve:

$$\begin{bmatrix} 1 & 1 & 1 & | & 30{,}000 \\ 0.07 & 0.09 & 0.11 & | & 2000 \end{bmatrix} \rightarrow \begin{bmatrix} 1 & 1 & 1 & | & 30{,}000 \\ 7 & 9 & 11 & | & 200{,}000 \end{bmatrix} \rightarrow \begin{bmatrix} 1 & 1 & 1 & | & 30{,}000 \\ 0 & 2 & 4 & | & -10{,}000 \end{bmatrix}$$

$$R_2 = 100r_2 \qquad\qquad R_1 = r_2 - 7r_1$$

$$\rightarrow \begin{bmatrix} 1 & 1 & 1 & | & 30{,}000 \\ 0 & 1 & 2 & | & -5000 \end{bmatrix} \rightarrow \begin{bmatrix} 1 & 0 & -1 & | & 35{,}000 \\ 0 & 1 & 2 & | & -5000 \end{bmatrix}$$

$$R_2 = \tfrac{1}{2}r_2 \qquad\qquad R_1 = r_1 - r_2$$

The matrix in the last step represents the system $\begin{cases} x - z = 35{,}000 \\ y + 2z = -5000 \end{cases}$

Therefore the solution is $x = 35{,}000 + z$; $y = -5000 - 2z$; z is any real number.
However, y and z cannot be negative. From $y = 5000 - 2z$, we must have $y = z = 0$.
One possible investment strategy

Amount invested at

7%	9%	11%
30,000	0	0

This will yield ($30,000)(0.07) = $2100, which is more than the required income.

(d)   Answers will vary.

87. Let x = the amount of liquid 1.
Let y = the amount of liquid 2.
Let z = the amount of liquid 3.

$$0.20x + 0.40y + 0.30z = 40 \quad \text{Vitamin C}$$

$$0.30x + 0.20y + 0.50z = 30 \quad \text{Vitamin D}$$

multiplying each equation by 10 yields

$$2x + 4y + 3z = 400$$
$$3x + 2y + 5z = 300$$

Set up a matrix and solve:

$$\begin{bmatrix} 2 & 4 & 3 & | & 400 \\ 3 & 2 & 5 & | & 300 \end{bmatrix} \rightarrow \begin{bmatrix} 1 & 2 & \frac{3}{2} & | & 200 \\ 3 & 2 & 5 & | & 300 \end{bmatrix} \rightarrow \begin{bmatrix} 1 & 2 & \frac{3}{2} & | & 200 \\ 0 & -4 & \frac{1}{2} & | & -300 \end{bmatrix}$$

$$R_1 = \tfrac{1}{2} r_1 \qquad\qquad R_2 = r_2 - 3r_1$$

$$\rightarrow \begin{bmatrix} 1 & 2 & \frac{3}{2} & | & 200 \\ 0 & 1 & -\frac{1}{8} & | & 75 \end{bmatrix} \rightarrow \begin{bmatrix} 1 & 0 & \frac{7}{4} & | & 50 \\ 0 & 1 & -\frac{1}{8} & | & 75 \end{bmatrix}$$

$$R_2 = -\tfrac{1}{4} r_2 \qquad\qquad R_1 = r_1 - 2r_2$$

The matrix in the last step represents the system $\begin{cases} x + \frac{7}{4}z = 50 \\ y - \frac{1}{8}z = 75 \end{cases}$

Therefore the solution is $x = 50 - \dfrac{7}{4}z$; $y = 75 + \dfrac{1}{8}z$; z is any real number

Possible combinations:

Liquid 1	Liquid 2	Liquid 3
50mg	75mg	0mg
36mg	76mg	8mg
22mg	77mg	16mg
8mg	78mg	24mg

89–91. Answers will vary.

Systems of Equations and Inequalities

10.3 Systems of Linear Equations: Determinants

1. determinants

3. False

5. $\begin{vmatrix} 3 & 1 \\ 4 & 2 \end{vmatrix} = 3(2) - 4(1) = 6 - 4 = 2$

7. $\begin{vmatrix} 6 & 4 \\ -1 & 3 \end{vmatrix} = 6(3) - (-1)(4) = 18 + 4 = 22$

9. $\begin{vmatrix} -3 & -1 \\ 4 & 2 \end{vmatrix} = -3(2) - 4(-1) = -6 + 4 = -2$

11. $\begin{vmatrix} 3 & 4 & 2 \\ 1 & -1 & 5 \\ 1 & 2 & -2 \end{vmatrix} = 3\begin{vmatrix} -1 & 5 \\ 2 & -2 \end{vmatrix} - 4\begin{vmatrix} 1 & 5 \\ 1 & -2 \end{vmatrix} + 2\begin{vmatrix} 1 & -1 \\ 1 & 2 \end{vmatrix}$

$$= 3\left[(-1)(-2) - 2(5)\right] - 4\left[1(-2) - 1(5)\right] + 2\left[1(2) - 1(-1)\right]$$

$$= 3(-8) - 4(-7) + 2(3)$$

$$= -24 + 28 + 6 = 10$$

13. $\begin{vmatrix} 4 & -1 & 2 \\ 6 & -1 & 0 \\ 1 & -3 & 4 \end{vmatrix} = 4\begin{vmatrix} -1 & 0 \\ -3 & 4 \end{vmatrix} - (-1)\begin{vmatrix} 6 & 0 \\ 1 & 4 \end{vmatrix} + 2\begin{vmatrix} 6 & -1 \\ 1 & -3 \end{vmatrix}$

$$= 4\left[-1(4) - 0(-3)\right] + 1\left[6(4) - 1(0)\right] + 2\left[6(-3) - 1(-1)\right]$$

$$= 4(-4) + 1(24) + 2(-17)$$

$$= -16 + 24 - 34$$

$$= -26$$

15. $\begin{cases} x + y = 8 \\ x - y = 4 \end{cases}$

$$D = \begin{vmatrix} 1 & 1 \\ 1 & -1 \end{vmatrix} = -1 - 1 = -2; \quad D_x = \begin{vmatrix} 8 & 1 \\ 4 & -1 \end{vmatrix} = -8 - 4 = -12; \quad D_y = \begin{vmatrix} 1 & 8 \\ 1 & 4 \end{vmatrix} = 4 - 8 = -4$$

Find the solutions by Cramer's Rule: $x = \dfrac{D_x}{D} = \dfrac{-12}{-2} = 6 \qquad y = \dfrac{D_y}{D} = \dfrac{-4}{-2} = 2$

17. $\begin{cases} 5x - y = 13 \\ 2x + 3y = 12 \end{cases}$

$$D = \begin{vmatrix} 5 & -1 \\ 2 & 3 \end{vmatrix} = 15 + 2 = 17$$

$$D_x = \begin{vmatrix} 13 & -1 \\ 12 & 3 \end{vmatrix} = 39 + 12 = 51; \qquad D_y = \begin{vmatrix} 5 & 13 \\ 2 & 12 \end{vmatrix} = 60 - 26 = 34$$

Find the solutions by Cramer's Rule: $x = \dfrac{D_x}{D} = \dfrac{51}{17} = 3 \qquad y = \dfrac{D_y}{D} = \dfrac{34}{17} = 2$

19. $\begin{cases} 3x \quad\;\; = 24 \\ x + 2y = \;\; 0 \end{cases}$

$$D = \begin{vmatrix} 3 & 0 \\ 1 & 2 \end{vmatrix} = 6 - 0 = 6$$

$$D_x = \begin{vmatrix} 24 & 0 \\ 0 & 2 \end{vmatrix} = 48 - 0 = 48; \qquad D_y = \begin{vmatrix} 3 & 24 \\ 1 & 0 \end{vmatrix} = 0 - 24 = -24$$

Find the solutions by Cramer's Rule: $x = \dfrac{D_x}{D} = \dfrac{48}{6} = 8 \qquad y = \dfrac{D_y}{D} = \dfrac{-24}{6} = -4$

21. $\begin{cases} 3x - 6y = 24 \\ 5x + 4y = 12 \end{cases}$

$$D = \begin{vmatrix} 3 & -6 \\ 5 & 4 \end{vmatrix} = 12 + 30 = 42$$

$$D_x = \begin{vmatrix} 24 & -6 \\ 12 & 4 \end{vmatrix} = 96 + 72 = 168; \qquad D_y = \begin{vmatrix} 3 & 24 \\ 5 & 12 \end{vmatrix} = 36 - 120 = -84$$

Find the solutions by Cramer's Rule: $x = \dfrac{D_x}{D} = \dfrac{168}{42} = 4 \qquad y = \dfrac{D_y}{D} = \dfrac{-84}{42} = -2$

23. $\begin{cases} 3x - 2y = 4 \\ 6x - 4y = 0 \end{cases}$

$$D = \begin{vmatrix} 3 & -2 \\ 6 & -4 \end{vmatrix} = -12 - (-12) = 0$$

Since $D = 0$, Cramer's Rule does not apply.

25. $\begin{cases} 2x - 4y = -2 \\ 3x + 2y = \;\; 3 \end{cases}$

$$D = \begin{vmatrix} 2 & -4 \\ 3 & 2 \end{vmatrix} = 4 + 12 = 16$$

$$D_x = \begin{vmatrix} -2 & -4 \\ 3 & 2 \end{vmatrix} = -4 + 12 = 8; \qquad D_y = \begin{vmatrix} 2 & -2 \\ 3 & 3 \end{vmatrix} = 6 + 6 = 12$$

Find the solutions by Cramer's Rule: $x = \dfrac{D_x}{D} = \dfrac{8}{16} = \dfrac{1}{2} \qquad y = \dfrac{D_y}{D} = \dfrac{12}{16} = \dfrac{3}{4}$

27. $\begin{cases} 2x - 3y = -1 \\ 10x + 10y = 5 \end{cases}$

$$D = \begin{vmatrix} 2 & -3 \\ 10 & 10 \end{vmatrix} = 20 + 30 = 50$$

$$D_x = \begin{vmatrix} -1 & -3 \\ 5 & 10 \end{vmatrix} = -10 + 15 = 5; \qquad D_y = \begin{vmatrix} 2 & -1 \\ 10 & 5 \end{vmatrix} = 10 + 10 = 20$$

Find the solutions by Cramer's Rule: $x = \dfrac{D_x}{D} = \dfrac{5}{50} = \dfrac{1}{10}$ $\qquad y = \dfrac{D_y}{D} = \dfrac{20}{50} = \dfrac{2}{5}$

29. $\begin{cases} 2x + 3y = 6 \\ x - y = \dfrac{1}{2} \end{cases}$

$$D = \begin{vmatrix} 2 & 3 \\ 1 & -1 \end{vmatrix} = -2 - 3 = -5$$

$$D_x = \begin{vmatrix} 6 & 3 \\ \frac{1}{2} & -1 \end{vmatrix} = -6 - \frac{3}{2} = -\frac{15}{2}; \qquad D_y = \begin{vmatrix} 2 & 6 \\ 1 & \frac{1}{2} \end{vmatrix} = 1 - 6 = -5$$

Find the solutions by Cramer's Rule: $x = \dfrac{D_x}{D} = \dfrac{-\frac{15}{2}}{-5} = \dfrac{3}{2}$ $\qquad y = \dfrac{D_y}{D} = \dfrac{-5}{-5} = 1$

31. $\begin{cases} 3x - 5y = 3 \\ 15x + 5y = 21 \end{cases}$

$$D = \begin{vmatrix} 3 & -5 \\ 15 & 5 \end{vmatrix} = 15 + 75 = 90$$

$$D_x = \begin{vmatrix} 3 & -5 \\ 21 & 5 \end{vmatrix} = 15 + 105 = 120; \qquad D_y = \begin{vmatrix} 3 & 3 \\ 15 & 21 \end{vmatrix} = 63 - 45 = 18$$

Find the solutions by Cramer's Rule: $x = \dfrac{D_x}{D} = \dfrac{120}{90} = \dfrac{4}{3}$ $\qquad y = \dfrac{D_y}{D} = \dfrac{18}{90} = \dfrac{1}{5}$

33. $\begin{cases} x + y - z = 6 \\ 3x - 2y + z = -5 \\ x + 3y - 2z = 14 \end{cases}$

$$D = \begin{vmatrix} 1 & 1 & -1 \\ 3 & -2 & 1 \\ 1 & 3 & -2 \end{vmatrix} = 1 \begin{vmatrix} -2 & 1 \\ 3 & -2 \end{vmatrix} - 1 \begin{vmatrix} 3 & 1 \\ 1 & -2 \end{vmatrix} + (-1) \begin{vmatrix} 3 & -2 \\ 1 & 3 \end{vmatrix}$$

$$= 1(4 - 3) - 1(-6 - 1) - 1(9 + 2) = 1 + 7 - 11 = -3$$

$$D_x = \begin{vmatrix} 6 & 1 & -1 \\ -5 & -2 & 1 \\ 14 & 3 & -2 \end{vmatrix} = 6 \begin{vmatrix} -2 & 1 \\ 3 & -2 \end{vmatrix} - 1 \begin{vmatrix} -5 & 1 \\ 14 & -2 \end{vmatrix} + (-1) \begin{vmatrix} -5 & -2 \\ 14 & 3 \end{vmatrix}$$

$$= 6(4 - 3) - 1(10 - 14) - 1(-15 + 28) = 6 + 4 - 13 = -3$$

$$D_y = \begin{vmatrix} 1 & 6 & -1 \\ 3 & -5 & 1 \\ 1 & 14 & -2 \end{vmatrix} = 1 \begin{vmatrix} -5 & 1 \\ 14 & -2 \end{vmatrix} - 6 \begin{vmatrix} 3 & 1 \\ 1 & -2 \end{vmatrix} + (-1) \begin{vmatrix} 3 & -5 \\ 1 & 14 \end{vmatrix}$$

$$= 1(10 - 14) - 6(-6 - 1) - 1(42 + 5) = -4 + 42 - 47 = -9$$

$$D_z = \begin{vmatrix} 1 & 1 & 6 \\ 3 & -2 & -5 \\ 1 & 3 & 14 \end{vmatrix} = 1 \begin{vmatrix} -2 & -5 \\ 3 & 14 \end{vmatrix} - 1 \begin{vmatrix} 3 & -5 \\ 1 & 14 \end{vmatrix} + 6 \begin{vmatrix} 3 & -2 \\ 1 & 3 \end{vmatrix}$$

$$= 1(-28 + 15) - 1(42 + 5) + 6(9 + 2) = -13 - 47 + 66 = 6$$

Find the solutions by Cramer's Rule:

$$x = \frac{D_x}{D} = \frac{-3}{-3} = 1 \qquad y = \frac{D_y}{D} = \frac{-9}{-3} = 3 \qquad z = \frac{D_z}{D} = \frac{6}{-3} = -2$$

35. $\begin{cases} x + 2y - z = -3 \\ 2x - 4y + z = -7 \\ -2x + 2y - 3z = 4 \end{cases}$

$$D = \begin{vmatrix} 1 & 2 & -1 \\ 2 & -4 & 1 \\ -2 & 2 & -3 \end{vmatrix} = 1 \begin{vmatrix} -4 & 1 \\ 2 & -3 \end{vmatrix} - 2 \begin{vmatrix} 2 & 1 \\ -2 & -3 \end{vmatrix} + (-1) \begin{vmatrix} 2 & -4 \\ -2 & 2 \end{vmatrix}$$

$$= 1(12 - 2) - 2(-6 + 2) - 1(4 - 8) = 10 + 8 + 4 = 22$$

$$D_x = \begin{vmatrix} -3 & 2 & -1 \\ -7 & -4 & 1 \\ 4 & 2 & -3 \end{vmatrix} = -3 \begin{vmatrix} -4 & 1 \\ 2 & -3 \end{vmatrix} - 2 \begin{vmatrix} -7 & 1 \\ 4 & -3 \end{vmatrix} + (-1) \begin{vmatrix} -7 & -4 \\ 4 & 2 \end{vmatrix}$$

$$= -3(12 - 2) - 2(21 - 4) - 1(-14 + 16) = -30 - 34 - 2 = -66$$

$$D_y = \begin{vmatrix} 1 & -3 & -1 \\ 2 & -7 & 1 \\ -2 & 4 & -3 \end{vmatrix} = 1 \begin{vmatrix} -7 & 1 \\ 4 & -3 \end{vmatrix} - (-3) \begin{vmatrix} 2 & 1 \\ -2 & -3 \end{vmatrix} + (-1) \begin{vmatrix} 2 & -7 \\ -2 & 4 \end{vmatrix}$$

$$= 1(21 - 4) + 3(-6 + 2) - 1(8 - 14) = 17 - 12 + 6 = 11$$

$$D_z = \begin{vmatrix} 1 & 2 & -3 \\ 2 & -4 & -7 \\ -2 & 2 & 4 \end{vmatrix} = 1\begin{vmatrix} -4 & -7 \\ 2 & 4 \end{vmatrix} - 2\begin{vmatrix} 2 & -7 \\ -2 & 4 \end{vmatrix} + (-3)\begin{vmatrix} 2 & -4 \\ -2 & 2 \end{vmatrix}$$

$$= 1(-16+14) - 2(8-14) - 3(4-8) = -2 + 12 + 12 = 22$$

Find the solutions by Cramer's Rule:

$$x = \frac{D_x}{D} = \frac{-66}{22} = -3 \qquad y = \frac{D_y}{D} = \frac{11}{22} = \frac{1}{2} \qquad z = \frac{D_z}{D} = \frac{22}{22} = 1$$

37. $\begin{cases} x - 2y + 3z = 1 \\ 3x + y - 2z = 0 \\ 2x - 4y + 6z = 2 \end{cases}$

$$D = \begin{vmatrix} 1 & -2 & 3 \\ 3 & 1 & -2 \\ 2 & -4 & 6 \end{vmatrix} = 1\begin{vmatrix} 1 & -2 \\ -4 & 6 \end{vmatrix} - (-2)\begin{vmatrix} 3 & -2 \\ 2 & 6 \end{vmatrix} + 3\begin{vmatrix} 3 & 1 \\ 2 & -4 \end{vmatrix}$$

$$= 1(6-8) + 2(18+4) + 3(-12-2) = -2 + 44 - 42 = 0$$

Since $D = 0$, Cramer's Rule does not apply.

39. $\begin{cases} x + 2y - z = 0 \\ 2x - 4y + z = 0 \\ -2x + 2y - 3z = 0 \end{cases}$

$$D = \begin{vmatrix} 1 & 2 & -1 \\ 2 & -4 & 1 \\ -2 & 2 & -3 \end{vmatrix} = 1\begin{vmatrix} -4 & 1 \\ 2 & -3 \end{vmatrix} - 2\begin{vmatrix} 2 & 1 \\ -2 & -3 \end{vmatrix} + (-1)\begin{vmatrix} 2 & -4 \\ -2 & 2 \end{vmatrix}$$

$$= 1(12-2) - 2(-6+2) - 1(4-8) = 10 + 8 + 4 = 22$$

$$D_x = \begin{vmatrix} 0 & 2 & -1 \\ 0 & -4 & 1 \\ 0 & 2 & -3 \end{vmatrix} = 0 \quad \text{(By Theorem 12)}$$

$$D_y = \begin{vmatrix} 1 & 0 & -1 \\ 2 & 0 & 1 \\ -2 & 0 & -3 \end{vmatrix} = 0 \quad \text{(By Theorem 12)}; \quad D_z = \begin{vmatrix} 1 & 2 & 0 \\ 2 & -4 & 0 \\ -2 & 2 & 0 \end{vmatrix} = 0 \quad \text{(By Theorem 12)}$$

Find the solutions by Cramer's Rule:

$$x = \frac{D_x}{D} = \frac{0}{22} = 0 \qquad y = \frac{D_y}{D} = \frac{0}{22} = 0 \qquad z = \frac{D_z}{D} = \frac{0}{22} = 0$$

41. $\begin{cases} x - 2y + 3z = 0 \\ 3x + y - 2z = 0 \\ 2x - 4y + 6z = 0 \end{cases}$

$$D = \begin{vmatrix} 1 & -2 & 3 \\ 3 & 1 & -2 \\ 2 & -4 & 6 \end{vmatrix} = 1\begin{vmatrix} 1 & -2 \\ -4 & 6 \end{vmatrix} - (-2)\begin{vmatrix} 3 & -2 \\ 2 & 6 \end{vmatrix} + 3\begin{vmatrix} 3 & 1 \\ 2 & -4 \end{vmatrix}$$

$$= 1(6 - 8) + 2(18 + 4) + 3(-12 - 2) = -2 + 44 - 42 = 0$$

Since $D = 0$, Cramer's Rule does not apply.

43. Solve for x:

$$\begin{vmatrix} x & x \\ 4 & 3 \end{vmatrix} = 5$$

$$3x - 4x = 5$$

$$-x = 5$$

$$x = -5$$

45. Solve for x:

$$\begin{vmatrix} x & 1 & 1 \\ 4 & 3 & 2 \\ -1 & 2 & 5 \end{vmatrix} = 2$$

$$x\begin{vmatrix} 3 & 2 \\ 2 & 5 \end{vmatrix} - 1\begin{vmatrix} 4 & 2 \\ -1 & 5 \end{vmatrix} + 1\begin{vmatrix} 4 & 3 \\ -1 & 2 \end{vmatrix} = 2$$

$$x(15 - 4) - (20 + 2) + (8 + 3) = 2$$

$$11x - 22 + 11 = 2$$

$$11x = 13$$

$$x = \frac{13}{11}$$

47. Solve for x:

$$\begin{vmatrix} x & 2 & 3 \\ 1 & x & 0 \\ 6 & 1 & -2 \end{vmatrix} = 7$$

$$x\begin{vmatrix} x & 0 \\ 1 & -2 \end{vmatrix} - 2\begin{vmatrix} 1 & 0 \\ 6 & -2 \end{vmatrix} + 3\begin{vmatrix} 1 & x \\ 6 & 1 \end{vmatrix} = 7$$

$$x(-2x) - 2(-2) + 3(1 - 6x) = 7$$

$$-2x^2 + 4 + 3 - 18x = 7$$

$$-2x^2 - 18x = 0$$

$$-2x(x + 9) = 0 \Rightarrow x = 0 \text{ or } x = -9$$

49. Let $\begin{vmatrix} x & y & z \\ u & v & w \\ 1 & 2 & 3 \end{vmatrix} = 4$

Then $\begin{vmatrix} 1 & 2 & 3 \\ u & v & w \\ x & y & z \end{vmatrix} = -4$ by Theorem 11

The value of the determinant changes sign when two rows are interchanged.

51. Let $\begin{vmatrix} x & y & z \\ u & v & w \\ 1 & 2 & 3 \end{vmatrix} = 4$

$$\begin{vmatrix} x & y & z \\ -3 & -6 & -9 \\ u & v & w \end{vmatrix} = -3\begin{vmatrix} x & y & z \\ 1 & 2 & 3 \\ u & v & w \end{vmatrix} = -3(-1)\begin{vmatrix} x & y & z \\ u & v & w \\ 1 & 2 & 3 \end{vmatrix} = 3(4) = 12$$

$$\qquad\qquad\quad \text{Theorem 14} \qquad\quad \text{Theorem 11}$$

53. Let $\begin{vmatrix} x & y & z \\ u & v & w \\ 1 & 2 & 3 \end{vmatrix} = 4$

$$\begin{vmatrix} 1 & 2 & 3 \\ x-3 & y-6 & z-9 \\ 2u & 2v & 2w \end{vmatrix} = 2\begin{vmatrix} 1 & 2 & 3 \\ x-3 & y-6 & z-9 \\ u & v & w \end{vmatrix} = 2(-1)\begin{vmatrix} x-3 & y-6 & z-9 \\ 1 & 2 & 3 \\ u & v & w \end{vmatrix}$$

$$\qquad\qquad\quad \text{Theorem 14} \qquad\qquad\qquad \text{Theorem 11}$$

$$= 2(-1)(-1)\begin{vmatrix} x-3 & y-6 & z-9 \\ u & v & w \\ 1 & 2 & 3 \end{vmatrix} = 2(-1)(-1)\begin{vmatrix} x & y & z \\ u & v & w \\ 1 & 2 & 3 \end{vmatrix}$$

$$\qquad\qquad\quad \text{Theorem 11} \qquad\qquad\qquad\qquad\qquad \text{Theorem 15 } (R_1 = -3r_3 + r_1)$$

$$= 2(-1)(-1)(4) = 8$$

55. Let $\begin{vmatrix} x & y & z \\ u & v & w \\ 1 & 2 & 3 \end{vmatrix} = 4$

$$\begin{vmatrix} 1 & 2 & 3 \\ 2x & 2y & 2z \\ u-1 & v-2 & w-3 \end{vmatrix} = 2\begin{vmatrix} 1 & 2 & 3 \\ x & y & z \\ u-1 & v-2 & w-3 \end{vmatrix} = 2(-1)\begin{vmatrix} x & y & z \\ 1 & 2 & 3 \\ u-1 & v-2 & w-3 \end{vmatrix}$$

$$\qquad\qquad\quad \text{Theorem 14} \qquad\qquad\qquad \text{Theorem 11}$$

$$= 2(-1)(-1)\begin{vmatrix} x & y & z \\ u-1 & v-2 & w-3 \\ 1 & 2 & 3 \end{vmatrix} = 2(-1)(-1)\begin{vmatrix} x & y & z \\ u & v & w \\ 1 & 2 & 3 \end{vmatrix}$$

$$\qquad\qquad\quad \text{Theorem 11} \qquad\qquad\qquad\qquad\qquad \text{Theorem 15 } (R_2 = -r_3 + r_2)$$

$$= 2(-1)(-1)(4) = 8$$

57.   Expanding the determinant:

$$\begin{vmatrix} x & y & 1 \\ x_1 & y_1 & 1 \\ x_2 & y_2 & 1 \end{vmatrix} = 0$$

$$x\begin{vmatrix} y_1 & 1 \\ y_2 & 1 \end{vmatrix} - y\begin{vmatrix} x_1 & 1 \\ x_2 & 1 \end{vmatrix} + 1\begin{vmatrix} x_1 & y_1 \\ x_2 & y_2 \end{vmatrix} = 0$$

$$x(y_1 - y_2) - y(x_1 - x_2) + (x_1 y_2 - x_2 y_1) = 0$$

$$x(y_1 - y_2) + y(x_2 - x_1) = x_2 y_1 - x_1 y_2$$

$$y(x_2 - x_1) = x_2 y_1 - x_1 y_2 + x(y_2 - y_1)$$

$$y(x_2 - x_1) - y_1(x_2 - x_1) = x_2 y_1 - x_1 y_2 + x(y_2 - y_1) - y_1(x_2 - x_1)$$

$$(x_2 - x_1)(y - y_1) = x(y_2 - y_1) + x_2 y_1 - x_1 y_2 - y_1 x_2 + y_1 x_1$$

$$(x_2 - x_1)(y - y_1) = (y_2 - y_1)x - (y_2 - y_1)x_1$$

$$(x_2 - x_1)(y - y_1) = (y_2 - y_1)(x - x_1)$$

$$(y - y_1) = \frac{(y_2 - y_1)}{(x_2 - x_1)}(x - x_1)$$

This is the 2-point form of the equation for a line.

59.   Expanding the determinant:

$$\begin{vmatrix} x^2 & x & 1 \\ y^2 & y & 1 \\ z^2 & z & 1 \end{vmatrix} = x^2\begin{vmatrix} y & 1 \\ z & 1 \end{vmatrix} - x\begin{vmatrix} y^2 & 1 \\ z^2 & 1 \end{vmatrix} + 1\begin{vmatrix} y^2 & y \\ z^2 & z \end{vmatrix}$$

$$= x^2(y - z) - x(y^2 - z^2) + 1(y^2 z - z^2 y)$$

$$= x^2(y - z) - x(y - z)(y + z) + yz(y - z)$$

$$= (y - z)\left[x^2 - xy - xz + yz \right]$$

$$= (y - z)\left[x(x - y) - z(x - y) \right] = (y - z)(x - y)(x - z)$$

61. Evaluating the determinant to show the relationship:

$$\begin{vmatrix} a_{13} & a_{12} & a_{11} \\ a_{23} & a_{22} & a_{21} \\ a_{33} & a_{32} & a_{31} \end{vmatrix} = a_{13}\begin{vmatrix} a_{22} & a_{21} \\ a_{32} & a_{31} \end{vmatrix} - a_{12}\begin{vmatrix} a_{23} & a_{21} \\ a_{33} & a_{31} \end{vmatrix} + a_{11}\begin{vmatrix} a_{23} & a_{22} \\ a_{33} & a_{32} \end{vmatrix}$$

$$= a_{13}(a_{22}a_{31} - a_{21}a_{32}) - a_{12}(a_{23}a_{31} - a_{21}a_{33}) + a_{11}(a_{23}a_{32} - a_{22}a_{33})$$

$$= a_{13}a_{22}a_{31} - a_{13}a_{21}a_{32} - a_{12}a_{23}a_{31} + a_{12}a_{21}a_{33} + a_{11}a_{23}a_{32} - a_{11}a_{22}a_{33}$$

$$= -a_{11}a_{22}a_{33} + a_{11}a_{23}a_{32} + a_{12}a_{21}a_{33} - a_{12}a_{23}a_{31} - a_{13}a_{21}a_{32} + a_{13}a_{22}a_{31}$$

$$= -a_{11}(a_{22}a_{33} - a_{23}a_{32}) + a_{12}(a_{21}a_{33} - a_{23}a_{31}) - a_{13}(a_{21}a_{32} - a_{22}a_{31})$$

$$= -a_{11}\begin{vmatrix} a_{22} & a_{23} \\ a_{32} & a_{33} \end{vmatrix} + a_{12}\begin{vmatrix} a_{21} & a_{23} \\ a_{31} & a_{33} \end{vmatrix} - a_{13}\begin{vmatrix} a_{21} & a_{22} \\ a_{31} & a_{32} \end{vmatrix}$$

$$= -\left(a_{11}\begin{vmatrix} a_{22} & a_{23} \\ a_{32} & a_{33} \end{vmatrix} - a_{12}\begin{vmatrix} a_{21} & a_{23} \\ a_{31} & a_{33} \end{vmatrix} + a_{13}\begin{vmatrix} a_{21} & a_{22} \\ a_{31} & a_{32} \end{vmatrix} \right)$$

$$= -\begin{vmatrix} a_{11} & a_{12} & a_{13} \\ a_{21} & a_{22} & a_{23} \\ a_{31} & a_{32} & a_{33} \end{vmatrix}$$

63. Set up a 3 by 3 determinant in which the first column and third column are the same and evaluate by expanding down column 2:

$$\begin{vmatrix} a & b & a \\ c & d & c \\ e & f & e \end{vmatrix} = -b\begin{vmatrix} c & c \\ e & e \end{vmatrix} + d\begin{vmatrix} a & a \\ e & e \end{vmatrix} - f\begin{vmatrix} a & a \\ c & c \end{vmatrix}$$

$$= -b(ce - ce) + d(ae - ae) - f(ac - ac)$$

$$= -b(0) + d(0) - f(0) = 0$$

Systems of Equations and Inequalities

10.4 Matrix Algebra

1. inverse

3. identity

5. False

7. $A + B = \begin{bmatrix} 0 & 3 & -5 \\ 1 & 2 & 6 \end{bmatrix} + \begin{bmatrix} 4 & 1 & 0 \\ -2 & 3 & -2 \end{bmatrix} = \begin{bmatrix} 0+4 & 3+1 & -5+0 \\ 1+(-2) & 2+3 & 6+(-2) \end{bmatrix} = \begin{bmatrix} 4 & 4 & -5 \\ -1 & 5 & 4 \end{bmatrix}$

9. $4A = 4\begin{bmatrix} 0 & 3 & -5 \\ 1 & 2 & 6 \end{bmatrix} = \begin{bmatrix} 4\cdot 0 & 4\cdot 3 & 4(-5) \\ 4\cdot 1 & 4\cdot 2 & 4\cdot 6 \end{bmatrix} = \begin{bmatrix} 0 & 12 & -20 \\ 4 & 8 & 24 \end{bmatrix}$

11. $3A - 2B = 3\begin{bmatrix} 0 & 3 & -5 \\ 1 & 2 & 6 \end{bmatrix} - 2\begin{bmatrix} 4 & 1 & 0 \\ -2 & 3 & -2 \end{bmatrix} = \begin{bmatrix} 0 & 9 & -15 \\ 3 & 6 & 18 \end{bmatrix} - \begin{bmatrix} 8 & 2 & 0 \\ -4 & 6 & -4 \end{bmatrix} = \begin{bmatrix} -8 & 7 & -15 \\ 7 & 0 & 22 \end{bmatrix}$

13. $AC = \begin{bmatrix} 0 & 3 & -5 \\ 1 & 2 & 6 \end{bmatrix} \cdot \begin{bmatrix} 4 & 1 \\ 6 & 2 \\ -2 & 3 \end{bmatrix} = \begin{bmatrix} 0(4)+3(6)+(-5)(-2) & 0(1)+3(2)+(-5)(3) \\ 1(4)+2(6)+6(-2) & 1(1)+2(2)+6(3) \end{bmatrix} = \begin{bmatrix} 28 & -9 \\ 4 & 23 \end{bmatrix}$

15. $CA = \begin{bmatrix} 4 & 1 \\ 6 & 2 \\ -2 & 3 \end{bmatrix} \cdot \begin{bmatrix} 0 & 3 & -5 \\ 1 & 2 & 6 \end{bmatrix} = \begin{bmatrix} 4(0)+1(1) & 4(3)+1(2) & 4(-5)+1(6) \\ 6(0)+2(1) & 6(3)+2(2) & 6(-5)+2(6) \\ -2(0)+3(1) & -2(3)+3(2) & -2(-5)+3(6) \end{bmatrix} = \begin{bmatrix} 1 & 14 & -14 \\ 2 & 22 & -18 \\ 3 & 0 & 28 \end{bmatrix}$

17. $C(A+B) = \begin{bmatrix} 4 & 1 \\ 6 & 2 \\ -2 & 3 \end{bmatrix} \left(\begin{bmatrix} 0 & 3 & -5 \\ 1 & 2 & 6 \end{bmatrix} + \begin{bmatrix} 4 & 1 & 0 \\ -2 & 3 & -2 \end{bmatrix} \right) = \begin{bmatrix} 4 & 1 \\ 6 & 2 \\ -2 & 3 \end{bmatrix} \cdot \begin{bmatrix} 4 & 4 & -5 \\ -1 & 5 & 4 \end{bmatrix} = \begin{bmatrix} 15 & 21 & -16 \\ 22 & 34 & -22 \\ -11 & 7 & 22 \end{bmatrix}$

19. $AC - 3I_2 = \begin{bmatrix} 0 & 3 & -5 \\ 1 & 2 & 6 \end{bmatrix} \cdot \begin{bmatrix} 4 & 1 \\ 6 & 2 \\ -2 & 3 \end{bmatrix} - 3\begin{bmatrix} 1 & 0 \\ 0 & 1 \end{bmatrix} = \begin{bmatrix} 28 & -9 \\ 4 & 23 \end{bmatrix} - \begin{bmatrix} 3 & 0 \\ 0 & 3 \end{bmatrix} = \begin{bmatrix} 25 & -9 \\ 4 & 20 \end{bmatrix}$

21.　$CA - CB = \begin{bmatrix} 4 & 1 \\ 6 & 2 \\ -2 & 3 \end{bmatrix} \cdot \begin{bmatrix} 0 & 3 & -5 \\ 1 & 2 & 6 \end{bmatrix} - \begin{bmatrix} 4 & 1 \\ 6 & 2 \\ -2 & 3 \end{bmatrix} \cdot \begin{bmatrix} 4 & 1 & 0 \\ -2 & 3 & -2 \end{bmatrix}$

$= \begin{bmatrix} 1 & 14 & -14 \\ 2 & 22 & -18 \\ 3 & 0 & 28 \end{bmatrix} - \begin{bmatrix} 14 & 7 & -2 \\ 20 & 12 & -4 \\ -14 & 7 & -6 \end{bmatrix} = \begin{bmatrix} -13 & 7 & -12 \\ -18 & 10 & -14 \\ 17 & -7 & 34 \end{bmatrix}$

23.　$\begin{bmatrix} 2 & -2 \\ 1 & 0 \end{bmatrix} \begin{bmatrix} 2 & 1 & 4 & 6 \\ 3 & -1 & 3 & 2 \end{bmatrix}$

$= \begin{bmatrix} 2(2)+(-2)(3) & 2(1)+(-2)(-1) & 2(4)+(-2)(3) & 2(6)+(-2)(2) \\ 1(2)+0(3) & 1(1)+0(-1) & 1(4)+0(3) & 1(6)+0(2) \end{bmatrix}$

$= \begin{bmatrix} -2 & 4 & 2 & 8 \\ 2 & 1 & 4 & 6 \end{bmatrix}$

25.　$\begin{bmatrix} 1 & 2 & 3 \\ 0 & -1 & 4 \end{bmatrix} \begin{bmatrix} 1 & 2 \\ -1 & 0 \\ 2 & 4 \end{bmatrix} = \begin{bmatrix} 1(1)+2(-1)+3(2) & 1(2)+2(0)+3(4) \\ 0(1)+(-1)(-1)+4(2) & 0(2)+(-1)(0)+4(4) \end{bmatrix} = \begin{bmatrix} 5 & 14 \\ 9 & 16 \end{bmatrix}$

27.　$\begin{bmatrix} 1 & 0 & 1 \\ 2 & 4 & 1 \\ 3 & 6 & 1 \end{bmatrix} \begin{bmatrix} 1 & 3 \\ 6 & 2 \\ 8 & -1 \end{bmatrix} = \begin{bmatrix} 1(1)+0(6)+1(8) & 1(3)+0(2)+1(-1) \\ 2(1)+4(6)+1(8) & 2(3)+4(2)+1(-1) \\ 3(1)+6(6)+1(8) & 3(3)+6(2)+1(-1) \end{bmatrix} = \begin{bmatrix} 9 & 2 \\ 34 & 13 \\ 47 & 20 \end{bmatrix}$

29.　Augment the matrix with the identity and use row operations to find the inverse:

$A = \begin{bmatrix} 2 & 1 \\ 1 & 1 \end{bmatrix} \rightarrow \left[\begin{array}{cc|cc} 2 & 1 & 1 & 0 \\ 1 & 1 & 0 & 1 \end{array} \right]$

$\rightarrow \left[\begin{array}{cc|cc} 1 & 1 & 0 & 1 \\ 2 & 1 & 1 & 0 \end{array} \right] \rightarrow \left[\begin{array}{cc|cc} 1 & 1 & 0 & 1 \\ 0 & -1 & 1 & -2 \end{array} \right] \rightarrow \left[\begin{array}{cc|cc} 1 & 1 & 0 & 1 \\ 0 & 1 & -1 & 2 \end{array} \right] \rightarrow \left[\begin{array}{cc|cc} 1 & 0 & 1 & -1 \\ 0 & 1 & -1 & 2 \end{array} \right]$

Interchange　　　$R_2 = -2r_1 + r_2$　　$R_2 = -r_2$　　　$R_1 = -r_2 + r_1$

r_1 and r_2

$A^{-1} = \begin{bmatrix} 1 & -1 \\ -1 & 2 \end{bmatrix}$

31. Augment the matrix with the identity and use row operations to find the inverse:

$$A = \begin{bmatrix} 6 & 5 \\ 2 & 2 \end{bmatrix} \rightarrow \begin{bmatrix} 6 & 5 & | & 1 & 0 \\ 2 & 2 & | & 0 & 1 \end{bmatrix}$$

$$\rightarrow \begin{bmatrix} 2 & 2 & | & 0 & 1 \\ 6 & 5 & | & 1 & 0 \end{bmatrix} \rightarrow \begin{bmatrix} 2 & 2 & | & 0 & 1 \\ 0 & -1 & | & 1 & -3 \end{bmatrix} \rightarrow \begin{bmatrix} 1 & 1 & | & 0 & \frac{1}{2} \\ 0 & 1 & | & -1 & 3 \end{bmatrix} \rightarrow \begin{bmatrix} 1 & 0 & | & 1 & -\frac{5}{2} \\ 0 & 1 & | & -1 & 3 \end{bmatrix}$$

Interchange　　　$R_2 = -3r_1 + r_2$　　$R_1 = \frac{1}{2}r_1$　　　　$R_1 = -r_2 + r_1$

r_1 and r_2　　　　　　　　　　　$R_2 = -r_2$

$$A^{-1} = \begin{bmatrix} 1 & -\frac{5}{2} \\ -1 & 3 \end{bmatrix}$$

33. Augment the matrix with the identity and use row operations to find the inverse:

$$A = \begin{bmatrix} 2 & 1 \\ a & a \end{bmatrix} \rightarrow \begin{bmatrix} 2 & 1 & | & 1 & 0 \\ a & a & | & 0 & 1 \end{bmatrix} \text{ where } a \neq 0.$$

$$\rightarrow \begin{bmatrix} 1 & \frac{1}{2} & | & \frac{1}{2} & 0 \\ a & a & | & 0 & 1 \end{bmatrix} \rightarrow \begin{bmatrix} 1 & \frac{1}{2} & | & \frac{1}{2} & 0 \\ 0 & \frac{1}{2}a & | & -\frac{1}{2}a & 1 \end{bmatrix} \rightarrow \begin{bmatrix} 1 & \frac{1}{2} & | & \frac{1}{2} & 0 \\ 0 & 1 & | & -1 & \frac{2}{a} \end{bmatrix} \rightarrow \begin{bmatrix} 1 & 0 & | & 1 & -\frac{1}{a} \\ 0 & 1 & | & -1 & \frac{2}{a} \end{bmatrix}$$

$R_1 = \frac{1}{2}r_1$　　　　$R_2 = -ar_1 + r_2$　　　$R_2 = \frac{2}{a}r_2$　　　　$R_1 = -\frac{1}{2}r_2 + r_1$

$$A^{-1} = \begin{bmatrix} 1 & -\frac{1}{a} \\ -1 & \frac{2}{a} \end{bmatrix}$$

35. Augment the matrix with the identity and use row operations to find the inverse:

$$A = \begin{bmatrix} 1 & -1 & 1 \\ 0 & -2 & 1 \\ -2 & -3 & 0 \end{bmatrix} \rightarrow \begin{bmatrix} 1 & -1 & 1 & | & 1 & 0 & 0 \\ 0 & -2 & 1 & | & 0 & 1 & 0 \\ -2 & -3 & 0 & | & 0 & 0 & 1 \end{bmatrix}$$

$$\rightarrow \begin{bmatrix} 1 & -1 & 1 & | & 1 & 0 & 0 \\ 0 & -2 & 1 & | & 0 & 1 & 0 \\ 0 & -5 & 2 & | & 2 & 0 & 1 \end{bmatrix} \rightarrow \begin{bmatrix} 1 & -1 & 1 & | & 1 & 0 & 0 \\ 0 & 1 & -\frac{1}{2} & | & 0 & -\frac{1}{2} & 0 \\ 0 & -5 & 2 & | & 2 & 0 & 1 \end{bmatrix} \rightarrow \begin{bmatrix} 1 & 0 & \frac{1}{2} & | & 1 & -\frac{1}{2} & 0 \\ 0 & 1 & -\frac{1}{2} & | & 0 & -\frac{1}{2} & 0 \\ 0 & 0 & -\frac{1}{2} & | & 2 & -\frac{5}{2} & 1 \end{bmatrix}$$

$R_3 = 2r_1 + r_3$　　　　　　　$R_2 = -\frac{1}{2}r_2$　　　　　　　$R_1 = r_2 + r_1$

$R_3 = 5r_2 + r_3$

$$\rightarrow \begin{bmatrix} 1 & 0 & \frac{1}{2} & | & 1 & -\frac{1}{2} & 0 \\ 0 & 1 & -\frac{1}{2} & | & 0 & -\frac{1}{2} & 0 \\ 0 & 0 & 1 & | & -4 & 5 & -2 \end{bmatrix} \rightarrow \begin{bmatrix} 1 & 0 & 0 & | & 3 & -3 & 1 \\ 0 & 1 & 0 & | & -2 & 2 & -1 \\ 0 & 0 & 1 & | & -4 & 5 & -2 \end{bmatrix}$$

$R_3 = -2r_3$　　　　　　　　$R_1 = -\frac{1}{2}r_3 + r_1$

$R_2 = \frac{1}{2}r_3 + r_2$

$$A^{-1} = \begin{bmatrix} 3 & -3 & 1 \\ -2 & 2 & -1 \\ -4 & 5 & -2 \end{bmatrix}$$

37. Augment the matrix with the identity and use row operations to find the inverse:

$$A = \begin{bmatrix} 1 & 1 & 1 \\ 3 & 2 & -1 \\ 3 & 1 & 2 \end{bmatrix} \rightarrow \left[\begin{array}{ccc|ccc} 1 & 1 & 1 & 1 & 0 & 0 \\ 3 & 2 & -1 & 0 & 1 & 0 \\ 3 & 1 & 2 & 0 & 0 & 1 \end{array}\right]$$

$$\rightarrow \left[\begin{array}{ccc|ccc} 1 & 1 & 1 & 1 & 0 & 0 \\ 0 & -1 & -4 & -3 & 1 & 0 \\ 0 & -2 & -1 & -3 & 0 & 1 \end{array}\right] \rightarrow \left[\begin{array}{ccc|ccc} 1 & 1 & 1 & 1 & 0 & 0 \\ 0 & 1 & 4 & 3 & -1 & 0 \\ 0 & -2 & -1 & -3 & 0 & 1 \end{array}\right] \rightarrow \left[\begin{array}{ccc|ccc} 1 & 0 & -3 & -2 & 1 & 0 \\ 0 & 1 & 4 & 3 & -1 & 0 \\ 0 & 0 & 7 & 3 & -2 & 1 \end{array}\right]$$

$$R_2 = -3r_1 + r_2 \qquad\qquad R_2 = -r_2 \qquad\qquad R_1 = -r_2 + r_1$$
$$R_3 = -3r_1 + r_3 \qquad\qquad\qquad\qquad\qquad\qquad R_3 = 2r_2 + r_3$$

$$\rightarrow \left[\begin{array}{ccc|ccc} 1 & 0 & -3 & -2 & 1 & 0 \\ 0 & 1 & 4 & 3 & -1 & 0 \\ 0 & 0 & 1 & \frac{3}{7} & -\frac{2}{7} & \frac{1}{7} \end{array}\right] \rightarrow \left[\begin{array}{ccc|ccc} 1 & 0 & 0 & -\frac{5}{7} & \frac{1}{7} & \frac{3}{7} \\ 0 & 1 & 0 & \frac{9}{7} & \frac{1}{7} & -\frac{4}{7} \\ 0 & 0 & 1 & \frac{3}{7} & -\frac{2}{7} & \frac{1}{7} \end{array}\right]$$

$$R_3 = \tfrac{1}{7}r_3 \qquad\qquad\qquad R_1 = 3r_3 + r_1$$
$$R_2 = -4r_3 + r_2$$

$$A^{-1} = \begin{bmatrix} -\frac{5}{7} & \frac{1}{7} & \frac{3}{7} \\ \frac{9}{7} & \frac{1}{7} & -\frac{4}{7} \\ \frac{3}{7} & -\frac{2}{7} & \frac{1}{7} \end{bmatrix}$$

39. Rewrite the system of equations in matrix form:

$$\begin{cases} 2x + y = 8 \\ x + y = 5 \end{cases} \qquad A = \begin{bmatrix} 2 & 1 \\ 1 & 1 \end{bmatrix}, \quad X = \begin{bmatrix} x \\ y \end{bmatrix}, \quad B = \begin{bmatrix} 8 \\ 5 \end{bmatrix}$$

Find the inverse of A and solve $X = A^{-1}B$:

From Problem 29, $A^{-1} = \begin{bmatrix} 1 & -1 \\ -1 & 2 \end{bmatrix}$ and $X = A^{-1}B = \begin{bmatrix} 1 & -1 \\ -1 & 2 \end{bmatrix}\begin{bmatrix} 8 \\ 5 \end{bmatrix} = \begin{bmatrix} 3 \\ 2 \end{bmatrix}$.

The solution is $x = 3$, $y = 2$.

41. Rewrite the system of equations in matrix form:

$$\begin{cases} 2x + y = 0 \\ x + y = 5 \end{cases} \qquad A = \begin{bmatrix} 2 & 1 \\ 1 & 1 \end{bmatrix}, \quad X = \begin{bmatrix} x \\ y \end{bmatrix}, \quad B = \begin{bmatrix} 0 \\ 5 \end{bmatrix}$$

Find the inverse of A and solve $X = A^{-1}B$:

From Problem 29, $A^{-1} = \begin{bmatrix} 1 & -1 \\ -1 & 2 \end{bmatrix}$ and $X = A^{-1}B = \begin{bmatrix} 1 & -1 \\ -1 & 2 \end{bmatrix}\begin{bmatrix} 0 \\ 5 \end{bmatrix} = \begin{bmatrix} -5 \\ 10 \end{bmatrix}$.

The solution is $x = -5$, $y = 10$.

43. Rewrite the system of equations in matrix form:

$$\begin{cases} 6x + 5y = 7 \\ 2x + 2y = 2 \end{cases} \qquad A = \begin{bmatrix} 6 & 5 \\ 2 & 2 \end{bmatrix}, \quad X = \begin{bmatrix} x \\ y \end{bmatrix}, \quad B = \begin{bmatrix} 7 \\ 2 \end{bmatrix}$$

Find the inverse of A and solve $X = A^{-1}B$:

From Problem 31, $A^{-1} = \begin{bmatrix} 1 & -\frac{5}{2} \\ -1 & 3 \end{bmatrix}$ and $X = A^{-1}B = \begin{bmatrix} 1 & -\frac{5}{2} \\ -1 & 3 \end{bmatrix}\begin{bmatrix} 7 \\ 2 \end{bmatrix} = \begin{bmatrix} 2 \\ -1 \end{bmatrix}$.

The solution is $x = 2,\ y = -1$.

45. Rewrite the system of equations in matrix form:

$$\begin{cases} 6x + 5y = 13 \\ 2x + 2y = 5 \end{cases} \qquad A = \begin{bmatrix} 6 & 5 \\ 2 & 2 \end{bmatrix}, \quad X = \begin{bmatrix} x \\ y \end{bmatrix}, \quad B = \begin{bmatrix} 13 \\ 5 \end{bmatrix}$$

Find the inverse of A and solve $X = A^{-1}B$:

From Problem 31, $A^{-1} = \begin{bmatrix} 1 & -\frac{5}{2} \\ -1 & 3 \end{bmatrix}$ and $X = A^{-1}B = \begin{bmatrix} 1 & -\frac{5}{2} \\ -1 & 3 \end{bmatrix}\begin{bmatrix} 13 \\ 5 \end{bmatrix} = \begin{bmatrix} \frac{1}{2} \\ 2 \end{bmatrix}$.

The solution is $x = \frac{1}{2},\ y = 2$.

47. Rewrite the system of equations in matrix form:

$$\begin{cases} 2x + y = -3 \\ ax + ay = -a \end{cases} \quad a \neq 0 \qquad A = \begin{bmatrix} 2 & 1 \\ a & a \end{bmatrix}, \quad X = \begin{bmatrix} x \\ y \end{bmatrix}, \quad B = \begin{bmatrix} -3 \\ -a \end{bmatrix}$$

Find the inverse of A and solve $X = A^{-1}B$:

From Problem 33, $A^{-1} = \begin{bmatrix} 1 & -\frac{1}{a} \\ -1 & \frac{2}{a} \end{bmatrix}$ and $X = A^{-1}B = \begin{bmatrix} 1 & -\frac{1}{a} \\ -1 & \frac{2}{a} \end{bmatrix}\begin{bmatrix} -3 \\ -a \end{bmatrix} = \begin{bmatrix} -2 \\ 1 \end{bmatrix}$.

The solution is $x = -2,\ y = 1$.

49. Rewrite the system of equations in matrix form:

$$\begin{cases} 2x + y = \dfrac{7}{a} \\ ax + ay = 5 \end{cases} \quad a \neq 0 \qquad A = \begin{bmatrix} 2 & 1 \\ a & a \end{bmatrix}, \quad X = \begin{bmatrix} x \\ y \end{bmatrix}, \quad B = \begin{bmatrix} \frac{7}{a} \\ 5 \end{bmatrix}$$

Find the inverse of A and solve $X = A^{-1}B$:

From Problem 33, $A^{-1} = \begin{bmatrix} 1 & -\frac{1}{a} \\ -1 & \frac{2}{a} \end{bmatrix}$ and $X = A^{-1}B = \begin{bmatrix} 1 & -\frac{1}{a} \\ -1 & \frac{2}{a} \end{bmatrix}\begin{bmatrix} \frac{7}{a} \\ 5 \end{bmatrix} = \begin{bmatrix} \frac{2}{a} \\ \frac{3}{a} \end{bmatrix}$.

The solution is $x = \dfrac{2}{a},\ y = \dfrac{3}{a}$.

51. Rewrite the system of equations in matrix form:

$$\begin{cases} x - y + z = 0 \\ -2y + z = -1 \\ -2x - 3y = -5 \end{cases} \quad A = \begin{bmatrix} 1 & -1 & 1 \\ 0 & -2 & 1 \\ -2 & -3 & 0 \end{bmatrix}, \quad X = \begin{bmatrix} x \\ y \\ z \end{bmatrix}, \quad B = \begin{bmatrix} 0 \\ -1 \\ -5 \end{bmatrix}$$

Find the inverse of A and solve $X = A^{-1}B$:

From Problem 35,

$$A^{-1} = \begin{bmatrix} 3 & -3 & 1 \\ -2 & 2 & -1 \\ -4 & 5 & -2 \end{bmatrix} \text{ and } X = A^{-1}B = \begin{bmatrix} 3 & -3 & 1 \\ -2 & 2 & -1 \\ -4 & 5 & -2 \end{bmatrix}\begin{bmatrix} 0 \\ -1 \\ -5 \end{bmatrix} = \begin{bmatrix} -2 \\ 3 \\ 5 \end{bmatrix}.$$

The solution is $x = -2$, $y = 3$, $z = 5$.

53. Rewrite the system of equations in matrix form:

$$\begin{cases} x - y + z = 2 \\ -2y + z = 2 \\ -2x - 3y = \dfrac{1}{2} \end{cases} \quad A = \begin{bmatrix} 1 & -1 & 1 \\ 0 & -2 & 1 \\ -2 & -3 & 0 \end{bmatrix}, \quad X = \begin{bmatrix} x \\ y \\ z \end{bmatrix}, \quad B = \begin{bmatrix} 2 \\ 2 \\ \frac{1}{2} \end{bmatrix}$$

Find the inverse of A and solve $X = A^{-1}B$:

From Problem 35,

$$A^{-1} = \begin{bmatrix} 3 & -3 & 1 \\ -2 & 2 & -1 \\ -4 & 5 & -2 \end{bmatrix} \text{ and } X = A^{-1}B = \begin{bmatrix} 3 & -3 & 1 \\ -2 & 2 & -1 \\ -4 & 5 & -2 \end{bmatrix}\begin{bmatrix} 2 \\ 2 \\ \frac{1}{2} \end{bmatrix} = \begin{bmatrix} \frac{1}{2} \\ -\frac{1}{2} \\ 1 \end{bmatrix}.$$

The solution is $x = \dfrac{1}{2}$, $y = -\dfrac{1}{2}$, $z = 1$.

55. Rewrite the system of equations in matrix form:

$$\begin{cases} x + y + z = 9 \\ 3x + 2y - z = 8 \\ 3x + y + 2z = 1 \end{cases} \quad A = \begin{bmatrix} 1 & 1 & 1 \\ 3 & 2 & -1 \\ 3 & 1 & 2 \end{bmatrix}, \quad X = \begin{bmatrix} x \\ y \\ z \end{bmatrix}, \quad B = \begin{bmatrix} 9 \\ 8 \\ 1 \end{bmatrix}$$

Find the inverse of A and solve $X = A^{-1}B$:

From Problem 37,

$$A^{-1} = \begin{bmatrix} -\frac{5}{7} & \frac{1}{7} & \frac{3}{7} \\ \frac{9}{7} & \frac{1}{7} & -\frac{4}{7} \\ \frac{3}{7} & -\frac{2}{7} & \frac{1}{7} \end{bmatrix} \text{ and } X = A^{-1}B = \begin{bmatrix} -\frac{5}{7} & \frac{1}{7} & \frac{3}{7} \\ \frac{9}{7} & \frac{1}{7} & -\frac{4}{7} \\ \frac{3}{7} & -\frac{2}{7} & \frac{1}{7} \end{bmatrix}\begin{bmatrix} 9 \\ 8 \\ 1 \end{bmatrix} = \begin{bmatrix} -\frac{34}{7} \\ \frac{85}{7} \\ \frac{12}{7} \end{bmatrix}.$$

The solution is $x = -\dfrac{34}{7}$, $y = \dfrac{85}{7}$, $z = \dfrac{12}{7}$.

57.   Rewrite the system of equations in matrix form:

$$\begin{cases} x + y + z = 2 \\ 3x + 2y - z = \dfrac{7}{3} \\ 3x + y + 2z = \dfrac{10}{3} \end{cases} \qquad A = \begin{bmatrix} 1 & 1 & 1 \\ 3 & 2 & -1 \\ 3 & 1 & 2 \end{bmatrix}, \quad X = \begin{bmatrix} x \\ y \\ z \end{bmatrix}, \quad B = \begin{bmatrix} 2 \\ \frac{7}{3} \\ \frac{10}{3} \end{bmatrix}$$

Find the inverse of A and solve $X = A^{-1}B$:

From Problem 37,

$$A^{-1} = \begin{bmatrix} -\frac{5}{7} & \frac{1}{7} & \frac{3}{7} \\ \frac{9}{7} & \frac{1}{7} & -\frac{4}{7} \\ \frac{3}{7} & -\frac{2}{7} & \frac{1}{7} \end{bmatrix} \text{ and } X = A^{-1}B = \begin{bmatrix} -\frac{5}{7} & \frac{1}{7} & \frac{3}{7} \\ \frac{9}{7} & \frac{1}{7} & -\frac{4}{7} \\ \frac{3}{7} & -\frac{2}{7} & \frac{1}{7} \end{bmatrix} \begin{bmatrix} 2 \\ \frac{7}{3} \\ \frac{10}{3} \end{bmatrix} = \begin{bmatrix} \frac{1}{3} \\ 1 \\ \frac{2}{3} \end{bmatrix}.$$

The solution is $x = \dfrac{1}{3}, y = 1, z = \dfrac{2}{3}$.

59.   Augment the matrix with the identity and use row operations to find the inverse:

$$A = \begin{bmatrix} 4 & 2 \\ 2 & 1 \end{bmatrix} \rightarrow \left[\begin{array}{cc|cc} 4 & 2 & 1 & 0 \\ 2 & 1 & 0 & 1 \end{array} \right] \rightarrow \left[\begin{array}{cc|cc} 4 & 2 & 1 & 0 \\ 0 & 0 & -\frac{1}{2} & 1 \end{array} \right] \rightarrow \left[\begin{array}{cc|cc} 1 & \frac{1}{2} & \frac{1}{4} & 0 \\ 0 & 0 & -\frac{1}{2} & 1 \end{array} \right]$$

$$R_2 = -\tfrac{1}{2}r_1 + r_2 \qquad R_1 = \tfrac{1}{4}r_1$$

There is no way to obtain the identity matrix on the left; thus there is no inverse.

61.   Augment the matrix with the identity and use row operations to find the inverse:

$$A = \begin{bmatrix} 15 & 3 \\ 10 & 2 \end{bmatrix} \rightarrow \left[\begin{array}{cc|cc} 15 & 3 & 1 & 0 \\ 10 & 2 & 0 & 1 \end{array} \right] \rightarrow \left[\begin{array}{cc|cc} 15 & 3 & 1 & 0 \\ 0 & 0 & -\frac{2}{3} & 1 \end{array} \right] \rightarrow \left[\begin{array}{cc|cc} 1 & \frac{1}{5} & \frac{1}{15} & 0 \\ 0 & 0 & -\frac{2}{3} & 1 \end{array} \right]$$

$$R_2 = -\tfrac{2}{3}r_1 + r_2 \qquad R_1 = \tfrac{1}{15}r_1$$

There is no way to obtain the identity matrix on the left; thus there is no inverse.

63.   Augment the matrix with the identity and use row operations to find the inverse:

$$A = \begin{bmatrix} -3 & 1 & -1 \\ 1 & -4 & -7 \\ 1 & 2 & 5 \end{bmatrix} \rightarrow \left[\begin{array}{ccc|ccc} -3 & 1 & -1 & 1 & 0 & 0 \\ 1 & -4 & -7 & 0 & 1 & 0 \\ 1 & 2 & 5 & 0 & 0 & 1 \end{array} \right]$$

$$\rightarrow \left[\begin{array}{ccc|ccc} 1 & 2 & 5 & 0 & 0 & 1 \\ 1 & -4 & -7 & 0 & 1 & 0 \\ -3 & 1 & -1 & 1 & 0 & 0 \end{array} \right] \rightarrow \left[\begin{array}{ccc|ccc} 1 & 2 & 5 & 0 & 0 & 1 \\ 0 & -6 & -12 & 0 & 1 & -1 \\ 0 & 7 & 14 & 1 & 0 & 3 \end{array} \right] \rightarrow \left[\begin{array}{ccc|ccc} 1 & 2 & 5 & 0 & 0 & 1 \\ 0 & 1 & 2 & 0 & -\frac{1}{6} & \frac{1}{6} \\ 0 & 7 & 14 & 1 & 0 & 3 \end{array} \right]$$

$$\text{Interchange } r_1 \text{ and } r_3 \qquad R_2 = -r_1 + r_2 \qquad\qquad R_2 = -\tfrac{1}{6}r_2$$

$$R_3 = 3r_1 + r_3$$

$$\rightarrow \begin{bmatrix} 1 & 0 & 1 & | & 0 & \frac{1}{3} & \frac{2}{3} \\ 0 & 1 & 2 & | & 0 & -\frac{1}{6} & \frac{1}{6} \\ 0 & 0 & 0 & | & 1 & \frac{7}{6} & \frac{11}{6} \end{bmatrix}$$

$$R_1 = -2r_2 + r_1$$
$$R_3 = -7r_2 + r_3$$

There is no way to obtain the identity matrix on the left; thus there is no inverse.

65. Use MATRIX EDIT to create the matrix

$$A = \begin{bmatrix} 25 & 61 & -12 \\ 18 & -2 & 4 \\ 8 & 35 & 21 \end{bmatrix}$$

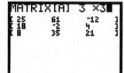

Then compute $[A]^{\wedge}(-1)$:

$$A^{-1} \approx \begin{bmatrix} 0.01 & 0.05 & -0.01 \\ 0.01 & -0.02 & 0.01 \\ -0.02 & 0.01 & 0.03 \end{bmatrix}$$

67. Use MATRIX EDIT to create the matrix

$$A = \begin{bmatrix} 44 & 21 & 18 & 6 \\ -2 & 10 & 15 & 5 \\ 21 & 12 & -12 & 4 \\ -8 & -16 & 4 & 9 \end{bmatrix}$$

Then compute $[A]^{\wedge}(-1)$:

$$A^{-1} = \begin{bmatrix} 0.02 & -0.04 & -0.01 & 0.01 \\ -0.02 & 0.05 & 0.03 & -0.03 \\ 0.02 & 0.01 & -0.04 & 0.00 \\ -0.02 & 0.06 & 0.07 & 0.06 \end{bmatrix}$$

69. Use MATRIX EDIT to create the matrices

$$A = \begin{bmatrix} 25 & 61 & -12 \\ 18 & -12 & 7 \\ 3 & 4 & -1 \end{bmatrix}; \quad B = \begin{bmatrix} 10 \\ -9 \\ 12 \end{bmatrix}$$

Then compute $\left([A]^{\wedge}(-1)\right)*[B]$:

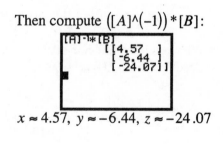

$$x \approx 4.57, \; y \approx -6.44, \; z \approx -24.07$$

71. Use MATRIX EDIT to create the matrices

$$A = \begin{bmatrix} 25 & 61 & -12 \\ 18 & -12 & 7 \\ 3 & 4 & -1 \end{bmatrix} ; \ B = \begin{bmatrix} 21 \\ 7 \\ -2 \end{bmatrix}$$

Then compute $([A]^\wedge(-1)) * [B]$:

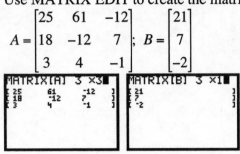

$x \approx -1.19, \ y \approx 2.46, \ z \approx 8.27$

73. (a) The rows of the 2 by 3 matrix represent stainless steel and aluminum. The columns represent 10-gallon, 5-gallon, and 1-gallon.

The 2 by 3 matrix is:

$$\begin{bmatrix} 500 & 350 & 400 \\ 700 & 500 & 850 \end{bmatrix}$$

The 3 by 2 matrix is:

$$\begin{bmatrix} 500 & 700 \\ 350 & 500 \\ 400 & 850 \end{bmatrix}$$

(b) The 3 by 1 matrix representing the amount of material is:

$$\begin{bmatrix} 15 \\ 8 \\ 3 \end{bmatrix}$$

(c) The days usage of materials is:

$$\begin{bmatrix} 500 & 350 & 400 \\ 700 & 500 & 850 \end{bmatrix} \cdot \begin{bmatrix} 15 \\ 8 \\ 3 \end{bmatrix} = \begin{bmatrix} 11{,}500 \\ 17{,}050 \end{bmatrix}$$

11,500 pounds of stainless steel and 17,050 pounds of aluminum were used that day.

(d) The 1 by 2 matrix representing cost is:

$$\begin{bmatrix} 0.10 & 0.05 \end{bmatrix}$$

(e) The total cost of the day's production was:

$$\begin{bmatrix} 0.10 & 0.05 \end{bmatrix} \cdot \begin{bmatrix} 11{,}500 \\ 17{,}050 \end{bmatrix} = \begin{bmatrix} 2002.50 \end{bmatrix}$$

The total cost of the day's production was $2002.50.

75. We need to consider 2 different cases.

Case 1: $a \neq 0$, $D = ad - bc \neq 0$.

$$A = \left[\begin{array}{cc|cc} a & b & 1 & 0 \\ c & d & 0 & 1 \end{array} \right] \longrightarrow \left[\begin{array}{cc|cc} 1 & \dfrac{b}{a} & \dfrac{1}{a} & 0 \\ c & d & 0 & 1 \end{array} \right] \longrightarrow \left[\begin{array}{cc|cc} 1 & \dfrac{b}{a} & \dfrac{1}{a} & 0 \\ c & d - \dfrac{cb}{a} & -\dfrac{c}{a} & 1 \end{array} \right] = \left[\begin{array}{cc|cc} 1 & \dfrac{b}{b} & \dfrac{1}{a} & 0 \\ 0 & \dfrac{a - cb}{a} & -\dfrac{c}{a} & 1 \end{array} \right]$$

$$R_1 = \tfrac{1}{a} \cdot r_1 \qquad\qquad R_2 = -c \cdot r_1 + r_2$$

$$\longrightarrow \begin{bmatrix} 1 & \dfrac{b}{a} & \bigg| & \dfrac{1}{a} & 0 \\[2mm] 0 & 1 & \bigg| & -\dfrac{c}{ad-bc} & \dfrac{a}{ad-bc} \end{bmatrix} \longrightarrow \begin{bmatrix} 1 & 0 & \bigg| & \dfrac{1}{a}+\dfrac{bc}{a(ad-bc)} & \dfrac{-b}{ad-bc} \\[2mm] 0 & 1 & \bigg| & -\dfrac{c}{ad-bc} & \dfrac{a}{ad-bc} \end{bmatrix}$$

$$R_2 = \left(\dfrac{a}{ad-bc}\right)\cdot r_2 \qquad\qquad R_1 = \left(-\dfrac{b}{a}\right)r_2 + r_1$$

Note that $\quad \dfrac{1}{a}+\dfrac{bc}{a(ad-bc)} = \dfrac{1\cdot(ad-bc)+bc}{a(ad-bc)} = \dfrac{ad-bc+bc}{a(ad-bc)} = \dfrac{ad}{a(ad-bc)} = \dfrac{d}{ad-bc}$

So, we have $\begin{bmatrix} 1 & 0 & \bigg| & \dfrac{1}{a}+\dfrac{bc}{a(ad-bc)} & \dfrac{-b}{ad-bc} \\[2mm] 0 & 1 & \bigg| & -\dfrac{c}{ad-bc} & \dfrac{a}{ad-bc} \end{bmatrix} = \begin{bmatrix} 1 & 0 & \bigg| & \dfrac{d}{ad-bc} & \dfrac{-b}{ad-bc} \\[2mm] 0 & 1 & \bigg| & -\dfrac{c}{ad-bc} & \dfrac{a}{ad-bc} \end{bmatrix}.$

Therefore $\quad A^{-1} = \begin{bmatrix} \dfrac{d}{ad-bc} & \dfrac{-b}{ad-bc} \\[2mm] -\dfrac{c}{ad-bc} & \dfrac{a}{ad-bc} \end{bmatrix} = \left(\dfrac{1}{ad-bc}\right)\begin{bmatrix} d & -b \\ -c & a \end{bmatrix} = \left(\dfrac{1}{D}\right)\begin{bmatrix} d & -b \\ -c & a \end{bmatrix},$

where $D = ad - bc$.

Case 2: $a = 0$, $D = ad - bc \neq 0$.

First note that $D = ad - bc = 0\cdot d - bc = -bc \neq 0 \Rightarrow b \neq 0$ and $c \neq 0$.

$$A = \begin{bmatrix} 0 & b & \big| & 1 & 0 \\ c & d & \big| & 0 & 1 \end{bmatrix} \longrightarrow \begin{bmatrix} c & d & \big| & 0 & 1 \\ 0 & b & \big| & 1 & 0 \end{bmatrix} \longrightarrow \begin{bmatrix} 1 & \dfrac{d}{c} & \bigg| & 0 & \dfrac{1}{c} \\[2mm] 0 & b & \big| & 1 & 0 \end{bmatrix} \longrightarrow \begin{bmatrix} 1 & \dfrac{d}{c} & \bigg| & 0 & \dfrac{1}{c} \\[2mm] 0 & 1 & \bigg| & \dfrac{1}{b} & 0 \end{bmatrix}$$

$$\text{interchange } r_1 \text{ and } r_2 \qquad R_1 = \dfrac{1}{c}\cdot r_1 \qquad\qquad R_2 = \dfrac{1}{b}\cdot r_2$$

$$\longrightarrow \begin{bmatrix} 1 & 0 & \bigg| & \dfrac{-d}{bc} & \dfrac{1}{c} \\[2mm] 0 & 1 & \bigg| & \dfrac{1}{b} & 0 \end{bmatrix}$$

$$R_1 = \left(\dfrac{-d}{c}\right)\cdot r_2 + r_1$$

Therefore, $A^{-1} = \begin{bmatrix} \dfrac{-d}{bc} & \dfrac{1}{c} \\[2mm] \dfrac{1}{b} & 0 \end{bmatrix} = \left(\dfrac{1}{-bc}\right)\begin{bmatrix} d & -b \\ -c & 0 \end{bmatrix} = \left(\dfrac{1}{D}\right)\begin{bmatrix} d & -b \\ -c & 0 \end{bmatrix},$

where $D = ad - bc = 0\cdot d - bc = -bc$.

Systems of Equations and Inequalities

10.5 Partial Fraction Decomposition

1. True

3. $3x^4 + 6x^3 + 3x^2 = 3x^2(x^2 + 2x + 1) = 3x^2(x+1)(x+1)$

5. The rational expression $\dfrac{x}{x^2 - 1}$ is proper, since the degree of the numerator is less than the degree of the denominator.

7. The rational expression $\dfrac{x^2 + 5}{x^2 - 4}$ is improper, so perform the division:

$$x^2 - 4 \overline{)\begin{array}{l} 1 \\ x^2 + 5 \\ \underline{x^2 - 4} \\ 9 \end{array}}$$

The proper rational expression is:
$$\frac{x^2 + 5}{x^2 - 4} = 1 + \frac{9}{x^2 - 4}$$

9. The rational expression $\dfrac{5x^3 + 2x - 1}{x^2 - 4}$ is improper, so perform the division:

$$x^2 - 4 \overline{)\begin{array}{l} 5x \\ 5x^3 + 2x - 1 \\ \underline{5x^3 - 20x} \\ 22x - 1 \end{array}}$$

The proper rational expression is:
$$\frac{5x^3 + 2x - 1}{x^2 - 4} = 5x + \frac{22x - 1}{x^2 - 4}$$

11. The rational expression $\dfrac{x(x-1)}{(x+4)(x-3)} = \dfrac{x^2 - x}{x^2 + x - 12}$ is improper, so perform the division:

$$x^2 + x - 12 \overline{)\begin{array}{l} 1 \\ x^2 - x + 0 \\ \underline{x^2 + x - 12} \\ -2x + 12 \end{array}}$$

The proper rational expression is:
$$\frac{x(x-1)}{(x+4)(x-3)} = 1 + \frac{-2x + 12}{x^2 + x - 12} = 1 + \frac{-2(x-6)}{(x+4)(x-3)}$$

13. Find the partial fraction decomposition:

$$\frac{4}{x(x-1)} = \frac{A}{x} + \frac{B}{x-1} \qquad \text{(Multiply both sides by } x(x-1).)$$

$$4 = A(x-1) + Bx$$

Let $x = 1$:   then $4 = A(0) + B \Rightarrow B = 4$

Let $x = 0$:   then $4 = A(-1) + B(0) \Rightarrow A = -4$

$$\frac{4}{x(x-1)} = \frac{-4}{x} + \frac{4}{x-1}$$

15. Find the partial fraction decomposition:

$$\frac{1}{x(x^2+1)} = \frac{A}{x} + \frac{Bx+C}{x^2+1} \qquad \text{(Multiply both sides by } x(x^2+1).)$$

$$1 = A(x^2+1) + (Bx+C)x$$

Let $x = 0$:   then $1 = A(1) + (B(0)+C)(0) \Rightarrow A = 1$

Let $x = 1$:   then $1 = A(1+1) + (B(1)+C)(1) \Rightarrow 1 = 2A + B + C$

$$\Rightarrow \quad 1 = 2(1) + B + C \Rightarrow -1 = B + C$$

Let $x = -1$:   then $1 = A(1+1) + (B(-1)+C)(-1) \Rightarrow 1 = 2A + B - C$

$$\Rightarrow \quad 1 = 2(1) + B - C \Rightarrow \ -1 = B - C$$

Solve the system of equations:

$$
\begin{array}{ll}
B + C = -1 & \\
\underline{B - C = -1} & \\
2B \quad\ = -2 & \qquad -1 + C = -1 \\
B \quad\ = -1 & \qquad C = 0
\end{array}
$$

$$\frac{1}{x(x^2+1)} = \frac{1}{x} + \frac{-x}{x^2+1}$$

17. Find the partial fraction decomposition:

$$\frac{x}{(x-1)(x-2)} = \frac{A}{x-1} + \frac{B}{x-2} \qquad \text{(Multiply both sides by } (x-1)(x-2).)$$

$$x = A(x-2) + B(x-1)$$

Let $x = 1$:   then $1 = A(1-2) + B(1-1) \Rightarrow 1 = -A \Rightarrow A = -1$

Let $x = 2$:   then $2 = A(2-2) + B(2-1) \Rightarrow 2 = B \Rightarrow B = 2$

$$\frac{x}{(x-1)(x-2)} = \frac{-1}{x-1} + \frac{2}{x-2}$$

19. Find the partial fraction decomposition:

$$\frac{x^2}{(x-1)^2(x+1)} = \frac{A}{x-1} + \frac{B}{(x-1)^2} + \frac{C}{x+1} \qquad \text{(Multiply both sides by } (x-1)^2(x+1).)$$

$$x^2 = A(x-1)(x+1) + B(x+1) + C(x-1)^2$$

Let $x = 1$:   then $1^2 = A(1-1)(1+1) + B(1+1) + C(1-1)^2$

$$\Rightarrow 1 = 2B \Rightarrow B = \frac{1}{2}$$

Let $x = -1$:   then $(-1)^2 = A(-1-1)(-1+1) + B(-1+1) + C(-1-1)^2$

$$\Rightarrow 1 = 4C \Rightarrow C = \frac{1}{4}$$

Let $x = 0$:   then $0^2 = A(0-1)(0+1) + B(0+1) + C(0-1)^2$

$$\Rightarrow 0 = -A + B + C \Rightarrow A = \frac{1}{2} + \frac{1}{4} = \frac{3}{4}$$

$$\frac{x^2}{(x-1)^2(x+1)} = \frac{(3/4)}{x-1} + \frac{(1/2)}{(x-1)^2} + \frac{(1/4)}{x+1}$$

21. Find the partial fraction decomposition:

$$\frac{1}{x^3 - 8} = \frac{1}{(x-2)(x^2 + 2x + 4)} = \frac{A}{x-2} + \frac{Bx + C}{x^2 + 2x + 4}$$

(Multiply both sides by $(x-2)(x^2 + 2x + 4)$.)

$$1 = A(x^2 + 2x + 4) + (Bx + C)(x - 2)$$

Let $x = 2$:   then $1 = A(2^2 + 2(2) + 4) + (B(2) + C)(2 - 2)$

$$\Rightarrow 1 = 12A \Rightarrow A = \frac{1}{12}$$

Let $x = 0$:   then $1 = A(0^2 + 2(0) + 4) + (B(0) + C)(0 - 2)$

$$\Rightarrow 1 = 4A - 2C \Rightarrow 1 = 4(1/12) - 2C \Rightarrow -2C = \frac{2}{3} \Rightarrow C = -\frac{1}{3}$$

Let $x = 1$:   then $1 = A(1^2 + 2(1) + 4) + (B(1) + C)(1 - 2)$

$$\Rightarrow 1 = 7A - B - C \Rightarrow 1 = 7(1/12) - B + \frac{1}{3} \Rightarrow B = -\frac{1}{12}$$

$$\frac{1}{x^3 - 8} = \frac{(1/12)}{x-2} + \frac{-(1/12)x - 1/3}{x^2 + 2x + 4} = \frac{(1/12)}{x-2} + \frac{-(1/12)(x+4)}{x^2 + 2x + 4}$$

23. Find the partial fraction decomposition:

$$\frac{x^2}{(x-1)^2(x+1)^2} = \frac{A}{x-1} + \frac{B}{(x-1)^2} + \frac{C}{x+1} + \frac{D}{(x+1)^2}$$

(Multiply both sides by $(x-1)^2(x+1)^2$.)

$$x^2 = A(x-1)(x+1)^2 + B(x+1)^2 + C(x-1)^2(x+1) + D(x-1)^2$$

Let $x = 1$:   then $1^2 = A(1-1)(1+1)^2 + B(1+1)^2 + C(1-1)^2(1+1) + D(1-1)^2$

$$\Rightarrow 1 = 4B \Rightarrow B = \frac{1}{4}$$

Let $x = -1$:   then

$$(-1)^2 = A(-1-1)(-1+1)^2 + B(-1+1)^2 + C(-1-1)^2(-1+1) + D(-1-1)^2$$

$$\Rightarrow 1 = 4D \Rightarrow D = \frac{1}{4}$$

Let $x = 0$:   then

$$0^2 = A(0-1)(0+1)^2 + B(0+1)^2 + C(0-1)^2(0+1) + D(0-1)^2$$

$$\Rightarrow 0 = -A + B + C + D \Rightarrow A - C = \frac{1}{4} + \frac{1}{4} = \frac{1}{2}$$

Let $x = 2$:   then

$$2^2 = A(2-1)(2+1)^2 + B(2+1)^2 + C(2-1)^2(2+1) + D(2-1)^2$$

$$\Rightarrow 4 = 9A + 9B + 3C + D \Rightarrow 9A + 3C = 4 - \frac{9}{4} - \frac{1}{4} = \frac{3}{2}$$

$$\Rightarrow 3A + C = \frac{1}{2}$$

Solve the system of equations:

$$A - C = \frac{1}{2}$$

$$3A + C = \frac{1}{2}$$

$$\overline{\phantom{3A + C = \frac{1}{2}}}$$

$$4A \quad = 1 \Rightarrow A = \frac{1}{4} \Rightarrow \frac{3}{4} + C = \frac{1}{2} \Rightarrow C = -\frac{1}{4}$$

$$\frac{x^2}{(x-1)^2(x+1)^2} = \frac{(1/4)}{x-1} + \frac{(1/4)}{(x-1)^2} + \frac{(-1/4)}{x+1} + \frac{(1/4)}{(x+1)^2}$$

25. Find the partial fraction decomposition:

$$\frac{x-3}{(x+2)(x+1)^2} = \frac{A}{x+2} + \frac{B}{x+1} + \frac{C}{(x+1)^2}$$

(Multiply both sides by $(x+2)(x+1)^2$.)

$$x - 3 = A(x+1)^2 + B(x+2)(x+1) + C(x+2)$$

Let $x = -2$:   then $-2 - 3 = A(-2+1)^2 + B(-2+2)(-2+1) + C(-2+2)$

$$\Rightarrow -5 = A \Rightarrow A = -5$$

Let $x = -1$:   then $-1 - 3 = A(-1+1)^2 + B(-1+2)(-1+1) + C(-1+2)$

$$\Rightarrow -4 = C \Rightarrow C = -4$$

Let $x = 0$:   then

$$0 - 3 = A(0+1)^2 + B(0+2)(0+1) + C(0+2) \Rightarrow -3 = A + 2B + 2C$$

$$\Rightarrow -3 = -5 + 2B + 2(-4) \Rightarrow 2B = 10 \Rightarrow B = 5$$

$$\frac{x-3}{(x+2)(x+1)^2} = \frac{-5}{x+2} + \frac{5}{x+1} + \frac{-4}{(x+1)^2}$$

27. Find the partial fraction decomposition:

$$\frac{x+4}{x^2(x^2+4)} = \frac{A}{x} + \frac{B}{x^2} + \frac{Cx+D}{x^2+4}$$ (Multiply both sides by $x^2(x^2+4)$.)

$$x + 4 = Ax(x^2+4) + B(x^2+4) + (Cx+D)x^2$$

Let $x = 0$:   then $0 + 4 = A(0)(0^2+4) + B(0^2+4) + (C(0)+D)(0)^2$

$$\Rightarrow 4 = 4B \Rightarrow B = 1$$

Let $x = 1$:   then $1 + 4 = A(1)(1^2 + 4) + B(1^2 + 4) + (C(1) + D)(1)^2$

$$\Rightarrow 5 = 5A + 5B + C + D \Rightarrow 5 = 5A + 5 + C + D$$

$$\Rightarrow 5A + C + D = 0$$

Let $x = -1$: then

$$-1 + 4 = A(-1)((-1)^2 + 4) + B((-1)^2 + 4) + (C(-1) + D)(-1)^2$$

$$\Rightarrow 3 = -5A + 5B - C + D \Rightarrow 3 = -5A + 5 - C + D$$

$$\Rightarrow -5A - C + D = -2$$

Let $x = 2$:   then $2 + 4 = A(2)(2^2 + 4) + B(2^2 + 4) + (C(2) + D)(2)^2$

$$\Rightarrow 6 = 16A + 8B + 8C + 4D \Rightarrow 6 = 16A + 8 + 8C + 4D$$

$$\Rightarrow 16A + 8C + 4D = -2$$

Solve the system of equations:

$$5A + C + D = 0$$
$$\underline{-5A - C + D = -2}$$
$$2D = -2 \qquad\qquad 5A + C - 1 = 0$$
$$D = -1 \qquad\qquad C = 1 - 5A$$
$$16A + 8(1 - 5A) + 4(-1) = -2$$
$$16A + 8 - 40A - 4 = -2 \qquad\qquad C = 1 - 5\left(\frac{1}{4}\right)$$
$$-24A = -6$$
$$A = \frac{1}{4} \qquad\qquad C = 1 - \frac{5}{4} = -\frac{1}{4}$$

$$\frac{x + 4}{x^2(x^2 + 4)} = \frac{(1/4)}{x} + \frac{1}{x^2} + \frac{(-(1/4)x - 1)}{x^2 + 4} = \frac{(1/4)}{x} + \frac{1}{x^2} + \frac{-(1/4)(x + 4)}{x^2 + 4}$$

29. Find the partial fraction decomposition:

$$\frac{x^2 + 2x + 3}{(x + 1)(x^2 + 2x + 4)} = \frac{A}{x + 1} + \frac{Bx + C}{x^2 + 2x + 4}$$

(Multiply both sides by $(x + 1)(x^2 + 2x + 4)$.)

$$x^2 + 2x + 3 = A(x^2 + 2x + 4) + (Bx + C)(x + 1)$$

Let $x = -1$: then

$$(-1)^2 + 2(-1) + 3 = A((-1)^2 + 2(-1) + 4) + (B(-1) + C)(-1 + 1)$$

$$\Rightarrow 2 = 3A \Rightarrow A = \frac{2}{3}$$

Let $x = 0$:   then $0^2 + 2(0) + 3 = A(0^2 + 2(0) + 4) + (B(0) + C)(0 + 1)$

$$\Rightarrow 3 = 4A + C \Rightarrow 3 = 4(2/3) + C \Rightarrow C = \frac{1}{3}$$

Let $x = 1$: then

$$1^2 + 2(1) + 3 = A(1^2 + 2(1) + 4) + (B(1) + C)(1 + 1)$$

$$\Rightarrow 6 = 7A + 2B + 2C \Rightarrow 6 = 7(2/3) + 2B + 2(1/3)$$

$$\Rightarrow 2B = 6 - \frac{14}{3} - \frac{2}{3} \Rightarrow 2B = \frac{2}{3} \Rightarrow B = \frac{1}{3}$$

$$\frac{x^2 + 2x + 3}{(x+1)(x^2 + 2x + 4)} = \frac{(2/3)}{x+1} + \frac{((1/3)x + 1/3)}{x^2 + 2x + 4} = \frac{(2/3)}{x+1} + \frac{(1/3)(x+1)}{x^2 + 2x + 4}$$

31. Find the partial fraction decomposition:

$$\frac{x}{(3x-2)(2x+1)} = \frac{A}{3x-2} + \frac{B}{2x+1} \quad \text{(Multiply both sides by } (3x-2)(2x+1).)$$

$$x = A(2x+1) + B(3x-2)$$

Let $x = -\frac{1}{2}$: then $-\frac{1}{2} = A(2(-1/2) + 1) + B(3(-1/2) - 2) \Rightarrow -\frac{1}{2} = -\frac{7}{2}B \Rightarrow B = \frac{1}{7}$

Let $x = \frac{2}{3}$: then $\frac{2}{3} = A(2(2/3) + 1) + B(3(2/3) - 2) \Rightarrow \frac{2}{3} = \frac{7}{3}A \Rightarrow A = \frac{2}{7}$

$$\frac{x}{(3x-2)(2x+1)} = \frac{(2/7)}{3x-2} + \frac{(1/7)}{2x+1}$$

33. Find the partial fraction decomposition:

$$\frac{x}{x^2 + 2x - 3} = \frac{x}{(x+3)(x-1)} = \frac{A}{x+3} + \frac{B}{x-1}$$

$$\text{(Multiply both sides by } (x+3)(x-1).)$$

$$x = A(x-1) + B(x+3)$$

Let $x = 1$: then $1 = A(1-1) + B(1+3) \Rightarrow 1 = 4B \Rightarrow B = \frac{1}{4}$

Let $x = -3$: then $-3 = A(-3-1) + B(-3+3) \Rightarrow -3 = -4A \Rightarrow A = \frac{3}{4}$

$$\frac{x}{x^2 + 2x - 3} = \frac{(3/4)}{x+3} + \frac{(1/4)}{x-1}$$

35. Find the partial fraction decomposition:

$$\frac{x^2 + 2x + 3}{(x^2 + 4)^2} = \frac{Ax + B}{x^2 + 4} + \frac{Cx + D}{(x^2 + 4)^2} \quad \text{(Multiply both sides by } (x^2 + 4)^2.)$$

$$x^2 + 2x + 3 = (Ax + B)(x^2 + 4) + Cx + D$$

$$x^2 + 2x + 3 = Ax^3 + Bx^2 + 4Ax + 4B + Cx + D$$

$$x^2 + 2x + 3 = Ax^3 + Bx^2 + (4A + C)x + 4B + D$$

$$A = 0; B = 1$$

$$4A + C = 2 \Rightarrow 4(0) + C = 2 \Rightarrow C = 2$$

$$4B + D = 3 \Rightarrow 4(1) + D = 3 \Rightarrow D = -1$$

$$\frac{x^2 + 2x + 3}{(x^2 + 4)^2} = \frac{1}{x^2 + 4} + \frac{2x - 1}{(x^2 + 4)^2}$$

37. Find the partial fraction decomposition:

$$\frac{7x+3}{x^3-2x^2-3x} = \frac{7x+3}{x(x-3)(x+1)} = \frac{A}{x} + \frac{B}{x-3} + \frac{C}{x+1}$$

(Multiply both sides by $x(x-3)(x+1)$.)

$$7x+3 = A(x-3)(x+1) + Bx(x+1) + Cx(x-3)$$

Let $x=0$: then $7(0)+3 = A(0-3)(0+1) + B(0)(0+1) + C(0)(0-3)$

$$\Rightarrow 3 = -3A \Rightarrow A = -1$$

Let $x=3$: then $7(3)+3 = A(3-3)(3+1) + B(3)(3+1) + C(3)(3-3)$

$$\Rightarrow 24 = 12B \Rightarrow B = 2$$

Let $x=-1$: then

$$7(-1)+3 = A(-1-3)(-1+1) + B(-1)(-1+1) + C(-1)(-1-3)$$

$$\Rightarrow -4 = 4C \Rightarrow C = -1$$

$$\frac{7x+3}{x^3-2x^2-3x} = \frac{-1}{x} + \frac{2}{x-3} + \frac{-1}{x+1}$$

39. Perform synthetic division to find a factor:

$$
\begin{array}{r|rrrr}
2) & 1 & -4 & 5 & -2 \\
   &   & 2 & -4 & 2 \\
\hline
   & 1 & -2 & 1 & 0
\end{array}
$$

$$x^3-4x^2+5x-2 = (x-2)(x^2-2x+1) = (x-2)(x-1)^2$$

Find the partial fraction decomposition:

$$\frac{x^2}{x^3-4x^2+5x-2} = \frac{x^2}{(x-2)(x-1)^2} = \frac{A}{x-2} + \frac{B}{x-1} + \frac{C}{(x-1)^2}$$

(Multiply both sides by $(x-2)(x-1)^2$.)

$$x^2 = A(x-1)^2 + B(x-2)(x-1) + C(x-2)$$

Let $x=2$: then $2^2 = A(2-1)^2 + B(2-2)(2-1) + C(2-2)$

$$\Rightarrow A = 4$$

Let $x=1$: then $1^2 = A(1-1)^2 + B(1-2)(1-1) + C(1-2)$

$$\Rightarrow 1 = -C \Rightarrow C = -1$$

Let $x=0$: then $0^2 = A(0-1)^2 + B(0-2)(0-1) + C(0-2)$

$$\Rightarrow 0 = A + 2B - 2C \Rightarrow 0 = 4 + 2B - 2(-1)$$

$$\Rightarrow 2B = -6 \Rightarrow B = -3$$

$$\frac{x^2}{x^3-4x^2+5x-2} = \frac{4}{x-2} + \frac{-3}{x-1} + \frac{-1}{(x-1)^2}$$

41. Find the partial fraction decomposition:

$$\frac{x^3}{(x^2+16)^3} = \frac{Ax+B}{x^2+16} + \frac{Cx+D}{(x^2+16)^2} + \frac{Ex+F}{(x^2+16)^3}$$

(Multiply both sides by $(x^2+16)^3$.)

$$x^3 = (Ax+B)(x^2+16)^2 + (Cx+D)(x^2+16) + Ex+F$$

$$x^3 = (Ax+B)(x^4+32x^2+256) + Cx^3+Dx^2+16Cx+16D+Ex+F$$

$$x^3 = Ax^5+Bx^4+32Ax^3+32Bx^2+256Ax+256B+Cx^3+Dx^2$$
$$+16Cx+16D+Ex+F$$

$$x^3 = Ax^5+Bx^4+(32A+C)x^3+(32B+D)x^2+(256A+16C+E)x$$
$$+(256B+16D+F)$$

$$A=0$$

$$B=0$$

$$32A+C=1 \;\Rightarrow\; 32(0)+C=1 \;\Rightarrow\; C=1$$

$$32B+D=0 \;\Rightarrow\; 32(0)+D=0 \;\Rightarrow\; D=0$$

$$256A+16C+E=0 \;\Rightarrow 256(0)+16(1)+E=0 \;\Rightarrow\; E=-16$$

$$256B+16D+F=0 \;\Rightarrow\; 256(0)+16(0)+F=0 \Rightarrow\; F=0$$

$$\frac{x^3}{(x^2+16)^3} = \frac{x}{(x^2+16)^2} + \frac{-16x}{(x^2+16)^3}$$

43. Find the partial fraction decomposition:

$$\frac{4}{2x^2-5x-3} = \frac{4}{(x-3)(2x+1)} = \frac{A}{x-3} + \frac{B}{2x+1}$$ (Multiply both sides by $(x-3)(2x+1)$.)

$$4 = A(2x+1) + B(x-3)$$

Let $x=-\dfrac{1}{2}$: then $4 = A\!\left(2(-1/2)+1\right) + B\!\left(-\dfrac{1}{2}-3\right) \Rightarrow 4 = -\dfrac{7}{2}B \Rightarrow B = -\dfrac{8}{7}$

Let $x=3$: then $4 = A(2(3)+1) + B(3-3) \;\Rightarrow\; 4 = 7A \;\Rightarrow\; A = \dfrac{4}{7}$

$$\frac{4}{2x^2-5x-3} = \frac{(4/7)}{x-3} + \frac{(-8/7)}{2x+1}$$

45. Find the partial fraction decomposition:

$$\frac{2x+3}{x^4-9x^2} = \frac{2x+3}{x^2(x-3)(x+3)} = \frac{A}{x} + \frac{B}{x^2} + \frac{C}{x-3} + \frac{D}{x+3}$$

(Multiply both sides by $x^2(x-3)(x+3)$.)

$$2x+3 = Ax(x-3)(x+3) + B(x-3)(x+3) + Cx^2(x+3) + Dx^2(x-3)$$

Let $x=0$: then

$$2\cdot0+3 = A\cdot0(0-3)(0+3) + B(0-3)(0+3) + C\cdot0^2(0+3) + D\cdot0^2(0-3)$$

$$\Rightarrow 3 = -9B \;\Rightarrow\; B = -\frac{1}{3}$$

Let $x = 3$:   then

$$2 \cdot 3 + 3 = A \cdot 3(3 - 3)(3 + 3) + B(3 - 3)(3 + 3) + C \cdot 3^2(3 + 3) + D \cdot 3^2(3 - 3)$$

$$\Rightarrow 9 = 54C \Rightarrow C = \frac{1}{6}$$

Let $x = -3$:   then

$$2(-3) + 3 = A(-3)(-3 - 3)(-3 + 3) + B(-3 - 3)(-3 + 3) + C(-3)^2(-3 + 3)$$
$$+ D(-3)^2(-3 - 3)$$

$$\Rightarrow -3 = -54D \Rightarrow D = \frac{1}{18}$$

Let $x = 1$:   then

$$2 \cdot 1 + 3 = A \cdot 1(1 - 3)(1 + 3) + B(1 - 3)(1 + 3) + C \cdot 1^2(1 + 3) + D \cdot 1^2(1 - 3)$$

$$\Rightarrow 5 = -8A - 8B + 4C - 2D$$

$$\Rightarrow 5 = -8A - 8(-1/3) + 4(1/6) - 2(1/18)$$

$$\Rightarrow 5 = -8A + \frac{8}{3} + \frac{2}{3} - \frac{1}{9} \Rightarrow -8A = \frac{16}{9} \Rightarrow A = -\frac{2}{9}$$

$$\frac{2x + 3}{x^4 - 9x^2} = \frac{(-2/9)}{x} + \frac{(-1/3)}{x^2} + \frac{(1/6)}{x - 3} + \frac{(1/18)}{x + 3}$$

Systems of Equations and Inequalities

10.6 Systems of Nonlinear Equations

1. $y = 3x + 2$
   The graph is a line.
   x-intercept:
   $$0 = 3x + 2$$
   $$3x = -2$$
   $$x = -\frac{2}{3}$$
   y-intercept: $y = 3(0) + 2 = 2$

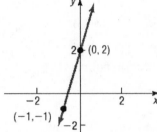

3. $y^2 = x^2 - 1 \Rightarrow x^2 - y^2 = 1$
   The graph is a hyperbola.
   x-intercepts:
   $$0^2 = x^2 - 1$$
   $$1 = x^2$$
   $$x = -1, x = 1$$
   y-intercepts:
   $$0 - y^2 = 1$$
   $$-y^2 = 1$$
   $$y^2 = -1, \text{ no real solution}$$
   There are no y-intercepts.

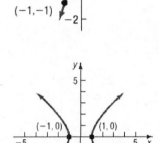

5. $\begin{cases} y = x^2 + 1 \\ y = x + 1 \end{cases}$

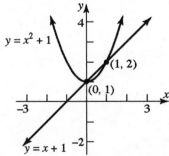

(0, 1) and (1, 2) are the intersection points.

Solve by substitution:
$$x^2 + 1 = x + 1$$
$$x^2 - x = 0$$
$$x(x - 1) = 0$$
$$x = 0 \quad \text{or} \quad x = 1$$
$$y = 1 \qquad y = 2$$
Solutions: (0, 1) and (1, 2)

7. $\begin{cases} y = \sqrt{36 - x^2} \\ y = 8 - x \end{cases}$

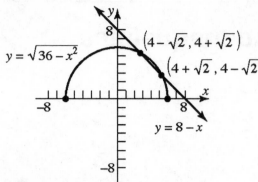

$y = \sqrt{36 - x^2}$

$\left(4 - \sqrt{2}, 4 + \sqrt{2}\right)$

$\left(4 + \sqrt{2}, 4 - \sqrt{2}\right)$

$y = 8 - x$

(2.59, 5.41) and (5.41, 2.59) are the intersection points.

Solve by substitution:

$$\sqrt{36 - x^2} = 8 - x$$
$$36 - x^2 = 64 - 16x + x^2$$
$$2x^2 - 16x + 28 = 0$$
$$x^2 - 8x + 14 = 0$$
$$x = \frac{8 \pm \sqrt{64 - 56}}{2}$$
$$= \frac{8 \pm 2\sqrt{2}}{2}$$
$$= 4 \pm \sqrt{2}$$

If $x = 4 + \sqrt{2}$, $y = 8 - \left(4 + \sqrt{2}\right) = 4 - \sqrt{2}$

If $x = 4 - \sqrt{2}$, $y = 8 - \left(4 - \sqrt{2}\right) = 4 + \sqrt{2}$

Solutions:

$\left(4 + \sqrt{2}, 4 - \sqrt{2}\right)$ and $\left(4 - \sqrt{2}, 4 + \sqrt{2}\right)$

9. $\begin{cases} y = \sqrt{x} \\ y = 2 - x \end{cases}$

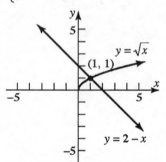

$y = \sqrt{x}$

(1, 1)

$y = 2 - x$

(1, 1) is the intersection point.

Solve by substitution:

$$\sqrt{x} = 2 - x$$
$$x = 4 - 4x + x^2$$
$$x^2 - 5x + 4 = 0$$
$$(x - 4)(x - 1) = 0$$
$$x = 4 \quad \text{or} \quad x = 1$$
$$y = -2 \quad \text{or} \quad y = 1$$

Eliminate (4, –2); it does not check.
Solution: (1, 1)

11. $\begin{cases} x = 2y \\ x = y^2 - 2y \end{cases}$

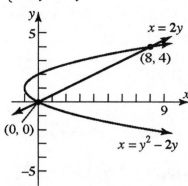

(0, 0) and (8, 4) are the intersection points.

Solve by substitution:
$$2y = y^2 - 2y$$
$$y^2 - 4y = 0$$
$$y(y - 4) = 0$$
$$y = 0 \quad \text{or} \quad y = 4$$
$$x = 0 \quad \text{or} \quad x = 8$$
Solutions: (0, 0) and (8, 4)

13. $\begin{cases} x^2 + y^2 = 4 \\ x^2 + 2x + y^2 = 0 \end{cases}$

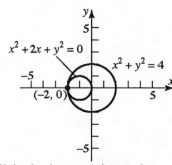

(–2, 0) is the intersection point.

Substitute 4 for $x^2 + y^2$ in the second equation:
$$2x + 4 = 0$$
$$2x = -4$$
$$x = -2$$
$$y = \sqrt{4 - (-2)^2} = 0$$
Solution: (–2, 0)

15. $\begin{cases} y = 3x - 5 \\ x^2 + y^2 = 5 \end{cases}$

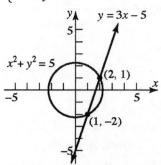

(1, –2) and (2, 1) are the intersection points.

Solve by substitution:
$$x^2 + (3x - 5)^2 = 5$$
$$x^2 + 9x^2 - 30x + 25 = 5$$
$$10x^2 - 30x + 20 = 0$$
$$x^2 - 3x + 2 = 0$$
$$(x - 1)(x - 2) = 0$$
$$x = 1 \quad \text{or} \quad x = 2$$
$$y = -2 \qquad y = 1$$
Solutions: (1, –2) and (2, 1)

17. $\begin{cases} x^2 + y^2 = 4 \\ y^2 - x = 4 \end{cases}$

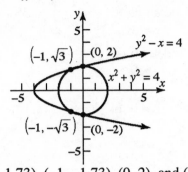

$(-1, 1.73)$, $(-1, -1.73)$, $(0, 2)$, and $(0, -2)$ are the intersection points.

Substitute $x + 4$ for y^2 in the first equation:

$$x^2 + x + 4 = 4$$
$$x^2 + x = 0$$
$$x(x + 1) = 0$$

$$x = 0 \quad \text{or} \quad x = -1$$
$$y^2 = 4 \qquad y^2 = 3$$
$$y = \pm 2 \qquad y^2 = \pm\sqrt{3}$$

Solutions:

$$(0, -2), (0, 2), \left(-1, \sqrt{3}\right), \left(-1, -\sqrt{3}\right)$$

19. $\begin{cases} xy = 4 \\ x^2 + y^2 = 8 \end{cases}$

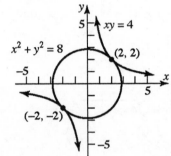

$(-2, -2)$ and $(2, 2)$ are the intersection points.

Solve by substitution:

$$x^2 + \left(\frac{4}{x}\right)^2 = 8$$
$$x^2 + \frac{16}{x^2} = 8$$
$$x^4 + 16 = 8x^2$$
$$x^4 - 8x^2 + 16 = 0$$
$$\left(x^2 - 4\right)^2 = 0$$
$$x^2 - 4 = 0$$
$$x^2 = 4$$
$$x = 2 \quad \text{or} \quad x = -2$$
$$y = 2 \qquad y = -2$$

Solutions: $(-2, -2)$ and $(2, 2)$

21. $\begin{cases} x^2 + y^2 = 4 \\ y = x^2 - 9 \end{cases}$

No solution; Inconsistent.

Solve by substitution:

$$x^2 + (x^2 - 9)^2 = 4$$
$$x^2 + x^4 - 18x^2 + 81 = 4$$
$$x^4 - 17x^2 + 77 = 0$$
$$x^2 = \frac{17 \pm \sqrt{289 - 4(77)}}{2}$$
$$= \frac{17 \pm \sqrt{-19}}{2}$$

There are no real solutions to this expression. Inconsistent.

23. $\begin{cases} y = x^2 - 4 \\ y = 6x - 13 \end{cases}$

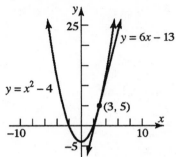

$y = 6x - 13$

$y = x^2 - 4$

(3, 5)

−10   10

−5

(3,5) is the intersection point.

Solve by substitution:

$$x^2 - 4 = 6x - 13$$

$$x^2 - 6x + 9 = 0$$

$$(x - 3)^2 = 0$$

$$x - 3 = 0$$

$$x = 3$$

$$y = 5$$

Solution: (3,5)

25. Solve the second equation for y, substitute into the first equation and solve:

$$\begin{cases} 2x^2 + y^2 = 18 \\ xy = 4 \implies y = \dfrac{4}{x} \end{cases}$$

$$2x^2 + \left(\frac{4}{x}\right)^2 = 18 \implies 2x^2 + \frac{16}{x^2} = 18$$

$$2x^4 + 16 = 18x^2 \implies 2x^4 - 18x^2 + 16 = 0 \implies x^4 - 9x^2 + 8 = 0 \implies (x^2 - 8)(x^2 - 1) = 0$$

$$x^2 = 8 \implies x = \pm\sqrt{8} = \pm 2\sqrt{2}$$

$$\text{or } x^2 = 1 \implies x = \pm 1$$

If $x = 2\sqrt{2}$: $y = \dfrac{4}{2\sqrt{2}} = \sqrt{2}$

If $x = -2\sqrt{2}$: $y = \dfrac{4}{-2\sqrt{2}} = -\sqrt{2}$

If $x = 1$: $y = \dfrac{4}{1} = 4$

If $x = -1$: $y = \dfrac{4}{-1} = -4$

Solutions: $\left(2\sqrt{2}, \sqrt{2}\right), \left(-2\sqrt{2}, -\sqrt{2}\right), (1, 4), (-1, -4)$

27. Substitute the first equation into the second equation and solve:

$$\begin{cases} y = 2x + 1 \\ 2x^2 + y^2 = 1 \end{cases}$$

$$2x^2 + (2x + 1)^2 = 1 \implies 2x^2 + 4x^2 + 4x + 1 = 1 \implies 6x^2 + 4x = 0 \implies 2x(3x + 2) = 0$$

$$\implies x = 0 \text{ or } x = -\frac{2}{3}$$

If $x = 0$: $y = 2(0) + 1 = 1$

If $x = -\dfrac{2}{3}$: $y = 2\left(-\dfrac{2}{3}\right) + 1 = -\dfrac{4}{3} + 1 = -\dfrac{1}{3}$

Solutions: $(0, 1), \left(-\dfrac{2}{3}, -\dfrac{1}{3}\right)$

29.   Solve the first equation for y, substitute into the second equation and solve:

$$\begin{cases} x + y + 1 = 0 \;\Rightarrow\; y = -x - 1 \\ x^2 + y^2 + 6y - x = -5 \end{cases}$$

$x^2 + (-x - 1)^2 + 6(-x - 1) - x = -5 \Rightarrow x^2 + x^2 + 2x + 1 - 6x - 6 - x = -5$

$2x^2 - 5x = 0 \Rightarrow x(2x - 5) = 0 \Rightarrow x = 0 \;$ or $\; x = \dfrac{5}{2}$

If $x = 0$: $y = -(0) - 1 = -1$

If $x = \dfrac{5}{2}$: $y = -\dfrac{5}{2} - 1 = -\dfrac{7}{2}$

Solutions: $(0, -1), \left(\dfrac{5}{2}, -\dfrac{7}{2}\right)$

31.   Solve the second equation for y, substitute into the first equation and solve:

$$\begin{cases} 4x^2 - 3xy + 9y^2 = 15 \\ 2x + 3y = 5 \;\Rightarrow\; y = -\dfrac{2}{3}x + \dfrac{5}{3} \end{cases}$$

$4x^2 - 3x\left(-\dfrac{2}{3}x + \dfrac{5}{3}\right) + 9\left(-\dfrac{2}{3}x + \dfrac{5}{3}\right)^2 = 15 \Rightarrow 4x^2 + 2x^2 - 5x + 4x^2 - 20x + 25 = 15$

$10x^2 - 25x + 10 = 0 \Rightarrow 2x^2 - 5x + 2 = 0$

$(2x - 1)(x - 2) = 0 \Rightarrow x = \dfrac{1}{2} \;$ or $\; x = 2$

If $x = \dfrac{1}{2}$: $y = -\dfrac{2}{3}\left(\dfrac{1}{2}\right) + \dfrac{5}{3} = \dfrac{4}{3}$

If $x = 2$: $y = -\dfrac{2}{3}(2) + \dfrac{5}{3} = \dfrac{1}{3}$

Solutions: $\left(\dfrac{1}{2}, \dfrac{4}{3}\right), \left(2, \dfrac{1}{3}\right)$

33.   Multiply each side of the second equation by 4 and add the equations to eliminate y:

$$\begin{cases} x^2 - 4y^2 = -7 \;\longrightarrow\; x^2 - 4y^2 = -7 \\ 3x^2 + y^2 = 31 \;\xrightarrow{\;4\;}\; \underline{12x^2 + 4y^2 = 124} \end{cases}$$

$13x^2 \qquad = 117 \Rightarrow x^2 = 9 \Rightarrow x = \pm 3$

If $x = 3$: $3(3)^2 + y^2 = 31 \Rightarrow y^2 = 4 \Rightarrow y = \pm 2$

If $x = -3$: $3(-3)^2 + y^2 = 31 \Rightarrow y^2 = 4 \Rightarrow y = \pm 2$

Solutions: $(3, 2), (3, -2), (-3, 2), (-3, -2)$

35. Multiply each side of the first equation by 5 and each side of the second equation by 3 and add the equations to eliminate y:

$$\begin{cases} 7x^2 - 3y^2 + 5 = 0 \Rightarrow 7x^2 - 3y^2 = -5 \xrightarrow{\;5\;} 35x^2 - 15y^2 = -25 \\ 3x^2 + 5y^2 = 12 \quad \Rightarrow 3x^2 + 5y^2 = 12 \xrightarrow{\;3\;} \underline{9x^2 + 15y^2 = 36} \end{cases}$$

$$44x^2 \qquad = \quad 11 \Rightarrow x^2 = \frac{1}{4} \Rightarrow x = \pm\frac{1}{2}$$

If $x = \dfrac{1}{2}$: $\quad 3\left(\dfrac{1}{2}\right)^2 + 5y^2 = 12 \Rightarrow 5y^2 = \dfrac{45}{4} \Rightarrow y^2 = \dfrac{9}{4} \Rightarrow y = \pm\dfrac{3}{2}$

If $x = -\dfrac{1}{2}$: $\quad 3\left(-\dfrac{1}{2}\right)^2 + 5y^2 = 12 \Rightarrow 5y^2 = \dfrac{45}{4} \Rightarrow y^2 = \dfrac{9}{4} \Rightarrow y = \pm\dfrac{3}{2}$

Solutions: $\left(\dfrac{1}{2}, \dfrac{3}{2}\right), \left(\dfrac{1}{2}, -\dfrac{3}{2}\right), \left(-\dfrac{1}{2}, \dfrac{3}{2}\right), \left(-\dfrac{1}{2}, -\dfrac{3}{2}\right)$

37. Multiply each side of the second equation by 2 and add the equations to eliminate xy:

$$\begin{cases} x^2 + 2xy = 10 \xrightarrow{\qquad} \quad x^2 + 2xy = 10 \\ 3x^2 - xy = 2 \xrightarrow{\;2\;} \underline{6x^2 - 2xy = 4} \end{cases}$$

$$7x^2 \qquad = 14 \Rightarrow x^2 = 2 \Rightarrow x = \pm\sqrt{2}$$

If $x = \sqrt{2}$: $\quad 3\left(\sqrt{2}\right)^2 - \sqrt{2}\cdot y = 2 \Rightarrow -\sqrt{2}\cdot y = -4 \Rightarrow y = \dfrac{4}{\sqrt{2}} \Rightarrow y = 2\sqrt{2}$

If $x = -\sqrt{2}$: $\quad 3\left(-\sqrt{2}\right)^2 - \left(-\sqrt{2}\right)y = 2 \Rightarrow \sqrt{2}\cdot y = -4 \Rightarrow y = \dfrac{-4}{\sqrt{2}} \Rightarrow y = -2\sqrt{2}$

Solutions: $\left(\sqrt{2}, 2\sqrt{2}\right), \left(-\sqrt{2}, -2\sqrt{2}\right)$

39. Multiply each side of the first equation by 2 and add the equations to eliminate y:

$$\begin{cases} 2x^2 + y^2 = 2 \Rightarrow 2x^2 + y^2 = 2 \xrightarrow{\;2\;} 4x^2 + 2y^2 = 4 \\ x^2 - 2y^2 + 8 = 0 \Rightarrow \quad x^2 - 2y^2 = -8 \xrightarrow{\qquad} \underline{x^2 - 2y^2 = -8} \end{cases}$$

$$5x^2 \qquad = -4 \Rightarrow x^2 = -\frac{4}{5}$$

No solution. The system is inconsistent.

41. Multiply each side of the second equation by 2 and add the equations to eliminate y:

$$\begin{cases} x^2 + 2y^2 = 16 \\ 4x^2 - y^2 = 24 \end{cases} \xrightarrow{} \begin{aligned} x^2 + 2y^2 &= 16 \\ \underline{8x^2 - 2y^2} &= \underline{48} \end{aligned}$$

$$9x^2 \qquad = 64 \Rightarrow x^2 = \frac{64}{9} \Rightarrow x = \pm\frac{8}{3}$$

If $x = \frac{8}{3}$: $\quad \left(\frac{8}{3}\right)^2 + 2y^2 = 16 \Rightarrow 2y^2 = \frac{80}{9} \Rightarrow y^2 = \frac{40}{9} \Rightarrow y = \pm\frac{2\sqrt{10}}{3}$

If $x = -\frac{8}{3}$: $\quad \left(-\frac{8}{3}\right)^2 + 2y^2 = 16 \Rightarrow 2y^2 = \frac{80}{9} \Rightarrow y^2 = \frac{40}{9} \Rightarrow y = \pm\frac{2\sqrt{10}}{3}$

Solutions: $\left(\frac{8}{3}, \frac{2\sqrt{10}}{3}\right), \left(\frac{8}{3}, -\frac{2\sqrt{10}}{3}\right), \left(-\frac{8}{3}, \frac{2\sqrt{10}}{3}\right), \left(-\frac{8}{3}, -\frac{2\sqrt{10}}{3}\right)$

43. Multiply each side of the second equation by 2 and add the equations to eliminate y:

$$\begin{cases} \dfrac{5}{x^2} - \dfrac{2}{y^2} + 3 = 0 \to \dfrac{5}{x^2} - \dfrac{2}{y^2} = -3 \longrightarrow \dfrac{5}{x^2} - \dfrac{2}{y^2} = -3 \\ \dfrac{3}{x^2} + \dfrac{1}{y^2} = 7 \to \dfrac{3}{x^2} + \dfrac{1}{y^2} = 7 \xrightarrow{2} \dfrac{6}{x^2} + \dfrac{2}{y^2} = 14 \end{cases}$$

$$\frac{11}{x^2} \qquad = 11 \Rightarrow 11 = 11x^2 \Rightarrow x^2 = 1 \Rightarrow x = \pm 1$$

If $x = 1$: $\quad \dfrac{3}{(1)^2} + \dfrac{1}{y^2} = 7 \Rightarrow \dfrac{1}{y^2} = 4 \Rightarrow y^2 = \dfrac{1}{4} \Rightarrow y = \pm\dfrac{1}{2}$

If $x = -1$: $\quad \dfrac{3}{(-1)^2} + \dfrac{1}{y^2} = 7 \Rightarrow \dfrac{1}{y^2} = 4 \Rightarrow y^2 = \dfrac{1}{4} \Rightarrow y = \pm\dfrac{1}{2}$

Solutions: $\left(1, \frac{1}{2}\right), \left(1, -\frac{1}{2}\right), \left(-1, \frac{1}{2}\right), \left(-1, -\frac{1}{2}\right)$

45. Multiply each side of the first equation by -2 and add the equations to eliminate x:

$$\begin{cases} \dfrac{1}{x^4} + \dfrac{6}{y^4} = 6 \xrightarrow{-2} -\dfrac{2}{x^4} - \dfrac{12}{y^4} = -12 \\ \dfrac{2}{x^4} - \dfrac{2}{y^4} = 19 \longrightarrow \dfrac{2}{x^4} - \dfrac{2}{y^4} = 19 \end{cases}$$

$$-\frac{14}{y^4} \qquad = 7 \Rightarrow -14 = 7y^4 \Rightarrow y^4 = -2$$

There are no real solutions. The system is inconsistent.

47. Factor the first equation, solve for x, substitute into the second equation and solve:
$$\begin{cases} x^2 - 3xy + 2y^2 = 0 \Rightarrow (x-2y)(x-y) = 0 \Rightarrow x = 2y \text{ or } x = y \\ \quad x^2 + xy = 6 \end{cases}$$

Substitute $x = 2y$ and solve:
$$x^2 + xy = 6$$
$$(2y)^2 + (2y)y = 6$$
$$4y^2 + 2y^2 = 6 \Rightarrow 6y^2 = 6$$
$$y^2 = 1 \Rightarrow y = \pm 1$$
If $y = 1$: $x = 2 \cdot 1 = 2$
If $y = -1$: $x = 2(-1) = -2$

Substitute $x = y$ and solve:
$$x^2 + xy = 6$$
$$y^2 + y \cdot y = 6$$
$$y^2 + y^2 = 6 \Rightarrow 2y^2 = 6$$
$$y^2 = 3 \Rightarrow y = \pm\sqrt{3}$$
If $y = \sqrt{3}$: $x = \sqrt{3}$
If $y = -\sqrt{3}$: $x = -\sqrt{3}$

Solutions: $(2, 1), (-2, -1), \left(\sqrt{3}, \sqrt{3}\right), \left(-\sqrt{3}, -\sqrt{3}\right)$

49. Multiply each side of the second equation by $-y$ and add the equations to eliminate y:
$$\begin{cases} y^2 + y + x^2 - x - 2 = 0 \longrightarrow \quad y^2 + y + x^2 - x - 2 = 0 \\ \quad y + 1 + \dfrac{x-2}{y} = 0 \xrightarrow{-y} \quad \underline{-y^2 - y \quad - x + 2 = 0} \end{cases}$$
$$x^2 - 2x = 0 \Rightarrow x(x-2) = 0$$
$$\Rightarrow x = 0 \text{ or } x = 2$$
If $x = 0$: $y^2 + y + 0^2 - 0 - 2 = 0 \Rightarrow y^2 + y - 2 = 0 \Rightarrow (y+2)(y-1) = 0$
$$\Rightarrow y = -2 \text{ or } y = 1$$
If $x = 2$: $y^2 + y + 2^2 - 2 - 2 = 0 \Rightarrow y^2 + y = 0 \Rightarrow y(y+1) = 0$
$$\Rightarrow y = 0 \text{ or } y = -1 \quad \text{Note: } y \neq 0 \text{ because of division by zero.}$$
Solutions: $(0, -2), (0, 1), (2, -1)$

51. Rewrite each equation in exponential form:
$$\begin{cases} \log_x y = 3 \rightarrow y = x^3 \\ \log_x(4y) = 5 \rightarrow 4y = x^5 \end{cases}$$
Substitute the first equation into the second and solve:
$$4x^3 = x^5$$
$$x^5 - 4x^3 = 0 \Rightarrow x^3(x^2 - 4) = 0 \Rightarrow x^3 = 0 \text{ or } x^2 = 4 \Rightarrow x = 0 \text{ or } x = \pm 2$$
The base of a logarithm must be positive, thus $x \neq 0$ and $x \neq -2$.
If $x = 2$: $y = 2^3 = 8$
Solution: $(2, 8)$

53.   Rewrite each equation in exponential form:

$$\begin{cases} \ln x = 4\ln y \;\Rightarrow\; x = e^{4\ln y} = e^{\ln y^4} = y^4 \\ \log_3 x = 2 + 2\log_3 y \;\Rightarrow\; x = 3^{2+2\log_3 y} = 3^2 \cdot 3^{2\log_3 y} = 3^2 \cdot 3^{\log_3 y^2} = 9y^2 \end{cases}$$

So we have the system $\begin{cases} x = y^4 \\ x = 9y^2 \end{cases}$

Therefore we have : $9y^2 = y^4 \Rightarrow 9y^2 - y^4 = 0 \Rightarrow y^2(9 - y^2) = 0$

$$y^2(3 + y)(3 - y) = 0 \Rightarrow y = 0 \;\text{ or }\; y = -3 \;\text{ or }\; y = 3$$

Since $\ln y$ is undefined when $y \le 0$, the only solution is $y = 3$.

        If $y = 3$: $x = y^4 \Rightarrow x = 3^4 = 81$

Solution: $(81, 3)$

55.   Solve the first equation for x, substitute into the second equation and solve:

$$\begin{cases} x + 2y = 0 \;\Rightarrow\; x = -2y \\ (x-1)^2 + (y-1)^2 = 5 \end{cases}$$

$$(-2y - 1)^2 + (y - 1)^2 = 5$$

$$4y^2 + 4y + 1 + y^2 - 2y + 1 = 5 \Rightarrow 5y^2 + 2y - 3 = 0$$

$$(5y - 3)(y + 1) = 0$$

$$y = \frac{3}{5} = 0.6 \;\text{ or }\; y = -1$$

$$x = -\frac{6}{5} = -1.2 \;\text{ or } x = 2$$

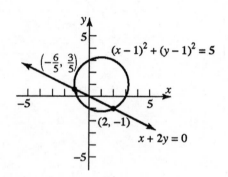

The points of intersection are $(-1.2, 0.6), (2, -1)$.

57.   Complete the square on the second equation, substitute into the first equation and solve:

$$\begin{cases} (x-1)^2 + (y+2)^2 = 4 \\ y^2 + 4y - x + 1 = 0 \;\Rightarrow\; y^2 + 4y + 4 = x - 1 + 4 \Rightarrow (y+2)^2 = x + 3 \end{cases}$$

$$(x - 1)^2 + x + 3 = 4$$

$$x^2 - 2x + 1 + x + 3 = 4$$

$$x^2 - x = 0$$

$$x(x - 1) = 0$$

$$x = 0 \;\text{ or } x = 1$$

If $x = 0$: $(y + 2)^2 = 0 + 3$

$$y + 2 = \pm\sqrt{3} \Rightarrow y = -2 \pm \sqrt{3}$$

If $x = 1$: $(y + 2)^2 = 1 + 3$

$$y + 2 = \pm 2 \Rightarrow y = -2 \pm 2$$

The points of intersection are:

$\left(0, -2 - \sqrt{3}\right), \left(0, -2 + \sqrt{3}\right), \left(1, -4\right), \left(1, 0\right)$.

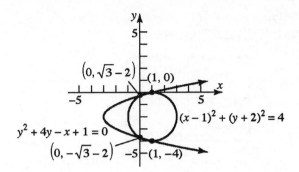

59.   Solve the first equation for x, substitute into the second equation and solve:

$$\begin{cases} y = \dfrac{4}{x-3} \;\Rightarrow\; x - 3 = \dfrac{4}{y} \;\Rightarrow\; x = \dfrac{4}{y} + 3 \\[2mm] x^2 - 6x + y^2 + 1 = 0 \end{cases}$$

$$\left(\frac{4}{y} + 3\right)^2 - 6\left(\frac{4}{y} + 3\right) + y^2 + 1 = 0$$

$$\frac{16}{y^2} + \frac{24}{y} + 9 - \frac{24}{y} - 18 + y^2 + 1 = 0$$

$$\frac{16}{y^2} + y^2 - 8 = 0$$

$$16 + y^4 - 8y^2 = 0$$

$$y^4 - 8y^2 + 16 = 0$$

$$(y^2 - 4)^2 = 0$$

$$y^2 - 4 = 0$$

$$y^2 = 4$$

$$y = \pm 2$$

If $y = 2$: $x = \dfrac{4}{2} + 3 = 5$

If $y = -2$: $x = \dfrac{4}{-2} + 3 = 1$

The points of intersection are: $(1, -2), (5, 2)$.

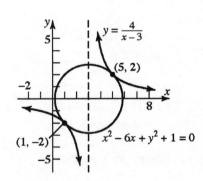

61.   Graph: $y_1 = x \wedge (2/3);\quad y_2 = e \wedge (-x)$
      Use INTERSECT to solve:

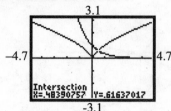

Solution: $(0.48, 0.62)$

63.   Graph: $y_1 = \sqrt[3]{2 - x^2};\quad y_2 = 4/x^3$
      Use INTERSECT to solve:

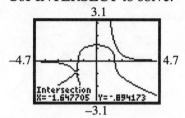

Solution: $(-1.65, -0.89)$

65.   Graph:
$$y_1 = \sqrt[4]{12 - x^4};\quad y_2 = -\sqrt[4]{12 - x^4};$$
$$y_3 = \sqrt{2/x};\quad y_4 = -\sqrt{2/x}$$
      Use INTERSECT to solve:

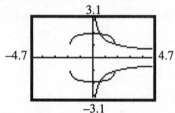

Solutions: $(0.58, 1.86),\ (1.81, 1.05),$
$\quad\quad\quad (1.81, -1.05),\ (0.58, -1.86)$

67.   Graph: $y_1 = 2/x;\quad y_2 = \ln x$
      Use INTERSECT to solve:

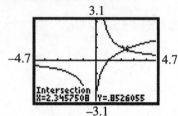

Solution: $(2.35, 0.85)$

69.   Let x and y be the two numbers. The system of equations is:
$$\begin{cases} x - y = 2 & \Rightarrow x = y + 2 \\ x^2 + y^2 = 10 \end{cases}$$
Solve the first equation for x, substitute into the second equation and solve:
$$(y + 2)^2 + y^2 = 10$$
$$y^2 + 4y + 4 + y^2 = 10$$
$$2y^2 + 4y - 6 = 0$$
$$y^2 + 2y - 3 = 0$$
$$(y + 3)(y - 1) = 0 \Rightarrow y = -3 \text{ or } y = 1$$
If $y = -3$: $x = -3 + 2 = -1$
If $y = 1$: $x = 1 + 2 = 3$
The two numbers are 1 and 3 or −1 and −3.

71. Let x and y be the two numbers. The system of equations is:

$$\begin{cases} xy = 4 \Rightarrow x = \dfrac{4}{y} \\ x^2 + y^2 = 8 \end{cases}$$

Solve the first equation for x, substitute into the second equation and solve:

$$\left(\frac{4}{y}\right)^2 + y^2 = 8$$

$$\frac{16}{y^2} + y^2 = 8$$

$$16 + y^4 = 8y^2$$

$$y^4 - 8y^2 + 16 = 0$$

$$(y^2 - 4)^2 = 0$$

$$y^2 - 4 = 0$$

$$y^2 = 4 \Rightarrow y = \pm 2$$

If $y = 2$: $x = \dfrac{4}{2} = 2$ If $y = -2$: $x = \dfrac{4}{-2} = -2$

The two numbers are 2 and 2 or –2 and –2.

73. Let x and y be the two numbers. The system of equations is:

$$\begin{cases} x - y = xy \\ \dfrac{1}{x} + \dfrac{1}{y} = 5 \end{cases}$$

Solve the first equation for x, substitute into the second equation and solve:

$$x - xy = y$$

$$x(1 - y) = y$$

$$x = \frac{y}{1 - y}$$

$$\frac{1}{\dfrac{y}{1-y}} + \frac{1}{y} = 5$$

$$\frac{1 - y}{y} + \frac{1}{y} = 5$$

$$\frac{2 - y}{y} = 5$$

$$2 - y = 5y$$

$$6y = 2 \Rightarrow y = \frac{1}{3} \Rightarrow x = \frac{(1/3)}{1 - \dfrac{1}{3}} = \frac{(1/3)}{(2/3)} = \frac{1}{2}$$

The two numbers are $\dfrac{1}{2}$ and $\dfrac{1}{3}$.

75. $\begin{cases} \dfrac{a}{b} = \dfrac{2}{3} \\ a + b = 10 \Rightarrow a = 10 - b \end{cases}$

Solve the second equation for a, substitute into the first equation and solve:

$$\frac{10 - b}{b} = \frac{2}{3}$$

$$3(10 - b) = 2b$$

$$30 - 3b = 2b$$

$$30 = 5b$$

$$b = 6 \Rightarrow a = 4$$

$$a + b = 10; \quad b - a = 2$$

The ratio of $a + b$ to $b - a$ is $\dfrac{10}{2} = 5$.

77. Let $x =$ the width of the rectangle.
Let $y =$ the length of the rectangle.

$$\begin{cases} 2x + 2y = 16 \\ xy = 15 \end{cases}$$

Solve the first equation for y, substitute into the second equation and solve.

$$2x + 2y = 16 \qquad\qquad x(8 - x) = 15$$

$$2y = 16 - 2x \qquad\qquad 8x - x^2 = 15$$

$$y = 8 - x \qquad\qquad x^2 - 8x + 15 = 0$$

$$(x - 5)(x - 3) = 0$$

$$x = 5 \ \text{ or } \ x = 3$$

$$y = 3 \qquad y = 5$$

The dimensions of the rectangle are 3 inches by 5 inches.

79. Let $x =$ the radius of the first circle.
Let $y =$ the radius of the second circle.

$$\begin{cases} 2\pi x + 2\pi y = 12\pi \\ \pi x^2 + \pi y^2 = 20\pi \end{cases}$$

Solve the first equation for y, substitute into the second equation and solve:

$$2\pi x + 2\pi y = 12\pi \qquad\qquad \pi x^2 + \pi y^2 = 20\pi$$

$$x + y = 6 \qquad\qquad x^2 + y^2 = 20$$

$$y = 6 - x \qquad\qquad x^2 + (6 - x)^2 = 20$$

$$x^2 + 36 - 12x + x^2 = 20 \Rightarrow 2x^2 - 12x + 16 = 0$$

$$x^2 - 6x + 8 = 0 \Rightarrow (x - 4)(x - 2) = 0$$

$$x = 4 \ \text{ or } \ x = 2$$

$$y = 2 \qquad y = 4$$

The radii of the circles are 2 centimeters and 4 centimeters.

81. The tortoise takes $9 + 3 = 12$ minutes or 0.2 hour longer to complete the race than the hare.
    Let r = the rate of the hare.
    Let t = the time for the hare to complete the race.
    Then $t + 0.2$ = the time for the tortoise and $r - 0.5$ = the rate for the tortoise.
    Since the length of the race is 21 meters, the distance equations are:

$$\begin{cases} rt = 21 \Rightarrow r = \dfrac{21}{t} \\ (r - 0.5)(t + 0.2) = 21 \end{cases}$$

Solve the first equation for r, substitute into the second equation and solve:

$$\left(\frac{21}{t} - 0.5\right)(t + 0.2) = 21$$

$$21 + \frac{4.2}{t} - 0.5t - 0.1 = 21$$

$$10t \cdot \left(21 + \frac{4.2}{t} - 0.5t - 0.1\right) = 10t \cdot (21)$$

$$210t + 42 - 5t^2 - t = 210t$$

$$5t^2 + t - 42 = 0$$

$$(5t - 14)(t + 3) = 0 \Rightarrow t = \frac{14}{5} = 2.8 \text{ or } t = -3$$

$t = -3$ makes no sense, since time cannot be negative.
Solve for r:

$$r = \frac{21}{2.8} = 7.5$$

The average speed of the hare is 7.5 meters per hour, and the average speed for the tortoise is 7 meters per hour.

83. Let x = the width of the cardboard. Let y = the length of the cardboard.
    The width of the box will be $x - 4$, the length of the box will be $y - 4$, and the height is 2.
    The volume is $V = (x - 4)(y - 4)(2)$.
    Solve the system of equations:

$$\begin{cases} xy = 216 \qquad \Rightarrow y = \dfrac{216}{x} \\ 2(x - 4)(y - 4) = 224 \end{cases}$$

Solve the first equation for y, substitute into the second equation and solve.

$$(2x - 8)\left(\frac{216}{x} - 4\right) = 224$$

$$432 - 8x - \frac{1728}{x} + 32 = 224$$

$$432x - 8x^2 - 1728 + 32x = 224x$$

$$-8x^2 + 240x - 1728 = 0$$

$$x^2 - 30x + 216 = 0$$
$$(x - 12)(x - 18) = 0$$
$$x = 12 \text{ or } x = 18$$
$$y = 18 \qquad y = 12$$

The cardboard should be 12 centimeters by 18 centimeters.

85. Find equations relating area and perimeter:
$$\begin{cases} x^2 + y^2 = 4500 \\ 3x + 3y + (x - y) = 300 \end{cases}$$

Solve the second equation for y, substitute into the first equation and solve:

$$4x + 2y = 300 \qquad\qquad x^2 + (150 - 2x)^2 = 4500$$
$$2y = 300 - 4x \qquad\qquad x^2 + 22{,}500 - 600x + 4x^2 = 4500$$
$$y = 150 - 2x \qquad\qquad 5x^2 - 600x + 18{,}000 = 0$$
$$x^2 - 120x + 3600 = 0$$
$$(x - 60)^2 = 0$$
$$x - 60 = 0$$
$$x = 60$$
$$y = 150 - 2(60) = 30$$

The sides of the squares are 30 feet and 60 feet.

87. Solve the system for l and w:
$$\begin{cases} 2l + 2w = P \\ lw = A \end{cases}$$

Solve the first equation for l, substitute into the second equation and solve.

$$2l = P - 2w$$
$$l = \frac{P}{2} - w$$
$$\left(\frac{P}{2} - w\right)w = A$$
$$\frac{P}{2}w - w^2 = A$$
$$w^2 - \frac{P}{2}w + A = 0$$

$$w = \frac{\dfrac{P}{2} \pm \sqrt{\dfrac{P^2}{4} - 4A}}{2} = \frac{\dfrac{P}{2} \pm \sqrt{\dfrac{P^2 - 16A}{4}}}{2}$$

$$= \frac{\dfrac{P}{2} \pm \dfrac{\sqrt{P^2 - 16A}}{2}}{2}$$

$$= \frac{P \pm \sqrt{P^2 - 16A}}{4}$$

If $w = \dfrac{P + \sqrt{P^2 - 16A}}{4}$ then $l = \dfrac{P}{2} - \dfrac{P + \sqrt{P^2 - 16A}}{4} = \dfrac{P - \sqrt{P^2 - 16A}}{4}$

If $w = \dfrac{P - \sqrt{P^2 - 16A}}{4}$ then $l = \dfrac{P}{2} - \dfrac{P - \sqrt{P^2 - 16A}}{4} = \dfrac{P + \sqrt{P^2 - 16A}}{4}$

If it is required that length be greater than width, then the solution is:

$$w = \frac{P - \sqrt{P^2 - 16A}}{4} \text{ and } l = \frac{P + \sqrt{P^2 - 16A}}{4}$$

89. Solve the equation: $m^2 - 4(2m - 4) = 0$

$$m^2 - 8m + 16 = 0$$

$$(m - 4)^2 = 0$$

$$m - 4 = 0 \Rightarrow m = 4$$

Use the point-slope equation with slope 4 and the point (2, 4) to obtain the equation of the tangent line: $y - 4 = 4(x - 2) \Rightarrow y - 4 = 4x - 8 \Rightarrow y = 4x - 4$

91. Solve the system:

$$\begin{cases} y = x^2 + 2 \\ y = mx + b \end{cases}$$

Solve the system by substitution:

$$x^2 + 2 = mx + b \Rightarrow x^2 - mx + 2 - b = 0$$

Note that the tangent line passes through (1, 3).

Find the relation between m and b: $3 = m(1) + b \Rightarrow b = 3 - m$

Substitute into the quadratic to eliminate b: $x^2 - mx + 2 - (3 - m) = 0 \Rightarrow x^2 - mx + (m - 1) = 0$

Find when the discriminant equals 0:

$$(-m)^2 - 4(1)(m - 1) = 0$$

$$m^2 - 4m + 4 = 0$$

$$(m - 2)^2 = 0$$

$$m - 2 = 0$$

$$m = 2 \Rightarrow b = 3 - 2 = 1$$

The equation of the tangent line is $y = 2x + 1$.

93. Solve the system:

$$\begin{cases} 2x^2 + 3y^2 = 14 \\ \qquad y = mx + b \end{cases}$$

Solve the system by substitution:

$$2x^2 + 3(mx + b)^2 = 14$$

$$2x^2 + 3m^2x^2 + 6mbx + 3b^2 = 14$$

$$(3m^2 + 2)x^2 + 6mbx + 3b^2 - 14 = 0$$

Note that the tangent line passes through (1, 2).

Find the relation between m and b: $2 = m(1) + b \Rightarrow b = 2 - m$

Substitute into the quadratic to eliminate b:

$$(3m^2 + 2)x^2 + 6m(2 - m)x + 3(2 - m)^2 - 14 = 0$$

$$(3m^2 + 2)x^2 + (12m - 6m^2)x + 12 - 12m + 3m^2 - 14 = 0$$

$$(3m^2 + 2)x^2 + (12m - 6m^2)x + (3m^2 - 12m - 2) = 0$$

Find when the discriminant equals 0:

$$(12m - 6m^2)^2 - 4(3m^2 + 2)(3m^2 - 12m - 2) = 0$$

$$144m^2 - 144m^3 + 36m^4 - 4(9m^4 - 36m^3 - 24m - 4) = 0$$

$$144m^2 - 144m^3 + 36m^4 - 36m^4 + 144m^3 + 96m + 16 = 0$$

$$144m^2 + 96m + 16 = 0$$

$$9m^2 + 6m + 1 = 0$$

$$(3m + 1)^2 = 0$$

$$3m + 1 = 0$$

$$m = -\frac{1}{3} \Rightarrow b = 2 - \left(-\frac{1}{3}\right) = \frac{7}{3}$$

The equation of the tangent line is $y = -\frac{1}{3}x + \frac{7}{3}$.

95. Solve the system:

$$\begin{cases} x^2 - y^2 = 3 \\ \quad\ y = mx + b \end{cases}$$

Solve the system by substitution:

$$x^2 - (mx + b)^2 = 3$$

$$x^2 - m^2x^2 - 2mbx - b^2 = 3$$

$$(1 - m^2)x^2 - 2mbx - b^2 - 3 = 0$$

Note that the tangent line passes through (2, 1).

Find the relation between m and b:

$$1 = m(2) + b \Rightarrow b = 1 - 2m$$

Substitute into the quadratic to eliminate b:

$$(1 - m^2)x^2 - 2m(1 - 2m)x - (1 - 2m)^2 - 3 = 0$$

$$(1 - m^2)x^2 + (-2m + 4m^2)x - 1 + 4m - 4m^2 - 3 = 0$$

$$(1 - m^2)x^2 + (-2m + 4m^2)x + (-4m^2 + 4m - 4) = 0$$

Find when the discriminant equals 0:

$$(-2m + 4m^2)^2 - 4(1 - m^2)(-4m^2 + 4m - 4) = 0$$

$$4m^2 - 16m^3 + 16m^4 - 4(4m^4 - 4m^3 + 4m - 4) = 0$$

$$4m^2 - 16m^3 + 16m^4 - 16m^4 + 16m^3 - 16m + 16 = 0$$

$$4m^2 - 16m + 16 = 0$$

$$m^2 - 4m + 4 = 0$$

$$(m - 2)^2 = 0$$

$$m - 2 = 0 \Rightarrow m = 2$$

$$b = 1 - 2(2) = -3$$

The equation of the tangent line is $y = 2x - 3$.

97. Solve for r_1 and r_2:

$$\begin{cases} r_1 + r_2 = -\dfrac{b}{a} \\ r_1 r_2 = \dfrac{c}{a} \end{cases}$$

Substitute and solve:

$$r_1 = -r_2 - \frac{b}{a}$$

$$\left(-r_2 - \frac{b}{a}\right) r_2 = \frac{c}{a}$$

$$-r_2^2 - \frac{b}{a} r_2 - \frac{c}{a} = 0$$

$$ar_2^2 + br_2 + c = 0$$

$$r_2 = \frac{-b \pm \sqrt{b^2 - 4ac}}{2a}$$

$$r_1 = -r_2 - \frac{b}{a} = -\left(\frac{-b \pm \sqrt{b^2 - 4ac}}{2a}\right) - \frac{2b}{2a}$$

$$= \frac{-b \mp \sqrt{b^2 - 4ac}}{2a}$$

The solutions are: $\dfrac{-b + \sqrt{b^2 - 4ac}}{2a}$ and $\dfrac{-b - \sqrt{b^2 - 4ac}}{2a}$.

99. Since the area of the square piece of sheet metal is 100 square feet, the sheet's dimensions are 10 feet by 10 feet. Let x = the length of the cut.

The dimensions of the box are length $= 10 - 2x$; width $= 10 - 2x$; height $= x$
Note that each of these expressions must be positive. So we must have
   $x > 0$ and $10 - 2x > 0 \Rightarrow x < 5$, that is, $0 < x < 5$.
So the volume of the box is given by

$$V = (\text{length}) \cdot (\text{width}) \cdot (\text{height}) = (10 - 2x)(10 - 2x)(x) = (10 - 2x)^2 (x)$$

(a) In order to get a volume equal to 9 cubic feet, we solve $(10 - 2x)^2 (x) = 9$.

$$(10 - 2x)^2 (x) = 9 \Rightarrow \left(100 - 40x + 4x^2\right)x = 9 \Rightarrow 100x - 40x^2 + 4x^3 = 9$$

So we need to solve the equation $4x^3 - 40x^2 + 100x - 9 = 0$.

Graphing the function $y_1 = 4x^3 - 40x^2 + 100x - 9$ on a calculator yields the graph

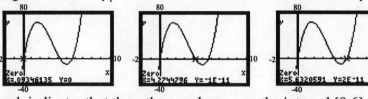

The graph indicates that there three real zeros on the interval [0,6].

Using the ZERO feature of a graphing calculator, we find that the three roots shown occur at $x \approx 0.09$, $x \approx 4.27$ and $x \approx 5.63$.

But we've already noted that we must have $0 < x < 5$, so the only practical values for the cut are $x \approx 0.09$ feet and $x \approx 4.27$ feet.

(b)   Answers will vary.

Systems of Equations and Inequalities

10.7 Systems of Inequalities

1. $3x + 4 < 8 - x$

$\quad 4x < 4$

$\quad\quad x < 1$

$\left\{x \mid x < 1\right\}$ or $(-\infty, 1)$

3. $x^2 + y^2 = 9$

The graph is a circle.

Center: $(0, 0)$; Radius: 3

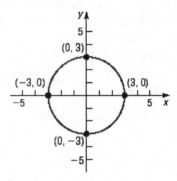

5. True

7. satisfied

9. False

11. $x \geq 0$

Graph the line $x = 0$. Use a solid line since the inequality uses $\geq$.

Choose a test point not on the line, such as $(2, 0)$.

Since $2 \geq 0$ is true, shade the side of the line containing $(2, 0)$.

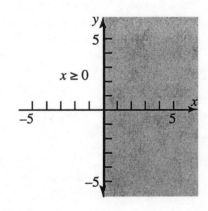

13. $x \geq 4$

Graph the line $x = 4$. Use a solid line since the
inequality uses $\geq$.
Choose a test point not on the line, such as $(5, 0)$.
Since $5 \geq 0$ is true, shade the side of the line
containing $(5, 0)$.

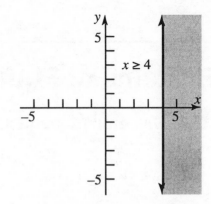

15. $2x + y \geq 6$

Graph the line $2x + y = 6$. Use a solid line since
the inequality uses $\geq$.
Choose a test point not on the line, such as $(0, 0)$.
Since $2(0) + 0 \geq 6$ is false, shade the opposite side
of the line from $(0, 0)$.

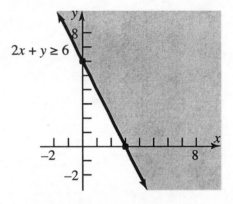

17. $x^2 + y^2 > 1$

Graph the circle $x^2 + y^2 = 1$. Use a dashed line
since the inequality uses $>$.
Choose a test point not on the circle, such as $(0, 0)$.
Since $0^2 + 0^2 > 1$ is false, shade the opposite side of
the circle from $(0, 0)$.

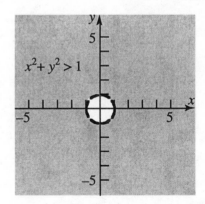

19. $y \le x^2 - 1$

Graph the parabola $y = x^2 - 1$. Use a solid line since the inequality uses $\le$.
Choose a test point not on the parabola, such as $(0, 0)$.
Since $0 \le 0^2 - 1$ is false, shade the opposite side of the parabola from $(0, 0)$.

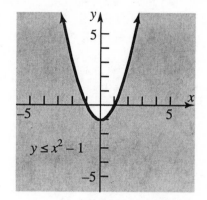

21. $xy \ge 4$

Graph the hyperbola $xy = 4$. Use a solid line since the inequality uses $\ge$.
Choose a test point not on the hyperbola, such as $(0, 0)$.
Since $0 \cdot 0 \ge 4$ is false, shade the opposite side of the hyperbola from $(0, 0)$.

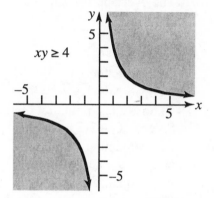

23. $\begin{cases} x + y \le 2 \\ 2x + y \ge 4 \end{cases}$

Graph the line $x + y = 2$. Use a solid line since the Inequality uses $\le$. Choose a test point not on the line, such as $(0, 0)$.
Since $0 + 0 \le 2$ is true, shade the side of the line containing $(0, 0)$.
Graph the line $2x + y = 4$. Use a solid line since the inequality uses $\ge$.
Choose a test point not on the line, such as $(0, 0)$.
Since $2(0) + 0 \ge 4$ is false, shade the opposite side of the line from $(0, 0)$.
The overlapping region is the solution.

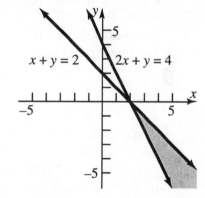

25. $\begin{cases} 2x - y \le 4 \\ 3x + 2y \ge -6 \end{cases}$

Graph the line $2x - y = 4$. Use a solid line since the inequality uses $\le$.
Choose a test point not on the line, such as $(0, 0)$.
Since $2(0) - 0 \le 4$ is true, shade the side of the line containing $(0, 0)$.
Graph the line $3x + 2y = -6$. Use a solid line since the inequality uses $\ge$.
Choose a test point not on the line, such as $(0, 0)$.
Since $3(0) + 2(0) \ge -6$ is true, shade the side of the line containing $(0, 0)$.
The overlapping region is the solution.

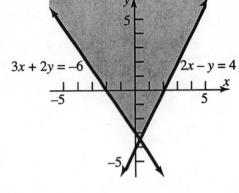

27. $\begin{cases} 2x - 3y \le 0 \\ 3x + 2y \le 6 \end{cases}$

Graph the line $2x - 3y = 0$. Use a solid line since the inequality uses $\le$.
Choose a test point not on the line, such as $(0, 3)$.
Since $2(0) - 3(3) \le 0$ is true, shade the side of the line containing $(0, 3)$.
Graph the line $3x + 2y = 6$. Use a solid line since the inequality uses $\le$.
Choose a test point not on the line, such as $(0, 0)$.
Since $3(0) + 2(0) \le 6$ is true, shade the side of the line containing $(0, 0)$.
The overlapping region is the solution.

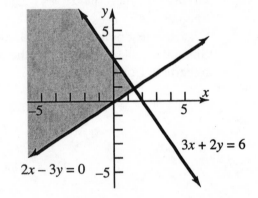

29. $\begin{cases} x^2 + y^2 \le 9 \\ x + y \ge 3 \end{cases}$

Graph the circle $x^2 + y^2 = 9$. Use a solid line since the inequality uses $\le$. Choose a test point not on the circle, such as $(0, 0)$. Since $0^2 + 0^2 \le 9$ is true, shade the same side of the circle as $(0, 0)$.
Graph the line $x + y = 3$. Use a solid line since the inequality uses $\ge$. Choose a test point not on the line, such as $(0, 0)$. Since $0 + 0 \ge 3$ is false, shade the opposite side of the line from $(0, 0)$.
The overlapping region is the solution.

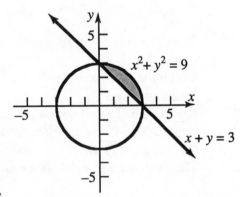

31.
$$\begin{cases} y \geq x^2 - 4 \\ y \leq x - 2 \end{cases}$$

Graph the parabola $y = x^2 - 4$. Use a solid line since the inequality uses $\geq$. Choose a test point not on the parabola, such as $(0, 0)$. Since $0 \geq 0^2 - 4$ is true, shade the same side of the parabola as $(0, 0)$.
Graph the line $y = x - 2$. Use a solid line since the inequality uses $\leq$. Choose a test point not on the line, such as $(0, 0)$. Since $0 \leq 0 - 2$ is false, shade the opposite side of the line from $(0, 0)$.
The overlapping region is the solution.

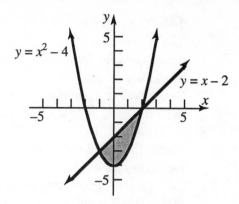

33.
$$\begin{cases} xy \geq 4 \\ y \geq x^2 + 1 \end{cases}$$

Graph the hyperbola $xy = 4$. Use a solid line since the inequality uses $\geq$. Choose a test point not on the parabola, such as $(0, 0)$. Since $0 \cdot 0 \geq 4$ is false, shade the opposite side of the hyperbola from $(0, 0)$.
Graph the parabola $y = x^2 + 1$. Use a solid line since the inequality uses $\geq$. Choose a test point not on the parabola, such as $(0, 0)$. Since $0 \geq 0^2 + 1$ is false, shade the opposite side of the parabola from $(0, 0)$.
The overlapping region is the solution.

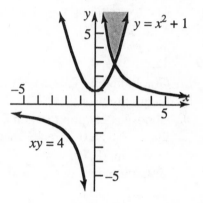

35.
$$\begin{cases} x - 2y \leq 6 \\ 2x - 4y \geq 0 \end{cases}$$

Graph the line $x - 2y = 6$. Use a solid line since the inequality uses $\leq$.
Choose a test point not on the line, such as $(0, 0)$. Since $0 - 2(0) \leq 6$ is true, shade the side of the line containing $(0, 0)$.
Graph the line $2x - 4y = 0$. Use a solid line since the inequality uses $\geq$. Choose a test point not on the line, such as $(0, 2)$. Since $2(0) - 4(2) \geq 0$ is false, shade the opposite side of the line from $(0, 2)$.
The overlapping region is the solution.

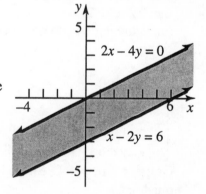

37. $\begin{cases} 2x + y \ge -2 \\ 2x + y \ge 2 \end{cases}$

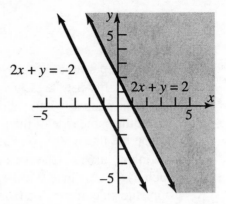

Graph the line $2x + y = -2$. Use a solid line since the inequality uses $\ge$.
Choose a test point not on the line, such as $(0, 0)$. Since $2(0) + 0 \ge -2$ is true, shade the side of the line containing $(0, 0)$.
Graph the line $2x + y = 2$. Use a solid line since the inequality uses $\ge$.
Choose a test point not on the line, such as $(0, 0)$. Since $2(0) + 0 \ge 2$ is false, shade the opposite side of the line from $(0, 0)$.
The overlapping region is the solution.

39. $\begin{cases} 2x + 3y \ge 6 \\ 2x + 3y \le 0 \end{cases}$

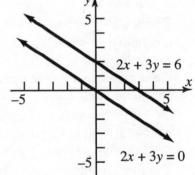

Graph the line $2x + 3y = 6$. Use a solid line since the inequality uses $\ge$.
Choose a test point not on the line, such as $(0, 0)$. Since $2(0) + 3(0) \ge 6$ is false, shade the opposite side of the line from $(0, 0)$.
Graph the line $2x + 3y = 0$. Use a solid line since the inequality uses $\le$.
Choose a test point not on the line, such as $(0, 2)$. Since $2(0) + 3(2) \le 0$ is false, shade the opposite side of the line from $(0, 2)$.
Since the regions do not overlap, the solution is an empty set.

41. $\begin{cases} x \ge 0 \\ y \ge 0 \\ 2x + y \le 6 \\ x + 2y \le 6 \end{cases}$

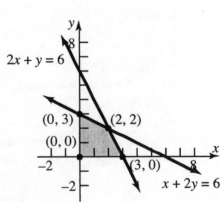

Graph $x \ge 0$; $y \ge 0$. Shaded region is the first quadrant.
Graph the line $2x + y = 6$. Use a solid line since the inequality uses $\le$.
Choose a test point not on the line, such as $(0, 0)$.
Since $2(0) + 0 \le 6$ is true, shade the side of the line containing $(0, 0)$.
Graph the line $x + 2y = 6$. Use a solid line since the inequality uses $\le$.
Choose a test point not on the line, such as $(0, 0)$.
Since $0 + 2(0) \le 6$ is true, shade the side of the line containing $(0, 0)$.
The overlapping region is the solution.
The graph is bounded.

Find the vertices:
The x-axis and y-axis intersect at $(0, 0)$.
The intersection of $x + 2y = 6$ and the y-axis is $(0, 3)$.
The intersection of $2x + y = 6$ and the x-axis is $(3, 0)$.
To find the intersection of $x + 2y = 6$ and $2x + y = 6$, solve the system:

$$\begin{cases} x + 2y = 6 \quad \Rightarrow \quad x = 6 - 2y \\ 2x + y = 6 \end{cases}$$

Substitute and solve:
$$2(6 - 2y) + y = 6$$
$$12 - 4y + y = 6$$
$$-3y = -6 \Rightarrow y = 2$$
$$x = 6 - 2(2) = 6 - 4 = 2$$

The point of intersection is $(2, 2)$.
The four corner points are $(0, 0)$, $(0, 3)$, $(3, 0)$, and $(2, 2)$.

43. $\begin{cases} x \geq 0 \\ y \geq 0 \\ x + y \geq 2 \\ 2x + y \geq 4 \end{cases}$

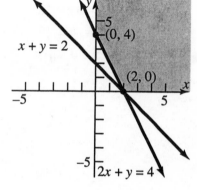

Graph $x \geq 0$; $y \geq 0$. Shaded region is the first quadrant.
Graph the line $x + y = 2$. Use a solid line since the
inequality uses $\geq$.
Choose a test point not on the line, such as $(0, 0)$.
Since $0 + 0 \geq 2$ is false, shade the opposite side of the
line from $(0, 0)$.
Graph the line $2x + y = 4$. Use a solid line since the inequality uses $\geq$.
Choose a test point not on the line, such as $(0, 0)$. Since $2(0) + 0 \geq 4$ is false, shade the
opposite side of the line from $(0, 0)$.
The overlapping region is the solution.
The graph is unbounded.
Find the vertices:
The intersection of $x + y = 2$ and the x-axis is $(2, 0)$.
The intersection of $2x + y = 4$ and the y-axis is $(0, 4)$.
The two corner points are $(2, 0)$, and $(0, 4)$.

45. $\begin{cases} x \geq 0 \\ y \geq 0 \\ x + y \geq 2 \\ 2x + 3y \leq 12 \\ 3x + y \leq 12 \end{cases}$

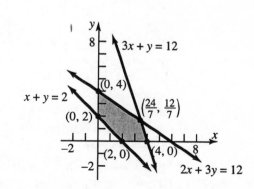

Graph $x \geq 0$; $y \geq 0$. Shaded region is the first quadrant.
Graph the line $x + y = 2$. Use a solid line since the
inequality uses $\geq$.

Choose a test point not on the line, such as $(0, 0)$. Since $0 + 0 \geq 2$ is false, shade the opposite side of the line from $(0, 0)$.

Graph the line $2x + 3y = 12$. Use a solid line since the inequality uses $\leq$.

Choose a test point not on the line, such as $(0, 0)$. Since $2(0) + 3(0) \leq 12$ is true, shade the side of the line containing $(0, 0)$.

Graph the line $3x + y = 12$. Use a solid line since the inequality uses $\leq$.

Choose a test point not on the line, such as $(0, 0)$. Since $3(0) + 0 \leq 12$ is true, shade the side of the line containing $(0, 0)$.

The overlapping region is the solution.

The graph is bounded.

Find the vertices:

The intersection of $x + y = 2$ and the y-axis is $(0, 2)$.

The intersection of $x + y = 2$ and the x-axis is $(2, 0)$.

The intersection of $2x + 3y = 12$ and the y-axis is $(0, 4)$.

The intersection of $3x + y = 12$ and the x-axis is $(4, 0)$.

To find the intersection of $2x + 3y = 12$ and $3x + y = 12$, solve the system:

$$\begin{cases} 2x + 3y = 12 \\ 3x + y = 12 \Rightarrow y = 12 - 3x \end{cases}$$

Substitute and solve:

$$2x + 3(12 - 3x) = 12$$
$$2x + 36 - 9x = 12$$
$$-7x = -24 \Rightarrow x = \frac{24}{7}$$
$$y = 12 - 3\left(\frac{24}{7}\right) = 12 - \frac{72}{2} = \frac{12}{7}$$

The point of intersection is $(24/7, 12/7)$.

The five corner points are $(0, 2)$, $(0, 4)$, $(2, 0)$, $(4, 0)$, and $(24/7, 12/7)$.

47. $\begin{cases} x \geq 0 \\ y \geq 0 \\ x + y \geq 2 \\ x + y \leq 8 \\ 2x + y \leq 10 \end{cases}$

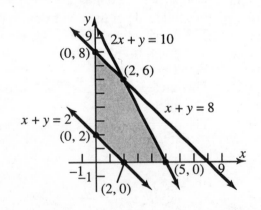

Graph $x \geq 0$; $y \geq 0$. Shaded region is the first quadrant.

Graph the line $x + y = 2$. Use a solid line since the inequality uses $\geq$.

Choose a test point not on the line, such as $(0, 0)$.

Since $0 + 0 \geq 2$ is false, shade the opposite side of the line from $(0, 0)$.

Graph the line $x + y = 8$. Use a solid line since the inequality uses $\leq$.

Choose a test point not on the line, such as $(0, 0)$. Since $0 + 0 \leq 8$ is true, shade the side of the line containing $(0, 0)$.

Graph the line $2x + y = 10$. Use a solid line since the inequality uses $\leq$.

Choose a test point not on the line, such as $(0, 0)$.

Since $2(0) + 0 \leq 10$ is true, shade the side of the line containing $(0, 0)$.
The overlapping region is the solution.
The graph is bounded.
Find the vertices:
The intersection of $x + y = 2$ and the y-axis is $(0, 2)$.
The intersection of $x + y = 2$ and the x-axis is $(2, 0)$.
The intersection of $x + y = 8$ and the y-axis is $(0, 8)$.
The intersection of $2x + y = 10$ and the x-axis is $(5, 0)$.
To find the intersection of $x + y = 8$ and $2x + y = 10$, solve the system:
$$\begin{cases} x + y = 8 \quad \Rightarrow \quad y = 8 - x \\ 2x + y = 10 \end{cases}$$
Substitute and solve:
$$2x + 8 - x = 10 \Rightarrow x = 2$$
$$y = 8 - 2 = 6$$
The point of intersection is $(2, 6)$.
The five corner points are $(0, 2)$, $(0, 8)$, $(2, 0)$, $(5, 0)$, and $(2, 6)$.

49.
$$\begin{cases} x \geq 0 \\ y \geq 0 \\ x + 2y \geq 1 \\ x + 2y \leq 10 \end{cases}$$

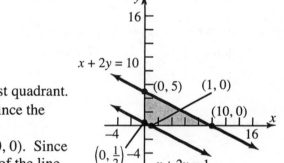

Graph $x \geq 0$; $y \geq 0$. Shaded region is the first quadrant.
Graph the line $x + 2y = 1$. Use a solid line since the
inequality uses $\geq$.
Choose a test point not on the line, such as $(0, 0)$. Since
$0 + 2(0) \geq 1$ is false, shade the opposite side of the line
from $(0, 0)$.
Graph the line $x + 2y = 10$. Use a solid line since the inequality uses $\leq$.
Choose a test point not on the line, such as $(0, 0)$. Since $0 + 2(0) \leq 10$ is true, shade the side
of the line containing $(0, 0)$.
The overlapping region is the solution.
The graph is bounded.
Find the vertices:
The intersection of $x + 2y = 1$ and the y-axis is $(0, 0.5)$.
The intersection of $x + 2y = 1$ and the x-axis is $(1, 0)$.
The intersection of $x + 2y = 10$ and the y-axis is $(0, 5)$.
The intersection of $x + 2y = 10$ and the x-axis is $(10, 0)$.
The four corner points are $(0, 0.5)$, $(0, 5)$, $(1, 0)$, and $(10, 0)$.

51.   The system of linear inequalities is:

$$\begin{cases} x \ge 0 \\ y \ge 0 \\ x \le 4 \\ x + y \le 6 \end{cases}$$

53.   The system of linear inequalities is:

$$\begin{cases} x \ge 0 \\ y \ge 15 \\ x \le 20 \\ x + y \le 50 \\ x - y \le 0 \end{cases}$$

55.   Above $y = x + 2$ $\begin{cases} y \ge x + 2 \\ y \le -x^2 + 4 \end{cases}$
      Below $y = -x^2 + 4$

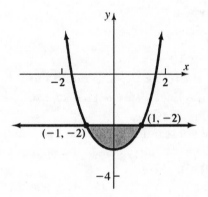

57.   Above $y = x^2 - 3$ $\begin{cases} y \ge x^2 - 3 \\ y \le -2 \end{cases}$
      Below $y = -2$

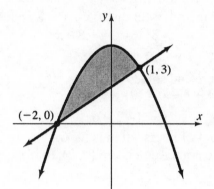

59.   Above $y = (x - 2)^2$ $\begin{cases} y \ge (x - 2)^2 \\ y \le -x^2 + 4 \end{cases}$
      Below $y = -x^2 + 4$

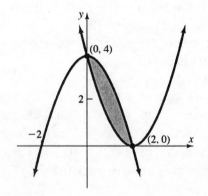

61.   (a)   Let x = the amount invested in Treasury bills.
           Let y = the amount invested in corporate bonds.
           The constraints are:
           $x \ge 0, y \ge 0$ A non-negative amount must be invested.
           $x + y \le 50,000$ Total investment cannot exceed \$50,000.

$y \le 10,000$ Amount invested in corporate bonds must not exceed \$10,000.
$x \ge 35,000$ Amount invested in Treasury bills must be at least \$35,000.

(b)   Graph the system.

$$\begin{cases} x \ge 0 \\ y \ge 0 \\ x + y \le 50,000 \\ y \le 10,000 \\ x \ge 35,000 \end{cases}$$

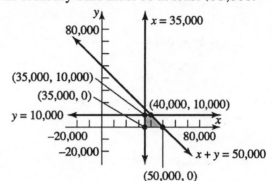

The corner points are $(35,000, 0)$, $(35,000, 10,000)$, $(40,000, 10,000)$, $(50,000, 0)$.

63.   (a)   Let x = the # of packages of the economy blend.
Let y = the # of packages of the superior blend.
The constraints are:

$x \ge 0, y \ge 0$ A non-negative # of packages must be produced.

$4x + 8y \le 75 \cdot 16$ Total amount of "A grade" coffee cannot exceed 75
    pounds. (Note: 75 pounds = (75)(16) ounces.)

$12x + 8y \le 120 \cdot 16$ Total amount of "B grade" coffee cannot exceed 120
    pounds. (Note: 120 pounds = (120)(16) ounces.)

We can simplify the equations

$\quad 4x + 8y \le 75 \cdot 16 \qquad\qquad\qquad 12x + 8y \le 120 \cdot 16$

$\quad\; x + 2y \le 75 \cdot 4 \qquad\qquad\qquad\;\; 3x + 2y \le 120 \cdot 4$

$\quad\; x + 2y \le 300 \qquad\qquad\qquad\quad\; 3x + 2y \le 480$

(b)   Graph the system.

$$\begin{cases} x \ge 0 \\ y \ge 0 \\ x + 2y \le 300 \\ 3x + 2y \le 480 \end{cases}$$

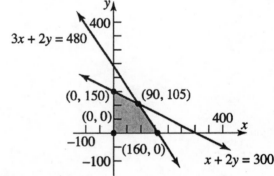

The corner points are $(0, 0)$, $(0, 150)$, $(90, 105)$, $(160, 0)$.

65.   (a)   Let x = the # of microwaves.
Let y = the # of printers.
The constraints are:

$x \ge 0, y \ge 0$ A non-negative # of items must be shipped.

$30x + 20y \le 1600$ Total cargo weight cannot exceed 1600 pounds.

$2x + 3y \le 150$ Total cargo volume cannot exceed 150 cubic feet.

Note: $30x + 20y \le 1600 \Rightarrow 3x + 2y \le 160$

(b) Graph the system.

$$\begin{cases} x \geq 0 \\ y \geq 0 \\ 3x + 2y \leq 160 \\ 2x + 3y \leq 150 \end{cases}$$

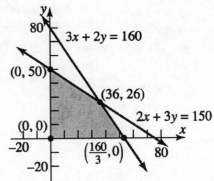

The corner points are $(0, 0)$, $(0, 50)$, $(36, 26)$, $(160/3, 0)$.

Systems of Equations and Inequalities

10.8 Linear Programming

1.   objective function

3. $z = x + y$

Vertex	Value of $z = x + y$
(0, 3)	$z = 0 + 3 = 3$
(0, 6)	$z = 0 + 6 = 6$
(5, 6)	$z = 5 + 6 = 11$
(5, 2)	$z = 5 + 2 = 7$
(4, 0)	$z = 4 + 0 = 4$

The maximum value is 11 at (5, 6), and the minimum value is 3 at (0, 3).

5. $z = x + 10y$

Vertex	Value of $z = x + 10y$
(0, 3)	$z = 0 + 10(3) = 30$
(0, 6)	$z = 0 + 10(6) = 60$
(5, 6)	$z = 5 + 10(6) = 65$
(5, 2)	$z = 5 + 10(2) = 25$
(4, 0)	$z = 4 + 10(0) = 4$

The maximum value is 65 at (5, 6), and the minimum value is 4 at (4, 0).

7. $z = 5x + 7y$

Vertex	Value of $z = 5x + 7y$
(0, 3)	$z = 5(0) + 7(3) = 21$
(0, 6)	$z = 5(0) + 7(6) = 42$
(5, 6)	$z = 5(5) + 7(6) = 67$
(5, 2)	$z = 5(5) + 7(2) = 39$
(4, 0)	$z = 5(4) + 7(0) = 20$

The maximum value is 67 at (5, 6), and the minimum value is 20 at (4, 0).

9.   Maximize $z = 2x + y$

Subject to $x \geq 0, \quad y \geq 0, \quad x + y \leq 6, \quad x + y \geq 1$

Graph the constraints.

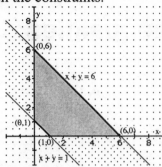

The corner points are $(0, 1)$, $(1, 0)$, $(0, 6)$, $(6, 0)$.

Evaluate the objective function:

Vertex	Value of $z = 2x + y$
$(0, 1)$	$z = 2(0) + 1 = 1$
$(0, 6)$	$z = 2(0) + 6 = 6$
$(1, 0)$	$z = 2(1) + 0 = 2$
$(6, 0)$	$z = 2(6) + 0 = 12$

The maximum value is 12 at $(6, 0)$.

11.   Minimize $z = 2x + 5y$

Subject to $x \geq 0, \quad y \geq 0, \quad x + y \geq 2, \quad x \leq 5, \quad y \leq 3$

Graph the constraints.

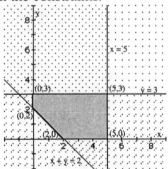

The corner points are $(0, 2)$, $(2, 0)$, $(0, 3)$, $(5, 0)$, $(5, 3)$.

Evaluate the objective function:

Vertex	Value of $z = 2x + 5y$
$(0, 2)$	$z = 2(0) + 5(2) = 10$
$(0, 3)$	$z = 2(0) + 5(3) = 15$
$(2, 0)$	$z = 2(2) + 5(0) = 4$
$(5, 0)$	$z = 2(5) + 5(0) = 10$
$(5, 3)$	$z = 2(5) + 5(3) = 25$

The minimum value is 4 at $(2, 0)$.

13. Maximize $z = 3x + 5y$

Subject to $x \geq 0, \quad y \geq 0, \quad x + y \geq 2, \quad 2x + 3y \leq 12, \quad 3x + 2y \leq 12$

Graph the constraints.

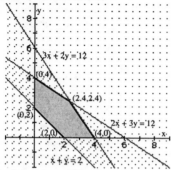

To find the intersection of $2x + 3y = 12$ and $3x + 2y = 12$, solve the system:

$$\begin{cases} 2x + 3y = 12 \\ 3x + 2y = 12 \quad \Rightarrow \quad y = 6 - \dfrac{3}{2}x \end{cases}$$

Substitute and solve:

$$2x + 3\left(6 - \frac{3}{2}x\right) = 12 \Rightarrow 2x + 18 - \frac{9}{2}x = 12 \Rightarrow -\frac{5}{2}x = -6$$

$$x = \frac{12}{5} \quad \Rightarrow \quad y = 6 - \frac{3}{2}\left(\frac{12}{5}\right) = 6 - \frac{18}{5} = \frac{12}{5}$$

The point of intersection is $(2.4, 2.4)$.

The corner points are (0, 2), (2, 0), (0, 4), (4, 0), (2.4, 2.4).

Evaluate the objective function:

Vertex	Value of $z = 3x + 5y$
(0, 2)	$z = 3(0) + 5(2) = 10$
(0, 4)	$z = 3(0) + 5(4) = 20$
(2, 0)	$z = 3(2) + 5(0) = 6$
(4, 0)	$z = 3(4) + 5(0) = 12$
(2.4, 2.4)	$z = 3(2.4) + 5(2.4) = 19.2$

The maximum value is 20 at (0, 4).

15. Minimize $z = 5x + 4y$

Subject to $x \geq 0, \quad y \geq 0, \quad x + y \geq 2, \quad 2x + 3y \leq 12, \quad 3x + y \leq 12$

Graph the constraints.

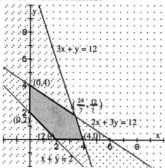

To find the intersection of $2x + 3y = 12$ and $3x + y = 12$, solve the system:

$$\begin{cases} 2x + 3y = 12 \\ 3x + y = 12 \quad \Rightarrow \quad y = 12 - 3x \end{cases}$$

Substitute and solve:

$$2x + 3(12 - 3x) = 12$$

$$2x + 36 - 9x = 12$$

$$-7x = -24 \Rightarrow x = \frac{24}{7}$$

$$y = 12 - 3\left(\frac{24}{7}\right) = 12 - \frac{72}{7} = \frac{12}{7}$$

The point of intersection is $\left(\frac{24}{7}, \frac{12}{7}\right)$.

The corner points are $(0, 2)$, $(2, 0)$, $(0, 4)$, $(4, 0)$, $\left(\frac{24}{7}, \frac{12}{7}\right)$.

Evaluate the objective function:

Vertex	Value of $z = 5x + 4y$
$(0, 2)$	$z = 5(0) + 4(2) = 8$
$(0, 4)$	$z = 5(0) + 4(4) = 16$
$(2, 0)$	$z = 5(2) + 4(0) = 10$
$(4, 0)$	$z = 5(4) + 4(0) = 20$
$\left(\frac{24}{7}, \frac{12}{7}\right)$	$z = 5\left(\frac{24}{7}\right) + 4\left(\frac{12}{7}\right) = \frac{120}{7} + \frac{48}{7} = \frac{168}{7} = 24$

The minimum value is 8 at $(0, 2)$.

17. Maximize $z = 5x + 2y$

Subject to $x \geq 0$, $y \geq 0$, $x + y \leq 10$, $2x + y \geq 10$, $x + 2y \geq 10$

Graph the constraints.

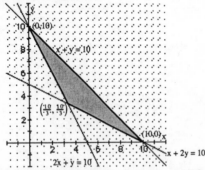

To find the intersection of $2x + y = 10$ and $x + 2y = 10$, solve the system:

$$\begin{cases} 2x + y = 10 \quad \Rightarrow \quad y = 10 - 2x \\ x + 2y = 10 \end{cases}$$

Substitute and solve:
$$x + 2(10 - 2x) = 10$$
$$x + 20 - 4x = 10$$
$$-3x = -10 \Rightarrow x = \frac{10}{3}$$
$$y = 10 - 2\left(\frac{10}{3}\right) = 10 - \frac{20}{3} = \frac{10}{3}$$

The point of intersection is $\left(\frac{10}{3}, \frac{10}{3}\right)$.

The corner points are $(0, 10)$, $(10, 0)$, $\left(\frac{10}{3}, \frac{10}{3}\right)$.

Evaluate the objective function:

Vertex	Value of $z = 5x + 2y$
$(0, 10)$	$z = 5(0) + 2(10) = 20$
$(10, 0)$	$z = 5(10) + 2(0) = 50$
$\left(\frac{10}{3}, \frac{10}{3}\right)$	$z = 5\left(\frac{10}{3}\right) + 2\left(\frac{10}{3}\right) = \frac{50}{3} + \frac{20}{3} = \frac{70}{3} = 23\frac{1}{3}$

The maximum value is 50 at $(10, 0)$.

19. Let x = the number of downhill skis produced.
Let y = the number of cross-country skis produced.
The total profit is: $P = 70x + 50y$. Profit is to be maximized; thus this is the objective function.
The constraints are:

$x \geq 0$, $y \geq 0$ A positive number of skis must be produced.
$2x + y \leq 40$ Only 40 hours of manufacturing time is available.
$x + y \leq 32$ Only 32 hours of finishing time is available.

Graph the constraints.

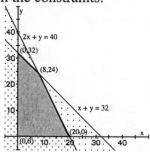

To find the intersection of $x + y = 32$ and $2x + y = 40$, solve the system:
$$\begin{cases} x + y = 32 & \Rightarrow \quad y = 32 - x \\ 2x + y = 40 \end{cases}$$

Substitute and solve:
$$2x + 32 - x = 40$$
$$x = 8$$
$$y = 32 - 8 = 24$$

The point of intersection is $(8, 24)$.

The corner points are $(0, 0)$, $(0, 32)$, $(20, 0)$, $(8, 24)$.
Evaluate the objective function:

Vertex	Value of $P = 70x + 50y$
$(0, 0)$	$P = 70(0) + 50(0) = 0$
$(0, 32)$	$P = 70(0) + 50(32) = 1600$
$(20, 0)$	$P = 70(20) + 50(0) = 1400$
$(8, 24)$	$P = 70(8) + 50(24) = 1760$

The maximum profit is \$1760, when 8 downhill skis and 24 cross-country skis are produced.

With the increase of the manufacturing time to 48 hours, we do the following:
The constraints are:

$x \geq 0$, $y \geq 0$ A positive number of skis must be produced.

$2x + y \leq 48$ Only 48 hours of manufacturing time is available.

$x + y \leq 32$ Only 32 hours of finishing time is available.

Graph the constraints.

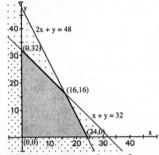

To find the intersection of $x + y = 32$ and $2x + y = 48$, solve the system:

$$\begin{cases} x + y = 32 \implies y = 32 - x \\ 2x + y = 48 \end{cases}$$

Substitute and solve:

$$2x + 32 - x = 48$$

$$x = 16$$

$$y = 32 - 16 = 16$$

The point of intersection is $(16, 16)$.
The corner points are $(0, 0)$, $(0, 32)$, $(24, 0)$, $(16, 16)$.
Evaluate the objective function:

Vertex	Value of $P = 70x + 50y$
$(0, 0)$	$P = 70(0) + 50(0) = 0$
$(0, 32)$	$P = 70(0) + 50(32) = 1600$
$(24, 0)$	$P = 70(24) + 50(0) = 1680$
$(16, 16)$	$P = 70(16) + 50(16) = 1920$

The maximum profit is \$1920, when 16 downhill skis and 16 cross-country skis are produced.

21. Let x = the number of acres of corn planted.
Let y = the number of acres of soybeans planted.
The total profit is: $P = 250x + 200y$. Profit is to be maximized; thus, this is the objective function.
The constraints are:

$x \geq 0, \ y \geq 0$ A non-negative number of acres must be planted.

$x + y \leq 100$ Acres available to plant.

$60x + 40y \leq 1800$ Money available for cultivation costs.

$60x + 60y \leq 2400$ Money available for labor costs.

Graph the constraints.

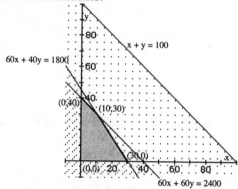

To find the intersection of $60x + 40y = 1800$ and $60x + 60y = 2400$, solve the system:

$$\begin{cases} 60x + 40y = 1800 \ \Rightarrow \ 60x = 1800 - 40y \\ 60x + 60y = 2400 \end{cases}$$

Substitute and solve:

$$1800 - 40y + 60y = 2400$$

$$20y = 600 \Rightarrow y = 30$$

$$60x = 1800 - 40(30)$$

$$60x = 600$$

$$x = 10$$

The point of intersection is (10, 30).
The corner points are (0, 0), (0, 40), (30, 0), (10, 30).
Evaluate the objective function:

Vertex	Value of $P = 250x + 200y$
(0, 0)	$P = 250(0) + 200(0) = 0$
(0, 40)	$P = 250(0) + 200(40) = 8000$
(30, 0)	$P = 250(30) + 200(0) = 7500$
(10, 30)	$P = 250(10) + 200(30) = 8500$

The maximum profit is $8500, when 10 acres of corn and 30 acres of soybeans are planted.

23. Let x = the number of hours that machine 1 is operated.
Let y = the number of hours that machine 2 is operated.
The total cost is: $C = 50x + 30y$. Cost is to be minimized; thus, this is the objective function.

The constraints are:

$x \geq 0$, $y \geq 0$ A positive number of hours must be used.

$x \leq 10$ 10 hours available on machine 1.

$y \leq 10$ 10 hours available on machine 2.

$60x + 40y \geq 240$ At least 240 8-inch pliers must be produced.

$70x + 20y \geq 140$ At least 140 6-inch pliers must be produced.

Graph the constraints.

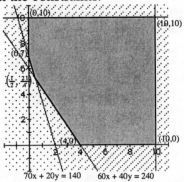

To find the intersection of $60x + 40y = 240$ and $70x + 20y = 140$, solve the system:

$$\begin{cases} 60x + 40y = 240 \\ 70x + 20y = 140 \quad \Rightarrow \quad 20y = 140 - 70x \end{cases}$$

Substitute and solve:

$$60x + 2(140 - 70x) = 240$$

$$60x + 280 - 140x = 240$$

$$-80x = -40$$

$$x = 0.5$$

$$20y = 140 - 70(0.5)$$

$$20y = 105$$

$$y = 5.25$$

The point of intersection is $(0.5, 5.25)$.

The corner points are (0, 7), (0, 10), (4, 0), (10, 0), (10, 10), $(0.5, 5.25)$.

Evaluate the objective function:

Vertex	Value of $C = 50x + 30y$
(0, 7)	$C = 50(0) + 30(7) = 210$
(0, 10)	$C = 50(0) + 30(10) = 300$
(4, 0)	$C = 50(4) + 30(0) = 200$
(10, 0)	$C = 50(10) + 30(0) = 500$
(10, 10)	$C = 50(10) + 30(10) = 800$
$(0.5, 5.25)$	$C = 50(0.5) + 30(5.25) = 182.50$

The minimum cost is $182.50, when machine 1 is used for 0.5 hour and machine 2 is used for 5.25 hours.

25. Let x = the number of pounds of ground beef.
     Let y = the number of pounds of ground pork.
     The total cost is: $C = 0.75x + 0.45y$.
     Cost is to be minimized; thus this is the objective function.
     The constraints are:
     $x \geq 0, \ y \geq 0$ A positive number of pounds must be used.
     $x \leq 200$ Only 200 pounds of ground beef are available.
     $y \geq 50$ At least 50 pounds of ground pork must be used.
     $0.75x + 0.60y \geq 0.70(x + y) \Rightarrow 0.05x \geq 0.10y$ Leanness condition to be met.
     Graph the constraints.

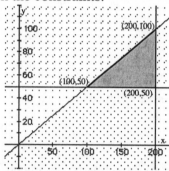

     The corner points are (100, 50), (200, 50), (200, 100).
     Evaluate the objective function:

Vertex	Value of $C = 0.75x + 0.45y$
(100, 50)	$C = 0.75(100) + 0.45(50) = 97.50$
(200, 50)	$C = 0.75(200) + 0.45(50) = 172.50$
(200, 100)	$C = 0.75(200) + 0.45(100) = 195.00$

     The minimum cost is $97.50, when 100 pounds of ground beef and 50 pounds of ground pork are used.

27. Let x = the number of racing skates manufactured.
     Let y = the number of figure skates manufactured.
     The total profit is: $P = 10x + 12y$. Profit is to be maximized; thus, this is the objective function.
     The constraints are:
     $x \geq 0, \ y \geq 0$ A positive number of skates must be manufactured.
     $6x + 4y \leq 120$ Only 120 hours are available for fabrication.
     $x + 2y \leq 40$ Only 40 hours are available for finishing.
     Graph the constraints.

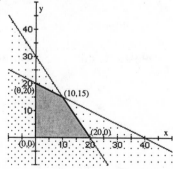

To find the intersection of $6x + 4y = 120$ and $x + 2y = 40$, solve the system:

$$\begin{cases} 6x + 4y = 120 \\ x + 2y = 40 \quad \Rightarrow \quad x = 40 - 2y \end{cases}$$

Substitute and solve:

$$6(40 - 2y) + 4y = 120 \Rightarrow 240 - 12y + 4y = 120$$

$$-8y = -120 \Rightarrow y = 15 \Rightarrow x = 40 - 2(15) = 10$$

The point of intersection is (10, 15).

The corner points are (0, 0), (0, 20), (20, 0), (10, 15).

Evaluate the objective function:

Vertex	Value of $P = 10x + 12y$
(0, 0)	$P = 10(0) + 12(0) = 0$
(0, 20)	$P = 10(0) + 12(20) = 240$
(20, 0)	$P = 10(20) + 12(0) = 200$
(10, 15)	$P = 10(10) + 12(15) = 280$

The maximum profit is $280, when 10 racing skates and 15 figure skates are produced.

29. Let x = the number of metal fasteners.

Let y = the number of plastic fasteners.

The total cost is: $C = 9x + 4y$. Cost is to be minimized; thus, this is the objective function.

The constraints are:

$x \geq 2, \; y \geq 2$ At least 2 of each fastener must be made.

$x + y \geq 6$ At least 6 fasteners are needed.

$4x + 2y \leq 24$ Only 24 hours are available.

Graph the constraints.

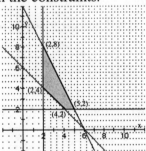

The corner points are (2, 4), (2, 8), (4, 2), (5, 2).

Evaluate the objective function:

Vertex	Value of $C = 9x + 4y$
(2, 4)	$C = 9(2) + 4(4) = 34$
(2, 8)	$C = 9(2) + 4(8) = 50$
(4, 2)	$C = 9(4) + 4(2) = 44$
(5, 2)	$C = 9(5) + 4(2) = 53$

The minimum cost is $34, when 2 metal fasteners and 4 plastic fasteners are ordered.

31. Let x = the number of first class seats.
Let y = the number of coach seats.
Using the hint, the revenue from x first class seats and y coach seats is $Fx + Cy$, where $F > C > 0$. Thus, $R = Fx + Cy$ is the objective function to be maximized.
The constraints are:

$8 \leq x \leq 16$ Restriction on first class seats.
$80 \leq y \leq 120$ Restriction on coach seats.

(a) $\dfrac{x}{y} \leq \dfrac{1}{12}$ Ratio of seats.

The constraints are:

$8 \leq x \leq 16$
$80 \leq y \leq 120$
$12x \leq y$

Graph the constraints.

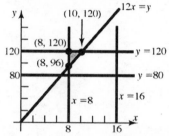

The corner points are $(8, 96)$, $(8, 120)$, and $(10, 120)$.
Evaluate the objective function:

Vertex	Value of $R = Fx + Cy$
$(8, 96)$	$R = 8F + 96C$
$(8, 120)$	$R = 8F + 120C$
$(10, 120)$	$R = 10F + 120C$

Since $C > 0$, $120C > 96C$, so $8F + 120C > 8F + 96C$.
Since $F > 0$, $10F > 8F$, so $10F + 120C > 8F + 120C$.
Thus, the maximum revenue occurs when the aircraft is configured with 10 first class seats and 120 coach seats.

(b) $\dfrac{x}{y} \leq \dfrac{1}{8}$

The constraints are:

$8 \leq x \leq 16$
$80 \leq y \leq 120$
$8x \leq y$

Graph the constraints.

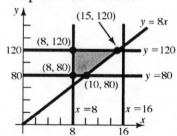

The corner points are $(8, 80)$, $(8, 120)$, $(15, 120)$, and $(10, 80)$.

Evaluate the objective function:

Vertex	Value of $R = Fx + Cy$
(8, 80)	$R = 8F + 80C$
(8, 120)	$R = 8F + 120C$
(15, 120)	$R = 15F + 120C$
(10, 80)	$R = 10F + 80C$

Since $F > 0$ and $C > 0$, $120C > 96C$, the maximum value of R occurs at (15, 120). The maximum revenue occurs when the aircraft is configured with 10 first class seats and 120 coach seats.

(c)   Answers will vary.

33. Answers will vary.

Systems of Equations and Inequalities

10.R Chapter Review

1. Solve the first equation for y, substitute into the second equation and solve:

$$\begin{cases} 2x - y = 5 & \Rightarrow \quad y = 2x - 5 \\ 5x + 2y = 8 \end{cases}$$

$$5x + 2(2x - 5) = 8 \Rightarrow 5x + 4x - 10 = 8$$

$$9x = 18 \Rightarrow x = 2 \Rightarrow y = 2(2) - 5 = 4 - 5 = -1$$

The solution is $x = 2, \ y = -1$.

3. Solve the second equation for x, substitute into the first equation and solve:

$$\begin{cases} 3x - 4y = 4 \\ x - 3y = \dfrac{1}{2} & \Rightarrow \quad x = 3y + \dfrac{1}{2} \end{cases}$$

$$3\left(3y + \frac{1}{2}\right) - 4y = 4 \Rightarrow 9y + \frac{3}{2} - 4y = 4 \Rightarrow 5y = \frac{5}{2} \Rightarrow y = \frac{1}{2} \Rightarrow x = 3\left(\frac{1}{2}\right) + \frac{1}{2} = 2$$

The solution is $x = 2, \ y = \dfrac{1}{2}$.

5. Solve the first equation for x, substitute into the second equation and solve:

$$\begin{cases} x - 2y - 4 = 0 & \Rightarrow \quad x = 2y + 4 \\ 3x + 2y - 4 = 0 \end{cases}$$

$$3(2y + 4) + 2y - 4 = 0 \Rightarrow 6y + 12 + 2y - 4 = 0$$

$$8y = -8 \Rightarrow y = -1 \Rightarrow x = 2(-1) + 4 = 2$$

The solution is $x = 2, \ y = -1$.

7. Substitute the first equation into the second equation and solve:

$$\begin{cases} y = 2x - 5 \\ x = 3y + 4 \end{cases}$$

$$x = 3(2x - 5) + 4 \Rightarrow x = 6x - 15 + 4$$

$$-5x = -11 \Rightarrow x = \frac{11}{5} \Rightarrow y = 2\left(\frac{11}{5}\right) - 5 = -\frac{3}{5}$$

The solution is $x = \dfrac{11}{5}, \ y = -\dfrac{3}{5}$.

9. Multiply each side of the first equation by 3 and each side of the second equation by −6 and add:

$$\begin{cases} x - 3y + 4 = 0 & \xrightarrow{\ 3\ } & 3x - 9y + 12 = 0 \\ \dfrac{1}{2}x - \dfrac{3}{2}y + \dfrac{4}{3} = 0 & \xrightarrow{\ -6\ } & \begin{array}{r} -3x + 9y - 8 = 0 \\ \hline 4 = 0 \end{array} \end{cases}$$

There is no solution to the system. The system is inconsistent.

11. Multiply each side of the first equation by 2 and each side of the second equation by 3 and add to eliminate y:

$$\begin{cases} 2x + 3y - 13 = 0 & \xrightarrow{\ 2\ } & 4x + 6y - 26 = 0 \\ 3x - 2y \quad\ = 0 & \xrightarrow{\ 3\ } & \begin{array}{r} 9x - 6y \quad\ = 0 \\ \hline 13x \quad\quad - 26 = 0 \\ 13x = 26 \\ x = 2 \end{array} \end{cases}$$

Substitute and solve for y:

$$3(2) - 2y = 0$$
$$-2y = -6$$
$$y = 3$$

The solution of the system is $x = 2,\ y = 3$.

13. Multiply each side of the second equation by −3 and add to eliminate x:

$$\begin{cases} 3x - 2y = 8 & \longrightarrow & 3x - 2y = \ 8 \\ x - \dfrac{2}{3}y = 12 & \xrightarrow{\ -3\ } & \begin{array}{r} -3x + 2y = -36 \\ \hline 0 = -28 \end{array} \end{cases}$$

The system has no solution, so the system is inconsistent.

15. Multiply each side of the first equation by −2 and add to the second equation to eliminate x; multiply each side of the first equation by −3 and add to the third equation to eliminate x:

$$\begin{cases} x + 2y - z = \ \ 6 & \xrightarrow{\ -2\ } & -2x - 4y + 2z = -12 \\ 2x - \ y + 3z = -13 & \longrightarrow & \begin{array}{r} 2x - \ y + 3z = -13 \\ \hline -5y + 5z = -25 \end{array} \xrightarrow{\ -1/5\ } y - z = 5 \\ 3x - 2y + 3z = -16 & \end{cases}$$

$$\xrightarrow{\ -3\ } \ -3x - 6y + 3z = -18$$
$$\longrightarrow \ \begin{array}{r} 3x - 2y + 3z = -16 \\ \hline -8y + 6z = -34 \end{array}$$

Multiply each side of the first result by 8 and add to the second result to eliminate y:

$$\begin{array}{r} y - z = \ \ 5 \quad \xrightarrow{\ 8\ } \quad 8y - 8z = \ \ 40 \\ -8y + 6z = -34 \quad \longrightarrow \quad \begin{array}{r} -8y + 6z = -34 \\ \hline -2z = \ \ 6 \\ z = -3 \end{array} \end{array}$$

Substituting and solving for the other variables:

$$y - (-3) = 5 \qquad\qquad x + 2(2) - (-3) = 6$$
$$y = 2 \qquad\qquad\qquad x + 4 + 3 = 6$$
$$x = -1$$

The solution is $x = -1$, $y = 2$, $z = -3$.

Multiply the first equation by -1 and the second equation by 2 and then add to eliminate x; multiply the second equation by -5 and add to the third equation to eliminate x:

$$\begin{cases} 2x - 4y + z = -15 \\ x + 2y - 4z = 27 \\ 5x - 6y - 2z = -3 \end{cases}$$

$$2x - 4y + z = -15 \xrightarrow{\ -1\ } -2x + 4y - z = 15$$
$$x + 2y - 4z = 27 \xrightarrow{\ 2\ } \underline{2x + 4y - 8z = 54}$$
$$8y - 9z = 69$$

$$x + 2y - 4z = 27 \xrightarrow{\ -5\ } -5x - 10y + 20z = -135$$
$$5x - 6y - 2z = -3 \longrightarrow \underline{5x - 6y - 2z \quad\;\; = -3}$$
$$-16y + 18z = -138$$

Now eliminate y in the resulting system:

$$8y - 9z = \quad 69 \xrightarrow{\ 2\ } 16y - 18z = 138$$
$$-16y + 18z = -138 \longrightarrow \underline{-16y + 18z = -138}$$
$$0 = 0$$

$$-16y + 18z = -138 \Rightarrow 18z + 138 = 16y \Rightarrow y = \frac{9}{8}z + \frac{69}{8}, \; z \text{ is any real number}$$

Substituting into the second equation and solving for x:

$$x + 2\left(\frac{9}{8}z + \frac{69}{8}\right) - 4z = 27$$

$$x + \frac{9}{4}z + \frac{69}{4} - 4z = 27 \Rightarrow x = \frac{7}{4}z + \frac{39}{4}, \; z \text{ is any real number}$$

The solution is $x = \frac{7}{4}z + \frac{39}{4}$, $y = \frac{9}{8}z + \frac{69}{8}$, z is any real number.

$$\begin{cases} 3x + 2y = 8 \\ x + 4y = -1 \end{cases}$$

$$A + C = \begin{bmatrix} 1 & 0 \\ 2 & 4 \\ -1 & 2 \end{bmatrix} + \begin{bmatrix} 3 & -4 \\ 1 & 5 \\ 5 & 2 \end{bmatrix} = \begin{bmatrix} 4 & -4 \\ 3 & 9 \\ 4 & 4 \end{bmatrix}$$

$$6A = 6 \cdot \begin{bmatrix} 1 & 0 \\ 2 & 4 \\ -1 & 2 \end{bmatrix} = \begin{bmatrix} 6 & 0 \\ 12 & 24 \\ -6 & 12 \end{bmatrix}$$

25. $AB = \begin{bmatrix} 1 & 0 \\ 2 & 4 \\ -1 & 2 \end{bmatrix} \cdot \begin{bmatrix} 4 & -3 & 0 \\ 1 & 1 & -2 \end{bmatrix} = \begin{bmatrix} 4 & -3 & 0 \\ 12 & -2 & -8 \\ -2 & 5 & -4 \end{bmatrix}$

27. $CB = \begin{bmatrix} 3 & -4 \\ 1 & 5 \\ 5 & 2 \end{bmatrix} \cdot \begin{bmatrix} 4 & -3 & 0 \\ 1 & 1 & -2 \end{bmatrix} = \begin{bmatrix} 8 & -13 & 8 \\ 9 & 2 & -10 \\ 22 & -13 & -4 \end{bmatrix}$

29. Augment the matrix with the identity and use row operations to find the inverse:

$$A = \begin{bmatrix} 4 & 6 \\ 1 & 3 \end{bmatrix} \rightarrow \left[\begin{array}{cc|cc} 4 & 6 & 1 & 0 \\ 1 & 3 & 0 & 1 \end{array} \right]$$

$$\rightarrow \left[\begin{array}{cc|cc} 1 & 3 & 0 & 1 \\ 4 & 6 & 1 & 0 \end{array} \right] \rightarrow \left[\begin{array}{cc|cc} 1 & 3 & 0 & 1 \\ 0 & -6 & 1 & -4 \end{array} \right] \rightarrow \left[\begin{array}{cc|cc} 1 & 3 & 0 & 1 \\ 0 & 1 & -\frac{1}{6} & \frac{2}{3} \end{array} \right] \rightarrow \left[\begin{array}{cc|cc} 1 & 0 & \frac{1}{2} & -1 \\ 0 & 1 & -\frac{1}{6} & \frac{2}{3} \end{array} \right]$$

      Interchange $R_2 = -4r_1 + r_2$ $R_2 = -\frac{1}{6}r_2$ $R_1 = -3r_2 + r_1$

    r_1 and r_2

$$A^{-1} = \begin{bmatrix} \frac{1}{2} & -1 \\ -\frac{1}{6} & \frac{2}{3} \end{bmatrix}$$

31. Augment the matrix with the identity and use row operations to find the inverse:

$$A = \begin{bmatrix} 1 & 3 & 3 \\ 1 & 2 & 1 \\ 1 & -1 & 2 \end{bmatrix} \rightarrow \left[\begin{array}{ccc|ccc} 1 & 3 & 3 & 1 & 0 & 0 \\ 1 & 2 & 1 & 0 & 1 & 0 \\ 1 & -1 & 2 & 0 & 0 & 1 \end{array} \right]$$

$$\rightarrow \left[\begin{array}{ccc|ccc} 1 & 3 & 3 & 1 & 0 & 0 \\ 0 & -1 & -2 & -1 & 1 & 0 \\ 0 & -4 & -1 & -1 & 0 & 1 \end{array} \right] \rightarrow \left[\begin{array}{ccc|ccc} 1 & 3 & 3 & 1 & 0 & 0 \\ 0 & 1 & 2 & 1 & -1 & 0 \\ 0 & -4 & -1 & -1 & 0 & 1 \end{array} \right] \rightarrow \left[\begin{array}{ccc|ccc} 1 & 0 & -3 & -2 & 3 & 0 \\ 0 & 1 & 2 & 1 & -1 & 0 \\ 0 & 0 & 7 & 3 & -4 & 1 \end{array} \right]$$

  $R_2 = -r_1 + r_2$ $R_2 = -r_2$ $R_1 = -3r_2 + r_1$

  $R_3 = -r_1 + r_3$ $R_3 = 4r_2 + r_3$

$$\rightarrow \left[\begin{array}{ccc|ccc} 1 & 0 & -3 & -2 & 3 & 0 \\ 0 & 1 & 2 & 1 & -1 & 0 \\ 0 & 0 & 1 & \frac{3}{7} & -\frac{4}{7} & \frac{1}{7} \end{array} \right] \rightarrow \left[\begin{array}{ccc|ccc} 1 & 0 & 0 & -\frac{5}{7} & \frac{9}{7} & \frac{3}{7} \\ 0 & 1 & 0 & \frac{1}{7} & \frac{1}{7} & -\frac{2}{7} \\ 0 & 0 & 1 & \frac{3}{7} & -\frac{4}{7} & \frac{1}{7} \end{array} \right]$$

      $R_3 = \frac{1}{7}r_3$ $R_1 = 3r_3 + r_1$

                                      $R_2 = -2r_3 + r_2$

$$A^{-1} = \begin{bmatrix} -\frac{5}{7} & \frac{9}{7} & \frac{3}{7} \\ \frac{1}{7} & \frac{1}{7} & -\frac{2}{7} \\ \frac{3}{7} & -\frac{4}{7} & \frac{1}{7} \end{bmatrix}$$

33. Augment the matrix with the identity and use row operations to find the inverse:

$$A = \begin{bmatrix} 4 & -8 \\ -1 & 2 \end{bmatrix} \rightarrow \begin{bmatrix} 4 & -8 & | & 1 & 0 \\ -1 & 2 & | & 0 & 1 \end{bmatrix}$$

$$\rightarrow \begin{bmatrix} -1 & 2 & | & 0 & 1 \\ 4 & -8 & | & 1 & 0 \end{bmatrix} \rightarrow \begin{bmatrix} -1 & 2 & | & 0 & 1 \\ 0 & 0 & | & 1 & 4 \end{bmatrix} \rightarrow \begin{bmatrix} 1 & -2 & | & 0 & -1 \\ 0 & 0 & | & 1 & 4 \end{bmatrix}$$

$$\begin{array}{ccc} \text{Interchange} & R_2 = 4r_1 + r_2 & R_1 = -r_1 \\ r_1 \text{ and } r_2 & & \end{array}$$

There is no inverse because there is no way to obtain the identity on the left side. The matrix is singular.

35. $\begin{cases} 3x - 2y = 1 \\ 10x + 10y = 5 \end{cases}$ Write the augmented matrix: $\begin{bmatrix} 3 & -2 & | & 1 \\ 10 & 10 & | & 5 \end{bmatrix}$

$$\rightarrow \begin{bmatrix} 3 & -2 & | & 1 \\ 1 & 16 & | & 2 \end{bmatrix} \rightarrow \begin{bmatrix} 1 & 16 & | & 2 \\ 3 & -2 & | & 1 \end{bmatrix} \rightarrow \begin{bmatrix} 1 & 16 & | & 2 \\ 0 & -50 & | & -5 \end{bmatrix} \rightarrow \begin{bmatrix} 1 & 16 & | & 2 \\ 0 & 1 & | & \frac{1}{10} \end{bmatrix} \rightarrow \begin{bmatrix} 1 & 0 & | & \frac{2}{5} \\ 0 & 1 & | & \frac{1}{10} \end{bmatrix}$$

$$\begin{array}{ccccc} R_2 = -3r_1 + r_2 & \text{Interchange} & R_2 = -3r_1 + r_2 & R_2 = -\frac{1}{50}r_2 & R_1 = -16r_2 + r_1 \\ & r_1 \text{ and } r_2 & & & \end{array}$$

The solution is $x = \dfrac{2}{5}, y = \dfrac{1}{10}$.

37. $\begin{cases} 5x + 6y - 3z = 6 \\ 4x - 7y - 2z = -3 \\ 3x + y - 7z = 1 \end{cases}$ Write the augmented matrix: $\begin{bmatrix} 5 & 6 & -3 & | & 6 \\ 4 & -7 & -2 & | & -3 \\ 3 & 1 & -7 & | & 1 \end{bmatrix}$

$$\rightarrow \begin{bmatrix} 1 & 13 & -1 & | & 9 \\ 4 & -7 & -2 & | & -3 \\ 3 & 1 & -7 & | & 1 \end{bmatrix} \rightarrow \begin{bmatrix} 1 & 13 & -1 & | & 9 \\ 0 & -59 & 2 & | & -39 \\ 0 & -38 & -4 & | & -26 \end{bmatrix} \rightarrow \begin{bmatrix} 1 & 13 & -1 & | & 9 \\ 0 & 1 & -\frac{2}{59} & | & \frac{39}{59} \\ 0 & -38 & -4 & | & -26 \end{bmatrix}$$

$$\begin{array}{ccc} R_1 = -r_2 + r_1 & R_2 = -4r_1 + r_2 & R_2 = -\frac{1}{59}r_2 \\ & R_3 = -3r_1 + r_3 & \end{array}$$

$$\rightarrow \begin{bmatrix} 1 & 0 & -\frac{33}{59} & | & \frac{24}{59} \\ 0 & 1 & -\frac{2}{59} & | & \frac{39}{59} \\ 0 & 0 & -\frac{312}{59} & | & -\frac{52}{59} \end{bmatrix} \rightarrow \begin{bmatrix} 1 & 0 & -\frac{33}{59} & | & \frac{24}{59} \\ 0 & 1 & -\frac{2}{59} & | & \frac{39}{59} \\ 0 & 0 & 1 & | & \frac{1}{6} \end{bmatrix} \rightarrow \begin{bmatrix} 1 & 0 & 0 & | & \frac{1}{2} \\ 0 & 1 & 0 & | & \frac{2}{3} \\ 0 & 0 & 1 & | & \frac{1}{6} \end{bmatrix}$$

$$\begin{array}{ccc} R_1 = -13r_2 + r_1 & R_3 = -\frac{59}{312}r_3 & R_1 = \frac{33}{59}r_3 + r_1 \\ R_3 = 38r_2 + r_3 & & R_2 = \frac{2}{59}r_3 + r_2 \end{array}$$

The solution is $x = \dfrac{1}{2}, y = \dfrac{2}{3}, z = \dfrac{1}{6}$.

39. $\begin{cases} x \quad\; -2z = 1 \\ 2x + 3y \quad\;\; = -3 \\ 4x - 3y - 4z = 3 \end{cases}$ Write the augmented matrix: $\begin{bmatrix} 1 & 0 & -2 & | & 1 \\ 2 & 3 & 0 & | & -3 \\ 4 & -3 & -4 & | & 3 \end{bmatrix}$

$\rightarrow \begin{bmatrix} 1 & 0 & -2 & | & 1 \\ 0 & 3 & 4 & | & -5 \\ 0 & -3 & 4 & | & -1 \end{bmatrix} \rightarrow \begin{bmatrix} 1 & 0 & -2 & | & 1 \\ 0 & 1 & \frac{4}{3} & | & -\frac{5}{3} \\ 0 & -3 & 4 & | & -1 \end{bmatrix} \rightarrow \begin{bmatrix} 1 & 0 & -2 & | & 1 \\ 0 & 1 & \frac{4}{3} & | & -\frac{5}{3} \\ 0 & 0 & 8 & | & -6 \end{bmatrix}$

$R_2 = -2r_1 + r_2 \qquad\quad R_2 = \frac{1}{3}r_2 \qquad\qquad R_3 = 3r_2 + r_3$

$R_3 = -4r_1 + r_3$

$\rightarrow \begin{bmatrix} 1 & 0 & -2 & | & 1 \\ 0 & 1 & \frac{4}{3} & | & -\frac{5}{3} \\ 0 & 0 & 1 & | & -\frac{3}{4} \end{bmatrix} \rightarrow \begin{bmatrix} 1 & 0 & 0 & | & -\frac{1}{2} \\ 0 & 1 & 0 & | & -\frac{2}{3} \\ 0 & 0 & 1 & | & -\frac{3}{4} \end{bmatrix}$

$R_3 = \frac{1}{8}r_3 \qquad\qquad R_1 = 2r_3 + r_1$

$\qquad\qquad\qquad R_2 = -\frac{4}{3}r_3 + r_2$

The solution is $x = -\dfrac{1}{2}, y = -\dfrac{2}{3}, z = -\dfrac{3}{4}$.

41. $\begin{cases} x - y + z = 0 \\ x - y - 5z = 6 \\ 2x - 2y + z = 1 \end{cases}$ Write the augmented matrix: $\begin{bmatrix} 1 & -1 & 1 & | & 0 \\ 1 & -1 & -5 & | & 6 \\ 2 & -2 & 1 & | & 1 \end{bmatrix}$

$\rightarrow \begin{bmatrix} 1 & -1 & 1 & | & 0 \\ 0 & 0 & -6 & | & 6 \\ 0 & 0 & -1 & | & 1 \end{bmatrix} \rightarrow \begin{bmatrix} 1 & -1 & 1 & | & 0 \\ 0 & 0 & 1 & | & -1 \\ 0 & 0 & -1 & | & 1 \end{bmatrix} \rightarrow \begin{bmatrix} 1 & -1 & 0 & | & 1 \\ 0 & 0 & 1 & | & -1 \\ 0 & 0 & 0 & | & 0 \end{bmatrix} \Rightarrow \begin{cases} x = y + 1 \\ z = -1 \end{cases}$

$R_2 = -r_1 + r_2 \qquad\quad R_2 = -\frac{1}{6}r_2 \qquad\qquad R_1 = -r_2 + r_1$

$R_3 = -2r_1 + r_3 \qquad\qquad\qquad\qquad\qquad R_3 = r_2 + r_3$

The solution is $x = y + 1, z = -1, y$ is any real number.

43. $\begin{cases} x - y - z - t = 1 \\ 2x + y - z + 2t = 3 \\ x - 2y - 2z - 3t = 0 \\ 3x - 4y + z + 5t = -3 \end{cases}$ Write the augmented matrix: $\begin{bmatrix} 1 & -1 & -1 & -1 & | & 1 \\ 2 & 1 & -1 & 2 & | & 3 \\ 1 & -2 & -2 & -3 & | & 0 \\ 3 & -4 & 1 & 5 & | & -3 \end{bmatrix}$

$\rightarrow \begin{bmatrix} 1 & -1 & -1 & -1 & | & 1 \\ 0 & 3 & 1 & 4 & | & 1 \\ 0 & -1 & -1 & -2 & | & -1 \\ 0 & -1 & 4 & 8 & | & -6 \end{bmatrix} \rightarrow \begin{bmatrix} 1 & -1 & -1 & -1 & | & 1 \\ 0 & -1 & -1 & -2 & | & -1 \\ 0 & 3 & 1 & 4 & | & 1 \\ 0 & -1 & 4 & 8 & | & -6 \end{bmatrix} \rightarrow \begin{bmatrix} 1 & -1 & -1 & -1 & | & 1 \\ 0 & 1 & 1 & 2 & | & 1 \\ 0 & 3 & 1 & 4 & | & 1 \\ 0 & -1 & 4 & 8 & | & -6 \end{bmatrix}$

$R_2 = -2r_1 + r_2 \qquad\qquad$ Interchange r_2 and $r_3 \qquad R_2 = -r_2$

$R_3 = -r_1 + r_3$

$R_4 = -3r_1 + r_4$

$$\rightarrow \begin{bmatrix} 1 & 0 & 0 & 1 & | & 2 \\ 0 & 1 & 1 & 2 & | & 1 \\ 0 & 0 & -2 & -2 & | & -2 \\ 0 & 0 & 5 & 10 & | & -5 \end{bmatrix} \rightarrow \begin{bmatrix} 1 & 0 & 0 & 1 & | & 2 \\ 0 & 1 & 1 & 2 & | & 1 \\ 0 & 0 & 1 & 1 & | & 1 \\ 0 & 0 & 1 & 2 & | & -1 \end{bmatrix} \rightarrow \begin{bmatrix} 1 & 0 & 0 & 1 & | & 2 \\ 0 & 1 & 0 & 1 & | & 0 \\ 0 & 0 & 1 & 1 & | & 1 \\ 0 & 0 & 0 & 1 & | & -2 \end{bmatrix}$$

$R_1 = r_2 + r_1$ $R_3 = -\frac{1}{2}r_3$ $R_2 = -r_3 + r_2$

$R_3 = -3r_2 + r_3$ $R_4 = \frac{1}{5}r_4$ $R_4 = -r_3 + r_4$

$R_4 = r_2 + r_4$

$$\rightarrow \begin{bmatrix} 1 & 0 & 0 & 0 & | & 4 \\ 0 & 1 & 0 & 0 & | & 2 \\ 0 & 0 & 1 & 0 & | & 3 \\ 0 & 0 & 0 & 1 & | & -2 \end{bmatrix}$$

$R_1 = -r_4 + r_1$

$R_2 = -r_4 + r_2$

$R_3 = -r_4 + r_3$

The solution is $x = 4$, $y = 2$, $z = 3$, $t = -2$.

45. $\begin{vmatrix} 3 & 4 \\ 1 & 3 \end{vmatrix} = 3(3) - 4(1) = 9 - 4 = 5$

47. $\begin{vmatrix} 1 & 4 & 0 \\ -1 & 2 & 6 \\ 4 & 1 & 3 \end{vmatrix} = 1 \begin{vmatrix} 2 & 6 \\ 1 & 3 \end{vmatrix} - 4 \begin{vmatrix} -1 & 6 \\ 4 & 3 \end{vmatrix} + 0 \begin{vmatrix} -1 & 2 \\ 4 & 1 \end{vmatrix}$

$= 1(6 - 6) - 4(-3 - 24) + 0(-1 - 8)$

$= 1(0) - 4(-27) + 0(-9) = 0 + 108 + 0$

$= 108$

49. $\begin{vmatrix} 2 & 1 & -3 \\ 5 & 0 & 1 \\ 2 & 6 & 0 \end{vmatrix} = 2 \begin{vmatrix} 0 & 1 \\ 6 & 0 \end{vmatrix} - 1 \begin{vmatrix} 5 & 1 \\ 2 & 0 \end{vmatrix} + (-3) \begin{vmatrix} 5 & 0 \\ 2 & 6 \end{vmatrix}$

$= 2(0 - 6) - 1(0 - 2) - 3(30 - 0) = -12 + 2 - 90$

$= -100$

51. Set up and evaluate the determinants to use Cramer's Rule:

$\begin{cases} x - 2y = 4 \\ 3x + 2y = 4 \end{cases}$

$D = \begin{vmatrix} 1 & -2 \\ 3 & 2 \end{vmatrix} = 2 + 6 = 8; \quad D_x = \begin{vmatrix} 4 & -2 \\ 4 & 2 \end{vmatrix} = 8 + 8 = 16; \quad D_y = \begin{vmatrix} 1 & 4 \\ 3 & 4 \end{vmatrix} = 4 - 12 = -8$

Find the solutions by Cramer's Rule: $x = \dfrac{D_x}{D} = \dfrac{16}{8} = 2 \qquad y = \dfrac{D_y}{D} = \dfrac{-8}{8} = -1$

53. Set up and evaluate the determinants to use Cramer's Rule:

$$\begin{cases} 2x + 3y - 13 = 0 \\ 3x - 2y = 0 \end{cases} \Rightarrow \begin{cases} 2x + 3y = 13 \\ 3x - 2y = 0 \end{cases}$$

$$D = \begin{vmatrix} 2 & 3 \\ 3 & -2 \end{vmatrix} = -4 - 9 = -13; \quad D_x = \begin{vmatrix} 13 & 3 \\ 0 & -2 \end{vmatrix} = -26 - 0 = -26; \quad D_y = \begin{vmatrix} 2 & 13 \\ 3 & 0 \end{vmatrix} = 0 - 39 = -39$$

Find the solutions by Cramer's Rule: $x = \dfrac{D_x}{D} = \dfrac{-26}{-13} = 2 \qquad y = \dfrac{D_y}{D} = \dfrac{-39}{-13} = 3$

55. Set up and evaluate the determinants to use Cramer's Rule:

$$\begin{cases} x + 2y - z = 6 \\ 2x - y + 3z = -13 \\ 3x - 2y + 3z = -16 \end{cases}$$

$$D = \begin{vmatrix} 1 & 2 & -1 \\ 2 & -1 & 3 \\ 3 & -2 & 3 \end{vmatrix} = 1 \begin{vmatrix} -1 & 3 \\ -2 & 3 \end{vmatrix} - 2 \begin{vmatrix} 2 & 3 \\ 3 & 3 \end{vmatrix} + (-1) \begin{vmatrix} 2 & -1 \\ 3 & -2 \end{vmatrix}$$

$$= 1(-3 + 6) - 2(6 - 9) - 1(-4 + 3) = 3 + 6 + 1 = 10$$

$$D_x = \begin{vmatrix} 6 & 2 & -1 \\ -13 & -1 & 3 \\ -16 & -2 & 3 \end{vmatrix} = 6 \begin{vmatrix} -1 & 3 \\ -2 & 3 \end{vmatrix} - 2 \begin{vmatrix} -13 & 3 \\ -16 & 3 \end{vmatrix} + (-1) \begin{vmatrix} -13 & -1 \\ -16 & -2 \end{vmatrix}$$

$$= 6(-3 + 6) - 2(-39 + 48) - 1(26 - 16) = 18 - 18 - 10 = -10$$

$$D_y = \begin{vmatrix} 1 & 6 & -1 \\ 2 & -13 & 3 \\ 3 & -16 & 3 \end{vmatrix} = 1 \begin{vmatrix} -13 & 3 \\ -16 & 3 \end{vmatrix} - 6 \begin{vmatrix} 2 & 3 \\ 3 & 3 \end{vmatrix} + (-1) \begin{vmatrix} 2 & -13 \\ 3 & -16 \end{vmatrix}$$

$$= 1(-39 + 48) - 6(6 - 9) - 1(-32 + 39) = 9 + 18 - 7 = 20$$

$$D_z = \begin{vmatrix} 1 & 2 & 6 \\ 2 & -1 & -13 \\ 3 & -2 & -16 \end{vmatrix} = 1 \begin{vmatrix} -1 & -13 \\ -2 & -16 \end{vmatrix} - 2 \begin{vmatrix} 2 & -13 \\ 3 & -16 \end{vmatrix} + 6 \begin{vmatrix} 2 & -1 \\ 3 & -2 \end{vmatrix}$$

$$= 1(16 - 26) - 2(-32 + 39) + 6(-4 + 3) = -10 - 14 - 6 = -30$$

Find the solutions by Cramer's Rule:

$$x = \dfrac{D_x}{D} = \dfrac{-10}{10} = -1 \qquad y = \dfrac{D_y}{D} = \dfrac{20}{10} = 2 \qquad z = \dfrac{D_z}{D} = \dfrac{-30}{10} = -3$$

57. Let $\begin{vmatrix} x & y \\ a & b \end{vmatrix} = 8$

Then $\begin{vmatrix} 2x & y \\ 2a & b \end{vmatrix} = 2(8) = 16$ by Theorem 14.

The value of the determinant is multiplied by k when the elements of a column are multiplied by k.

59. Find the partial fraction decomposition:

$$\frac{6}{x(x-4)} = \frac{A}{x} + \frac{B}{x-4} \quad \text{(Multiply both sides by } x(x-4).)$$

$$6 = A(x-4) + Bx$$

Let $x = 4$: then $6 = A(4-4) + B(4) \Rightarrow 4B = 6 \Rightarrow B = \frac{3}{2}$

Let $x = 0$: then $6 = A(0-4) + B(0) \Rightarrow -4A = 6 \Rightarrow A = -\frac{3}{2}$

$$\frac{6}{x(x-4)} = \frac{(-3/2)}{x} + \frac{(3/2)}{x-4}$$

61. Find the partial fraction decomposition:

$$\frac{x-4}{x^2(x-1)} = \frac{A}{x} + \frac{B}{x^2} + \frac{C}{x-1} \text{(Multiply both sides by } x^2(x-1).)$$

$$x-4 = Ax(x-1) + B(x-1) + Cx^2$$

Let $x = 1$: then $1-4 = A(1)(1-1) + B(1-1) + C(1)^2 \Rightarrow -3 = C \Rightarrow C = -3$

Let $x = 0$: then $0-4 = A(0)(0-1) + B(0-1) + C(0)^2 \Rightarrow -4 = -B \Rightarrow B = 4$

Let $x = 2$: then $2-4 = A(2)(2-1) + B(2-1) + C(2)^2 \Rightarrow -2 = 2A + B + 4C$

$$\Rightarrow 2A = -2 - 4 - 4(-3) \Rightarrow 2A = 6 \Rightarrow A = 3$$

$$\frac{x-4}{x^2(x-1)} = \frac{3}{x} + \frac{4}{x^2} + \frac{-3}{x-1}$$

63. Find the partial fraction decomposition:

$$\frac{x}{(x^2+9)(x+1)} = \frac{A}{x+1} + \frac{Bx+C}{x^2+9} \quad \text{(Multiply both sides by } (x+1)(x^2+9).)$$

$$x = A(x^2+9) + (Bx+C)(x+1)$$

Let $x = -1$: then $-1 = A\left((-1)^2+9\right) + \left(B(-1)+C\right)(-1+1)$

$$\Rightarrow -1 = A(10) + (-B+C)(0) \Rightarrow -1 = 10A \Rightarrow A = -\frac{1}{10}$$

Let $x = 1$: then $1 = A(1^2+9) + (B(1)+C)(1+1) \Rightarrow 1 = 10A + 2B + 2C$

$$\Rightarrow 1 = 10\left(-\frac{1}{10}\right) + 2B + 2C \Rightarrow 2 = 2B + 2C \Rightarrow B + C = 1$$

Let $x = 0$: then $0 = A(0^2+9) + (B(0)+C)(0+1) \Rightarrow 0 = 9A + C$

$$\Rightarrow 0 = 9\left(-\frac{1}{10}\right) + C \Rightarrow C = \frac{9}{10}$$

$$B = 1 - C \Rightarrow B = 1 - \frac{9}{10} \Rightarrow B = \frac{1}{10}$$

$$\frac{x}{(x^2+9)(x+1)} = \frac{-(1/10)}{x+1} + \frac{((1/10)x + 9/10)}{x^2+9}$$

65. Find the partial fraction decomposition:

$$\frac{x^3}{(x^2+4)^2} = \frac{Ax+B}{x^2+4} + \frac{Cx+D}{(x^2+4)^2} \quad \text{(Multiply both sides by } (x^2+4)^2.\text{)}$$

$$x^3 = (Ax+B)(x^2+4) + Cx+D$$

$$x^3 = Ax^3 + Bx^2 + 4Ax + 4B + Cx + D \Rightarrow x^3 = Ax^3 + Bx^2 + (4A+C)x + 4B + D$$

$$A = 1; \ B = 0$$

$$4A + C = 0 \ \Rightarrow \ 4(1) + C = 0 \ \Rightarrow \ C = -4$$

$$4B + D = 0 \ \Rightarrow \ 4(0) + D = 0 \ \Rightarrow \ D = 0$$

$$\frac{x^3}{(x^2+4)^2} = \frac{x}{x^2+4} + \frac{-4x}{(x^2+4)^2}$$

67. Find the partial fraction decomposition:

$$\frac{x^2}{(x^2+1)(x^2-1)} = \frac{x^2}{(x^2+1)(x-1)(x+1)} = \frac{A}{x-1} + \frac{B}{x+1} + \frac{Cx+D}{x^2+1}$$

$$\text{(Multiply both sides by } (x-1)(x+1)(x^2+1).\text{)}$$

$$x^2 = A(x+1)(x^2+1) + B(x-1)(x^2+1) + (Cx+D)(x-1)(x+1)$$

Let $x = 1$: then $\quad 1^2 = A(1+1)(1^2+1) + B(1-1)(1^2+1) + (C(1)+D)(1-1)(1+1)$

$$\Rightarrow 1 = 4A \Rightarrow A = \frac{1}{4}$$

Let $x = -1$: then

$$(-1)^2 = A(-1+1)((-1)^2+1) + B(-1-1)((-1)^2+1) + (C(-1)+D)(-1-1)(-1+1)$$

$$\Rightarrow 1 = -4B \Rightarrow B = -\frac{1}{4}$$

Let $x = 0$: then

$$0^2 = A(0+1)(0^2+1) + B(0-1)(0^2+1) + (C(0)+D)(0-1)(0+1)$$

$$\Rightarrow 0 = A - B - D \Rightarrow 0 = \frac{1}{4} - \left(-\frac{1}{4}\right) - D \Rightarrow D = \frac{1}{2}$$

Let $x = 2$: then

$$2^2 = A(2+1)(2^2+1) + B(2-1)(2^2+1) + (C(2)+D)(2-1)(2+1)$$

$$\Rightarrow 4 = 15A + 5B + 6C + 3D \Rightarrow 4 = 15\left(\frac{1}{4}\right) + 5\left(-\frac{1}{4}\right) + 6C + 3\left(\frac{1}{2}\right)$$

$$\Rightarrow 6C = 4 - \frac{15}{4} + \frac{5}{4} - \frac{3}{2} \Rightarrow 6C = 0 \Rightarrow C = 0$$

$$\frac{x^2}{(x^2+1)(x^2-1)} = \frac{(1/4)}{x-1} + \frac{-(1/4)}{x+1} + \frac{(1/2)}{x^2+1}$$

Solve the first equation for y, substitute into the second equation and solve:

$$\begin{cases} 2x + y + 3 = 0 \Rightarrow y = -2x - 3 \\ x^2 + y^2 = 5 \end{cases}$$

$$x^2 + (-2x - 3)^2 = 5 \Rightarrow x^2 + 4x^2 + 12x + 9 = 5$$

$$5x^2 + 12x + 4 = 0 \Rightarrow (5x + 2)(x + 2) = 0 \Rightarrow x = -\frac{2}{5} \quad \text{or} \quad x = -2$$

$$y = -\frac{11}{5} \qquad y = 1$$

Solutions: $\left(-\frac{2}{5}, -\frac{11}{5}\right)$, $(-2, 1)$.

Multiply each side of the second equation by 2 and add the equations to eliminate xy:

$$\begin{cases} 2xy + y^2 = 10 \longrightarrow \quad 2xy + y^2 = 10 \\ -xy + 3y^2 = 2 \xrightarrow{\ 2\ } -2xy + 6y^2 = 4 \end{cases}$$

$$7y^2 = 14 \Rightarrow y^2 = 2 \Rightarrow y = \pm\sqrt{2}$$

If $y = \sqrt{2}$: $\quad 2x(\sqrt{2}) + (\sqrt{2})^2 = 10 \Rightarrow 2\sqrt{2}x = 8 \Rightarrow x = \frac{8}{2\sqrt{2}} = 2\sqrt{2}$

If $y = -\sqrt{2}$: $\quad 2x(-\sqrt{2}) + (-\sqrt{2})^2 = 10 \Rightarrow -2\sqrt{2}x = 8 \Rightarrow x = \frac{8}{-2\sqrt{2}} = -2\sqrt{2}$

Solutions: $(2\sqrt{2}, \sqrt{2}), (-2\sqrt{2}, -\sqrt{2})$

Substitute the second equation into the first equation and solve:

$$\begin{cases} x^2 + y^2 = 6y \\ x^2 = 3y \end{cases}$$

$$3y + y^2 = 6y \Rightarrow y^2 - 3y = 0 \Rightarrow y(y - 3) = 0 \Rightarrow y = 0 \text{ or } y = 3$$

If $y = 0$: $\quad x^2 = 3(0) \Rightarrow x^2 = 0 \Rightarrow x = 0$

If $y = 3$: $\quad x^2 = 3(3) \Rightarrow x^2 = 9 \Rightarrow x = \pm 3$

Solutions: $(0, 0), (-3, 3), (3, 3)$.

Factor the second equation, solve for x, substitute into the first equation and solve:

$$\begin{cases} 3x^2 + 4xy + 5y^2 = 8 \\ x^2 + 3xy + 2y^2 = 0 \Rightarrow (x + 2y)(x + y) = 0 \Rightarrow x = -2y \text{ or } x = -y \end{cases}$$

Substitute $x = -2y$ and solve:

$$3x^2 + 4xy + 5y^2 = 8$$
$$3(-2y)^2 + 4(-2y)y + 5y^2 = 8$$
$$12y^2 - 8y^2 + 5y^2 = 8$$
$$9y^2 = 8$$
$$y^2 = \frac{8}{9} \Rightarrow y = \pm\frac{2\sqrt{2}}{3}$$

Substitute $x = -y$ and solve:

$$3x^2 + 4xy + 5y^2 = 8$$
$$3(-y)^2 + 4(-y)y + 5y^2 = 8$$
$$3y^2 - 4y^2 + 5y^2 = 8$$
$$4y^2 = 8$$
$$y^2 = 2 \Rightarrow y = \pm\sqrt{2}$$

If $y = \frac{2\sqrt{2}}{3}$: $x = -2\left(\frac{2\sqrt{2}}{3}\right) = \frac{-4\sqrt{2}}{3}$ If $y = \sqrt{2}$: $x = -\sqrt{2}$

If $y = -\sqrt{2}$: $x = \sqrt{2}$

If $y = \frac{-2\sqrt{2}}{3}$: $x = -2\left(\frac{-2\sqrt{2}}{3}\right) = \frac{4\sqrt{2}}{3}$

Solutions: $\left(-\frac{4\sqrt{2}}{3}, \frac{2\sqrt{2}}{3}\right), \left(\frac{4\sqrt{2}}{3}, -\frac{2\sqrt{2}}{3}\right), \left(-\sqrt{2}, \sqrt{2}\right), \left(\sqrt{2}, -\sqrt{2}\right)$

77.   Multiply each side of the second equation by $-y$ and add the equations to eliminate y:
$$\begin{cases} x^2 - 3x + y^2 + y = -2 & \longrightarrow & x^2 - 3x + y^2 + y = -2 \\ \frac{x^2 - x}{y} + y + 1 = 0 & \xrightarrow{-y} & -x^2 + x - y^2 - y = 0 \end{cases}$$

$$-2x = -2 \Rightarrow x = 1$$

If $x = 1$: $1^2 - 3(1) + y^2 + y = -2 \Rightarrow y^2 + y = 0$

$\Rightarrow y(y + 1) = 0$

$\Rightarrow y = 0$ or $y = -1$

Note that $y \neq 0$ because $y = 0$ would cause division by zero in the original system.
Solution: $(1, -1)$

79. $3x + 4y \leq 12$

Graph the line $3x + 4y = 12$. Use a solid line since the inequality uses $\leq$.
Choose a test point not on the line, such as $(0, 0)$.
Since $3(0) + 4(0) \leq 12$ is true, shade the side of the line containing $(0, 0)$.

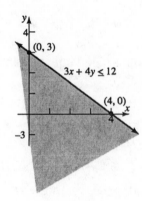

81.   Graph the system of linear inequalities:
$$\begin{cases} -2x + y \leq 2 \\ x + y \geq 2 \end{cases}$$
Graph the line $-2x + y = 2$. Use a solid line since the inequality uses $\leq$.
Choose a test point not on the line, such as $(0, 0)$. Since $-2(0) + 0 \leq 2$ is true, shade the side of the line containing $(0, 0)$.
Graph the line $x + y = 2$. Use a solid line since the inequality uses $\geq$. Choose a test point not on the line, such as $(0, 0)$. Since $0 + 0 \geq 2$ is false, shade the opposite side of the line from $(0, 0)$.
The overlapping region is the solution

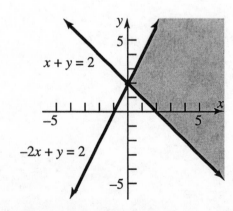

The graph is unbounded.
Find the vertices:
To find the intersection of $x + y = 2$ and $-2x + y = 2$, solve the system:
$$\begin{cases} x + y = 2 & \Rightarrow \quad x = 2 - y \\ -2x + y = 2 \end{cases}$$
Substitute and solve:
$$-2(2 - y) + y = 2 \Rightarrow -4 + 2y + y = 2 \Rightarrow 3y = 6 \Rightarrow y = 2$$
$$x = 2 - 2 = 0$$
The point of intersection is $(0, 2)$.
The corner point is $(0, 2)$.

83. Graph the system of linear inequalities:
$$\begin{cases} x \geq 0 \\ y \geq 0 \\ x + y \leq 4 \\ 2x + 3y \leq 6 \end{cases}$$

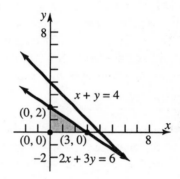

Graph $x \geq 0; y \geq 0$. Shaded region is the first quadrant.
Graph the line $x + y = 4$. Use a solid line since the
inequality uses $\leq$. Choose a test point not on the line,
such as $(0, 0)$. Since $0 + 0 \leq 4$ is true, shade the side of
the line containing $(0, 0)$.
Graph the line $2x + 3y = 6$. Use a solid line since
the inequality uses $\leq$. Choose a test point not on the
line, such as $(0, 0)$. Since $2(0) + 3(0) \leq 6$ is true, shade
the side of the line containing $(0, 0)$.
The overlapping region is the solution.
The graph is bounded.
Find the vertices:
The x-axis and y-axis intersect at $(0, 0)$.
The intersection of $2x + 3y = 6$ and the y-axis is $(0, 2)$.
The intersection of $2x + 3y = 6$ and the x-axis is $(3, 0)$.
The three corner points are $(0, 0)$, $(0, 2)$, and $(3, 0)$.

85. Graph the system of linear inequalities:

$$\begin{cases} x \geq 0 \\ y \geq 0 \\ 2x + y \leq 8 \\ x + 2y \geq 2 \end{cases}$$

Graph $x \geq 0$; $y \geq 0$. Shaded region is the first quadrant.

Graph the line $2x + y = 8$. Use a solid line since the inequality uses $\leq$. Choose a test point not on the line, such as $(0, 0)$. Since $2(0) + 0 \leq 8$ is true, shade the side of the line containing $(0, 0)$.

Graph the line $x + 2y = 2$. Use a solid line since the inequality uses $\geq$. Choose a test point not on the line, such as $(0, 0)$. Since $0 + 2(0) \geq 2$ is false, shade the opposite side of the line from $(0, 0)$.

The overlapping region is the solution.

The graph is bounded

Find the vertices:

The intersection of $x + 2y = 2$ and the y-axis is $(0, 1)$.

The intersection of $x + 2y = 2$ and the x-axis is $(2, 0)$.

The intersection of $2x + y = 8$ and the y-axis is $(0, 8)$.

The intersection of $2x + y = 8$ and the x-axis is $(4, 0)$.

The four corner points are $(0, 1)$, $(0, 8)$, $(2, 0)$, and $(4, 0)$.

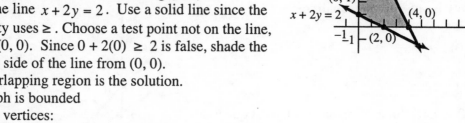

87. Graph the system of inequalities:

$$\begin{cases} x^2 + y^2 \leq 16 \\ x + y \geq 2 \end{cases}$$

Graph the circle $x^2 + y^2 = 16$. Use a solid line since the inequality uses $\leq$. Choose a test point not on the circle, such as $(0, 0)$. Since $0^2 + 0^2 \leq 16$ is true, shade the side of the circle containing $(0, 0)$.

Graph the line $x + y = 2$. Use a solid line since the inequality uses $\geq$. Choose a test point not on the line, such as $(0, 0)$. Since $0 + 0 \geq 2$ is false, shade the opposite side of the line from $(0, 0)$.

The overlapping region is the solution.

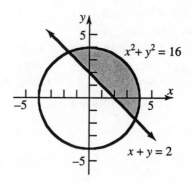

89. Graph the system of inequalities:
$$\begin{cases} y \le x^2 \\ xy \le 4 \end{cases}$$

Graph the parabola $y = x^2$. Use a solid line since the inequality uses $\le$. Choose a test point not on the parabola, such as (1, 2). Since $2 \le 1^2$ is false, shade the opposite side of the parabola from (1, 2).
Graph the hyperbola $xy = 4$. Use a solid line since the inequality uses $\le$. Choose a test point not on the hyperbola, such as (1, 2). Since $1 \cdot 2 \le 4$ is true, shade the same side of the hyperbola as (1, 2).
The overlapping region is the solution.

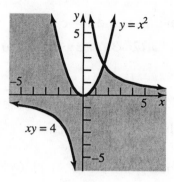

91. Maximize $z = 3x + 4y$ Subject to $x \ge 0$, $y \ge 0$, $3x + 2y \ge 6$, $x + y \le 8$
Graph the constraints.

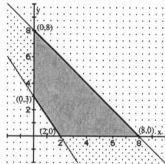

The corner points are (0, 3), (2, 0), (0, 8), (8, 0).
Evaluate the objective function:

Vertex	Value of $z = 3x + 4y$
(0, 3)	$z = 3(0) + 4(3) = 12$
(0, 8)	$z = 3(0) + 4(8) = 32$
(2, 0)	$z = 3(2) + 4(0) = 6$
(8, 0)	$z = 3(8) + 4(0) = 24$

The maximum value is 32 at (0, 8).

93. Minimize $z = 3x + 5y$
Subject to $x \ge 0$, $y \ge 0$, $x + y \ge 1$, $3x + 2y \le 12$, $x + 3y \le 12$
Graph the constraints.

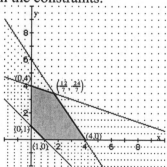

To find the intersection of $3x + 2y = 12$ and $x + 3y = 12$, solve the system:

$$\begin{cases} 3x + 2y = 12 \\ x + 3y = 12 \quad \Rightarrow \quad x = 12 - 3y \end{cases}$$

Substitute and solve:

$$3(12 - 3y) + 2y = 12 \Rightarrow 36 - 9y + 2y = 12 \Rightarrow -7y = -24 \Rightarrow y = \frac{24}{7}$$

$$x = 12 - 3\left(\frac{24}{7}\right) = 12 - \frac{72}{7} = \frac{12}{7}$$

The point of intersection is $\left(\frac{12}{7}, \frac{24}{7}\right)$.

The corner points are $(0, 1)$, $(1, 0)$, $(0, 4)$, $(4, 0)$, $\left(\frac{12}{7}, \frac{24}{7}\right)$.

Evaluate the objective function:

Vertex	Value of $z = 3x + 5y$
$(0, 1)$	$z = 3(0) + 5(1) = 5$
$(0, 4)$	$z = 3(0) + 5(4) = 20$
$(1, 0)$	$z = 3(1) + 5(0) = 3$
$(4, 0)$	$z = 3(4) + 5(0) = 12$
$\left(\dfrac{12}{7}, \dfrac{24}{7}\right)$	$z = 3\left(\dfrac{12}{7}\right) + 5\left(\dfrac{24}{7}\right) = \dfrac{36}{7} + \dfrac{120}{7} = \dfrac{156}{7} \approx 22.3$

The minimum value is 3 at $(1, 0)$.

95.   Multiply each side of the first equation by -2 and eliminate x:

$$\begin{cases} 2x + 5y = 5 \quad \xrightarrow{\ -2\ } \quad -4x - 10y = -10 \\ 4x + 10y = A \quad \xrightarrow{\qquad} \quad \underline{\ 4x + 10y = \quad A\ } \\ \qquad\qquad\qquad\qquad\qquad 0 = A - 10 \end{cases}$$

If there are to be infinitely many solutions, the result of elimination should be $0 = 0$. Therefore, $A - 10 = 0$ or $A = 10$.

97. $y = ax^2 + bx + c$

At $(0, 1)$ the equation becomes:

$$1 = a(0)^2 + b(0) + c$$
$$c = 1$$

At $(1, 0)$ the equation becomes:

$$0 = a(1)^2 + b(1) + c$$
$$0 = a + b + c$$
$$a + b + c = 0$$

At $(-2, 1)$ the equation becomes:

$$1 = a(-2)^2 + b(-2) + c = 4a - 2b + c \Rightarrow 4a - 2b + c = 1$$

The system of equations is:

$$\begin{cases} a + b + c = 0 \\ 4a - 2b + c = 1 \\ \qquad\quad c = 1 \end{cases}$$

Substitute $c = 1$ into the first and second equations and simplify:

$$\begin{cases} a + b + 1 = 0 \quad \Rightarrow \quad a + b = -1 \quad \Rightarrow \quad a = -b - 1 \\ 4a - 2b + 1 = 1 \quad \Rightarrow \quad 4a - 2b = 0 \end{cases}$$

Solve the first equation for a, substitute into the second equation and solve:

$$4(-b-1)-2b=0 \Rightarrow -4b-4-2b=0$$

$$-6b=4 \Rightarrow b=-\frac{2}{3} \Rightarrow a=\frac{2}{3}-1=-\frac{1}{3}$$

The quadratic function is $y=-\frac{1}{3}x^2-\frac{2}{3}x+1$.

99. Let x = the number of pounds of coffee that costs \$3.00 per pound.
     Let y = the number of pounds of coffee that costs \$6.00 per pound.
     Then $x+y=100$ represents the total amount of coffee in the blend.
     The value of the blend will be represented by the equation: $3x+6y=3.90(100)$.
     Solve the system of equations:

     $$\begin{cases} x+y=100 \Rightarrow y=100-x \\ 3x+6y=390 \end{cases}$$

     Solve by substitution:

     $$3x+6(100-x)=390 \Rightarrow 3x+600-6x=390$$

     $$-3x=-210 \Rightarrow x=70$$

     $$y=100-70=30$$

     The blend is made up of 70 pounds of the \$3-per pound coffee and 30 pounds of the \$6-per pound coffee.

101. Let x = the number of small boxes.
     Let y = the number of medium boxes.
     Let z = the number of large boxes.
     Oatmeal raisin equation: $x+2y+2z=15$
     Chocolate chip equation: $x+y+2z=10$
     Shortbread equation: $y+3z=11$
     Multiply each side of the second equation by -1 and add to the first equation to eliminate x:

     $$\begin{cases} x+2y+2z=15 \\ x+y+2z=10 \\ y+3z=11 \end{cases} \quad \xrightarrow{} \quad \xrightarrow{-1} \quad \begin{array}{rcl} x+2y+2z &=& 15 \\ -x-y-2z &=& -10 \\ \hline y &=& 5 \end{array}$$

     Substituting and solving for the other variables:

     $$\begin{array}{ll} 5+3z=11 & x+5+2(2)=10 \\ 3z=6 & x+9=10 \\ z=2 & x=1 \end{array}$$

     1 small box, 5 medium boxes, and 2 large boxes of cookies should be purchased.

103. Let x = the speed of the boat in still water.
     Let y = the speed of the river current.
     The distance from Chiritza to the Flotel Orellana is 100 kilometers.

	Rate	Time	Distance
trip downstream	$x+y$	$5/2$	100
trip downstream	$x-y$	3	100

The system of equations is:

$$\begin{cases} (x+y)(5/2) = 100 & \Rightarrow \quad 5x + 5y = 200 \\ (x-y)(3) = d & \Rightarrow \qquad 3x - 3y = 100 \end{cases}$$

$$5x + 5y = 200 \xrightarrow{\;3\;} 15x + 15y = 600$$

$$3x - 3y = 100 \xrightarrow{\;5\;} \underline{15x - 15y = 500}$$

$$30x = 1100$$

$$x = \frac{1100}{30} = \frac{110}{3}$$

$$\Rightarrow 3(110/3) - 3y = 100 \Rightarrow 110 - 3y = 100 \Rightarrow 10 = 3y \Rightarrow y = \frac{10}{3}$$

The speed of the boat $= 110/3 \approx 36.67$ km/hr; the speed of the current $= 10/3 \approx 3.33$ km/hr.

105. Let x = the number of hours for Bruce to do the job alone.
   Let y = the number of hours for Bryce to do the job alone.
   Let z = the number of hours for Marty to do the job alone.
   Then $1/x$ represents the fraction of the job that Bruce does in one hour.
   $\qquad$ $1/y$ represents the fraction of the job that Bryce does in one hour.
   $\qquad$ $1/z$ represents the fraction of the job that Marty does in one hour.
   The equation representing Bruce and Bryce working together is:

$$\frac{1}{x} + \frac{1}{y} = \frac{1}{(4/3)} = \frac{3}{4} = 0.75$$

The equation representing Bryce and Marty working together is:

$$\frac{1}{y} + \frac{1}{z} = \frac{1}{(8/5)} = \frac{5}{8} = 0.625$$

The equation representing Bruce and Marty working together is:

$$\frac{1}{x} + \frac{1}{z} = \frac{1}{(8/3)} = \frac{3}{8} = 0.375$$

Solve the system of equations: $\qquad\qquad$ Let $u = x^{-1},\ v = y^{-1},\ w = z^{-1}$

$$\begin{cases} x^{-1} + y^{-1} = 0.75 \\ y^{-1} + z^{-1} = 0.625 \\ x^{-1} + z^{-1} = 0.375 \end{cases} \qquad \begin{cases} u + v = 0.75 & \Rightarrow \quad u = 0.75 - v \\ v + w = 0.625 & \Rightarrow \quad w = 0.625 - v \\ u + w = 0.375 \end{cases}$$

Substitute into the third equation and solve:

$$0.75 - v + 0.625 - v = 0.375 \Rightarrow -2v = -1 \Rightarrow v = 0.5$$

$$u = 0.75 - 0.5 = 0.25$$

$$w = 0.625 - 0.5 = 0.125$$

Solve for x, y, and z: $\quad x = 4,\ y = 2,\ z = 8$ (reciprocals)
Bruce can do the job in 4 hours, Bryce in 2 hours, and Marty in 8 hours.

107. Let x = the number of gasoline engines produced each week.
Let y = the number of diesel engines produced each week.
The total cost is: $C = 450x + 550y$.
Cost is to be minimized; thus, this is the objective function.
The constraints are:

$20 \leq x \leq 60$ number of gasoline engines needed and capacity each week.
$15 \leq y \leq 40$ number of diesel engines needed and capacity each week.
$x + y \geq 50$ number of engines produced to prevent layoffs.

Graph the constraints.

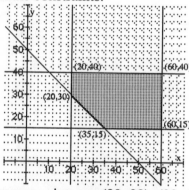

The corner points are (20, 30), (20, 40), (35, 15), (60, 15), (60, 40)
Evaluate the objective function:

Vertex	Value of $C = 450x + 550y$
(20, 30)	$C = 450(20) + 550(30) = 25,500$
(20, 40)	$C = 450(20) + 550(40) = 31,000$
(35, 15)	$C = 450(35) + 550(15) = 24,000$
(60, 15)	$C = 450(60) + 550(15) = 35,250$
(60, 40)	$C = 450(60) + 550(40) = 49,000$

The minimum cost is $24,000, when 35 gasoline engines and 15 diesel engines are produced.
The excess capacity is 15 gasoline engines, since only 20 gasoline engines had to be delivered.

Systems of Equations and Inequalities

10.CR   **Cumulative Review**

1. $2x^2 - x = 0$

    $x(2x - 1) = 0$

    $x = 0$ or $2x - 1 = 0$

    $\qquad\qquad 2x = 1$

    $\qquad\qquad x = \dfrac{1}{2}$

    The solution set is $\left\{ 0, \dfrac{1}{2} \right\}$.

3. $2x^3 - 3x^2 - 8x - 3 = 0$

    The graph of $y_1 = 2x^3 - 3x^2 - 8x - 3$
    appears to have an x-intercept at $x = 3$.

    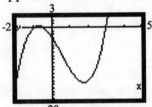

    Using synthetic division:

    $$
    \begin{array}{r|rrrr}
    3 & 2 & -3 & -8 & -3 \\
      &   & 6 & 9 & 3 \\
    \hline
      & 2 & 3 & 1 & 0
    \end{array}
    $$

    Therefore,

    $2x^3 - 3x^2 - 8x - 3 = 0$

    $(x - 3)(2x^2 + 3x + 1) = 0$

    $(x - 3)(2x + 1)(x + 1) = 0$

    $x = 3$ or $x = -\dfrac{1}{2}$ or $x = -1$

    The solution set is $\left\{ -1, -\dfrac{1}{2}, 3 \right\}$.

5. $\log_3(x - 1) + \log_3(2x + 1) = 2$

    $\log_3\big((x - 1)(2x + 1)\big) = 2$

    $\qquad (x - 1)(2x + 1) = 3^2$

    $\qquad\qquad 2x^2 - x - 1 = 9$

    $\qquad\qquad 2x^2 - x - 10 = 0$

    $\qquad\qquad (2x - 5)(x + 2) = 0$

    $\qquad\qquad x = \dfrac{5}{2}$ or $x = -2$

    Since $x = -2$ makes the original
    logarithms undefined, the solution
    set is $\left\{ \dfrac{5}{2} \right\}$.

7. $g(x) = \dfrac{2x^3}{x^4 + 1}$

$g(-x) = \dfrac{2(-x)^3}{(-x)^4 + 1} = \dfrac{-2x^3}{x^4 + 1} = -g(x),$

therefore g is an odd function and its graph is symmetric with respect to the origin.

9. $f(x) = 3^{x-2} + 1$

Using the graph of $y = 3^x$, shift the graph horizontally 2 units to the right, then shift the graph vertically upward 1 unit.
Domain: $(-\infty, \infty)$
Range: $(1, \infty)$
Horizontal Asymptote: $y = 1$

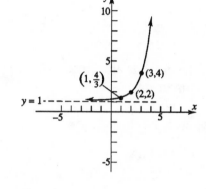

11. (a) $y = 3x + 6$

The graph is a line.
x-intercept:
  $0 = 3x + 6$

$3x = -6$

$x = -2$
y-intercept: $y = 3(0) + 6 = 6$

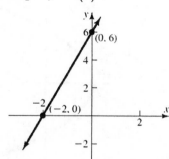

(b) $x^2 + y^2 = 4$

The graph is a circle.
Center: $(0, 0)$; Radius: 2

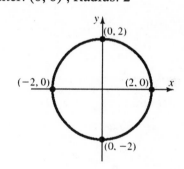

(c) $x^2 - y^2 = 4$
The graph is a hyperbola.
x-intercepts:
$$x^2 - 0^2 = 4$$
$$x^2 = 4$$
$$x = -2, x = 2$$
y-intercepts:

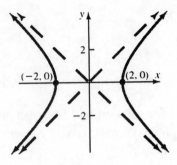

(d) $9x^2 + y^2 = 9$
$$x^2 + \frac{y^2}{9} = 1$$
The graph is an ellipse.
x-intercepts:
$$9x^2 + 0^2 = 9$$
$$9x^2 = 9$$
$$x^2 = 1$$
$$x = -1, x = 1$$
y-intercepts:
$$9(0)^2 + y^2 = 9$$
$$y^2 = 9$$
$$y^2 = 9$$
$$y = -3, y = 3$$

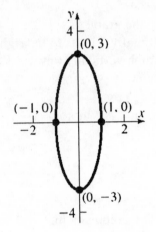

(e) $y^2 = 3x + 6$
The graph is a parabola.
x-intercept:
$$0^2 = 3x + 6$$
$$-3x = 6$$
$$x = -2$$
y-intercepts:
$$y^2 = 3(0) + 6$$
$$y^2 = 6$$
$$y = -\sqrt{6}, y = \sqrt{6}$$

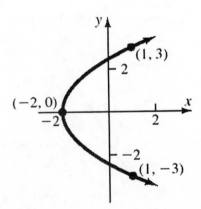

(f) $y = x^3$

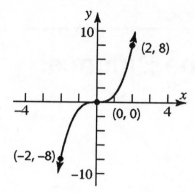

(g) $y = \dfrac{1}{x}$

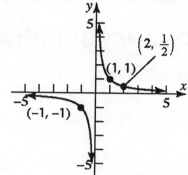

(h) $y = \sqrt{x}$

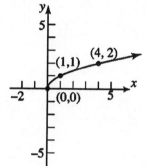

(i) $y = e^x$

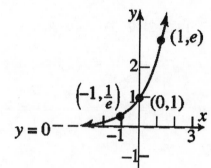

(j) $y = \ln x$

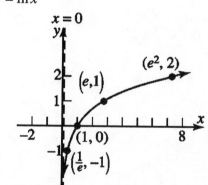

(k) $y = \sin x$

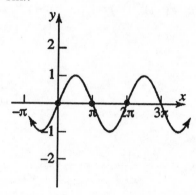

(l) $y = \cos x$

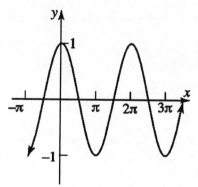

693

Chapter 11

Sequences; Induction; the Binomial Theorem

11.1 Sequences

1. $f(2) = \dfrac{2-1}{2} = \dfrac{1}{2}$; $f(3) = \dfrac{3-1}{3} = \dfrac{2}{3}$

3. sequence

5. $2 \cdot 1 + 2 \cdot 2 + 2 \cdot 3 + 2 \cdot 4 = 20$

7. True

9. $a_1 = 1$, $a_2 = 2$, $a_3 = 3$, $a_4 = 4$, $a_5 = 5$

11. $a_1 = \dfrac{1}{1+2} = \dfrac{1}{3}$, $a_2 = \dfrac{2}{2+2} = \dfrac{2}{4} = \dfrac{1}{2}$, $a_3 = \dfrac{3}{3+2} = \dfrac{3}{5}$, $a_4 = \dfrac{4}{4+2} = \dfrac{4}{6} = \dfrac{2}{3}$,

$a_5 = \dfrac{5}{5+2} = \dfrac{5}{7}$

13. $a_1 = (-1)^{1+1}(1^2) = 1$, $a_2 = (-1)^{2+1}(2^2) = -4$, $a_3 = (-1)^{3+1}(3^2) = 9$,

$a_4 = (-1)^{4+1}(4^2) = -16$, $a_5 = (-1)^{5+1}(5^2) = 25$

15. $a_1 = \dfrac{2^1}{3^1+1} = \dfrac{2}{4} = \dfrac{1}{2}$, $a_2 = \dfrac{2^2}{3^2+1} = \dfrac{4}{10} = \dfrac{2}{5}$, $a_3 = \dfrac{2^3}{3^3+1} = \dfrac{8}{28} = \dfrac{2}{7}$,

$a_4 = \dfrac{2^4}{3^4+1} = \dfrac{16}{82} = \dfrac{8}{41}$, $a_5 = \dfrac{2^5}{3^5+1} = \dfrac{32}{244} = \dfrac{8}{61}$

17. $a_1 = \dfrac{(-1)^1}{(1+1)(1+2)} = \dfrac{-1}{2 \cdot 3} = -\dfrac{1}{6}$, $a_2 = \dfrac{(-1)^2}{(2+1)(2+2)} = \dfrac{1}{3 \cdot 4} = \dfrac{1}{12}$,

$a_3 = \dfrac{(-1)^3}{(3+1)(3+2)} = \dfrac{-1}{4 \cdot 5} = -\dfrac{1}{20}$, $a_4 = \dfrac{(-1)^4}{(4+1)(4+2)} = \dfrac{1}{5 \cdot 6} = \dfrac{1}{30}$,

$a_5 = \dfrac{(-1)^5}{(5+1)(5+2)} = \dfrac{-1}{6 \cdot 7} = -\dfrac{1}{42}$

19. $a_1 = \dfrac{1}{e^1} = \dfrac{1}{e}$, $a_2 = \dfrac{2}{e^2}$, $a_3 = \dfrac{3}{e^3}$, $a_4 = \dfrac{4}{e^4}$, $a_5 = \dfrac{5}{e^5}$

21. $\dfrac{n}{n+1}$

23. $\dfrac{1}{2^{n-1}}$

25. $(-1)^{n+1}$

27. $(-1)^{n+1} n$

29. $a_1 = 2,\ a_2 = 3 + 2 = 5,\ a_3 = 3 + 5 = 8,\ a_4 = 3 + 8 = 11,\ a_5 = 3 + 11 = 14$

31. $a_1 = -2,\ a_2 = 2 + (-2) = 0,\ a_3 = 3 + 0 = 3,\ a_4 = 4 + 3 = 7,\ a_5 = 5 + 7 = 12$

33. $a_1 = 5,\ a_2 = 2 \cdot 5 = 10,\ a_3 = 2 \cdot 10 = 20,\ a_4 = 2 \cdot 20 = 40,\ a_5 = 2 \cdot 40 = 80$

35. $a_1 = 3,\ a_2 = \dfrac{3}{2},\ a_3 = \dfrac{\frac{3}{2}}{3} = \dfrac{1}{2},\ a_4 = \dfrac{\frac{1}{2}}{4} = \dfrac{1}{8},\ a_5 = \dfrac{\frac{1}{8}}{5} = \dfrac{1}{40}$

37. $a_1 = 1,\ a_2 = 2,\ a_3 = 2 \cdot 1 = 2,\ a_4 = 2 \cdot 2 = 4,\ a_5 = 4 \cdot 2 = 8$

39. $a_1 = A,\ a_2 = A + d,\ a_3 = (A + d) + d = A + 2d,\ a_4 = (A + 2d) + d = A + 3d,$
$$a_5 = (A + 3d) + d = A + 4d$$

41. $a_1 = \sqrt{2},\ a_2 = \sqrt{2 + \sqrt{2}},\ a_3 = \sqrt{2 + \sqrt{2 + \sqrt{2}}},\ a_4 = \sqrt{2 + \sqrt{2 + \sqrt{2 + \sqrt{2}}}},$
$$a_5 = \sqrt{2 + \sqrt{2 + \sqrt{2 + \sqrt{2 + \sqrt{2}}}}}$$

43. $\displaystyle\sum_{k=1}^{10} 5 = \underbrace{5 + 5 + 5 + \cdots + 5}_{10\ \text{times}} = 50$

45. $\displaystyle\sum_{k=1}^{6} k = 1 + 2 + 3 + 4 + 5 + 6 = 21$

47. $\displaystyle\sum_{k=1}^{5} (5k + 3) = (5 \cdot 1 + 3) + (5 \cdot 2 + 3) + (5 \cdot 3 + 3) + (5 \cdot 4 + 3) + (5 \cdot 5 + 3) = 8 + 13 + 18 + 23 + 28 = 90$

49. $\displaystyle\sum_{k=1}^{3} \left(k^2 + 4\right) = \left(1^2 + 4\right) + \left(2^2 + 4\right) + \left(3^2 + 4\right) = 5 + 8 + 13 = 26$

51. $\displaystyle\sum_{k=1}^{6} (-1)^k 2^k = (-1)^1 \cdot 2^1 + (-1)^2 \cdot 2^2 + (-1)^3 \cdot 2^3 + (-1)^4 \cdot 2^4 + (-1)^5 \cdot 2^5 + (-1)^6 \cdot 2^6$
$$= -2 + 4 - 8 + 16 - 32 + 64 = 42$$

53. $\displaystyle\sum_{k=1}^{4} \left(k^3 - 1\right) = \left(1^3 - 1\right) + \left(2^3 - 1\right) + \left(3^3 - 1\right) + \left(4^3 - 1\right) = 0 + 7 + 26 + 63 = 96$

55. $\displaystyle\sum_{k=1}^{n} (k + 2) = 3 + 4 + 5 + 6 + 7 + \cdots + (n + 2)$

57. $\displaystyle\sum_{k=1}^{n} \dfrac{k^2}{2} = \dfrac{1}{2} + 2 + \dfrac{9}{2} + 8 + \dfrac{25}{2} + 18 + \dfrac{49}{2} + 32 + \cdots + \dfrac{n^2}{2}$

59. $\displaystyle\sum_{k=0}^{n} \dfrac{1}{3^k} = 1 + \dfrac{1}{3} + \dfrac{1}{9} + \dfrac{1}{27} + \cdots + \dfrac{1}{3^n}$

61. $\displaystyle\sum_{k=0}^{n-1} \dfrac{1}{3^{k+1}} = \dfrac{1}{3} + \dfrac{1}{9} + \dfrac{1}{27} + \cdots + \dfrac{1}{3^n}$

63. $\displaystyle\sum_{k=2}^{n}(-1)^k \ln k = \ln 2 - \ln 3 + \ln 4 - \ln 5 + \cdots + (-1)^n \ln n$

65. $1 + 2 + 3 + \cdots + 20 = \displaystyle\sum_{k=1}^{20} k$

67. $\dfrac{1}{2} + \dfrac{2}{3} + \dfrac{3}{4} + \cdots + \dfrac{13}{13+1} = \displaystyle\sum_{k=1}^{13} \dfrac{k}{k+1}$

69. $1 - \dfrac{1}{3} + \dfrac{1}{9} - \dfrac{1}{27} + \cdots + (-1)^6 \left(\dfrac{1}{3^6}\right) = \displaystyle\sum_{k=0}^{6}(-1)^k \left(\dfrac{1}{3^k}\right)$

71. $3 + \dfrac{3^2}{2} + \dfrac{3^3}{3} + \cdots + \dfrac{3^n}{n} = \displaystyle\sum_{k=1}^{n} \dfrac{3^k}{k}$

73. $a + (a+d) + (a+2d) + \cdots + (a+nd) = \displaystyle\sum_{k=0}^{n}(a+kd)$

75. $B_1 = 1.01B_0 - 100 = 1.01(3000) - 100 = \$2930$

77. $p_1 = 1.03p_0 + 20 = 1.03(2000) + 20 = 2080$
    $p_2 = 1.03p_1 + 20 = 1.03(2080) + 20 = 2162.4$
    There are approximately 2162 trout in the pond after 2 months.

79. $a_1 = 1,\ a_2 = 1,\ a_3 = 2,\ a_4 = 3,\ a_5 = 5,\ a_6 = 8,\ a_7 = 13,\ a_8 = 21,\ a_n = a_{n-1} + a_{n-2}$
    $a_8 = a_7 + a_6 = 13 + 8 = 21$
    After 7 months there are 21 pairs of mature rabbits.

81. 1, 1, 2, 3, 5, 8, 13   This is the Fibonacci sequence.

83. To show that $1 + 2 + 3 + \cdots + (n-1) + n = \dfrac{n(n+1)}{2}$

    Let $S = 1 + 2 + 3 + \cdots + (n-1) + n$, we can reverse the order to get
    $S = n + (n-1) + (n-2) + \cdots + 2 + 1$. Now add these two lines to get
    $S = 1 + 2 + 3 + \cdots + (n-1) + n$

    $\underline{+S = n + (n-1) + (n-2) + \cdots + 2 + 1}$

    $2S = [1+n] + [2+(n-1)] + [3+(n-2)] + \cdots + [(n-1)+2] + [n+1]$

    n terms

    So we have
    $2S = [1+n] + [1+n] + [1+n] + \cdots + [n+1] + [n+1] = n \cdot [n+1]$

    n terms

    Therefore, $S = \dfrac{n \cdot (n+1)}{2}$.

Chapter 11

Sequences; Induction; the Binomial Theorem

11.2 Arithmetic Sequences

1. arithmetic

3. $d = a_{n+1} - a_n = (n+1+4) - (n+4) = n+5-n-4 = 1$
   $a_1 = 1+4 = 5, \ a_2 = 2+4 = 6, \ a_3 = 3+4 = 7, \ a_4 = 4+4 = 8$

5. $d = a_{n+1} - a_n = (2(n+1)-5) - (2n-5) = 2n+2-5-2n+5 = 2$
   $a_1 = 2 \cdot 1 - 5 = -3, \ a_2 = 2 \cdot 2 - 5 = -1, \ a_3 = 2 \cdot 3 - 5 = 1, \ a_4 = 2 \cdot 4 - 5 = 3$

7. $d = a_{n+1} - a_n = (6 - 2(n+1)) - (6 - 2n) = 6 - 2n - 2 - 6 + 2n = -2$
   $a_1 = 6 - 2 \cdot 1 = 4, \ a_2 = 6 - 2 \cdot 2 = 2, \ a_3 = 6 - 2 \cdot 3 = 0, \ a_4 = 6 - 2 \cdot 4 = -2$

9. $d = a_{n+1} - a_n = \left(\dfrac{1}{2} - \dfrac{1}{3}(n+1)\right) - \left(\dfrac{1}{2} - \dfrac{1}{3}n\right) = \dfrac{1}{2} - \dfrac{1}{3}n - \dfrac{1}{3} - \dfrac{1}{2} + \dfrac{1}{3}n = -\dfrac{1}{3}$
   $a_1 = \dfrac{1}{2} - \dfrac{1}{3} \cdot 1 = \dfrac{1}{6}, \ a_2 = \dfrac{1}{2} - \dfrac{1}{3} \cdot 2 = -\dfrac{1}{6}, \ a_3 = \dfrac{1}{2} - \dfrac{1}{3} \cdot 3 = -\dfrac{1}{2}, \ a_4 = \dfrac{1}{2} - \dfrac{1}{3} \cdot 4 = -\dfrac{5}{6}$

11. $d = a_{n+1} - a_n = \ln\left(3^{n+1}\right) - \ln\left(3^n\right) = (n+1)\ln(3) - n\ln(3) = \ln 3(n+1-n) = \ln(3)$
    $a_1 = \ln\left(3^1\right) = \ln(3), \ a_2 = \ln\left(3^2\right) = 2\ln(3), \ a_3 = \ln\left(3^3\right) = 3\ln(3), \ a_4 = \ln\left(3^4\right) = 4\ln(3)$

13. $a_n = a + (n-1)d = 2 + (n-1)3 = 2 + 3n - 3 = 3n - 1$
    $a_5 = 3 \cdot 5 - 1 = 14$

15. $a_n = a + (n-1)d = 5 + (n-1)(-3) = 5 - 3n + 3 = 8 - 3n$
    $a_5 = 8 - 3 \cdot 5 = -7$

17. $a_n = a + (n-1)d = 0 + (n-1)\dfrac{1}{2} = \dfrac{1}{2}n - \dfrac{1}{2} = \dfrac{1}{2}(n-1)$
    $a_5 = \dfrac{1}{2}\left(5 - \dfrac{1}{2}\right) = 2$

19. $a_n = a + (n-1)d = \sqrt{2} + (n-1)\sqrt{2} = \sqrt{2} + \sqrt{2}n - \sqrt{2} = \sqrt{2}n$
    $a_5 = 5\sqrt{2}$

21. $a_1 = 2, \ d = 2, \quad a_n = a + (n-1)d$
$a_{12} = 2 + (12-1)2 = 2 + 11(2) = 2 + 22 = 24$

23. $a_1 = 1, \ d = -2 - 1 = -3, \quad a_n = a + (n-1)d$
$a_{10} = 1 + (10-1)(-3) = 1 + 9(-3) = 1 - 27 = -26$

25. $a_1 = a, \ d = (a+b) - a = b, \quad a_n = a + (n-1)d$
$a_8 = a + (8-1)b = a + 7b$

27. $a_8 = a + 7d = 8 \qquad a_{20} = a + 19d = 44$
Solve the system of equations by subtracting the first equation from the second:
$\qquad 12d = 36 \Rightarrow d = 3$
$\qquad a = 8 - 7(3) = 8 - 21 = -13$
Recursive formula: $\quad a_1 = -13 \qquad a_n = a_{n-1} + 3$

29. $a_9 = a + 8d = -5 \qquad a_{15} = a + 14d = 31$
Solve the system of equations by subtracting the first equation from the second:
$\qquad 6d = 36 \Rightarrow d = 6$
$\qquad a = -5 - 8(6) = -5 - 48 = -53$
Recursive formula: $\quad a_1 = -53 \qquad a_n = a_{n-1} + 6$

31. $a_{15} = a + 14d = 0 \qquad a_{40} = a + 39d = -50$
Solve the system of equations by subtracting the first equation from the second:
$\qquad 25d = -50 \Rightarrow d = -2$
$\qquad a = -14(-2) = 28$
Recursive formula: $\quad a_1 = 28 \qquad a_n = a_{n-1} - 2$

33. $a_{14} = a + 13d = -1 \qquad a_{18} = a + 17d = -9$
Solve the system of equations by subtracting the first equation from the second:
$\qquad 4d = -8 \Rightarrow d = -2$
$\qquad a = -1 - 13(-2) = -1 + 26 = 25$
Recursive formula: $\quad a_1 = 25 \qquad a_n = a_{n-1} - 2$

35. $S_n = \dfrac{n}{2}(a + a_n) = \dfrac{n}{2}\big(1 + (2n-1)\big) = \dfrac{n}{2}(2n) = n^2$

37. $S_n = \dfrac{n}{2}(a + a_n) = \dfrac{n}{2}\big(7 + (2 + 5n)\big) = \dfrac{n}{2}(9 + 5n)$

39. $a_1 = 2, \ d = 4 - 2 = 2, \quad a_n = a + (n-1)d$
$\qquad 70 = 2 + (n-1)2 \Rightarrow 70 = 2 + 2n - 2 \Rightarrow 70 = 2n \Rightarrow n = 35$
$\qquad S_n = \dfrac{n}{2}(a + a_n) = \dfrac{35}{2}(2 + 70) = \dfrac{35}{2}(72) = 35(36) = 1260$

41. $a_1 = 5,\ d = 9 - 5 = 4,\ \ a_n = a + (n-1)d$

$49 = 5 + (n-1)4 \Rightarrow 49 = 5 + 4n - 4 \Rightarrow 48 = 4n \Rightarrow n = 12$

$S_n = \dfrac{n}{2}(a + a_n) = \dfrac{12}{2}(5 + 49) = 6(54) = 324$

43.   Using the sum of the sequence feature:

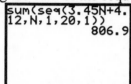

45. $d = 5.2 - 2.8 = 2.4$

$a = 2.8$

$36.4 = 2.8 + (n-1)2.4$

$36.4 = 2.8 + 2.4n - 2.4$

$36 = 2.4n$

$n = 15$

$a_n = 2.8 + (n-1)2.4 = 2.8 + 2.4n - 2.4 = 2.4n + 0.4$

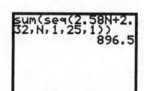

47. $d = 7.48 - 4.9 = 2.58$

$a = 4.9$

$66.82 = 4.9 + (n-1)2.58$

$66.82 = 4.9 + 2.58n - 2.58$

$64.5 = 2.58n$

$n = 25$

$a_n = 4.9 + (n-1)2.58 = 4.9 + 2.58n - 2.58$

$a_n = 2.58n + 2.32$

49.   Find the common difference of the terms and solve the system of equations:

$(2x + 1) - (x + 3) = d \ \Rightarrow\ x - 2 = d$

$(5x + 2) - (2x + 1) = d \ \Rightarrow\ 3x + 1 = d$

$3x + 1 = x - 2$

$2x = -3$

$x = -1.5$

51.   The total number of seats is: $S = 25 + 26 + 27 + \cdots + \big(25 + 29(1)\big)$

This is the sum of an arithmetic sequence with $d = 1,\ a = 25,$ and $n = 30$.

Find the sum of the sequence: $S_{30} = \dfrac{30}{2}[2(25) + (30 - 1)(1)] = 15(50 + 29) = 15(79) = 1185$

There are 1185 seats in the theater.

53. The lighter colored tiles have 20 tiles in the bottom row and 1 tile in the top row. The number decreases by 1 as we move up the triangle. This is an arithmetic sequence with $a_1 = 20$, $d = -1$, and $n = 20$. Find the sum:

$S = \dfrac{20}{2}[2(20) + (20 - 1)(-1)] = 10(40 - 19) = 10(21) = 210$ lighter tiles.

The darker colored tiles have 19 tiles in the bottom row and 1 tile in the top row. The number decreases by 1 as we move up the triangle. This is an arithmetic sequence with $a_1 = 19$, $d = -1$, and $n = 19$. Find the sum:

$S = \dfrac{19}{2}[2(19) + (19 - 1)(-1)] = \dfrac{19}{2}(38 - 18) = \dfrac{19}{2}(20) = 190$ darker tiles.

55. Find n in an arithmetic sequence with $a_1 = 10$, $d = 4$, $S_n = 2040$.

$$S_n = \frac{n}{2}[2a_1 + (n-1)d]$$

$$2040 = \frac{n}{2}[2(10) + (n-1)4]$$

$$4080 = n[20 + 4n - 4]$$

$$4080 = n(4n + 16)$$

$$4080 = 4n^2 + 16n$$

$$1020 = n^2 + 4n$$

$$n^2 + 4n - 1020 = 0$$

$$(n + 34)(n - 30) = 0$$

$$n = -34 \text{ or } n = 30$$

There are 30 rows in the corner section of the stadium.

57.   Answers will vary.

Sequences; Induction; the Binomial Theorem

11.3 Geometric Sequences; Geometric Series

1. $A = 1000\left(1 + \dfrac{0.04}{2}\right)^{2 \cdot 2} = \$1082.43$

3. geometric

5. annuity

7. False

9. $r = \dfrac{3^{n+1}}{3^n} = 3^{n+1-n} = 3$

    $a_1 = 3^1 = 3, \ a_2 = 3^2 = 9, \ a_3 = 3^3 = 27, \ a_4 = 3^4 = 81$

11. $r = \dfrac{-3\left(\dfrac{1}{2}\right)^{n+1}}{-3\left(\dfrac{1}{2}\right)^n} = \left(\dfrac{1}{2}\right)^{n+1-n} = \dfrac{1}{2}$

    $a_1 = -3\left(\dfrac{1}{2}\right)^1 = -\dfrac{3}{2}, \ a_2 = -3\left(\dfrac{1}{2}\right)^2 = -\dfrac{3}{4}, \ a_3 = -3\left(\dfrac{1}{2}\right)^3 = -\dfrac{3}{8}, \ a_4 = -3\left(\dfrac{1}{2}\right)^4 = -\dfrac{3}{16}$

13. $r = \dfrac{\dfrac{2^{n+1-1}}{4}}{\dfrac{2^{n-1}}{4}} = \dfrac{2^n}{2^{n-1}} = 2^{n-(n-1)} = 2$

    $a_1 = \dfrac{2^{1-1}}{4} = \dfrac{2^0}{2^2} = 2^{-2} = \dfrac{1}{4}, \ a_2 = \dfrac{2^{2-1}}{4} = \dfrac{2^1}{2^2} = 2^{-1} = \dfrac{1}{2}, \ a_3 = \dfrac{2^{3-1}}{4} = \dfrac{2^2}{2^2} = 1,$

        $a_4 = \dfrac{2^{4-1}}{4} = \dfrac{2^3}{2^2} = 2$

15. $r = \dfrac{2^{\left(\frac{n+1}{3}\right)}}{2^{\left(\frac{n}{3}\right)}} = 2^{\left(\frac{n+1}{3} - \frac{n}{3}\right)} = 2^{1/3}$

    $a_1 = 2^{1/3}, \ a_2 = 2^{2/3}, \ a_3 = 2^{3/3} = 2, \ a_4 = 2^{4/3}$

17. $r = \dfrac{\dfrac{3^{n+1-1}}{2^{n+1}}}{\dfrac{3^{n-1}}{2^n}} = \dfrac{3^n}{3^{n-1}} \cdot \dfrac{2^n}{2^{n+1}} = 3^{n-(n-1)} \cdot 2^{n-(n+1)} = 3 \cdot 2^{-1} = \dfrac{3}{2}$

$a_1 = \dfrac{3^{1-1}}{2^1} = \dfrac{3^0}{2} = \dfrac{1}{2}, \; a_2 = \dfrac{3^{2-1}}{2^2} = \dfrac{3^1}{2^2} = \dfrac{3}{4}, \; a_3 = \dfrac{3^{3-1}}{2^3} = \dfrac{3^2}{2^3} = \dfrac{9}{8},$

$a_4 = \dfrac{3^{4-1}}{2^4} = \dfrac{3^3}{2^4} = \dfrac{27}{16}$

19. $\{n+2\}$ Arithmetic
$d = (n+1+2) - (n+2) = n+3-n-2 = 1$

21. $\{4n^2\}$ Examine the terms of the sequence: 4, 16, 36, 64, 100, ...
There is no common difference; there is no common ratio; neither.

23. $\left\{3 - \dfrac{2}{3}n\right\}$ Arithmetic
$d = \left(3 - \dfrac{2}{3}(n+1)\right) - \left(3 - \dfrac{2}{3}n\right) = 3 - \dfrac{2}{3}n - \dfrac{2}{3} - 3 + \dfrac{2}{3}n = -\dfrac{2}{3}$

25. 1, 3, 6, 10, ...   Neither
There is no common difference or common ratio.

27. $\left\{\left(\dfrac{2}{3}\right)^n\right\}$ Geometric $r = \dfrac{\left(\dfrac{2}{3}\right)^{n+1}}{\left(\dfrac{2}{3}\right)^n} = \left(\dfrac{2}{3}\right)^{n+1-n} = \dfrac{2}{3}$

29. $-1, -2, -4, -8, ...$ Geometric $r = \dfrac{-2}{-1} = \dfrac{-4}{-2} = \dfrac{-8}{-4} = 2$

31. $\{3^{n/2}\}$ Geometric $r = \dfrac{3^{\left(\frac{n+1}{2}\right)}}{3^{\left(\frac{n}{2}\right)}} = 3^{\left(\frac{n+1}{2} - \frac{n}{2}\right)} = 3^{1/2}$

33. $a_5 = 2 \cdot 3^{5-1} = 2 \cdot 3^4 = 2 \cdot 81 = 162$ $a_n = 2 \cdot 3^{n-1}$

35. $a_5 = 5(-1)^{5-1} = 5(-1)^4 = 5 \cdot 1 = 5$ $a_n = 5 \cdot (-1)^{n-1}$

37. $a_5 = 0 \cdot \left(\dfrac{1}{2}\right)^{5-1} = 0 \cdot \left(\dfrac{1}{2}\right)^4 = 0$ $a_n = 0 \cdot \left(\dfrac{1}{2}\right)^{n-1} = 0$

39. $a_5 = \sqrt{2} \cdot \left(\sqrt{2}\right)^{5-1} = \sqrt{2} \cdot \left(\sqrt{2}\right)^4 = \sqrt{2} \cdot 4 = 4\sqrt{2}$ $a_n = \sqrt{2} \cdot \left(\sqrt{2}\right)^{n-1} = \left(\sqrt{2}\right)^n$

41. $a=1$, $r=\dfrac{1}{2}$, $n=7$ $a_7 = 1\cdot\left(\dfrac{1}{2}\right)^{7-1} = \left(\dfrac{1}{2}\right)^6 = \dfrac{1}{64}$

43. $a=1$, $r=-1$, $n=9$ $a_9 = 1\cdot(-1)^{9-1} = (-1)^8 = 1$

45. $a=0.4$, $r=0.1$, $n=8$ $a_8 = 0.4\cdot(0.1)^{8-1} = 0.4(0.1)^7 = 0.00000004$

47. $a=\dfrac{1}{4}$, $r=2$ $S_n = a\left(\dfrac{1-r^n}{1-r}\right) = \dfrac{1}{4}\left(\dfrac{1-2^n}{1-2}\right) = -\dfrac{1}{4}(1-2^n)$

49. $a=\dfrac{2}{3}$, $r=\dfrac{2}{3}$ $S_n = a\left(\dfrac{1-r^n}{1-r}\right) = \dfrac{2}{3}\left(\dfrac{1-\left(\dfrac{2}{3}\right)^n}{1-\dfrac{2}{3}}\right) = \dfrac{2}{3}\left(\dfrac{1-\left(\dfrac{2}{3}\right)^n}{\dfrac{1}{3}}\right) = 2\left(1-\left(\dfrac{2}{3}\right)^n\right)$

51. $a=-1$, $r=2$ $S_n = a\left(\dfrac{1-r^n}{1-r}\right) = -1\left(\dfrac{1-2^n}{1-2}\right) = 1-2^n$

53. Using the sum of the sequence feature:
```
sum(seq(2^N/4,N,
0,14,1))
            8191.75
```

55. Using the sum of the sequence feature:
```
sum(seq((2/3)^N,
N,1,15,1))
         1.995432683
```

57. Using the sum of the sequence feature:
```
sum(seq(-1*2^N,N
,0,14,1))
            -32767
```

59. $a=1$, $r=\dfrac{1}{3}$ Since $|r|<1$, $S = \dfrac{a}{1-r} = \dfrac{1}{1-\dfrac{1}{3}} = \dfrac{1}{\dfrac{2}{3}} = \dfrac{3}{2}$

61. $a=8$, $r=\dfrac{1}{2}$ Since $|r|<1$, $S = \dfrac{a}{1-r} = \dfrac{8}{1-\dfrac{1}{2}} = \dfrac{8}{\dfrac{1}{2}} = 16$

63. $a=2$, $r=-\dfrac{1}{4}$ Since $|r|<1$, $S = \dfrac{a}{1-r} = \dfrac{2}{1-\left(-\dfrac{1}{4}\right)} = \dfrac{2}{\dfrac{5}{4}} = \dfrac{8}{5}$

65. $a = 5$, $r = \dfrac{1}{4}$ Since $|r| < 1$, $S = \dfrac{a}{1-r} = \dfrac{5}{1 - \dfrac{1}{4}} = \dfrac{5}{\dfrac{3}{4}} = \dfrac{20}{3}$

67. $a = 6$, $r = -\dfrac{2}{3}$ Since $|r| < 1$, $S = \dfrac{a}{1-r} = \dfrac{6}{1 - \left(-\dfrac{2}{3}\right)} = \dfrac{6}{\dfrac{5}{3}} = \dfrac{18}{5}$

69. Find the common ratio of the terms and
     solve the system of equations:
     $$\dfrac{x+2}{x} = r, \quad \dfrac{x+3}{x+2} = r$$
     $$\dfrac{x+2}{x} = \dfrac{x+3}{x+2}$$
     $$x^2 + 4x + 4 = x^2 + 3x$$
     $$x = -4$$

71. This is a geometric series with $a = \$18{,}000$, $r = 1.05$, $n = 5$. Find the 5th term:
     $$a_5 = 18{,}000(1.05)^{5-1} = 18{,}000(1.05)^4 \approx \$21{,}879.11$$

73. (a) Find the 10th term of the geometric sequence:
         $a = 2$, $r = 0.9$, $n = 10$ $a_{10} = 2(0.9)^{10-1} = 2(0.9)^9 \approx 0.775$ feet
     (b) Find n when $a_n < 1$:
         $$2(0.9)^{n-1} < 1$$
         $$(0.9)^{n-1} < 0.5$$
         $$(n-1)\log(0.9) < \log(0.5)$$
         $$n - 1 > \dfrac{\log(0.5)}{\log(0.9)}$$
         $$n > \dfrac{\log(0.5)}{\log(0.9)} + 1 \approx 7.58$$
         On the 8th swing the arc is less than 1 foot.
     (c) Find the sum of the first 15 swings:
         $$S_{15} = 2\left(\dfrac{1 - (0.9)^{15}}{1 - 0.9}\right) = 2\left(\dfrac{1 - (0.9)^{15}}{0.1}\right) = 20\left(1 - (0.9)^{15}\right) \approx 15.88 \text{ feet}$$
     (d) Find the sum of the geometric series:
         $$S = \dfrac{2}{1 - 0.9} = \dfrac{2}{0.1} = 20 \text{ feet}$$

75. This is an ordinary annuity with $P = \$100$ and $n = (12)(30) = 360$ payment periods.

The interest rate per period is $\dfrac{0.12}{12} = 0.01$. Thus,

$$A = 100\left[\dfrac{[1+0.01]^{360}-1}{0.01}\right] \approx \$349,496.41$$

77. This is an ordinary annuity with $P = \$500$ and $n = (4)(20) = 80$ payment periods.

The interest rate per period is $\dfrac{0.08}{4} = 0.02$. Thus,

$$A = 500\left[\dfrac{[1+0.02]^{80}-1}{0.02}\right] \approx \$96,885.98$$

79. This is an ordinary annuity with $A = \$50,000$ and $n = (12)(10) = 120$ payment

periods. The interest rate per period is $\dfrac{0.06}{12} = 0.005$. Thus,

$$50,000 = P\left[\dfrac{[1+0.005]^{120}-1}{0.005}\right]$$

$$P = 50,000\left[\dfrac{0.005}{[1+0.005]^{120}-1}\right] \approx \$305.10$$

81. Both options are geometric sequences:
Option A: $a = \$20,000$; $r = 1.06$; $n = 5$

$a_5 = 20,000(1.06)^{5-1} = 20,000(1.06)^4 \approx \$25,250$

$$S_5 = 20,000\left(\dfrac{1-(1.06)^5}{1-1.06}\right) \approx \$112,742$$

Option B: $a = \$22,000$; $r = 1.03$; $n = 5$

$a_5 = 22,000(1.03)^{5-1} = 22,000(1.03)^4 \approx \$24,761$

$$S_5 = 22,000\left(\dfrac{1-(1.03)^5}{1-1.03}\right) \approx \$116,801$$

Option A provides more money in the 5th year, while Option B provides the greatest total amount of money over the 5-year period.

83. Option 1: Total Salary $= \$2,000,000(7) + \$100,000(7) = \$14,700,000$

Option 2: Geometric series with: $a = \$2,000,000$, $r = 1.045$, $n = 7$
Find the sum of the geometric series:

$$S = 2,000,000\left(\dfrac{1-(1.045)^7}{1-1.045}\right) \approx \$16,038,304$$

Option 3:   Arithmetic series with: $a = \$2,000,000$, $d = \$95,000$, $n = 7$

Find the sum of the arithmetic series:

$$S_7 = \frac{7}{2}(2(2,000,000) + (7-1)(95,000)) = \$15,995,000$$

Option 2 provides the most money; Option 1 provides the least money.

85.   This is a geometric sequence with $a = 1$, $r = 2$, $n = 64$.

Find the sum of the geometric series:

$$S_{64} = 1\left(\frac{1-2^{64}}{1-2}\right) = \frac{1-2^{64}}{-1} = 2^{64} - 1 \approx 1.845 \times 10^{19} \text{ grains}$$

87.   The common ratio, $r = 0.90 < 1$. The sum is: $S = \dfrac{1}{1-0.9} = \dfrac{1}{0.10} = 10$.

The multiplier is 10.

89.   This is an infinite geometric series with $a = 4$, and $r = \dfrac{1.03}{1.09}$.

Find the sum: Price $= \dfrac{4}{1 - \dfrac{1.03}{1.09}} \approx \$72.67$.

91.   Yes, a sequence can be both arithmetic and geometric. For example, the constant sequence $3, 3, 3, 3, \ldots$ can be viewed as an arithmetic sequence with $a = 3$ and $d = 0$. Alternatively, the same sequence can be viewed as a geometric sequence with $a = 3$ and $r = 1$.

93.   Answers will vary.

Chapter 11

Sequences; Induction; the Binomial Theorem

11.4 Mathematical Induction

1. **I:** $n = 1$: $2 \cdot 1 = 2$ and $1(1+1) = 2$

   II: If $2 + 4 + 6 + \cdots + 2k = k(k+1)$

   then $2 + 4 + 6 + \cdots + 2k + 2(k+1)$

   $$= \left[2 + 4 + 6 + \cdots + 2k\right] + 2(k+1) = k(k+1) + 2(k+1)$$
   $$= (k+1)(k+2)$$

   Conditions I and II are satisfied; the statement is true.

3. **I:** $n = 1$: $1 + 2 = 3$ and $\dfrac{1}{2} \cdot 1(1+5) = 3$

   II: If $3 + 4 + 5 + \cdots + (k+2) = \dfrac{1}{2} \cdot k(k+5)$

   then $3 + 4 + 5 + \cdots + (k+2) + [(k+1) + 2]$

   $$= \left[3 + 4 + 5 + \cdots + (k+2)\right] + (k+3) = \frac{1}{2} \cdot k(k+5) + (k+3)$$
   $$= \frac{1}{2}k^2 + \frac{5}{2}k + k + 3 = \frac{1}{2}k^2 + \frac{7}{2}k + 3 = \frac{1}{2} \cdot \left(k^2 + 7k + 6\right)$$
   $$= \frac{1}{2} \cdot (k+1)(k+6)$$

   Conditions I and II are satisfied; the statement is true.

5. **I:** $n = 1$: $3 \cdot 1 - 1 = 2$ and $\dfrac{1}{2} \cdot 1(3 \cdot 1 + 1) = 2$

   II: If $2 + 5 + 8 + \cdots + (3k - 1) = \dfrac{1}{2} \cdot k(3k+1)$

   then $2 + 5 + 8 + \cdots + (3k - 1) + [3(k+1) - 1]$

   $$= \left[2 + 5 + 8 + \cdots + (3k - 1)\right] + (3k + 2) = \frac{1}{2} \cdot k(3k+1) + (3k+2)$$
   $$= \frac{3}{2}k^2 + \frac{1}{2}k + 3k + 2 = \frac{3}{2}k^2 + \frac{7}{2}k + 2 = \frac{1}{2} \cdot \left(3k^2 + 7k + 4\right)$$
   $$= \frac{1}{2} \cdot (k+1)(3k+4)$$

   Conditions I and II are satisfied; the statement is true.

7. I: $n = 1$: $2^{1-1} = 1$ and $2^1 - 1 = 1$

    II: If $1 + 2 + 2^2 + \cdots + 2^{k-1} = 2^k - 1$

          then $1 + 2 + 2^2 + \cdots + 2^{k-1} + 2^{k+1-1}$

$$= \left[1 + 2 + 2^2 + \cdots + 2^{k-1}\right] + 2^k = 2^k - 1 + 2^k$$

$$= 2 \cdot 2^k - 1 = 2^{k+1} - 1$$

Conditions I and II are satisfied; the statement is true.

9. I: $n = 1$: $4^{1-1} = 1$ and $\dfrac{1}{3} \cdot \left(4^1 - 1\right) = 1$

    II: If $1 + 4 + 4^2 + \cdots + 4^{k-1} = \dfrac{1}{3} \cdot \left(4^k - 1\right)$

          then $1 + 4 + 4^2 + \cdots + 4^{k-1} + 4^{k+1-1}$

$$= \left[1 + 4 + 4^2 + \cdots + 4^{k-1}\right] + 4^k = \frac{1}{3} \cdot \left(4^k - 1\right) + 4^k$$

$$= \frac{1}{3} \cdot 4^k - \frac{1}{3} + 4^k = \frac{4}{3} \cdot 4^k - \frac{1}{3} = \frac{1}{3}\left(4 \cdot 4^k - 1\right) = \frac{1}{3} \cdot \left(4^{k+1} - 1\right)$$

Conditions I and II are satisfied; the statement is true.

11. I: $n = 1$: $\dfrac{1}{1(1+1)} = \dfrac{1}{2}$ and $\dfrac{1}{1+1} = \dfrac{1}{2}$

     II: If $\dfrac{1}{1 \cdot 2} + \dfrac{1}{2 \cdot 3} + \dfrac{1}{3 \cdot 4} + \cdots + \dfrac{1}{k(k+1)} = \dfrac{k}{k+1}$

           then $\dfrac{1}{1 \cdot 2} + \dfrac{1}{2 \cdot 3} + \dfrac{1}{3 \cdot 4} + \cdots + \dfrac{1}{k(k+1)} + \dfrac{1}{(k+1)(k+1+1)}$

$$= \left[\frac{1}{1 \cdot 2} + \frac{1}{2 \cdot 3} + \frac{1}{3 \cdot 4} + \cdots + \frac{1}{k(k+1)}\right] + \frac{1}{(k+1)(k+2)}$$

$$= \frac{k}{k+1} + \frac{1}{(k+1)(k+2)} = \frac{k}{k+1} \cdot \frac{k+2}{k+2} + \frac{1}{(k+1)(k+2)}$$

$$= \frac{k^2 + 2k + 1}{(k+1)(k+2)} = \frac{(k+1)(k+1)}{(k+1)(k+2)} = \frac{k+1}{k+2}$$

Conditions I and II are satisfied; the statement is true.

13. I: $n=1$: $1^2=1$ and $\frac{1}{6}\cdot 1(1+1)(2\cdot 1+1)=1$

II: If $1^2+2^2+3^2+\cdots+k^2=\frac{1}{6}\cdot k(k+1)(2k+1)$

then $1^2+2^2+3^2+\cdots+k^2+(k+1)^2$

$$=\left[1^2+2^2+3^2+\cdots+k^2\right]+(k+1)^2=\frac{1}{6}k(k+1)(2k+1)+(k+1)^2$$

$$=(k+1)\left[\frac{1}{6}k(2k+1)+k+1\right]=(k+1)\left[\frac{1}{3}k^2+\frac{1}{6}k+k+1\right]$$

$$=(k+1)\left[\frac{1}{3}k^2+\frac{7}{6}k+1\right]=\frac{1}{6}(k+1)\left[2k^2+7k+6\right]$$

$$=\frac{1}{6}\cdot(k+1)(k+2)(2k+3)$$

Conditions I and II are satisfied; the statement is true.

15. I: $n=1$: $5-1=4$ and $\frac{1}{2}\cdot 1(9-1)=4$

II: If $4+3+2+\cdots+(5-k)=\frac{1}{2}\cdot k(9-k)$

then $4+3+2+\cdots+(5-k)+\left(5-(k+1)\right)$

$$=\left[4+3+2+\cdots+(5-k)\right]+(4-k)=\frac{1}{2}k(9-k)+(4-k)$$

$$=\frac{9}{2}k-\frac{1}{2}k^2+4-k=-\frac{1}{2}k^2+\frac{7}{2}k+4=-\frac{1}{2}\cdot\left[k^2-7k-8\right]$$

$$=-\frac{1}{2}\cdot(k+1)(k-8)=\frac{1}{2}\cdot(k+1)(8-k)=\frac{1}{2}\cdot(k+1)[9-(k+1)]$$

Conditions I and II are satisfied; the statement is true.

17. I: $n=1$: $1(1+1)=2$ and $\frac{1}{3}\cdot 1(1+1)(1+2)=2$

II: If $1\cdot 2+2\cdot 3+3\cdot 4+\cdots+k(k+1)=\frac{1}{3}\cdot k(k+1)(k+2)$

then $1\cdot 2+2\cdot 3+3\cdot 4+\cdots+k(k+1)+(k+1)(k+1+1)$

$$=\left[1\cdot 2+2\cdot 3+3\cdot 4+\cdots+k(k+1)\right]+(k+1)(k+2)$$

$$=\frac{1}{3}\cdot k(k+1)(k+2)+(k+1)(k+2)=(k+1)(k+2)\left[\frac{1}{3}k+1\right]$$

$$=\frac{1}{3}\cdot(k+1)(k+2)(k+3)$$

Conditions I and II are satisfied; the statement is true.

19.   I: $n = 1$: $1^2 + 1 = 2$ is divisible by 2
     II: If $k^2 + k$ is divisible by 2
            then $(k+1)^2 + (k+1) = k^2 + 2k + 1 + k + 1 = (k^2 + k) + (2k + 2)$
            Since $k^2 + k$ is divisible by 2 and $2k + 2$ is divisible by 2, then $(k+1)^2 + (k+1)$
            is divisible by 2.
     Conditions I and II are satisfied; the statement is true.

21.   I: $n = 1$: $1^2 - 1 + 2 = 2$ is divisible by 2
     II: If $k^2 - k + 2$ is divisible by 2
            then $(k+1)^2 - (k+1) + 2 = k^2 + 2k + 1 - k - 1 + 2 = (k^2 - k + 2) + (2k)$
            Since $k^2 - k + 2$ is divisible by 2 and $2k$ is divisible by 2, then
            $(k+1)^2 - (k+1) + 2$ is divisible by 2.
     Conditions I and II are satisfied; the statement is true.

23.   I: $n = 1$:   If $x > 1$ then $x^1 = x > 1$.
     II: Assume, for any natural number k, that if $x > 1$, then $x^k > 1$.
            Show that if $x^k > 1$, then $x^{k+1} > 1$:
$$x^{k+1} = x^k \cdot x > 1 \cdot x = x > 1$$
$$\uparrow$$
$$(x^k > 1)$$
     Conditions I and II are satisfied; the statement is true.

25.   I: $n = 1$: $a - b$ is a factor of $a^1 - b^1 = a - b$.
     II: If $a - b$ is a factor of $a^k - b^k$
            Show that $a - b$ is a factor of $a^{k+1} - b^{k+1} = a \cdot a^k - b \cdot b^k$.
$$a^{k+1} - b^{k+1} = a \cdot a^k - b \cdot b^k$$
$$= a \cdot a^k - a \cdot b^k + a \cdot b^k - b \cdot b^k = a\left(a^k - b^k\right) + b^k(a - b)$$
            Since $a - b$ is a factor of $a^k - b^k$ and $a - b$ is a factor of $a - b$, then
            $a - b$ is a factor of $a^{k+1} - b^{k+1}$.
     Conditions I and II are satisfied; the statement is true.

27. $n = 1$: $1^2 - 1 + 41 = 41$ is a prime number.
       $n = 41$: $41^2 - 41 + 41 = 41^2$ is not a prime number.

29. **I:** $n=1$: $ar^{1-1}=a$ and $a\left(\dfrac{1-r^1}{1-r}\right)=a$

II: If $a+ar+ar^2+\cdots+ar^{k-1}=a\left(\dfrac{1-r^k}{1-r}\right)$

then $a+ar+ar^2+\cdots+ar^{k-1}+ar^{k+1-1}$

$$=\left[a+ar+ar^2+\cdots+ar^{k-1}\right]+ar^k=a\left(\dfrac{1-r^k}{1-r}\right)+ar^k$$

$$=\dfrac{a(1-r^k)+ar^k(1-r)}{1-r}=\dfrac{a-ar^k+ar^k-ar^{k+1}}{1-r}=a\left(\dfrac{1-r^{k+1}}{1-r}\right)$$

Conditions I and II are satisfied; the statement is true.

31. **I:** $n=4$: The number of diagonals of a quadrilateral is $\dfrac{1}{2}\cdot 4(4-3)=2$.

II: Assume that for any integer k, the number of diagonals of a convex polygon with k sides (k vertices) is $\dfrac{1}{2}\cdot k(k-3)$. A convex polygon with $k+1$ sides ($k+1$ vertices) consists of a convex polygon with k sides (k vertices) plus a triangle, for a total of ($k+1$) vertices. The diagonals of this $k+1$-sided convex polygon are formed by the k original diagonals, plus $k-1$ additional diagonals.
Thus, we have the equation:

$$\dfrac{1}{2}\cdot k(k-3)+(k-1)=\dfrac{1}{2}k^2-\dfrac{3}{2}k+k-1$$

$$=\dfrac{1}{2}k^2-\dfrac{1}{2}k-1$$

$$=\dfrac{1}{2}\cdot\left(k^2-k-2\right)$$

$$=\dfrac{1}{2}\cdot(k+1)(k-2)$$

Conditions I and II are satisfied; the statement is true.

33. Answers will vary.

Sequences; Induction; the Binomial Theorem

11.5 The Binomial Theorem

1. Pascal Triangle

3. False

5. $\dbinom{5}{3} = \dfrac{5!}{3!\,2!} = \dfrac{5\cdot4\cdot3\cdot2\cdot1}{3\cdot2\cdot1\cdot2\cdot1} = \dfrac{5\cdot4}{2\cdot1} = 10$

7. $\dbinom{7}{5} = \dfrac{7!}{5!\,2!} = \dfrac{7\cdot6\cdot5\cdot4\cdot3\cdot2\cdot1}{5\cdot4\cdot3\cdot2\cdot1\cdot2\cdot1} = \dfrac{7\cdot6}{2\cdot1} = 21$

9. $\dbinom{50}{49} = \dfrac{50!}{49!\,1!} = \dfrac{50\cdot49!}{49!\cdot1} = \dfrac{50}{1} = 50$

11. $\dbinom{1000}{1000} = \dfrac{1000!}{1000!\,0!} = \dfrac{1}{1} = 1$

13. $\dbinom{55}{23} = \dfrac{55!}{23!\,32!} \approx 1.866442159\times10^{15}$

15. $\dbinom{47}{25} = \dfrac{47!}{25!\,22!} \approx 1.483389769\times10^{13}$

17. $(x+1)^5 = \dbinom{5}{0}x^5 + \dbinom{5}{1}x^4 + \dbinom{5}{2}x^3 + \dbinom{5}{3}x^2 + \dbinom{5}{4}x^1 + \dbinom{5}{5}x^0$

    $= x^5 + 5x^4 + 10x^3 + 10x^2 + 5x + 1$

19. $(x-2)^6 = \dbinom{6}{0}x^6 + \dbinom{6}{1}x^5(-2) + \dbinom{6}{2}x^4(-2)^2 + \dbinom{6}{3}x^3(-2)^3 + \dbinom{6}{4}x^2(-2)^4$

    $\qquad + \dbinom{6}{5}x(-2)^5 + \dbinom{6}{6}x^0(-2)^6$

    $= x^6 + 6x^5(-2) + 15x^4\cdot4 + 20x^3(-8) + 15x^2\cdot16 + 6x\cdot(-32) + 64$

    $= x^6 - 12x^5 + 60x^4 - 160x^3 + 240x^2 - 192x + 64$

21. $(3x+1)^4 = \dbinom{4}{0}(3x)^4 + \dbinom{4}{1}(3x)^3 + \dbinom{4}{2}(3x)^2 + \dbinom{4}{3}(3x) + \dbinom{4}{4}$

    $= 81x^4 + 4\cdot27x^3 + 6\cdot9x^2 + 4\cdot3x + 1 = 81x^4 + 108x^3 + 54x^2 + 12x + 1$

23. $\left(x^2+y^2\right)^5 = \binom{5}{0}\left(x^2\right)^5 + \binom{5}{1}\left(x^2\right)^4 y^2 + \binom{5}{2}\left(x^2\right)^3\left(y^2\right)^2 + \binom{5}{3}\left(x^2\right)^2\left(y^2\right)^3$

$$+ \binom{5}{4}x^2\left(y^2\right)^4 + \binom{5}{5}\left(y^2\right)^5$$

$$= x^{10} + 5x^8 y^2 + 10x^6 y^4 + 10x^4 y^6 + 5x^2 y^8 + y^{10}$$

25. $\left(\sqrt{x}+\sqrt{2}\right)^6 = \binom{6}{0}\left(\sqrt{x}\right)^6 + \binom{6}{1}\left(\sqrt{x}\right)^5\left(\sqrt{2}\right)^1 + \binom{6}{2}\left(\sqrt{x}\right)^4\left(\sqrt{2}\right)^2 + \binom{6}{3}\left(\sqrt{x}\right)^3\left(\sqrt{2}\right)^3$

$$\binom{6}{4}\left(\sqrt{x}\right)^2\left(\sqrt{2}\right)^4 + \binom{6}{5}\left(\sqrt{x}\right)\left(\sqrt{2}\right)^5 + \binom{6}{6}\left(\sqrt{2}\right)^6$$

$$= x^3 + 6\sqrt{2}x^{5/2} + 15\cdot 2x^2 + 20\cdot 2\sqrt{2}x^{3/2} + 15\cdot 4x + 6\cdot 4\sqrt{2}x^{1/2} + 8$$

$$= x^3 + 6\sqrt{2}x^{5/2} + 30x^2 + 40\sqrt{2}x^{3/2} + 60x + 24\sqrt{2}x^{1/2} + 8$$

27. $\left(ax+by\right)^5 = \binom{5}{0}\left(ax\right)^5 + \binom{5}{1}\left(ax\right)^4 \cdot by + \binom{5}{2}\left(ax\right)^3\left(by\right)^2 + \binom{5}{3}\left(ax\right)^2\left(by\right)^3$

$$+ \binom{5}{4}ax\left(by\right)^4 + \binom{5}{5}\left(by\right)^5$$

$$= a^5 x^5 + 5a^4 x^4 by + 10a^3 x^3 b^2 y^2 + 10a^2 x^2 b^3 y^3 + 5axb^4 y^4 + b^5 y^5$$

29. $n=10,\ j=4,\ x=x,\ a=3$

$$\binom{10}{4}x^6 \cdot 3^4 = \frac{10!}{4!\,6!}\cdot 81x^6 = \frac{10\cdot 9\cdot 8\cdot 7}{4\cdot 3\cdot 2\cdot 1}\cdot 81x^6 = 17{,}010x^6$$

The coefficient of x^6 is 17,010.

31. $n=12,\ j=5,\ x=2x,\ a=-1$

$$\binom{12}{5}(2x)^7 \cdot (-1)^5 = \frac{12!}{5!\,7!}\cdot 128x^7(-1) = \frac{12\cdot 11\cdot 10\cdot 9\cdot 8}{5\cdot 4\cdot 3\cdot 2\cdot 1}\cdot(-128)x^7 = -101{,}376x^7$$

The coefficient of x^7 is $-101{,}376$.

33. $n=9,\ j=2,\ x=2x,\ a=3$

$$\binom{9}{2}(2x)^7 \cdot 3^2 = \frac{9!}{2!\,7!}\cdot 128x^7(9) = \frac{9\cdot 8}{2\cdot 1}\cdot 128x^7\cdot 9 = 41{,}472x^7$$

The coefficient of x^7 is $41{,}472$.

35. $n=7,\ j=4,\ x=x,\ a=3$

$$\binom{7}{4}x^3 \cdot 3^4 = \frac{7!}{4!\,3!}\cdot 81x^3 = \frac{7\cdot 6\cdot 5}{3\cdot 2\cdot 1}\cdot 81x^3 = 2835x^3$$

37. $n = 9, \; j = 2, \; x = 3x, \; a = -2$

$$\binom{9}{2}(3x)^7 \cdot (-2)^2 = \frac{9!}{2! \, 7!} \cdot 2187x^7 \cdot 4 = \frac{9 \cdot 8}{2 \cdot 1} \cdot 8748x^7 = 314{,}928x^7$$

39. The constant term in $\displaystyle\sum_{j=0}^{12}\binom{12}{j}\left(x^2\right)^{12-j}\left(\frac{1}{x}\right)^j$ occurs when:

$$2(12 - j) = j$$
$$24 - 2j = j$$
$$3j = 24$$
$$j = 8$$

Evaluate the 9th term:

$$\binom{12}{8}\left(x^2\right)^4 \cdot \left(\frac{1}{x}\right)^8 = \frac{12!}{8! \, 4!}x^8 \cdot \frac{1}{x^8} = \frac{12 \cdot 11 \cdot 10 \cdot 9}{4 \cdot 3 \cdot 2 \cdot 1}x^0 = 495$$

41. The x^4 term in $\displaystyle\sum_{j=0}^{10}\binom{10}{j}(x)^{10-j}\left(\frac{-2}{\sqrt{x}}\right)^j$ occurs when:

$$10 - j - \frac{1}{2}j = 4$$
$$-\frac{3}{2}j = -6$$
$$j = 4$$

Evaluate the 5th term:

$$\binom{10}{4}(x)^6 \cdot \left(\frac{-2}{\sqrt{x}}\right)^4 = \frac{10!}{6! \, 4!}x^6 \cdot \frac{16}{x^2} = \frac{10 \cdot 9 \cdot 8 \cdot 7}{4 \cdot 3 \cdot 2 \cdot 1} \cdot 16x^4 = 3360x^4$$

The coefficient is 3360.

43. $(1.001)^5 = \left(1 + 10^{-3}\right)^5 = \binom{5}{0} \cdot 1^5 + \binom{5}{1} \cdot 1^4 \cdot 10^{-3} + \binom{5}{2} \cdot 1^3 \cdot \left(10^{-3}\right)^2 + \binom{5}{3} \cdot 1^2 \cdot \left(10^{-3}\right)^3 + \cdots$

$$= 1 + 5(0.001) + 10(0.000001) + 10(0.000000001) + \cdots$$
$$= 1 + 0.005 + 0.000010 + 0.000000010 + \cdots$$
$$= 1.00501 \quad \text{(correct to 5 decimal places)}$$

45. $\dbinom{n}{n-1} = \dfrac{n!}{(n-1)!(n-(n-1))!} = \dfrac{n!}{(n-1)!(1)!} = \dfrac{n(n-1)!}{(n-1)!} = n$

$\dbinom{n}{n} = \dfrac{n!}{n!(n-n)!} = \dfrac{n!}{n! \, 0!} = \dfrac{n!}{n! \cdot 1} = \dfrac{n!}{n!} = 1$

47. Show that
$$\binom{n}{0} + \binom{n}{1} + \cdots + \binom{n}{n} = 2^n$$

$$2^n = (1+1)^n$$

$$= \binom{n}{0} \cdot 1^n + \binom{n}{1} \cdot 1^{n-1} \cdot 1 + \binom{n}{2} \cdot 1^{n-2} \cdot 1^2 + \cdots + \binom{n}{n} \cdot 1^{n-n} \cdot 1^n$$

$$= \binom{n}{0} + \binom{n}{1} + \cdots + \binom{n}{n}$$

49. $$\binom{5}{0}\left(\frac{1}{4}\right)^5 + \binom{5}{1}\left(\frac{1}{4}\right)^4\left(\frac{3}{4}\right) + \binom{5}{2}\left(\frac{1}{4}\right)^3\left(\frac{3}{4}\right)^2 + \binom{5}{3}\left(\frac{1}{4}\right)^2\left(\frac{3}{4}\right)^3$$
$$+ \binom{5}{4}\left(\frac{1}{4}\right)\left(\frac{3}{4}\right)^4 + \binom{5}{5}\left(\frac{3}{4}\right)^5$$

$$= \left(\frac{1}{4} + \frac{3}{4}\right)^5 = (1)^5 = 1$$

Chapter 11

Sequences; Induction; the Binomial Theorem

11.R Chapter Review

1. $a_1 = (-1)^1 \dfrac{1+3}{1+2} = -\dfrac{4}{3}, \ a_2 = (-1)^2 \dfrac{2+3}{2+2} = \dfrac{5}{4}, \ a_3 = (-1)^3 \dfrac{3+3}{3+2} = -\dfrac{6}{5},$

$a_4 = (-1)^4 \dfrac{4+3}{4+2} = \dfrac{7}{6}, \ a_5 = (-1)^5 \dfrac{5+3}{5+2} = -\dfrac{8}{7}$

3. $a_1 = \dfrac{2^1}{1^2} = \dfrac{2}{1} = 2, \ a_2 = \dfrac{2^2}{2^2} = \dfrac{4}{4} = 1, \ a_3 = \dfrac{2^3}{3^2} = \dfrac{8}{9}, \ a_4 = \dfrac{2^4}{4^2} = \dfrac{16}{16} = 1, \ a_5 = \dfrac{2^5}{5^2} = \dfrac{32}{25}$

5. $a_1 = 3, \ a_2 = \dfrac{2}{3} \cdot 3 = 2, \ a_3 = \dfrac{2}{3} \cdot 2 = \dfrac{4}{3}, \ a_4 = \dfrac{2}{3} \cdot \dfrac{4}{3} = \dfrac{8}{9}, \ a_5 = \dfrac{2}{3} \cdot \dfrac{8}{9} = \dfrac{16}{27}$

7. $a_1 = 2, \ a_2 = 2 - 2 = 0, \ a_3 = 2 - 0 = 2, \ a_4 = 2 - 2 = 0, \ a_5 = 2 - 0 = 2$

9. $\displaystyle\sum_{k=1}^{4} (4k + 2) = (4 \cdot 1 + 2) + (4 \cdot 2 + 2) + (4 \cdot 3 + 2) + (4 \cdot 4 + 2) = 6 + 10 + 14 + 18 = 48$

11. $1 - \dfrac{1}{2} + \dfrac{1}{3} - \dfrac{1}{4} + \cdots + \dfrac{1}{13} = \displaystyle\sum_{k=1}^{13} (-1)^{k+1} \left(\dfrac{1}{k} \right)$

13. $\{n + 5\}$ Arithmetic $\quad d = (n + 1 + 5) - (n + 5) = n + 6 - n - 5 = 1$

$S_n = \dfrac{n}{2} [6 + n + 5] = \dfrac{n}{2}(n + 11)$

15. $\{2n^3\}$ Examine the terms of the sequence: 2, 16, 54, 128, 250, ...
Neither. There is no common difference, no common ratio.

17. $\{2^{3n}\}$ Geometric $\quad r = \dfrac{2^{3(n+1)}}{2^{3n}} = \dfrac{2^{3n+3}}{2^{3n}} = 2^{3n+3-3n} = 2^3 = 8$

$S_n = 8 \left(\dfrac{1 - 8^n}{1 - 8} \right) = 8 \left(\dfrac{1 - 8^n}{-7} \right) = \dfrac{8}{7} (8^n - 1)$

19. 0, 4, 8, 12, ...   Arithmetic $d = 4 - 0 = 4$

$$S_n = \frac{n}{2}(2(0) + (n-1)4) = \frac{n}{2}(4(n-1)) = 2n(n-1)$$

21. $3, \dfrac{3}{2}, \dfrac{3}{4}, \dfrac{3}{5}, \dfrac{3}{16}, \ldots$ Geometric $r = \dfrac{\frac{3}{2}}{3} = \dfrac{3}{2} \cdot \dfrac{1}{3} = \dfrac{1}{2}$

$$S_n = 3\left(\frac{1 - \left(\frac{1}{2}\right)^n}{1 - \frac{1}{2}}\right) = 3\left(\frac{1 - \left(\frac{1}{2}\right)^n}{\frac{1}{2}}\right) = 6\left(1 - \left(\frac{1}{2}\right)^n\right)$$

23. $\dfrac{2}{3}, \dfrac{3}{4}, \dfrac{4}{5}, \dfrac{5}{6}, \ldots$ Neither. There is no common difference, no common ratio.

25. $\displaystyle\sum_{k=1}^{5}(k^2 + 12) = 13 + 16 + 21 + 28 + 37 = 115$

27. $\displaystyle\sum_{k=1}^{10}(3k - 9) = \sum_{k=1}^{10}3k - \sum_{k=1}^{10}9 = 3\sum_{k=1}^{10}k - \sum_{k=1}^{10}9 = 3\left(\frac{10(10+1)}{2}\right) - 10(9) = 165 - 90 = 75$

29. $\displaystyle\sum_{k=1}^{7}\left(\frac{1}{3}\right)^k = \frac{1}{3}\left(\frac{1 - \left(\frac{1}{3}\right)^7}{1 - \frac{1}{3}}\right) = \frac{1}{3}\left(\frac{1 - \left(\frac{1}{3}\right)^7}{\frac{2}{3}}\right) = \frac{1}{2}\left(1 - \frac{1}{2187}\right) = \frac{1}{2} \cdot \frac{2186}{2187} = \frac{1093}{2187}$

31. Arithmetic $a_1 = 3, \ d = 4, \ a_n = a + (n-1)d$
$a_9 = 3 + (9-1)4 = 3 + 8(4) = 3 + 32 = 35$

33. Geometric $a = 1, \ r = \dfrac{1}{10}, \ a_n = ar^{n-1}$ $a_{11} = 1 \cdot \left(\dfrac{1}{10}\right)^{11-1} = \left(\dfrac{1}{10}\right)^{10} = \dfrac{1}{10,000,000,000}$

35. Arithmetic $a_1 = \sqrt{2}, \ d = \sqrt{2}, \ a_n = a + (n-1)d$
$a_9 = \sqrt{2} + (9-1)\sqrt{2} = \sqrt{2} + 8\sqrt{2} = 9\sqrt{2}$

37. $a_7 = a + 6d = 31$ $a_{20} = a + 19d = 96$
Solve the system of equations:
$31 - 6d + 19d = 96$

$\begin{aligned}13d &= 65 \\ d &= 5 \\ a = 31 - 6(5) &= 31 - 30 = 1\end{aligned}$ $\begin{aligned}a_n &= 1 + (n-1)(5) = 1 + 5n - 5 \\ &= 5n - 4 \\ \text{General formula: } & \{5n - 4\}\end{aligned}$

39. $a_{10} = a + 9d = 0$ $a_{18} = a + 17d = 8$
Solve the system of equations:
$$-9d + 17d = 8$$
$$8d = 8$$
$$d = 1$$
$$a = -9(1) = -9$$

$a_n = -9 + (n-1)(1) = -9 + n - 1$
$$= n - 10$$
General formula: $\{n - 10\}$

41. $a = 3,\ r = \dfrac{1}{3}$

Since $|r| < 1,\ \ S = \dfrac{a}{1-r} = \dfrac{3}{1 - \dfrac{1}{3}} = \dfrac{3}{\dfrac{2}{3}} = \dfrac{9}{2}$

43. $a = 2,\ r = -\dfrac{1}{2}$

Since $|r| < 1,\ \ S = \dfrac{a}{1-r} = \dfrac{2}{1 - \left(-\dfrac{1}{2}\right)} = \dfrac{2}{\dfrac{3}{2}} = \dfrac{4}{3}$

45. $a = 4,\ r = \dfrac{1}{2}$

Since $|r| < 1,\ \ S = \dfrac{a}{1-r} = \dfrac{4}{1 - \dfrac{1}{2}} = \dfrac{4}{\dfrac{1}{2}} = 8$

47.   I: $n = 1$: $3 \cdot 1 = 3$ and $\dfrac{3 \cdot 1}{2}(1+1) = 3$

II: If $3 + 6 + 9 + \cdots + 3k = \dfrac{3k}{2}(k+1)$
then $3 + 6 + 9 + \cdots + 3k + 3(k+1)$
$$= \left[3 + 6 + 9 + \cdots + 3k\right] + 3(k+1) = \dfrac{3k}{2}(k+1) + 3(k+1)$$
$$= (k+1)\left(\dfrac{3k}{2} + 3\right) = \dfrac{3}{2}(k+1)(k+2)$$

Conditions I and II are satisfied; the statement is true.

49.   I: $n = 1$: $2 \cdot 3^{1-1} = 2$ and $3^1 - 1 = 2$
II: If $2 + 6 + 18 + \cdots + 2 \cdot 3^{k-1} = 3^k - 1$
then $2 + 6 + 18 + \cdots + 2 \cdot 3^{k-1} + 2 \cdot 3^{k+1-1}$
$$= \left[2 + 6 + 18 + \cdots + 2 \cdot 3^{k-1}\right] + 2 \cdot 3^k = 3^k - 1 + 2 \cdot 3^k$$
$$= 3 \cdot 3^k - 1 = 3^{k+1} - 1$$

Conditions I and II are satisfied; the statement is true.

51. I: $n = 1$: $(3 \cdot 1 - 2)^2 = 1$ and $\dfrac{1}{2} \cdot 1(6 \cdot 1^2 - 3 \cdot 1 - 1) = 1$

    II: If $1^2 + 4^2 + 7^2 + \cdots + (3k - 2)^2 = \dfrac{1}{2} \cdot k\left(6k^2 - 3k - 1\right)$

       then $1^2 + 4^2 + 7^2 + \cdots + (3k - 2)^2 + \left(3(k + 1) - 2\right)^2$

$$= \left[1^2 + 4^2 + 7^2 + \cdots + (3k - 2)^2\right] + (3k + 1)^2$$

$$= \frac{1}{2} \cdot k\left(6k^2 - 3k - 1\right) + (3k + 1)^2$$

$$= \frac{1}{2} \cdot \left[6k^3 - 3k^2 - k + 18k^2 + 12k + 2\right] = \frac{1}{2} \cdot \left[6k^3 + 15k^2 + 11k + 2\right]$$

$$= \frac{1}{2} \cdot (k + 1)\left[6k^2 + 9k + 2\right] = \frac{1}{2} \cdot (k + 1)\left[6k^2 + 12k + 6 - 3k - 3 - 1\right]$$

$$= \frac{1}{2} \cdot (k + 1)\left[6(k^2 + 2k + 1) - 3(k + 1) - 1\right]$$

$$= \frac{1}{2} \cdot (k + 1)\left[6(k + 1)^2 - 3(k + 1) - 1\right]$$

Conditions I and II are satisfied; the statement is true.

53. $\dbinom{5}{2} = \dfrac{5!}{2! \, 3!} = \dfrac{5 \cdot 4 \cdot 3 \cdot 2 \cdot 1}{2 \cdot 1 \cdot 3 \cdot 2 \cdot 1} = \dfrac{5 \cdot 4}{2 \cdot 1} = 10$

55. $(x + 2)^5 = \dbinom{5}{0}x^5 + \dbinom{5}{1}x^4 \cdot 2 + \dbinom{5}{2}x^3 \cdot 2^2 + \dbinom{5}{3}x^2 \cdot 2^3 + \dbinom{5}{4}x^1 \cdot 2^4 + \dbinom{5}{5} \cdot 2^5$

$$= x^5 + 5 \cdot 2x^4 + 10 \cdot 4x^3 + 10 \cdot 8x^2 + 5 \cdot 16x + 32$$

$$= x^5 + 10x^4 + 40x^3 + 80x^2 + 80x + 32$$

57. $(2x + 3)^5 = \dbinom{5}{0}(2x)^5 + \dbinom{5}{1}(2x)^4 \cdot 3 + \dbinom{5}{2}(2x)^3 \cdot 3^2 + \dbinom{5}{3}(2x)^2 \cdot 3^3$

$$+ \dbinom{5}{4}(2x)^1 \cdot 3^4 + \dbinom{5}{5} \cdot 3^5$$

$$= 32x^5 + 5 \cdot 16x^4 \cdot 3 + 10 \cdot 8x^3 \cdot 9 + 10 \cdot 4x^2 \cdot 27 + 5 \cdot 2x \cdot 81 + 243$$

$$= 32x^5 + 240x^4 + 720x^3 + 1080x^2 + 810x + 243$$

59. $n = 9, \quad j = 2, \quad x = x, \quad a = 2$

$$\dbinom{9}{2}x^7 \cdot 2^2 = \frac{9!}{2! \, 7!} \cdot 4x^7 = \frac{9 \cdot 8}{2 \cdot 1} \cdot 4x^7 = 144x^7$$

The coefficient of x^7 is 144.

61. $n = 7, \ j = 5, \ x = 2x, \ a = 1$

$$\binom{7}{5}(2x)^2 \cdot 1^5 = \frac{7!}{5! \, 2!} \cdot 4x^2(1) = \frac{7 \cdot 6}{2 \cdot 1} \cdot 4x^2 = 84x^2$$

The coefficient of x^2 is 84.

63.   This is an arithmetic sequence with $a = 80, \ d = -3, \ n = 25$

(a) $a_{25} = 80 + (25 - 1)(-3) = 80 - 72 = 8$ bricks

(b) $S_{25} = \dfrac{25}{2}(80 + 8) = 25(44) = 1100$ bricks

1100 bricks are needed to build the steps.

65.   This is a geometric sequence with $a = 20, \ r = \dfrac{3}{4}$.

(a)   After striking the ground the third time, the height is $20\left(\dfrac{3}{4}\right)^3 = \dfrac{135}{16} \approx 8.44$ feet.

(b)   After striking the ground the nth time, the height is $20\left(\dfrac{3}{4}\right)^n$ feet.

(c)   If the height is less than 6 inches, that is, 0.5 feet, then:

$$0.5 > 20\left(\frac{3}{4}\right)^n$$

$$0.025 > \left(\frac{3}{4}\right)^n$$

$$\log 0.025 > n \log \frac{3}{4}$$

$$n > \frac{\log 0.025}{\log \dfrac{3}{4}} \approx 12.82$$

The height is less than 6 inches after the 13th strike.

(d)   Since this is a geometric sequence with $|r| < 1$, the distance is the sum of the two infinite geometric series–the distances going down plus the distances going up.

Distance going down: $S_{down} = \dfrac{20}{1 - \dfrac{3}{4}} = \dfrac{20}{\dfrac{1}{4}} = 80$ feet.

Distance going up: $S_{up} = \dfrac{15}{1 - \dfrac{3}{4}} = \dfrac{15}{\dfrac{1}{4}} = 60$ feet.

The total distance traveled is $80 + 60 = 140$ feet.

67.   This is an ordinary annuity with $P = \$500$ and $n = (4)(30) = 120$ payment periods.

The interest rate per period is $\dfrac{0.08}{4} = 0.02$. Thus,

$$A = 500\left[\frac{[1 + 0.02]^{120} - 1}{0.02}\right] \approx \$244{,}129.08$$

Sequences; Induction; the Binomial Theorem

11.CR   Cumulative Review

1. $\left|x^2\right| = 9$

   $x^2 = 9 \quad$ or $\quad x^2 = -9$

   $x = \pm 3 \quad$ or $\quad x = \pm 3i$

   The solution set is $\{-3i, \ 3i, \ -3, \ 3\}$.

3. $2e^x = 5$

   $e^x = \dfrac{5}{2}$

   $\ln e^x = \ln \dfrac{5}{2} \Rightarrow x = \ln \dfrac{5}{2}$

   The solution set is $\left\{\ln \dfrac{5}{2}\right\}$.

5. Given a circle with center $(-1,2)$ and containing the point $(3,5)$, we first use the distance formula to determine the radius.

$$r = \sqrt{\left(3-(-1)\right)^2 + \left(5-2\right)^2}$$
$$= \sqrt{4^2 + 3^2} = \sqrt{16+9}$$
$$= \sqrt{25}$$
$$= 5$$

   Therefore, the equation of the circle is given by

$$\left(x-(-1)\right)^2 + \left(y-2\right)^2 = 5^2$$
$$\left(x+1\right)^2 + \left(y-2\right)^2 = 5^2$$
$$x^2 + 2x + 1 + y^2 - 4y + 4 = 25$$
$$x^2 + y^2 + 2x + y^2 - 4y - 20 = 0$$

7. Center: $(0, 0)$; Focus: $(0, 3)$;

   Vertex: $(0, 4)$;

   Major axis is the y-axis; $a = 4$; $c = 3$.

   Find b:

   $b^2 = a^2 - c^2 = 16 - 9 = 7 \Rightarrow b = \sqrt{7}$

   Write the equation:

$$\frac{\left(x-0\right)^2}{\left(\sqrt{7}\right)^2} + \frac{\left(y-0\right)^2}{\left(4\right)^2} = 1$$

$$\frac{x^2}{7} + \frac{y^2}{16} = 1$$

9. A circle with center point $(0,4)$, passing through the pole has radius 4.

   Using rectangular coordinates:

$$\left(x-h\right)^2 + \left(y-k\right)^2 = r^2$$
$$\left(x-0\right)^2 + \left(y-4\right)^2 = 4^2$$
$$x^2 + y^2 - 8y + 16 = 16$$
$$x^2 + y^2 - 8y = 0$$

   Converting to polar coordinates:

$$r^2 - 8r\sin\theta = 0$$
$$r^2 = 8r\sin\theta$$
$$r = 8\sin\theta$$

11. $\cos^{-1}(-0.5)$

We are finding the angle θ, $-\pi \leq \theta \leq \pi$, whose cosine equals -0.5.

$$\cos\theta = -0.5 \quad -\pi \leq \theta \leq \pi$$

$$\theta = \frac{2\pi}{3}$$

$$\cos^{-1}(-0.5) = \frac{2\pi}{3}$$

Chapter 12

Counting and Probability

12.1 Sets and Counting

1. union

3. True

5. $A \cup B = \{1, 3, 5, 7, 9\} \cup \{1, 5, 6, 7\} = \{1, 3, 5, 6, 7, 9\}$

7. $A \cap B = \{1, 3, 5, 7, 9\} \cap \{1, 5, 6, 7\} = \{1, 5, 7\}$

9. $(A \cup B) \cap C = (\{1, 3, 5, 7, 9\} \cup \{1, 5, 6, 7\}) \cap \{1, 2, 4, 6, 8, 9\}$
   $= \{1, 3, 5, 6, 7, 9\} \cap \{1, 2, 4, 6, 8, 9\}$
   $= \{1, 6, 9\}$

11. $(A \cap B) \cup C = (\{1, 3, 5, 7, 9\} \cap \{1, 5, 6, 7\}) \cup \{1, 2, 4, 6, 8, 9\}$
    $= \{1, 5, 7\} \cup \{1, 2, 4, 6, 8, 9\}$
    $= \{1, 2, 4, 5, 6, 7, 8, 9\}$

13. $(A \cup C) \cap (B \cup C)$
    $= (\{1, 3, 5, 7, 9\} \cup \{1, 2, 4, 6, 8, 9\}) \cap (\{1, 5, 6, 7\} \cup \{1, 2, 4, 6, 8, 9\})$
    $= \{1, 2, 3, 4, 5, 6, 7, 8, 9\} \cap \{1, 2, 4, 5, 6, 7, 8, 9\}$
    $= \{1, 2, 4, 5, 6, 7, 8, 9\}$

15. $\overline{A} = \{0, 2, 6, 7, 8\}$

17. $\overline{A \cap B} = \overline{\{1, 3, 4, 5, 9\} \cap \{2, 4, 6, 7, 8\}} = \overline{\{4\}} = \{0, 1, 2, 3, 5, 6, 7, 8, 9\}$

19. $\overline{A} \cup \overline{B} = \{0, 2, 6, 7, 8\} \cup \{0, 1, 3, 5, 9\} = \{0, 1, 2, 3, 5, 6, 7, 8, 9\}$

21. $\overline{A} \cap \overline{C} = \overline{\{1, 3, 4, 5, 9\}} \cap \overline{\{0, 2, 5, 7, 8, 9\}} = \overline{\{5, 9\}} = \{0, 1, 2, 3, 4, 6, 7, 8\}$

23. $\overline{A \cup B \cup C} = \overline{\{1, 3, 4, 5, 9\} \cup \{2, 4, 6, 7, 8\} \cup \{1, 3, 4, 6\}}$
    $= \overline{\{1, 2, 3, 4, 5, 6, 7, 8, 9\}} = \{0\}$

25. $\{a\}, \{b\}, \{c\}, \{d\}, \{a, b\}, \{a, c\}, \{a, d\}, \{b, c\}, \{b, d\}, \{c, d\}, \{a, b, c\}, \{a, b, d\}, \{a, c, d\}, \{b, c, d\}, \{a, b, c, d\}, \varnothing$

27. $n(A) = 15$, $n(B) = 20$, $n(A \cap B) = 10$

$$n(A \cup B) = n(A) + n(B) - n(A \cap B) = 15 + 20 - 10 = 25$$

29. $n(A \cup B) = 50$, $n(A \cap B) = 10$, $n(B) = 20$

$$n(A \cup B) = n(A) + n(B) - n(A \cap B)$$
$$50 = n(A) + 20 - 10$$
$$40 = n(A)$$

31.   From the figure: $n(A) = 15 + 3 + 5 + 2 = 25$

33.   From the figure: $n(A \text{ or } B) = n(A \cup B) = 15 + 2 + 5 + 3 + 10 + 2 = 37$

35.   From the figure: $n(A \text{ but not } C) = n(A) - n(A \cap C) = 25 - 7 = 18$

37.   From the figure: $n(A \text{ and } B \text{ and } C) = n(A \cap B \cap C) = 5$

39.   Let $A = \{\text{those who will purchase a major appliance}\}$

$B = \{\text{those who will buy a car}\}$

$n(U) = 500$, $n(A) = 200$, $n(B) = 150$, $n(A \cap B) = 25$

$n(A \cup B) = n(A) + n(B) - n(A \cap B) = 200 + 150 - 25 = 325$

$n(\text{purchase neither}) = 500 - 325 = 175$

$n(\text{purchase only a car}) = 150 - 25 = 125$

41.   Construct a Venn diagram:

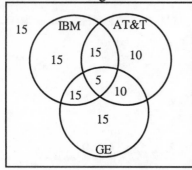

     (a)   15
     (b)   15
     (c)   15
     (d)   25
     (e)   40

43.   (a) $n(\text{married}) = n(\text{married, spouse present}) + n(\text{married, spouse absent})$
$$= 54{,}654 + 3232 = 57{,}886 \text{ thousand}$$
     (b) $n(\text{widowed or divorced}) = n(\text{widowed}) + n(\text{divorced})$
$$= 2686 + 8208 = 10{,}894 \text{ thousand}$$
     (c) $n(\text{married, spouse absent, widowed or divorced})$
$$= n(\text{married, spouse absent}) + n(\text{widowed}) + n(\text{divorced})$$
$$= 3232 + 2686 + 8208$$
$$= 14{,}126 \text{ thousand}$$

45.   Answers will vary.

Counting and Probability

12.2 Permutations and Combinations

1. 1, 1

3. permutation

5. True

7. $P(6, 2) = \dfrac{6!}{(6-2)!} = \dfrac{6!}{4!} = \dfrac{6 \cdot 5 \cdot 4!}{4!} = 30$

9. $P(4, 4) = \dfrac{4!}{(4-4)!} = \dfrac{4!}{0!} = \dfrac{4 \cdot 3 \cdot 2 \cdot 1}{1} = 24$

11. $P(7, 0) = \dfrac{7!}{(7-0)!} = \dfrac{7!}{7!} = 1$

13. $P(8, 4) = \dfrac{8!}{(8-4)!} = \dfrac{8!}{4!} = \dfrac{8 \cdot 7 \cdot 6 \cdot 5 \cdot 4!}{4!} = 1680$

15. $C(8, 2) = \dfrac{8!}{(8-2)!\, 2!} = \dfrac{8!}{6!\, 2!} = \dfrac{8 \cdot 7 \cdot 6!}{6! \cdot 2 \cdot 1} = 28$

17. $C(7, 4) = \dfrac{7!}{(7-4)!\, 4!} = \dfrac{7!}{3!\, 4!} = \dfrac{7 \cdot 6 \cdot 5 \cdot 4!}{4! \cdot 3 \cdot 2 \cdot 1} = 35$

19. $C(15, 15) = \dfrac{15!}{(15-15)!\, 15!} = \dfrac{15!}{0!\, 15!} = \dfrac{15!}{15! \cdot 1} = 1$

21. $C(26, 13) = \dfrac{26!}{(26-13)!\, 13!} = \dfrac{26!}{13!\, 13!} = 10,400,600$

23. {*abc, abd, abe, acb, acd, ace, adb, adc, ade, aeb, aec, aed, bac, bad, bae, bca,*
bcd, bce, bda, bdc, bde, bea, bec, bed, cab, cad, cae, cba, cbd, cbe, cda, cdb,
cde, cea, ceb, ced, dab, dac, dae, dba, dbc, dbe, dca, dcb, dce, dea, deb, dec,
eab, eac, ead, eba, ebc, ebd, eca, ecb, ecd, eda, edb, edc}
$$P(5,3) = \dfrac{5!}{(5-3)!} = \dfrac{5!}{2!} = \dfrac{5 \cdot 4 \cdot 3 \cdot 2!}{2!} = 60$$

25. {123, 124, 132, 134, 142, 143, 213, 214, 231, 234, 241, 243, 312, 314, 321, 324,
341, 342, 412, 413, 421, 423, 431, 432}
$$P(4,3) = \dfrac{4!}{(4-3)!} = \dfrac{4!}{1!} = \dfrac{4 \cdot 3 \cdot 2 \cdot 1}{1} = 24$$

27. {*abc, abd, abe, acd, ace, ade, bcd, bce, bde, cde*}
$$C(5,3) = \dfrac{5!}{(5-3)!\, 3!} = \dfrac{5 \cdot 4 \cdot 3!}{2 \cdot 1 \cdot 3!} = 10$$

29.   {123, 124, 134, 234} $C(4,3) = \dfrac{4!}{(4-3)!\,3!} = \dfrac{4 \cdot 3!}{1!\,3!} = 4$

31.   There are 5 choices of shirts and 3 choices of ties; there are (5)(3) = 15 combinations.

33.   There are 4 choices for the first letter in the code and 4 choices for the second letter in the code; there are (4)(4) = 16 possible two-letter codes.

35.   There are two choices for each of three positions; there are (2)(2)(2) = 8 possible three-digit numbers.

37.   To line up the four people, there are 4 choices for the first position, 3 choices for the second position, 2 choices for the third position, and 1 choice for the fourth position. Thus there are (4)(3)(2)(1) = 24 possible ways four people can be lined up.

39.   Since no letter can be repeated, there are 5 choices for the first letter, 4 choices for the second letter, and 3 choices for the third letter. Thus, there are (5)(4)(3) = 60 possible three-letter codes.

41.   There are 26 possible one-letter names. There are (26)(26) = 676 possible two-letter names. There are (26)(26)(26) = 17,576 possible three-letter names. Thus, there are 26 + 676 + 17,576 = 18,278 possible companies that can be listed on the New York Stock Exchange.

43.   A committee of 4 from a total of 7 students is given by:
$$C(7,4) = \dfrac{7!}{(7-4)!\,4!} = \dfrac{7!}{3!\,4!} = \dfrac{7 \cdot 6 \cdot 5 \cdot 4!}{3 \cdot 2 \cdot 1 \cdot 4!} = 35$$
35 committees are possible.

45.   There are 2 possible answers for each question. Therefore, there are $2^{10} = 1024$ different possible arrangements of the answers.

47.   There are 9 choices for the first digit, and 10 choices for each of the other three digits. Thus, there are (9)(10)(10)(10) = 9000 possible four-digit numbers.

49.   There are 5 choices for the first position, 4 choices for the second position, 3 choices for the third position, 2 choices for the fourth position, and 1 choice for the fifth position. Thus, there are (5)(4)(3)(2)(1) = 120 possible arrangements of the books.

51.   There are 8 choices for the DOW stocks, 15 choices for the NASDAQ stocks, and 4 choices for the global stocks. Thus, there are (8)(15)(4) = 480 different portfolios.

53.   The first person can have any of 365 days, the second person can have any of the remaining 364 days. Thus, there are (365)(364) = 132,860 possible ways two people can have different birthdays.

55. Choosing 2 boys from the 4 boys can be done $C(4,2)$ ways, and choosing 3 girls from the 8 girls can be done in $C(8,3)$ ways. Thus, there are a total of:

$$C(4,2)\cdot C(8,3) = \frac{4!}{(4-2)!\,2!}\cdot\frac{8!}{(8-3)!\,3!} = \frac{4!}{2!\,2!}\cdot\frac{8!}{5!\,3!}$$

$$= \frac{4\cdot 3!}{2\cdot 1\cdot 2\cdot 1}\cdot\frac{8\cdot 7\cdot 6\cdot 5!}{5!\,3!} = 336$$

57. This is a permutation with repetition. There are $\dfrac{9!}{2!\,2!} = 90{,}720$ different words.

59. (a) $C(7,2)\cdot C(3,1) = 21\cdot 3 = 63$
     (b) $C(7,3)\cdot C(3,0) = 35\cdot 1 = 35$
     (c) $C(3,3)\cdot C(7,0) = 1\cdot 1 = 1$

61. There are $C(100, 22)$ ways to form the first committee. There are 78 senators left, so there are $C(78, 13)$ ways to form the second committee. There are $C(65, 10)$ ways to form the third committee. There are $C(55, 5)$ ways to form the fourth committee. There are $C(50, 16)$ ways to form the fifth committee. There are $C(34, 17)$ ways to form the sixth committee. There are $C(17, 17)$ ways to form the seventh committee.
     The total number of committees

$$= C(100,22)\cdot C(78,13)\cdot C(65,10)\cdot C(55,5)\cdot C(50,16)\cdot C(34,17)\cdot C(17,17)$$

$$\approx 1.1568\times 10^{76}$$

63. There are 9 choices for the first position, 8 choices for the second position, 7 for the third position, etc. There are $9\cdot 8\cdot 7\cdot 6\cdot 5\cdot 4\cdot 3\cdot 2\cdot 1 = 9! = 362{,}880$ possible batting orders.

65. The team must have 1 pitcher and 8 position players (non-pitchers). For pitcher, choose 1 player from a group of 4 players, i.e., $C(4, 1)$. For position players, choose 8 players from a group of 11 players, i.e., $C(11, 8)$. Therefore, the number different teams possible is $C(4,1)\cdot C(11,8) = 4\cdot 165 = 660$.

67. Choose 2 players from a group of 6 players. Therefore, there are $C(6,2) = 15$ different teams possible.

69–71.   Answers will vary.

Chapter 12

Counting and Probability

12.3 Probability

1. equally likely

3. False

5. Probabilities must be between 0 and 1, inclusive. Thus, 0, 0.01, 0.35, and 1 could be probabilities.

7. All the probabilities are between 0 and 1.
The sum of the probabilities is $0.2 + 0.3 + 0.1 + 0.4 = 1$.
This is a probability model.

9. All the probabilities are between 0 and 1.
The sum of the probabilities is $0.3 + 0.2 + 0.1 + 0.3 = 0.9$.
This is not a probability model.

11. The sample space is: $S = \{HH, HT, TH, TT\}$.
Each outcome is equally likely to occur; so $P(E) = \dfrac{n(E)}{n(S)}$.
The probabilities are: $P(HH) = \dfrac{1}{4}$, $P(HT) = \dfrac{1}{4}$, $P(TH) = \dfrac{1}{4}$, $P(TT) = \dfrac{1}{4}$.

13. The sample space of tossing two fair coins and a fair die is:
$$S = \{HH1, HH2, HH3, HH4, HH5, HH6, HT1, HT2, HT3, HT4, HT5,$$
$$HT6, TH1, TH2, TH3, TH4, TH5, TH6, TT1, TT2, TT3, TT4, TT5, TT6\}$$
There are 24 equally likely outcomes and the probability of each is $\dfrac{1}{24}$.

15. The sample space for tossing three fair coins is:
$$S = \{HHH, HHT, HTH, THH, HTT, THT, TTH, TTT\}$$
There are 8 equally likely outcomes and the probability of each is $\dfrac{1}{8}$.

17. The sample space is:
$$S = \{1 \text{ Yellow}, 1 \text{ Red}, 1 \text{ Green}, 2 \text{ Yellow}, 2 \text{ Red}, 2 \text{ Green}, 3 \text{ Yellow}, 3 \text{ Red},$$
$$3 \text{ Green}, 4 \text{ Yellow}, 4 \text{ Red}, 4 \text{ Green}\}$$
There are 12 equally likely events and the probability of each is $\dfrac{1}{12}$. The probability of getting a 2 or 4 followed by a Red is $P(2 \text{ Red}) + P(4 \text{ Red}) = \dfrac{1}{12} + \dfrac{1}{12} = \dfrac{1}{6}$.

19. The sample space is:

S = {1 Yellow Forward, 1 Yellow Backward, 1 Red Forward, 1 Red Backward, 1 Green Forward, 1 Green Backward, 2 Yellow Forward, 2 Yellow Backward, 2 Red Forward, 2 Red Backward, 2 Green Forward, 2 Green Backward, 3 Yellow Forward, 3 Yellow Backward, 3 Red Forward, 3 Red Backward, 3 Green Forward, 3 Green Backward, 4 Yellow Forward, 4 Yellow Backward, 4 Red Forward, 4 Red Backward, 4 Green Forward, 4 Green Backward}

There are 24 equally likely events and the probability of each is $\frac{1}{24}$.

The probability of getting a 1, followed by a Red or Green, followed by a Backward is

$P(1 \text{ Red Backward}) + P(1 \text{ Green Backward}) = \frac{1}{24} + \frac{1}{24} = \frac{1}{12}$.

21. The sample space is:

S = {1 1 Yellow, 1 1 Red, 1 1 Green, 1 2 Yellow, 1 2 Red, 1 2 Green, 1 3 Yellow, 1 3 Red, 1 3 Green, 1 4 Yellow, 1 4 Red, 1 4 Green, 2 1 Yellow, 2 1 Red, 2 1 Green, 2 2 Yellow, 2 2 Red, 2 2 Green, 2 3 Yellow, 2 3 Red, 2 3 Green, 2 4 Yellow, 2 4 Red, 2 4 Green, 3 1 Yellow, 3 1 Red, 3 1 Green, 3 2 Yellow, 3 2 Red, 3 2 Green, 3 3 Yellow, 3 3 Red, 3 3 Green, 3 4 Yellow, 3 4 Red, 3 4 Green, 4 1 Yellow, 4 1 Red, 4 1 Green, 4 2 Yellow, 4 2 Red, 4 2 Green, 4 3 Yellow, 4 3 Red, 4 3 Green, 4 4 Yellow, 4 4 Red, 4 4 Green}

There are 48 equally likely events and the probability of each is $\frac{1}{48}$.

The probability of getting a 2, followed by a 2 or 4, followed by a Red or Green is

$P(2\ 2\text{ Red}) + P(2\ 4\text{ Red}) + P(2\ 2\text{ Green}) + P(2\ 4\text{ Green}) = \frac{1}{48} + \frac{1}{48} + \frac{1}{48} + \frac{1}{48} = \frac{1}{12}$

23. A, B, C, F   25. B

27. Let $P(\text{tails}) = x$, then $P(\text{heads}) = 4x$

$$x + 4x = 1$$
$$5x = 1$$
$$x = \frac{1}{5}$$
$$P(\text{tails}) = \frac{1}{5}, \quad P(\text{heads}) = \frac{4}{5}$$

29. $P(2) = P(4) = P(6) = x \qquad P(1) = P(3) = P(5) = 2x$

$$P(1) + P(2) + P(3) + P(4) + P(5) + P(6) = 1$$
$$2x + x + 2x + x + 2x + x = 1$$
$$9x = 1$$
$$x = \frac{1}{9}$$
$$P(2) = P(4) = P(6) = \frac{1}{9} \qquad P(1) = P(3) = P(5) = \frac{2}{9}$$

31. $P(E) = \dfrac{n(E)}{n(S)} = \dfrac{n\{1,2,3\}}{10} = \dfrac{3}{10}$

33. $P(E) = \dfrac{n(E)}{n(S)} = \dfrac{n\{2,4,6,8,10\}}{10} = \dfrac{5}{10} = \dfrac{1}{2}$

35. $P(\text{white}) = \dfrac{n(\text{white})}{n(S)} = \dfrac{5}{5+10+8+7} = \dfrac{5}{30} = \dfrac{1}{6}$

37. The sample space is: $S = \{BBB, BBG, BGB, GBB, BGG, GBG, GGB, GGG\}$

   $P(3 \text{ boys}) = \dfrac{n(3 \text{ boys})}{n(S)} = \dfrac{1}{8}$

39. The sample space is:
   $S = \{BBBB, BBBG, BBGB, BGBB, GBBB, BBGG, BGBG, GBBG, BGGB, GBGB, GGBB,$
   $BGGG, GBGG, GGBG, GGGB, GGGG\}$

   $P(1 \text{ girl, } 3 \text{ boys}) = \dfrac{n(1 \text{ girl, } 3 \text{ boys})}{n(S)} = \dfrac{4}{16} = \dfrac{1}{4}$

41. $P(\text{sum of two dice is } 7) = \dfrac{n(\text{sum of two dice is } 7)}{n(S)}$

   $= \dfrac{n\{1,6 \text{ or } 2,5 \text{ or } 3,4 \text{ or } 4,3 \text{ or } 5,2 \text{ or } 6,1\}}{n(S)} = \dfrac{6}{36} = \dfrac{1}{6}$

43. $P(\text{sum of two dice is } 3) = \dfrac{n(\text{sum of two dice is } 3)}{n(S)} = \dfrac{n\{1,2 \text{ or } 2,1\}}{n(S)} = \dfrac{2}{36} = \dfrac{1}{18}$

45. $P(A \cup B) = P(A) + P(B) - P(A \cap B) = 0.25 + 0.45 - 0.15 = 0.55$

46. $P(A \cap B) = P(A) + P(B) - P(A \cup B) = 0.25 + 0.45 - 0.6 = 0.1$

47. $P(A \cup B) = P(A) + P(B) = 0.25 + 0.45 = 0.70$

49. $P(A \cup B) = P(A) + P(B) - P(A \cap B)$
    $0.85 = 0.60 + P(B) - 0.05$
    $P(B) = 0.85 - 0.60 + 0.05 = 0.30$

51. $P(\text{not victim}) = 1 - P(\text{victim}) = 1 - 0.265 = 0.735$

53. $P(\text{not in } 70\text{'s}) = 1 - P(\text{in } 70\text{'s}) = 1 - 0.3 = 0.7$

55. $P(\text{white or green}) = P(\text{white}) + P(\text{green}) = \dfrac{n(\text{white}) + n(\text{green})}{n(S)} = \dfrac{9+8}{9+8+3} = \dfrac{17}{20}$

57. $P(\text{not white}) = 1 - P(\text{white}) = 1 - \dfrac{n(\text{white})}{n(S)} = 1 - \dfrac{9}{20} = \dfrac{11}{20}$

59. $P(\text{strike or one}) = P(\text{strike}) + P(\text{one}) = \dfrac{n(\text{strike}) + n(\text{one})}{n(S)} = \dfrac{3+1}{8} = \dfrac{4}{8} = \dfrac{1}{2}$

61.   There are 30 households out of 100 with an income of \$30,000 or more.

$$P(E) = \frac{n(E)}{n(S)} = \frac{n(30{,}000 \text{ or more})}{n(\text{total households})} = \frac{30}{100} = \frac{3}{10} = 0.30$$

63.   There are 40 households out of 100 with an income of less than \$20,000.

$$P(E) = \frac{n(E)}{n(S)} = \frac{n(\text{less than } \$20{,}000)}{n(\text{total households})} = \frac{40}{100} = \frac{2}{5} = 0.40$$

65.   (a) $P(1 \text{ or } 2) = P(1) + P(2) = 0.24 + 0.33 = 0.57$

(b) $P(1 \text{ or more}) = P(1) + P(2) + P(3) + P(4 \text{ or more})$

$$= 0.24 + 0.33 + 0.21 + 0.17 = 0.95$$

(c) $P(3 \text{ or fewer}) = P(0) + P(1) + P(2) + P(3) = 0.05 + 0.24 + 0.33 + 0.21 = 0.83$

(d) $P(3 \text{ or more}) = P(3) + P(4 \text{ or more}) = 0.21 + 0.17 = 0.38$

(e) $P(\text{fewer than } 2) = P(0) + P(1) = 0.05 + 0.24 = 0.29$

(f) $P(\text{fewer than } 1) = P(0) = 0.05$

(g) $P(1, 2, \text{ or } 3) = P(1) + P(2) + P(3) = 0.24 + 0.33 + 0.21 = 0.78$

(h) $P(2 \text{ or more}) = P(2) + P(3) + P(4 \text{ or more}) = 0.33 + 0.21 + 0.17 = 0.71$

67.   (a) $P(\text{freshman or female}) = P(\text{freshman}) + P(\text{female}) - P(\text{freshman and female})$

$$= \frac{n(\text{freshman}) + n(\text{female}) - n(\text{freshman and female})}{n(S)}$$

$$= \frac{18 + 15 - 8}{33} = \frac{25}{33}$$

(b) $P(\text{sophomore or male}) = P(\text{sophomore}) + P(\text{male}) - P(\text{sophomore and male})$

$$= \frac{n(\text{sophomore}) + n(\text{male}) - n(\text{sophomore and male})}{n(S)}$$

$$= \frac{15 + 18 - 8}{33} = \frac{25}{33}$$

69. $P(\text{at least 2 with same birthday}) = 1 - P(\text{none with same birthday})$

$$= 1 - \frac{n(\text{different birthdays})}{n(S)}$$

$$= 1 - \frac{365 \cdot 364 \cdot 363 \cdot 362 \cdot 361 \cdot 360 \cdots 354}{365^{12}}$$

$$\approx 1 - 0.833$$

$$= 0.167$$

71.   The sample space for picking 5 out of 10 numbers in a particular order contains

$$P(10,5) = \frac{10!}{(10-5)!} = \frac{10!}{5!} = 30{,}240 \text{ possible outcomes.}$$

One of these is the desired outcome. Thus, the probability of winning is:

$$P(E) = \frac{n(E)}{n(S)} = \frac{n(\text{winning})}{n(\text{total possible outcomes})} = \frac{1}{30{,}240} \approx 0.000033069$$

Chapter 12

Counting and Probability

12.R Chapter Review

1. {Dave}, {Joanne}, {Erica},
   {Dave, Joanne}, {Dave, Erica}, {Joanne, Erica},
   {Dave, Joanne, Erica}, $\varnothing$

3. $A \cup B = \{1, 3, 5, 7\} \cup \{3, 5, 6, 7, 8\} = \{1, 3, 5, 6, 7, 8\}$

5. $A \cap C = \{1, 3, 5, 7\} \cap \{2, 3, 7, 8, 9\} = \{3, 7\}$

7. $\overline{A} \cup \overline{B} = \overline{\{1, 3, 5, 7\}} \cup \overline{\{3, 5, 6, 7, 8\}} = \{2, 4, 6, 8, 9\} \cup \{1, 2, 4, 9\} = \{1, 2, 4, 6, 8, 9\}$

9. $\overline{B \cap C} = \overline{\{3, 5, 6, 7, 8\} \cap \{2, 3, 7, 8, 9\}} = \overline{\{3, 7, 8\}} = \{1, 2, 4, 5, 6, 9\}$

11. $n(A) = 8, n(B) = 12, n(A \cap B) = 3$
    $n(A \cup B) = n(A) + n(B) - n(A \cap B) = 8 + 12 - 3 = 17$

13. From the figure: $n(A) = 20 + 2 + 6 + 1 = 29$

15. From the figure: $n(A \text{ and } C) = n(A \cap C) = 1 + 6 = 7$

17. From the figure: $n(\text{neither in } A \text{ nor in } C) = n(\overline{A \cup C}) = 20 + 5 = 25$

19. $P(8,3) = \dfrac{8!}{(8-3)!} = \dfrac{8!}{5!} = \dfrac{8 \cdot 7 \cdot 6 \cdot 5!}{5!} = 336$

21. $C(8,3) = \dfrac{8!}{(8-3)!\,3!} = \dfrac{8!}{5!\,3!} = \dfrac{8 \cdot 7 \cdot 6 \cdot 5!}{5! \cdot 3 \cdot 2 \cdot 1} = 56$

23. There are 2 choices of material, 3 choices of color, and 10 choices of size. The complete assortment would have: $2 \cdot 3 \cdot 10 = 60$ suits.

25. There are two possible outcomes for each game or
    $2 \cdot 2 \cdot 2 \cdot 2 \cdot 2 \cdot 2 \cdot 2 = 2^7 = 128$ outcomes for 7 games.

27. Since order is significant, this is a permutation.
    $P(9,4) = \dfrac{9!}{(9-4)!} = \dfrac{9!}{5!} = \dfrac{9 \cdot 8 \cdot 7 \cdot 6 \cdot 5!}{5!} = 3024$ ways to seat 4 people in 9 seats.

29. Choose 4 runners –order is not significant:

$$C(8,4) = \frac{8!}{(8-4)!\,4!} = \frac{8!}{4!\,4!} = \frac{8\cdot7\cdot6\cdot5\cdot4!}{4\cdot3\cdot2\cdot1\cdot4!} = 70 \text{ ways a squad can be chosen.}$$

31. Choose 2 teams from 14–order is not significant:

$$C(14,2) = \frac{14!}{(14-2)!\,2!} = \frac{14!}{12!\,2!} = \frac{14\cdot13\cdot12!}{12!\cdot2\cdot1} = 91 \text{ ways to choose 2 teams.}$$

33. There are $8\cdot10\cdot10\cdot10\cdot10\cdot2 = 1{,}600{,}000$ possible phone numbers.

35. There are $24\cdot9\cdot10\cdot10\cdot10 = 216{,}000$ possible license plates.

37. Since there are repeated letters:

$$\frac{7!}{2!\cdot2!} = \frac{7\cdot6\cdot5\cdot4\cdot3\cdot2\cdot1}{2\cdot1\cdot2\cdot1} = 1260 \text{ different words can be formed.}$$

39. (a) $C(9,4)\cdot C(9,3)\cdot C(9,2) = 126\cdot84\cdot36 = 381{,}024$ committees can be formed.
    (b) $C(9,4)\cdot C(5,3)\cdot C(2,2) = 126\cdot10\cdot1 = 1260$ committees can be formed.

41. (a) $365\cdot364\cdot363\cdot362\cdots348 = 8.634628387\times10^{45}$

    (b) $P(\text{no one has same birthday}) = \dfrac{365\cdot364\cdot363\cdot362\cdots348}{365^{18}} \approx 0.6531$

    (c) $P(\text{at least 2 have same birthday}) = 1 - P(\text{no one has same birthday})$
    $$= 1 - 0.6531 = 0.3469$$

43. (a) $P(\text{unemployed}) = 0.058$
    (b) $P(\text{not unemployed}) = 1 - P(\text{unemployed}) = 1 - 0.058 = 0.942$

45. $P(\$1 \text{ bill}) = \dfrac{n(\$1 \text{ bill})}{n(S)} = \dfrac{4}{9}$

47. Let S be all possible selections, let D be a card that is divisible by 5, and let PN be a 1 or a prime number.

   $n(S) = 100$

   $n(D) = 20$ (There are 20 numbers divisible by 5 between 1 and 100.)

   $n(PN) = 26$ (There are 25 prime numbers less than or equal to 100.)

   $$P(D) = \frac{n(D)}{n(S)} = \frac{20}{100} = \frac{1}{5} = 0.2$$

   $$P(PN) = \frac{n(PN)}{n(S)} = \frac{26}{100} = \frac{13}{50} = 0.26$$

Chapter 12

Counting and Probability

12.CR   Cumulative Review

1. $3x^2 - 2x = -1 \Rightarrow 3x^2 - 2x + 1 = 0$

$$x = \frac{-b \pm \sqrt{b^2 - 4ac}}{2a} = \frac{-(-2) \pm \sqrt{(-2)^2 - 4(3)(1)}}{2(3)} = \frac{2 \pm \sqrt{4 - 12}}{6}$$

$$= \frac{2 \pm \sqrt{-8}}{6} = \frac{2 \pm 2\sqrt{2}i}{6} = \frac{1 \pm \sqrt{2}i}{3}$$

The solution set is $\left\{ \dfrac{1 - \sqrt{2}i}{3}, \dfrac{1 + \sqrt{2}i}{3} \right\}$.

3. $y = 2(x + 1)^2 - 4$

Using the graph of $y = x^2$, horizontally shift to the left 1 unit, vertically stretch by a factor of 2, and vertically shift down 4 units.

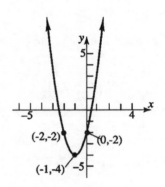

5. $f(x) = 5x^4 - 9x^3 - 7x^2 - 31x - 6$

Step 1: $f(x)$ has at most 4 real zeros.

Step 2: Possible rational zeros:
$p = \pm 1, \pm 2, \pm 3, \pm 6; \quad q = \pm 1, \pm 5;$

$$\frac{p}{q} = \pm 1, \pm \frac{1}{5}, \pm 2, \pm \frac{2}{5}, \pm 3, \pm \frac{4}{5}, \pm 6, \pm \frac{6}{5}$$

Step 3: Using the Bounds on Zeros Theorem:
$$f(x) = 5\left(x^4 - 1.8x^3 - 1.4x^2 - 6.2x - 1.2\right)$$

$a_3 = -1.8, \quad a_2 = -1.4, \quad a_1 = -6.2, \quad a_0 = -1.2$

$\text{Max}\left\{1, |-1.2| + |-6.2| + |-1.4| + |-1.8|\right\} = \text{Max}\{1, 10.6\} = 10.6$

$1 + \text{Max}\left\{|-1.2|, |-6.2|, |-1.4|, |-1.8|\right\} = 1 + 6.2 = 7.2$

Graphing using the bounds: (Second graph has a better window.)

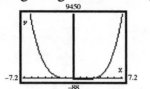 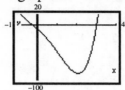

Step 4:   From the graph it appears that there are x-intercepts at -0.2 and 3.
Using synthetic division with 3:

$$
3\overline{)\begin{array}{ccccc} 5 & -9 & -7 & -31 & -6 \\ & 15 & 18 & 33 & 6 \\ \hline 5 & 6 & 11 & 2 & 0 \end{array}}
$$

Since the remainder is 0, $x-3$ is a factor. The other factor is the quotient: $5x^3 + 6x^2 + 11x + 2$.
Using synthetic division with 2 on the quotient:

$$
-0.2\overline{)\begin{array}{cccc} 5 & 6 & 11 & 2 \\ & -1 & -1 & -2 \\ \hline 5 & 5 & 10 & 0 \end{array}}
$$

Since the remainder is 0, $x-(-0.2) = x + 0.2$ is a factor. The other factor is the quotient: $5x^2 + 5x + 10 = 5\left(x^2 + x + 2\right)$.

Factoring, $f(x) = 5(x^2 + x + 2)(x - 3)(x + 0.2)$
The real zeros are 3 and -0.2.
The complex zeros come from solving $x^2 + x + 2 = 0$.

$$
x = \frac{-b \pm \sqrt{b^2 - 4ac}}{2a} = \frac{-1 \pm \sqrt{1^2 - 4(1)(2)}}{2(1)} = \frac{-1 \pm \sqrt{1 - 8}}{2} = \frac{-1 \pm \sqrt{-7}}{2} = \frac{-1 \pm \sqrt{7}i}{2}
$$

Therefore, the over the set of complex numbers, $f(x) = 5x^4 - 9x^3 - 7x^2 - 31x - 6$ has zeros -0.2, 3, $\dfrac{-1 - \sqrt{7}i}{2}, \dfrac{-1 + \sqrt{7}i}{2}$.

7. $\log_3(9) = \log_3\left(3^2\right) = 2$

9.   Multiply each side of the first equation by -3 and add to the second equation to eliminate x;
multiply each side of the first equation by 2 and add to the third equation to eliminate x:

$$
\begin{cases} x - 2y + z = 15 \\ 3x + y - 3z = -8 \\ -2x + 4y - z = -27 \end{cases}
$$

$$
\begin{array}{l}
x - 2y + z = 15 \xrightarrow{-3} -3x + 6y - 3z = -45 \\
\phantom{x - 2y + z = 15 \xrightarrow{-3}} \underline{3x + y - 3z = -8} \\
\phantom{x - 2y + z = 15 \xrightarrow{-3}} 7y - 6z = -53
\end{array}
$$

$$
\begin{array}{l}
x - 2y + z = 15 \xrightarrow{2} 2x - 4y + 2z = 30 \\
-2x + 4y - z = -27 \longrightarrow \underline{-2x + 4y - z = -27} \\
 z = 3
\end{array}
$$

735

Substituting and solving for the other variables:
$$z = 3 \Rightarrow 7y - 6(3) = -53$$
$$7y = -35$$
$$y = -5$$

$$z = 3, y = -5 \Rightarrow x - 2(-5) + 3 = 15$$
$$x + 10 + 3 = 15 \Rightarrow x = 2$$

The solution is $x = 2$, $y = -5$, $z = 3$.

11. $y = 3\sin(2x + \pi) = 3\sin\left(2\left(x + \dfrac{\pi}{2}\right)\right)$

Amplitude : $|A| = |3| = 3$

Period : $T = \dfrac{2\pi}{2} = \pi$

Phase Shift : $\dfrac{\phi}{\omega} = \dfrac{-\pi}{2} = -\dfrac{\pi}{2}$

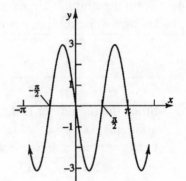

A Preview of Calculus: The Limit, Derivative, and Integral of a Function

13.1 Finding Limits Using Tables and Graphs

1. $f(x) = \begin{cases} 3x - 2 & \text{if } x \neq 2 \\ 3 & \text{if } x = 2 \end{cases}$

3. $\lim\limits_{x \to c} f(x)$

5. True

7. $\lim\limits_{x \to 2} \left(4x^3 \right)$

X	Y1
1.99	31.522
1.999	31.952
1.9999	31.995
2.0001	32.005
2.001	32.048
2.01	32.482

Y1◻4X^3

$\lim\limits_{x \to 2} \left(4x^3 \right) = 32$

9. $\lim\limits_{x \to 0} \left(\dfrac{x+1}{x^2+1} \right)$

X	Y1
-.01	.9899
-.001	.999
-1E-4	.9999
1E-4	1.0001
.001	1.001
.01	1.0099

Y1◻(X+1)/(X²+1)

$\lim\limits_{x \to 0} \left(\dfrac{x+1}{x^2+1} \right) = 1$

11. $\lim\limits_{x \to 4} \left(\dfrac{x^2 - 4x}{x - 4} \right)$

X	Y1
3.99	3.99
3.999	3.999
3.9999	3.9999
4.0001	4.0001
4.001	4.001
4.01	4.01

Y1◻(X²-4X)/(X-4)

$\lim\limits_{x \to 4} \left(\dfrac{x^2 - 4x}{x - 4} \right) = 4$

Chapter 13 A Preview of Calculus:
The Limit, Derivative, and Integral of a Function

13. $\lim\limits_{x\to 0}\left(e^x+1\right)$

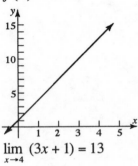

$\lim\limits_{x\to 0}\left(e^x+1\right)=2$

15. $\lim\limits_{x\to 0}\left(\dfrac{\cos x-1}{x}\right)$

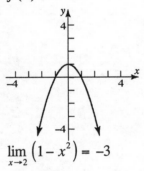

$\lim\limits_{x\to 0}\left(\dfrac{\cos x-1}{x}\right)=0$

17. $\lim\limits_{x\to 2}f(x)=3$

19. $\lim\limits_{x\to 2}f(x)=4$

21. $\lim\limits_{x\to 3}f(x)$ does not exist because as x gets closer to 3, but is less than 3, $f(x)$ gets closer to 3. However, as x gets closer to 3, but is greater than 3, $f(x)$ gets closer to 6.

23. $f(x)=3x+1$

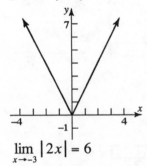

$\lim\limits_{x\to 4}(3x+1)=13$

25. $f(x)=1-x^2$

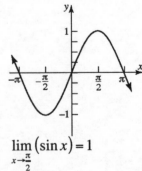

$\lim\limits_{x\to 2}\left(1-x^2\right)=-3$

27. $f(x)=|2x|$

$\lim\limits_{x\to -3}|2x|=6$

29. $f(x)=\sin x$

$\lim\limits_{x\to \frac{\pi}{2}}(\sin x)=1$

738

31. $f(x) = e^x$

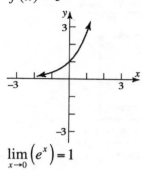

$$\lim_{x \to 0} \left(e^x \right) = 1$$

33. $f(x) = \dfrac{1}{x}$

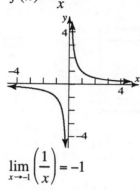

$$\lim_{x \to -1} \left(\dfrac{1}{x} \right) = -1$$

35. $f(x) = \begin{cases} x^2 & x \geq 0 \\ 2x & x < 0 \end{cases}$

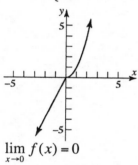

$$\lim_{x \to 0} f(x) = 0$$

37. $f(x) = \begin{cases} 3x & x \leq 1 \\ x+1 & x > 1 \end{cases}$

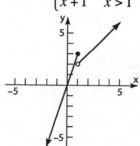

$$\lim_{x \to 1} f(x) \text{ does not exist}$$

39. $f(x) = \begin{cases} x & x < 0 \\ 1 & x = 0 \\ 3x & x > 0 \end{cases}$

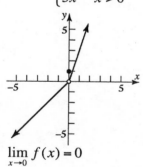

$$\lim_{x \to 0} f(x) = 0$$

41. $f(x) = \begin{cases} \sin x & x \leq 0 \\ x^2 & x > 0 \end{cases}$

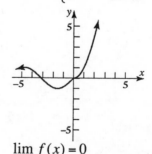

$$\lim_{x \to 0} f(x) = 0$$

43. $\lim\limits_{x \to 1} \left(\dfrac{x^3 - x^2 + x - 1}{x^4 - x^3 + 2x - 2} \right)$

X	Y1
.99	.66663
.999	.66667
.9999	.66667
1	ERROR
1.0001	.66667
1.001	.66667
1.01	.66663

Y1■(X^3–X²+X–1)...

$\lim\limits_{x \to 1} \left(\dfrac{x^3 - x^2 + x - 1}{x^4 - x^3 + 2x - 2} \right) \approx 0.67$

45. $\lim\limits_{x \to 2} \left(\dfrac{x^3 - 2x^2 + 4x - 8}{x^2 + x - 6} \right)$

X	Y1
1.99	1.5952
1.999	1.5995
1.9999	1.6
2	ERROR
2.0001	1.6
2.001	1.6005
2.01	1.6048

Y1■(X^3–2X²+4X–...

$\lim\limits_{x \to 2} \left(\dfrac{x^3 - 2x^2 + 4x - 8}{x^2 + x - 6} \right) = 1.60$

47. $\lim\limits_{x \to -1} \left(\dfrac{x^3 + 2x^2 + x}{x^4 + x^3 + 2x + 2} \right)$

X	Y1
-1.01	.01042
-1.001	.001
-1	ERROR
-.9999	-1E-4
-.999	-1E-3
-.99	-.0096

Y1■(X^3+2X²+X)/...

$\lim\limits_{x \to -1} \left(\dfrac{x^3 + 2x^2 + x}{x^4 + x^3 + 2x + 2} \right) = 0.00$

A Preview of Calculus: The Limit, Derivative, and Integral of a Function

13.2 Algebra Techniques for Finding Limits

1. product

3. 1

5. False

7. $\lim\limits_{x \to 1} (5) = 5$

9. $\lim\limits_{x \to 4} (x) = 4$

11. $\lim\limits_{x \to 2} (3x + 2) = 3(2) + 2 = 8$

13. $\lim\limits_{x \to -1} (3x^2 - 5x) = 3(-1)^2 - 5(-1) = 8$

15. $\lim\limits_{x \to 1} (5x^4 - 3x^2 + 6x - 9) = 5(1)^4 - 3(1)^2 + 6(1) - 9 = 5 - 3 + 6 - 9 = -1$

17. $\lim\limits_{x \to 1} (x^2 + 1)^3 = \left(\lim\limits_{x \to 1} (x^2 + 1) \right)^3 = \left(1^2 + 1\right)^3 = 2^3 = 8$

19. $\lim\limits_{x \to 1} \sqrt{5x + 4} = \sqrt{\lim\limits_{x \to 1} (5x + 4)} = \sqrt{5(1) + 4} = \sqrt{9} = 3$

21. $\lim\limits_{x \to 0} \left(\dfrac{x^2 - 4}{x^2 + 4} \right) = \dfrac{\lim\limits_{x \to 0}(x^2 - 4)}{\lim\limits_{x \to 0}(x^2 + 4)} = \dfrac{0^2 - 4}{0^2 + 4} = \dfrac{-4}{4} = -1$

23. $\lim\limits_{x \to 2} (3x - 2)^{5/2} = \left(\lim\limits_{x \to 2} (3x - 2) \right)^{5/2} = \left(3(2) - 2\right)^{5/2} = 4^{5/2} = 32$

25. $\lim\limits_{x \to 2} \left(\dfrac{x^2 - 4}{x^2 - 2x} \right) = \lim\limits_{x \to 2} \left(\dfrac{(x - 2)(x + 2)}{x(x - 2)} \right) = \lim\limits_{x \to 2} \left(\dfrac{x + 2}{x} \right) = \dfrac{2 + 2}{2} = \dfrac{4}{2} = 2$

27. $\lim\limits_{x \to -3} \left(\dfrac{x^2 - x - 12}{x^2 - 9} \right) = \lim\limits_{x \to -3} \left(\dfrac{(x - 4)(x + 3)}{(x - 3)(x + 3)} \right) = \lim\limits_{x \to -3} \left(\dfrac{x - 4}{x - 3} \right) = \dfrac{-3 - 4}{-3 - 3} = \dfrac{-7}{-6} = \dfrac{7}{6}$

29. $\lim\limits_{x \to 1} \left(\dfrac{x^3 - 1}{x - 1} \right) = \lim\limits_{x \to 1} \left(\dfrac{(x - 1)(x^2 + x + 1)}{x - 1} \right) = \lim\limits_{x \to 1} \left(x^2 + x + 1 \right) = 1^2 + 1 + 1 = 3$

31. $\lim\limits_{x \to -1}\left(\dfrac{(x+1)^2}{x^2-1}\right) = \lim\limits_{x \to -1}\left(\dfrac{(x+1)^2}{(x-1)(x+1)}\right) = \lim\limits_{x \to -1}\left(\dfrac{x+1}{x-1}\right) = \dfrac{-1+1}{-1-1} = \dfrac{0}{-2} = 0$

33. $\lim\limits_{x \to 1}\left(\dfrac{x^3-x^2+x-1}{x^4-x^3+2x-2}\right) = \lim\limits_{x \to 1}\left(\dfrac{x^2(x-1)+1(x-1)}{x^3(x-1)+2(x-1)}\right) = \lim\limits_{x \to 1}\left(\dfrac{(x-1)(x^2+1)}{(x-1)(x^3+2)}\right) = \lim\limits_{x \to 1}\left(\dfrac{x^2+1}{x^3+2}\right)$

$$= \dfrac{1^2+1}{1^3+2} = \dfrac{2}{3}$$

35. $\lim\limits_{x \to 2}\left(\dfrac{x^3-2x^2+4x-8}{x^2+x-6}\right) = \lim\limits_{x \to 2}\left(\dfrac{x^2(x-2)+4(x-2)}{(x+3)(x-2)}\right) = \lim\limits_{x \to 2}\left(\dfrac{(x-2)(x^2+4)}{(x+3)(x-2)}\right) = \lim\limits_{x \to 2}\left(\dfrac{x^2+4}{x+3}\right)$

$$= \dfrac{2^2+4}{2+3} = \dfrac{8}{5}$$

37. $\lim\limits_{x \to -1}\left(\dfrac{x^3+2x^2+x}{x^4+x^3+2x+2}\right) = \lim\limits_{x \to -1}\left(\dfrac{x(x^2+2x+1)}{x^3(x+1)+2(x+1)}\right) = \lim\limits_{x \to -1}\left(\dfrac{x(x+1)^2}{(x+1)(x^3+2)}\right)$

$$= \lim\limits_{x \to -1}\left(\dfrac{x(x+1)}{x^3+2}\right) = \dfrac{-1(-1+1)}{(-1)^3+2} = \dfrac{-1(0)}{-1+2} = \dfrac{0}{1} = 0$$

39. $\lim\limits_{x \to 2}\left(\dfrac{f(x)-f(2)}{x-2}\right) = \lim\limits_{x \to 2}\left(\dfrac{(5x-3)-7}{x-2}\right) = \lim\limits_{x \to 2}\left(\dfrac{5x-10}{x-2}\right) = \lim\limits_{x \to 2}\left(\dfrac{5(x-2)}{x-2}\right) = \lim\limits_{x \to 2}(5) = 5$

41. $\lim\limits_{x \to 3}\left(\dfrac{f(x)-f(3)}{x-3}\right) = \lim\limits_{x \to 3}\left(\dfrac{x^2-9}{x-3}\right) = \lim\limits_{x \to 3}\left(\dfrac{(x-3)(x+3)}{x-3}\right) = \lim\limits_{x \to 3}(x+3) = 3+3 = 6$

43. $\lim\limits_{x \to -1}\left(\dfrac{f(x)-f(-1)}{x-(-1)}\right) = \lim\limits_{x \to -1}\left(\dfrac{x^2+2x-(-1)}{x+1}\right) = \lim\limits_{x \to -1}\left(\dfrac{x^2+2x+1}{x+1}\right) = \lim\limits_{x \to -1}\left(\dfrac{(x+1)^2}{x+1}\right)$

$$= \lim\limits_{x \to -1}(x+1) = -1+1 = 0$$

45. $\lim\limits_{x \to 0}\left(\dfrac{f(x)-f(0)}{x-0}\right) = \lim\limits_{x \to 0}\left(\dfrac{3x^3-2x^2+4-4}{x}\right) = \lim\limits_{x \to 0}\left(\dfrac{3x^3-2x^2}{x}\right) = \lim\limits_{x \to 0}(3x^2-2x) = 0$

47. $\lim\limits_{x \to 1}\left(\dfrac{f(x)-f(1)}{x-1}\right) = \lim\limits_{x \to 1}\left(\dfrac{\frac{1}{x}-1}{x-1}\right) = \lim\limits_{x \to 1}\left(\dfrac{\frac{1-x}{x}}{x-1}\right) = \lim\limits_{x \to 1}\left(\dfrac{-1(x-1)}{x(x-1)}\right) = \lim\limits_{x \to 1}\left(\dfrac{-1}{x}\right) = \dfrac{-1}{1} = -1$

49. $\lim\limits_{x\to 0}\left(\dfrac{\tan x}{x}\right)=\lim\limits_{x\to 0}\left(\dfrac{\frac{\sin x}{\cos x}}{x}\right)=\lim\limits_{x\to 0}\left(\dfrac{\sin x}{x}\cdot\dfrac{1}{\cos x}\right)=\left(\lim\limits_{x\to 0}\left(\dfrac{\sin x}{x}\right)\right)\cdot\left(\lim\limits_{x\to 0}\left(\dfrac{1}{\cos x}\right)\right)$

$$=1\cdot\left(\dfrac{\lim\limits_{x\to 0}1}{\lim\limits_{x\to 0}(\cos x)}\right)=1\cdot\dfrac{1}{1}=1$$

51. $\lim\limits_{x\to 0}\left(\dfrac{3\sin x+\cos x-1}{4x}\right)=\lim\limits_{x\to 0}\left(\dfrac{3\sin x}{4x}+\dfrac{\cos x-1}{4x}\right)=\lim\limits_{x\to 0}\left(\dfrac{3\sin x}{4x}\right)+\lim\limits_{x\to 0}\left(\dfrac{\cos x-1}{4x}\right)$

$$=\dfrac{3}{4}\lim\limits_{x\to 0}\left(\dfrac{\sin x}{x}\right)+\dfrac{1}{4}\lim\limits_{x\to 0}\left(\dfrac{\cos x-1}{x}\right)=\dfrac{3}{4}\cdot 1+\dfrac{1}{4}\cdot 0=\dfrac{3}{4}$$

A Preview of Calculus: The Limit, Derivative, and Integral of a Function

13.3 One-Sided Limits; Continuous Functions

1. $f(0) = 0^2 = 0,\ f(2) = 5 - 2 = 3$ 3. True

5. True   7. one-sided

9. continuous, c 11. True

13. Domain: $\left\{ x \mid -8 \le x < -6 \text{ or } -6 < x < 4 \text{ or } 4 < x \le 6 \right\}$

15. x-intercepts: $-8, -5, -3$ 17. $f(-8) = 0;\ f(-4) = 2$

19. $\displaystyle\lim_{x \to -6^-} f(x) = \infty$ 21. $\displaystyle\lim_{x \to -4^-} f(x) = 2$

23. $\displaystyle\lim_{x \to 2^-} f(x) = 1$

25. $\displaystyle\lim_{x \to 4} f(x)$ does exist. $\displaystyle\lim_{x \to 4} f(x) = 0$ since $\displaystyle\lim_{x \to 4^-} f(x) = \lim_{x \to 4^+} f(x) = 0$

27. f is not continuous at -6 because $f(-6)$ does not exist.

29. f is continuous at 0 because $f(0) = \displaystyle\lim_{x \to 0^-} f(x) = \lim_{x \to 0^+} f(x) = 3$

31. f is not continuous at 4 because $f(4)$ does not exist.

33. $\displaystyle\lim_{x \to 1^+} (2x + 3) = 2(1) + 3 = 5$

35. $\displaystyle\lim_{x \to 1^-} \left(2x^3 + 5x \right) = 2(1)^3 + 5(1) = 2 + 5 = 7$

37. $\displaystyle\lim_{x \to \frac{\pi}{2}^+} (\sin x) = \sin \frac{\pi}{2} = 1$

39. $\displaystyle\lim_{x \to 2^+} \left(\frac{x^2 - 4}{x - 2} \right) = \lim_{x \to 2^+} \left(\frac{(x+2)(x-2)}{x-2} \right) = \lim_{x \to 2^+} (x+2) = 2 + 2 = 4$

41. $\displaystyle\lim_{x \to -1^-} \left(\frac{x^2 - 1}{x^3 + 1} \right) = \lim_{x \to -1^-} \left(\frac{(x+1)(x-1)}{(x+1)(x^2 - x + 1)} \right) = \lim_{x \to -1^-} \left(\frac{x-1}{x^2 - x + 1} \right) = \frac{-1-1}{(-1)^2 - (-1) + 1} = -\frac{2}{3}$

43. $\displaystyle\lim_{x \to -2^+} \left(\frac{x^2 + x - 2}{x^2 + 2x} \right) = \lim_{x \to -2^+} \left(\frac{(x+2)(x-1)}{x(x+2)} \right) = \lim_{x \to -2^+} \left(\frac{x-1}{x} \right) = \frac{-2-1}{-2} = \frac{-3}{-2} = \frac{3}{2}$

45. $f(x) = x^3 - 3x^2 + 2x - 6; \quad c = 2$

    1. $f(2) = 2^3 - 3 \cdot 2^2 + 2 \cdot 2 - 6 = -6$

    2. $\displaystyle\lim_{x \to 2^-} f(x) = 2^3 - 3 \cdot 2^2 + 2 \cdot 2 - 6 = -6$

    3. $\displaystyle\lim_{x \to 2^+} f(x) = 2^3 - 3 \cdot 2^2 + 2 \cdot 2 - 6 = -6$

Thus, $f(x)$ is continuous at $c = 2$.

47. $f(x) = \dfrac{x^2 + 5}{x - 6}; \quad c = 3$

    1. $f(3) = \dfrac{3^2 + 5}{3 - 6} = \dfrac{14}{-3} = -\dfrac{14}{3}$

    2. $\displaystyle\lim_{x \to 3^-} f(x) = \dfrac{3^2 + 5}{3 - 6} = \dfrac{14}{-3} = -\dfrac{14}{3}$

    3. $\displaystyle\lim_{x \to 3^+} f(x) = \dfrac{3^2 + 5}{3 - 6} = \dfrac{14}{-3} = -\dfrac{14}{3}$

Thus, $f(x)$ is continuous at $c = 3$.

49. $f(x) = \dfrac{x + 3}{x - 3}; \quad c = 3$

Since $f(x)$ is not defined at $c = 3$, the function is not continuous at $c = 3$.

51. $f(x) = \dfrac{x^3 + 3x}{x^2 - 3x}; \quad c = 0$

Since $f(x)$ is not defined at $c = 0$, the function is not continuous at $c = 0$.

53. $f(x) = \begin{cases} \dfrac{x^3 + 3x}{x^2 - 3x} & \text{if } x \neq 0 \\ 1 & \text{if } x = 0 \end{cases}; \quad c = 0$

    1. $f(0) = 1$

    2. $\displaystyle\lim_{x \to 0^-} f(x) = \lim_{x \to 0^-} \left(\frac{x^3 + 3x}{x^2 - 3x} \right) = \lim_{x \to 0^-} \left(\frac{x(x^2 + 3)}{x(x - 3)} \right) = \lim_{x \to 0^-} \left(\frac{x^2 + 3}{x - 3} \right) = \frac{3}{-3} = -1$

Since $\displaystyle\lim_{x \to 0^-} f(x) \neq f(0)$, the function is not continuous at $c = 0$.

55. $f(x) = \begin{cases} \dfrac{x^3 + 3x}{x^2 - 3x} & \text{if } x \neq 0 \\ -1 & \text{if } x = 0 \end{cases}$; $c = 0$

    1. $f(0) = -1$

    2. $\lim\limits_{x \to 0^-} f(x) = \lim\limits_{x \to 0^-} \left(\dfrac{x^3 + 3x}{x^2 - 3x} \right) = \lim\limits_{x \to 0^-} \left(\dfrac{x(x^2 + 3)}{x(x - 3)} \right) = \lim\limits_{x \to 0^-} \left(\dfrac{x^2 + 3}{x - 3} \right) = \dfrac{3}{-3} = -1$

    3. $\lim\limits_{x \to 0^+} f(x) = \lim\limits_{x \to 0^+} \left(\dfrac{x^3 + 3x}{x^2 - 3x} \right) = \lim\limits_{x \to 0^+} \left(\dfrac{x(x^2 + 3)}{x(x - 3)} \right) = \lim\limits_{x \to 0^+} \left(\dfrac{x^2 + 3}{x - 3} \right) = \dfrac{3}{-3} = -1$

The function is continuous at $c = 0$.

57. $f(x) = \begin{cases} \dfrac{x^3 - 1}{x^2 - 1} & \text{if } x < 1 \\ 2 & \text{if } x = 1 \\ \dfrac{3}{x + 1} & \text{if } x > 1 \end{cases}$; $c = 1$

    1. $f(1) = 2$

    2. $\lim\limits_{x \to 1^-} f(x) = \lim\limits_{x \to 1^-} \left(\dfrac{x^3 - 1}{x^2 - 1} \right) = \lim\limits_{x \to 1^-} \left(\dfrac{(x - 1)(x^2 + x + 1)}{(x - 1)(x + 1)} \right) = \lim\limits_{x \to 1^-} \left(\dfrac{x^2 + x + 1}{x + 1} \right) = \dfrac{3}{2}$

Since $\lim\limits_{x \to 1^-} f(x) \neq f(1)$, the function is not continuous at $c = 1$.

59. $f(x) = \begin{cases} 2e^x & \text{if } x < 0 \\ 2 & \text{if } x = 0 \\ \dfrac{x^3 + 2x^2}{x^2} & \text{if } x > 0 \end{cases}$; $c = 0$

    1. $f(0) = 2$

    2. $\lim\limits_{x \to 0^-} f(x) = \lim\limits_{x \to 0^-} \left(2e^x \right) = 2e^0 = 2 \cdot 1 = 2$

    3. $\lim\limits_{x \to 0^+} f(x) = \lim\limits_{x \to 0^+} \left(\dfrac{x^3 + 2x^2}{x^2} \right) = \lim\limits_{x \to 0^+} \left(\dfrac{x^2(x + 2)}{x^2} \right) = \lim\limits_{x \to 0^+} (x + 2) = 0 + 2 = 2$

The function is continuous at $c = 0$.

61. The domain of $f(x) = 2x + 3$ is all real numbers, and $f(x)$ is a polynomial function. Therefore, $f(x)$ is continuous everywhere.

63. The domain of $f(x) = 3x^2 + x$ is all real numbers, and $f(x)$ is a polynomial function. Therefore, $f(x)$ is continuous everywhere.

65. The domain of $f(x) = 4\sin x$ is all real numbers, and trigonometric functions are continuous at every point in their domains. Therefore, $f(x)$ is continuous everywhere.

67. The domain of $f(x) = 2\tan x$ is all real numbers except odd integer multiples of $\frac{\pi}{2}$, and trigonometric functions are continuous at every point in their domains. Therefore, $f(x)$ is continuous everywhere except where $x = \frac{k\pi}{2}$ where k is an odd integer. $f(x)$ is discontinuous at $x = \frac{k\pi}{2}$ where k is an odd integer.

69. $f(x) = \dfrac{2x+5}{x^2-4} = \dfrac{2x+5}{(x-2)(x+2)}$. The domain of $f(x)$ is all real numbers except $x = 2$ and $x = -2$, and $f(x)$ is a rational function. Therefore, $f(x)$ is continuous everywhere except at $x = 2$ and $x = -2$. $f(x)$ is discontinuous at $x = 2$ and $x = -2$.

71. $f(x) = \dfrac{x-3}{\ln x}$. The domain of $f(x)$ is $(0, 1)$ or $(1, \infty)$. Thus, $f(x)$ is continuous on the interval $(0, \infty)$ except at $x = 1$. $f(x)$ is discontinuous at $x = 1$.

73. $R(x) = \dfrac{x-1}{x^2-1} = \dfrac{x-1}{(x-1)(x+1)}$. The domain of R is $\left\{ x \mid x \neq -1, \ x \neq 1 \right\}$. Thus R is discontinuous at both -1 and 1. $\displaystyle\lim_{x\to-1^-} R(x) = \lim_{x\to-1^-} \left(\dfrac{x-1}{(x-1)(x+1)} \right) = \lim_{x\to-1^-} \left(\dfrac{1}{x+1} \right) = -\infty$ since when $x < -1$, $\dfrac{1}{x+1} < 0$, and as x approaches -1, $\dfrac{1}{x+1}$ becomes unbounded.

$\displaystyle\lim_{x\to-1^+} R(x) = \lim_{x\to-1^+} \left(\dfrac{1}{x+1} \right) = \infty$ since when $x > -1$, $\dfrac{1}{x+1} > 0$, and as x approaches -1, $\dfrac{1}{x+1}$ becomes unbounded. $\displaystyle\lim_{x\to 1} R(x) = \lim_{x\to 1} \left(\dfrac{1}{x+1} \right) = \dfrac{1}{2}$. Note there is a hole in the graph at $\left(1, \dfrac{1}{2} \right)$.

75. $R(x) = \dfrac{x^2+x}{x^2-1} = \dfrac{x(x+1)}{(x-1)(x+1)}$. The domain of R is $\left\{ x \mid x \neq -1, \ x \neq 1 \right\}$. Thus R is discontinuous at both -1 and 1. $\displaystyle\lim_{x\to 1^-} R(x) = \lim_{x\to 1^-} \left(\dfrac{x(x+1)}{(x-1)(x+1)} \right) = \lim_{x\to 1^-} \left(\dfrac{x}{x-1} \right) = -\infty$ since when $x < 1$, $\dfrac{x}{x-1} < 0$, and as x approaches 1, $\dfrac{x}{x-1}$ becomes unbounded.

$\lim\limits_{x\to1^+} R(x) = \lim\limits_{x\to1^+} \dfrac{x}{x-1} = \infty$ since when $x > 1$, $\dfrac{x}{x-1} > 0$, and as x approaches 1, $\dfrac{x}{x-1}$

becomes unbounded. $\lim\limits_{x\to-1} R(x) = \lim\limits_{x\to-1}\left(\dfrac{x}{x-1}\right) = \dfrac{-1}{-2} = \dfrac{1}{2}$. Note there is a hole in the graph

at $\left(-1, \dfrac{1}{2}\right)$.

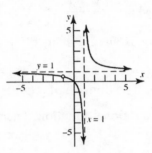

77. $R(x) = \dfrac{x^3 - x^2 + x - 1}{x^4 - x^3 + 2x - 2} = \dfrac{x^2(x-1) + 1(x-1)}{x^3(x-1) + 2(x-1)} = \dfrac{(x-1)(x^2+1)}{(x-1)(x^3+2)} = \dfrac{x^2+1}{x^3+2}$, $x \ne 1$

There is a vertical asymptote where $x^3 + 2 = 0$. $x = -\sqrt[3]{2}$ is a vertical asymptote. There is
a hole in the graph at $x = 1$.

79. $R(x) = \dfrac{x^3 - 2x^2 + 4x - 8}{x^2 + x - 6} = \dfrac{x^2(x-2) + 4(x-2)}{(x+3)(x-2)} = \dfrac{(x-2)(x^2+4)}{(x+3)(x-2)} = \dfrac{x^2+4}{x+3}$, $x \ne 2$

There is a vertical asymptote where $x + 3 = 0$. $x = -3$ is a vertical asymptote. There is a
hole in the graph at $x = 2$.

81. $R(x) = \dfrac{x^3 + 2x^2 + x}{x^4 + x^3 + 2x + 2} = \dfrac{x(x^2 + 2x + 1)}{x^3(x+1) + 2(x+1)} = \dfrac{x(x+1)^2}{(x+1)(x^3+2)} = \dfrac{x(x+1)}{x^3+2}$, $x \ne -1$

There is a vertical asymptote where $x^3 + 2 = 0$. $x = -\sqrt[3]{2}$ is a vertical asymptote. There is
a hole in the graph at $x = -1$.

83. $R(x) = \dfrac{x^3 - x^2 + x - 1}{x^4 - x^3 + 2x - 2}$

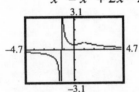

85. $R(x) = \dfrac{x^3 - 2x^2 + 4x - 8}{x^2 + x - 6}$

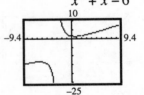

87. $R(x) = \dfrac{x^3 + 2x^2 + x}{x^4 + x^3 + 2x + 2}$

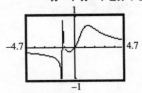

89.   Answers will vary. Three possible
      function are:
      $f(x) = x^2$; $g(x) = \sin(x)$; $h(x) = e^x$.

A Preview of Calculus:
The Limit, Derivative, and Integral of a Function

13.4 The Tangent Problem; The Derivative

1. Slope 5; containing point $(2,-4)$

$$y - y_1 = m(x - x_1)$$
$$y - (-4) = 5(x - 2)$$
$$y + 4 = 5x - 10$$
$$y = 5x - 14$$

3. tangent line

5. velocity

7. True

9. $f(x) = 3x + 5$ at $(1, 8)$

$$m_{\tan} = \lim_{x \to 1}\left(\frac{f(x) - f(1)}{x - 1}\right) = \lim_{x \to 1}\left(\frac{3x + 5 - 8}{x - 1}\right)$$

$$= \lim_{x \to 1}\left(\frac{3x - 3}{x - 1}\right) = \lim_{x \to 1}\left(\frac{3(x - 1)}{x - 1}\right) = \lim_{x \to 1}(3) = 3$$

Tangent Line: $y - 8 = 3(x - 1)$
$$y - 8 = 3x - 3$$
$$y = 3x + 5$$

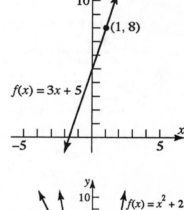

11. $f(x) = x^2 + 2$ at $(-1, 3)$

$$m_{\tan} = \lim_{x \to -1}\left(\frac{f(x) - f(-1)}{x + 1}\right) = \lim_{x \to -1}\left(\frac{x^2 + 2 - 3}{x + 1}\right)$$

$$= \lim_{x \to -1}\left(\frac{x^2 - 1}{x + 1}\right) = \lim_{x \to -1}\left(\frac{(x + 1)(x - 1)}{x + 1}\right)$$

$$= \lim_{x \to -1}(x - 1) = -1 - 1 = -2$$

Tangent Line: $y - 3 = -2(x - (-1))$
$$y - 3 = -2x - 2$$
$$y = -2x + 1$$

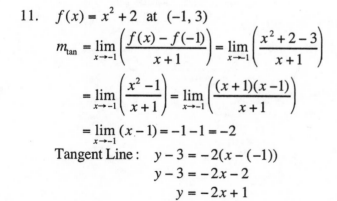

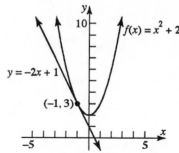

13. $f(x) = 3x^2$ at $(2, 12)$

$m_{\tan} = \lim\limits_{x \to 2} \left(\dfrac{f(x) - f(2)}{x - 2} \right) = \lim\limits_{x \to 2} \left(\dfrac{3x^2 - 12}{x - 2} \right)$

$\qquad = \lim\limits_{x \to 2} \dfrac{3(x^2 - 4)}{x - 2} = \lim\limits_{x \to 2} \left(\dfrac{3(x + 2)(x - 2)}{x - 2} \right)$

$\qquad = \lim\limits_{x \to 2} \left(3(x + 2) \right) = 3(2 + 2) = 12$

Tangent Line: $y - 12 = 12(x - 2)$

$\qquad\qquad\quad y - 12 = 12x - 24$

$\qquad\qquad\qquad\quad y = 12x - 12$

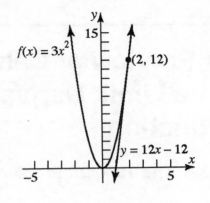

15. $f(x) = 2x^2 + x$ at $(1, 3)$

$m_{\tan} = \lim\limits_{x \to 1} \left(\dfrac{f(x) - f(1)}{x - 1} \right) = \lim\limits_{x \to 1} \left(\dfrac{2x^2 + x - 3}{x - 1} \right)$

$\qquad = \lim\limits_{x \to 1} \left(\dfrac{(2x + 3)(x - 1)}{x - 1} \right) = \lim\limits_{x \to 1} (2x + 3)$

$\qquad = 2(1) + 3 = 5$

Tangent Line: $y - 3 = 5(x - 1)$

$\qquad\qquad\quad y - 3 = 5x - 5$

$\qquad\qquad\qquad y = 5x - 2$

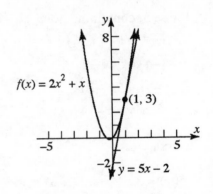

17. $f(x) = x^2 - 2x + 3$ at $(-1, 6)$

$m_{\tan} = \lim\limits_{x \to -1} \dfrac{f(x) - f(-1)}{x + 1} = \lim\limits_{x \to -1} \dfrac{x^2 - 2x + 3 - 6}{x + 1}$

$\qquad = \lim\limits_{x \to -1} \dfrac{x^2 - 2x - 3}{x + 1} = \lim\limits_{x \to -1} \dfrac{(x + 1)(x - 3)}{x + 1}$

$\qquad = \lim\limits_{x \to -1} (x - 3) = -1 - 3 = -4$

Tangent Line: $y - 6 = -4(x - (-1))$

$\qquad\qquad\quad y - 6 = -4x - 4$

$\qquad\qquad\qquad y = -4x + 2$

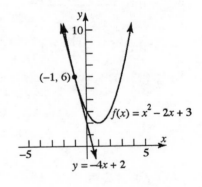

19. $f(x) = x^3 + x$ at $(2, 10)$

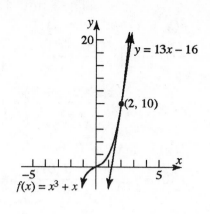

$$m_{\tan} = \lim_{x \to 2}\left(\frac{f(x) - f(2)}{x - 2}\right) = \lim_{x \to 2}\left(\frac{x^3 + x - 10}{x - 2}\right)$$

$$= \lim_{x \to 2}\left(\frac{x^3 - 8 + x - 2}{x - 2}\right)$$

$$= \lim_{x \to 2}\left(\frac{(x - 2)(x^2 + 2x + 4) + x - 2}{x - 2}\right)$$

$$= \lim_{x \to 2}\left(\frac{(x - 2)(x^2 + 2x + 4 + 1)}{x - 2}\right)$$

$$= \lim_{x \to 2}(x^2 + 2x + 5) = 4 + 4 + 5 = 13$$

Tangent Line : $y - 10 = 13(x - 2)$
$$y - 10 = 13x - 26$$
$$y = 13x - 16$$

21. $f(x) = -4x + 5$ at 3

$$f'(3) = \lim_{x \to 3}\left(\frac{f(x) - f(3)}{x - 3}\right) = \lim_{x \to 3}\left(\frac{-4x + 5 - (-7)}{x - 3}\right) = \lim_{x \to 3}\left(\frac{-4x + 12}{x - 3}\right) = \lim_{x \to 3}\left(\frac{-4(x - 3)}{x - 3}\right)$$

$$= \lim_{x \to 3}(-4) = -4$$

23. $f(x) = x^2 - 3$ at 0

$$f'(0) = \lim_{x \to 0}\left(\frac{f(x) - f(0)}{x - 0}\right) = \lim_{x \to 0}\left(\frac{x^2 - 3 - (-3)}{x}\right) = \lim_{x \to 0}\left(\frac{x^2}{x}\right) = \lim_{x \to 0}(x) = 0$$

25. $f(x) = 2x^2 + 3x$ at 1

$$f'(1) = \lim_{x \to 1}\left(\frac{f(x) - f(1)}{x - 1}\right) = \lim_{x \to 1}\left(\frac{2x^2 + 3x - 5}{x - 1}\right) = \lim_{x \to 1}\left(\frac{(2x + 5)(x - 1)}{x - 1}\right) = \lim_{x \to 1}(2x + 5) = 7$$

27. $f(x) = x^3 + 4x$ at -1

$$f'(-1) = \lim_{x \to -1}\left(\frac{f(x) - f(-1)}{x - (-1)}\right) = \lim_{x \to -1}\left(\frac{x^3 + 4x - (-5)}{x + 1}\right) = \lim_{x \to -1}\left(\frac{x^3 + 1 + 4x + 4}{x + 1}\right)$$

$$= \lim_{x \to -1}\left(\frac{(x + 1)(x^2 - x + 1) + 4(x + 1)}{x + 1}\right) = \lim_{x \to -1}\left(\frac{(x + 1)(x^2 - x + 1 + 4)}{x + 1}\right)$$

$$= \lim_{x \to -1}(x^2 - x + 5) = (-1)^2 - (-1) + 5 = 7$$

29. $f(x) = x^3 + x^2 - 2x$ at 1

$$f'(1) = \lim_{x \to 1}\left(\frac{f(x) - f(1)}{x - 1}\right) = \lim_{x \to 1}\left(\frac{x^3 + x^2 - 2x - 0}{x - 1}\right) = \lim_{x \to 1}\left(\frac{x(x^2 + x - 2)}{x - 1}\right)$$

$$= \lim_{x \to 1}\left(\frac{x(x + 2)(x - 1)}{x - 1}\right) = \lim_{x \to 1}\left(x(x + 2)\right) = 1(1 + 2) = 3$$

31. $f(x) = \sin x$ at 0

$$f'(0) = \lim_{x \to 0}\left(\frac{f(x) - f(0)}{x - 0}\right) = \lim_{x \to 0}\left(\frac{\sin x - 0}{x - 0}\right) = \lim_{x \to 0}\left(\frac{\sin x}{x}\right) = 1$$

33. Use NDeriv:

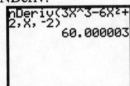

35. Use NDeriv:

37. Use NDeriv:

39. Use NDeriv:

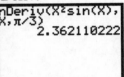

41. Use NDeriv:

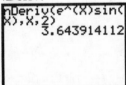

43. $V(r) = 3\pi r^2$ at $r = 3$

$$V'(3) = \lim_{r \to 3}\left(\frac{V(r) - V(3)}{r - 3}\right) = \lim_{r \to 3}\left(\frac{3\pi r^2 - 27\pi}{r - 3}\right) = \lim_{r \to 3}\left(\frac{3\pi(r^2 - 9)}{r - 3}\right)$$

$$= \lim_{r \to 3}\left(\frac{3\pi(r - 3)(r + 3)}{r - 3}\right) = \lim_{r \to 3}\left(3\pi(r + 3)\right) = 3\pi(3 + 3) = 18\pi$$

At the instant $r = 3$ feet, the volume of the cylinder is increasing at a rate of 18π cubic feet.

45. $V(r) = \frac{4}{3}\pi r^3$ at $r = 2$

$$V'(2) = \lim_{r \to 2}\left(\frac{V(r) - V(2)}{r - 2}\right) = \lim_{r \to 2}\left(\frac{\frac{4}{3}\pi r^3 - \frac{32}{3}\pi}{r - 2}\right) = \lim_{r \to 2}\left(\frac{\left(\frac{4}{3}\pi(r^3 - 8)\right)}{r - 2}\right)$$

$$= \lim_{r \to 2}\left(\frac{\left(\frac{4}{3}\pi(r - 2)(r^2 + 2r + 4)\right)}{r - 2}\right) = \lim_{r \to 2}\left(\frac{4}{3}\pi(r^2 + 2r + 4)\right) = \frac{4}{3}\pi(4 + 4 + 4) = 16\pi$$

At the instant $r = 2$ feet, the volume of the sphere is increasing at a rate of 16π cubic feet.

47. (a) $-16t^2 + 96t = 0$
      $-16t(t - 6) = 0$
          $t = 0$ or $t = 6$
      The ball strikes the ground after 6 seconds.

   (b) $\dfrac{\Delta s}{\Delta t} = \dfrac{s(2) - s(0)}{2 - 0} = \dfrac{-16(2)^2 + 96(2) - 0}{2} = \dfrac{128}{2} = 64$ feet / sec

   (c) $s'(t_0) = \lim_{t \to t_0}\left(\dfrac{s(t) - s(t_0)}{t - t_0}\right) = \lim_{t \to t_0}\left(\dfrac{-16t^2 + 96t - \left(-16t_0^2 + 96t_0\right)}{t - t_0}\right)$

   $$= \lim_{t \to t_0}\left(\dfrac{-16t^2 + 16t_0^2 + 96t - 96t_0}{t - t_0}\right) = \lim_{t \to t_0}\left(\dfrac{-16\left(t^2 - t_0^2\right) + 96(t - t_0)}{t - t_0}\right)$$

   $$= \lim_{t \to t_0}\left(\dfrac{-16(t - t_0)(t + t_0) + 96(t - t_0)}{t - t_0}\right) = \lim_{t \to t_0}\left(\dfrac{(t - t_0)(-16(t + t_0) + 96)}{t - t_0}\right)$$

   $$= \lim_{t \to t_0}\left(-16(t + t_0) + 96\right) = \left(-16(t_0 + t_0) + 96\right) = -32t_0 + 96$$

   (d) $s'(2) = -32(2) + 96 = -64 + 96 = 32$ feet/sec
   (e) $s'(t) = 0$
       $-32t + 96 = 0$
          $-32t = -96$
              $t = 3$ seconds
   (f) $s(3) = -16(3)^2 + 96(3) = -144 + 288 = 144$ feet
   (g) $s'(6) = -32(6) + 96 = -192 + 96 = -96$ feet/sec

49. (a) $\dfrac{\Delta s}{\Delta t} = \dfrac{s(4) - s(1)}{4 - 1} = \dfrac{917 - 987}{3} = \dfrac{-70}{3} = -23\dfrac{1}{3}$ feet/sec

   (b) $\dfrac{\Delta s}{\Delta t} = \dfrac{s(3) - s(1)}{3 - 1} = \dfrac{945 - 987}{2} = \dfrac{-42}{2} = -21$ feet/sec

   (c) $\dfrac{\Delta s}{\Delta t} = \dfrac{s(2) - s(1)}{2 - 1} = \dfrac{969 - 987}{1} = \dfrac{-18}{1} = -18$ feet/sec

   (d) $s(t) = -2.631t^2 - 10.269t + 999.933$

(e) $s'(1) = \lim_{t \to 1}\left(\dfrac{s(t) - s(1)}{t-1}\right) = \lim_{t \to 1}\left(\dfrac{-2.631t^2 - 10.269t + 999.933 - \left(-2.631(1)^2 - 10.269(1) + 999\right.}{t-1}\right.$

$= \lim_{t \to 1}\left(\dfrac{-2.631\,t^2 - 10.269t + 12.9}{t-1}\right)$

$= \lim_{t \to 1}\left(\dfrac{-2.631t^2 + 2.631t - 12.9t + 12.9}{t-1}\right)$

$= \lim_{t \to 1}\left(\dfrac{-2.631\,t(t-1) - 12.9(t-1)}{t-1}\right) = \lim_{t \to 1}\left(\dfrac{(-2.631t - 12.9)(t-1)}{t-1}\right)$

$= \lim_{t \to 1}(-2.631t - 12.9) = -2.631(1) - 12.9 = -15.531 \text{ feet/sec}$

At the instant when $t = 1$, the instantaneous speed of the ball is -15.531 feet / sec.

A Preview of Calculus:
The Limit, Derivative, and Integral of a Function

13.5 The Area Problem; The Integral

1. $A = lw$

3. $\displaystyle\int_a^b f(x)\,dx$

5. $A \approx f(1)\cdot 1 + f(2)\cdot 1 = 1\cdot 1 + 2\cdot 1 = 1 + 2 = 3$

7. $A \approx f(0)\cdot 2 + f(2)\cdot 2 + f(4)\cdot 2 + f(6)\cdot 2 = 10\cdot 2 + 6\cdot 2 + 7\cdot 2 + 5\cdot 2$
     $\quad = 20 + 12 + 14 + 10 = 56$

9.   (a)   Graph $f(x) = 3x$:

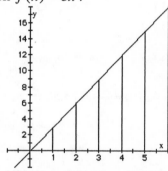

   (b) $A \approx f(0)(2) + f(2)(2) + f(4)(2) = 0(2) + 6(2) + 12(2) = 0 + 12 + 24 = 36$
   (c) $A \approx f(2)(2) + f(4)(2) + f(6)(2) = 6(2) + 12(2) + 18(2) = 12 + 24 + 36 = 72$
   (d) $A \approx f(0)(1) + f(1)(1) + f(2)(1) + f(3)(1) + f(4)(1) + f(5)(1)$
         $\quad = 0(1) + 3(1) + 6(1) + 9(1) + 12(1) + 15(1) = 0 + 3 + 6 + 9 + 12 + 15 = 45$
   (e) $A \approx f(1)(1) + f(2)(1) + f(3)(1) + f(4)(1) + f(5)(1) + f(6)(1)$
         $\quad = 3(1) + 6(1) + 9(1) + 12(1) + 15(1) + 18(1) = 3 + 6 + 9 + 12 + 15 + 18 = 63$

   (f)   The actual area is the area of a triangle: $A = \dfrac{1}{2}(6)(18) = 54$

11.   (a)   Graph $f(x) = -3x + 9$:

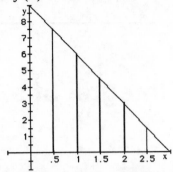

(b) $A \approx f(0)(1) + f(1)(1) + f(2)(1) = 9(1) + 6(1) + 3(1) = 9 + 6 + 3 = 18$

(c) $A \approx f(1)(1) + f(2)(1) + f(3)(1) = 6(1) + 3(1) + 0(1) = 6 + 3 + 0 = 9$

(d) $A \approx f(0)\left(\dfrac{1}{2}\right) + f\left(\dfrac{1}{2}\right)\left(\dfrac{1}{2}\right) + f(1)\left(\dfrac{1}{2}\right) + f\left(\dfrac{3}{2}\right)\left(\dfrac{1}{2}\right) + f(2)\left(\dfrac{1}{2}\right) + f\left(\dfrac{5}{2}\right)\left(\dfrac{1}{2}\right)$

$= 9\left(\dfrac{1}{2}\right) + \dfrac{15}{2}\left(\dfrac{1}{2}\right) + 6\left(\dfrac{1}{2}\right) + \dfrac{9}{2}\left(\dfrac{1}{2}\right) + 3\left(\dfrac{1}{2}\right) + \dfrac{3}{2}\left(\dfrac{1}{2}\right)$

$= \dfrac{9}{2} + \dfrac{15}{4} + 3 + \dfrac{9}{4} + \dfrac{3}{2} + \dfrac{3}{4} = \dfrac{63}{4} = 15.75$

(e) $A \approx f\left(\dfrac{1}{2}\right)\left(\dfrac{1}{2}\right) + f(1)\left(\dfrac{1}{2}\right) + f\left(\dfrac{3}{2}\right)\left(\dfrac{1}{2}\right) + f(2)\left(\dfrac{1}{2}\right) + f\left(\dfrac{5}{2}\right)\left(\dfrac{1}{2}\right) + f(3)\left(\dfrac{1}{2}\right)$

$= \dfrac{15}{2}\left(\dfrac{1}{2}\right) + 6\left(\dfrac{1}{2}\right) + \dfrac{9}{2}\left(\dfrac{1}{2}\right) + 3\left(\dfrac{1}{2}\right) + \dfrac{3}{2}\left(\dfrac{1}{2}\right) + 0\left(\dfrac{1}{2}\right)$

$= \dfrac{15}{4} + 3 + \dfrac{9}{4} + \dfrac{3}{2} + \dfrac{3}{4} + 0 = \dfrac{45}{4} = 11.25$

(f)   The actual area is the area of a triangle: $A = \dfrac{1}{2}(3)(9) = \dfrac{27}{2} = 13.5$

13.   (a)   Graph $f(x) = x^2 + 2$, $[0, 4]$:

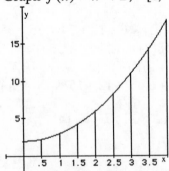

(b) $A \approx f(0)(1) + f(1)(1) + f(2)(1) + f(3)(1) = 2(1) + 3(1) + 6(1) + 11(1)$

$\quad = 2 + 3 + 6 + 11 = 22$

(c) $A \approx f(0)\left(\dfrac{1}{2}\right) + f\left(\dfrac{1}{2}\right)\left(\dfrac{1}{2}\right) + f(1)\left(\dfrac{1}{2}\right) + f\left(\dfrac{3}{2}\right)\left(\dfrac{1}{2}\right) + f(2)\left(\dfrac{1}{2}\right) + f\left(\dfrac{5}{2}\right)\left(\dfrac{1}{2}\right)$

$\qquad\qquad\qquad\qquad\qquad + f(3)\left(\dfrac{1}{2}\right) + f\left(\dfrac{7}{2}\right)\left(\dfrac{1}{2}\right)$

$$= 2\left(\frac{1}{2}\right) + \frac{9}{4}\left(\frac{1}{2}\right) + 3\left(\frac{1}{2}\right) + \frac{17}{4}\left(\frac{1}{2}\right) + 6\left(\frac{1}{2}\right) + \frac{33}{4}\left(\frac{1}{2}\right) + 11\left(\frac{1}{2}\right) + \frac{57}{4}\left(\frac{1}{2}\right)$$

$$= 1 + \frac{9}{8} + \frac{3}{2} + \frac{17}{8} + 3 + \frac{33}{8} + \frac{11}{2} + \frac{57}{8} = \frac{51}{2}$$

(d) $A = \displaystyle\int_0^4 (x^2 + 2)\,dx$

(e)   Use fnInt function:

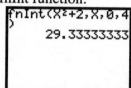

15. (a)   Graph $f(x) = x^3$, $[0, 4]$:

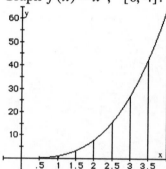

(b) $A \approx f(0)(1) + f(1)(1) + f(2)(1) + f(3)(1) = 0(1) + 1(1) + 8(1) + 27(1)$
        $= 0 + 1 + 8 + 27 = 36$

(c) $A \approx f(0)\left(\frac{1}{2}\right) + f\left(\frac{1}{2}\right)\left(\frac{1}{2}\right) + f(1)\left(\frac{1}{2}\right) + f\left(\frac{3}{2}\right)\left(\frac{1}{2}\right) + f(2)\left(\frac{1}{2}\right) + f\left(\frac{5}{2}\right)\left(\frac{1}{2}\right)$

$$+ f(3)\left(\frac{1}{2}\right) + f\left(\frac{7}{2}\right)\left(\frac{1}{2}\right)$$

$$= 0\left(\frac{1}{2}\right) + \frac{1}{8}\left(\frac{1}{2}\right) + 1\left(\frac{1}{2}\right) + \frac{27}{8}\left(\frac{1}{2}\right) + 8\left(\frac{1}{2}\right) + \frac{125}{8}\left(\frac{1}{2}\right) + 27\left(\frac{1}{2}\right) + \frac{343}{8}\left(\frac{1}{2}\right)$$

$$= 0 + \frac{1}{16} + \frac{1}{2} + \frac{27}{16} + 4 + \frac{125}{16} + \frac{27}{2} + \frac{343}{16} = 49$$

(d) $A = \displaystyle\int_0^4 x^3\,dx$

(e)   Use fnInt function:

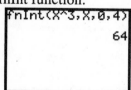

17. (a) Graph $f(x) = \dfrac{1}{x}$, $[1, 5]$:

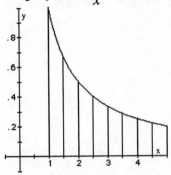

(b) $A \approx f(1)(1) + f(2)(1) + f(3)(1) + f(4)(1) = 1(1) + \dfrac{1}{2}(1) + \dfrac{1}{3}(1) + \dfrac{1}{4}(1)$

$= 1 + \dfrac{1}{2} + \dfrac{1}{3} + \dfrac{1}{4} = \dfrac{25}{12}$

(c) $A \approx f(1)\left(\dfrac{1}{2}\right) + f\left(\dfrac{3}{2}\right)\left(\dfrac{1}{2}\right) + f(2)\left(\dfrac{1}{2}\right) + f\left(\dfrac{5}{2}\right)\left(\dfrac{1}{2}\right) + f(3)\left(\dfrac{1}{2}\right) + f\left(\dfrac{7}{2}\right)\left(\dfrac{1}{2}\right)$

$+ f(4)\left(\dfrac{1}{2}\right) + f\left(\dfrac{9}{2}\right)\left(\dfrac{1}{2}\right)$

$= 1\left(\dfrac{1}{2}\right) + \dfrac{2}{3}\left(\dfrac{1}{2}\right) + \dfrac{1}{2}\left(\dfrac{1}{2}\right) + \dfrac{2}{5}\left(\dfrac{1}{2}\right) + \dfrac{1}{3}\left(\dfrac{1}{2}\right) + \dfrac{2}{7}\left(\dfrac{1}{2}\right) + \dfrac{1}{4}\left(\dfrac{1}{2}\right) + \dfrac{2}{9}\left(\dfrac{1}{2}\right)$

$= \dfrac{1}{2} + \dfrac{1}{3} + \dfrac{1}{4} + \dfrac{1}{5} + \dfrac{1}{6} + \dfrac{1}{7} + \dfrac{1}{8} + \dfrac{1}{9} = \dfrac{4609}{2520} \approx 1.829$

(d) $A = \displaystyle\int_{1}^{5} \dfrac{1}{x}\,dx$

(e) Use fnInt function:

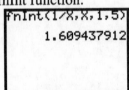

19. (a) Graph $f(x) = e^{x}$, $[-1, 3]$:

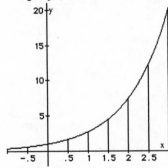

(b) $A \approx \left(f(-1) + f(0) + f(1) + f(2)\right)(1) \approx (0.3679 + 1 + 2.7183 + 7.3891)(1)$

≈ 11.48

(c) $A \approx \left(f(-1) + f\left(-\dfrac{1}{2}\right) + f(0) + f\left(\dfrac{1}{2}\right) + f(1) + f\left(\dfrac{3}{2}\right) + f(2) + f\left(\dfrac{5}{2}\right) \right) \cdot \left(\dfrac{1}{2}\right)$

$\approx (0.3679 + 0.6065 + 1 + 1.6487 + 2.7183 + 4.4817 + 7.3891 + 12.1825)(0.5)$

$= 30.3947(0.5) \approx 15.20$

(d) $A = \int_{-1}^{3} e^{x}\, dx$

(e)   Use fnInt function:

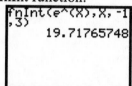

21.   (a)   Graph $f(x) = \sin x,\quad [0,\ \pi]$:

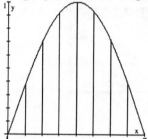

(b) $A \approx \left(f(0) + f\left(\dfrac{\pi}{4}\right) + f\left(\dfrac{\pi}{2}\right) + f\left(\dfrac{3\pi}{4}\right) \right)\left(\dfrac{\pi}{4}\right) = \left(0 + \dfrac{\sqrt{2}}{2} + 1 + \dfrac{\sqrt{2}}{2} \right)\left(\dfrac{\pi}{4}\right)$

$= \left(1 + \sqrt{2}\right)\left(\dfrac{\pi}{4}\right) \approx 1.90$

(c) $A \approx \left(f(0) + f\left(\dfrac{\pi}{8}\right) + f\left(\dfrac{\pi}{4}\right) + f\left(\dfrac{3\pi}{8}\right) + f\left(\dfrac{\pi}{2}\right) + f\left(\dfrac{5\pi}{8}\right) + f\left(\dfrac{3\pi}{4}\right) + f\left(\dfrac{7\pi}{8}\right) \right)\left(\dfrac{\pi}{8}\right)$

$\approx (0 + 0.3827 + 0.7071 + 0.9239 + 1 + 0.9239 + 0.7071 + 0.3827)\left(\dfrac{\pi}{8}\right)$

$= 5.0274\left(\dfrac{\pi}{8}\right) \approx 1.97$

(d) $A = \displaystyle\int_{0}^{\pi} \sin x\, dx$

(e)   Use fnInt function:

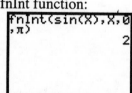

23. (a) The integral represents the area under the graph of $f(x) = 3x + 1$ from $x = 0$ to $x = 4$.

(b)

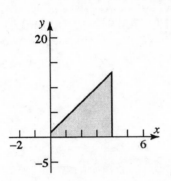

(c)

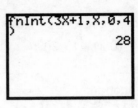

25. (a) The integral represents the area under the graph of $f(x) = x^2 - 1$ from $x = 2$ to $x = 5$.

(b)

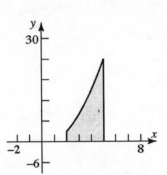

(c)

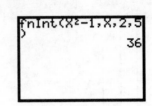

27. (a) The integral represents the area under the graph of $f(x) = \sin x$ from $x = 0$ to $x = \dfrac{\pi}{2}$.

(b)

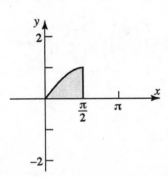

(c)

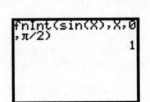

29. (a)   The integral represents the area under the graph of $f(x) = e^x$ from $x = 0$ to $x = 2$.

(b)

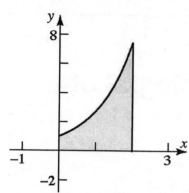

(c)

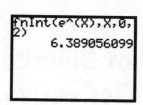

31. Using left endpoints:
$n = 2$: $0 + 0.5 = 0.5$

$n = 4$: $0 + 0.125 + 0.25 + 0.375 = 0.75$

$n = 10$: $0 + 0.02 + 0.04 + 0.06 + \cdots + 0.18 = \dfrac{10}{2}(0 + 0.18) = 0.9$

$n = 100$: $0 + 0.0002 + 0.0004 + 0.0006 + \cdots + 0.0198 = \dfrac{100}{2}(0 + 0.0198) = 0.99$

Using right endpoints:
$n = 2$: $0.5 + 1 = 1.5$

$n = 4$: $0.125 + 0.25 + 0.375 + 0.5 = 1.25$

$n = 10$: $0.02 + 0.04 + 0.06 + \cdots + 0.20 = \dfrac{10}{2}(0.02 + 0.20) = 1.1$

$n = 100$: $0.0002 + 0.0004 + 0.0006 + \cdots + 0.02 = \dfrac{100}{2}(0.0002 + 0.02) = 1.01$

A Preview of Calculus: The Limit, Derivative, and Integral of a Function

13.R Chapter Review

1. $\lim\limits_{x \to 2} \left(3x^2 - 2x + 1\right) = 3(2)^2 - 2(2) + 1 = 12 - 4 + 1 = 9$

3. $\lim\limits_{x \to -2} \left(x^2 + 1\right)^2 = \left(\lim\limits_{x \to -2} \left(x^2 + 1\right)\right)^2 = \left((-2)^2 + 1\right)^2 = 5^2 = 25$

5. $\lim\limits_{x \to 3} \sqrt{x^2 + 7} = \sqrt{\lim\limits_{x \to 3} (x^2 + 7)} = \sqrt{3^2 + 7} = \sqrt{16} = 4$

7. $\lim\limits_{x \to 1^-} \sqrt{1 - x^2} = \sqrt{\lim\limits_{x \to 1^-} (1 - x^2)} = \sqrt{1 - 1^2} = \sqrt{0} = 0$

9. $\lim\limits_{x \to 2} (5x + 6)^{3/2} = \left(\lim\limits_{x \to 2} (5x + 6)\right)^{3/2} = \left(5(2) + 6\right)^{3/2} = 16^{3/2} = 64$

11. $\lim\limits_{x \to -1} \left(\dfrac{x^2 + x + 2}{x^2 - 9}\right) = \dfrac{\lim\limits_{x \to -1}(x^2 + x + 2)}{\lim\limits_{x \to -1}(x^2 - 9)} = \dfrac{(-1)^2 + (-1) + 2}{(-1)^2 - 9} = \dfrac{2}{-8} = -\dfrac{1}{4}$

13. $\lim\limits_{x \to 1} \left(\dfrac{x - 1}{x^3 - 1}\right) = \lim\limits_{x \to 1} \left(\dfrac{x - 1}{(x - 1)(x^2 + x + 1)}\right) = \lim\limits_{x \to 1} \left(\dfrac{1}{x^2 + x + 1}\right) = \dfrac{1}{1^2 + 1 + 1} = \dfrac{1}{3}$

15. $\lim\limits_{x \to -3} \left(\dfrac{x^2 - 9}{x^2 - x - 12}\right) = \lim\limits_{x \to -3} \left(\dfrac{(x - 3)(x + 3)}{(x - 4)(x + 3)}\right) = \lim\limits_{x \to -3} \left(\dfrac{x - 3}{x - 4}\right) = \dfrac{-3 - 3}{-3 - 4} = \dfrac{-6}{-7} = \dfrac{6}{7}$

17. $\lim\limits_{x \to -1^-} \left(\dfrac{x^2 - 1}{x^3 - 1}\right) = \lim\limits_{x \to -1^-} \left(\dfrac{(x + 1)(x - 1)}{(x - 1)(x^2 + x + 1)}\right) = \lim\limits_{x \to -1^-} \left(\dfrac{x + 1}{x^2 + x + 1}\right) = \dfrac{-1 + 1}{(-1)^2 + (-1) + 1} = \dfrac{0}{1} = 0$

19. $\lim\limits_{x \to 2} \left(\dfrac{x^3 - 8}{x^3 - 2x^2 + 4x - 8}\right) = \lim\limits_{x \to 2} \left(\dfrac{(x - 2)(x^2 + 2x + 4)}{x^2(x - 2) + 4(x - 2)}\right) = \lim\limits_{x \to 2} \left(\dfrac{(x - 2)(x^2 + 2x + 4)}{(x - 2)(x^2 + 4)}\right)$

$= \lim\limits_{x \to 2} \left(\dfrac{x^2 + 2x + 4}{x^2 + 4}\right) = \left(\dfrac{2^2 + 2(2) + 4}{2^2 + 4}\right) = \dfrac{12}{8} = \dfrac{3}{2}$

21. $\displaystyle\lim_{x\to3}\left(\frac{x^4-3x^3+x-3}{x^3-3x^2+2x-6}\right)=\lim_{x\to3}\left(\frac{x^3(x-3)+1(x-3)}{x^2(x-3)+2(x-3)}\right)=\lim_{x\to3}\left(\frac{(x-3)(x^3+1)}{(x-3)(x^2+2)}\right)=\lim_{x\to3}\left(\frac{x^3+1}{x^2+2}\right)$

$$=\frac{3^3+1}{3^2+2}=\frac{28}{11}$$

23. $f(x)=3x^4-x^2+2;\ c=5$

    1. $f(5)=3(5)^4-5^2+2=1852$

    2. $\displaystyle\lim_{x\to5^-}f(x)=3(5)^4-5^2+2=1852$

    3. $\displaystyle\lim_{x\to5^+}f(x)=3(5)^4-5^2+2=1852$

Thus, $f(x)$ is continuous at $c=5$.

25. $f(x)=\dfrac{x^2-4}{x+2};\ c=-2$

Since $f(x)$ is not defined at $c=-2$, the function is not continuous at $c=-2$.

27. $f(x)=\begin{cases}\dfrac{x^2-4}{x+2} & \text{if } x\neq-2 \\ 4 & \text{if } x=-2\end{cases};\quad c=-2$

    1. $f(-2)=4$

    2. $\displaystyle\lim_{x\to-2^-}f(x)=\lim_{x\to-2^-}\left(\frac{x^2-4}{x+2}\right)=\lim_{x\to-2^-}\left(\frac{(x-2)(x+2)}{x+2}\right)=\lim_{x\to-2^-}(x-2)=-4$

Since $\displaystyle\lim_{x\to-2^-}f(x)\neq f(-2)$, the function is not continuous at $c=-2$.

29. $f(x)=\begin{cases}\dfrac{x^2-4}{x+2} & \text{if } x\neq-2 \\ -4 & \text{if } x=-2\end{cases};\quad c=-2$

    1. $f(-2)=-4$

    2. $\displaystyle\lim_{x\to-2^-}f(x)=\lim_{x\to-2^-}\left(\frac{x^2-4}{x+2}\right)=\lim_{x\to-2^-}\left(\frac{(x-2)(x+2)}{x+2}\right)=\lim_{x\to-2^-}(x-2)=-4$

    3. $\displaystyle\lim_{x\to-2^+}f(x)=\lim_{x\to-2^+}\left(\frac{x^2-4}{x+2}\right)=\lim_{x\to-2^+}\left(\frac{(x-2)(x+2)}{x+2}\right)=\lim_{x\to-2^+}(x-2)=-4$

The function is continuous at $c=-2$.

31. Domain: $\left\{x\,\middle|\,-6\leq x<2 \text{ or } 2<x<5 \text{ or } 5<x\leq6\right\}$

33. x-intercepts: $1,6$ 35. $f(-6)=2;\ f(-4)=1$ 37. $\displaystyle\lim_{x\to-4^-}f(x)=4$

39. $\displaystyle\lim_{x\to-2^-}f(x)=-2$ 41. $\displaystyle\lim_{x\to2^-}f(x)=-\infty$

43. $\displaystyle\lim_{x\to0}f(x)$ does not exist because $\displaystyle\lim_{x\to0^-}f(x)=4\neq\lim_{x\to0^+}f(x)=1$

45. f is not continuous at -2 because $\lim\limits_{x \to -2^-} f(x) \neq \lim\limits_{x \to -2^+} f(x)$

47. f is not continuous at 0 because $\lim\limits_{x \to 0^-} f(x) \neq \lim\limits_{x \to 0^+} f(x)$

49. f is continuous at 4 because $f(4) = \lim\limits_{x \to 4^-} f(x) = \lim\limits_{x \to 4^+} f(x)$

51. $R(x) = \dfrac{x+4}{x^2-16} = \dfrac{x+4}{(x-4)(x+4)}$. The domain of R is $\left\{ x \mid x \neq -4,\ x \neq 4 \right\}$. Thus R is

discontinuous at both -4 and 4. $\lim\limits_{x \to 4^-} R(x) = \lim\limits_{x \to 4^-} \left(\dfrac{x+4}{(x-4)(x+4)} \right) = \lim\limits_{x \to 4^-} \left(\dfrac{1}{x-4} \right) = -\infty$ since

when $x < 4$, $\dfrac{1}{x-4} < 0$, and as x approaches 4, $\dfrac{1}{x-4}$ becomes unbounded.

$\lim\limits_{x \to 4^+} R(x) = \lim\limits_{x \to 4^+} \left(\dfrac{1}{x-4} \right) = \infty$ since when $x > 4$, $\dfrac{1}{x-4} > 0$, and as x approaches 4, $\dfrac{1}{x-4}$

becomes unbounded. $\lim\limits_{x \to -4} R(x) = \lim\limits_{x \to -4} \left(\dfrac{1}{x-4} \right) = -\dfrac{1}{8}$. Note there is a hole in the

graph at $\left(-4, -\dfrac{1}{8} \right)$.

53. $R(x) = \dfrac{x^3 - 2x^2 + 4x - 8}{x^2 - 11x + 18} = \dfrac{x^2(x-2) + 4(x-2)}{(x-9)(x-2)} = \dfrac{(x-2)(x^2+4)}{(x-9)(x-2)} = \dfrac{x^2+4}{x-9},\ x \neq 2$

Undefined at $x = 2$ and $x = 9$. There is a vertical asymptote where $x - 9 = 0$. $x = 9$ is a
vertical asymptote. There is a hole in the graph at $x = 2$.

55. $f(x) = 2x^2 + 8x$ at $(1, 10)$

$m_{\tan} = \lim\limits_{x \to 1} \left(\dfrac{f(x) - f(1)}{x-1} \right) = \lim\limits_{x \to 1} \left(\dfrac{2x^2 + 8x - 10}{x-1} \right)$

$= \lim\limits_{x \to 1} \left(\dfrac{2(x+5)(x-1)}{x-1} \right) = \lim\limits_{x \to 1} \left(2(x+5) \right)$

$= 2(1+5) = 12$

Tangent Line : $y - 10 = 12(x-1)$

$y - 10 = 12x - 12$

$y = 12x - 2$

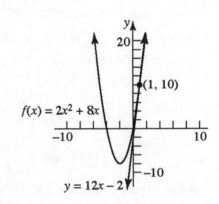

57. $f(x) = x^2 + 2x - 3$ at $(-1, -4)$

$$m_{\tan} = \lim_{x \to -1}\left(\frac{f(x) - f(-1)}{x + 1}\right) = \lim_{x \to -1}\left(\frac{x^2 + 2x - 3 - (-4)}{x + 1}\right)$$

$$= \lim_{x \to -1}\left(\frac{x^2 + 2x + 1}{x + 1}\right) = \lim_{x \to -1}\left(\frac{(x+1)^2}{x + 1}\right) = \lim_{x \to -1}(x + 1)$$

$$= -1 + 1 = 0$$

Tangent Line: $\quad y - (-4) = 0(x - (-1))$

$$y + 4 = 0$$

$$y = -4$$

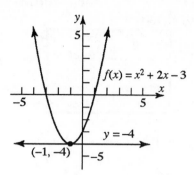

59. $f(x) = x^3 + x^2$ at $(2, 12)$

$$m_{\tan} = \lim_{x \to 2}\left(\frac{f(x) - f(2)}{x - 2}\right) = \lim_{x \to 2}\left(\frac{x^3 + x^2 - 12}{x - 2}\right)$$

$$= \lim_{x \to 2}\left(\frac{x^3 - 2x^2 + 3x^2 - 12}{x - 2}\right)$$

$$= \lim_{x \to 2}\left(\frac{x^2(x - 2) + 3(x - 2)(x + 2)}{x - 2}\right)$$

$$= \lim_{x \to 2}\left(\frac{(x - 2)(x^2 + 3x + 6)}{x - 2}\right)$$

$$= \lim_{x \to 2}(x^2 + 3x + 6) = 4 + 6 + 6 = 16$$

Tangent Line: $\quad y - 12 = 16(x - 2)$

$$y - 12 = 16x - 32$$

$$y = 16x - 20$$

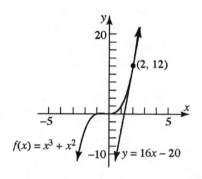

61. $f(x) = -4x^2 + 5$ at 3

$$f'(3) = \lim_{x \to 3}\left(\frac{f(x) - f(3)}{x - 3}\right) = \lim_{x \to 3}\left(\frac{-4x^2 + 5 - (-31)}{x - 3}\right) = \lim_{x \to 3}\left(\frac{-4x^2 + 36}{x - 3}\right)$$

$$= \lim_{x \to 3}\left(\frac{-4(x^2 - 9)}{x - 3}\right) = \lim_{x \to 3}\left(\frac{-4(x - 3)(x + 3)}{x - 3}\right) = \lim_{x \to 3}((-4)(x + 3)) = -4(6) = -24$$

63. $f(x) = x^2 - 3x$ at 0

$$f'(0) = \lim_{x \to 0}\left(\frac{f(x) - f(0)}{x - 0}\right) = \lim_{x \to 0}\left(\frac{x^2 - 3x - 0}{x}\right) = \lim_{x \to 0}\left(\frac{x(x - 3)}{x}\right) = \lim_{x \to 0}(x - 3) = -3$$

65. $f(x) = 2x^2 + 3x + 2$ at 1

$$f'(1) = \lim_{x \to 1}\left(\frac{f(x) - f(1)}{x - 1}\right) = \lim_{x \to 1}\left(\frac{2x^2 + 3x + 2 - 7}{x - 1}\right) = \lim_{x \to 1}\left(\frac{2x^2 + 3x - 5}{x - 1}\right)$$

$$= \lim_{x \to 1}\left(\frac{(2x + 5)(x - 1)}{x - 1}\right) = \lim_{x \to 1}(2x + 5) = 7$$

67. Use NDeriv:

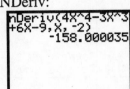

69. Use NDeriv:

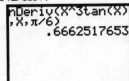

71. (a) $-16t^2 + 96t + 112 = 0$

$-16(t^2 - 6t - 7) = 0$

$-16(t + 1)(t - 7) = 0$

$t = -1$ or $t = 7$

The ball strikes the ground after 7 seconds in the air.

(b) $-16t^2 + 96t + 112 = 112$

$-16t^2 + 96t = 0$

$-16t(t - 6) = 0$

$t = 0$ or $t = 6$

The ball passes the rooftop after 6 seconds.

(c) $\dfrac{\Delta s}{\Delta t} = \dfrac{s(2) - s(0)}{2 - 0} = \dfrac{-16(2)^2 + 96(2) + 112 - 112}{2} = \dfrac{128}{2} = 64$ feet/sec

(d) $s'(t_0) = \lim\limits_{t \to t_0}\left(\dfrac{s(t) - s(t_0)}{t - t_0}\right) = \lim\limits_{t \to t_0}\left(\dfrac{-16t^2 + 96t + 112 - \left(-16t_0^2 + 96t_0 + 112\right)}{t - t_0}\right)$

$= \lim\limits_{t \to t_0}\left(\dfrac{-16t^2 + 16t_0^2 + 96t - 96t_0}{t - t_0}\right) = \lim\limits_{t \to t_0}\left(\dfrac{-16\left(t^2 - t_0^2\right) + 96(t - t_0)}{t - t_0}\right)$

$= \lim\limits_{t \to t_0}\left(\dfrac{-16(t - t_0)(t + t_0) + 96(t - t_0)}{t - t_0}\right) = \lim\limits_{t \to t_0}\left(\dfrac{(t - t_0)(-16(t + t_0) + 96)}{t - t_0}\right)$

$= \lim\limits_{t \to t_0}(-16(t + t_0) + 96) = (-16(t_0 + t_0) + 96) = -32t_0 + 96$

(e) $s'(2) = -32(2) + 96 = -64 + 96 = 32$ feet/sec

(f) $s'(t) = 0$

$-32t + 96 = 0$

$-32t = -96$

$t = 3$ seconds

(g) $s'(6) = -32(6) + 96 = -192 + 96 = -96$ feet/sec

(h) $s'(7) = -32(7) + 96 = -224 + 96 = -128$ feet/sec

73. (a) $\dfrac{\Delta R}{\Delta x} = \dfrac{8775 - 2340}{130 - 25} = \dfrac{6435}{105} \approx \$61.29/\text{watch}$

(b) $\dfrac{\Delta R}{\Delta x} = \dfrac{6975 - 2340}{90 - 25} = \dfrac{4635}{65} \approx \$71.31/\text{watch}$

(c) $\dfrac{\Delta R}{\Delta x} = \dfrac{4375 - 2340}{50 - 25} = \dfrac{2035}{25} \approx \$81.40/\text{watch}$

(d) $R(x) = -0.25x^2 + 100.01x - 1.24$

(e) $R'(25) = \lim_{x \to 25}\left(\dfrac{R(x) - R(25)}{x - 25}\right) = \lim_{x \to 25}\left(\dfrac{-0.25x^2 + 100.01x - 1.24 - \left(-0.25(25)^2 + 100.01(25) - 1.24\right)}{x - 25}\right)$

$\qquad = \lim_{x \to 25}\left(\dfrac{-0.25x^2 + 100.014x - 2344}{x - 25}\right)$

$\qquad = \lim_{x \to 25}\left(\dfrac{(x - 25)(-0.25x + 93.76)}{x - 25}\right)$

$\qquad = \lim_{x \to 25}(-0.25x + 93.76) = -0.25(25) + 93.76 = \$87.51/\text{ watch}$

75. (a) Graph $f(x) = 2x + 3$:

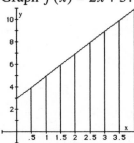

(b) $A \approx \left(f(0) + f(1) + f(2) + f(3)\right)(1) = (3 + 5 + 7 + 9)(1) = 24(1) = 24$

(c) $A \approx \left(f(1) + f(2) + f(3) + f(4)\right)(1) = (5 + 7 + 9 + 11)(1) = 32(1) = 32$

(d) $A \approx \left(f(0) + f\left(\dfrac{1}{2}\right) + f(1) + f\left(\dfrac{3}{2}\right) + f(2) + f\left(\dfrac{5}{2}\right) + f(3) + f\left(\dfrac{7}{2}\right)\right) \cdot \left(\dfrac{1}{2}\right)$

$\qquad = (3 + 4 + 5 + 6 + 7 + 8 + 9 + 10)\left(\dfrac{1}{2}\right) = 52\left(\dfrac{1}{2}\right) = 26$

(e) $A \approx \left(f\left(\dfrac{1}{2}\right) + f(1) + f\left(\dfrac{3}{2}\right) + f(2) + f\left(\dfrac{5}{2}\right) + f(3) + f\left(\dfrac{7}{2}\right) + f(4)\right) \cdot \left(\dfrac{1}{2}\right)$

$\qquad = (4 + 5 + 6 + 7 + 8 + 9 + 10 + 11)\left(\dfrac{1}{2}\right) = 60\left(\dfrac{1}{2}\right) = 30$

(f) The actual area is the area of a trapezoid: $A = \dfrac{1}{2}(3 + 11)(4) = \dfrac{56}{2} = 28$

77. (a) Graph $f(x) = 4 - x^2$, $[-1, 2]$:

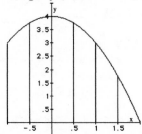

(b) $A \approx \left(f(-1) + f(0) + f(1)\right)(1) = (3 + 4 + 3)(1) = 10(1) = 10$

(c) $A \approx \left(f(-1) + f\left(-\dfrac{1}{2}\right) + f(0) + f\left(\dfrac{1}{2}\right) + f(1) + f\left(\dfrac{3}{2}\right) \right) \cdot \left(\dfrac{1}{2}\right)$

$= \left(3 + \dfrac{15}{4} + 4 + \dfrac{15}{4} + 3 + \dfrac{7}{4} \right)\left(\dfrac{1}{2}\right) = \dfrac{77}{4}\left(\dfrac{1}{2}\right) = \dfrac{77}{8} = 9.625$

(d) $A = \int_{-1}^{2} \left(4 - x^2 \right) dx$

(e)   Use fnInt function:

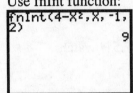

79.   (a)   Graph $f(x) = \dfrac{1}{x^2}$, $[1, 4]$:

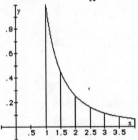

(b) $A \approx \left(f(1) + f(2) + f(3) \right)(1) = \left(1 + \dfrac{1}{4} + \dfrac{1}{9} \right)(1) = \dfrac{49}{36}(1) = \dfrac{49}{36} = 1.36$

(c) $A \approx \left(f(1) + f\left(\dfrac{3}{2}\right) + f(2) + f\left(\dfrac{5}{2}\right) + f(3) + f\left(\dfrac{7}{2}\right) \right) \cdot \left(\dfrac{1}{2}\right)$

$= \left(1 + \dfrac{4}{9} + \dfrac{1}{4} + \dfrac{4}{25} + \dfrac{1}{9} + \dfrac{4}{49} \right)\left(\dfrac{1}{2}\right) = 1.02$

(d) $A = \int_{1}^{4} \left(\dfrac{1}{x^2} \right) dx$

(e)   Use fnInt function:

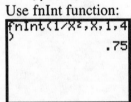

81. (a) The integral represents the area under the graph of $f(x) = 9 - x^2$ from $x = -1$ to $x = 3$.

(b)

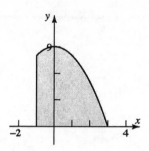

(c)

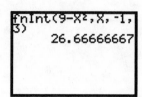

83. (a) The integral represents the area under the graph of $f(x) = e^x$ from $x = -1$ to $x = 1$.

(b)

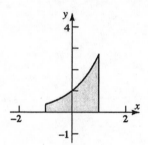

(c)

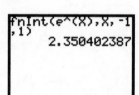

Appendix A

Review

A.1 Algebra Review

1. variable

3. strict

5. $|-5| = 5$

7. False

9. $x = -2, y = 3 \Rightarrow x + 2y = -2 + 2 \cdot 3 = -2 + 6 = 4$

11. $x = -2, y = 3 \Rightarrow 5xy + 2 = 5(-2)(3) + 2 = -30 + 2 = -28$

13. $x = -2, y = 3 \Rightarrow \dfrac{2x}{x-y} = \dfrac{2(-2)}{-2-3} = \dfrac{-4}{-5} = \dfrac{4}{5}$

15. $x = -2, y = 3 \Rightarrow \dfrac{3x+2y}{2+y} = \dfrac{3(-2)+2(3)}{2+3} = \dfrac{-6+6}{5} = \dfrac{0}{5} = 0$

17. $\dfrac{x^2-1}{x}$ Value (c) must be excluded.

The value $x = 0$ must be excluded from the domain because it causes division by 0.

19. $\dfrac{x}{x^2-9} = \dfrac{x}{(x-3)(x+3)}$ Value (a) must be excluded.

The values $x = -3$ and $x = 3$ must be excluded from the domain because they cause division by 0.

21. $\dfrac{x^2}{x^2+1}$ None of the given values are excluded. The domain is all real numbers.

23. $\dfrac{x^2+5x-10}{x^3-x} = \dfrac{x^2+5x-10}{x(x-1)(x+1)}$ Values (b), (c), and (d) must be excluded.

The values $x = 0$, $x = 1$, and $x = -1$ must be excluded from the domain because they cause division by 0.

25. $\dfrac{4}{x-5}$ Domain $= \{x \mid x \neq 5\}$

27. $\dfrac{x}{x+4}$ Domain $= \{x \mid x \neq -4\}$

29.

31. $\dfrac{1}{2} > 0$ 33. $-1 > -2$ 35. $\pi > 3.14$

37. $\dfrac{1}{2} = 0.5$ 39. $\dfrac{2}{3} < 0.67$ 41. $x > 0$

43. $x < 2$ 45. $x \le 1$

47. $x \ge -2$ 49. $x > -1$

51. $x = 3, y = -2 \Rightarrow |x + y| = |3 + (-2)| = |1| = 1$

53. $x = 3, y = -2 \Rightarrow |x| + |y| = |3| + |-2| = 3 + 2 = 5$

55. $x = 3, y = -2 \Rightarrow \dfrac{|x|}{x} = \dfrac{|3|}{3} = \dfrac{3}{3} = 1$

57. $x = 3, y = -2 \Rightarrow -1|4x - 5y| = |4(3) - 5(-2)| = |12 + 10| = |22| = 22$

59. $x = 3, y = -2 \Rightarrow \big||4x| - |5y|\big| = \big||4(3)| - |5(-2)|\big| = \big||12| - |-10|\big| = |12 - 10| = |2| = 2$

61. $d(C, D) = d(0,1) = |1 - 0| = |1| = 1$ 63. $d(D, E) = d(1,3) = |3 - 1| = |2| = 2$

65. $d(A, E) = d(-3,3) = |3 - (-3)| = |6| = 6$ 67. $(-4)^2 = (-4)(-4) = 16$

69. $4^{-2} = \dfrac{1}{4^2} = \dfrac{1}{16}$ 71. $3^{-6} \cdot 3^4 = 3^{-6+4} = 3^{-2} = \dfrac{1}{3^2} = \dfrac{1}{9}$

73. $\left(3^{-2}\right)^{-1} = 3^{(-2)(-1)} = 3^2 = 9$ 75. $\sqrt{25} = \sqrt{5^2} = |5| = 5$

77. $\sqrt{(-4)^2} = |-4| = 4$ 79. $\left(8x^3\right)^2 = 8^2 \cdot x^{3 \cdot 2} = 64x^6$

81. $\left(x^2 y^{-1}\right)^2 = \left(\dfrac{x^2}{y}\right)^2 = \dfrac{x^{2 \cdot 2}}{y^{1 \cdot 2}} = \dfrac{x^4}{y^2}$

83. $\dfrac{x^{-2}y^3}{xy^4} = \dfrac{x^{-2}}{x} \cdot \dfrac{y^3}{y^4} = x^{-2-1}y^{3-4} = x^{-3}y^{-1} = \dfrac{1}{x^3} \cdot \dfrac{1}{y} = \dfrac{1}{x^3 y}$

85. $\dfrac{(-2)^3 x^4 (yz)^2}{3^2 x y^3 z} = \dfrac{-8x^4 y^2 z^2}{9xy^3 z} = \dfrac{-8}{9} x^{4-1} y^{2-3} z^{2-1} = \dfrac{-8}{9} x^3 y^{-1} z^1 = \dfrac{-8}{9} x^3 \cdot \dfrac{1}{y} \cdot z = -\dfrac{8x^3 z}{9y}$

87. $\left(\dfrac{3x^{-1}}{4y^{-1}}\right)^{-2} = \left(\dfrac{3y^1}{4x^1}\right)^{-2} = \dfrac{1}{\left(\dfrac{3y^1}{4x^1}\right)^2} = \dfrac{1}{\dfrac{3^2 y^{1\cdot 2}}{4^2 x^{1\cdot 2}}} = \dfrac{1}{\dfrac{9y^2}{16x^2}} = \dfrac{16x^2}{9y^2}$

89. $A = l \cdot w$ 91. $C = \pi \cdot d$

93. $A = \dfrac{\sqrt{3}}{4} \cdot x^2$ 95. $V = \dfrac{4}{3} \cdot \pi \cdot r^3$

97. $V = x^3$

99. $|x - 115| \le 5$
   (a) $|x - 115| = |113 - 115| = |-2| = 2 \le 5$ 113 volts is acceptable.
   (b) $|x - 115| = |109 - 115| = |-6| = 6 > 5$ 109 volts is not acceptable.

101. $|x - 3| \le 0.01$
   (a) $|x - 3| = |2.999 - 3| = |-0.001| = 0.001 \le 0.01$
        A radius of 2.999 centimeters is acceptable.
   (b) $|x - 3| = |2.89 - 3| = |-0.11| = 0.11 > 0.01$
        A radius of 2.89 centimeters is _not_ acceptable.

103. $\dfrac{1}{3} = 0.333333... > 0.333 \Rightarrow \dfrac{1}{3}$ is larger by approximately $0.0003333...$

105. No. Suppose that c = the number closest to 0, with $c > 0$. Then $c/2 < c$, and $c/2$ is half-way between 0 and c. Therefore, the number c can't be the number closest to 0.

107. Answers will vary.

Appendix A

Review

A.2 Geometry Review

1. right, hypotenuse   3. $C = 2\pi r$ 5. True

7. $a = 5,\ b = 12,\ c^2 = a^2 + b^2 = 5^2 + 12^2 = 25 + 144 = 169 \Rightarrow c = 13$

9. $a = 10,\ b = 24,\ c^2 = a^2 + b^2 = 10^2 + 24^2 = 100 + 576 = 676 \Rightarrow c = 26$

11. $a = 7,\ b = 24,\ c^2 = a^2 + b^2 = 7^2 + 24^2 = 49 + 576 = 625 \Rightarrow c = 25$

13. $5^2 = 3^2 + 4^2 \Rightarrow 25 = 9 + 16 \Rightarrow 25 = 25$
The given triangle is a right triangle. The length of the hypotenuse is 5.

15. $6^2 = 4^2 + 5^2 \Rightarrow 36 = 16 + 25 \Rightarrow 36 \neq 41$
The given triangle is not a right triangle.

17. $25^2 = 7^2 + 24^2 \Rightarrow 625 = 49 + 576 \Rightarrow 625 = 625$
The given triangle is a right triangle. The length of the hypotenuse is 25.

19. $6^2 = 3^2 + 4^2 \Rightarrow 36 = 9 + 16 \Rightarrow 36 \neq 25$
The given triangle is not a right triangle.

21. $A = lw = 4 \cdot 2 = 8$ in^2 23. $A = \frac{1}{2}bh = \frac{1}{2}(2)(4) = 4$ in^2

25. $A = \pi r^2 = \pi(5)^2 = 25\pi$ m^2 $C = 2\pi r = 2\pi(5) = 10\pi$ m

27. $V = lwh = 8 \cdot 4 \cdot 7 = 224$ ft^3 $S = 2(lw + lh + wh) = 2 \cdot (8 \cdot 4 + 8 \cdot 7 + 4 \cdot 7) = 232$ ft^2

29. $V = \frac{4}{3}\pi r^3 = \frac{4}{3}\pi \cdot 4^3 = \frac{256}{3}\pi$ cm^3 $S = 4\pi r^2 = 4\pi \cdot 4^2 = 64\pi$ cm^2

31. $V = \pi r^2 h = \pi(9)^2(8) = 648\pi$ in^3
$S = 2\pi r^2 + 2\pi rh = 2\pi \cdot 9^2 + 2\pi \cdot 9 \cdot 8 = 306\pi$ in^2

33. The diameter of the circle is 2, so its radius is 1. $A = \pi r^2 = \pi(1)^2 = \pi$ square units

35.   The diameter of the circle is the length of the diagonal of the square.
$$d^2 = 2^2 + 2^2 = 4 + 4 = 8 \Rightarrow d = \sqrt{8} = 2\sqrt{2} \Rightarrow r = \sqrt{2}$$
The area of the circle is: $A = \pi r^2 = \pi\left(\sqrt{2}\right)^2 = 2\pi$ square units

37.   The total distance traveled is 4 times the circumference of the wheel.
$$\text{Total distance} = 4C = 4(\pi d) = 4\pi \cdot 16 = 64\pi \approx 201.1 \text{ inches} \approx 16.8 \text{ feet}$$

39.   Area of the border = area of $EFGH$ – area of $ABCD$ = $10^2 - 6^2 = 100 - 36 = 64 \text{ ft}^2$

41.   Area of the window = area of the rectangle + area of the semicircle.
$$A = (6)(4) + \frac{1}{2}\pi \cdot 2^2 = 24 + 2\pi \approx 30.28 \text{ ft}^2$$
Perimeter of the window = 2 heights + width + one-half the circumference.
$$P = 2(6) + 4 + \frac{1}{2}\pi(4) = 12 + 4 + 2\pi = 16 + 2\pi \approx 22.28 \text{ feet}$$

43.   Convert 20 feet to miles, and use the Pythagorean theorem to find the distance:
$$20 \text{ feet} = 20 \text{ feet} \cdot \frac{1 \text{ mile}}{5280 \text{ feet}} \approx 0.003788 \text{ miles}$$
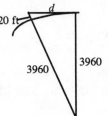
$$d^2 \approx (3960 + 0.003788)^2 - 3960^2 \approx 30$$
$$d \approx 5.477 \text{ miles}$$

45.   Convert 100 feet to miles, and use the Pythagorean theorem to find the distance:
$$100 \text{ feet} = 100 \text{ feet} \cdot \frac{1 \text{ mile}}{5280 \text{ feet}} \approx 0.018939 \text{ miles}$$
$$d^2 \approx (3960 + 0.018939)^2 - 3960^2 \approx 150 \Rightarrow d \approx 12.247 \text{ miles}$$
Convert 150 feet to miles, and solve the Pythagorean theorem to find the distance:
$$150 \text{ feet} = 150 \text{ feet} \cdot \frac{1 \text{ mile}}{5280 \text{ feet}} \approx 0.028409 \text{ miles}$$
$$d^2 \approx (3960 + 0.028409)^2 - 3960^2 \approx 225 \Rightarrow d \approx 15 \text{ miles}$$

47.   Given a rectangle with perimeter = 1000 feet, the largest area will be enclosed by a square with dimensions 250 by 250 feet. Then, the area = $250^2 = 62,500$ square feet.

A circular pool with circumference = 1000 feet yields the equation : $2\pi r = 1000 \Rightarrow r = \dfrac{500}{\pi}$

The area enclosed by the circular pool is:
$$A = \pi r^2 = \pi\left(\frac{500}{\pi}\right)^2 = \frac{500^2}{\pi} \approx 79,577.47 \text{ square feet}$$
Therefore, a circular pool will enclose the most area.

Appendix A

Review

A.3 Polynomials and Rational Expressions

1. $4, 3$

3. $x^3 - 8$

5. True

7. $3x(x^2 - 4) = 3x(x - 2)(x + 2)$

9. True

11. in lowest terms

13. True

15. $(10x^5 - 8x^2) + (3x^3 - 2x^2 + 6) = 10x^5 + 3x^3 - 10x^2 + 6$

17. $(x + a)^2 - x^2 = x^2 + xa + ax + a^2 - x^2 = 2ax + a^2$

19. $(x + 8)(2x + 1) = 2x^2 + x + 16x + 8 = 2x^2 + 17x + 8$

21. $(x^2 + x - 1)(x^2 - x + 1) = x^4 - x^3 + x^2 + x^3 - x^2 + x - x^2 + x - 1 = x^4 - x^2 + 2x - 1$

23. $(x + 1)^3 - (x - 1)^3$ can be factored as a difference of two cubes.
$$(x + 1)^3 - (x - 1)^3 = \left[(x + 1) - (x - 1)\right]\left[(x + 1)^2 + (x + 1)(x - 1) + (x - 1)^2\right]$$
$$= \left[x + 1 - x + 1\right]\left[x^2 + 2x + 1 + x^2 - 1 + x^2 - 2x + 1\right]$$
$$= [2]\left[3x^2 + 1\right]$$
$$= 6x^2 + 2$$

25. $x^2 - 36 = (x - 6)(x + 6)$

27. $1 - 4x^2 = (1 - 2x)(1 + 2x)$

29. $x^2 + 7x + 10 = (x + 2)(x + 5)$

31. $x^2 - 2x + 8 = (x + 2)(x - 4)$

33. $x^2 + 4x + 16$ is prime because there are no factors of 16 whose sum is 4.

35. $15 + 2x - x^2 = (3 + x)(5 - x)$

37. $3x^2 - 12x - 36 = 3(x^2 - 4x - 12) = 3(x - 6)(x + 2)$

39. $y^4 + 11y^3 + 30y^2 = y^2(y^2 + 11y + 30) = y^2(y + 6)(y + 5)$

41. $4x^2 + 12x + 9 = (2x + 3)(2x + 3) = (2x + 3)^2$

43. $3x^2 + 4x + 1 = (3x + 1)(x + 1)$

45. $x^4 - 81 = \left(x^2 + 9\right)\left(x^2 - 9\right) = \left(x^2 + 9\right)(x - 3)(x + 3)$

47. $x^6 - 2x^3 + 1 = \left(x^3 - 1\right)\left(x^3 - 1\right) = \left(x^3 - 1\right)^2 = \left[(x - 1)\left(x^2 + x + 1\right)\right]^2 = (x - 1)^2\left(x^2 + x + 1\right)^2$

49. $x^7 - x^5 = x^5\left(x^2 - 1\right) = x^5(x - 1)(x + 1)$

51. $5 + 16x - 16x^2 = (5 - 4x)(1 + 4x)$ 53. $4y^2 - 16y + 15 = (2y - 3)(2y - 5)$

55. $1 - 8x^2 - 9x^4 = -(9x^4 + 8x^2 - 1) = -(9x^2 - 1)(x^2 + 1) = -(3x - 1)(3x + 1)(x^2 + 1)$

57. $x(x + 3) - 6(x + 3) = (x + 3)(x - 6)$

59. $(x + 2)^2 - 5(x + 2) = (x + 2)\left[(x + 2) - 5\right] = (x + 2)(x - 3)$

61. $6x(2 - x)^4 - 9x^2(2 - x)^3 = 3x(2 - x)^3\left[2(2 - x) - 3x\right] = 3x(2 - x)^3\left[4 - 2x - 3x\right]$
$$= 3x(2 - x)^3(4 - 5x)$$

63. $x^3 + 2x^2 - x - 2 = x^2(x + 2) - (x + 2) = (x + 2)(x^2 - 1) = (x + 2)(x - 1)(x + 1)$

65. $x^4 - x^3 + x - 1 = x^3(x - 1) + (x - 1) = (x - 1)(x^3 + 1) = (x - 1)(x + 1)(x^2 - x + 1)$

67. $\dfrac{3x - 6}{5x} \cdot \dfrac{x^2 - x - 6}{x^2 - 4} = \dfrac{3(x - 2)}{5x} \cdot \dfrac{(x - 3)(x + 2)}{(x - 2)(x + 2)} = \dfrac{3(x - 3)}{5x}$

69. $\dfrac{4x^2 - 1}{x^2 - 16} \cdot \dfrac{x^2 - 4x}{2x + 1} = \dfrac{(2x - 1)(2x + 1)}{(x - 4)(x + 4)} \cdot \dfrac{x(x - 4)}{2x + 1} = \dfrac{x(2x - 1)}{x + 4}$

71. $\dfrac{x}{x^2 - 7x + 6} - \dfrac{x}{x^2 - 2x - 24} = \dfrac{x}{(x - 6)(x - 1)} - \dfrac{x}{(x - 6)(x + 4)}$
$$= \dfrac{x(x + 4)}{(x - 6)(x - 1)(x + 4)} - \dfrac{x(x - 1)}{(x - 6)(x + 4)(x - 1)}$$
$$= \dfrac{x^2 + 4x - x^2 + x}{(x - 6)(x + 4)(x - 1)}$$
$$= \dfrac{5x}{(x - 6)(x + 4)(x - 1)}$$

73. $\dfrac{4x}{x^2-4} - \dfrac{2}{x^2+x-6} = \dfrac{4x}{(x-2)(x+2)} - \dfrac{2}{(x+3)(x-2)}$

$$= \dfrac{4x(x+3)}{(x-2)(x+2)(x+3)} - \dfrac{2(x+2)}{(x+3)(x-2)(x+2)}$$

$$= \dfrac{4x^2+12x-2x-4}{(x-2)(x+2)(x+3)} = \dfrac{4x^2+10x-4}{(x-2)(x+2)(x+3)}$$

$$= \dfrac{2(2x^2+5x-2)}{(x-2)(x+2)(x+3)}$$

75. $\dfrac{1}{x} - \dfrac{2}{x^2+x} + \dfrac{3}{x^3-x^2} = \dfrac{1}{x} - \dfrac{2}{x(x+1)} + \dfrac{3}{x^2(x-1)} = \dfrac{x(x+1)(x-1)-2x(x-1)+3(x+1)}{x^2(x+1)(x-1)}$

$$= \dfrac{x(x^2-1)-2x^2+2x+3x+3}{x^2(x+1)(x-1)} = \dfrac{x^3-x-2x^2+5x+3}{x^2(x+1)(x-1)}$$

$$= \dfrac{x^3-2x^2+4x+3}{x^2(x+1)(x-1)}$$

77. $\dfrac{1}{h}\left[\dfrac{1}{x+h} - \dfrac{1}{x}\right] = \dfrac{1}{h}\left[\dfrac{x}{(x+h)x} - \dfrac{x+h}{x(x+h)}\right] = \dfrac{1}{h}\left[\dfrac{x-x-h}{x(x+h)}\right] = \dfrac{-h}{hx(x+h)} = -\dfrac{1}{x(x+h)}$

79. $2(3x+4)^2 + (2x+3)\cdot 2(3x+4)\cdot 3$

$= 2(3x+4)\left[(3x+4)+(2x+3)\cdot 3\right]$

$= 2(3x+4)\left[3x+4+6x+9\right]$

$= 2(3x+4)(9x+13)$

81. $2x(2x+5) + x^2\cdot 2 = 2x\left[2x+5+x\right] = 2x(3x+5)$

83. $2(x+3)(x-2)^3 + (x+3)^2\cdot 3(x-2)^2 = (x+3)(x-2)^2\left[2(x-2)+(x+3)\cdot 3\right]$

$= (x+3)(x-2)^2\left[2x-4+3x+9\right]$

$= (x+3)(x-2)^2\left[5x+5\right]$

$= 5(x+3)(x-2)^2(x+1)$

85. $(4x-3)^2 + x\cdot 2(4x-3)\cdot 4 = (4x-3)\left[(4x-3)+x\cdot 2\cdot 4\right]$

$= (4x-3)\left[4x-3+8x\right]$

$= (4x-3)\left[12x-3\right]$

$= 3(4x-3)(4x-1)$

87. $2(3x-5)\cdot 3(2x+1)^3 + (3x-5)^2\cdot 3(2x+1)^2\cdot 2 = 6(3x-5)(2x+1)^2\left[(2x+1)+(3x-5)\right]$

$= 6(3x-5)(2x+1)^2(5x-4)$

89. $$\frac{(2x+3)\cdot 3-(3x-5)\cdot 2}{(3x-5)^2}=\frac{6x+9-6x+10}{(3x-5)^2}=\frac{19}{(3x-5)^2}$$

91. $$\frac{x\cdot 2x-(x^2+1)\cdot 1}{(x^2+1)^2}=\frac{2x^2-x^2-1}{(x^2+1)^2}=\frac{x^2-1}{(x^2+1)^2}=\frac{(x+1)(x-1)}{(x^2+1)^2}$$

93. $$\frac{(3x+1)\cdot 2x-x^2\cdot 3}{(3x+1)^2}=\frac{6x^2+2x-3x^2}{(3x+1)^2}=\frac{3x^2+2x}{(3x+1)^2}=\frac{x(3x+2)}{(3x+1)^2}$$

95. $$\frac{(x^2+1)\cdot 3-(3x+4)\cdot 2x}{(x^2+1)^2}=\frac{3x^2+3-6x^2-8x}{(x^2+1)^2}=\frac{-3x^2-8x+3}{(x^2+1)^2}$$
$$=\frac{-(3x^2+8x-3)}{(x^2+1)^2}$$
$$=-\frac{(3x-1)(x+3)}{(x^2+1)^2}$$

Appendix A

Review

A.4 Polynomial Division; Synthetic Division

1. quotient, divisor, remainder

3. True

5.

$$
\begin{array}{r}
4x^2 - 11x + 23 \\
x+2\overline{)4x^3 - 3x^2 + x + 1} \\
\underline{4x^3 + 8x^2} \\
-11x^2 + x \\
\underline{-11x^2 - 22x} \\
23x + 1 \\
\underline{23x + 46} \\
-45
\end{array}
$$

Check:

$(x+2)(4x^2 - 11x + 23) + (-45)$
$= 4x^3 - 11x^2 + 23x + 8x^2 - 22x + 46 - 45$
$= 4x^3 - 3x^2 + x + 1$

The quotient is $4x^2 - 11x + 23$; the remainder is -45.

7.

$$
\begin{array}{r}
4x - 3 \\
x^2\overline{)4x^3 - 3x^2 + x + 1} \\
\underline{4x^3} \\
-3x^2 \\
\underline{-3x^2} \\
x + 1
\end{array}
$$

Check :

$(x^2)(4x - 3) + (x+1) = 4x^3 - 3x^2 + x + 1$

The quotient is $4x - 3$; the remainder is $x + 1$.

9.

$$
\begin{array}{r}
5x^2 - 13 \\
x^2 + 2\overline{)5x^4 - 3x^2 + x + 1} \\
\underline{5x^4 + 10x^2} \\
-13x^2 + x + 1 \\
\underline{-13x^2 - 26} \\
x + 27
\end{array}
$$

Check :

$(x^2 + 2)(5x^2 - 13) + (x + 27)$
$= 5x^4 - 13x^2 + 10x^2 - 26 + x + 27$
$= 5x^4 - 3x^2 + x + 1$

The quotient is $5x^2 - 13$; the remainder is $x + 27$.

11.

$$
\begin{array}{r}
2x^2 \\
2x^3 - 1\overline{)4x^5 - 3x^2 + x + 1} \\
\underline{4x^5 - 2x^2} \\
-x^2 + x + 1
\end{array}
$$

Check :

$(2x^3 - 1)(2x^2) + (-x^2 + x + 1)$
$= 4x^5 - 2x^2 - x^2 + x + 1$
$= 4x^5 - 3x^2 + x + 1$

The quotient is $2x^2$; the remainder is $-x^2 + x + 1$.

13.

$$\begin{array}{r} x^2 - 2x + \dfrac{1}{2} \\ 2x^2 + x + 1{\overline{\smash{\big)}\,2x^4 - 3x^3 + x + 1}} \end{array}$$

$$\underline{2x^4 + x^3 + x^2}$$
$$-4x^3 - x^2 + x$$
$$\underline{-4x^3 - 2x^2 - 2x}$$
$$x^2 + 3x + 1$$
$$\underline{x^2 + \dfrac{1}{2}x + \dfrac{1}{2}}$$
$$\dfrac{5}{2}x + \dfrac{1}{2}$$

Check :

$$\left(x^2 - 2x + \dfrac{1}{2}\right)\!\left(2x^2 + x + 1\right) + \left(\dfrac{5}{2}x + \dfrac{1}{2}\right)$$

$$= 2x^4 + x^3 + x^2 - 4x^3 - 2x^2 - 2x + x^2 + \dfrac{1}{2}x$$

$$+ \dfrac{1}{2} + \dfrac{5}{2}x + \dfrac{1}{2}$$

$$= 2x^4 - 3x^3 + x + 1$$

The quotient is $x^2 - 2x + \dfrac{1}{2}$; the remainder is $\dfrac{5}{2}x + \dfrac{1}{2}$.

15.

$$\begin{array}{r} -4x^2 - 3x - 3 \\ x - 1{\overline{\smash{\big)}\,-4x^3 + x^2 + - 4}} \end{array}$$

$$\underline{-4x^3 + 4x^2}$$
$$-3x^2$$
$$\underline{-3x^2 + 3x}$$
$$-3x - 4$$
$$\underline{-3x + 3}$$
$$-7$$

Check :

$$(x - 1)(-4x^2 - 3x - 3) + (-7)$$

$$= -4x^3 - 3x^2 - 3x + 4x^2 + 3x + 3 - 7$$

$$= -4x^3 + x^2 - 4$$

The quotient is $-4x^2 - 3x - 3$; the remainder is -7.

17.

$$\begin{array}{r} x^2 - x - 1 \\ x^2 + x + 1{\overline{\smash{\big)}\,x^4 - x^2 + 1}} \end{array}$$

$$\underline{x^4 + x^3 + x^2}$$
$$-x^3 - 2x^2$$
$$\underline{-x^3 - x^2 - x}$$
$$-x^2 + x + 1$$
$$\underline{-x^2 - x - 1}$$
$$2x + 2$$

Check :

$$\left(x^2 + x + 1\right)\!\left(x^2 - x - 1\right) + \left(2x + 2\right)$$

$$= x^4 - x^3 - x^2 + x^3 - x^2 - x + x^2 - x - 1 + 2x + 2$$

$$= x^4 - x^2 + 1$$

The quotient is $x^2 - x - 1$; the remainder is $2x + 2$.

19.

$$x - a \overline{)\begin{array}{c} x^2 + ax + a^2 \\ x^3 \qquad\qquad - a^3 \end{array}}$$

$$\underline{x^3 - ax^2}$$

$$ax^2$$

$$\underline{ax^2 - a^2 x}$$

$$a^2 x - a^3$$

$$\underline{a^2 x - a^3}$$

$$0$$

Check :

$(x - a)(x^2 + ax + a^2) + 0$

$= x^3 + ax^2 + a^2 x - ax^2 - a^2 x - a^3$

$= x^3 - a^3$

The quotient is $x^2 + ax + a^2$; the remainder is 0.

21.
$$2\overline{)\begin{array}{cccc} 1 & -1 & 2 & 4 \\ & 2 & 2 & 8 \\ \hline 1 & 1 & 4 & 12 \end{array}}$$

Quotient: $x^2 + x + 4$

Remainder: 12

23.
$$3\overline{)\begin{array}{cccc} 3 & 2 & -1 & 3 \\ & 9 & 33 & 96 \\ \hline 3 & 11 & 32 & 99 \end{array}}$$

Quotient: $3x^2 + 11x + 32$

Remainder: 99

25.
$$-3\overline{)\begin{array}{cccccc} 1 & 0 & -4 & 0 & 1 & 0 \\ & -3 & 9 & -15 & 45 & -138 \\ \hline 1 & -3 & 5 & -15 & 46 & -138 \end{array}}$$

Quotient : $x^4 - 3x^3 + 5x^2 - 15x + 46$

Remainder : -138

27.
$$1\overline{)\begin{array}{ccccccc} 4 & 0 & -3 & 0 & 1 & 0 & 5 \\ & 4 & 4 & 1 & 1 & 2 & 2 \\ \hline 4 & 4 & 1 & 1 & 2 & 2 & 7 \end{array}}$$

Quotient : $4x^5 + 4x^4 + x^3 + x^2 + 2x + 2$

Remainder : 7

29.
$$-1.1\overline{)\begin{array}{cccc} 0.1 & 0 & 0.2 & 0 \\ & -0.11 & 0.121 & -0.3531 \\ \hline 0.1 & -0.11 & 0.321 & -0.3531 \end{array}}$$

Quotient: $0.1x^2 - 0.11x + 0.321$

Remainder: -0.3531

31.
$$1\overline{)\begin{array}{cccccc} 1 & 0 & 0 & 0 & 0 & -1 \\ & 1 & 1 & 1 & 1 & 1 \\ \hline 1 & 1 & 1 & 1 & 1 & 0 \end{array}}$$

Quotient: $x^4 + x^3 + x^2 + x + 1$

Remainder: 0

33.
$$2\overline{)\begin{array}{cccc} 4 & -3 & -8 & 4 \\ & 8 & 10 & 4 \\ \hline 4 & 5 & 2 & 8 \end{array}}$$

Remainder = 8 ≠ 0;
therefore $x - 2$ is not a factor of the given polynomial.

35.
$$2 \overline{)\begin{array}{rrrrr} 3 & -6 & 0 & -5 & 10 \\ & 6 & 0 & 0 & -10 \\ \hline 3 & 0 & 0 & -5 & 0 \end{array}}$$

Remainder = 0;
therefore $x - 2$ is a factor of the given polynomial.

37.
$$-3 \overline{)\begin{array}{rrrrrrr} 3 & 0 & 0 & 82 & 0 & 0 & 27 \\ & -9 & 27 & -81 & -3 & 9 & -27 \\ \hline 3 & -9 & 27 & 1 & -3 & 9 & 0 \end{array}}$$

Remainder = 0;
therefore $x + 3$ is a factor of the given polynomial.

39.
$$-4 \overline{)\begin{array}{rrrrrrr} 4 & 0 & -64 & 0 & 1 & 0 & -15 \\ & -16 & 64 & 0 & 0 & -4 & 16 \\ \hline 4 & -16 & 0 & 0 & 1 & -4 & 1 \end{array}}$$

Remainder = $1 \neq 0$;
therefore $x + 4$ is not a factor of the given polynomial.

41.
$$\frac{1}{2} \overline{)\begin{array}{rrrrr} 2 & -1 & 0 & 2 & -1 \\ & 1 & 0 & 0 & 1 \\ \hline 2 & 0 & 0 & 2 & 0 \end{array}}$$

Remainder = 0;
therefore $x - \dfrac{1}{2}$ is a factor of the given polynomial.

43. $\dfrac{x^3 - 2x^2 + 3x + 5}{x + 2} = ax^2 + bx + c + \dfrac{d}{x + 2}$

In order to find $a + b + c + d$, we do the long division and then look at the coefficients of the quotient and remainder.

Divide:

$$\begin{array}{r} x^2 - 4x + 11 \\ x + 2 \overline{)x^3 - 2x^2 + 3x + 5} \\ \underline{x^3 + 2x^2} \\ -4x^2 + 3x \\ \underline{-4x^2 - 8x} \\ 11x + 5 \\ \underline{11x + 22} \\ -17 \end{array}$$

Therefore,

$$\frac{x^3 - 2x^2 + 3x + 5}{x + 2} = x^2 - 4x + 11 + \frac{-17}{x + 2}$$

$$a + b + c + d = 1 - 4 + 11 - 17$$
$$= -9$$

Appendix A

Review

A.5   Solving Equations

1. $x^2 - 5x - 6 = (x-6)(x+1)$

3. $\left\{-\dfrac{5}{3}, 3\right\}$

5. equivalent

7. False

9. add, $\left(\dfrac{5}{2}\right)^2 = \dfrac{25}{4}$

11. False

13.
$$3x = 21$$
$$\frac{3x}{3} = \frac{21}{3}$$
$$x = 7$$

15.
$$5x + 15 = 0$$
$$5x + 15 - 15 = 0 - 15$$
$$5x = -15$$
$$\frac{5x}{5} = \frac{-15}{5}$$
$$x = -3$$

17.
$$2x - 3 = 5$$
$$2x - 3 + 3 = 5 + 3$$
$$2x = 8$$
$$\frac{2x}{2} = \frac{8}{2}$$
$$x = 4$$

19.
$$\frac{1}{3}x = \frac{5}{12}$$
$$(3)\left(\frac{1}{3}x\right) = \left(\frac{5}{12}\right)(3)$$
$$x = \frac{5}{4}$$

21.
$$6 - x = 2x + 9$$
$$6 - x - 6 = 2x + 9 - 6$$
$$-x = 2x + 3$$
$$-x - 2x = 2x + 3 - 2x$$
$$-3x = 3$$
$$\frac{-3x}{-3} = \frac{3}{-3}$$
$$x = -1$$

23.
$$2(3 + 2x) = 3(x - 4)$$
$$6 + 4x = 3x - 12$$
$$6 + 4x - 6 = 3x - 12 - 6$$
$$4x = 3x - 18$$
$$4x - 3x = 3x - 18 - 3x$$
$$x = -18$$

25. $8x - (2x + 1) = 3x - 10$

$8x - 2x - 1 = 3x - 10$

$6x - 1 = 3x - 10$

$6x - 1 + 1 = 3x - 10 + 1$

$6x = 3x - 9$

$6x - 3x = 3x - 9 - 3x$

$3x = -9$

$\dfrac{3x}{3} = \dfrac{-9}{3}$

$x = -3$

27. $\dfrac{1}{2}x - 4 = \dfrac{3}{4}x$

$\dfrac{1}{2}x - 4 + 4 = \dfrac{3}{4}x + 4$

$\dfrac{1}{2}x = \dfrac{3}{4}x + 4$

$\dfrac{1}{2}x - \dfrac{3}{4}x = \dfrac{3}{4}x + 4 - \dfrac{3}{4}x$

$\dfrac{2}{4}x - \dfrac{3}{4}x = 4$

$-\dfrac{1}{4}x = 4$

$(-4)\left(-\dfrac{1}{4}x\right) = (4)(-4)$

$x = -16$

29. $0.9t = 0.4 + 0.1t$

$0.9t - 0.1t = 0.4 + 0.1t - 0.1t$

$0.8t = 0.4$

$\dfrac{0.8t}{0.8} = \dfrac{0.4}{0.8}$

$t = 0.5$

31. $\dfrac{2}{y} + \dfrac{4}{y} = 3$

$y\left(\dfrac{2}{y} + \dfrac{4}{y}\right) = y(3)$

$2 + 4 = 3y$

$6 = 3y$

$\dfrac{6}{3} = \dfrac{3y}{3}$

$y = 2$

Since $y = 2$ does not cause a denominator to equal zero, the solution set is {2}.

33. $(x + 7)(x - 1) = (x + 1)^2$

$x^2 + 6x - 7 = x^2 + 2x + 1$

$x^2 + 6x - 7 - x^2 = x^2 + 2x + 1 - x^2$

$6x - 7 = 2x + 1$

$6x - 7 + 7 = 2x + 1 + 7$

$6x = 2x + 8$

$6x - 2x = 2x + 8 - 2x$

$4x = 8$

$\dfrac{4x}{4} = \dfrac{8}{4}$

$x = 2$

35. $z(z^2 + 1) = 3 + z^3$

$z^3 + z = 3 + z^3$

$z^3 + z - z^3 = 3 + z^3 - z^3$

$z = 3$

37.
$$x^2 = 9x$$
$$x^2 - 9x = 0$$
$$x(x-9) = 0$$
$$x = 0$$
or $x - 9 = 0 \Rightarrow x = 9$
The solution set is $\{0, 9\}$.

39.
$$t^3 - 9t^2 = 0$$
$$t^2(t-9) = 0$$
$$t^2 = 0 \Rightarrow t = 0$$
or $t - 9 = 0 \Rightarrow t = 9$
The solution set is $\{0, 9\}$.

41.
$$\frac{3}{2x-3} = \frac{2}{x+5}$$
$$(2x-3)(x+5)\left(\frac{3}{2x-3}\right) = \left(\frac{2}{x+5}\right)(2x-3)(x+5)$$
$$(x+5)(3) = (2)(2x-3)$$
$$3x + 15 = 4x - 6$$
$$3x + 15 - 4x = 4x - 6 - 4x$$
$$15 - x = -6$$
$$15 - x - 15 = -6 - 15$$
$$-x = -21$$
$$x = 21$$
Since $x = 21$ does not cause any denominator to equal zero, the solution set is $\{21\}$.

43.
$$(x+2)(3x) = (x+2)(6)$$
$$3x^2 + 6x = 6x + 12$$
$$3x^2 + 6x - 6x = 6x + 12 - 6x$$
$$3x^2 = 12$$
$$3x^2 - 12 = 12 - 12$$
$$3x^2 - 12 = 0$$
$$3(x^2 - 4) = 0$$
$$3(x-2)(x+2) = 0$$
$$x - 2 = 0 \Rightarrow x = 2$$
or $x + 2 = 0 \Rightarrow x = -2$
The solution set is $\{-2, 2\}$.

45.
$$\frac{2}{x-2}=\frac{3}{x+5}+\frac{10}{(x+5)(x-2)}$$

$$(x-2)(x+5)\left(\frac{2}{x-2}\right)=\left(\frac{3}{x+5}+\frac{10}{(x+5)(x-2)}\right)(x-2)(x+5)$$

$$(x+5)(2)=(3)(x-2)+10$$

$$2x+10=3x-6+10$$

$$2x+10=3x+4$$

$$2x+10-10=3x+4-10$$

$$2x=3x-6$$

$$2x-3x=3x-6-3x$$

$$-x=-6$$

$$x=6$$

Since $x=6$ does not cause any denominator to equal zero, the solution set is $\{6\}$.

47. $|2x|=6$

$2x=6$ or $2x=-6$

$x=3$ or $x=-3$

The solution set is $\{-3, 3\}$.

49. $|2x+3|=5$

$2x+3=5$ or $2x+3=-5$

$2x=2$ or $2x=-8$

$x=1$ or $x=-4$

The solution set is $\{-4, 1\}$.

51. $|1-4t|=5$

$1-4t=5$ or $1-4t=-5$

$-4t=4$ or $-4t=-6$

$t=-1$ or $t=\frac{3}{2}$

The solution set is $\left\{-1, \frac{3}{2}\right\}$.

53. $|-2x|=8$

$-2x=8$ or $-2x=-8$

$x=-4$ or $x=4$

The solution set is $\{-4, 4\}$.

55. $|-2|x=4$

$2x=4$

$x=2$

The solution set is $\{2\}$.

57. $|x-2|=-\frac{1}{2}$

The equation has no solution, since absolute value always yields a non-negative number.

59. $\left| x^2 - 4 \right| = 0$

$$x^2 - 4 = 0$$
$$x^2 = 4$$
$$x = \pm 2$$

The solution set is $\{-2, 2\}$.

61. $\left| x^2 - 2x \right| = 3$

$$x^2 - 2x = 3 \quad \text{or} \quad x^2 - 2x = -3$$
$$x^2 - 2x - 3 = 0 \quad \text{or} \quad x^2 - 2x + 3 = 0$$
$$(x - 3)(x + 1) = 0 \Rightarrow x = 3, x = -1$$
$$\text{or} \quad x = \frac{2 \pm \sqrt{4 - 12}}{2}$$
$$= \frac{2 \pm \sqrt{-8}}{2} \Rightarrow \text{no real solution}$$

The solution set is $\{-1, 3\}$.

63. $\left| x^2 + x - 1 \right| = 1$

$$x^2 + x - 1 = 1 \quad \text{or} \quad x^2 + x - 1 = -1$$
$$x^2 + x - 2 = 0 \quad \text{or} \quad x^2 + x = 0$$
$$(x - 1)(x + 2) = 0 \quad \text{or} \quad x(x + 1) = 0$$
$$x = 1, x = -2 \quad \text{or} \quad x = 0, x = -1$$

The solution set is $\{-2, -1, 0, 1\}$.

65. $x^2 = 4x$

$$x^2 - 4x = 0$$
$$x(x - 4) = 0$$
$$x = 0$$
$$\text{or } x - 4 = 0 \Rightarrow x = 4$$

The solution set is $\{0, 4\}$.

67. $z^2 + 4z - 12 = 0$

$$(z + 6)(z - 2) = 0$$
$$z + 6 = 0 \Rightarrow z = -6$$
$$\text{or } z - 2 = 0 \Rightarrow z = 2$$

The solution set is $\{-6, 2\}$.

69. $2x^2 - 5x - 3 = 0$

$$(2x + 1)(x - 3) = 0$$
$$2x + 1 = 0$$
$$2x = -1 \Rightarrow x = -\frac{1}{2}$$
$$\text{or } x - 3 = 0 \Rightarrow x = 3$$

The solution set is $\left\{ -\frac{1}{2}, 3 \right\}$.

71. $x(x - 7) + 12 = 0$

$$x^2 - 7x + 12 = 0$$
$$(x - 4)(x - 3) = 0$$
$$x - 4 = 0 \Rightarrow x = 4$$
$$\text{or } x - 3 = 0 \Rightarrow x = 3$$

The solution set is $\{3, 4\}$.

73. $4x^2 + 9 = 12x$

$$4x^2 - 12x + 9 = 0$$
$$(2x - 3)^2 = 0 \Rightarrow x = \frac{3}{2}$$

The solution set is $\left\{ \frac{3}{2} \right\}$.

75.

$$6x - 5 = \frac{6}{x}$$

$$6x^2 - 5x = 6$$

$$6x^2 - 5x - 6 = 0$$

$$(3x + 2)(2x - 3) = 0$$

$$3x + 2 = 0 \Rightarrow x = -\frac{2}{3}$$

$$\text{or } 2x - 3 = 0 \Rightarrow x = \frac{3}{2}$$

Since neither of these values causes a denominator to equal zero, the solution set is $\left\{-\frac{2}{3}, \frac{3}{2}\right\}$.

77.

$$\frac{4(x-2)}{x-3} + \frac{3}{x} = \frac{-3}{x(x-3)}$$

$$x(x-3)\left(\frac{4(x-2)}{x-3} + \frac{3}{x}\right) = \left(\frac{-3}{x(x-3)}\right)x(x-3)$$

$$x(4(x-2)) + (x-3)(3) = -3$$

$$4x^2 - 8x + 3x - 9 = -3$$

$$4x^2 - 5x - 6 = 0$$

$$(4x + 3)(x - 2) = 0$$

$$4x + 3 = 0 \Rightarrow x = -\frac{3}{4}$$

$$\text{or } x - 2 = 0 \Rightarrow x = 2$$

Since neither of these values causes a denominator to equal zero, the solution set is $\left\{-\frac{3}{4}, 2\right\}$.

79. $x^2 = 25$

$$x = \pm\sqrt{25}$$

$$x = \pm 5$$

The solution set is $\{-5, 5\}$.

81. $(x - 1)^2 = 4$

$$x - 1 = \pm\sqrt{4}$$

$$x - 1 = \pm 2$$

$$x - 1 = 2 \Rightarrow x = 3$$

$$\text{or } x - 1 = -2 \Rightarrow x = -1$$

The solution set is $\{-1, 3\}$.

83. $(2x+3)^2 = 9$

$$2x+3 = \pm\sqrt{9}$$
$$2x+3 = \pm 3$$
$$2x+3 = 3$$
$$2x = 0 \Rightarrow x = 0$$
$$\text{or } 2x+3 = -3$$
$$2x = -6 \Rightarrow x = -3$$

The solution set is $\{-3, 0\}$.

85. $\left(\dfrac{8}{2}\right)^2 = 4^2 = 16$

87. $\left(\dfrac{\left(\dfrac{1}{2}\right)}{2}\right)^2 = \left(\dfrac{1}{4}\right)^2 = \dfrac{1}{16}$

89. $\left(\dfrac{\left(-\dfrac{2}{3}\right)}{2}\right)^2 = \left(-\dfrac{1}{3}\right)^2 = \dfrac{1}{9}$

91. $x^2 + 4x = 21$

$$x^2 + 4x + 4 = 21 + 4$$
$$(x+2)^2 = 25$$
$$x+2 = \pm\sqrt{25}$$
$$x+2 = \pm 5$$
$$x+2 = 5 \Rightarrow x = 3$$
$$\text{or } x+2 = -5 \Rightarrow x = -7$$

The solution set is $\{-7, 3\}$.

93. $x^2 - \dfrac{1}{2}x - \dfrac{3}{16} = 0$

$$x^2 - \frac{1}{2}x = \frac{3}{16}$$
$$x^2 - \frac{1}{2}x + \frac{1}{16} = \frac{3}{16} + \frac{1}{16}$$
$$\left(x - \frac{1}{4}\right)^2 = \frac{1}{4}$$
$$x - \frac{1}{4} = \pm\sqrt{\frac{1}{4}}$$
$$x - \frac{1}{4} = \pm\frac{1}{2}$$
$$x - \frac{1}{4} = \frac{1}{2} \Rightarrow x = \frac{3}{4}$$
$$\text{or } x - \frac{1}{4} = -\frac{1}{2} \Rightarrow x = -\frac{1}{4}$$

The solution set is $\left\{-\dfrac{1}{4}, \dfrac{3}{4}\right\}$.

95.
$$3x^2 + x - \frac{1}{2} = 0$$

$$x^2 + \frac{1}{3}x - \frac{1}{6} = 0$$

$$x^2 + \frac{1}{3}x = \frac{1}{6}$$

$$x^2 + \frac{1}{3}x + \frac{1}{36} = \frac{1}{6} + \frac{1}{36}$$

$$\left(x + \frac{1}{6}\right)^2 = \frac{7}{36}$$

$$x + \frac{1}{6} = \pm\sqrt{\frac{7}{36}}$$

$$x + \frac{1}{6} = \pm\frac{\sqrt{7}}{6}$$

$$x + \frac{1}{6} = \frac{\sqrt{7}}{6} \Rightarrow x = -\frac{1}{6} + \frac{\sqrt{7}}{6}$$

$$\text{or } x + \frac{1}{6} = -\frac{\sqrt{7}}{6} \Rightarrow x = -\frac{1}{6} - \frac{\sqrt{7}}{6}$$

The solution set is $\left\{ -\frac{1}{6} - \frac{\sqrt{7}}{6},\ -\frac{1}{6} + \frac{\sqrt{7}}{6} \right\}$.

97.
$$x^2 - 4x + 2 = 0$$
$$a = 1, \quad b = -4, \quad c = 2$$

$$x = \frac{-(-4) \pm \sqrt{(-4)^2 - 4(1)(2)}}{2(1)}$$

$$= \frac{4 \pm \sqrt{16 - 8}}{2} = \frac{4 \pm \sqrt{8}}{2}$$

$$= \frac{4 \pm 2\sqrt{2}}{2} = 2 \pm \sqrt{2}$$

The solution set is $\left\{2 - \sqrt{2}, 2 + \sqrt{2}\right\}$.

99.
$$x^2 - 5x - 1 = 0$$
$$a = 1, \quad b = -5, \quad c = -1$$

$$x = \frac{-(-5) \pm \sqrt{(-5)^2 - 4(1)(-1)}}{2(1)}$$

$$= \frac{5 \pm \sqrt{25 + 4}}{2} = \frac{5 \pm \sqrt{29}}{2}$$

The solution set is $\left\{ \frac{5 - \sqrt{29}}{2}, \frac{5 + \sqrt{29}}{2} \right\}$.

101.
$$2x^2 - 5x + 3 = 0$$
$$a = 2, \quad b = -5, \quad c = 3$$

$$x = \frac{-(-5) \pm \sqrt{(-5)^2 - 4(2)(3)}}{2(2)}$$

$$= \frac{5 \pm \sqrt{25 - 24}}{4} = \frac{5 \pm 1}{4}$$

The solution set is $\left\{1, \frac{3}{2}\right\}$.

103.
$$4y^2 - y + 2 = 0$$
$$a = 4, \quad b = -1, \quad c = 2$$

$$y = \frac{-(-1) \pm \sqrt{(-1)^2 - 4(4)(2)}}{2(4)}$$

$$= \frac{1 \pm \sqrt{1 - 32}}{8} = \frac{1 \pm \sqrt{-31}}{8}$$

No real solution.

105. $4x^2 = 1 - 2x \Rightarrow 4x^2 + 2x - 1 = 0$
$a = 4, \quad b = 2, \quad c = -1$

$$x = \frac{-2 \pm \sqrt{2^2 - 4(4)(-1)}}{2(4)}$$

$$= \frac{-2 \pm \sqrt{4 + 16}}{8} = \frac{-2 \pm \sqrt{20}}{8}$$

$$= \frac{-2 \pm 2\sqrt{5}}{8} = \frac{-1 \pm \sqrt{5}}{4}$$

The solution set is $\left\{\dfrac{-1 - \sqrt{5}}{4}, \dfrac{-1 + \sqrt{5}}{4}\right\}$.

107. $x^2 + \sqrt{3}x - 3 = 0$
$a = 1, \quad b = \sqrt{3}, \quad c = -3$

$$x = \frac{-\left(\sqrt{3}\right) \pm \sqrt{\left(\sqrt{3}\right)^2 - 4(1)(-3)}}{2(1)}$$

$$= \frac{-\sqrt{3} \pm \sqrt{3 + 12}}{2} = \frac{-\sqrt{3} \pm \sqrt{15}}{2}$$

The solution set is $\left\{\dfrac{-\sqrt{3} - \sqrt{15}}{2}, \dfrac{-\sqrt{3} + \sqrt{15}}{2}\right\}$.

109. $x^2 - 5x + 7 = 0$
$a = 1, \quad b = -5, \quad c = 7$

$b^2 - 4ac = (-5)^2 - 4(1)(7)$

$\qquad = 25 - 28 = -3$

Since the discriminant is negative, there are no real solutions.

111. $9x^2 - 30x + 25 = 0$
$a = 9, \quad b = -30, \quad c = 25$

$b^2 - 4ac = (-30)^2 - 4(9)(25)$

$\qquad = 900 - 900 = 0$

Since the discriminant is zero, there is one repeated real solution.

113. $3x^2 + 5x - 8 = 0$
$a = 3, \quad b = 5, \quad c = -8$

$b^2 - 4ac = 5^2 - 4(3)(-8)$

$\qquad = 25 + 96 = 121$

Since the discriminant is positive, there are two unequal real solutions.

115.
$$ax - b = c, \quad a \neq 0$$
$$ax - b + b = c + b$$
$$ax = c + b$$
$$\frac{ax}{a} = \frac{c + b}{a}$$
$$x = \frac{c + b}{a}$$

117.
$$\frac{x}{a} + \frac{x}{b} = c, \quad a \neq 0, b \neq 0, a \neq -b$$

$$ab\left(\frac{x}{a} + \frac{x}{b}\right) = ab \cdot c$$

$$bx + ax = abc$$

$$x(a + b) = abc$$

$$\frac{x(a + b)}{a + b} = \frac{abc}{a + b}$$

$$x = \frac{abc}{a + b}$$

119.

$$\frac{1}{x-a} + \frac{1}{x+a} = \frac{2}{x-1}$$

$$(x-a)(x+a)(x-1)\left(\frac{1}{x-a} + \frac{1}{x+a}\right) = \left(\frac{2}{x-1}\right)(x-a)(x+a)(x-1)$$

$$(x+a)(x-1)(1) + (x-a)(x-1)(1) = (2)(x-a)(x+a)$$

$$x^2 - x + ax - a + x^2 - x - ax + a = 2x^2 - 2a^2$$

$$2x^2 - 2x = 2x^2 - 2a^2$$

$$-2x = -2a^2$$

$$\frac{-2x}{-2} = \frac{-2a^2}{-2}$$

$$x = a^2$$

such that $x \neq \pm a, x \neq 1$

121. Solving for R:

$$\frac{1}{R} = \frac{1}{R_1} + \frac{1}{R_2}$$

$$RR_1R_2\left(\frac{1}{R}\right) = RR_1R_2\left(\frac{1}{R_1} + \frac{1}{R_2}\right)$$

$$R_1R_2 = RR_2 + RR_1$$

$$R_1R_2 = R(R_2 + R_1)$$

$$\frac{R_1R_2}{R_2 + R_1} = \frac{R(R_2 + R_1)}{R_2 + R_1}$$

$$\frac{R_1R_2}{R_2 + R_1} = R$$

123. Solving for R:

$$F = \frac{mv^2}{R}$$

$$RF = R\left(\frac{mv^2}{R}\right)$$

$$RF = mv^2$$

$$\frac{RF}{F} = \frac{mv^2}{F} \Rightarrow R = \frac{mv^2}{F}$$

125. Solving for r:

$$S = \frac{a}{1-r}$$

$$S(1-r) = \left(\frac{a}{1-r}\right)(1-r)$$

$$S - Sr = a$$

$$S - Sr - S = a - S$$

$$-Sr = a - S$$

$$\frac{-Sr}{-S} = \frac{a-S}{-S}$$

$$r = \frac{S-a}{S}$$

127. Given a quadratic equation $ax^2 + bx + c = 0$,

the solutions are given by $x = \dfrac{-b + \sqrt{b^2 - 4ac}}{2a}$ and $x = \dfrac{-b - \sqrt{b^2 - 4ac}}{2a}$.

Adding these two values we get

$$\dfrac{-b + \sqrt{b^2 - 4ac}}{2a} + \dfrac{-b - \sqrt{b^2 - 4ac}}{2a} = \dfrac{-b + \sqrt{b^2 - 4ac} - b - \sqrt{b^2 - 4ac}}{2a} = \dfrac{-2b}{2a} = -\dfrac{b}{a}$$

129. The quadratic equation $kx^2 + x + k = 0$ will have a repeated solution provided the discriminant is 0. That is, we need $b^2 - 4ac = 0$ in the given quadratic equation.

$$b^2 - 4ac = (1)^2 - 4(k)(k) = 1 - 4k^2$$

So we solve

$$1 - 4k^2 = 0$$

$$1 = 4k^2$$

$$\dfrac{1}{4} = k^2$$

$$\pm\sqrt{\dfrac{1}{4}} = k \Rightarrow \pm\dfrac{1}{2} = k$$

131. The quadratic equation $ax^2 + bx + c = 0$ has solutions given by

$$x_1 = \dfrac{-b + \sqrt{b^2 - 4ac}}{2a} \quad \text{and} \quad x_2 = \dfrac{-b - \sqrt{b^2 - 4ac}}{2a}.$$

The quadratic equation $ax^2 - bx + c = 0$ has solutions given by

$$x_3 = \dfrac{-(-b) + \sqrt{(-b)^2 - 4ac}}{2a} = \dfrac{b + \sqrt{b^2 - 4ac}}{2a} = -x_2$$

and

$$x_4 = \dfrac{-(-b) - \sqrt{(-b)^2 - 4ac}}{2a} = \dfrac{b - \sqrt{b^2 - 4ac}}{2a} = -x_1$$

So we have the negatives of the first pair of solutions.

133–139.   Answers will vary.

Review

A.6 Complex Numbers; Quadratic Equations in the Complex Number System

1. True

3. False

5. $\{-2i, 2i\}$

7. True

9. $(2 - 3i) + (6 + 8i) = (2 + 6) + (-3 + 8)i = 8 + 5i$

11. $(-3 + 2i) - (4 - 4i) = (-3 - 4) + (2 - (-4))i = -7 + 6i$

13. $(2 - 5i) - (8 + 6i) = (2 - 8) + (-5 - 6)i = -6 - 11i$

15. $3(2 - 6i) = 6 - 18i$

17. $2i(2 - 3i) = 4i - 6i^2 = 4i - 6(-1) = 6 + 4i$

19. $(3 - 4i)(2 + i) = 6 + 3i - 8i - 4i^2 = 6 - 5i - 4(-1) = 10 - 5i$

21. $(-6 + i)(-6 - i) = 36 + 6i - 6i - i^2 = 36 - (-1) = 37$

23. $\dfrac{10}{3 - 4i} = \dfrac{10}{3 - 4i} \cdot \dfrac{3 + 4i}{3 + 4i} = \dfrac{30 + 40i}{9 + 16} = \dfrac{30 + 40i}{25} = \dfrac{30}{25} + \dfrac{40}{25}i = \dfrac{6}{5} + \dfrac{8}{5}i$

25. $\dfrac{2 + i}{i} = \dfrac{2 + i}{i} \cdot \dfrac{-i}{-i} = \dfrac{-2i - i^2}{-i^2} = \dfrac{-2i - (-1)}{-(-1)} = \dfrac{1 - 2i}{1} = 1 - 2i$

27. $\dfrac{6 - i}{1 + i} = \dfrac{6 - i}{1 + i} \cdot \dfrac{1 - i}{1 - i} = \dfrac{6 - 7i + (-1)}{1 + 1} = \dfrac{5 - 7i}{2} = \dfrac{5}{2} - \dfrac{7}{2}i$

29. $\left(\dfrac{1}{2} + \dfrac{\sqrt{3}}{2}i\right)^2 = \dfrac{1}{4} + 2\left(\dfrac{1}{2}\right)\left(\dfrac{\sqrt{3}}{2}i\right) + \dfrac{3}{4}i^2 = \dfrac{1}{4} + \dfrac{\sqrt{3}}{2}i + \dfrac{3}{4}(-1) = -\dfrac{1}{2} + \dfrac{\sqrt{3}}{2}i$

31. $(1 + i)^2 = 1 + 2i + i^2 = 1 + 2i + (-1) = 2i$

33. $i^{23} = i^{20+3} = i^{20} \cdot i^3 = \left(i^4\right)^5 \cdot i^3 = (1)^5(-i) = -i$

35. $i^{-15} = \dfrac{1}{i^{15}} = \dfrac{1}{i^{12+3}} = \dfrac{1}{i^{12} \cdot i^3} = \dfrac{1}{\left(i^4\right)^3 \cdot i^3} = \dfrac{1}{(1)^3(-i)} = \dfrac{1}{(1)(-i)} = \dfrac{1}{-i} \cdot \dfrac{i}{i} = \dfrac{i}{-i^2} = \dfrac{i}{-(-1)} = i$

37. $i^6 - 5 = i^4 \cdot i^2 - 5 = (1)(-1) - 5 = -1 - 5 = -6$

39. $6i^3 - 4i^5 = i^3\left(6 - 4i^2\right) = i^2 \cdot i\left(6 - 4(-1)\right) = -1 \cdot i(10) = -10i$

41. $(1+i)^3 = (1+i)(1+i)(1+i) = \left(1 + 2i + i^2\right)(1+i) = (1 + 2i - 1)(1+i) = 2i(1+i)$

    $= 2i + 2i^2 = 2i + 2(-1) = -2 + 2i$

43. $i^7\left(1 + i^2\right) = i^7\left(1 + (-1)\right) = i^7(0) = 0$

45. $i^6 + i^4 + i^2 + 1 = \left(i^2\right)^3 + \left(i^2\right)^2 + i^2 + 1 = (-1)^3 + (-1)^2 + (-1) + 1 = -1 + 1 - 1 + 1 = 0$

47. $\sqrt{-4} = 2i$ 49. $\sqrt{-25} = 5i$

51. $\sqrt{(3 + 4i)(4i - 3)} = \sqrt{12i - 9 + 16i^2 - 12i} = \sqrt{-9 + 16(-1)} = \sqrt{-25} = 5i$

53. $x^2 + 4 = 0$ 55. $x^2 - 16 = 0$

    $x^2 = -4$ $x^2 = 16$

    $x = \pm\sqrt{-4} = \pm 2i$ $x = \pm\sqrt{16} = \pm 4$

    The solution set is $\{\pm 2i\}$.   The solution set is $\{\pm 4\}$.

57. $x^2 - 6x + 13 = 0$

    $a = 1, b = -6, c = 13, \quad b^2 - 4ac = (-6)^2 - 4(1)(13) = 36 - 52 = -16$

    $x = \dfrac{-(-6) \pm \sqrt{-16}}{2(1)} = \dfrac{6 \pm 4i}{2} = 3 \pm 2i$

    The solution set is $\{3 - 2i, 3 + 2i\}$.

59. $x^2 - 6x + 10 = 0$

    $a = 1, b = -6, c = 10, \quad b^2 - 4ac = (-6)^2 - 4(1)(10) = 36 - 40 = -4$

    $x = \dfrac{-(-6) \pm \sqrt{-4}}{2(1)} = \dfrac{6 \pm 2i}{2} = 3 \pm i$

    The solution set is $\{3 - i, 3 + i\}$.

61. $8x^2 - 4x + 1 = 0$
    $a = 8, b = -4, c = 1, \quad b^2 - 4ac = (-4)^2 - 4(8)(1) = 16 - 32 = -16$

$$x = \frac{-(-4) \pm \sqrt{-16}}{2(8)} = \frac{4 \pm 4i}{16} = \frac{1}{4} \pm \frac{1}{4}i$$

The solution set is $\left\{ \dfrac{1}{4} - \dfrac{1}{4}i, \ \dfrac{1}{4} + \dfrac{1}{4}i \right\}$.

63. $5x^2 + 1 = 2x \Rightarrow 5x^2 - 2x + 1 = 0$
    $a = 5, b = -2, c = 1, \quad b^2 - 4ac = (-2)^2 - 4(5)(1) = 4 - 20 = -16$

$$x = \frac{-(-2) \pm \sqrt{-16}}{2(5)} = \frac{2 \pm 4i}{10} = \frac{1}{5} \pm \frac{2}{5}i$$

The solution set is $\left\{ \dfrac{1}{5} - \dfrac{2}{5}i, \ \dfrac{1}{5} + \dfrac{2}{5}i \right\}$.

65. $x^2 + x + 1 = 0$
    $a = 1, b = 1, c = 1, \quad b^2 - 4ac = 1^2 - 4(1)(1) = 1 - 4 = -3$

$$x = \frac{-1 \pm \sqrt{-3}}{2(1)} = \frac{-1 \pm \sqrt{3}i}{2} = -\frac{1}{2} \pm \frac{\sqrt{3}}{2}i$$

The solution set is $\left\{ -\dfrac{1}{2} - \dfrac{\sqrt{3}}{2}i, \ -\dfrac{1}{2} + \dfrac{\sqrt{3}}{2}i \right\}$.

67. $x^3 - 8 = 0$
    $(x - 2)(x^2 + 2x + 4) = 0$

$$x - 2 = 0 \Rightarrow x = 2$$

$$x^2 + 2x + 4 = 0$$

$a = 1, b = 2, c = 4, \quad b^2 - 4ac = 2^2 - 4(1)(4) = 4 - 16 = -12$

$$x = \frac{-2 \pm \sqrt{-12}}{2(1)} = \frac{-2 \pm 2\sqrt{3}i}{2} = -1 \pm \sqrt{3}i$$

The solution set is $\left\{ 2, \ -1 - \sqrt{3}i, \ -1 + \sqrt{3}i \right\}$.

69. $x^4 = 16 \Rightarrow x^4 - 16 = 0$
$$(x^2 - 4)(x^2 + 4) = 0$$
$$(x - 2)(x + 2)(x^2 + 4) = 0$$
$$x - 2 = 0 \Rightarrow x = 2$$
$$x + 2 = 0 \Rightarrow x = -2$$
$$x^2 + 4 = 0 \Rightarrow x^2 = -4 \Rightarrow x = \pm\sqrt{-4} \Rightarrow x = \pm 2i$$
The solution set is $\{-2, \ 2, \ -2i, \ 2i\}$.

71. $x^4 + 13x^2 + 36 = 0$
$$(x^2 + 9)(x^2 + 4) = 0$$
$$x^2 + 9 = 0 \Rightarrow x^2 = -9 \Rightarrow x = \pm\sqrt{-9} \Rightarrow x = \pm 3i$$
$$x^2 + 4 = 0 \Rightarrow x^2 = -4 \Rightarrow x = \pm\sqrt{-4} \Rightarrow x = \pm 2i$$
The solution set is $\{-3i, \ 3i, \ -2i, \ 2i\}$.

73. $3x^2 - 3x + 4 = 0$
$a = 3, b = -3, c = 4, \quad b^2 - 4ac = (-3)^2 - 4(3)(4) = 9 - 48 = -39$
The equation has two complex conjugate solutions.

75. $2x^2 + 3x - 4 = 0$
$a = 2, b = 3, c = -4, \quad b^2 - 4ac = 3^2 - 4(2)(-4) = 9 + 32 = 41$
The equation has two unequal real-number solutions.

77. $9x^2 - 12x + 4 = 0$
$a = 9, b = -12, c = 4, \quad b^2 - 4ac = (-12)^2 - 4(9)(4) = 144 - 144 = 0$
The equation has a repeated real-number solution.

79. The other solution is the conjugate of $2 + 3i$, that is, $2 - 3i$.

81. $z + \bar{z} = 3 - 4i + \overline{3 - 4i} = 3 - 4i + 3 + 4i = 6$

83. $z\bar{z} = (3 - 4i)(\overline{3 - 4i}) = (3 - 4i)(3 + 4i) = 9 + 12i - 12i - 16i^2 = 9 - 16(-1) = 25$

85. $z + \bar{z} = a + bi + \overline{a + bi} = a + bi + a - bi = 2a$
$z - \bar{z} = a + bi - (\overline{a + bi}) = a + bi - (a - bi) = a + bi - a + bi = 2bi$

87. $\overline{z + w} = \overline{(a + bi) + (c + di)} = \overline{(a + c) + (b + d)i} = (a + c) - (b + d)i$
$$= (a - bi) + (c - di) = \overline{a + bi} + \overline{c + di} = \bar{z} + \bar{w}$$

89. Answers will vary.

Review

A.7   Setting Up Equations; Applications

1.   mathematical modeling   3.   uniform motion   5.   True

7.   Let A represent the area of the circle and r the radius.
     The area of a circle is the product of the number π and the square of the radius. $A = \pi r^2$

9.   Let A represent the area of the square and s the length of a side.
     The area of the square is the square of the length of a side. $A = s^2$

11. Let F represent the force, m the mass, and a the acceleration.
     Force equals the product of mass and acceleration. $F = ma$

13. Let W represent the work, F the force, and d the distance.
     Work equals force times distance. $W = Fd$

15. Let C represent the total variable cost and x the number of dishwashers manufactured.
     $C = 150x$

17. Let x represent the amount of money invested in bonds.
     Then $x - 3000$ represents the amount of money invested in CD's

Amount in Bonds	Amount in CD's	Total
x	$x - 3000$	20,000

$x + x - 3000 = 20{,}000$

$2x - 3000 = 20{,}000$

$2x = 23{,}000$

$x = 11{,}500$

$11,500 will be invested in bonds. $8500 will be invested in CD's.

19.   Let x represent the score on the final exam and construct a table.

	Test1	Test2	Test3	Test4	Test5	Final Exam	Final Exam
score	80	83	71	61	95	x	x
weight	1/7	1/7	1/7	1/7	1/7	1/7	1/7

Compute the final average and set equal to 80.

$$\left(\frac{1}{7}\right)(80 + 83 + 71 + 61 + 95 + x + x) = 80$$

Now solve for x:

$$\left(\frac{1}{7}\right)(390 + 2x) = 80$$

$$390 + 2x = 560$$

$$2x = 170$$

$$x = 85$$

Brooke needs to score an 85 on the final exam to get an average of 80 in the course.

21. l = length, w = width

$2l + 2w = 60$ Perimeter $= 2l + 2w$

$l = w + 8$ The length is 8 more than the width.

$2(w + 8) + 2w = 60$

$2w + 16 + 2w = 60$

$4w + 16 = 60$

$4w = 44$

$w = 11$ feet, $l = 11 + 8 = 19$ feet

23.   Let x represent the amount of money invested in bonds.
Then $50,000 - x$ represents the amount of money invested in CD's.

	Principal	Rate	Time (yrs)	Interest
Bonds	x	0.15	1	$0.15x$
CD's	$50,000 - x$	0.07	1	$0.07(50,000 - x)$

Since the total interest is to be $6000, we have:

$$0.15x + 0.07(50,000 - x) = 6000$$

$$(100)(0.15x + 0.07(50,000 - x)) = (6000)(100)$$

$$15x + 7(50,000 - x) = 600,000$$

$$15x + 350,000 - 7x = 600,000$$

$$8x + 350,000 = 600,000$$

$$8x = 250,000 \Rightarrow x = 31,250$$

$31,250 should be invested in bonds at 15% and $18,750 should be invested in CD's at 7%.

25. Let x represent the amount of money loaned at 8%.
    Then $12,000 - x$ represents the amount of money loaned at 18%.

	Principal	Rate	Time (yrs)	Interest
Loan at 8%	x	0.08	1	$0.08x$
Loan at 18%	$12,000 - x$	0.18	1	$0.18(12,000 - x)$

Since the total interest is to be $1000, we have:
$$0.08x + 0.18(12,000 - x) = 1000$$

$$(100)(0.08x + 0.18(12,000 - x)) = (1000)(100)$$

$$8x + 18(12,000 - x) = 100,000 \Rightarrow 8x + 216,000 - 18x = 100,000$$

$$-10x + 216,000 = 100,000 \Rightarrow -10x = -116,000 \Rightarrow x = 11,600$$

$11,600 was loaned at 8%.

27. Let x represent the number of pounds of Earl Grey tea.
    Then $100 - x$ represents the number of pounds of Orange Pekoe tea.

	No. of pounds	Price per pound	Total Value
Earl Grey	x	$5.00	$5x$
Orange Pekoe	$100 - x$	$3.00	$3(100 - x)$
Blend	100	$4.50	$4.50(100)$

$$5x + 3(100 - x) = 4.50(100)$$

$$5x + 300 - 3x = 450$$

$$2x + 300 = 450$$

$$2x = 150$$

$$x = 75$$

75 pounds of Earl Grey tea must be blended with 25 pounds of Orange Pekoe tea.

29. Let x represent the number of pounds of cashews.
    Then $x + 60$ represents the number of pounds in the mixture.

	No. of pounds	Price per pound	Total Value
cashews	x	$4.00	$4x$
peanuts	60	$1.50	$1.50(60)$
mixture	$x + 60$	$2.50	$2.50(x + 60)$

$$4x + 1.50(60) = 2.50(x + 60)$$

$$4x + 90 = 2.50x + 150$$

$$1.5x = 60$$

$$x = 40$$

40 pounds of cashews must be added to the 60 pounds of peanuts.

31.   Let r represent the speed of the current.

	Rate	Time	Distance
Upstream	$16 - r$	$\dfrac{20}{60} = \dfrac{1}{3}$	$\dfrac{16-r}{3}$
Downstream	$16 + r$	$\dfrac{15}{60} = \dfrac{1}{4}$	$\dfrac{16+r}{4}$

Since the distance is the same in each direction:

$$\frac{16 - r}{3} = \frac{16 + r}{4}$$

$$4(16 - r) = 3(16 + r)$$

$$64 - 4r = 48 + 3r$$

$$16 = 7r$$

$$r = \frac{16}{7} \approx 2.286$$

The speed of the current is approximately 2.286 miles per hour.

33.   Let r represent the rate of the slower train. Then $r + 50$ represents the rate of the faster train.

	Rate	Time	Distance
Slower Train	r	3	$3r$
Faster Train	$r + 50$	1	$(1)(r + 50)$

The distance that the faster train has traveled is 10 miles less than the distance than the slower train has traveled.

$$3r - 10 = r + 50$$

$$3r = r + 60$$

$$2r = 60$$

$$r = 30$$

The Metra train is averaging 30 miles per hour; the Amtrak train is averaging $30 + 50 = 80$ miles per hour..

35.   Let t represent the time it takes to do the job together.

	Time to do job	Part of job done in one minute
Trent	30	$\dfrac{1}{30}$
Lois	20	$\dfrac{1}{20}$
Together	t	$\dfrac{1}{t}$

$$\frac{1}{30} + \frac{1}{20} = \frac{1}{t}$$

$$2t + 3t = 60$$

$$5t = 60 \Rightarrow t = 12$$

Working together, the job can be done in 12 minutes.

37.   Let x represent the length of the side of the sheet metal.

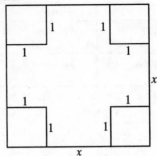

$$(x-2)(x-2)(1) = 4$$
$$x^2 - 4x + 4 = 4$$
$$x^2 - 4x = 0$$
$$x(x-4) = 0 \Rightarrow x = 0 \text{ or } x = 4$$

Since the side cannot be 0 feet long, the length of a side of the sheet metal is 4 feet. Hence, the dimensions of the sheet metal before the cut are 4 feet by 4 feet.

39.   Let r represent the speed of the current.

	Rate	Time	Distance
Upstream	$15 - r$	$\dfrac{10}{15 - r}$	10
Downstream	$15 + r$	$\dfrac{10}{15 + r}$	10

Since the total time is 1.5 hours, we have:

$$\frac{10}{15 - r} + \frac{10}{15 + r} = 1.5$$
$$10(15 + r) + 10(15 - r) = 1.5(15 - r)(15 + r)$$
$$150 + 10r + 150 - 10r = 1.5(225 - r^2)$$
$$300 = 1.5(225 - r^2)$$
$$200 = 225 - r^2$$
$$r^2 - 25 = 0$$
$$(r - 5)(r + 5) = 0 \Rightarrow r = 5 \text{ or } r = -5, \text{ which is not practical}$$

The speed of the current is 5 miles per hour.

41.   Let w represent the width of window.
Then $l = w + 2$ represents the length of the window.
Since the area is 143 square feet, we have: $w(w + 2) = 143$
$$w^2 + 2w - 143 = 0$$
$$(w - 11)(w + 13) = 0 \Rightarrow w = 11 \text{ or } w = -13, \text{ which is not practical}$$
The width of the rectangular window is 11 feet and the
length is $11 + 2 = 13$ feet.

43. Let l represent the length of the rectangle.
Let w represent the width of the rectangle.
The perimeter is 26 meters and the area is 40 square meters.

$$2l + 2w = 26 \implies l + w = 13 \implies w = 13 - l$$

$$lw = 40$$

$$l(13 - l) = 40$$

$$13l - l^2 = 40$$

$$l^2 - 13l + 40 = 0 \implies (l - 8)(l - 5) = 0$$

$$l = 8 \qquad \text{or} \qquad l = 5$$

$$w = \frac{40}{8} = 5 \qquad w = \frac{40}{5} = 8$$

The dimensions are 5 meters by 8 meters.

45. (a) $s = 96 + 80t - 16t^2$. The ball hits the ground when the height is 0.
So we solve $0 = 96 + 80t - 16t^2$.

$$0 = -16t^2 + 80t + 96$$

$$\frac{0}{-16} = \frac{-16t^2 + 80t + 96}{-16}$$

$$0 = t^2 - 5t - 6$$

$$0 = (t - 6)(t + 1) \implies t = 6 \text{ or } t = -1$$

Discard the negative value since t represents elapsed time. Therefore, the ball hits the ground after 6 seconds.

(b) $s = 96 + 80t - 16t^2$. The ball passes the top of the building when the height is 96.
So we solve $96 = 96 + 80t - 16t^2$.

$$96 = 96 + 80t - 16t^2$$

$$0 = 80t - 16t^2$$

$$0 = 16t(5 - t) \implies t = 0 \text{ or } t = 5$$

We know that the ball starts ($t = 0$) at a height of 96 feet. Therefore, the ball passes the top of the building on its way down after 5 seconds.

47. l = length of the garden
w = width of the garden
(a)   The length of the garden is to be twice its width. Thus, $l = 2w$.
The dimensions of the fenced area are $l + 4$ and $w + 4$.
The perimeter is 46 feet, so:

$$2(l + 4) + 2(w + 4) = 46 \implies 2(2w + 4) + 2(w + 4) = 46$$

$$4w + 8 + 2w + 8 = 46 \implies 6w + 16 = 46 \implies 6w = 30 \implies w = 5$$

The dimensions of the garden are 5 feet by 10 feet.
(b)   Area = $l \cdot w = 5 \cdot 10 = 50$ square feet.

(c)   If the dimensions of the garden are the same, then the length and width of the fence are also the same $(l+4)$. The perimeter is 46 feet, so:

$$2(l+4)+2(l+4)=46 \Rightarrow 2l+8+2l+8=46$$

$$4l+16=46 \Rightarrow 4l=30 \Rightarrow l=7.5$$

The dimensions of the garden are 7.5 feet by 7.5 feet.

(d)   Area $= l \cdot w = 7.5(7.5) = 56.25$ square feet.

49.   Let x represent the width of the border measured in feet.

The radius of the pool is 5 feet.

Then $x+5$ represents the radius of the circle, including both the pool and the border.

The total area of the pool and border is $A_T = \pi(x+5)^2$.

The area of the pool is $A_P = \pi(5)^2 = 25\pi$.

The area of the border is $A_B = A_T - A_P = \pi(x+5)^2 - 25\pi$.

Since the concrete is 3 inches or 0.25 feet thick, the volume of the concrete in the border is

$$0.25 A_B = 0.25\left(\pi(x+5)^2 - 25\pi\right)$$

Since the volume of the border is to be 27 cubic feet, we solve

$$0.25\left(\pi(x+5)^2 - 25\pi\right) = 27$$

$$\pi\left(x^2 + 10x + 25 - 25\right) = 108$$

$$\pi x^2 + 10\pi x - 108 = 0$$

$$x = \frac{-10\pi \pm \sqrt{(10\pi)^2 - 4(\pi)(-108)}}{2(\pi)}$$

$$\approx \frac{-31.42 \pm \sqrt{2344.1285}}{6.28} \approx \frac{-31.42 \pm 48.42}{6.28} \approx 2.71 \text{ or } -12.71$$

The width of the border is approximately 2.71 feet.

51.   Let x represent the width of the border measured in feet.

The total area is $A_T = (6+2x)(10+2x)$.

The area of the garden is $A_G = 6 \cdot 10 = 60$.

The area of the border is $A_B = A_T - A_G = (6+2x)(10+2x) - 60$.

Since the concrete is 3 inches or 0.25 feet thick, the volume of the concrete in the border is

$$0.25 A_B = 0.25((6+2x)(10+2x) - 60)$$

Since the volume of the border is to be 27 cubic feet, we solve

$$0.25((6+2x)(10+2x) - 60) = 27$$

$$60 + 32x + 4x^2 - 60 = 108$$

$$4x^2 + 32x - 108 = 0$$

$$x^2 + 8x - 27 = 0$$

$$x = \frac{-8 \pm \sqrt{8^2 - 4(1)(-27)}}{2(1)} = \frac{-8 \pm \sqrt{172}}{2}$$

$$\approx \frac{-8 \pm 13.11}{2} \approx 2.56 \text{ or } -10.56$$

The width of the border is approximately 2.56 feet

53. Let x represent the amount of water in gallons.
    Then $x + 1$ represents the number of gallons in the 60% solution.

	No. of gallons	Conc. of Antifreeze	Pure Antifreeze
water	x	0	0
100% antifreeze	1	1.00	$1(1)$
60% antifreeze	$x + 1$	0.60	$0.60(x + 1)$

$$0 + 1(1) = 0.60(x + 1)$$
$$1 = 0.6x + 0.6$$
$$0.4 = 0.6x$$
$$x = \frac{4}{6} = \frac{2}{3}$$

2/3 gallon of water should be added.

55. Let x represent the number of ounces of water to be evaporated.

	No. of ounces	Conc. of Salt	Pure Salt
Water	x	0.00	0
4% Salt	32	0.04	$0.04(32)$
6% Salt	$32 - x$	0.06	$0.06(32 - x)$

$$0 + 0.04(32) = 0.06(32 - x) \Rightarrow 1.28 = 1.92 - 0.06x \Rightarrow 0.06x = 0.64 \Rightarrow x = \frac{0.64}{0.06} = \frac{32}{3}$$

32/3 ounces of water need to be evaporated.

57. Let x represent the number of grams of pure gold.
    Then $60 - x$ represents the number of grams of 12 karat gold to be used.

	No. of grams	Conc. of gold	Pure gold
Pure gold	x	1.00	x
12-karat gold	$60 - x$	$\frac{1}{2}$	$\frac{1}{2}(60 - x)$
16-karat gold	60	$\frac{2}{3}$	$\frac{2}{3}(60)$

$$x + \frac{1}{2}(60 - x) = \frac{2}{3}(60)$$
$$x + 30 - 0.5x = 40$$
$$0.5x = 10$$
$$x = 20$$

20 grams of pure gold should be mixed with 40 grams of 12-karat gold.

59. Let t represent the time it takes for Mike to catch up with Dan.

	Time to run mile	Time	Part of mile run in one minute	Distance
Mike	6	t	$\dfrac{1}{6}$	$\dfrac{1}{6}t$
Dan	9	$t+1$	$\dfrac{1}{9}$	$\dfrac{1}{9}(t+1)$

Since the distances are the same, we have:

$$\frac{1}{6}t = \frac{1}{9}(t+1)$$
$$3t = 2t + 2$$
$$t = 2$$

Mike will pass Dan after 2 minutes, which is a distance of $\dfrac{1}{3}$ mile.

61. Let t represent the time the auxiliary pump needs to run.

	Time to do job alone	Part of job done in one hour	Time on Job	Part of total job done by each pump
Main Pump	4	$\dfrac{1}{4}$	3	$\dfrac{3}{4}$
Auxiliary Pump	9	$\dfrac{1}{9}$	t	$\dfrac{1}{9}t$

Since the two pumps are emptying one tanker, we have:

$$\frac{3}{4} + \frac{1}{9}t = 1$$
$$27 + 4t = 36$$
$$4t = 9$$
$$t = \frac{9}{4} = 2.25$$

The auxiliary pump must run for 2.25 hours. It should be started at 9:45 A.M.

63. Let t represent the time for the tub to fill with the faucets open and the stopper removed.

	Time to do job alone	Part of job done in one minute	Time on Job	Part of total job done by each
Faucets open	15	$\dfrac{1}{15}$	t	$\dfrac{t}{15}$
Stopper removed	20	$-\dfrac{1}{20}$	t	$-\dfrac{t}{20}$

Since one tub is being filled, we have:

$$\frac{t}{15} + \left(-\frac{t}{20}\right) = 1 \Rightarrow 4t - 3t = 60 \Rightarrow t = 60$$

Therefore, 60 minutes are required to fill the tub.

65. Burke's rate is $\dfrac{100}{12}$ meters/sec.

In 9.99 seconds, Burke will run $\dfrac{100}{12}(9.99) = 83.25$ meters.

Lewis would win by 16.75 meters.

67. Let x represent the number of highway miles.

Then $30,000 - x$ represents the number of city miles.

	No. of miles	Gallons per mile	Gallons used
highway	x	$\dfrac{1}{40}$	$\dfrac{1}{40}x$
city	$30,000 - x$	$\dfrac{1}{25}$	$(30,000 - x)\left(\dfrac{1}{25}\right)$

$$\frac{1}{40}x + (30,000 - x)\left(\frac{1}{25}\right) = 900 \Rightarrow \frac{1}{40}x + 1200 - \frac{1}{25}x = 900$$

$$\frac{1}{40}x - \frac{1}{25}x = -300$$

$$\frac{25 - 40}{(40)(25)}x = -300 \Rightarrow \frac{-15}{1000}x = -300$$

$$x = -300\left(\frac{1000}{-15}\right) = 20,000$$

Therese is allowed 20,000 miles as a business expense.

69. Solve the equation $\dfrac{1}{2}n(n - 3) = 65$.

$$2\left(\frac{1}{2}\right)n(n - 3) = (65)(2)$$

$$n(n - 3) = 130$$

$$n^2 - 3n = 130$$

$$n^2 - 3n - 130 = 0$$

$$(n + 10)(n - 13) = 0 \Rightarrow n = -10 \text{ or } n = 13$$

Since n must be a positive integer (it represents the number of sides of the polygon), we discard the negative value. Therefore, a polygon with 65 diagonals has 13 sides.

Solve the equation $\dfrac{1}{2}n(n - 3) = 80$.

$$2\left(\frac{1}{2}\right)n(n - 3) = (80)(2) \Rightarrow n(n - 3) = 160$$

$$n^2 - 3n = 160 \Rightarrow n^2 - 3n - 160 = 0$$

The solutions to this equation are $n = \dfrac{3 \pm \sqrt{9 + 640}}{2} = \dfrac{3 \pm \sqrt{649}}{2}$, but neither of these

answers is a whole number. Therefore, there is no polygon having 80 diagonals.

71. The time traveled with the tail wind was: $919 = 550t \Rightarrow t = \dfrac{919}{550} \approx 1.671$ hours

Since they were 20 minutes early, the time in still air would have been:
1.671 hours $+ 20$ minutes $= 1.671 + 0.333 \approx 2$ hours
Thus the rate in still air is: $919 = r(2) \Rightarrow r \approx 460$ nautical miles/hour
The tail wind was about $550 - 460 = 90$ nautical miles per hour.

73. Answers will vary.

Review

A.8 Interval Notation; Solving Inequalities

1. $x \geq -2$

3. 2

5. negative

7. $\{-5, 5\}$

9. True

11. $[0, 2]$ $0 \leq x \leq 2$

13. $(-1, 2)$ $-1 < x < 2$

15. $[0, 3)$ $0 \leq x < 3$

17. $[0, 4]$

19. $[4, 6)$

21. $[4, \infty)$

23. $(-\infty, -4)$

25. $2 \leq x \leq 5$

27. $-3 < x < -2$

29. $x \geq 4$

31. $x < -3$

33. (a) $6 < 8$
    (b) $-2 < 0$
    (c) $9 < 15$
    (d) $-6 > -10$

35. (a) $7 > 0$
    (b) $-1 > -8$
    (c) $12 > -9$
    (d) $-8 < 6$

37. (a) $2x + 4 < 5$
    (b) $2x - 4 < -3$
    (c) $6x + 3 < 6$
    (d) $-4x - 2 > -4$

39. If $x < 5$, then $x - 5 < 0$.

41. If $x > -4$, then $x + 4 > 0$.

43. If $x \geq -4$, then $3x \geq -12$.

45. If $x > 6$, then $-2x < -12$.

47. If $x \geq 5$, then $-4x \leq -20$.

49. If $2x > 6$, then $x > 3$.

51. If $-\dfrac{1}{2}x \leq 3$, then $x \geq -6$.

53. $x + 1 < 5$

$\quad x < 4$

$\left\{x \mid x < 4\right\}$ or $(-\infty, 4)$

4

55. $1 - 2x \leq 3$

$\quad -2x \leq 2 \Rightarrow x \geq -1$

$\left\{x \mid x \geq -1\right\}$ or $[-1, \infty)$

-1

57. $3x - 7 > 2$

$\quad 3x > 9 \Rightarrow x > 3$

$\left\{x \mid x > 3\right\}$ or $(3, \infty)$

3

59. $3x - 1 \geq 3 + x$

$\quad 2x \geq 4 \Rightarrow x \geq 2$

$\left\{x \mid x \geq 2\right\}$ or $[2, \infty)$

2

61. $-2(x + 3) < 8$

$\quad x + 3 > -4$

$\quad x > -7$

$\left\{x \mid x > -7\right\}$ or $(-7, \infty)$

-7

63. $4 - 3(1 - x) \leq 3$

$\quad 4 - 3 + 3x \leq 3$

$\quad 3x + 1 \leq 3$

$\quad 3x \leq 2 \Rightarrow x \leq \dfrac{2}{3}$

$\left\{x \mid x \leq \dfrac{2}{3}\right\}$ or $\left(-\infty, \dfrac{2}{3}\right]$

$\dfrac{2}{3}$

65. $\dfrac{1}{2}(x - 4) > x + 8$

$\quad \dfrac{1}{2}x - 2 > x + 8$

$\quad -\dfrac{1}{2}x > 10 \Rightarrow x < -20$

$\left\{x \mid x < -20\right\}$ or $(-\infty, -20)$

-20

67. $\dfrac{x}{2} \geq 1 - \dfrac{x}{4}$

$\quad 2x \geq 4 - x$

$\quad 3x \geq 4 \Rightarrow x \geq \dfrac{4}{3}$

$\left\{x \mid x \geq \dfrac{4}{3}\right\}$ or $\left[\dfrac{4}{3}, \infty\right)$

$\dfrac{4}{3}$

69. $0 \le 2x - 6 \le 4$

    $6 \le 2x \le 10$

    $3 \le x \le 5$

    $\{x \mid 3 \le x \le 5\}$ or $[3, 5]$

71. $-5 \le 4 - 3x \le 2$

    $-9 \le -3x \le -2$

    $3 \ge x \ge \dfrac{2}{3}$

    $\dfrac{2}{3} \le x \le 3$

    $\left\{x \mid \dfrac{2}{3} \le x \le 3\right\}$ or $\left[\dfrac{2}{3}, 3\right]$

73. $-3 < \dfrac{2x - 1}{4} < 0$

    $-12 < 2x - 1 < 0$

    $-11 < 2x < 1 \Rightarrow -\dfrac{11}{2} < x < \dfrac{1}{2}$

    $\left\{x \mid -\dfrac{11}{2} < x < \dfrac{1}{2}\right\}$ or $\left(-\dfrac{11}{2}, \dfrac{1}{2}\right)$

75. $1 < 1 - \dfrac{1}{2}x < 4$

    $0 < -\dfrac{1}{2}x < 3$

    $0 > x > -6 \Rightarrow -6 < x < 0$

    $\{x \mid -6 < x < 0\}$ or $(-6, 0)$

77. $(4x + 2)^{-1} < 0$

    $\dfrac{1}{4x + 2} < 0$

    $4x + 2 < 0$

    $x < -\dfrac{1}{2}$

    $\left\{x \mid x < -\dfrac{1}{2}\right\}$ or $\left(-\infty, -\dfrac{1}{2}\right)$

79. $0 < \dfrac{2}{x} < \dfrac{3}{5} \Rightarrow 0 < \dfrac{2}{x}$ and $\dfrac{2}{x} < \dfrac{3}{5}$

$0 < \dfrac{2}{x} \Rightarrow x > 0$ therefore $\dfrac{2}{x} < \dfrac{3}{5} \Rightarrow 10 < 3x \Rightarrow \dfrac{10}{3} < x$

$\left\{ x \,\middle|\, x > \dfrac{10}{3} \right\}$ or $\left(\dfrac{10}{3}, \infty \right)$

10/3

81. $|2x| < 8$

$-8 < 2x < 8$

$-4 < x < 4$

$\left\{ x \,\middle|\, -4 < x < 4 \right\}$ or $(-4, 4)$

-4 4

83. $|3x| > 12$

$3x < -12$ or $3x > 12$

$x < -4$ or $x > 4$

$\left\{ x \,\middle|\, x < -4 \text{ or } x > 4 \right\}$ or

$(-\infty, -4)$ or $(4, \infty)$

-4 4

85. $|2x - 1| \le 1$

$-1 \le 2x - 1 \le 1$

$0 \le \quad 2x \quad \le 2$

$0 \le \quad x \quad \le 1$

$\left\{ x \,\middle|\, 0 \le x \le 1 \right\}$ or $[0, 1]$

0 1

87. $|1 - 2x| > 3$

$1 - 2x < -3$ or $1 - 2x > 3$

$-2x < -4$ or $-2x > 2$

$x > 2$ or $x < -1$

$\left\{ x \,\middle|\, x < -1 \text{ or } x > 2 \right\}$

or $(-\infty, -1)$ or $(2, \infty)$

-1 2

89. $|-4x| + |-5| \le 9$

$|4x| + 5 \le 9$

$|4x| \le 4$

$-4 \le 4x \le 4$

$-1 \le x \le 1$

$\left\{ x \,\middle|\, -1 \le x \le 1 \right\}$ or $[-1, 1]$

-1 1

91. $|-2x| > |-4|$

$|2x| > 4$

$2x < -4$ or $2x > 4$

$x < -2$ or $x > 2$

$\left\{ x \,\middle|\, x < -2 \text{ or } x > 2 \right\}$

or $(-\infty, -2)$ or $(2, \infty)$

-2 2

93. x differs from 2 by less than $\frac{1}{2}$

$$|x-2| < \frac{1}{2}$$

$$-\frac{1}{2} < x-2 < \frac{1}{2}$$

$$\frac{3}{2} < x < \frac{5}{2}$$

$$\left\{ x \left| \frac{3}{2} < x < \frac{5}{2} \right. \right\}$$

95. x differs from -3 by more than 2

$$|x-(-3)| > 2$$

$$x+3 < -2 \text{ or } x+3 > 2$$

$$x < -5 \text{ or } \quad x > -1$$

$$\left\{ x \,|\, x < -5 \text{ or } x > -1 \right\}$$

97. $21 <$ young adult's age < 30

99. A temperature x that differs from $98.6°F$ by at least $1.5°$

$$|x - 98.6°| \geq 1.5°$$

$$x - 98.6° \leq -1.5° \text{ or } x - 98.6° \geq 1.5°$$

$$x \leq 97.1° \text{ or } \qquad x \geq 100.1°$$

The temperatures that are considered unhealthy are those that are less than $97.1°F$ or greater than $100.1°F$, inclusive.

101. (a) An average 25-year-old male can expect to live at least 48.4 more years. $25 + 48.4 = 73.4$. Therefore, the average life expectancy of a 25-year-old male will be ≥ 73.4.

   (b) An average 25-year-old female can expect to live at least 54.7 more years. $25 + 54.7 = 79.7$. Therefore, the average life expectancy of a 25-year-old female will be ≥ 79.7.

   (c) A female can expect to live 6.3 years longer.

103. Let P represent the selling price and C represent the commission.
Calculating the commission:
$C = 45,000 + 0.25(P - 900,000)$

$$= 45,000 + 0.25P - 225,000$$

$$= 0.25P - 180,000$$

Calculate the commission range, given the price range:

$$900,000 \leq \qquad P \qquad \leq 1,100,000$$

$$0.25(900,000) \leq \quad 0.25P \quad \leq 0.25(1,100,000)$$

$$225,000 \leq \quad 0.25P \quad \leq 275,000$$

$$225,000 - 180,000 \leq 0.25P - 180,000 \leq 275,000 - 180,000$$

$$45,000 \leq \qquad C \qquad \leq 95,000$$

The agent's commission ranges from \$45,000 to \$95,000, inclusive.

$$\frac{45,000}{900,000} = 0.05 = 5\% \text{ to } \frac{95,000}{1,100,000} \approx 0.086 \approx 8.6\%, \text{ inclusive.}$$

As a percent of selling price, the commission ranges from 5% to about 8.6%.

105. Let W represent the weekly wage and T represent the tax withheld.
In terms of wages the tax is
$$T = 74.35 + 0.25(W - 592) = 74.35 + 0.25W - 148 = 0.25W - 73.65$$
Now calculate the tax range given a wage range of $600 to $800, inclusive.
$$600 \le \qquad W \qquad \le 800$$
$$(0.25)600 \le \qquad 0.25W \qquad \le (0.25)800$$
$$150 \le \qquad 0.25W \qquad \le 200$$
$$150 - 73.65 \le 0.25W - 73.65 \le 200 - 73.65$$
$$76.35 \le 0.25W - 73.65 \le 126.35$$
The amount of tax withheld ranges from $76.35 to $126.35, inclusive.

107. Let K represent the monthly usage in kilowatt-hours.
Let C represent the monthly customer bill.
Calculating the bill:
$$C = 0.08275K + 7.58$$
Calculating the range of kilowatt-hours, given the range of bills:
$$63.47 \le \qquad C \qquad \le 214.53$$
$$63.47 \le 0.08275K + 7.58 \le 214.53$$
$$55.89 \le \qquad 0.08275K \qquad \le 206.95$$
$$675.41 \le \qquad K \qquad \le 2500.91$$
The range of usage in kilowatt-hours varied from 675.41 to 2500.91.

109. Let C represent the dealer's cost and M represent the markup over dealer's cost.
If the price is $8800, then $8800 = C + MC = C(1 + M)$.
Solving for C: $\quad C = \dfrac{8800}{1 + M}$
Calculating the range of dealer costs, given the range of markups:
$$0.12 \le \quad M \quad \le 0.18$$
$$1.12 \le 1 + M \le 1.18$$
$$\frac{1}{1.12} \ge \frac{1}{1 + M} \ge \frac{1}{1.18}$$
$$\frac{8800}{1.12} \ge \frac{8800}{1 + M} \ge \frac{8800}{1.18}$$
$$7857.14 \ge \quad C \quad \ge 7457.63$$
The dealer's cost ranged from $7457.63 to $7857.14.

111. Let T represent the score on the last test and G represent the course grade.
Calculating the course grade and solving for the last test:
$$G = \frac{68 + 82 + 87 + 89 + T}{5} = \frac{326 + T}{5} \Rightarrow T = 5G - 326$$
Calculating the range of scores on the last test, given the grade range:
$$80 \le \quad G \quad < 90$$
$$400 \le \quad 5G \quad < 450$$
$$74 \le 5G - 326 < 124$$
$$74 \le \quad T \quad < 124$$
The fifth test score must be greater than or equal to 74.

113. Since $a < b$

$$\frac{a}{2} < \frac{b}{2} \qquad\qquad \frac{a}{2} < \frac{b}{2}$$

$$\frac{a}{2} + \frac{a}{2} < \frac{a}{2} + \frac{b}{2} \qquad \frac{a}{2} + \frac{b}{2} < \frac{b}{2} + \frac{b}{2}$$

$$a < \frac{a+b}{2} \qquad\qquad \frac{a+b}{2} < b$$

Thus, $a < \dfrac{a+b}{2} < b$

115. If $0 < a < b$, then $0 < a^2 < ab$ and $0 < ab < b^2$.

$$0 < a^2 < ab \qquad\qquad 0 < ab < b^2$$

$$0 < a^2 < \left(\sqrt{ab}\right)^2 \qquad 0 < \left(\sqrt{ab}\right)^2 < b^2$$

$$0 < a < \sqrt{ab} \qquad\qquad 0 < \sqrt{ab} < b$$

Thus, $a < \sqrt{ab} < b$.

117. For $0 < a < b$, $\quad \dfrac{1}{h} = \dfrac{1}{2}\left(\dfrac{1}{a} + \dfrac{1}{b}\right)$

$$h \cdot \frac{1}{h} = \frac{1}{2}\left(\frac{b+a}{ab}\right) \cdot h \implies 1 = \frac{1}{2}\left(\frac{b+a}{ab}\right) \cdot h \implies \frac{2ab}{a+b} = h$$

$$h - a = \frac{2ab}{a+b} - a \qquad\qquad b - h = b - \frac{2ab}{a+b}$$

$$= \frac{2ab - a(a+b)}{a+b} \qquad\qquad = \frac{b(a+b) - 2ab}{a+b}$$

$$= \frac{2ab - a^2 - ab}{a+b} \qquad\qquad = \frac{ab + b^2 - 2ab}{a+b}$$

$$= \frac{ab - a^2}{a+b} \qquad\qquad = \frac{b^2 - ab}{a+b}$$

$$= \frac{a(b-a)}{a+b} > 0 \qquad\qquad = \frac{b(b-a)}{a+b} > 0$$

Therefore, $h > a$. $\qquad\qquad$ Therefore, $h < b$.

Thus, $a < h < b$.

119–121. Answers will vary.

Appendix A

Review

A.9 *n*th Roots; Rational Exponents; Radical Equations

1. False

3. index

5. True

7. $\sqrt[3]{27} = \sqrt[3]{3^3} = 3$

9. $\sqrt[3]{-8} = \sqrt[3]{(-2)^3} = -2$

11. $\sqrt{8} = \sqrt{4 \cdot 2} = 2\sqrt{2}$

13. $\sqrt[3]{-8x^4} = \sqrt[3]{-8 \cdot x^3 \cdot x} = -2x\sqrt[3]{x}$

15. $\sqrt[4]{x^{12}y^8} = \sqrt[4]{\left(x^3\right)^4 \left(y^2\right)^4} = x^3 y^2$

17. $\sqrt[4]{\dfrac{x^9 y^7}{x\, y^3}} = \sqrt[4]{x^8 y^4} = x^2 y$

19. $\sqrt{36x} = 6\sqrt{x}$

21. $\sqrt{3x^2}\sqrt{12x} = \sqrt{36x^2 \cdot x} = 6x\sqrt{x}$

23. $\left(\sqrt{5}\,\sqrt[3]{9}\right)^2 = 5\left(\sqrt[3]{81}\right) = 5\left(\sqrt[3]{27 \cdot 3}\right) = 5 \cdot 3\left(\sqrt[3]{3}\right) = 15\sqrt[3]{3}$

25. $\left(3\sqrt{6}\right)\left(2\sqrt{2}\right) = 6\sqrt{12} = 6\sqrt{4 \cdot 3} = 12\sqrt{3}$

27. $\left(\sqrt{3} + 3\right)\left(\sqrt{3} - 1\right) = \left(\sqrt{3}\right)^2 - \sqrt{3} + 3\sqrt{3} - 3 = 3 + 2\sqrt{3} - 3 = 2\sqrt{3}$

29. $\left(\sqrt{x} - 1\right)^2 = \left(\sqrt{x}\right)^2 - 2\sqrt{x} + 1 = x - 2\sqrt{x} + 1$

31. $3\sqrt{2} - 4\sqrt{8} = 3\sqrt{2} - 4\sqrt{4 \cdot 2} = 3\sqrt{2} - 8\sqrt{2} = -5\sqrt{2}$

33. $\sqrt[3]{16x^4} - \sqrt[3]{2x} = \sqrt[3]{8 \cdot 2 \cdot x^3 \cdot x} - \sqrt[3]{2x} = \sqrt[3]{8 \cdot x^3 \cdot 2 \cdot x} - \sqrt[3]{2x}$
$$= \sqrt[3]{\left(8x^3\right) \cdot 2x} - \sqrt[3]{2x}$$
$$= \sqrt[3]{\left(8x^3\right)} \cdot \sqrt[3]{2x} - \sqrt[3]{2x}$$
$$= 2x\sqrt[3]{2x} - \sqrt[3]{2x}$$
$$= (2x - 1)\sqrt[3]{2x}$$

35. $\dfrac{1}{\sqrt{2}}\cdot\dfrac{\sqrt{2}}{\sqrt{2}}=\dfrac{\sqrt{2}}{2}$

37. $\dfrac{-\sqrt{3}}{\sqrt{5}}\cdot\dfrac{\sqrt{5}}{\sqrt{5}}=\dfrac{-\sqrt{15}}{5}=-\dfrac{\sqrt{15}}{5}$

39. $\dfrac{\sqrt{3}}{5-\sqrt{2}}\cdot\dfrac{5+\sqrt{2}}{5+\sqrt{2}}=\dfrac{\sqrt{3}\left(5+\sqrt{2}\right)}{25-2}=\dfrac{\sqrt{3}\left(5+\sqrt{2}\right)}{23}$

41. $\dfrac{2-\sqrt{5}}{2+3\sqrt{5}}\cdot\dfrac{2-3\sqrt{5}}{2-3\sqrt{5}}=\dfrac{4-6\sqrt{5}-2\sqrt{5}+3\left(\sqrt{5}\right)^2}{4-9\left(\sqrt{5}\right)^2}=\dfrac{4-8\sqrt{5}+3\cdot5}{4-9\cdot5}$

$=\dfrac{4-8\sqrt{5}+15}{4-45}=\dfrac{19-8\sqrt{5}}{-41}$

$=\dfrac{-19+8\sqrt{5}}{41}$

43. $\dfrac{5}{\sqrt[3]{2}}\cdot\dfrac{\left(\sqrt[3]{2}\right)^2}{\left(\sqrt[3]{2}\right)^2}=\dfrac{5\left(\sqrt[3]{2}\right)^2}{\left(\sqrt[3]{2}\right)^3}=\dfrac{5\left(\sqrt[3]{2^2}\right)}{2}=\dfrac{5\sqrt[3]{4}}{2}$

45. $\dfrac{\sqrt{x+h}-\sqrt{x}}{\sqrt{x+h}+\sqrt{x}}\cdot\dfrac{\sqrt{x+h}-\sqrt{x}}{\sqrt{x+h}-\sqrt{x}}=\dfrac{\left(\sqrt{x+h}\right)^2-2\sqrt{x}\cdot\sqrt{x+h}+\left(\sqrt{x}\right)^2}{\left(\sqrt{x+h}\right)^2-\left(\sqrt{x}\right)^2}$

$=\dfrac{x+h-2\sqrt{x(x+h)}+x}{x+h-x}$

$=\dfrac{2x+h-2\sqrt{x(x+h)}}{h}$

47. $\sqrt[3]{2t-1}=2$

$\left(\sqrt[3]{2t-1}\right)^3=2^3$

$2t-1=8$

$2t=9$

$t=\dfrac{9}{2}$

Check $t=\dfrac{9}{2}$:

$\sqrt[3]{2\left(\dfrac{9}{2}\right)-1}=2$

$\sqrt[3]{9-1}=2$

$\sqrt[3]{8}=2$

$2=2$

The solution is $t=\dfrac{9}{2}$.

49.
$$\sqrt{15-2x} = x$$
$$\left(\sqrt{15-2x}\right)^2 = x^2$$
$$15-2x = x^2$$
$$0 = x^2 + 2x - 15$$
$$0 = (x+5)(x-3)$$
$$x = -5, x = 3$$

Check $x = -5$:
$$\sqrt{15-2(-5)} = -5$$
$$\sqrt{15+10} = -5$$
$$\sqrt{25} = -5$$
$$5 \neq -5$$
Check $x = 3$:
$$\sqrt{15-2(3)} = 3$$
$$\sqrt{15-6} = 3$$
$$\sqrt{9} = 3$$
$$3 = 3$$
The solution is $x = 3$.

51. $8^{2/3} = \left(2^3\right)^{2/3} = 2^2 = 4$

53. $(-27)^{1/3} = \left((-3)^3\right)^{1/3} = -3$

55. $16^{3/2} = \left(4^2\right)^{3/2} = 4^3 = 64$

57. $9^{-3/2} = \left(3^2\right)^{-3/2} = 3^{-3} = \dfrac{1}{3^3} = \dfrac{1}{27}$

59. $\left(\dfrac{9}{8}\right)^{3/2} = \dfrac{9^{3/2}}{8^{3/2}} = \dfrac{\left(3^2\right)^{3/2}}{\left(2^3\right)^{3/2}} = \dfrac{3^{6/2}}{2^{9/2}} = \dfrac{3^3}{2^4 \cdot 2^{1/2}} = \dfrac{27}{16 \cdot \sqrt{2}} \cdot \dfrac{\sqrt{2}}{\sqrt{2}} = \dfrac{27\sqrt{2}}{32}$

61. $\left(\dfrac{8}{9}\right)^{-3/2} = \left(\dfrac{9}{8}\right)^{3/2} = \left(\dfrac{3^2}{2^3}\right)^{3/2} = \dfrac{3^3}{2^{9/2}} = \dfrac{27}{2^4 \cdot 2^{1/2}} = \dfrac{27}{16 \cdot \sqrt{2}} \cdot \dfrac{\sqrt{2}}{\sqrt{2}} = \dfrac{27\sqrt{2}}{16 \cdot 2} = \dfrac{27\sqrt{2}}{32}$

63. $x^{3/4} \cdot x^{1/3} \cdot x^{-1/2} = x^{3/4 + 1/3 - 1/2} = x^{(9+4-6)/12} = x^{7/12}$

65. $\left(x^3 y^6\right)^{1/3} = \left(x^3\right)^{1/3}\left(y^6\right)^{1/3} = x y^2$

67. $\left(x^2 y\right)^{1/3}\left(x y^2\right)^{2/3} = x^{2/3} y^{1/3} x^{2/3} y^{4/3} = x^{4/3} y^{5/3}$

69. $\left(16 x^2 y^{-1/3}\right)^{3/4} = \left(2^4 x^2 y^{-1/3}\right)^{3/4} = 2^{4 \cdot 3/4} x^{2 \cdot 3/4} y^{(-1/3)(3/4)} = 2^3 x^{3/2} y^{-1/4} = \dfrac{8 x^{3/2}}{y^{1/4}}$

71. $\dfrac{x}{(1+x)^{1/2}} + 2(1+x)^{1/2} = \dfrac{x + 2(1+x)^{1/2}(1+x)^{1/2}}{(1+x)^{1/2}} = \dfrac{x + 2(1+x)}{(1+x)^{1/2}} = \dfrac{x + 2 + 2x}{(1+x)^{1/2}} = \dfrac{3x+2}{(1+x)^{1/2}}$

73.

$$2x\left(x^2+1\right)^{1/2} + x^2 \cdot \frac{1}{2}\left(x^2+1\right)^{-1/2} \cdot 2x = 2x\left(x^2+1\right)^{1/2} + \frac{x^3}{\left(x^2+1\right)^{1/2}}$$

$$= \frac{2x\left(x^2+1\right)^{1/2}\cdot\left(x^2+1\right)^{1/2}+x^3}{\left(x^2+1\right)^{1/2}} = \frac{2x\left(x^2+1\right)+x^3}{\left(x^2+1\right)^{1/2}}$$

$$= \frac{2x^3+2x+x^3}{\left(x^2+1\right)^{1/2}} = \frac{3x^3+2x}{\left(x^2+1\right)^{1/2}}$$

$$= \frac{x\left(3x^2+2\right)}{\left(x^2+1\right)^{1/2}}$$

75.

$$\sqrt{4x+3}\cdot\frac{1}{2\sqrt{x-5}}+\sqrt{x-5}\cdot\frac{1}{5\sqrt{4x+3}} = \frac{\sqrt{4x+3}}{2\sqrt{x-5}}+\frac{\sqrt{x-5}}{5\sqrt{4x+3}}$$

$$= \frac{\sqrt{4x+3}\cdot5\cdot\sqrt{4x+3}+\sqrt{x-5}\cdot2\cdot\sqrt{x-5}}{10\sqrt{x-5}\sqrt{4x+3}}$$

$$= \frac{5(4x+3)+2(x-5)}{10\sqrt{(x-5)(4x+3)}} = \frac{20x+15+2x-10}{10\sqrt{(x-5)(4x+3)}}$$

$$= \frac{22x+5}{10\sqrt{(x-5)(4x+3)}}$$

77.

$$\frac{\sqrt{1+x}-x\cdot\dfrac{1}{2\sqrt{1+x}}}{1+x} = \frac{\sqrt{1+x}-\dfrac{x}{2\sqrt{1+x}}}{1+x} = \frac{\dfrac{2\sqrt{1+x}\sqrt{1+x}-x}{2\sqrt{1+x}}}{1+x}$$

$$= \frac{2(1+x)-x}{2(1+x)^{1/2}}\cdot\frac{1}{1+x}$$

$$= \frac{2+x}{2(1+x)^{3/2}}$$

79.

$$\frac{\left(x+4\right)^{1/2}-2x\left(x+4\right)^{-1/2}}{x+4} = \frac{\left(x+4\right)^{1/2}-\dfrac{2x}{\left(x+4\right)^{1/2}}}{x+4} = \frac{\dfrac{\left(x+4\right)^{1/2}\left(x+4\right)^{1/2}}{\left(x+4\right)^{1/2}}-\dfrac{2x}{\left(x+4\right)^{1/2}}}{x+4}$$

$$= \frac{\dfrac{x+4-2x}{\left(x+4\right)^{1/2}}}{x+4} = \frac{-x+4}{\left(x+4\right)^{1/2}}\cdot\frac{1}{x+4} = \frac{-x+4}{\left(x+4\right)^{3/2}}$$

$$= \frac{4-x}{\left(x+4\right)^{3/2}}$$

81. $\dfrac{\dfrac{x^2}{\left(x^2-1\right)^{1/2}}-\left(x^2-1\right)^{1/2}}{x^2}=\dfrac{\dfrac{x^2-\left(x^2-1\right)^{1/2}\cdot\left(x^2-1\right)^{1/2}}{\left(x^2-1\right)^{1/2}}}{x^2}$

$$=\dfrac{x^2-\left(x^2-1\right)^{1/2}\cdot\left(x^2-1\right)^{1/2}}{\left(x^2-1\right)^{1/2}}\cdot\dfrac{1}{x^2}$$

$$=\dfrac{x^2-\left(x^2-1\right)}{\left(x^2-1\right)^{1/2}}\cdot\dfrac{1}{x^2}=\dfrac{x^2-x^2+1}{\left(x^2-1\right)^{1/2}}\cdot\dfrac{1}{x^2}$$

$$=\dfrac{1}{x^2\left(x^2-1\right)^{1/2}}$$

83. $\dfrac{\dfrac{1+x^2}{2\sqrt{x}}-2x\sqrt{x}}{\left(1+x^2\right)^2}=\dfrac{\dfrac{1+x^2-\left(2\sqrt{x}\right)\left(2x\sqrt{x}\right)}{2\sqrt{x}}}{\left(1+x^2\right)^2}$

$$=\dfrac{1+x^2-\left(2\sqrt{x}\right)\left(2x\sqrt{x}\right)}{2\sqrt{x}}\cdot\dfrac{1}{\left(1+x^2\right)^2}$$

$$=\dfrac{1+x^2-4x^2}{2\sqrt{x}}\cdot\dfrac{1}{\left(1+x^2\right)^2}$$

$$=\dfrac{1-3x^2}{2\sqrt{x}\left(1+x^2\right)^2}$$

85. $(x+1)^{3/2}+x\cdot\dfrac{3}{2}(x+1)^{1/2}=(x+1)^{1/2}\left(x+1+\dfrac{3}{2}x\right)=(x+1)^{1/2}\left(\dfrac{5}{2}x+1\right)=\dfrac{1}{2}(x+1)^{1/2}(5x+2)$

87. $6x^{1/2}\left(x^2+x\right)-8x^{3/2}-8x^{1/2}=2x^{1/2}\left(3(x^2+x)-4x-4\right)=2x^{1/2}\left(3x^2-x-4\right)$

$$=2x^{1/2}(3x-4)(x+1)$$

89. $3\left(x^2+4\right)^{4/3}+x\cdot4\left(x^2+4\right)^{1/3}\cdot2x=\left(x^2+4\right)^{1/3}\left[3\left(x^2+4\right)+8x^2\right]$

$$=\left(x^2+4\right)^{1/3}\left[3x^2+12+8x^2\right]$$

$$=\left(x^2+4\right)^{1/3}\left(11x^2+12\right)$$

91. $4(3x+5)^{1/3}(2x+3)^{3/2}+3(3x+5)^{4/3}(2x+3)^{1/2}$

$$=(3x+5)^{1/3}(2x+3)^{1/2}\left[4(2x+3)+3(3x+5)\right]=(3x+5)^{1/3}(2x+3)^{1/2}(8x+12+9x+15)$$

$$=(3x+5)^{1/3}(2x+3)^{1/2}(17x+27)$$

93. $3x^{-1/2} + \dfrac{3}{2}x^{1/2} = \dfrac{3}{x^{1/2}} + \dfrac{3}{2}x^{1/2} = \dfrac{3 \cdot 2 + 3x^{1/2} \cdot x^{1/2}}{2x^{1/2}} = \dfrac{6 + 3x}{2x^{1/2}}$

$= \dfrac{3(2 + x)}{2x^{1/2}}$

95. $x\left(\dfrac{1}{2}\right)\left(8 - x^2\right)^{-1/2}(-2x) + \left(8 - x^2\right)^{1/2} = \dfrac{-x^2}{\left(8 - x^2\right)^{1/2}} + \left(8 - x^2\right)^{1/2}$

$= \dfrac{-x^2 + \left(8 - x^2\right)^{1/2} \cdot \left(8 - x^2\right)^{1/2}}{\left(8 - x^2\right)^{1/2}}$

$= \dfrac{-x^2 + 8 - x^2}{\left(8 - x^2\right)^{1/2}} = \dfrac{8 - 2x^2}{\left(8 - x^2\right)^{1/2}} = \dfrac{2\left(4 - x^2\right)}{\left(8 - x^2\right)^{1/2}}$

$= \dfrac{2(2 - x)(2 + x)}{\left(8 - x^2\right)^{1/2}}$

Appendix B

Graphing Utilities

B.1 The Viewing Rectangle

1. $(-1, 4)$

3. $(3, 1)$

5. $X\min = -6$
   $X\max = 6$
   $X\operatorname{scl} = 2$
   $Y\min = -4$
   $Y\max = 4$
   $Y\operatorname{scl} = 2$

7. $X\min = -6$
   $X\max = 6$
   $X\operatorname{scl} = 2$
   $Y\min = -1$
   $Y\max = 3$
   $Y\operatorname{scl} = 1$

9. $X\min = 3$
   $X\max = 9$
   $X\operatorname{scl} = 1$
   $Y\min = 2$
   $Y\max = 10$
   $Y\operatorname{scl} = 2$

11. $X\min = -11$
    $X\max = 5$
    $X\operatorname{scl} = 1$
    $Y\min = -3$
    $Y\max = 6$
    $Y\operatorname{scl} = 1$

13. $X\min = -30$
    $X\max = 50$
    $X\operatorname{scl} = 10$
    $Y\min = -90$
    $Y\max = 50$
    $Y\operatorname{scl} = 10$

15. $X\min = -10$
    $X\max = 110$
    $X\operatorname{scl} = 10$
    $Y\min = -10$
    $Y\max = 160$
    $Y\operatorname{scl} = 10$

17. $P_1 = (1,3);\ P_2 = (5,15)$
$$d(P_1,P_2) = \sqrt{(5-1)^2 + (15-3)^2}$$
$$= \sqrt{(4)^2 + (12)^2}$$
$$= \sqrt{16 + 144}$$
$$= \sqrt{160} = 2\sqrt{10}$$

19. $P_1 = (-4,6);\ P_2 = (4,-8)$
$$d(P_1,P_2) = \sqrt{(4-(-4))^2 + (-8-6)^2}$$
$$= \sqrt{(8)^2 + (-14)^2}$$
$$= \sqrt{64 + 196}$$
$$= \sqrt{260} = 2\sqrt{65}$$

Graphing Utilities

B.2 Using a Graphing Utility to Graph Equations

1.   (a)

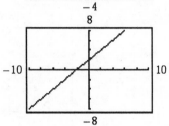

            (b)

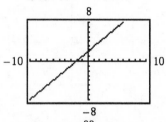

   (c)

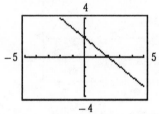

            (d)

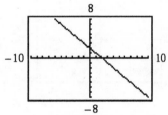

3.   (a)

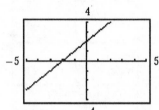

            (b)

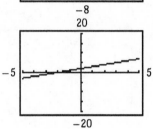

   (c)

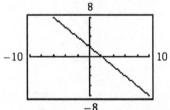

            (d)

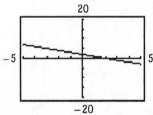

5. (a)

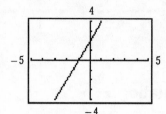

(b)

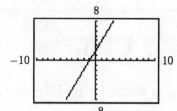

(c)

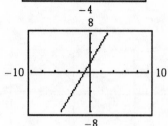

(d)

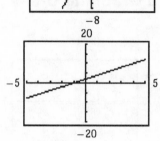

7. (a)

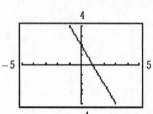

(b)

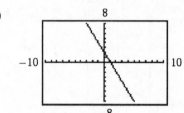

(c)

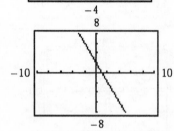

(d)

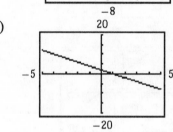

9. (a)

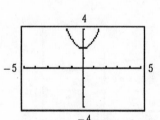

(b)

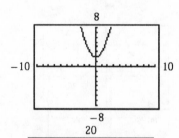

(c)

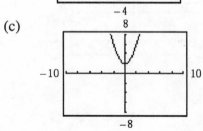

(d)

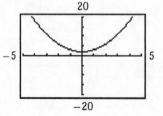

11.   (a)

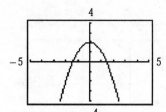

(b)

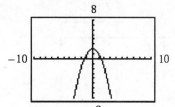

(c)

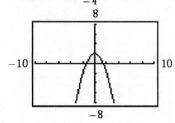

(d)

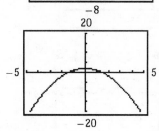

13. $3x + 2y = 6 \Rightarrow 2y = -3x + 6 \Rightarrow y = -\dfrac{3}{2}x + 3$

(a)

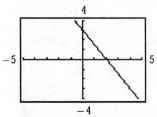

(b)

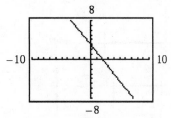

(c)

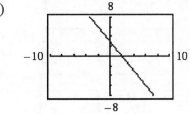

(d)

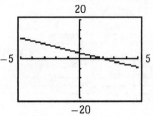

15. $-3x + 2y = 6 \Rightarrow 2y = 3x + 6 \Rightarrow y = \dfrac{3}{2}x + 3$

(a)

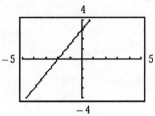

(b)

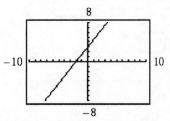

(c)

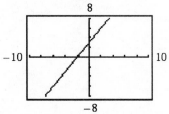

(d)

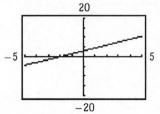

17. $y = x + 2;\quad -3 \le x \le 3$

X	Y1	
-3	-1	
-2	0	
-1	1	
0	2	
1	3	
2	4	
3	5	

X=3

19. $y = -x + 2;\quad -3 \le x \le 3$

X	Y1	
-3	5	
-2	4	
-1	3	
0	2	
1	1	
2	0	
3	-1	

X=3

21. $y = 2x + 2;\quad -3 \le x \le 3$

X	Y1	
-3	-4	
-2	-2	
-1	0	
0	2	
1	4	
2	6	
3	8	

X=3

23. $y = -2x + 2;\quad -3 \le x \le 3$

X	Y1	
-3	8	
-2	6	
-1	4	
0	2	
1	0	
2	-2	
3	-4	

X=3

25. $y = x^2 + 2;\quad -3 \le x \le 3$

X	Y1	
-3	11	
-2	6	
-1	3	
0	2	
1	3	
2	6	
3	11	

X=3

27. $y = -x^2 + 2;\quad -3 \le x \le 3$

X	Y1	
-3	-7	
-2	-2	
-1	1	
0	2	
1	1	
2	-2	
3	-7	

X=3

29. $3x + 2y = 6;\quad -3 \le x \le 3$

X	Y1	
-3	7.5	
-2	6	
-1	4.5	
0	3	
1	1.5	
2	0	
3	-1.5	

X=3

31. $-3x + 2y = 6;\quad -3 \le x \le 3$

X	Y1	
-3	-1.5	
-2	0	
-1	1.5	
0	3	
1	4.5	
2	6	
3	7.5	

X=3

Appendix B

Graphing Utilities

B.3 **Using a Graphing Utility to Locate Intercepts and Check for Symmetry**

1.

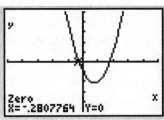

The smaller x-intercept is $x \approx -3.41$.

3.

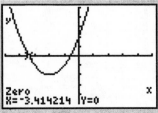

The smaller x-intercept is $x \approx -1.71$.

5.

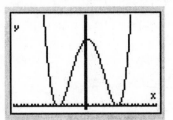

The smaller x-intercept is $x \approx -0.28$.

7.

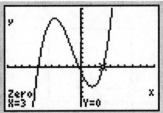

The positive x-intercept is $x = 3.00$.

We zoom in on the positive x-intercept:

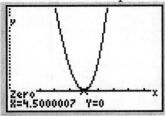

The positive x-intercept is $x \approx 4.50$.

9.

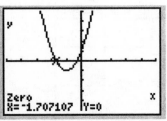

11.

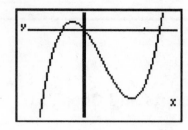

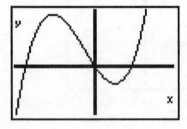

The largest positive x-intercept is
$x \approx 12.30$.

We zoom in on the positive x-intercept:

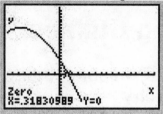

The smallest positive x-intercept is
$x \approx 0.32$.

13.

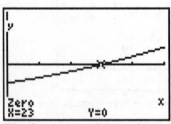

We zoom in on the positive x-intercept:

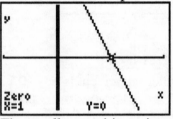

The smallest positive x-intercept is
$x = 1.00$.

The largest positive x-intercept is
$x = 23.00$.

Graphing Utilities

B.5   Square Screens

1. Yes
    X min = −3

    X max = 3

    X scl = 2

    Y min = −2

    Y max = 2

    Y scl = 2

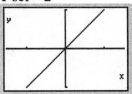

3. Yes
    X min = 0

    X max = 9

    X scl = 3

    Y min = −2

    Y max = 4

    Y scl = 2

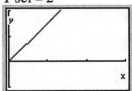

5. No
    X min = −6

    X max = 6

    X scl = 1

    Y min = −2

    Y max = 2

    Y scl = 0.5

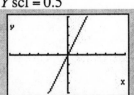

7. Yes
    X min = 0

    X max = 9

    X scl = 1

    Y min = −2

    Y max = 4

    Y scl = 1

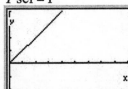

9. One possible answer:
    X min = −4

    X max = 8

    X scl = 1

    Y min = 4

    Y max = 12

    Y scl = 1